LabVIEW based Advanced Instrumentation Systems

S. Sumathi and P. Surekha

LabVIEW based Advanced Instrumentation Systems

With 488 Figures and 34 Tables

Dr. S. Sumathi
Assistant Professor
Department of Electrical
and Electronics Engineering
PSG College of Technology
Coimbatore 641 004
Tamil Nadu, India
E-mail: ss_eeein@yahoo.com

Prof. Surekha. P
Programmer Analyst
Cognizant Technology Solutions
5/535, old Mahabalipuram Road
Okkiyam Thoraipakkam
Chennai - 600 096
E-mail: surekha_3000@yahoo.com

Library of Congress Control Number: 2006936972

ISBN-10 3-540-48500-7 Springer Berlin Heidelberg New York
ISBN-13 978-3-540-48500-1 Springer Berlin Heidelberg New York

Springer is a part of Springer Science+Business Media.

springer.com

Typesetting by the authors and SPi
Cover design: KünkelLopka GmbH

Printed on acid-free paper SPIN 11803485 89/3100/SPi 5 4 3 2 1 0

Preface

Information is a valuable resource to an organization. User-friendly, computer-controlled instrumentation and data analysis techniques are revolutionizing the way measurements are being made, allowing nearly instantaneous comparison between theoretical predictions, simulations, and actual experimental results. This book provides comprehensive coverage of fundamentals of advanced instrumentation systems based on LabVIEW concepts. This book is for those who wish a better understanding of virtual instrumentation concepts, its purpose, its nature, and the applications developed using the National Instrument's LabVIEW software.

The evolution and pervasiveness of PCs as cost-effective computing platforms, recently joined by workstations with more powerful software tools, has resulted in a virtual explosion in data acquisition, signal processing and control systems from laboratory to industry including field applications. An ever-increasing array of industry-standard design and simulation tools provides the opportunity to fully integrate the use of computers directly in the laboratory. Advanced techniques in instrumentation and control such as Distributed Automation and SCADA are dealt in this book.

The current trends in instrumentation like Fiber optic instrumentation, LASER instrumentation, Smart Transmitters, and CASE have made virtual instrumentation to support high availability, and increase in popularity.

This text discusses a number of new technologies and challenges of virtual instrumentation systems in terms of applications in the areas including control systems, power systems, networking, robotics, communication, and artificial intelligence.

About the Book

The book is meant for wide range of readers from College, University Students wishing to learn basic as well as advanced concepts in virtual instrumentation system. It can also be meant for the programmers who may be involved in the programming based on the LabVIEW and virtual instrumentation applications.

Virtual Instrumentation System, at present is a well developed field, among academicians as well as between program developers. The various approaches to data transmission, the common interface buses and standards of instrumentation are given in detail.

The solutions to the problems in instrumentation are programmed using LabVIEW and the results are given. An overview of LabVIEW with examples is provided for easy reference of the students and professionals. This book also provides research projects, LabVIEW tools, and glossary of virtual instrumentation terms in appendix.

The book also presents Application Case Studies on a wide range of connected fields to facilitate the reader for better understanding. This book can be used from Under Graduation to Post-Graduate Level. We hope that the reader will find this book a truly helpful guide and a valuable source of information about the advanced instrumentation principles for their numerous practical applications.

Salient Features

The salient features of this book includes:

- Detailed description on virtual instrumentation system concepts.
- Worked out examples using LabVIEW software.
- Application case studies based on LabVIEW in various fields like Instrumentation and Control, Power Systems, Robotics, Networking and Communication, and Artificial Intelligence.
- LabVIEW Tools, Research Projects, and Glossary.

Organization of the Book

The book starts with the introduction to virtual instrumentation and covers in detail on the advanced virtual instrumentation concepts, techniques, and applications.

- Chapter 1 presents an introduction to virtual instrumentation concepts, architecture of a virtual instrumentation system and the role of various components in the architecture. It introduces the concept of distributed virtual instrumentation systems and conventional virtual instrumentation systems. The advantages of virtual instrumentation is discussed and compared with the conventional virtual instrumentation systems.
- Chapter 2 provides an overview of virtual instruments such as the front panel and the block diagram in virtual instrumentation software, LabVIEW. It discusses the menus used by the virtual instruments, 'G' Programming concepts, Data flow model in the block diagram, and the data types and its representation. The VI libraries and creation of a SubVI are also investigated here.

- Chapter 3 describes the structures available in LabVIEW such as For loop, While loop, Case structures and Sequence structure. This chapter also addresses issues related to arrays, clusters and formula node. Besides these aspects, data displaying elements on the front panel such as waveform charts, waveform graphs, XY graphs, and intensity plots are also illustrated with suitable examples.
- Chapter 4 deals with the components of a typical measurement system, origin of signals and the various types of signal acquiring devices such as sensors and transducers. The concepts of signal conditioning and the SCXI, a signal conditioning and front end fore plug in DAQ boards are also discussed. The output of the sensors are in analog form, hence to process them analog to digital converters are used. Conversions back to analog signals are accomplished using digital-to-analog converters.
- Chapter 5 describes the operation and characteristic feature of serial communication devices such as 4–20, 60 mA current loops along with the RS232C standard. The IEEE standard GPIB is also detailed in later part of this chapter. VISA, which is a high level API is capable of controlling VXI, GPIB, or Serial instruments is also delineated.
- Chapter 6 focusses on the most common and latest PC interface buses such as USB, PCI, PXI, PCMCIA, VXI, and LXI. Modern computer buses can use both parallel and bit-serial connections, and can be wired in either a multidrop or daisy chain topology, or connected by switched hubs.
- Chapter 7 touches the aspects related with signal grounding and digital I/O techniques. The approach of data acquisition in LabVIEW is elaborated with the DAQ components and the Hardware and Software configuration.
- Chapter 8 encompasses the operation and characteristic features of data transmission such as pulse codes, analog and digital modulation, wireless communication, RF analyser, distributed automation, and SCADA. Data transmission plays a very important role in all kind of digital equipments as it is the responsibility of these devices to transmit the data without being lost.
- Chapter 9 elaborates on the current trends in instrumentation such as fiber optic and laser instrumentation. The various types of fiber optic sensors including voltage, pressure, stress, temperature, and laser sensors including velocity, distance, and length are also discussed. The later sections of this chapter presents on the concepts of smart transmitter and CASE.
- Chapter 10 presents the recent approaches of LabVIEW in Virtual instrumentation with application examples in the areas of instrumentation, control systems, power systems, and robotics.
- Chapter 11 illustrates the application examples based on LabVIEW in the areas of communication, networking, artificial intelligence, and biomedical instrumentation.

About the Authors

S. Sumathi completed B.E. (Electronics and Communication Engineering), M.E. (Applied Electronics) at Government College of Technology, Coimbatore, Tamil Nadu, and Ph.D. in data mining. Currently, working as Assistant Professor in the Department of Electrical and Electronics Engineering, PSG College of Technology, Coimbatore with teaching and research experience of 16 years. She received the prestigious gold medal from the Institution of Engineers Journal Computer Engineering Division for the research paper titled, "Development of New Soft Computing Models for Data Mining" and also best project award for UG Technical Report titled, "Self-Organized Neural Network Schemes: As a Data mining tool"; Dr. R. Sundramoorthy award for Outstanding Academic of PSG College of Technology in the year 2006. She has guided a project which received Best M. Tech Thesis award from Indian Society for Technical Education, New Delhi. In appreciation of publishing various technical articles she has received National and International Journal Publication Awards. She has prepared manuals for Electronics and Instrumentation Lab and Electrical and Electronics Lab of EEE Department, PSG College of Technology, Coimbatore and also organized second National Conference on Intelligent and Efficient Electrical Systems in the year 2005 and conducted Short-Term Courses on "Neuro Fuzzy System Principles and Data Mining Applications." She has published several research articles in National and International Journals/Conferences and guided many UG and PG projects. She has also published three books on, "Introduction to Neural Networks with MATLAB," "Introduction to Fuzzy Systems with MATLAB" and "Introduction to Data mining and its Applications." She reviewed papers in National/International Journals and Conferences. The Research interests include Neural Networks, Fuzzy Systems and Genetic Algorithms, Pattern Recognition and Classification, Data Warehousing and Data Mining, Operating systems and Parallel Computing, etc.

Surekha P. completed her B.E. Degree in Electrical and Electronics Engineering in PARK College of Engineering and Technology, Coimbatore, Tamil Nadu, and Masters Degree in Control Systems at PSG College of Technology, Coimbatore, Tamil Nadu. She was a Rank Holder in both B.E. and M.E. programmes. She has received Alumni Award for best performance in curricular and cocurricular activities during her Masters Degree programme. She has presented papers in National Conferences and Journals. She is currently working as a programmer analyst in Cognizant Technology Solutions, Chennai, Tamil Nadu. Her research areas include robotics, virtual instrumentation, neural network, fuzzy logic theory and genetic algorithm.

Acknowledgement

The authors are always thankful to the Almighty for perseverance and achievements. They wish to thank Mr. G. Rangaswamy, Managing Trustee, PSG Institutions and Dr. R. Rudramoorthy, Principal, PSG College of Technology, Coimbatore, for their whole-hearted cooperation and great encouragement given in this successful endeavour. They also appreciate and acknowledge very much to Mr. K.K.N. Anburajan, Lab-in-Charges of EEE Department, PSG College of Technology, Coimbatore who have been with them in all their endeavours with their excellent, unforgettable help, and assistance in the successful execution of the work. Sumathi owes much to her daughter, Priyanka, who has helped and to the support rendered by her husband, brother, and family. Surekha would like to thank her parents, brother, and husband who shouldered a lot of extra responsibilities during the months this was being written. They did this with the long-term vision, depth of character, and positive outlook that are truly befitting of their name.

Contents

1 Introduction to Virtual Instrumentation 1
1.1 Introduction 1
1.2 History of Instrumentation Systems 2
1.3 Evolution of Virtual Instrumentation 4
1.4 Premature Challenges 5
1.5 Virtual Instrumentation 5
1.5.1 Definition 6
1.5.2 Architecture of Virtual Instrumentation 6
1.5.3 Presentation and Control 10
1.5.4 Functional Integration 11
1.6 Programming Requirements 12
1.7 Drawbacks of Recent Approaches 13
1.8 Conventional Virtual Instrumentation 13
1.9 Distributed Virtual Instrumentation 14
1.10 Virtual Instruments Versus Traditional Instruments 17
1.11 Advantages of VI 18
1.11.1 Performance 18
1.11.2 Platform-Independent Nature 19
1.11.3 Flexibility 19
1.11.4 Lower Cost 19
1.11.5 Plug-In and Networked Hardware 19
1.11.6 The Costs of a Measurement Application 20
1.11.7 Reducing System Specification Time Cost 20
1.11.8 Lowering the Cost of Hardware and Software 20
1.11.9 Minimising Set-Up and Configuration Time Costs 20
1.11.10 Decreasing Application Software Development Time Costs 21
1.12 Evolution of LabVIEW 21
1.13 Creating Virtual Instruments Using LabVIEW 22
1.13.1 Connectivity and Instrument Control 23
1.13.2 Open Environment 24

1.13.3 Reduces Cost and Preserves Investment 24
1.13.4 Multiple Platforms 24
1.13.5 Distributed Development 25
1.13.6 Analysis Capabilities 25
1.13.7 Visualization Capabilities 25
1.13.8 Flexibility and Scalability 26
1.14 Advantages of LabVIEW 26
1.14.1 Easy to Learn 26
1.14.2 Easy to Use 26
1.14.3 Complete Functionality 27
1.14.4 Modular Development 27
1.15 Virtual Instrumentation in the Engineering Process 27
1.15.1 Research and Design 28
1.15.2 Development Test and Validation 28
1.15.3 Manufacturing Test 28
1.15.4 Manufacturing Design 29
1.16 Virtual Instruments Beyond the Personal Computer 29

2 Programming Techniques 33
2.1 Introduction 33
2.2 Virtual Instruments 34
2.2.1 Front Panel 34
2.2.2 Block Diagram 39
2.3 LabVIEW Environment 42
2.3.1 Startup Menu 44
2.3.2 Shortcut Menu 44
2.3.3 Pull-Down Menu 45
2.3.4 Pop-Up Menu 50
2.3.5 Palletes 56
2.4 Dataflow Programming 61
2.5 'G' Programming 62
2.5.1 Data Types and Conversion 63
2.5.2 Representation and Precision 64
2.5.3 Creating and Saving VIs 66
2.5.4 Wiring, Editing, and Debugging 68
2.5.5 Creating SubVIs 73
2.5.6 VI Libraries 77

3 Programming Concepts of VI 81
3.1 Introduction 81
3.2 Control Structures 82
3.2.1 The For Loop 82
3.2.2 The While Loop 88
3.2.3 Shift Registers 95
3.2.4 Feedback Nodes 98

3.3 Selection Structures 100
3.3.1 Case Structures 101
3.3.2 Sequence Structures (Flat and Stacked Structures) . . 107
3.4 The Formula Node 111
3.5 Arrays 112
3.5.1 Single and Multidimensional Arrays 113
3.5.2 Autoindexing 115
3.5.3 Functions for Manipulating Arrays 117
3.5.4 Polymorphism 125
3.6 Clusters 126
3.6.1 Creating Cluster Controls and Indicators 128
3.6.2 Cluster Functions 130
3.6.3 Error Handling 136
3.7 Waveform Charts 138
3.7.1 Chart Components 142
3.7.2 Mechanical Action of Boolean Switches 145
3.8 Waveform Graphs 146
3.8.1 Single-Plot Waveform Graphs 147
3.8.2 Multiple-Plot Waveform Graphs 147
3.9 XY Graphs 148
3.10 Strings 155
3.10.1 Creating String Controls and Indicators 155
3.10.2 String Functions 157
3.11 Tables 161
3.12 List Boxes 163
3.13 File Input/Output 163
3.13.1 File I/O VIs and Functions 163
3.13.2 File I/O Express VIs 165

4 Inputs and Outputs 173
4.1 Introduction 173
4.2 Components of Measuring System 174
4.3 Origin of Signals 178
4.3.1 Transducers and Sensors 178
4.3.2 Acquiring the Signal 179
4.3.3 Sampling Theorem 180
4.3.4 Filtering and Averaging 180
4.3.5 Triggering 183
4.3.6 Throughput 184
4.4 Transducer 184
4.4.1 Selecting a Transducer 185
4.4.2 Electrical Transducer 186
4.5 Sensors 196
4.5.1 The Nose as a Sensor 198
4.5.2 Sensors and Biosensors: Definitions 199

4.5.3 Differences Between Chemical Sensors, Physical Sensors, and Biosensors . . . 200
4.5.4 Thermocouples . . . 201
4.5.5 RTD: Resistance Temperature Detector . . . 203
4.5.6 Strain Gauges . . . 204
4.6 General Signal Conditioning Functions . . . 206
4.6.1 Amplification . . . 206
4.6.2 Filtering and Averaging . . . 207
4.6.3 Isolation . . . 207
4.6.4 Multiplexing . . . 207
4.6.5 Digital Signal Conditioning . . . 208
4.6.6 Pulse Operation . . . 208
4.6.7 Signal Conditioning Systems for PC-Based DAQ Systems . . . 208
4.6.8 Signal Conditioning with SCXI . . . 209
4.7 Analog-to-Digital Control . . . 210
4.7.1 Understanding Integrating ADCs . . . 210
4.7.2 Understanding SAR ADC . . . 214
4.7.3 Understanding Flash ADCs . . . 218
4.7.4 Understanding Pipelined ADCs . . . 225
4.8 Digital-to-Analog Control . . . 231

5 Common Instrument Interfaces . . . 239
5.1 Introduction . . . 239
5.2 4–20 mA Current Loop . . . 239
5.2.1 Basic 2-wire Circuit . . . 241
5.2.2 4–20 mA Equation . . . 242
5.2.3 3 V/5 V DACs Support Intelligent Current Loop . . . 245
5.2.4 Basic Requirements for 4–20 mA Transducers . . . 245
5.2.5 Digitally Controlled 4–20 mA Current Loops . . . 245
5.3 60 mA Current Loop . . . 247
5.4 RS232 . . . 249
5.5 RS422 and RS485 . . . 253
5.6 GPIB . . . 254
5.6.1 History and Concept . . . 255
5.6.2 Types of GPIB Messages . . . 257
5.6.3 Physical Bus Structure . . . 257
5.6.4 Physical Standards . . . 261
5.6.5 IEEE 488.2 STANDARD . . . 261
5.6.6 Advantages of GPIB . . . 264
5.7 VISA . . . 264
5.7.1 Supported Platforms and Environments . . . 265
5.7.2 VISA Programming . . . 265
5.7.3 DEFAULT Resource Manager, Session, and Instrument Descriptors . . . 266

5.7.4 VISAIC and Message-Based Combination 271
5.7.5 Message-Based Communication 272
5.7.6 Register-Based Communication 274
5.7.7 VISA Attributes 275
5.7.8 Advantages of VISA 278

6 Interface Buses 281
6.1 Introduction 281
6.2 USB 282
6.2.1 Architecture of USB 282
6.2.2 Need for USB 283
6.2.3 Power Cables 284
6.2.4 Data Formats 286
6.2.5 Speed 288
6.2.6 Electrical Specifications 289
6.2.7 Suspend Mode 291
6.2.8 Cables or Pipes 291
6.2.9 USB Functions 292
6.2.10 USB Descriptor 293
6.2.11 Advantages of USB 295
6.3 PCI 295
6.3.1 A 32-Bit Bus 296
6.3.2 History 296
6.3.3 Architecture of PCI with Two Faces 297
6.3.4 Features of PCI 297
6.3.5 Low Profile PCI 300
6.3.6 PCI-X 300
6.3.7 PCI for Data Communication 301
6.3.8 PCI IDE Bus Mastering 302
6.3.9 PCI Internal Interrupts 303
6.3.10 PCI Bus Performance 303
6.3.11 PCI Expansion Slots 304
6.3.12 Standardization 305
6.3.13 Using PCI 305
6.4 PCI Express 306
6.4.1 Need for PCI Express 306
6.4.2 Types of PCI Express Architecture 306
6.4.3 Performance 307
6.4.4 Express Card 307
6.5 PXI 308
6.5.1 PXI Architecture 308
6.5.2 Interoperability with Compact PCI 311
6.5.3 Electrical Architecture Overview 312
6.5.4 Software Architecture Overview 315
6.6 PCMCIA 315

6.6.1 Features of PCMCIA . . . 316
6.6.2 Specifications . . . 317
6.6.3 Board Layout and Jumper Settings . . . 317
6.6.4 Types of PC Cards . . . 317
6.6.5 Features of PC Card Technology . . . 318
6.6.6 Utilities of PCMCIA Card in the Networking Category . . . 318
6.7 SCXI . . . 319
6.7.1 SCXI Hardware and Software . . . 319
6.7.2 Analog Input Signal Connections . . . 320
6.7.3 SCXI Software-Configurable Settings . . . 322
6.7.4 Theory of Operation . . . 325
6.7.5 Typical Program Flowchart . . . 328
6.8 VXI . . . 332
6.8.1 Need for VXI . . . 332
6.8.2 Features of VXI . . . 333
6.8.3 VXI Bus Mechanical Configuration . . . 333
6.8.4 Noise Incurred in VXI . . . 334
6.8.5 Hardware Registers . . . 335
6.8.6 Register-Based Devices . . . 336
6.8.7 Message-Based Communication and Serial Protocol . . . 336
6.8.8 Commander/Servant Hierarchies . . . 336
6.8.9 Three Ways to Control a VXI System . . . 338
6.8.10 Software Standards . . . 338
6.9 LXI . . . 338
6.9.1 LXI Modular Switching Chassis . . . 339
6.9.2 LXI/PXI Module Selection . . . 340

7 Hardware Aspects . . . 345
7.1 Introduction . . . 345
7.2 Signal Grounding . . . 346
7.2.1 Single-Ended Inputs . . . 346
7.2.2 Differential Inputs . . . 347
7.2.3 System Ground and Isolation . . . 348
7.2.4 Wiring Configurations . . . 349
7.3 Digital I/O Techniques . . . 352
7.3.1 Pull-Up and Pull-Down Resistors . . . 352
7.3.2 TTL to Solid-State Relays . . . 353
7.3.3 Voltage Dividers . . . 354
7.4 Data Acquisition in LabVIEW . . . 355
7.5 Hardware Installation and Configuration . . . 355
7.5.1 Buffers . . . 356
7.5.2 Triggering . . . 356

7.6 Components of DAQ . . . 357
7.6.1 System Components . . . 357
7.6.2 NI-DAQ . . . 358
7.7 DAQ Signal Accessory . . . 359
7.7.1 Function Generator . . . 361
7.7.2 Microphone . . . 362
7.7.3 Thermocouple and IC Temperature Sensor . . . 363
7.7.4 Noise Generator . . . 363
7.7.5 Digital Trigger . . . 363
7.7.6 Counter/Timers . . . 363
7.7.7 Quadrature Encoder . . . 363
7.8 DAQ Assistant . . . 364
7.8.1 MAX-Based Tasks . . . 366
7.8.2 Steps to Create a MAX-Based Task . . . 366
7.8.3 Project-Based Tasks . . . 367
7.8.4 Steps to Create a Project-Based Task . . . 367
7.8.5 Project-Based and MAX-Based Tasks . . . 369
7.8.6 Edit a Task . . . 371
7.8.7 Copy a MAX Task to Project . . . 372
7.9 DAQ Hardware . . . 372
7.9.1 Windows Configuration Manager . . . 372
7.9.2 Channel and Task Configuration . . . 373
7.9.3 Hardware Triggering . . . 373
7.9.4 Analog Input . . . 374
7.9.5 Analog Output . . . 375
7.9.6 Digital Output . . . 376
7.9.7 Counters and Timers . . . 378
7.10 DAQ Software . . . 378

8 Data Transmission Concepts . . . 381
8.1 Introduction . . . 381
8.2 Pulse Codes . . . 382
8.2.1 RZ and RB Recording . . . 382
8.2.2 NRZ Recording . . . 385
8.2.3 Phase Encoding . . . 388
8.3 Analog and Digital Modulation Techniques . . . 390
8.3.1 Amplitude Modulation . . . 392
8.3.2 Frequency Modulation (FM) . . . 394
8.3.3 Phase Modulation . . . 396
8.3.4 Need for Digital Modulation . . . 397
8.3.5 Digital Modulation and their Types . . . 398
8.3.6 Applications of Digital Modulation . . . 401
8.4 Wireless Communication . . . 401
8.4.1 Background . . . 402
8.4.2 Wireless Data . . . 403

8.4.3 Trends in Wireless Communication ... 404
8.4.4 Software Defined Radio ... 405
8.5 RF Network Analyser ... 407
8.6 Distributed Automation and Control Systems ... 413
8.6.1 Distributed Control Systems ... 413
8.6.2 Computers in Industrial Control ... 414
8.6.3 Applications of Computers in Process Industry ... 415
8.6.4 Direct Digital and Supervisory Control ... 416
8.6.5 Architecture of Distributed Control Systems ... 417
8.6.6 Advantages of Distributed Control Systems ... 420
8.6.7 CORBA-Based Automation Systems ... 422
8.7 SCADA ... 423
8.7.1 Architecture ... 424
8.7.2 Security Concerns ... 430
8.7.3 Analysis of the Vulnerabilities of SCADA Systems ... 431
8.7.4 Security Recommendations ... 433

9 Current Trends in Instrumentation ... 437
9.1 Introduction ... 437
9.2 Fiber-Optic Instrumentation ... 438
9.2.1 Fiber-Optic Sensors ... 438
9.2.2 Fiber-Optic Pressure Sensors ... 441
9.2.3 Fiber-Optic Voltage Sensor ... 442
9.2.4 Fiber-Optic Liquid Level Monitoring ... 444
9.2.5 Optical Fiber Temperature Sensors ... 447
9.2.6 Fiber-Optic Stress Sensor ... 449
9.2.7 Fiber-Optic Gyroscope: Polarization Maintaining ... 456
9.2.8 Gratings in Fiber ... 462
9.2.9 Advantages of Fiber Optics ... 464
9.3 Laser Instrumentation ... 465
9.3.1 Measurement of Velocity, Distance, and Length ... 465
9.3.2 LASER Heating, Welding, Melting, and Trimming ... 474
9.3.3 Laser Trimming and Melting ... 480
9.4 Smart Instruments ... 483
9.4.1 Smart Intelligent Transducers ... 483
9.4.2 Smart Transmitter with HART Communicator ... 491
9.5 Computer-Aided Software Engineering ... 495
9.5.1 The TEXspecTool for Computer-Aided Software Engineering ... 498

10 VI Applications: Part I ... 507
10.1 Fiber-Optic Component Inspection Using Integrated Vision and Motion Components ... 507
10.1.1 Fiber Basics ... 508
10.1.2 Fiber-Optic Inspection Platform Overview ... 509

10.1.3 Inspection Measurements . . . 509
10.1.4 Optical Inspection Overview . . . 509
10.1.5 Real Measurements . . . 509
10.1.6 IMAQ Vision Functions . . . 510
10.1.7 Motion Control . . . 512
10.2 Data Acquisition and User Interface of Beam Instrumentation System at SRRC . . . 514
10.2.1 Introduction to SRRC . . . 514
10.2.2 Outline of the Control and Beam Instrumentation System . . . 514
10.2.3 Specific Applications . . . 515
10.3 VISCP: A Virtual Instrumentation and CAD Tool for Electronic Engineering Learning . . . 519
10.3.1 Schematic Capture Program . . . 520
10.3.2 Netlist Generation Tool: Simulation . . . 521
10.3.3 Virtual Instrumentation . . . 522
10.3.4 User Interface . . . 523
10.3.5 Available Virtual Instruments . . . 524
10.3.6 Hardware . . . 525
10.4 Distributed Multiplatform Control System with LabVIEW . . 526
10.4.1 Overview . . . 526
10.4.2 The Software Structure . . . 527
10.4.3 Software Portability . . . 528
10.4.4 The New ODCS with the LabVIEW VI Server ODCS on Unix . . . 528
10.5 The Virtual Instrument Control System . . . 530
10.5.1 Introduction . . . 531
10.5.2 System Structure . . . 531
10.5.3 Applications . . . 533
10.6 Controller Design Using the Maple Professional Math Toolbox for LabVIEW . . . 536
10.6.1 The Two Tank System . . . 537
10.6.2 Controller Parameter Tuning . . . 539
10.6.3 Deployment of the Controller Parameters . . . 541
10.7 Embedding Remote Experimentation in Power Engineering Education . . . 542
10.7.1 Virtual Laboratories in Power Engineering . . . 543
10.7.2 Remote Experiments Over the Internet . . . 544
10.8 Design of an Automatic System for the Electrical Quality Assurance during the Assembly of the Electrical Circuits of the LHC . . . 549
10.8.1 Methodology of Verification . . . 550
10.8.2 Technical Design . . . 552
10.9 Internet-Ready Power Network Analyzer for Power Quality Measurements and Monitoring . . . 555

10.9.1 Computer-Based Power Analyzer . . . 556
10.9.2 Instruments Implemented in the Analyzer . . . 556
10.9.3 Measured Data Analysis . . . 559
10.9.4 Supervising Module . . . 559
10.9.5 Hardware Platforms for the Virtual Analyzer . . . 560
10.9.6 Advantages of the Virtual Analyzer . . . 560
10.9.7 Future Vision . . . 561
10.10 Application of Virtual Instrumentation in a Power Engineering Laboratory . . . 561
10.10.1 Lab Capabilities . . . 561
10.10.2 Single and Three phase Transformers . . . 562
10.10.3 DC Generator Characteristics . . . 565
10.10.4 Synchronous Machine . . . 567
10.10.5 Induction Machine . . . 569

11 VI Applications: Part II . . . 571
11.1 Implementation of a Virtual Factory Communication System Using the Manufacturing Message Specification Standard . . . 571
11.1.1 MMS on Top of TCP/IP . . . 572
11.1.2 Virtual Factory Communication System . . . 574
11.1.3 MMS Internet Monitoring System . . . 578
11.2 Developing Remote Front Panel LabVIEW Applications . . . 580
11.2.1 Reducing the Amount of Data Sent . . . 581
11.2.2 Reducing the Update Rate of the Data . . . 581
11.2.3 Minimizing the Amount of Advanced Communication . . . 583
11.2.4 Functionality to Avoid with Web Applications . . . 584
11.2.5 Security Issues . . . 585
11.3 Using the Timed Loop to Write Multirate Applications in LabVIEW . . . 586
11.3.1 Timed Loops . . . 587
11.3.2 Configuring Timed Loops . . . 588
11.3.3 Selecting Timing Sources . . . 588
11.3.4 Setting the Period and the Offset . . . 588
11.3.5 Setting Priorities . . . 589
11.3.6 Naming Timed Loops . . . 590
11.3.7 Timed Loop Modes . . . 591
11.3.8 Configuring Modes Using the Loop Configuration Dialog Box . . . 592
11.3.9 Configuring Modes Using the Input Node . . . 592
11.3.10 Changing Timed Loop Input Node Values Dynamically. . . . 593
11.3.11 Aborting a Timed Loop Execution . . . 593
11.3.12 Synchronizing Timed Loops . . . 594
11.3.13 Timed Loop Execution Overview . . . 595

11.4 Client–Server Applications in LabVIEW 595
11.4.1 Interprocess Communication 596
11.4.2 A Simple Read-Only Server 597
11.4.3 Two Way Communication: A Read–Write Server 598
11.4.4 The VI-Reference Server Process 600
11.4.5 The VI-Reference Client 600
11.4.6 Further Thoughts 601
11.5 Web-Based Matlab and Controller Design Learning with LabVIEW 601
11.5.1 Introduction to Web-Based MATLAB 602
11.5.2 Learning of MATLAB 602
11.5.3 Learning of Controller Design 603
11.6 Neural Networks for Measurement and Instrumentation in Virtual Environments 605
11.6.1 Modeling Natural Objects, Processes, and Behaviors for Real-Time Virtual Environment Applications 607
11.6.2 Hardware NN Architectures for Real-Time Modeling Applications 609
11.6.3 Case Study: NN Modeling of Electromagnetic Radiation for Virtual Prototyping Environments 614
11.7 LabVIEW Interface for School-Network DAQ Card 623
11.7.1 The WALTA LabVIEW Interface 625
11.8 PC and LabVIEW-Based Robot Control System 627
11.8.1 Introduction to Robot Control System 627
11.8.2 The Robot and the Control System 628
11.8.3 PCL-832 Servomotor Control Card 629
11.8.4 Digital Differential Analysis (DDA) 629
11.8.5 Closed-Loop Position Control of the Control Card . . 630
11.8.6 Modified Closed-Loop Position Control of the Control Card 631
11.8.7 Programming of the Control Card 631
11.8.8 Optimal Cruising Trajectory Planning Method 633
11.9 Mobile Robot Miniaturization: A Tool for Investigation in Control Algorithms 634
11.9.1 Hardware 635
11.9.2 Software 639
11.9.3 Experimentation Environment 640
11.9.4 Experimentation in Distributed Adaptive Control . . . 644
11.10 A Steady-Hand Robotic System for Microsurgical Augmentation 646
11.10.1 Robotically Assisted Micromanipulation 647
11.10.2 A Robotic System for Steady-Hand Micromanipulation 649
11.10.3 Current Status 654

A LabVIEW Research Projects 657
A.1 An Optical Fibre Sensor Based on Neural Networks for Online Detection of Process Water Contamination 657
A.2 An Intelligent Optical Fibre-Based Sensor System for Monitoring Food Quality 657
A.3 Networking Automatic Test Equipment Environments 658
A.4 Using LabVIEW to Prototype an Industrial-Quality Real-Time Solution for the Titan Outdoor 4WD Mobile Robot Controller 658
A.5 Intelligent Material Handling: Development and Implementation of a Matrix-Based Discrete-Event Controller 659
A.6 Curve Tracer with a Personal Computer and LabVIEW 659
A.7 Secure Two-Way Transfer of Measurement Data 659
A.8 Development of a LabVIEW-Based Test Facility for Standalone PV Systems 660
A.9 Semantic Virtual Environments with Adaptive Multimodal Interfaces 660
A.10 A Dynamic Compilation Framework for Controlling Microprocessor Energy and Performance 661
A.11 A Method to Record, Store, and Analyze Multiple Physiologic Pressure Recordings 661
A.12 Characterization of a Pseudorandom Testing Technique for Analog and Mixed-Signal Built-in Self-Test 662
A.13 Power-Aware Network Swapping for Wireless Palmtop PCs .. 662
A.14 Reducing Jitter in Embedded Systems Employing a Time-Triggered Software Architecture and Dynamic Voltage Scaling 663
A.15 End-to-End Testing for Boards and Systems Using Boundary Scan 663
A.16 An Approach to the Equivalent-Time Sampling Technique for Pulse Transient Measurements 664
A.17 Reactive Types for Dataflow-Oriented Software Architectures 664
A.18 Improving the Steering Efficiency of 1x4096 Opto-VLSI Processor Using Direct Power Measurement Method 665
A.19 Experimental Studies of the 2.4-GHz IMS Wireless Indoor Channel 665
A.20 Virtual Instrument for Condition Monitoring of On-Load Tap Change 665
A.21 Toward Evolvable Hardware Chips: Experiments with a Programmable Transistor Array 665
A.22 Remote Data Acquisition, Control and Analysis Using LabVIEW Front Panel and Real-Time Engine 666

B LabVIEW Tools 667
B.1 DIAdem 667
B.2 Electronics Workbench 667
B.3 DSC Module 668
B.4 Vision Development Module 668
B.5 FPGA 669
B.6 LabWindows/CVI 670
B.7 NI MATRIXx 670
B.8 Measurement Studio 671
B.9 VI Logger 672
B.10 Motion Control 672
B.11 TestStand 672
B.12 SignalExpress 673

C Glossary 675

Bibliography 711

1

Introduction to Virtual Instrumentation

Learning Objectives. On completion of this chapter the reader will have a knowledge on:

- History of Instrumentation Systems
- Evolution of Virtual Instrumentation
- Premature Challenges of VI
- Definition of Virtual Instrumentation
- Architecture of Virtual Instrumentation
- Programming Requirements of VI
- Conventional Virtual Instrumentation
- Distributed Virtual Instrumentation
- Virtual Instruments Versus Traditional Instruments
- Advantages of VI
- Evolution of LabVIEW
- Creating Virtual Instruments using LabVIEW
- Advantages of LabVIEW
- Virtual Instrumentation in the Engineering Process
- Virtual Instruments Beyond the Personal Computer

1.1 Introduction

An instrument is a device designed to collect data from an environment, or from a unit under test, and to display information to a user based on the collected data. Such an instrument may employ a transducer to sense changes in a physical parameter, such as temperature or pressure, and to convert the sensed information into electrical signals, such as voltage or frequency variations. The term instrument may also be defined as a physical software device that performs an analysis on data acquired from another instrument and then outputs the processed data to display or recording devices. This second category of recording instruments may include oscilloscopes, spectrum analyzers, and digital millimeters. The types of source data collected and analyzed by instruments may thus vary widely, including both physical parameters such as temperature, pressure, distance, frequency and amplitudes of light and sound, and also electrical parameters including voltage, current, and frequency.

Virtual instrumentation is an interdisciplinary field that merges sensing, hardware, and software technologies in order to create flexible and sophisticated instruments for control and monitoring applications. The concept of *virtual instrumentation* was born in late 1970s, when microprocessor technology enabled a machine's function to be more easily changed by changing its software. The flexibility is possible as the capabilities of a virtual instrument depend very little on dedicated hardware – commonly, only application-specific signal conditioning module and the analog-to-digital converter used as interface to the external world. Therefore, simple use of computers or specialized onboard processors in instrument control and data acquisition cannot be defined as virtual instrumentation. Increasing number of biomedical applications use virtual instrumentation to improve insights into the underlying nature of complex phenomena and reduce costs of medical equipment and procedures.

In this chapter, we describe basic concepts of virtual instrumentation. In Sect. 2 we give a brief history of virtual instrumentation. The architecture of a virtual instrument along with the definition and contemporary development tools are described in Sect. 3. In Sect. 4 we describe the organization of the distributed virtual instrumentation.

1.2 History of Instrumentation Systems

Historically, instrumentation systems originated in the distant past, with measuring rods, thermometers, and scales. In modern times, instrumentation systems have generally consisted of individual instruments, for example, an electromechanical pressure gauge comprising a sensing transducer wired to signal conditioning circuitry, outputs a processed signal to a display panel and perhaps also to a line recorder, in which a trace of changing conditions is linked onto a rotating drum by a mechanical arm, creating a time record of pressure changes. Complex systems such as chemical process control applications employed until the 1980s consisted of sets of individual physical instruments wired to a central control panel that comprised an array of physical data display devices such as dials and counters, together with sets of switches, knobs, and buttons for controlling the instruments.

A history of virtual instrumentation is characterized by continuous increase of flexibility and scalability of measurement equipment. Starting from first manual-controlled vendor-defined electrical instruments, the instrumentation field has made a great progress toward contemporary computer-controlled, user-defined, sophisticated measuring equipment. Instrumentation had the following phases:

- Analog measurement devices
- Data acquisition and processing devices

– Digital processing based on general purpose computing platform
– Distributed virtual instrumentation

The first phase is represented by early "pure" analog measurement devices, such as oscilloscopes or EEG recording systems. They were completely closed dedicated systems, which included power suppliers, sensors, translators, and displays. They required manual settings, presenting results on various counters, gauges, CRT displays, or on the paper. Further use of data was not part of the instrument package, and an operator had to physically copy data to a paper notebook or a data sheet. Performing complex or automated test procedures was rather complicated or impossible, as everything had to be set manually.

Second phase started in 1950s, as a result of demands from the industrial control field. Instruments incorporated rudiment control systems, with relays, rate detectors, and integrators. That led to creation of proportional–integral–derivative (PID) control systems, which allowed greater flexibility of test procedures and automation of some phases of measuring process. Instruments started to digitalize measured signals, allowing digital processing of data, and introducing more complex control or analytical decisions. However, real-time digital processing requirements were too high for any but an onboard special purpose computer or digital signal processor (DSP). The instruments still were standalone vendor defined boxes.

In the third phase, measuring instruments became computer based. They began to include interfaces that enabled communication between the instrument and the computer. This relationship started with the general-purpose interface bus (GPIB) originated in 1960s by Hewlett-Packard (HP), then called HPIB, for purpose of instrument control by HP computers. Initially, computers were primarily used as off-line instruments. They were further processing the data after first recording the measurements on disk or type. As the speed and capabilities of general-purpose computers advanced exponentially general-purpose computers became fast enough for complex real-time measurements. It soon became possible to adapt standard, by now high-speed computers, to the online applications required in real-time measurement and control. New general-purpose computers from most manufactures incorporated all the hardware and much of the general software required by the instruments for their specific purposes. The main advantages of standard personal computers are low price driven by the large market, availability, and standardization. Although computers' performance soon became high enough, computers were still not easy to use for experimentalists.

Nearly all of the early instrument control programs were written in BASIC, because it had been the dominant language used with dedicated instrument controllers. It required engineers and other users to become programmers before becoming instrument users, so it was hard for them to exploit potential that computerized instrumentation could bring. Therefore, an important milestone in the history of virtual instrumentation happened in 1986, when

National Instruments introduced LabVIEW 1.0 on a PC platform. LabVIEW introduced graphical user interfaces and visual programming into computerized instrumentation, joining simplicity of a user interface operation with increased capabilities of computers. Today, the PC is the platform on which most measurements are made, and the graphical user interface has made measurements user-friendlier. As a result, virtual instrumentation made possible decrease in price of an instrument. As the virtual instrument depends very little on dedicated hardware, a customer could now use his own computer, while an instrument manufactures could supply only what the user could not get in the general market.

The fourth phase became feasible with the development of local and global networks of general purpose computers. Since most instruments were already computerized, advances in telecommunications and network technologies made possible physical distribution of virtual instrument components into telemedical systems to provide medical information and services at a distance. Possible infrastructure for distributed virtual instrumentation includes the Internet, private networks, and cellular networks, where the interface between the components can be balanced for price and performance.

The introduction of computers into the field of instrumentation began as a way to couple an individual instrument, such as a pressure sensor, to a computer, and enable the display of measurement data on virtual instrument panel on the computer screen using appropriate software. The instrument also contained buttons for controlling the operation of the sensor. Thus, such instrumentation software enabled the creation of a simulated physical instrument, having the capability to control physical sensing components.

1.3 Evolution of Virtual Instrumentation

Virtual instrumentation combines mainstream commercial technologies, such as the PC, with flexible software and a wide variety of measurement and control hardware, so engineers and scientists can create user-defined systems that meet their exact application needs. With virtual instrumentation, engineers and scientists reduce development time, design higher quality products, and lower their design costs. A large variety of data collection instruments designed specifically for computerized control and operation were developed and made available on the commercial market, creating the field now called "virtual instrumentation."

Thus, virtual instrumentation refers to the use of general purpose computers and workstations, in combination with data collection hardware devices and virtual instrumentation software, to construct an integrated instrumentation system. In such a system, the data collection hardware devices are used to incorporate sensing elements for detecting changes in the conditions of test subjects. These hardware devices are intimately coupled to the computer, whereby the operations of the sensors are controlled by the computer software and the output of the data collection devices are displayed on the computer screen with the use of displays simulating in appearance of the physical dials, meters, and other data visualization devices of traditional instruments. Virtual instrumentation systems also comprise pure software "instruments," such as oscilloscopes and spectrum analyzers, for processing the collected sensor data and "messaging" it such that the users can make full use of the data.

1.4 Premature Challenges

Virtual instrumentation is necessary because it delivers instrumentation with the rapid adaptability required for today's concept, product, and process design, development, and delivery. Only with virtual instrumentation can engineers and scientists create the user-defined instruments required to keep up with the world's demands.

The early development of virtual instrumentation systems faced challenging and technical difficulties. Major obstacles included many types of electronic interfaces by which external data collection devices can be coupled to a computer, and a variety of "command sets" used by different hardware device vendors to control their respective products. Also, data collecting hardware devices differ in their internal structures and functions, enabling virtual instrumentation systems to take these differences into account.

To meet the ever-increasing demand to innovate and deliver ideas and products faster, scientists and engineers are turning to advanced electronics, processors, and software. Consider a modern cell phone. Most contain the latest features of the last generation, including audio, a phone book, and text messaging capabilities. New versions include a camera, MP3 player, and Bluetooth networking, and Internet browsing.

Some data acquisition devices are so-called "register-based" instruments since they are controlled by streams of 1s and 0s sent directly to control the components within the instruments. Other devices include "message-based" instruments which are controlled by "strings" of ASCII characters, effectively constituting written instructions that must be decoded within the instrument. In turn, different instruments use different protocols to output data, some as electrical frequencies and others as variations in a base voltage, etc. Thus, any virtual instrumentation system intended for connection to a typical variety of commercially available data collection hardware devices must accordingly comprise software tools capable of communicating effectively with the disparate types of hardware devices.

1.5 Virtual Instrumentation

Virtual instrumentation achieved mainstream adoption by providing a new model for building measurement and automation systems. Keys to its success include rapid PC advancement; explosive low-cost, high-performance data converter (semiconductor) development; and system design software emergence. These factors make virtual instrumentation systems accessible to a very broad base of users.

1.5.1 Definition

A virtual instrumentation system is a software that is used by the user to develop a computerized test and measurement system, for controlling an external measurement hardware device from a desktop computer, and for displaying test or measurement data on panels in the computer screen. The test and measurement data are collected by the external device interfaced with the desktop computer. Virtual instrumentation also extends to computerized systems for controlling processes based on the data collected and processed by a PC based instrumentation system.

There are several definitions of a virtual instrument available in the open literature. Santori defines a virtual instrument as "an instrument whose general function and capabilities are determined in software." Goldberg describes that "a virtual instrument is composed of some specialized subunits, some general-purpose computers, some software, and a little know-how". Although informal, these definition capture the basic idea of virtual instrumentation and virtual concepts in general – provided with sufficient resources, "any computer can simulate any other if we simply load it with software simulating the other computer." This universality introduces one of the basic properties of a virtual instrument – its ability to change form through software, enabling a user to modify its function at will to suit a wide range of applications.

1.5.2 Architecture of Virtual Instrumentation

A virtual instrument is composed of the following blocks:

- Sensor module
- Sensor interface
- Information systems interface
- Processing module
- Database interface
- User interface

Figure 1.1 shows the general architecture of a virtual instrument. The sensor module detects physical signal and transforms it into electrical form, conditions the signal, and transforms it into a digital form for further manipulation. Through a sensor interface, the sensor module communicates with a computer. Once the data are in a digital form on a computer, they can be processed, mixed, compared, and otherwise manipulated, or stored in a database. Then, the data may be displayed, or converted back to analog form for further process control. Virtual instruments are often integrated with some other information systems. In this way, the configuration settings and the

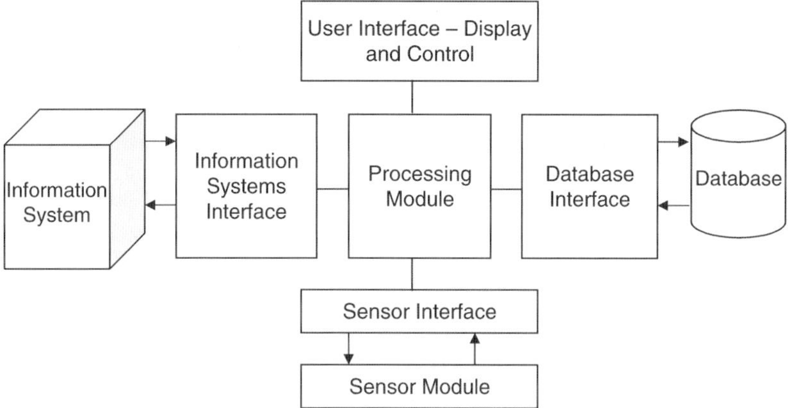

Fig. 1.1. Architecture of a virtual instrument

data measured may be stored and associated with the available records. In following sections each of the virtual instruments modules are described in more detail.

Sensor Module

The sensor module performs signal conditioning and transforms it into a digital form for further manipulation. Once the data are in a digital form on a computer, they can be displayed, processed, mixed, compared, stored in a database, or converted back to analog form for further process control. The database can also store configuration settings and signal records. The sensor module interfaces a virtual instrument to the external, mostly analog world transforming measured signals into computer readable form. A sensor module principally consists of three main parts:

- The sensor
- The signal conditioning part
- The A/D converter

The sensor detects physical signals from the environment. If the parameter being measured is not electrical, the sensor must include a transducer to convert the information to an electrical signal, for example, when measuring blood pressure.

The signal-conditioning module performs (usually analog) signal conditioning prior to AD conversion. This module usually does the amplification, transducer excitation, linearization, isolation, or filtering of detected signals. The A/D converter changes the detected and conditioned voltage into a digital value. The converter is defined by its resolution and sampling frequency. The converted data must be precisely time-stamped to allow later sophisticated analyses. Although most biomedical sensors are specialized in processing of certain signals, it is possible to use generic measurement components, such as data acquisition (DAQ), or image acquisition (IMAQ) boards, which may be applied to broader class of signals. Creating generic measuring board, and incorporating the most important components of different sensors into one unit, it is possible to perform the functions of many medical instruments on the same computer.

Sensor Interface

There are many interfaces used for communication between sensors modules and the computer. According to the type of connection, sensor interfaces can be classified as *wired* and *wireless.*

- *Wired Interfaces* are usually standard parallel interfaces, such as GPIB, Small Computer Systems Interface (SCSI), system buses (PCI eXtension for Instrumentation PXI or VME Extensions for Instrumentation (VXI), or serial buses (RS232 or USB interfaces).
- *Wireless Interfaces* are increasingly used because of convenience. Typical interfaces include 802.11 family of standards, Bluetooth, or GPRS/GSM interface. Wireless communication is especially important for implanted sensors where cable connection is impractical or not possible. In addition, standards, such as Bluetooth, define a self-identification protocol, allowing the network to configure dynamically and describe itself. In this way, it is possible to reduce installation cost and create plug-and-play like networks of sensors. Device miniaturization allowed development of Personal Area Networks (PANs) of intelligent sensors Communication with medical devices is also standardized with the IEEE 1073 family of standards. This interface is intended to be highly robust in an environment where devices are frequently connected to and disconnected from the network.

Processing Module

Integration of the general purpose microprocessors/microcontrollers allowed flexible implementation of sophisticated processing functions. As the functionality of a virtual instrument depends very little on dedicated hardware, which principally does not perform any complex processing, functionality and appearance of the virtual instrument may be completely changed utilizing different processing functions. Broadly speaking, processing function used in virtual instrumentation may be classified as *analytic processing* and *artificial intelligence techniques.*

- *Analytic processing.* Analytic functions define clear functional relations among input parameters. Some of the common analyses used in virtual instrumentation include spectral analysis, filtering, windowing, transforms, peak detection, or curve fitting. Virtual instruments often use various statistics function, such as, random assignment and biostatistical analyses. Most of those functions can nowadays be performed in real-time.

– *Artificial intelligence techniques.* Artificial intelligence technologies could be used to enhance and improve the efficiency, the capability, and the features of instrumentation in application areas related to measurement, system identification, and control. These techniques exploit the advanced computational capabilities of modern computing systems to manipulate the sampled input signals and extract the desired measurements. Artificial intelligence technologies, such as neural networks, fuzzy logic and expert systems, are applied in various applications, including sensor fusion to high-level sensors, system identification, prediction, system control, complex measurement procedures, calibration, and instrument fault detection and isolation. Various nonlinear signal processing, including fuzzy logic and neural networks, are also common tools in analysis of biomedical signals. Using artificial intelligence it is even possible to add medical intelligence to ordinary user interface devices. For example, several artificial intelligence techniques, such as pattern recognition and machine learning, were used in a software-based visual-field testing system.

Database Interface

Computerized instrumentation allows measured data to be stored for off-line processing, or to keep records as a part of the patient record. There are several currently available database technologies that can be used for this purpose. Simple usage of file systems interface leads to creation of many proprietary formats, so the interoperability may be a problem. The eXtensible Markup Language (XML) may be used to solve interoperability problem by providing universal syntax. The XML is a standard for describing document structure and content. It organizes data using markup tags, creating self-describing documents, as tags describe the information it contains. Contemporary database management systems such SQL Server and Oracle support XML import and export of data. Many virtual instruments use DataBase Management Systems (DBMSs). They provide efficient management of data and standardized insertion, update, deletion, and selection. Most of these DBMSs provided Structured Query Language (SQL) interface, enabling transparent execution of the same programs over database from different vendors. Virtual instruments use these DMBSs using some of programming interfaces, such as ODBC, JDBC, ADO, and DAO.

Information System Interface

Virtual instruments are increasingly integrated with other medical information systems, such as hospital information systems. They can be used to create executive dashboards, supporting decision support, real time alerts, and predictive warnings. Some virtual interfaces toolkits, such as LabVIEW, provide mechanisms for customized components, such as ActiveX objects, that allows communication with other information system, hiding details of the communication from virtual interface code. In Web based applications this integration is usually implemented using Unified Resource Locators (URLs). Each virtual instrument is identified with its URL, receiving configuration settings via parameters. The virtual instrument then can store the results of the processing into a database identified with its URL.

1.5.3 Presentation and Control

An effective user interface for presentation and control of a virtual instrument affects efficiency and precision of an operator do the measurements and facilitates result interpretation. Since computer's user interfaces are much easier shaped and changed than conventional instrument's user interfaces, it is possible to employ more presentation effects and to customize the interface for each user. According to presentation and interaction capabilities, we can classify interfaces used in virtual instrumentation in four groups:

- Terminal user interfaces
- Graphical user interfaces
- Multimodal user interfaces and
- Virtual and augmented reality interfaces

Terminal User Interfaces

First programs for instrumentation control and data acquisition had character-oriented terminal user interfaces. This was necessary as earlier general-purpose computers were not capable of presenting complex graphics. As terminal user interfaces require little of system resources, they were implemented on many platforms. In this interfaces, communication between a user and a computer is purely textual. The user sends requests to the computer typing commands, and receives response in a form of textual messages. Presentation is usually done on a screen with fixed resolution, for example 25 rows and 80 columns on ordinary PC, where each cell presents one of the characters from a fixed character set, such as the ASCII set. Additional effects, such as text and background color or blinking, are possible on most terminal user interfaces. Even with the limited set of characters, more sophisticated effects in a form of character graphics are possible. Although terminal user interfaces are not any more widely use on desktop PCs, they have again become important in a wide range of new pervasive devices, such as cellular phones or low-end personal digital assistants (PDAs). As textual services, such as Short Message System (SMS), require small presentation and network resources they are broadly supported and available on almost all cellular phone devices. These services may be very important in distributed virtual instrumentation, and for emergency alerts.

Graphical User Interfaces

Graphical user interfaces (GUIs) enabled more intuitive human–computer interaction, making virtual instrumentation more accessible. Simplicity of interaction and high intuitiveness of graphical user interface operations made possible creation of user-friendlier virtual instruments. GUIs allowed creation of many sophisticated graphical widgets such as graphs, charts, tables, gauges, or meters, which can easily be created with many user interface tools. In addition, improvements in presentation capabilities of personal computers allowed for development of various sophisticated 2D and 3D medical imaging technologies. Computer graphics extended the functionality of conventional medical diagnostic imaging in many ways, for example, by adding the visual tool of color. For instance, interpretation of radiographs, which are black and white images, requires lots of training, but, with color, it is possible to highlight problems clearly. In addition, improvements in presentation capabilities of personal computers allowed for development of various sophisticated 2D and 3D medical imaging technologies.

Multimodal Presentation

In addition to graphical user interfaces that improve visualization, contemporary personal computers are capable of presenting other modalities such as sonification or haptic rendering. Multimodal combinations of complementary modalities can greatly improve the perceptual quality of user interfaces. Sonification is the second most important presentation modality. Relationship between visualization and sonification is itself a complex design problem, due to the nature of the cognitive information processing. Efficiency of sonification, as acoustic presentation modality, depends on other presentation modalities. Sonification is effectively used in various biomedical applications, for example, in virtual instruments for EEG analysis.

Virtual and Augmented Reality

Virtual environments will most likely pervade the medical practice of the future. Many of the goals of virtual reality technology developers actually mirror those involved in virtual instrumentation work. Although virtual reality systems do not necessarily involve the use of virtual instrumentation, they nonetheless drive the development of new conditions under which physicians will need access to data in radically different forms. A combination of virtual presentation with real world objects creates augment reality interfaces.

1.5.4 Functional Integration

Functional integration of modules governs flexibility of a virtual instrument. The simplest, and the least flexible way, is to create a virtual instrument as a single, monolithic application with all software modules of the virtual instruments logically and physically integrated. This approach can achieve the best performance, but makes difficult maintenance and customization. Therefore, it is more convenient to use modular organization. An object-oriented method was identified as natural approach in modeling and design of instruments. Each module of a virtual instrument is then implemented as an object with clearly defined interface, integrated with other objects using message interchange. Similar approach is component-oriented approach, where, in addition to logical separation of components into objects, they are physically placed into different unit to allow reuse. Another approach, similar in its basic idea to the object-oriented approach, is a structural coupling paradigm for nonconventional controllers that define the layered approach to functional integration of sensor modules. This sensor model was applied in many domains, including electrophysiological interaction systems with sensors for human physiological signals. In this sensor interaction model, a stream of raw data from the sensing hardware, for example electroencephalogram (EEG) data, passes through up to two levels of signal preprocessing before it is either passed to an application or presented directly to a subject. Second command layer, which is optional, allows more flexible organization of data processing and plug-and-play like integration of complex processing mechanisms into a virtual instrument solution.

1.6 Programming Requirements

Until the 1990s, the programming of virtual instrumentation systems was a task strictly for professional programmers, who wrote the required software programs using "textual" programming languages such as BASIC, C + +, or PASCAL. In addition to the factors previously mentioned, the great variety in possible applications also called for professional expertise, in that one of customer's measurement applications was rarely suitable for another customer. For example, we customers may have needed only to collect a single value, outside temperature, once an hour, and to have all collected values stored in a file. The next customer possibly required that several related process temperatures, in a rubber-curing process, be monitored continuously, and that a shut-off valve be activated in the event that the relative temperature between two process steps should vary by more than 7° for a time period of 3 s or more, in any 15-min period, and to store only data concerning such episodes.

The development of instrumentation systems software by professionals using textual programming languages such as C++ is very time consuming and tedious, and it typically results in the production of a program consisting of many pages of source code, written in a computer language that is virtually unreadable by nonprogrammers, and which thus cannot be modified by the typical users of such programs.

1.7 Drawbacks of Recent Approaches

In the last ten years, there have appeared several commercial software products for the development of virtual instrumentation systems using purely graphical programming methods. Each of these products provides users, who are not skilled software programmers, with a "graphical development environment" within which a custom virtual instrumentation system is present. A "design desktop" environment is presented which has a look-and-feel environment familiar to users of Windows-based graphical applications, in which a variety of software options and "tools" are accessible from toolbars and dialog boxes featuring drop-down menus, and may be accessed by manipulating an on-screen cursor using the computer mouse.

However, these older software packages used for developing virtual instrumentation systems, using graphical programming means, provide the user with tools for designing so-called "data flow" diagrams. The user of these software packages is thus required to place icons representing desired system components onto a design desktop and then to do "wiring" connections between components. In order to design a dataflow diagram that corresponds to a workable measurement system application, the user is required to have comparably deep knowledge and understanding of the specific data paths and element combinations that will be required to attain the user's objective, which is a "solution" to the user's measurement requirements. User designed systems developed using this type of software are also prone to errors, because they generally allow the user to unwittingly wire together components that are functionally incompatible.

Wide range of applications in the subject of an instrumentation system spans the range of human activity. Therefore, a software development system that aims to provide a large cross-section of potential users with the tools to design their own customized instrumentation system must provide the user with a wide range of development tools and options which may or may not be mutually incompatible in a given application. Ideally, such a system should also organize the software tools provided to the user in a way that enables the user to select the particular tools best suited for the pertinent application. Older software packages lack built-in safeguards against the construction of unworkable combinations of components, and provide inadequate intuitive guidance to the user seeking to develop a measurement application.

1.8 Conventional Virtual Instrumentation

The rapid adoption of the PC in the last 20 years catalyzed a revolution in instrumentation for test, measurement, and automation. One major development resulting from the ubiquity of the PC is the concept of virtual instrumentation, which offers several benefits to engineers and scientists who require increased productivity, accuracy, and performance. A virtual instrument consists of an industry-standard computer or workstation equipped with powerful application software, cost-effective hardware such as plug-in boards, and driver software, which together perform the functions of traditional instruments. Virtual instruments represent a fundamental shift from traditional hardware-centered instrumentation systems to software-centered systems that exploit the computing power, productivity, display, and connectivity capabilities of popular desktop computers and workstations.

Although the PC and integrated circuit technology have experienced significant advances in the last two decades, it is software that truly provides the leverage to build on this powerful hardware foundation to create virtual instruments, providing better ways to innovate, and significantly reduces cost. With virtual instruments, engineers and scientists build measurement and automation systems that suit their needs exactly (user-defined) instead of being limited by traditional fixed-function instruments (vendor-defined). This section describes powerful programming tools, flexible acquisition hardware, and the personal computer, which are the essential components for virtual instrumentation. The synergy between them offers advantages that cannot be matched by traditional instrumentation.

1.9 Distributed Virtual Instrumentation

Advances in telecommunications and network technologies made possible physical distribution of virtual instrument components into telemedical systems to provide medical information and services at a distance. Distributed virtual instruments are naturally integrated into telemedical systems. Figure 1.2 illustrates possible infrastructure for distributed virtual instrumentation, where the interface between the components can be balanced for price and performance.

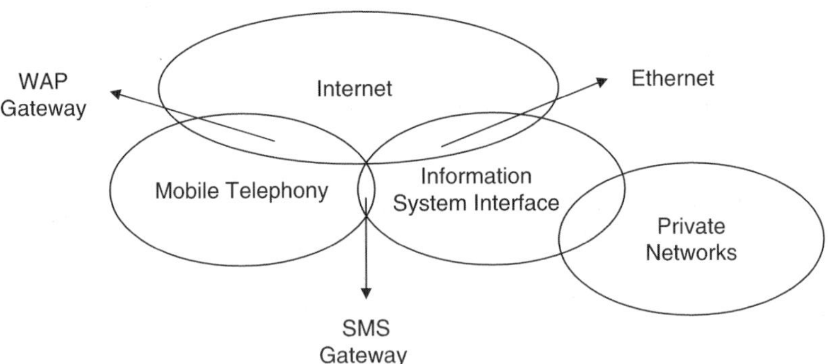

Fig. 1.2. Infrastructure of Distributed Virtual Instrumentation

Information System Networks and Private Networks

Information systems are usually integrated as Intranets using Local Area Network (LAN). Historically, information systems were firstly interconnected using various private networks, starting form point-to-point communication with fax and modems connected to analog telephone lines operating at speeds up to 56 kbps, ISDN lines of up to 128 kbps, T-1 lines having a capacity of 1.544 Mbps, and satellite links of 100 Mbps. Advanced virtual instrumentation solutions could be implemented using existing local and private networks. For example, the EVAC project demonstrated a prototype system for using virtual environments to control remote instrumentation, illustrating the potential of a virtual laboratory over high speed networks. Although private networks improve the performance, reliability and security, they are usually very expensive to develop and maintain.

The Internet

The Internet has enormous potential for distributed virtual instrumentation. Various remote devices, such as telerobots or remote experimental apparatus, can be directly controlled from the Internet. There are a great number of research activities that explore how the Internet can be applied to medicine. In addition, many of virtual instrumentation development tools, such as LabVIEW, directly support integration of virtual instruments in the Internet environment. The Web technologies make possible creation of sophisticated client-server applications on various platforms, using interoperable technologies such as HTML, Java Applets, Virtual Reality Modeling Language, and multimedia support. Although the Internet is already enabling technology for many biomedical applications, a recent United States study of healthcare applications in relation to Internet capabilities found clear requirements for the Internet's evolutionary development. However, the rapid progress of the Internet will probably very soon enable its usage in complex real-time applications.

Cellular Networks

Various mobile devices, such as mobile phones or PDAs, are commonplace today. Moreover, the underlying telecommunication infrastructure of these devices, primarily cellular networks, provides sophisticated data services that can be exploited for distributed applications. The most common data service on cellular networks is exchange of simple textual message. Most of mobile phone devices support simple data communication using standard SMS. Although simple, this system allows the various modes of communication such as:

– Wireless Access Protocol (WAP) is platform-independent wireless technology, which enables mobile devices to effectively access Internet content

and services, as well as to communicate with each other. WAP manages communication by exchanging messages written in Wireless Markup Language (WML). The WAP and the Internet can support new kinds of applications, such as remote monitoring using a wireless personal monitor and cellular phone link connected on request in the case of medical emergencies. The interface allows the following modes of communications:
  - Emergency WAP push, which sends WML messages to physicians or medical call enter in case of medical emergency
  - WML browsing, allows a participant to browse through information in medical information systems or in monitoring system
  - Data distribution WAP, which periodically sends messages to physicians. These data could be simple text or some 2D graphics with wireless bitmap (WBMP).

Distributed Integration

When the components are distributed, efficient communication mechanisms are needed. According to a conceptual model and abstractions they utilize, we can identify four approaches to distributed communication:

- Message passing systems
- Remote procedure calling (RPC) systems
- Distributed object systems, and
- Agent-based systems.

The message passing model allows communication between programs by exchange of messages or packets over the network. It supports a variety of communication patterns, such as pier-to-pier, group, broadcast and collective communication. For example, in virtual instrumentation application the data acquisition part could be a server for other units, sending messages with measured data at request or periodically to processing clients. Data processing clients may themselves be servers for data presentation devices. In a distributed environment there may be many interconnected servers and clients, each dedicated to one of the virtual instrument functions. Remote procedure call (RPC) is an abstraction on top of message passing architectures. RPC brings procedural programming paradigm to network programming, adding the abstraction of the function call to distributed systems. In RPC, communication between programs is accomplished by calling a function on that other computer's machine creating the illusion that communication with a remote program is not different than communication with a local program. Distributed object systems extend the idea of RPC with the object-oriented abstraction on top of procedure calls. Distributed object systems supply programs with references to remote objects, allowing the program to control, call methods, and store the remote object in the same way as a local object. The major standard in distributed objects is OMG CORBA, a language-neutral specification for communicating object systems. Many standards have been defined on top of CORBA, such as *CORBAmed* that defines standardized interfaces for healthcare objects. Competitors to CORBA include Microsoft's DCOM architecture and the various distributed object systems layered on top of Java. Agent based integration is potentially very effective distributed virtual instrument integration mechanism. Agent based systems add concepts of autonomity and proactivity to distributed object systems. Agent-oriented approach is well suited for developing complex, distributed systems. Agents can react asynchronously and autonomously to unexpected situations, increasing robustness and fault-tolerance that is very important in the case of fragile network connections, and for mobile devices. As an example of an agent-based distributed integration, we can present a Virtual Medical Device (VMD) agent framework with four types of agents: *data agents*, *processing agents*, *presentation agents*, and *monitoring agents* for distributed EEG monitoring. In this framework, data agents abstract data source, creating uniform view on different types of data, independent of data acquisition device. Processing agents produce derived data, such us power spectrum from raw data provided by the data agents. Presentation agents supply user interface components using a variety of user data views. User interface components are based on HTTP, SMS, and WAP protocols. Monitoring agents collaborate with data and processing agents providing support for data mining operations, and search for relevant patterns.

1.10 Virtual Instruments Versus Traditional Instruments

Stand-alone traditional instruments such as oscilloscopes and waveform generators are very powerful, expensive, and designed to perform one or more specific tasks defined by the vendor. However, the user generally cannot extend or customize them. The knobs and buttons on the instrument, the built-in circuitry, and the functions available to the user, are all specific to the nature of the instrument. In addition, special technology and costly components must be developed to build these instruments, making them very expensive and slow to adapt. Table 1.1 compares the traditional instruments with the virtual instruments based on different aspects.

Virtual instruments, by virtue of being PC-based, inherently take advantage of the benefits from the latest technology incorporated into off-the-shelf PCs. These advances in technology and performance, which are quickly closing the gap between stand-alone instruments and PCs, include powerful processors such as the Pentium 4 and operating systems and technologies such as Microsoft Windows XP, .NET, and Apple Mac OS X. In addition to incorporating powerful features, these platforms also offer easy access to powerful tools such as the Internet.

Table 1.1. Traditional versus virtual instruments

Traditional instruments	Virtual instruments
Vendor-defined	User-defined
Function-specific, stand-alone with limited connectivity	Application-oriented system with connectivity to networks, peripherals, and applications
Hardware is the key	Software is the key
Expensive	Low-cost, reusable
Closed, fixed functionality	Open, flexible functionality leveraging off familiar computer technology
Slow turn on technology (5–10 year life cycle)	Fast turn on technology (1–2 year life cycle)
Minimal economics of scale	Maximum economics of scale
High development and maintenance costs	Software minimizes development and maintenance costs

1.11 Advantages of VI

The virtual instruments running on notebook automatically incorporate their portable nature to the Engineers and scientists whose needs, applications and requirements change very quickly, need flexibility to create their own solutions. We can adapt a virtual instrument to our particular needs without having to replace the entire device because of the application software installed on the PC and the wide range of available plug-in hardware.

1.11.1 Performance

In terms of performance, LabVIEW includes a compiler that produces native code for the CPU platform. The graphical code is translated into executable machine code by interpreting the syntax and by compilation. The LabVIEW syntax is strictly enforced during the editing process and compiled into the executable machine code when requested to run or upon saving. In the latter case, the executable and the source code are merged into a single file. The executable runs with the help of the LabVIEW run-time engine, which contains some precompiled code to perform common tasks that are defined by the G language. The run-time engine reduces compile time and also provides a consistent interface to various operating systems, graphic systems, hardware components, etc. The run-time environment makes the code portable across platforms. Generally, LV code can be slower than equivalent compiled C code, although the differences often lie more with program optimization than inherent execution speed. Many libraries with a large number of functions for data acquisition, signal generation, mathematics, statistics, signal conditioning, analysis, etc., along with numerous graphical interface elements are provided in several LabVIEW package options. The LabVIEW Professional Development System allows creating stand-alone executables and the resultant executable can be distributed an unlimited number of times. The run-time engine and its libraries can be provided freely along with the executable.

1.11.2 Platform-Independent Nature

A benefit of the LabVIEW environment is the platform-independent nature of the G-code, which is (with the exception of a few platform specific functions) portable between the different LabVIEW systems for different operating systems (Windows, MacOSX, and Linux). National Instruments is increasingly focusing on the capability of deploying LabVIEW code onto an increasing number of targets including devices like Pharlap OS-based LabVIEW real-time controllers, PocketPCs, PDAs, Fieldpoint modules, and into FPGAs on special boards.

1.11.3 Flexibility

Except for the specialized components and circuitry found in traditional instruments, the general architecture of stand-alone instruments is very similar to that of a PC-based virtual instrument. Both require one or more microprocessors, communication ports (for example, serial and GPIB), and display capabilities, as well as data acquisition modules. These devices differ from one another in their flexibility and the fact that these devices can be modified and adapted to the particular needs. A traditional instrument might contain an integrated circuit to perform a particular set of data processing functions; in a virtual instrument, these functions would be performed by software running on the PC processor. We can extend the set of functions easily, limited only by the power of the software used.

1.11.4 Lower Cost

By employing virtual instrumentation solutions, lower capital costs, system development costs, and system maintenance costs are reduced, increasing the time to market and improving the quality of our own products.

1.11.5 Plug-In and Networked Hardware

There is a wide variety of available hardware that can either be plugged into the computer or accessed through a network. These devices offer a wide range of data acquisition capabilities at a significantly lower cost than that of dedicated devices. As integrated circuit technology advances, and off-the-shelf components become cheaper and more powerful, so do the boards that use them. With these advances in technology there arises an increase in data acquisition rates, measurement accuracy, precision, and better signal isolation. Depending on the particular application, the hardware we choose might include analog input or output, digital input or output, counters, timers, filters, simultaneous sampling, and waveform generation capabilities. The wide gamut of boards and hardware could include any one of these features or a combination of them.

1.11.6 The Costs of a Measurement Application

The costs involved in the development of a measurement application can be divided into five distinct areas composed of hardware and software prices and several time costs. The price of the hardware and software was considered as the single largest cost of their most recent test or measurement system. However, the cumulative time costs in the other areas make up the largest portion of the total application cost.

1.11.7 Reducing System Specification Time Cost

Deciding the types of measurements to take and the types of analysis to perform takes time. Once the user has set the measurement specifications, the user must then determine exactly the method to implement the measurement system. The time taken to perform these two steps equals the system specification time.

1.11.8 Lowering the Cost of Hardware and Software

The price of measurement hardware and software is undoubtedly the most visible cost of a data-acquisition system. Many people attempt to save money in this area without considering the effect on the total development cost. The price of hardware and software, on an average, makes up about 36 per cent of the total application cost. However, the time and money spent in other areas of the development process make up the majority of the total application cost. Virtual instrumentation lowers not only time costs but it also helps to minimize hardware and software costs.

1.11.9 Minimising Set-Up and Configuration Time Costs

Once the users have specified and purchased measurement hardware, the real task of developing the application begins. However, the user must first install the hardware and software, configure any necessary setting and ensure that all pieces of the system function properly. This set-up and configuration stage also includes any necessary fixturing, such as sensor connection, wiring and any preparation of the unit under test. Therefore, saving time in set-up and configuration can result in a large reduction in the total application cost. Virtual instrumentation again delivers this cost reduction with a combination of off-the-shelf technology and intelligent software.

1.11.10 Decreasing Application Software Development Time Costs

Developing the measurement application code typically consumes about 30%, which can be sometimes as high as 50% of the overall measurement application cost. Time-saving application development tools can often greatly reduce costs in this stage. NI data-acquisition software and its tight integration with the NI LabVIEW development environment have delivered results for numerous customers. NI-DAQ software saves significant time in the software development stage.

1.12 Evolution of LabVIEW

At the heart of any virtual instrument is flexible software, and National Instruments invented one of the world's best virtual instrumentation software platforms – LabVIEW expanded as *Lab*oratory *V*irtual *I*nstrumentation *E*ngineering *W*orkbench. LabVIEW is a powerful graphical development environment for signal acquisition, measurement analysis, and data presentation, giving the flexibility of a programming language without the complexity of traditional development tools. Since 1986, when National Instruments introduced LabVIEW for the Macintosh, it has quickly and consistently attracted engineers and scientists looking for a productive, powerful programming language to use in test, control and design applications. Today, LabVIEW is the preferred graphical development environment for thousands of engineers and scientists. With the right software tool, engineers and scientists can efficiently create their own applications, by designing and integrating the routines that a particular process requires. They can also create an appropriate user interface that best suits the purpose of the application and those who will interact with it. They can define how and when the application acquires data from the device, how it processes, manipulates and stores the data, and how the results are presented to the user. With powerful software, we can build intelligence and decision-making capabilities into the instrument so that it adapts when measured signals change inadvertently or when more or less processing power is required.

An important advantage that software provides is modularity. When dealing with a large project, engineers and scientists generally approach the task by breaking it down into functional solvable units. These subtasks are more manageable and easier to test, given the reduced dependencies that might cause unexpected behavior. We can design a virtual instrument to solve each of these subtasks, and then join them into a complete system to solve the larger task. The ease with which we can accomplish this division of tasks depends greatly on the underlying architecture of the software.

The premier virtual instrumentation graphical development environment, *LabVIEW*, attracts thousands of engineers and scientists building virtual instruments. By understanding customer project development needs, there has been a consistent improvement in significant software innovations, including Express technology, the LabVIEW Real-Time Module and LabVIEW PDA Module, and SignalExpress:

Express Technology. National Instruments created Express technology for LabVIEW, LabWindow/CVI, and Measurement Studio in 2003 to reduce code complexity while preserving power and functionality. Today, more than 50% of data acquisition customers use DAQ Assistant to simplify data acquisition tasks.

The LabVIEW Real-Time Module and LabVIEW PDA Modules. LabVIEW was extended for deterministic execution using the Real-Time Module and developed matching hardware platforms to make embedded application deployment a reality. The LabVIEW PDA Module extended virtual instrumentation and the LabVIEW platform to handheld devices.

SignalExpress. Design and test engineers required a virtual instrumentation software that interactively measures and analyzes data and hence SignalExpress was created. This tool had a drag-and-drop feature with *no-programming-required* environment ideal for exploratory measurements.

1.13 Creating Virtual Instruments Using LabVIEW

LabVIEW is an integral part of virtual instrumentation because it provides an easy-to-use application development environment designed specifically with the needs of engineers and scientists in mind. LabVIEW offers powerful features that make it easy to connect to a wide variety of hardware and other software. Graphical Programming is one of the most powerful feature that LabVIEW offers engineers and scientists. With LabVIEW, the user can design custom virtual instruments by creating a graphical user interface on the computer screen through which one can:

- Operate the instrumentation program
- Control selected hardware
- Analyze acquired data
- Display results

The user can customize front panels with knobs, buttons, dials, and graphs to emulate control panels of traditional instruments, create custom test panels, or visually represent the control and operation of processes. The similarity between standard flow charts and graphical programs shortens the learning curve associated with traditional text-based languages.

The user can determine the behavior of the virtual instruments by connecting icons together to create block diagram and front panel as shown in Fig 1.3. These diagrams depict natural design notations for scientists and engineers. With graphical programming, the user can develop systems more rapidly

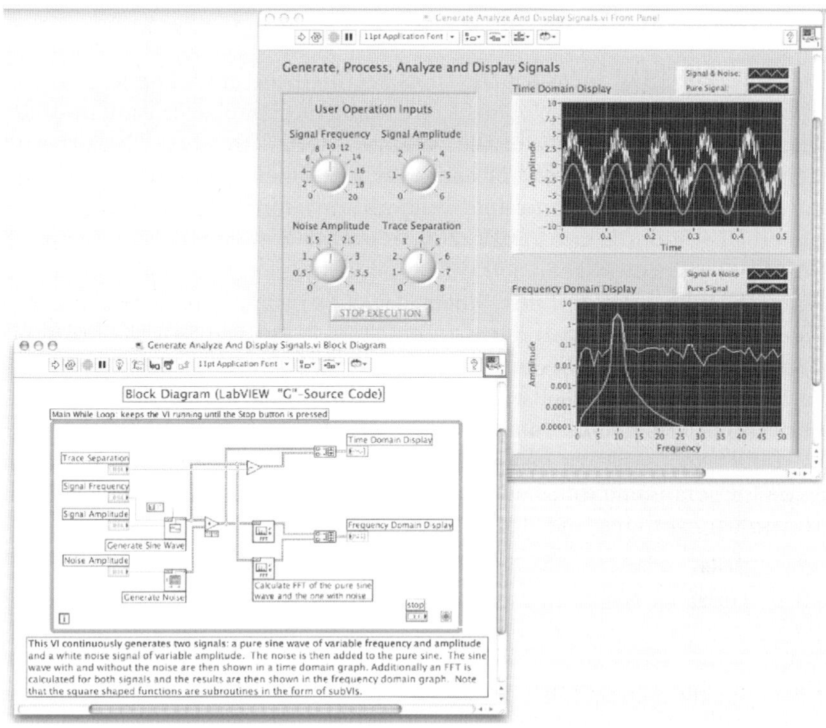

Fig. 1.3. LabVIEW front panel and block diagram

than with conventional programming languages, while retaining the power and flexibility needed to create a variety of applications.

1.13.1 Connectivity and Instrument Control

Virtual instrumentation software productivity has grown in advance since it includes built-in knowledge of hardware integration. This software is mainly designed to create test, measurement, and control systems. The software also includes extensive functionality for I/O of almost any kind. LabVIEW has ready-to-use libraries for integrating stand-alone instruments, data acquisition devices, motion control and vision products, GPIB/IEEE 488, serial/RS-232 devices, and PLCs. These tools are manly used to build a complete measurement and automation system. LabVIEW also incorporates major instrumentation standards such as:

- VISA, an interoperable standard for GPIB, serial, and VXI instrumentation
- PXI and software and hardware based on the PXI Systems Alliance Compact PCI standard
- IVI interchangeable virtual instrument drivers
- VXI plug and play, a driver standard for VXI instruments.

1.13.2 Open Environment

Although LabVIEW provides the tools required for most applications, LabVIEW is also an open development environment. Standardization of software relies greatly on the ability of the package selected to work with other software, measurement and control hardware, and open standards, which define interoperability between multiple vendors. In addition, conforming to open commercial standards LabVIEW reduces overall system cost. A large number of third-party hardware and software vendors develop and maintain hundreds of LabVIEW libraries and instrument drivers to enable the user to easily use their products with LabVIEW. However, this is not the only way to provide connectivity to LabVIEW-based applications. LabVIEW offers simple ways to incorporate ActiveX software, dynamic link libraries (DLLs), and shared libraries from other tools. In addition, the user can share LabVIEW code as a DLL, built executable, or using ActiveX. LabVIEW also offers a full range of options for communications and data standards, such as TCP/IP, OPC, SQL database connectivity, and XML data formats.

1.13.3 Reduces Cost and Preserves Investment

A single computer equipped with LabVIEW can be used for countless applications and purposes, hence it is considered as a versatile product. It is not only versatile but also extremely cost-effective. Virtual instrumentation with LabVIEW proves to be economical, not only in the reduced development costs but also in its preservation of capital investment over a long period of time. When the user needs a change, the system can be modified easily without the need to buy new equipment. The user can also create complete instrumentation libraries for less than the cost of a single traditional, commercial instrument.

1.13.4 Multiple Platforms

The majority of computer systems use some variation of the Microsoft Windows operating system. Nevertheless, other options offer clear advantages for certain types of applications. Real-time and embedded development continues to grow rapidly in most industries, as computing power is packaged into smaller and more specialized packages. Minimizing losses resulting from changing to new platforms is important and choosing the right software for this purpose is a key factor. LabVIEW minimizes this concern, because it runs on Windows 2000, NT, XP, Me, 98, 95, and NT embedded, as well as Mac OS, Sun Solaris, and Linux. LabVIEW also compiles code to run on the VenturCom ETS real-time operating system through the LabVIEW Real-Time Module.

Given the importance of legacy systems, National Instruments continues to make available of the older versions of LabVIEW for Windows, Mac OS, and Sun operating systems. LabVIEW is platform-independent; virtual instruments that are developed in one platform can transparently be ported to any other LabVIEW platform by simply opening the virtual instrument. Because LabVIEW applications are portable across platforms, it is assured that the work completed today will be applicable in the future. As new computer technologies emerge, LabVIEW offers easy migration of applications to new platforms and new operating systems. By creating platform-independent virtual instruments and by porting applications between platforms, the development time and other inconveniences related to platform portability are minimized.

1.13.5 Distributed Development

The user can easily develop distributed applications with LabVIEW across different platforms. With easy-to-use server tools, the user can offload processor-intensive routines to other machines for faster execution, or create remote monitoring and control applications. Powerful server technology can simplify the task of developing large, multicomputer applications. In addition, LabVIEW includes standard networking technologies such as TCP/IP, and incorporates robust publish and subscribe protocols.

1.13.6 Analysis Capabilities

Virtual instrumentation software requires comprehensive analysis and signal processing tools, because the application does not just stop when the data is collected. High-speed measurement applications in machine monitoring and control systems usually require order analysis for accurate vibration data. Closed-loop, embedded control systems might need point-by-point averaging for control algorithms to maintain stability. In addition to the advanced analysis libraries already included in LabVIEW, National Instruments provides add-on software such as the LabVIEW Signal Processing Toolset, the LabVIEW Sound and Vibration Toolkit, and the LabVIEW Order Analysis Toolkit to complement analysis offerings.

1.13.7 Visualization Capabilities

LabVIEW includes a wide array of built-in visualization tools to present data on the user interface of the virtual instrument – for charting and graphing as well as 2D and 3D visualization. We can instantly reconfigure attributes of the data presentation, such as colors, font size, graph types, and more, as well as dynamically rotate, zoom, and pan these graphs with the mouse. Rather than programming graphics and all custom attributes from scratch, we can simply drag-and-drop these objects onto the instrument front panels.

1.13.8 Flexibility and Scalability

Engineers and scientists have needs and requirements that change rapidly. They also need to have maintainable, extensible solutions that can be used for a long time. By creating virtual instruments based on powerful development software such as LabVIEW, we inherently design an open framework that seamlessly integrates software and hardware. This ensures that applications not only work well today but that it can be easily integrated with new technologies in the future as they become available, or extend solutions beyond the original scope, as new requirements are identified. Moreover, every application has its own unique requirements that require a broad range of solutions.

1.14 Advantages of LabVIEW

The choice of software is important because software is generally the central component that ties the entire system together. Choosing the right software can maximize productivity, while a software package that does not fit the needs could drain time and productivity. LabVIEW helps the user to do more projects done in less time, by streamlining the process from inception through completion. There are several areas in LabVIEW that contribute to a significant gain in productivity when compared to other development software.

1.14.1 Easy to Learn

The one factor that greatly contributes to starting the application faster is the initial learning curve generally associated with any software tool. A tool can be used to its full potential only when the user becomes proficient and understand the intricacies associated with it. LabVIEW significantly reduces the learning curve, by providing an intuitive interface and a set of tools that encapsulate functionality, while abstracting the underlying enabling technology. One feature that enhances the initial experience of users with LabVIEW is the fact that the development environment relies almost exclusively on a drag-and-drop interface, which is second nature to users of computer technology.

1.14.2 Easy to Use

LabVIEW continuously improves based on usability research, customer feedback, and better technology. LabVIEW simplifies even the most challenging tasks, such as multithreaded parallel execution, through the LabVIEW patented Dataflow technology. For example, the user can program ActiveX through simple drag-and-drop interactions, and publish to the Web with just a few mouse clicks.

1.14.3 Complete Functionality

Ease of use is not the only factor that provides the best productivity. The user has to be sure that the products and tools satisfy all the requirements and those of the project. With LabVIEW, the user can easily scale to meet current and future requirements, whether it is required to create a small and easy-to-use instrument control application or a solution that monitors and controls the entire factory. The user can select from different development packages with varying degrees of functionality. Even if a basic LabVIEW package is chosen initially, the user can easily upgrade to the more complete versions to increase power and flexibility. The user can build a complete solution with LabVIEW, as it offers all the necessary tools and features for any measurement and automation application:

- *Powerful Built-In Functionality.* LabVIEW development systems include a complete set of tools geared specifically for data acquisition, instrument control, and other hardware. Other tools include Web and network connectivity, visualization, storage, report generation, measurement analysis, and simulation.
- *Add-On Software for Special Needs.* Powerful LabVIEW toolsets have been developed by National Instruments and other companies. These toolsets expand and enhance the core functionality of LabVIEW for specialized tasks.
- *Configuration-Based Tools.* National Instruments offers tools that satisfy specific needs in a nonprogrammatic configuration-based environment. For example, NI DIAdem offers simple ways to transfer LabVIEW-acquired data for the purpose of interactive post acquisition analysis and report generation.
- *Third-Party Tools.* Many companies continuously develop and sell tools that work with or use LabVIEW for industry-specific needs such as biomedical applications, telecommunications, automotive, and others.

1.14.4 Modular Development

A large project can be broken down into functional solvable units to simplify the programming complexity. The programmer can easily manage and test these subtasks to reduce dependencies that might cause unexpected behavior. Once these subtasks are solved, the user can put them together to complete the system. The ease with which the user can accomplish this division of tasks depends greatly on the underlying architecture of the software.

1.15 Virtual Instrumentation in the Engineering Process

Virtual instruments provide significant advantages in every stage of the engineering process, from research and design to manufacturing test as shown in Fig 1.4.

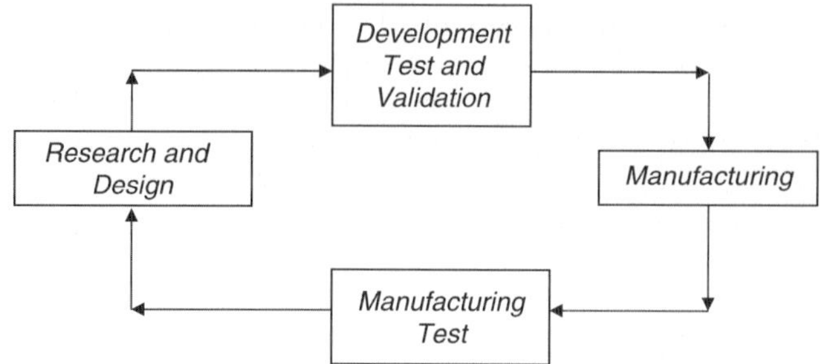

Fig. 1.4. Evolution of engineering process

1.15.1 Research and Design

In research and design, engineers and scientists demand rapid development and prototyping capabilities. With virtual instruments, we can quickly develop a program, take measurements from an instrument to test a prototype, and analyze results, all in a fraction of the time required to build tests with traditional instruments. When we need flexibility, a scalable open platform is essential, from the desktop, to embedded systems, to distributed networks. The demanding requirements of research and development (R&D) applications require seamless software and hardware integration. Either to interface stand-alone instruments using GPIB or directly acquire signals into the computer with a data acquisition board and signal conditioning hardware, LabVIEW makes integration simple. With virtual instruments, the user can also automate a testing procedure, eliminating the possibility of human error and ensuring the consistency of the results by not introducing unknown or unexpected variables.

1.15.2 Development Test and Validation

With the flexibility and power of virtual instruments, the user can easily build complex test procedures. For automated design verification testing, the user can create test routines in LabVIEW and integrate software such as National Instruments Test Stand, which offers powerful test management capabilities. One of the many advantages these tools offer across the organization is code reuse. Code can be developed in the design process and then plug the same programs into functional tools for validation, test, or manufacturing.

1.15.3 Manufacturing Test

Decreasing test time and simplifying development of test procedures are primary goals in manufacturing test. Virtual instruments based on LabVIEW combined with powerful test management software such as Test Stand deliver high performance to meet those needs. These tools meet rigorous throughput requirements with a high-speed, multithreaded engine for running multiple test sequences in parallel. Test Stand easily manages test sequencing, execution, and reporting based on routines written in LabVIEW. Test Stand integrates the creation of test code in LabVIEW. Test Stand can also reuse code created in R&D or design and validation. If we have manufacturing test applications, we can take full advantage of the work already done in the product life cycle.

1.15.4 Manufacturing Design

Manufacturing applications require software to be reliable, high in performance, and interoperable. Virtual instruments based on LabVIEW offer all these advantages, by integrating features such as alarm management, historical data trending, security, networking, industrial I/O, and enterprise connectivity. With this functionality, the user can easily connect many types of industrial devices such as PLCs, industrial networks, distributed I/O, and plug-in data acquisition boards. By sharing code across the enterprise, manufacturing can use the same LabVIEW applications developed in R&D or validation, and integrate seamlessly with manufacturing test processes.

1.16 Virtual Instruments Beyond the Personal Computer

Recently, commercial PC technologies have been migrating into embedded systems. Examples include Windows CE, Intel x86-based processors, PCI and Compact PCI buses, and Ethernet for embedded development. Because virtual instrumentation relies so heavily on commercial technologies for cost and performance advantages, it also has expanded to encompass more embedded and real-time capabilities. For example, LabVIEW runs on Linux as well as the embedded ETS real-time operating system from VenturCom on specific embedded targets. The option of using virtual instrumentation as a scalable framework that extends from the desktop to embedded devices should be considered a tool in the complete toolbox of an embedded systems developer. A dramatic technology change example that affects embedded systems development is networking and the Web. With the ubiquity of PCs, Ethernet now dominates as the standard network infrastructure for companies worldwide.

In addition, the popularity of the Web interface in the PC world has overflowed into the development of cell phones PDAs, and now industrial data acquisition and control systems. Embedded systems at one time meant standalone operation, or at most interfacing at a low level with a real-time bus to peripheral components. Now, the increased demand for information at all levels of the enterprise (and in consumer products) requires one to network embedded systems while continuing to guarantee reliable and often real-time operation. Because virtual instrumentation software can combine one development environment for both desktop and real-time systems using cross-platform compiled technology, we can capitalize on the built-in Web servers and easy-to-use networking functionality of desktop software and target it to real-time and embedded systems.

For example, we could use LabVIEW to simply configure a built-in Web server to export an application interface to defined secure machines on the network on Windows, and then download that application to run on a headless embedded system that can fit in the user's hand. This procedure happens with no additional programming required on the embedded system. Then the embedded system can be powered and connected to the application from a remote secure machine via Ethernet, and interface a standard Web browser. For more sophisticated networking applications, we can graphically program TCP/IP or other methods which are already familiar in LabVIEW and then run them in the embedded system. Embedded systems development is one of the fastest growing segments of engineering, and will continue to be for the foreseeable future as consumers demand smarter cars, appliances, homes, and so on.

The evolution of these commercial technologies will propel virtual instrumentation into being more applicable to a growing number of applications. Leading companies that provide virtual instrumentation software and hardware tools need to invest in expertise and product development to serve this growing set of applications. For example, for its flagship virtual instrumentation software platform, LabVIEW, National Instruments has described a vision that includes the ability to scale from development for desktop operating systems, to embedded real-time systems, to handheld personal digital assistant targets, to FPGA-based hardware, and even to enabling smart sensors. Next-generation virtual instrumentation tools need to include networking technology for quick and easy integration of Bluetooth, wireless Ethernet, and other standards. In addition to using these technologies, virtual instrumentation software needs a better way to describe and design timing and synchronization relationships between distributed systems in an intuitive way to help faster development and control of these often embedded systems.

The virtual instrumentation concepts of integrated software and hardware, flexible modular tools, and the use of commercial technologies combine to create a framework upon which we can rapidly complete ones systems development and also maintain them for the long term. Because virtual instrumentation offers so many options and capabilities in embedded development, it makes sense for embedded developers to understand and review these tools.

Summary

Virtual instrumentation is fueled by ever-advancing computer technology and offers us the power to create and define a system based on an open framework. This concept not only ensures that a work will be usable in the future but also provides the flexibility to adapt and extend as needs change. LabVIEW was designed with scientists and engineers in mind, providing powerful tools and a familiar development environment created specifically for the design of virtual instruments. Finally, system design software that provides an intuitive interface for designing custom instrumentation systems furthers virtual instrumentation. LabVIEW is an example of such software. The LabVIEW graphical development environment offers the performance and flexibility of a programming language, as well as high-level functionality and configuration utilities designed specifically for measurement and automation applications.

Review Questions

1. Give an overview of virtual instrument and LabVIEW.
2. Explain the concept of virtual instrumentation.
3. Define the Terms: virtual instrument and LabVIEW.
4. What is virtual instrument and why do we need it?
5. What are the rules for writing virtual instrumentation programs?
6. What is the necessity for multiprocessing machines?
7. Explain the stages involved in engineering of products using virtual instrument with a neat schematic diagram.
8. What are the various techniques adapted for LabVIEW program?
9. Compare and contrast virtual instruments versus traditional instruments.
10. State the principles of LabVIEW.
11. Explain the performance and merits of contemporary virtual instruments.
12. How does the LabVIEW methodology work.
13. Give some of the recent instrument technologies used for instrumentation.

2

Programming Techniques

Learning Objectives. On completion of this chapter the reader will have a knowledge on:

- Programming Techniques
- Front Panel and Block Diagram of Virtual Instruments
- LabVIEW Environment and its Menus
- Palletes of LabVIEW
- Data Flow Programming
- 'G' Programming Concepts
- Creating and Saving VIs
- Wiring, Editing, and Debugging VIs
- Creating SubVIs

2.1 Introduction

LabVIEW programs are called virtual instruments, or VIs, because their appearance and operation imitate physical instruments, such as oscilloscopes and millimeters. LabVIEW contains a comprehensive set of tools for acquiring, analyzing, displaying, and storing data, as well as tools to troubleshoot the code. LabVIEW VIs contain three components – the front panel, the block diagram, and the icon and connector pane. This section describes the front panel and the block diagram along with the controls and indicators used to create the user interface and functions to create the block diagram. Dataflow programming and 'G' programming are also discussed along with the data types and operations that can be performed in LabVIEW. The menus available on the two windows front panel and block diagram are illustrated to enable the user understand the basic operation of LabVIEW. The section also focuses on creating a VI, subVI along with the libraries available.

2.2 Virtual Instruments

Virtual Instruments are front panel and block diagram. The front panel or user interface is built with controls and indicators. Controls are knobs, push buttons, dials, and other input devices. Indicators are graphs, LEDs, and other displays. After the user interface is built, codes can be added using VIs and structures to control the front panel objects. The block diagram contains this code. In some ways, the block diagram resembles a flowchart.

2.2.1 Front Panel

When a blank VI is opened, an untitled front panel window appears. This window displays the front panel and is one of the two LabVIEW windows used to build a VI. The *front panel* is the window through which the user interacts with the program. The input data to the executing program is fed through the front panel and the output can also be viewed on the front panel, thus making it indispensable.

Front Panel Toolbar

The toolbar buttons are shown in Fig. 2.1. The toolbar buttons are individually explained in Table 2.1. The virtual instruments such as the front panel and block diagram of a temperature example are shown in Fig. 2.2 along with their toolbars.

Controls and Indicators

The front panel is primarily a combination of *controls* and *indicators*, which are the interactive input and output terminals of the VI, respectively. Controls are knobs, push buttons, dials, and other input devices. Indicators are graphs, LEDs, and other displays. Controls simulate instrument input devices and supply data to the block diagram of the VI. Indicators simulate instrument output devices and display data the block diagram acquires or generates. A simple way to understand about controls and indicators is illustrated as follows:

Controls* = *Inputs from the User* = *Source or "Terminals"
Indicators* = *Outputs to the User* = *Destinations or "Sinks"

Fig. 2.1. Front panel toolbar

Table 2.1. Front panel tool bar menus

S. No	Icon	Name	Meaning
1		**Run**	Used to run a VI. The VI runs if the **Run** button appears as a solid white arrow as shown at the left. The solid white arrow also indicates the VI can be used as a subVI if a connector pane for the VI is created.
2		**Run**	While the VI runs, the **Run** button appears as shown at left if the VI is a top-level VI, meaning it has no callers and therefore is not a subVI.
3		**Run**	If the VI that is running is a subVI, the **Run** button appears as shown at left.
4		**Run**	The **Run** button appears broken, shown at left, when the VI that is created and edited contains errors. If the **Run** button still appears broken after wiring of the block diagram is completed, then the VI is broken and cannot run. On clicking this button, an **Error list** window is displayed, which lists all errors and warnings.
5		**Run Continuously**	The **Run Continuously** button, shown at left, is used to run the VI until one abortion or pause execution.
6		**Abort Execution**	While the VI runs, the **Abort Execution** button, shown at left, appears. Click this button to stop the VI immediately if there is no other way to stop the VI. If more than one running top-level VI uses the VI, the button is dimmed.
7		**Pause**	Click the **Pause** button, shown at left, to pause a running VI. Upon clicking the **Pause** button, LabVIEW highlights on the block diagram the location where the execution was paused, and the **Pause** button appears red. Click the button again to continue running the VI.

Table 2.1. *Continued*

S. No	Icon	Name	Meaning
8	13pt Application Font	**Text Settings**	Select the **Text Settings** pull-down menu, shown at left, to change the font settings for the selected portions of the VI, including size, style, and color
9		**Align Objects**	Select the **Align Objects** pull-down menu, shown at left, to align objects along axes, including vertical, top edge, left, and so on.
10		**Distribute Objects**	Select the **Distribute Objects** pull-down menu, shown at left, to space objects evenly, including gaps, compression, and so on.
11		**Resize Objects**	Select the **Resize Objects** pull-down menu, shown at left, to resize multiple front panel objects to the same size.
12		**Reorder**	Select the **Reorder** pull-down menu, shown at left, when objects overlap each other and to define which one is in front or back of another. Select one of the objects with the Positioning tool and then select from **Move Forward**, **Move Backward**, **Move To Front**, and **Move To Back**.
13		**Context Help**	Select the **Show Context Help Window** button, shown at left, to toggle the display of the **Context Help** window.
14		**Type**	**Type** appears to remind the user that a new value is available to replace an old value. The Enter button disappears when clicked, press the <Enter> key, or click the front panel or block diagram workspace.

Controls and Indicators are generally not interchangeable; the difference should be clear among the user. The user can "drop" controls and indicators onto the front panel by selecting them from a subpalette of the floating **Controls** palette window shown in Fig. 2.3 and placing them in a desired spot.

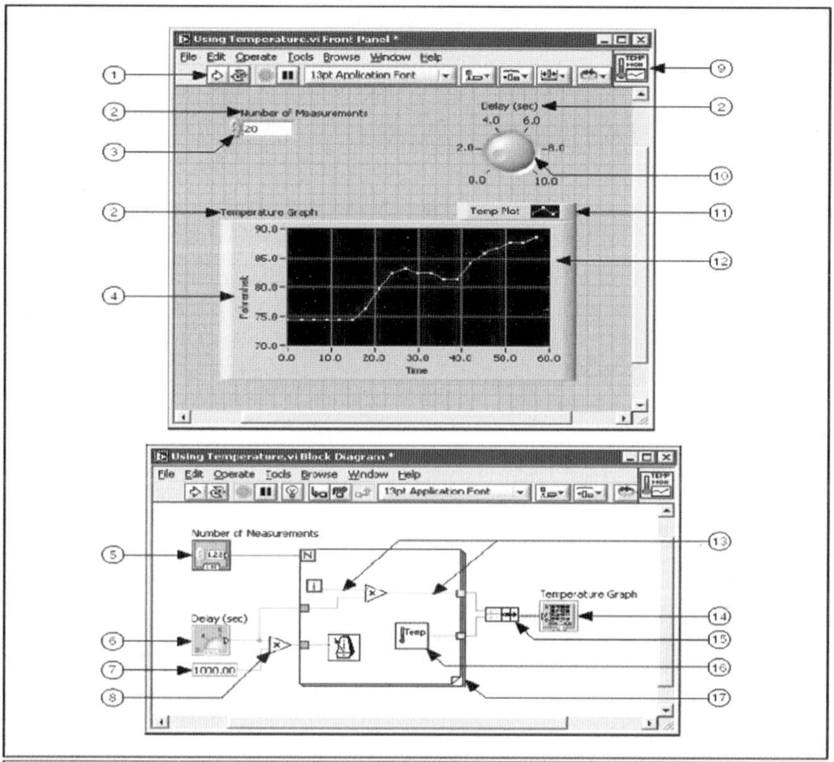

1 Toolbar
2 Owned Label
3 Numeric Control
4 Free Label
5 Numeric Control Terminal
6 Knob Terminal
7 Numeric Constant
8 Multiply Function
9 Icon
10 Knob Control
11 Plot Legend
12 XY Graph
13 Wire Data Path
14 XY Graph Terminal
15 Bundle Function
16 SubVI
17 For Loop Structure

Source: National Instruments

Fig. 2.2. Front panel and block diagram. *Source*: National Instruments

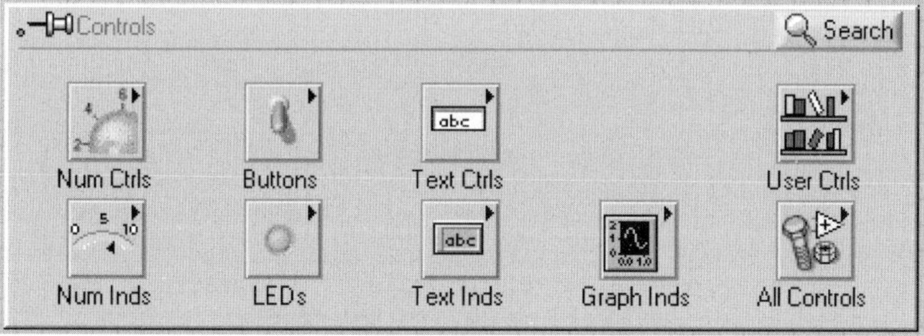

Fig. 2.3. Controls and indicators

Once an object is on the front panel, the user can easily adjust its size, shape, position, color, and other attributes.

Controls and indicators can be broadly classified as:

1. Numeric controls and indicators
2. Boolean controls and indicators

Numeric Controls and Indicators

The two most commonly used numeric objects are the numeric control and the numeric indicator, as shown in Fig. 2.4. The values in a numeric control can be entered or changed by selecting the increment and decrement buttons with the Operating tool or double-clicking the number with either the Labeling tool or the Operating tool.

Boolean Controls and Indicators

The Boolean controls and indicators (Fig. 2.5) are used to enter and display Boolean (True or false) values. Boolean objects simulate switches, push

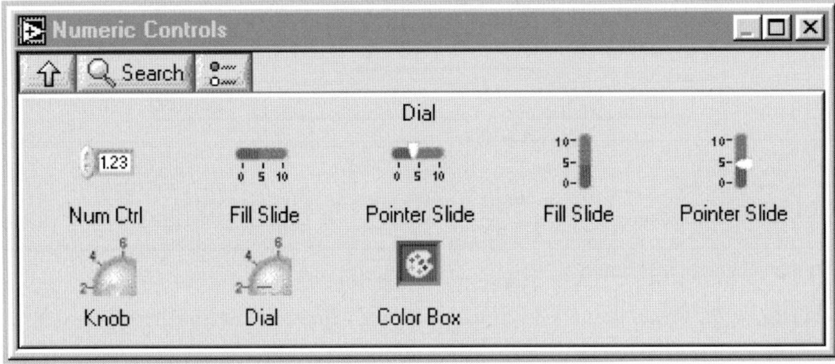

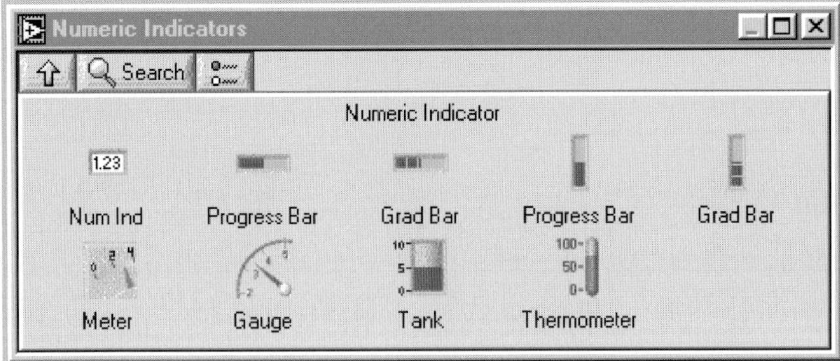

Fig. 2.4. Numeric controls and indicators

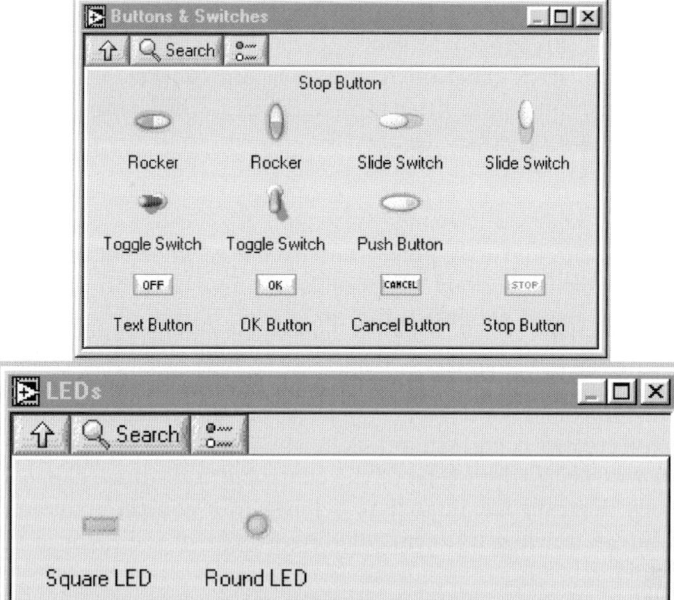

Fig. 2.5. Boolean controls and indicators

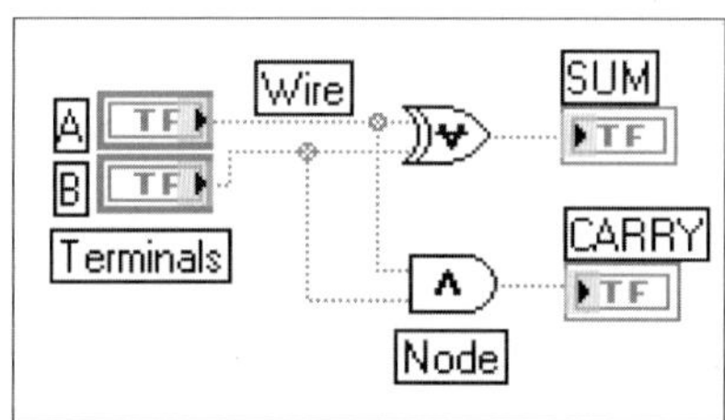

Fig. 2.6. Illustration of components in block diagram

buttons, and LEDs. The most common Boolean objects are the vertical toggle switch and the round LED.

2.2.2 Block Diagram

The block diagram window holds the graphical source code of a LabVIEW's block diagram corresponds to the lines of text found in a more conventional language like C or BASIC – it is the actual executable code. The block diagram can be constructed with the basic blocks such as: terminals, nodes, and wires.

The simple VI shown in Fig. 2.6 is an example of a digital Half Adder circuit. Its block diagram in shows examples of terminals, nodes, and wires. After the front panel is built, the user can add code using graphical representations of functions to control the front panel objects. The block diagram contains this graphical source code. Front panel objects appear as terminals, on the block diagram. Block diagram objects include terminals, subVIs, functions, constants, structures, and wires, which transfer data among other block diagram objects.

Terminals

Terminals are entry and exit ports that exchange information between the front panel and block diagram. Terminals are analogous to parameters and constants in text-based programming languages. Types of terminals include control or indicator terminals and node terminals. Control and indicator terminals belong to front panel controls and indicators.

When a control or indicator is placed on the front panel, LabVIEW automatically creates a corresponding terminal on the block diagram. The terminals represent the data type of the control or indicator. The user cannot delete a block diagram terminal that belongs to a control or indicator. The terminal disappears when its corresponding control or indicator is deleted on the front panel.

The front panel controls or indicators can be configured to appear as icon or data type terminals on the block diagram. By default, front panel objects appear as icon terminals. For example, a knob icon terminal, shown at left, represents a knob on the front panel. The DBL at the bottom of the terminal represents a data type of double-precision, floating-point numeric. To display a terminal as a data type on the block diagram, right-click the terminal and select **View as Icon** from the shortcut menu to remove the checkmark. A DBL data type terminal, represents a double-precision, floating-point numeric control or indicator.

Control terminals have thick borders, while indicator terminal borders are thin. It is very important to distinguish between the two since they are not functionally equivalent i.e., control = input, indicator = output, and so they are not interchangeable.

Data entered into the front panel controls enter the block diagram through the control terminals. In Fig. 2.6 the data enters the AND and XOR functions, when the AND and XOR functions complete their internal calculations, they produce new data values. The data then flow to the indicator terminals, where they exit the block diagram, re-enter the front panel, and appear in front panel indicators. The terminals in Fig. 2.6 belong to four front panel controls and indicators. On right-clicking the function node and selecting **Visible Items≫Terminals** from the shortcut menu the terminals of the function on the block diagram are displayed.

Nodes

A node is just a fancy word for a program execution element. Nodes are analogous to statements, operators, functions, and subroutines in standard programming languages. The AND and XOR functions represent one type of node. A structure is another type of node. Structures can execute code repeatedly or conditionally, similar to loops and Case statements in traditional programming languages. LabVIEW also has special nodes, called Formula Nodes, which are useful for evaluating mathematical formulas or expressions.

Wires

Wires connecting nodes and terminals hold a LabVIEW VI together. Wires are the data paths between source and destination terminals, they deliver data from one source terminal to one or more destination terminals. Wires are analogous to variables in text-based programming languages. If more than one source or no source at all is connected to a wire, LabVIEW disagrees and the wire will appear broken. This principle of wires connecting source and destination terminals explains why controls and indicators are not interchangeable; controls are source terminals, whereas indicators are destinations, or sinks. Each wire has a different style or color, depending on the data type that flows through the wire. Table 2.2 shows a few wires and their corresponding types.

By default, basic wire styles are used in block diagrams. To avoid confusion among the data types, the colors and styles are simply matched. In LabVIEW, wires are used to connect multiple terminals together to pass data in a VI. The wires must be connected to inputs and outputs that are compatible with the data that is transferred with the wire. For example, an array output cannot be wired to a numeric input. In addition, the direction of the wires must be correct. The wires must be connected to only one input and at least one output. For example, two indicators cannot be wired together. The components that determine wiring compatibility include the data type of the control and/or indicator and the data type of the terminal.

Block Diagram Toolbar

When a VI is run, buttons appear on the block diagram toolbar that can be used to debug the VI. The following toolbar shown in Fig. 2.7 appears on the block diagram. Table 2.3 explains the icons available on the block diagram toolbar.

Table 2.2. Data type and representation

	Scalar	1D array	2D array	Color
Floating point number				Orange
Integer number				Blue
Boolean				Green
String				Pink

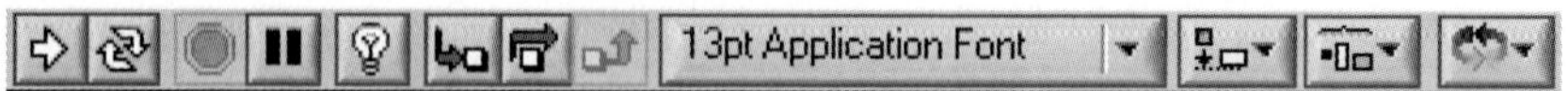

Fig. 2.7. Block diagram toolbar

Table 2.3. Block diagram toolbar menus

S. No	Icon	Name	Meaning
1		Highlight Execution	**Highlight Execution** button, shown at left, is used to display an animation of the block diagram execution when the **Run** button is selected. Used to see the flow of data through the block diagram. The same button is used to disable execution highlighting.
2		Step Into	**Step Into** button, shown at left, is used to open a node and pause. When clicked, it executes the first action and pauses at the next action of the subVI or structure. This function can also be enabled by selecting the <Ctrl> and down arrow key. Single-stepping through a VI steps through the VI node by node. Each node blinks to denote when it is ready to execute.
3		Step Over	**Step Over** button, shown at left, is used to execute a node and pause at the next node. This function can also be selected by using the <Ctrl> and right arrow keys. By stepping over the node, the node can be executed without single-stepping through the node.
4		Step Out	**Step Out** button, shown at left, is used to finish executing the current node and pause. When the VI finishes executing, the **Step Out** button becomes dimmed. This function can also be selected by using the <Ctrl> and up arrow keys. By stepping out of a node, single-stepping through the node is completed and execution goes to the next node.
5		Warning	The **Warning** button, shown at left, appears if a VI includes a warning and if a checkmark in the **Show Warnings** checkbox in the **Error List** window is placed. A warning indicates there is a potential problem with the block diagram, but it does not stop the VI from running.

2.3 LabVIEW Environment

When LabVIEW is launched, the following navigation dialog box appears that includes introductory material and common commands as shown in Fig. 2.8. The commands are briefly explained in Table 2.4.

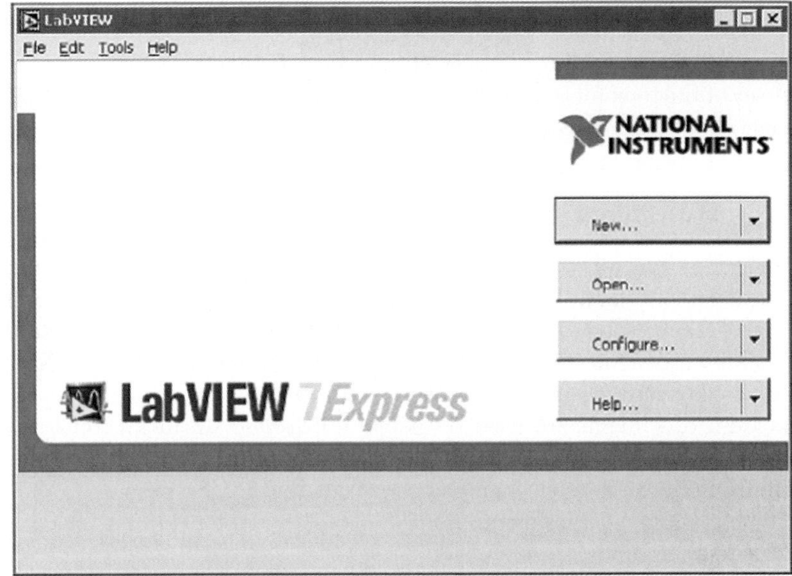

Source: National Instruments

Fig. 2.8. Navigation dialog box

Table 2.4. Dialog box menus

S No	Icon	Name	Meaning
1	New...	New	The **New** button is used to create a new VI. The arrow on the **New** button is used to open a blank VI or to open the **New** dialog box.
2	Open...	Open	The **Open** button is used to open an existing VI. The arrow on the **Open** button is used to open recent files.
3	Configure...	Configure	The **Configure** button is used to configure the data acquisition devices. The arrow on the **Configure** button is used to configure LabVIEW.
4	Help...	Help	The **Help** button is used to launch the *LabVIEW Help*. The arrow on the **Help** button is used for other Help options, including the NI Example Finder.

The **LabVIEW** dialog box includes the following components:

- A menu with standard items such as **File≫Exit**.
- A set of buttons for creating and opening VIs, configuring data acquisition devices, and finding helpful information.

2.3.1 Startup Menu

The menus at the top of a VI window contain items common to other applications, such as **Open**, **Save**, **Copy**, and **Paste**, and other items specific to LabVIEW. Some menu items also list shortcut key combinations **(Mac OS)**. The menus appear at the top of the screen **(Windows and UNIX)**. The menus display only the most recently used items by default. The arrows at the bottom of a menu are used to display all items. All menu items can be displayed by default by selecting **Tools≫Options** and selecting **Miscellaneous** from the pull-down menu as shown in Fig. 2.9. Some menu items are unavailable while a VI is in run mode.

2.3.2 Shortcut Menu

The most often-used menu is the object shortcut menu. All LabVIEW objects and empty space on the front panel and block diagram have associated

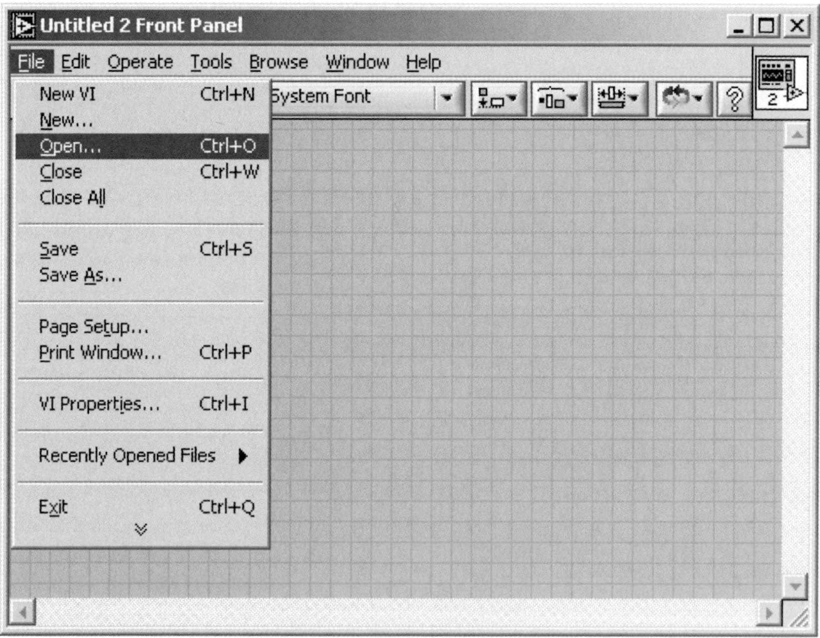

Fig. 2.9. Startup menu

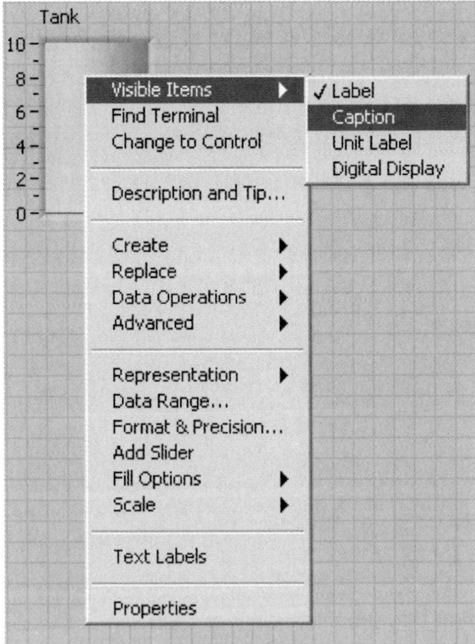

Fig. 2.10. Shortcut menu for tank

shortcut menus. The shortcut menu items are used to change the look or behavior of front panel and block diagram objects. To access the shortcut menu, right-click the object, front panel, or block diagram. The shortcut menu for a tank available on the front panel is shown in Fig. 2.10.

Property Dialog Boxes

Front panel objects also have property dialog boxes that can be used to change the look or behavior of front panel objects. Right-clicking the front panel object and selecting Properties from the shortcut menu can access the property dialog box for an object. Figure 2.11 shows the property dialog box for the tank. The options available on the property dialog box for an object are similar to the options available on the shortcut menu for that object.

2.3.3 Pull-Down Menu

The user should understand clearly that LabVIEW's capabilities are many and varied. The menu bar at the top of a VI window contains several pull-down menus. When an item on the menu bar is clicked, a menu appears below the bar. The pull-down menu contains items common to many applications, such as Open, Save, Copy, and Paste and many other functions particular to Lab View. Some of the basic pulls-down menu functions are discussed in

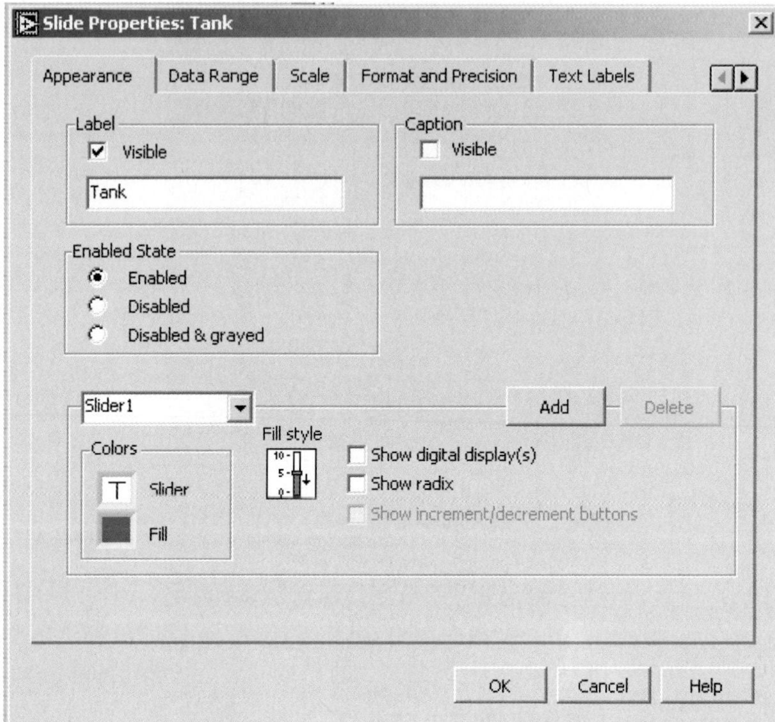

Fig. 2.11. Property dialog box for tank

the following sections. Many menus also list shortcut keyboard combinations for the user to use. To use keyboard shortcuts, press the appropriate key in conjunction with the <control> key on PCs, the <command> key on Macs, the <meta> key on Suns, and the <alt> key on HP machines. Many of the menu items show keyboard shortcuts to the right of their corresponding commands.

File Menu

Full down the File menu, which contains commands common to many applications, such as Save and Print. New VIs can be created or existing VIs can be opened from the File menu as shown in Fig. 2.12.

Edit Menu

Take a look at the Edit menu. It has some universal commands, like Undo, Cut, Copy, and Paste. The Find command can be used to search objects and remove bad wires form the block diagram as shown in Fig. 2.13.

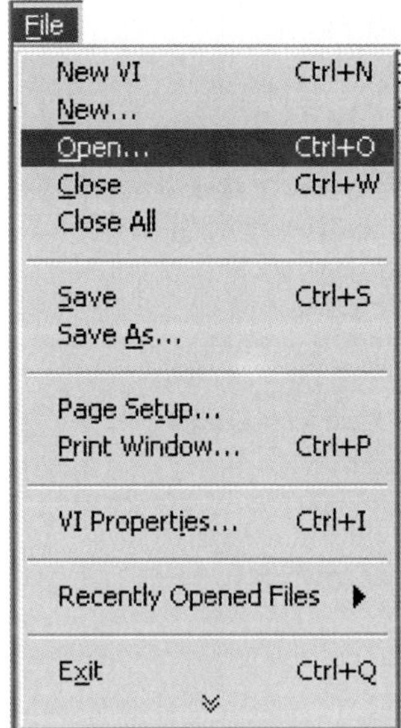

Fig. 2.12. Snapshot of File menu

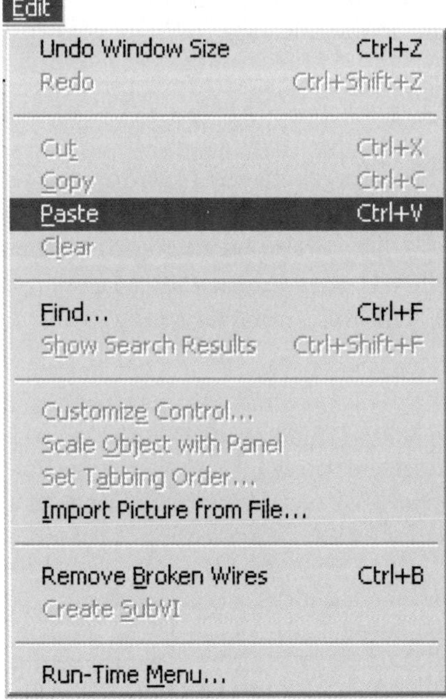

Fig. 2.13. Snapshot of Edit menu

Fig. 2.14. Snapshot of Operate menu

Operate Menu

The user can run or stop the program from the Operate menu. The user can also change a VI's default values, control print and log at completion features, and switch between run mode and edit mode as shown in Fig. 2.14.

Tool Menu

The tools menu lets the user to access built-in and add-on tools, and utilities that work with lab VIEW, such as the Measurement and Automation Explorer, where the DAQ devices are configured, or the Web Publishing Tool is used for creating HTML pages from Lab VIEW. The user can view and change the Lab VIEW Options as shown in Fig. 2.15.

Browse Menu

The Browse menu contains features for simple navigation among large number of VIs. The user can view VI hierarchy, determine VI's entire submenu and view debugging break points as shown in Fig. 2.16.

Windows Menu

The Windows menu in Fig. 2.17 is used to perform the following functions:

- Toggle between the panel and diagram windows
- Show the error list and the clipboard

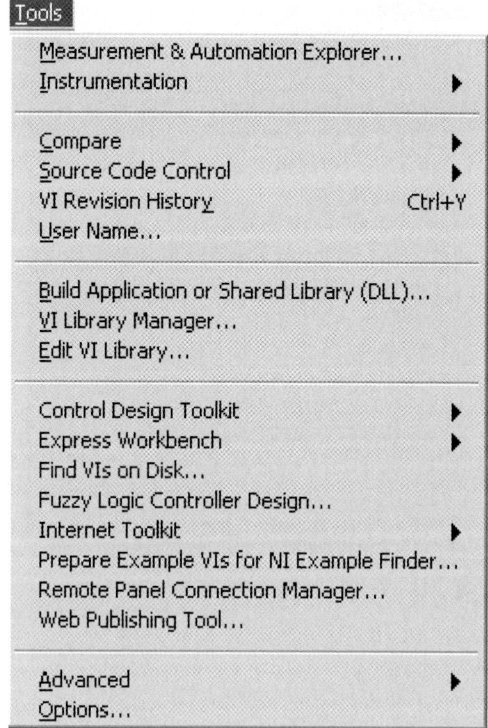

Fig. 2.15. Snapshot of Tools menu

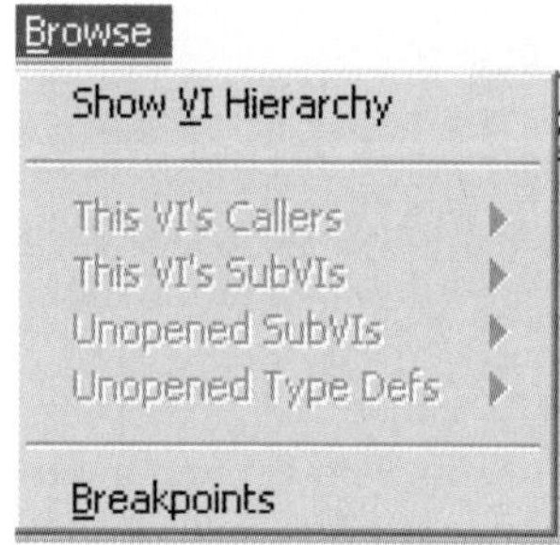

Fig. 2.16. Snapshot of Browse menu

- Tile both window so that the user can see them at the same time
- Switch between open VIs

The user can also bring up floating palettes if closed. In addition, the user can show VI information and development history from this menu.

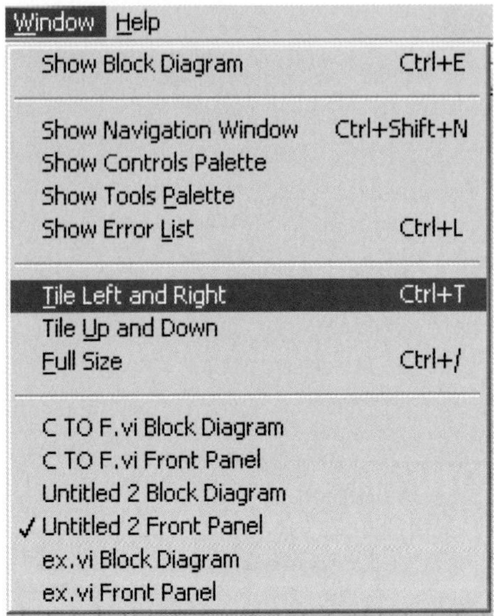

Fig. 2.17. Snapshot of Window menu

Help Menu

The Help Window contents can be shown, hidden, or locked using the Help menu. The user can also access LabView's online information and view the About Lab View information window as shown in Fig. 2.18.

2.3.4 Pop-Up Menu

Pop-up menus are more frequently used than any other LabVIEW menu. To pop-up, the cursor is positioned over the object whose menu is desired; then the mouse is right clicked on Windows and UNIX machines, or the <command> key is held and clicked on the MAC.

Virtually every LabVIEW object has a Pop-up menu of options and commands. Options available in this Pop-up menu depend on the kind of object, and they are different when the VI is in edit mode or run mode. For example, a numeric control will have a very different Pop-up menu than a graph indicator. If pop up function is accessed on an empty space in the front panel or block diagram, the **Controls** or **Functions** palette appears.

Pop-Up Menu Features

Many pop-up menu items expand into submenus called hierarchical menus, denoted by a right arrowhead. Hierarchical menus sometimes have a selection of mutually exclusive options. The currently selected option is denoted by a

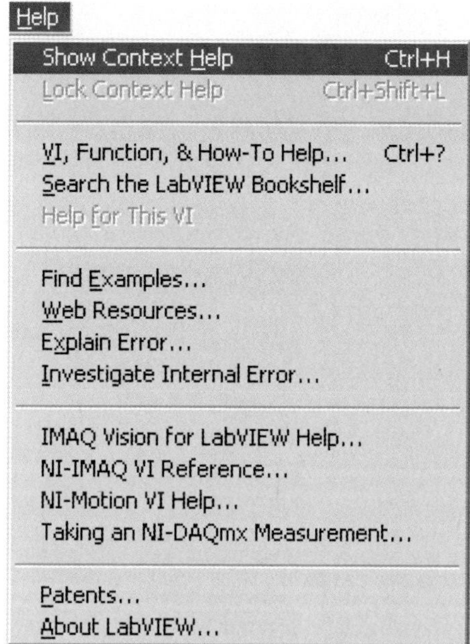

Fig. 2.18. Snapshot of Help menu

check mark for text-displayed options or is surrounded by a box for graphical options.

Some menu items Pop-up dialog boxes containing options for the user to configure. Menu items leading to dialog boxes are denoted by ellipses. Menu items without right arrowheads or ellipses are usually commands that execute immediately upon selection. A command usually appears in verb form, such as **Change to Indicator**. When selected, some commands are replaced in the menu by their inverse commands. For example, when the **Change to Indicator** is chosen, the menu selection becomes **Change to Control**. Sometimes different parts of an object have different Pop-up menus. For example, if an object's label is popped on, the menu contains only a Size to Text option. Popping up elsewhere on the object gives a full menu of options shown in Fig. 2.19.

Visible Items

Many items have **Visible Items** menu with which the user can show or hide certain cosmetic features like labels, captions, scrollbars, or wiring terminals. If **Visible Items** is selected, the user will get another menu off to the side, listing options of what can be shown. If an option has a check next to it, that option is currently visible; if it has no check, it is hidden. The status of an option can be toggled by releasing the mouse as shown in Fig. 2.20.

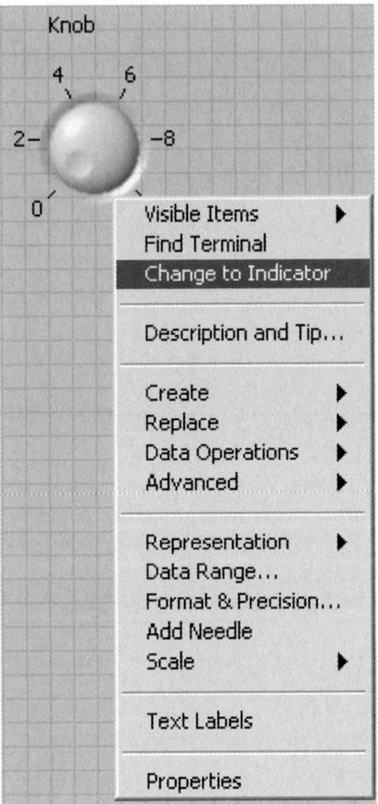

Fig. 2.19. Pop-up menu of Knob in Controls Palette

Find Terminal and Find Control/Indicator

If **Find Terminal** is selected from a front panel Pop-up menu, LabVIEW will locate and highlight its corresponding block diagram terminal. If **Find Control/Indicator** is selected from the block diagram Pop-up menu, LabVIEW will show its corresponding front panel object as shown in Fig. 2.21.

Change to Control and Change to Indicator

By selecting **Change to Indicator**, an existing control (an input object) can be changed into an indicator (an output object), or vice versa if **Change to Control** is selected. When an object is a control, its Pop-up menu contains the option to **Change to Indicator**. When it is an indicator, the Pop-up menu reads **Change to Control**. Since **Change to Control/Indicator** is an option in the Pop-up menu, it is easy to accidentally select it without realizing what is done. Because controls and indicators are not functionally interchangeable in a block diagram, the resulting errors may befuddle the

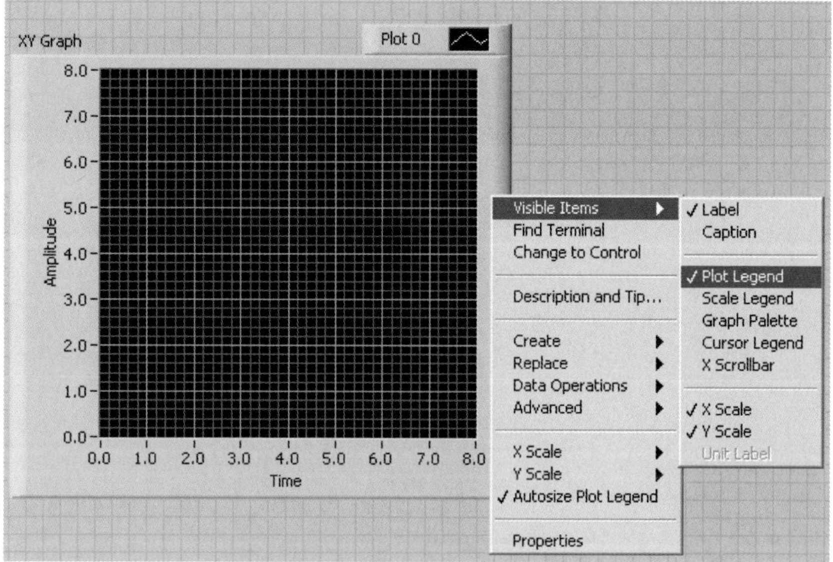

Fig. 2.20. Visible Items in XY graph

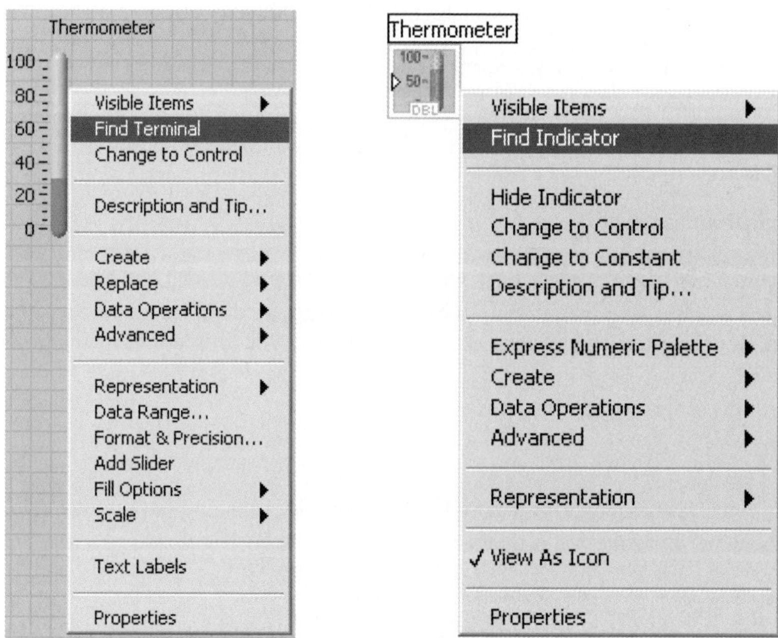

Fig. 2.21. Find Terminal and Find Indicator in Pop-up menu of thermometer in front panel and block diagram

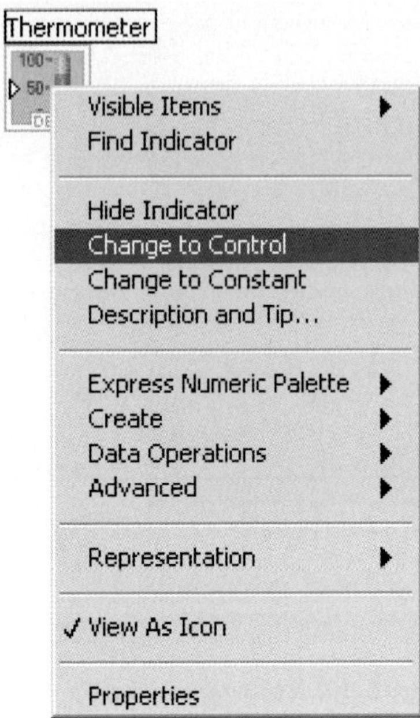

Fig. 2.22. Thermometer's icon in block diagram

user. A control terminal in the block diagram has a thicker border than an indicator terminal as shown in Fig. 2.22 for a thermometer icon.

Description and Tip

Selecting this option will allow the user to enter a description and a "tip". The description will appear in the Help window for that control, and the tip will show up when the mouse cursor is placed over this control as shown in Fig. 2.23.

Create

The Create option is an easy way for the user to create a property node, local variable, or reference for a given object as shown in Fig. 2.24.

Replace

The Replace option is extremely useful. It helps access to the controls or Functions palette and allows the user to replace the object popped-up with one of the user's choice, where possible, wires will remain intact as shown in Fig. 2.25.

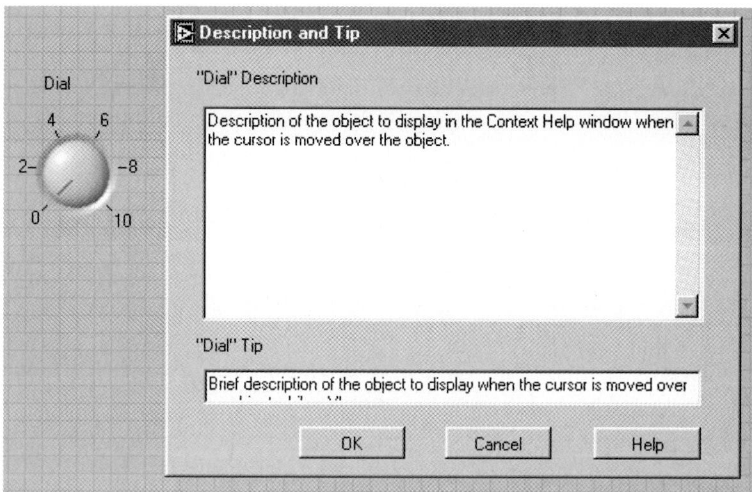

Fig. 2.23. Descriptions and tip for Knob Control

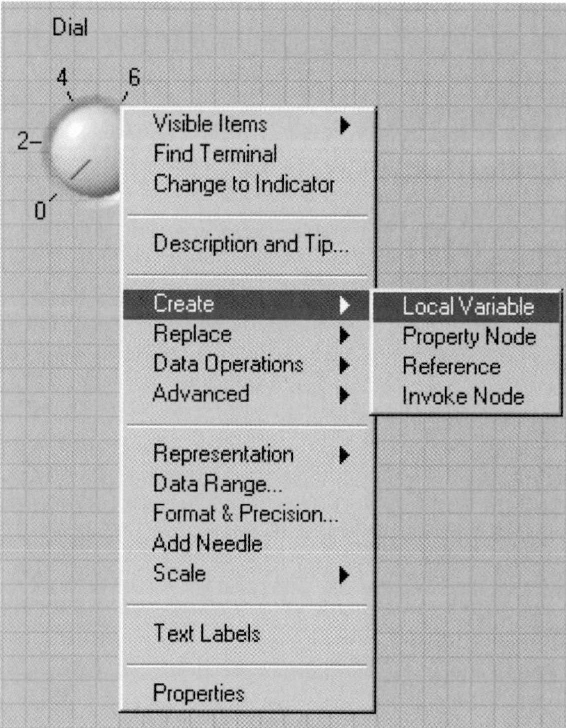

Fig. 2.24. Create for Knob Control

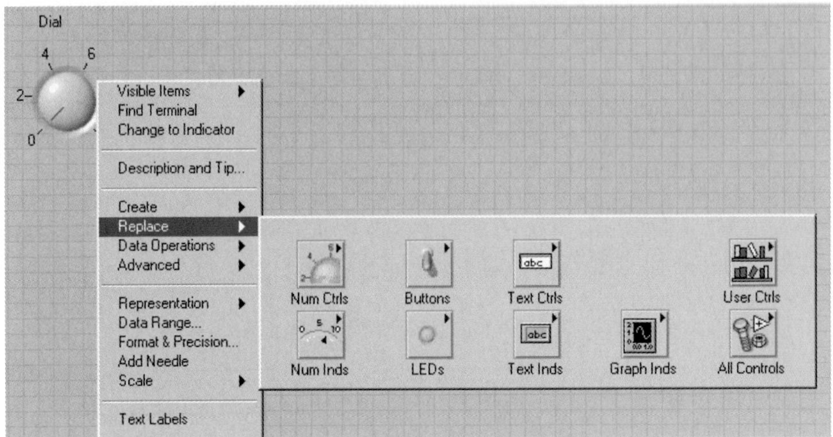

Fig. 2.25. Replace for Knob Control

2.3.5 Palletes

LabVIEW has graphical; floating palettes to help the user to create and run VIs. LabVIEW has three often-used floating palettes that can be placed in a convenient spot on the screen: the **Tools** palette, the **Controls** palette, and the **Functions** palette.

Controls and Functions Palettes

The **Controls** palette will often be used, since the controls and indicators that are required on the front panel are available. The user will probably use the Functions palette even more often, since it contains the functions and structures used to build a VI. The Controls and Functions palettes are unique in several ways. Most importantly the Controls palette is only visible when the front panel window is active, and the Functions palette is only visible when the block diagram window is active. Both palettes have subpalettes containing the objects to be accessed. As the cursor is passed over each subpalette button in the Controls and Functions palettes, the subpalette's name appears at the top of the window.

If a button is clicked, the associated subpalette appears and replaces the previous active palette. To select an object in the subpalette, the mouse button is clicked over the object, and then clicked on the front panel or block diagram to place it wherever desired. Like palette button names, subpalette object name appear when the cursor is run over them. To return to the previous ("owning") palette, the top-left arrow on each palette is selected. Clicking on the spyglass icon the user can search for a specific item in a palette, and then the user can edit palettes by clicking the options buttons.

There is another way to navigate palettes that some people find a little easier. Instead of having each subpalette replace the current palette, the user can pass through subpalettes in a hierarchical manner without them replacing their parent palettes. Now that some subpalettes have subpalettes containing more objects these are denoted by a little triangle in the upper-right corner of the icon and give the user a raised appearance.

Controls Palette

The **Controls** palette can be displayed by selecting **Window≫Show Controls Palette** or right-clicking the front panel workspace. The **Controls** palette can also be tacked down by clicking the thumbtack on the top left corner of the palette. By default, the **Controls** palette starts in the Express view. The Express palette view includes subpalettes on the top level of the **Controls** and **Functions** palettes. The **All Controls** and **All Functions** subpalettes contain the complete set of built-in controls, indicators, VIs, and functions. The **Advanced** palette view includes subpalettes on the top level of the **Controls** and **Functions** palettes that contain the complete set of built-in controls, indicators, VIs, and functions. The **Express** subpalettes contain Express VIs and other objects required to build common measurement applications. Click the **Options** button on the **Controls** or **Functions** palette to change to another palette view or format as shown in Fig. 2.26.

Functions Palette

The **Functions** palette is available only on the block diagram. The **Functions** palette contains the VIs and functions used to build the block diagram. The **Functions Palette** can be displayed by selecting the **Windows≫Show** or right-clicking the block diagram workspace to display the **Functions** palette. The **Functions** palette can also be displayed by clicking the thumbtack on the top left corner of the palette. The **Functions** palette starts in the Express view by default.

The **Functions** palette, shown in Fig. 2.27, is available only on the block diagram. The **Functions** palette contains the VIs and functions used to build

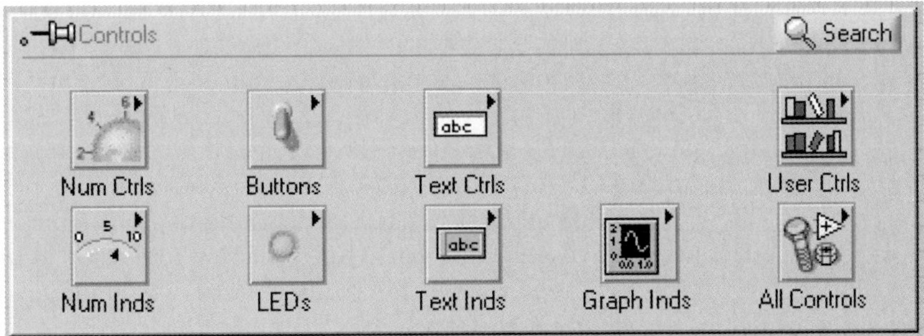

Fig. 2.26. Controls Palette

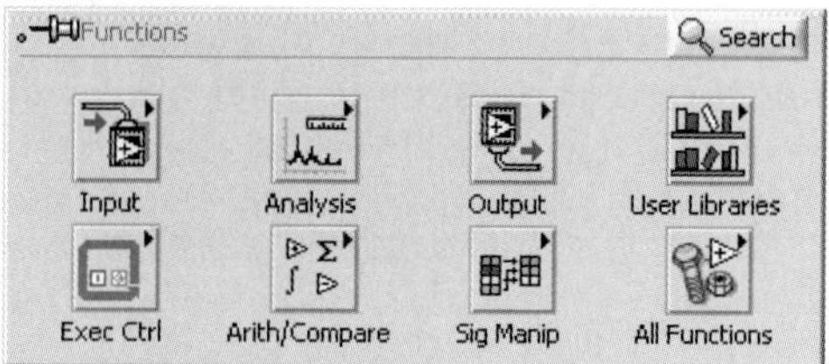

Fig. 2.27. Functions palette

the block diagram. Refer to the Block Diagram section of this chapter for more information about using the **Functions** palette on the block diagram. The VIs and functions located on the **Functions** palette depend on the palette view currently selected. The VIs and functions are located on subpalettes based on the types of VIs and functions.

Customizable Palettes

If LabVIEW's default organization of the **Controls** and **Functions** palettes does not fit the user's needs, they can be customized according to the choice. Access the menu editor by clicking on the "options" icon for a palette. From here, the user can create own palettes and customize existing views by adding new subpalettes, hiding items, or moving them from one palette to another. For example, if the user creates a VI using trigonometric functions, it can be placed in the existing **Trigonometric** subpalette for easy access. Editing the palettes is handy for placing most frequently used functions at the top level for easy access and burying those pesky functions.

Tools Palette

A *tool* is a special operating mode of the mouse cursor. Tools are used to perform specific editing and operation functions, similar to that used in a standard paint program. The **Tools** palette shown in Fig. 2.28 is available on both the front panel and the block diagram. Like the **Controls** and **Functions** palettes, the **Tools** palette window can be relocated or closed. To select a tool, the appropriate button on the **Tools** palette can be selected and the mouse cursor will change accordingly. By selecting **Window≫Show Tools Palette** the **Tools** palette can be displayed. If automatic tool selection is enabled and when the cursor is moved over objects on the front panel or block diagram, LabVIEW automatically selects the corresponding tool from the **Tools** palette. The user can disable automatic tool selection and select a tool manually by clicking the tool required on the **Tools** palette. The functions of the menus available in the **Tools Palette** are explained in Table 2.5.

Fig. 2.28. Tools palette

Table 2.5. Tools palette menus

S. No	Icon	Name	Meaning
1		Operating Tool	The Operating Tool is used to change values of front panel controls and indicators. Used to operate knobs, switches, and other objects with the Operating tool – hence the name. It is the only front panel tool available when the VI is running or in run mode.
2		Positioning tool	The Positioning tool selects, moves, and resizes objects.
3		Labeling tool	The Labeling tool creates and edits text labels.
4		Wiring tool	The Wiring tool wires objects together on the block diagram. It is also used to assign controls and indicators on the front panel to terminals on the VI's connector.
5		Color tool	The Color tool brightens objects and background by allowing the user to choose from a multitude of hues. Both foreground and background colors can be selected by clicking on the appropriate color area in the Tools palette. If an object is popped up with the Color tool, a hue from the color palette appears and the required color can be chosen.

Table 2.5. *Continued*

S. No	Icon	Name	Meaning
6		Pop-up tool	The Pop-up tool opens an object's Pop-up menu when the user clicks on the object with it. This tool can be used to access Pop-up menus instead of the standard method for popping up (right-clicking under Windows and Unix and <command>-clicking on MacOS).
7		Scroll tool	The Scroll tool lets the user to scroll in the active window.
8		Breakpoint tool	The Breakpoint tool sets breakpoints on VI diagrams to help in debugging the code. It causes execution to suspend so that the user can see what is going on and change input values if required.
9		Probe tool	The Probe tool creates probes on wires so that the user can view the data traveling through them while the VI is running.
10		Color Copy tool	Use the Color Copy tool to pick up a color from an existing object, and then use the Color tool to paste that color onto other objects. This technique is very useful if the user needs to duplicate an exact shade but can't remember which one it was. The user can also access the Color Copy tool when the Color tool is active by holding down the <control> key on Windows, <option> on MacOS, <meta> on Sun, and <alt> on Linux and HP-UX.
11		Automatic Tool Selection	The **Automatic Tool Selection** button on the **Tools** palette can be selected to disable automatic tool selection. Press the <Shift-Tab> keys or click the **Automatic Tool Selection** button to enable automatic tool selection again. Press the <Tab> or <Shift-Tab> keys or clicks the **Automatic Tool Selection** button on the **Tools** palette to enable automatic tool selection again. If automatic tool selection is disabled, the user can press the spacebar to switch to the next most useful tool.

2.4 Dataflow Programming

LabVIEW follows a dataflow model for running VIs. A block diagram node executes when all its inputs are available. When a node completes execution, it supplies data to its output terminals and passes the output data to the next node in the dataflow path. Visual Basic, C + +, JAVA, and most other text-based programming languages follow a control flow model of program execution. In control flow, the sequential order of program elements determines the execution order of a program. For a dataflow-programming example, consider a block diagram that adds two numbers and then subtracts 50.00 from the result of the addition as illustrated in Fig. 2.29a. In this case, the block diagram executes from left to right, not because the objects are placed in that order, but because the Subtract function cannot execute until the Add function finishes executing and passes the data to the Subtract function. Remember that a node executes only when data are available at all of its input terminals, and it supplies data to its output terminals only when it finishes execution. In the following example in Fig. 2.29b, consider which code segment would execute first – the Add, Random Number, or Divide function. In a situation where one code segment must execute before another and no data dependency exists between the functions, the user can use other programming methods, such as error clusters, to force the order of execution.

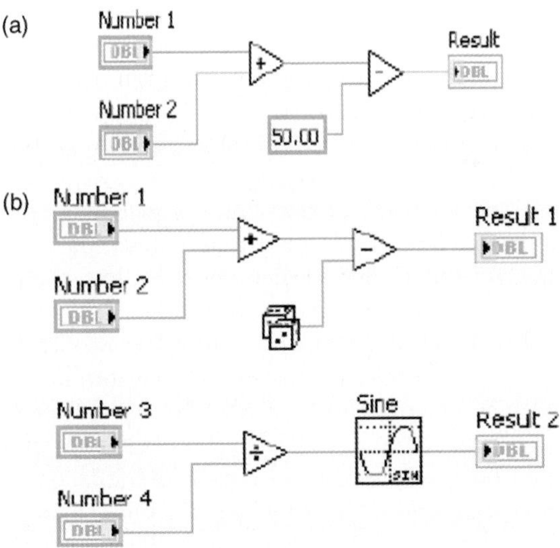

Fig. 2.29. Dataflow programming example

2.5 'G' Programming

The 'G' sequence structure is used to control execution order when natural data dependency does not exist. The user also can create an artificial data dependency in which the receiving node does not actually use the data received. Instead, the receiving node uses the arrival of data to trigger its execution.

The programming language used in LabVIEW, called "G", is a dataflow language. Execution is determined by the structure of a graphical block diagram (the LV-source code) on which the programmer connects different function-nodes by drawing wires. These wires propagate variables and any node can execute as soon as all its input data become available. Since this might be the case for multiple nodes simultaneously, "G" is inherently capable of parallel execution. Multiprocessing and multi-threading hardware is automatically exploited by the built-in scheduler, which multiplexes multiple OS threads over the nodes ready for execution.

Programmers with a background in conventional programming often show a certain reluctance to adopt the LabVIEW dataflow scheme, claiming that LabVIEW is prone to race conditions. In reality, this stems from a misunderstanding of the data-flow paradigm. The afore-mentioned data-flow (which can be "forced," typically by linking inputs and outputs of nodes) completely defines the execution sequence, and that can be fully controlled by the programmer.

Thus the execution sequence of the LabVIEW graphical syntax is as well-defined as with any textually coded language such as C, Visual BASIC, or Python, etc. Furthermore, LabVIEW does not require type definition of the variables; the wire type is defined by the data-supplying node. LabVIEW supports polymorphism in that wires automatically adjust to various types of data.

LabVIEW ties the creation of user interfaces (called front panels) into the development cycle. LabVIEW programs/subroutines are called virtual instruments (VIs). As mentioned earlier, each VI has three components: a block diagram, a front panel and a connector pane. The latter may represent the VI as a subVI in block diagrams of calling VIs. Controls and indicators on the front panel allow an operator to input data into or extract data from a running virtual instrument. However, the front panel can also serve as a programmatic interface. Thus a virtual instrument can either be run as a program, with the front panel serving as a user interface, or, when dropped as a node onto the block diagram, the front panel defines the inputs and outputs for the given node through the connector pane. This implies each VI can be easily tested before being embedded as a subroutine into a larger program.

The graphical approach also allows nonprogrammers to build programs by simply dragging and dropping virtual representations of the lab equipment with which they are already familiar. The LabVIEW programming environment, with the included examples and the documentation, makes it easy to create small applications. This is a benefit on one side but there is also a certain danger of underestimating the expertise needed for good quality "G" programming. For complex algorithms or large-scale code it is important that the programmer understands the special LabVIEW syntax and the topology of its memory management well. The most advanced LabVIEW development systems offer the possibility of building stand alone applications. Furthermore, it is possible to create distributed applications which communicate by a server/client scheme, which the inherently parallel nature of G-code makes easy.

2.5.1 Data Types and Conversion

Data types indicate the type of objects, inputs, and outputs that can be wired together. For example, a switch has a green border so the user can wire a switch to any input with a green label on an Express VI. A knob has an orange border so the user can wire a knob to any input with an orange label. However, an orange knob cannot be wired to an input with a green label. The dynamic data type stores the information generated or acquired by an Express VI. The dynamic data type appears as a dark blue terminal.

Most Express VIs accept and/or return the dynamic data type. The dynamic data type can be wired to any indicator or input that accepts numeric, waveform, or Boolean data. Wire the dynamic data type to an indicator that can best present the data. Indicators include a graph, chart, or numeric indicator. Most other VIs and functions in LabVIEW do not accept the dynamic data type. To use a built-in VI or function to analyze or process the data the dynamic data type must be converted to another form.

To perform the conversion the **Convert from Dynamic Data** Express VI can be used to convert the dynamic data type to numeric, Boolean, waveform, and array data types for use with other VIs and functions. When the **Convert from Dynamic Data** Express VI is placed on the block diagram, the **Configure Convert from Dynamic Data** dialog box appears. The **Configure Convert from Dynamic Data** dialog box displays options to format the data that the **Convert from Dynamic Data** Express VI returns.

When a dynamic data type is wired to an array indicator, LabVIEW automatically places the Convert from Dynamic Data Express VI on the block diagram. By double-clicking the Convert from Dynamic Data Express VI, the **Configure Convert from Dynamic Data** dialog box opens which is used to control the appearance of data in the array.

The Convert to Dynamic Data Express VI can also be used to convert numeric, Boolean, waveform, and array data types to the dynamic data type for use with Express VIs. When the Convert to Dynamic Data Express VI is placed on the block diagram, the **Configure Convert to Dynamic Data** dialog box appears. This dialog box is used to select the kind of data to convert to the dynamic data type.

Data Operations

The data operations Pop-up menu has several handy options to let one manipulate the data in a control or indicator. Some of the features are discussed as follows:

- **Reinitialize to Default** returns an object to its default value, while **Make Current Value Default** sets the default value to whatever data are currently there.
- Use **Cut Data**, **Copy Data**, and **Paste Data** to take data out or put data into a control or indicator.
- Data socket connection brings up a dialog box where the user can configure this control to be connected to a Data socket URL.

Advanced

The advanced Pop-up option gives access to less-frequently used feature that let the user fine-tune the appearance and behavior of the control or indicator:

- **Use Key navigation** is used to associate a keyboard key combination with a front panel object. When a user enters that key combination while a VI is running, LabVIEW acts as if the user had clicked on that object, and the object becomes the key focus.
- **Synchronous Display** is a selectable option that forces LabVIEW to refresh the display of this control or indicator on every update.
- **Customize** will bring up the Control Editor to allow the user to customize the graphical appearance of the control.
- **Hide control/indicator** is used to hide a front panel object. To view the front panel object again, select show control/indicator on the Pop-up menu of the block diagram terminal.
- **Enabled State** allows the user to set a control's state as **enabled, disabled,** or **disabled & grayed**.

2.5.2 Representation and Precision

The appearance of numeric terminals on the block diagram depends on the *representation* of the data. The different representations provide alternative methods of storing data, to help use memory more effectively. Different numeric representations may use a different number of bytes of memory to store data, or they may view data as *signed* (having only zero or positive values). Block diagram terminals are blue for integer data and orange for floating-point data (integer data have no digits to the right of the decimal point). The terminals contain a few letters describing the data type, such as "DBL" for double-precision floating-point data. The numeric data representations available in LabVIEW are shown in Table 2.6, along with their size in bytes.

Table 2.6. Numeric data representations

Representation	Abbreviation	Size (bytes)
Byte	18	1
Unsigned byte	U8	1
Word	I16	2
Unsigned word	U16	2
Long	i32	4
Unsigned long	U32	4
Single precision	SGL	4
Double precision	DBL	8
Extended precision	EXT	10[a]/12[b]/16[c]
Complex single	CSG	8
Complex double	CDB	16
Complex extended	CXT	20[a]/24[b]/32[c]

[a] Windows
[b] MacOS
[c] Unix

Popping up on the object and selecting **Representation≫** can change the representation of numeric constants, controls, and indicators. To be concerned about memory requirements, especially if we are is using larger structures such as arrays, the smallest representation that will hold the data without losing information is used. **Adapt to Source** automatically assigns the representation of the source data to the indicator. LabVIEW also contains functions that convert one data type to another.

Precision

LabVIEW lets the user select whether the digital displays are formatted for numeric values or for time and date. If numeric, the user can choose whether the notation is floating point, scientific, engineering, or relative time in seconds. The user can also choose the *precision* of the display, which refers to the number of digits to the right of the decimal point, from 0 to 20. The precision affects only the display of the value; the internal accuracy still depends on the representation.

Selecting **Format & Precision** from the object's Pop-up menu can modify the format and precision. The dialog box appears. To show time and date, **Time & Date** from the **Format** ring is chosen and the dialog box will change accordingly.

Sometimes it can be hard to discern exact values from graphical controls and indicators like graphs and thermometers. Use the Pop-up option to **Visible items≫Digital Display** to bring up a digital window next to the object and display the precise numeric value. This dialog display is part of the object itself and will not have a block diagram terminal.

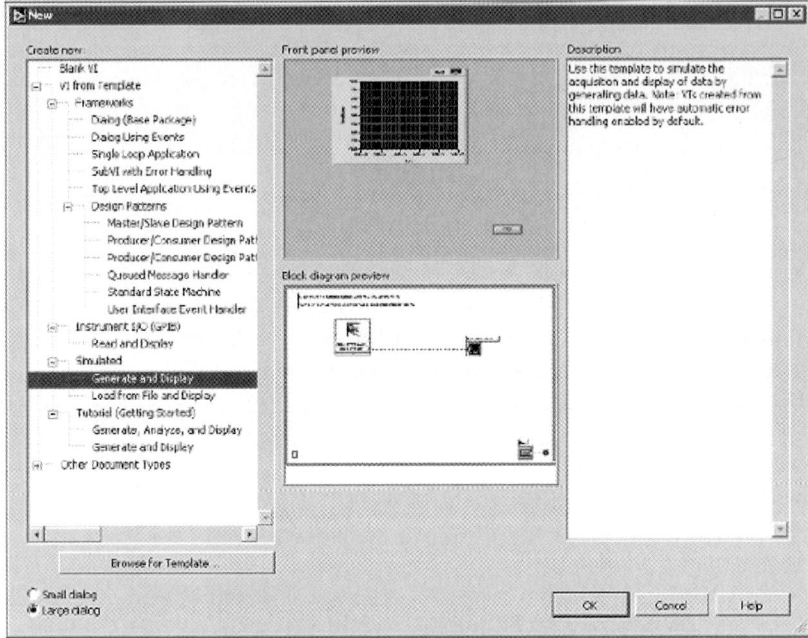

Fig. 2.30. New dialog box

2.5.3 Creating and Saving VIs

The **New** dialog box appears when the **New** button in the **LabVIEW** dialog box is selected. This dialog box can also be selected from **File≫New**. When a template in the **Create new** list is selected, previews of the VI appear in **Front Panel Preview** and **Block Diagram Preview** sections, and a description of the template appears in the **Description** section. Figure 2.30 shows the **New** dialog box and the subVI with Error Handling VI template.

If no template is available for the task to be created, the user can start with a blank VI and create a VI to accomplish the specific task. In the **LabVIEW** dialog box, click the arrow on the **New** button and select **Blank VI** from the shortcut menu or press the <Ctrl-N> keys to open a blank VI.

Open/Templates

Using the **New** dialog box, different components in LabVIEW can be created to build an application. The user can start with a blank VI to write a VI from scratch, or start with a template to simplify the programming. The **New** dialog box includes the following components:

- **Create new**. Displays a set of templates to start building VIs and other LabVIEW documents. The user can select from the following templates to start building a VI or other LabVIEW document. The templates are:

1. **Blank VI** – opens a blank front panel and blank block diagram.
2. **VI from Template** – opens a front panel and block diagram with components required to build different types of VIs.

- **DAQ**. Opens a front panel and a block diagram with the components required to measure or generates signals using the DAQ Assistant Express VI and NI-DAQmx.
- **Frameworks**. Opens a front panel and block diagram with the components and settings required to build VIs that include a specific type of functionality.
- **Instrument I/O**. Opens a front panel and block diagram with the components required to communicate with an external instrument attached to the computer through a port, such as a serial or GPIB-enabled device.
- **Simulated**. Opens a front panel and block diagram with the components required to simulate acquired data from a device.
- **Tutorial (Getting Started)**. Opens a front panel and block diagram with the components required to build the VIs
- **Browse for Template**. Displays the **Browse** dialog box so that the user can navigate to a VI, control, or template. If the user has previously browsed for and selected a template from this dialog box, use the pull-down menu of the **Browse** button to select a template to reopen it.
- **Front panel preview**. Displays the front panel for the VI template selected in the **Create new** list.
- **Block diagram preview**. Displays the block diagram for the VI template selected in the **Create new** list.
- **Description**. Displays a description of the template selected in the **Create new** list if the template includes a description.

Opening an Existing VI

A VI can be loaded into memory by selecting **File≫Open**. In the **Choose the VI to open** dialog box that appears, navigate the VI to be opened. As the VI loads, a status dialog box similar to the following example might appear.

The **Loading** section lists the subVIs of the VI as they are loaded into memory. **Number Loaded** is the number of subVIs loaded into memory so far. The file loading can be cancelled at any time by clicking the **Stop** button.

If LabVIEW cannot immediately locate a subVI, it begins searching through all directories specified by the VI Search Path. The user can edit the VI Search Path by selecting **Tools≫Options** and selecting **Paths** from the top pull-down menu. The **Searching** section lists directories or VIs as LabVIEW searches through them. SubVIs can be ignored by clicking the **Ignore SubVI** button, or by clicking the **Browse** button to search for the missing subVI.

Saving VIs

Select **Save**, **Save As**, **Save All**, or **Save with Options** from the **File** menu to save VIs as individual files or group several VIs together and save them in a VI library. VI library files end with the extension .llb. LabVIEW uses the native file dialog boxes so they act similar to other applications on the computer. Selecting Tools Options and selecting Miscellaneous from the top pull-down menu can disable this feature. If native file dialogs are disabled, LabVIEW uses its own platform-independent file dialog boxes with some convenient features, such as providing a list of recent paths and reducing the steps necessary to save VIs in VI libraries as shown in Fig. 2.31.

2.5.4 Wiring, Editing, and Debugging

Automatically Wiring Objects

LabVIEW automatically wires objects as they are placed on the block diagram. LabVIEW connects the terminals that best match and leaves terminals that do not match unconnected. As the user moves a selected object close to other objects on the block diagram, LabVIEW draws temporary wires to show the valid connections. When the mouse button is released, LabVIEW automatically connects the wires.

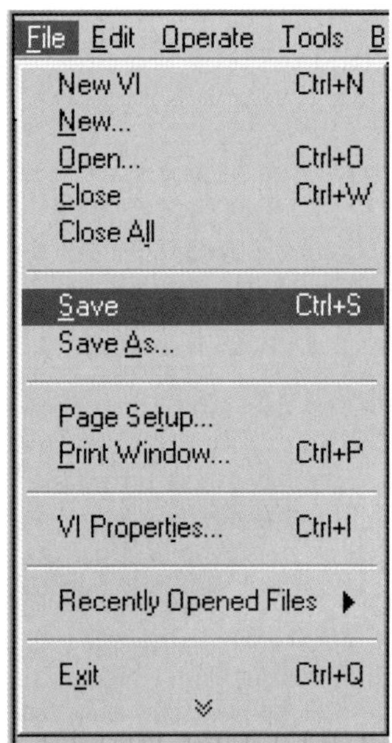

Fig. 2.31. File menu

Toggle automatic wiring by pressing the spacebar while an object is moved using the Positioning tool. The automatic wiring settings can be adjusted by selecting **Tools≫Options** and selecting **Block Diagram** from the top pulls-down menu.

Manually Wiring Objects

When the wiring tool is moved over a terminal, a tip strip appears with the name of the terminal. In addition, the terminal blinks in the **Context Help** window and on the icon to help the user verify that the objects are wired to the correct terminal.

Debugging Techniques

LabVIEW has many built in debugging features to help the user develop VIs. This section explains how to use the conveniences to the best advantage.

Fixing a Broken VI

A *broken VI* is a VI that cannot complete or run. The **Run** button appears as a broken arrow to indicate that the VI has a problem. It's perfectly normal for a VI to be broken while the user is creating or editing it, until all the icons are wired in the diagram. Sometimes it may be required to **Remove Broken Wires** (found in **Edit** menu) to clean up loose wires.

To find out why a VI is broken, click on the broken **Run** button or select **Show Error List** from the **Windows** menu. An information box titled "Error List" appears listing all errors for the VI. The user can choose to see the error list for other open VIs using a menu ring at the top of the window. To find out more about a particular error, click on it. The Error List window will display more information. To locate a particular error in the VI, double-click on the error in the list or highlight it and press the **Find** button. LabVIEW bring the relevant window to the front and highlights the object causing the error.

Finding Errors

By clicking the broken **Run** button or selecting **Windows≫Show Error List** the **Error list** window is displayed which lists all the errors. The relevant block diagram or front panel can be displayed by double-clicking an error description and highlight the object that contains the error.

Execution Highlighting

Sometimes it's nice to see exactly where the data is and the operational status of the data. In LabVIEW, the user can view an animation of VI block diagram execution. Clicking on the Execution Highlighting button in the Toolbar enables this mode.

As data is passed from one node to another, bubbles moving along the wires indicate the movement of data. The user will notice that highlighting greatly reduces the performance of a VI. Click again on the Execution Highlighting button to resume normal execution.

Node values are automatically shown during execution highlighting, if **Auto probe is selected** from the **Debugging** menu of the **Options** dialog. Execution highlighting is commonly used in conjunction with single-step mode to gain an understanding of how data flow through nodes. When these two nodes are used together, execution glyphs on the subVI's icon indicate the VIs that are running and are waiting to run.

Single-Stepping

Single-step through a VI is used to view each action of the VI on the block diagram as the VI runs. For debugging purposes, the user may want to execute a block diagram by node. *Nodes* include subVIs, functions, structures, code interface nodes (CINs), formula nodes, and property nodes. The single-stepping buttons affect execution only in a VI or subVI in single-step mode. Enter single-step mode by clicking the **Step Over** or **Step Into** button. The cursor can be moved over the **Step Over**, **Step Into**, or **Step Out** button to view a tip strip that describes the next step.

The user can execute by using single-step through subVIs or run them normally. If the single-step through a VI with execution highlighting is on, an execution glyph, appears on the icons of the subVIs that are currently running. While in *single-step mode*, any of the three step buttons that are active can be selected to precede to the next step. The step button pressed determines how the next step will be executed.

Step Into button is pressed to execute the first step of a subVI or structure and then pause at the next step of the subVI or structure. The Keyboard shortcut for this function is: down arrow key in conjunction with <control> under Windows and <command> on MACS (<meta> on Suns and <alt> on HP workstations).

Step Over button is pressed to execute a structure (sequence, loop, etc.) or a subVI and then pause at the next node. The Keyboard shortcut for this function is: right arrow key in conjunction with <control> under Windows and <command> on MACS.

Step Out button is pressed to finish executing the current block diagram, structure, or VI and then pause. The Keyboard shortcut for this function is: up arrow key in conjunction with <control> under Windows and <command> on MACS.

Probes

The *probe* is used to check intermediate values in a VI that executes but produces questionable or unexpected results. When execution pauses at a

Fig. 2.32. Custom Probe

node because of single-stepping or a breakpoint, we can also probe the wire that just executed to see the value that flowed through that wire. We can also create a custom probe to specify which indicator is used to view the probed data. For example, if a numeric data is viewed, the user can choose to see that data in a chart within the probe. A custom probe can be created by right-clicking a wire and selecting **Custom Probe≫New** from the shortcut menu as shown in Fig. 2.32.

Assume we have a program with a series of operations, any one of which may be the cause of correct output data. To fix it, we could create an indicator to display intermediate results on a wire, or we can leave the VI running and simply use a probe. To access the probe, the Probe tool from the **Tools** palette can be selected and its cursor can be slicked on a wire, or pop-uped on the wire and select Probe. The probe display, which is a floating window, first appears empty if VI is not running. When the VI is running, the probe display shows the value carried by its associated wire.

We can use the probe with execution highlighting and single-step mode to view values more easily. Each probe and the wire are arithmetically numbered by LabVIEW to keep track of the operation. A probe's number will not be visible if the name of the object probe is longer than the Probe window itself. If the user loses track of which probe goes with which wire, then the user can pop up on a probe or a wire and select **find Wire** or **Find Probe**, respectively, to highlight the corresponding object. The user cannot change data with the probe.

We can also probe using any conventional indicator by selecting **Custom Probe>** from the wire's Pop-up menu, and then choosing the desired indicator with which to probe. For example, we could use a chart to show the progress of a variable in a loop, since it displays past values as well as current ones. LabVIEW does not allow the user to select an indicator of a different data type than the wire.

Breakpoints

The Breakpoint tool is used to place a breakpoint on a VI, node, or wire on the block diagram and to pause execution at that location. When a breakpoint is set on a wire, execution pauses after data pass through the wire.

A breakpoint can be placed on the block diagram workspace to pause execution after all nodes on the block diagram execute. When a VI pauses at a breakpoint, LabVIEW brings the block diagram to the front and uses a marquee to highlight the node or wire that contains the breakpoint. LabVIEW highlights breakpoints with red borders for nodes and block diagrams and red bullets for wires. When we move the cursor over an existing breakpoint, the black area of the Breakpoint tool cursor appears white. Use the Breakpoint tool to click an existing breakpoint to remove it.

A breakpoint can be set by clicking on a block diagram object with the Breakpoint tool from the **Tools** palette. Clicking again on the object clears the breakpoint. The appearance of the breakpoint cursor indicates whether a breakpoint will be set or cleared.

Depending on where they are placed, breakpoints behave differently.

- If the breakpoint is set on a *block diagram*, a red border appears around the diagram and the pause will occur when the block diagram completes.
- If the breakpoint is set on a *node*, a red border frames the node and execution pauses just before the node executes.
- If the breakpoint is set on a *wire*, a red bullet appears on the wire and any attached probe will be surrounded by a red border. The pause will occur after data have passed through the wire.

When a VI pauses because of a breakpoint, the block diagram comes to the front, and the object causing the break is highlighted with a marquee. Breakpoints are saved with a VI are active only during execution.

Suspending Execution

We can also enable and disable breakpoints with the **Suspend when Called** option, found in the **Execution** menu of **VI Properties** (from the icon pane Pop-up menu in a VI's front panel). **Suspend when Called** causes the breakpoint to occur at all calls to the VI on which it's set. If a subVI is called from two locations in a block diagram, the subVI breakpoint suspends execution at both calls.

If the user wants to set a breakpoint to suspend execution only at a particular call to the subVI, the breakpoint can be set using the **SubVI Node Setup** option. This option can be accessed by popping up on the subVI icon (in the block diagram of the calling VI).

Warnings

To obtain extra debugging help, the user can choose **Show Warnings** in the Error List window by clicking in the appropriate box. A warning is something that is not illegal and would not cause a broken run arrow but does not make sense to LabVIEW, such as a control terminal that is not wired to anything. If **Show Warnings** is checked and have any outstanding warnings, the Warning button can be viewed on the Toolbar. By clicking on the Warning button the Error List window describes the warning.

The user can also configure LabVIEW's options to show warnings by default. The Debugging menu in the **Options** dialog box (accessed by selecting **Tools≫Options. . .**) is used to check the **Show warnings in error box by a fault box**.

Most Common Mistakes

Certain mistakes that are made more frequently are explained later. If run button is broken, one of the following describes the problem:

- A function terminal requiring an input is unwired. The user cannot leave unwired functions on the diagram while we run a VI to try out different algorithms.
- The block diagram contains a bad wire due to a data-type mismatch or a loose, unconnected end, which may be hidden under something or so tiny that cannot be seen. The **Remove Broken Wires** command from the **Edit** menu eliminates the bad wires.
- A subVI is broken, or edited its connector after placing its icon on the diagram. The **Replace** or **Relink to subVI** Pop-up option to a relink to the subVI.
- When two controls are wired together to the same indicator. The Error List window will bear the message, "Signal: has multiple sources," for this problem. This can be solved by changing one of the controls to an indicator.

2.5.5 Creating SubVIs

Much of LabVIEW's power and convenience stems arise from its modularity. The parts of the program can be built into one complete module at a time by creating subVIs. A subVI is simply a VI used in (or called by) another VI. A subVI node (comprised of icon/connector in a calling VI block diagram) is analogous to a subroutine call in a main program. A block diagram can contain several identical subVI nodes that call the same subVI several times.

Any VI can be used as a subVI in the block diagram of another VI, provided its icon has been created and its connector assigned. The existing VIs on a block diagram can be used as subVIs with the **Select a VI** button in the **Functions** palette. Choosing this option produces a file dialog box from which we can select any VI in the system; its icon will appear on the diagram.

A VI cannot be call itself directly as a subVI. This can be done by recursion in which a VI can call itself indirectly using a VI reference.

Creating a SubVI from a VI

Before a VI is used as a subVI, the user must supply a way for the VI to receive data from and pass data to the calling VI. To perform this, the user needs to assign the VI controls and indicators to terminals on its connector pane, and the user must create an icon to represent the VI.

Designing the Icon

Every subVI must have an icon to represent it in the block diagram of a calling VI; the icon to represent it in the block of a calling VI; the icon is in its graphical symbol. An icon is a graphical representation of a VI. It can contain text, images, or a combination of both. If we use a VI as a subVI, the icon identifies the subVI on the block diagram of the VI.

The default icon contains a number that indicates new VIs that have been opened since launching LabVIEW. The icon can be created by selecting **Edit Icon** from the Pop-up menu of the icon pane in the upper right-hand corner of the front panel. The user must be in edit mode to operate upon this menu.

The *Icon Editor* can be accessed by double clicking on the icon in the icon pane. The Icon Editor window appears on the front panel. The icons can also be edited by selecting **File≫VI Properties**, selecting **General** from the **Category** pull-down menu, and clicking the **Edit Icon** button. The tools on the left side of the **Icon Editor** Dialog box are used to create the icon design in the editing area. The normal image size of the icon appears in the appropriate box to the right of the editing area. Depending on the type of monitor used, the user can design a separate icon for monochrome, 16-color, and 256-color mode. LabVIEW uses the monochrome icon for printing unless we have a color printer.

The **Edit** menu is used to cut, copy, and paste images from and to the icon. When the user selects a portion of the icon and paste an image, LabVIEW resizes the image to fit into the selection area. A graphic can also be dragged from anywhere in the file system and dropped in the upper right corner of the front panel or block diagram. LabVIEW converts the graphic to a 32 pixel icon.

The **Copy from** option on the right side of the **Icon Editor** dialog box is used to copy from a color icon to a black-and-white icon and vice versa. After selecting a **Copy from** option, the user can click the **OK** button to complete the change.

The tools on the left side of the **Icon Editor** Dialog box are used to create the icon design in the editing area. The normal image size of the icon appears in the appropriate box to the right of the editing area.

The following tasks can be performed with the tools available on the icon editor:

- The Pencil tool to draw and erase pixel by pixel.
- The Line tool to draw straight lines. To draw horizontal, vertical, and diagonal lines, press the <Shift> key while this tool is used to drag the cursor.
- The Color Copy tool to copy the foreground color from an element in the icon.
- The Fill tool to fill an outlined area with the foreground color.
- The Rectangle tool to draw a rectangular border in the foreground color. This tool can be double-clicked to frame the icon in the foreground color.
- The Filled Rectangle tool to draw a rectangle with a foreground color frame and filled with the background color. This tool can be double-clicked to frame the icon in the foreground color and fill it with the background color.
- The Select tool to select an area of the icon to cut, copy, move, or make other changes. This tool can be double-clicked with the <Delete> key pressed to delete the entire icon.
- The Text tool to enter text into the icon. This tool can be double-clicked to select a different font. **(Windows)** The **Small Fonts** option works well in icons.
- The Foreground/Background tool to display the current foreground and background colors. Each rectangle can be clicked to display a color palette from which the user can select new colors.

The options on the right-hand side of the editing area is used to perform the following tasks:

- **Show Terminals**. Displays the terminal pattern of the connector pane.
- **OK**. Saves the drawing as the icon and returns to the front panel.
- **Cancel**. Returns to the front panel without saving any changes.

The menu bar in the **Icon Editor** dialog box contains more editing options such as **Undo**, **Redo**, **Cut**, **Copy**, **Paste**, and **Clear**.

Assigning the Connector

Before using a VI as a subVI, connector terminals have to be assigned, similar to the defining parameters for a subroutine in a conventional language. The connector pane is a set of terminals that corresponds to the controls and indicators of that VI, similar to the parameter list of a function call in text-based programming languages. The connector is a LabVIEW's way of passing data into and out of a subVI. The connector of a VI assigns the VIs control and indicators to input and output terminals.

To define the connector, the icon pane can be popped and **Show Connector** from Shortcut menu is selected. To view the icon again, the connector is popped up and **Show Icon** from Shortcut menu is selected. LabVIEW chooses a default connector based on the number of controls and indicators on the front panel. If the user wants a different one, it can be chosen from the Patterns menu, obtained by popping up on the connector. The user can also rotate and flip the connector if it doesn't have a convenient orientation, using commands in the connector Pop-up menu. The connector pane replaces the icon.

Assigning Terminals to Controls and Indicators

After selecting a pattern to use for the connector pane, the user should define connections by assigning a front panel control or indicator to each of the connector pane terminals. When there is a link between controls and indicators to the connector pane, the inputs are placed on the left and outputs are placed on the right to prevent complicated, unclear wiring patterns in the VIs. To assign a terminal to a front panel control or indicator, a terminal of the connector pane is selected first and the front panel control or indicator to be assigned to that terminal is selected. The terminal changes to the data type color of the control to indicate that the terminal has been connected. The user can also select the control or indicator first and then select the terminal. Each rectangle on the connector pane represents a terminal. These rectangles are used to assign inputs and outputs. The number of terminals LabVIEW displays on the connector pane depends on the number of controls and indicators on the front panel. The following front panel has four controls and one indicator, so LabVIEW displays four input terminals and one output terminal on the connector pane.

The following steps are used to assign a terminal to a control or indicator:

1. Click on a terminal in the connector. The cursor automatically changes to the Wiring tool, and the terminal turns black.
2. Click on the control or indicator that the terminal should represent. A moving dotted line frames the control or indicator.
3. Click in an open area on the front panel. The dotted line disappears and the selected terminal dims as pictured, indicating that the control or indicator has been assigned to that terminal.

The order of the first two steps can be reversed. If the terminal is white or black, it indicates that the connection is improper. To solve this, the previous steps are repeated. The user can have up to 28 connector terminals for each VI. If a mistake occurs, the user can **disconnect** a particular terminal or **Disconnect All** by selecting the appropriate action from the connector's Pop-up menu.

Selecting and Modifying Terminal Patterns

A different terminal pattern for a VI can be selected by right-clicking the connector pane and selecting **Patterns** from the shortcut menu. A connector pane pattern with extra terminals can also be selected. These extra terminals can be left unconnected until the user requires them. This flexibility enables us to make changes with minimal effect on the hierarchy of the VIs. There can also be more front panel controls or indicators than terminals. A solid border highlights the pattern currently associated with the icon.

The maximum number of terminals available for a subVI is 28. This pattern is used as a standard to assist in simplifying wiring. The top inputs and outputs are commonly used for passing references and the bottom inputs and outputs are used for error handling. To change the spatial arrangement of the connector pane patterns, right-click the connector pane and select **Flip Horizontal**, **Flip Vertical**, or **Rotate 90 Degrees** from the shortcut menu.

Creating SubVIs from a Block Diagram Selection

Fortunately, subVIs can also be created by converting a part of the code in an existing VI. The Positioning tool is used to select the section of the VI to be replaced with a subVI. Choosing **Create SubVI** from the **Edit** menu creates a subVI with correct wiring and an icon. The user can double-click on the new subVI to view its front panel edit its icon, look at its connector, and save it under a new name.

2.5.6 VI Libraries

VI libraries are special LabVIEW files that have the same load, save, and open capabilities as directories and folders within the LabVIEW environment. Several VIs can be grouped together and saved as a VI library. VI libraries offer several advantages and disadvantages; for example, they can contain only compressed versions of VIs, not data or other files. In addition, the operating system sees VI libraries as single files, and the user can access their contents only from LabVIEW.

Reasons for Using VI Libraries

- The user can use up to 255 characters to name the files.
- The user can transfer a VI library to other platforms more readily than transferring multiple individual VIs.
- The user can slightly reduce the file size of the project because VI libraries are compressed to reduce disk space requirements.

Reasons to Save VIs as Individual Files

- The user can use the file system to manage the individual files (e.g., copy, move, rename, backup) without having to go through LabVIEW.
- The user can use subdirectories.

- The user can store VIs and controls in individual files more robustly than storing the entire project in the same file.
- The user can use the Professional Development System built-in source code control tools.

How to Use VIs Libraries

A VI library can be created from the **Save** or **Save As** Dialog box by clicking on the **New VI Library** button under **Windows** or the **New** button on MacOS. The name of the new library is entered in the dialog box that appears, and appends an .llb extension. Then the **VI Library** button is clicked and the library is created. If the user does not include the .llb extension, LabVIEW adds it by default. Usually a VI library is created when a VI is saved, so after the library is created, a dialog box appears to allow the user name the VI and save it in the new library. Creating a VI library, saving VIs in it and accessing them through LabVIEW is similar to creating a directory or folder, but the individual VIs cannot be seen from the operating system.

The Edit VI Library Dialog

The contents of a VI library cannot be edited through the operating system; it can be done only with the use of the **Edit VI Library** dialog box. The **Edit VI Library** dialog box, available from the **Tools** menu, initially displays a list of the files in the VI library. As moving through the list, the creation and last modification dates for the selected file are shown at the bottom of the dialog box.

If VI marked as Top **Level**, it will load automatically when VI library is opened. More than one top-level VI in a library can be present. Top-level VI names will also appear in a separate section at the top of the Load dialog, making it easier for determining which VIs are main.

The VI Library Manager

The **VI Library Manager** (from the **Tools menu**) is used to simplify copying, renaming, and deleting files within VI libraries as well as the file system. This is also used to create new VI libraries and directories and convert VI libraries to and from directories if the user needs to manage the VIs with source code control tools.

Summary

- VIs contain three main components – the front panel, the block diagram, and the icon and connector pane.
- The front panel is the user interface of a VI and specifies the inputs and displays the outputs of the VI.

- The block diagram contains the graphical source code composed of nodes, terminals, and wires.
- The **Tools** palette is used to create, modify, and debug VIs. By pressing the <Shift> key and right clicking the cursor a temporary version of the **Tools** palette at the location of the cursor can be displayed.
- The **Controls** palette is used to place controls and indicators on the front panel. By right-clicking an open space on the front panel the **Controls** palette is displayed.
- The **Functions** palette is used to place VIs and functions on the block diagram. By right clicking an open space on the block diagram the **Functions** palette is displayed.
- The **Search** button on the **Controls** and **Functions** palettes is used to search for controls, VIs, and functions.
- All LabVIEW objects and empty space on the front panel and block diagram have associated shortcut menus, which is accessed by right-clicking an object, the front panel, or the block diagram.
- The **Help** menu is used to display the **Context Help** window and the *LabVIEW Help*, which describes most palettes, menus, tools, VIs, functions, and features.
- The user can build the front panel with controls and indicators, which are the interactive input and output terminals of the VI, respectively.
- Control terminals have a thicker border than indicator terminals. To change a control to an indicator or to change an indicator to a control, right-click the object and select **Change to Indicator** or **Change to Control** from the shortcut menu.

Review Questions

1. What are the recent research achievements in programming techniques?
2. Define the terms: front panel and block diagram.
3. Write about the various trends in the current scenario that effect LabVIEW.
4. Give some of the LabVIEW techniques.
5. Explain about numeric and Boolean controls and indicators with an example.
6. Define the LabVIEW environment.
7. Explain in detail about the process involved in LabVIEW environment.
8. What are the major essential elements in LabVIEW environment?
9. Define dataflow programming.
10. What factors influence the success of dataflow programming?
11. Explain dataflow programming with suitable example.
12. What are the components of dataflow programming?
13. Define G programming.

14. Give details on multidimensional hierarchies used in G programming.
15. State the G programming rules.
16. Explain about business information technology using G programming.
17. Write short notes on dataflow programming and G programming.
18. Compare and contrast G programming and dataflow programming.

3

Programming Concepts of VI

Learning Objectives. On completion of this chapter the reader will have a knowledge on:

- Control Structures such as the For Loop and the While Loop
- Shift Registers and Their function
- Feedback Nodes
- Selection Structures, Case Structures, and Sequence Structures
- The Formula Node
- Single and Multidimensional Arrays
- Autoindexing of Arrays
- Functions for Manipulating Arrays
- Creating Cluster Controls and Indicators
- Cluster functions
- Error Handling
- Waveform Charts, Components and Mechanical Actions of Switches
- Single Plot and Multiple Plot Waveform Graphs
- XY Graphs and Intensity Graphs
- Strings, Creating String Controls and Indicators, String Functions
- Tables and List Boxes
- File Input/Output, File I/O VIs and Functions

3.1 Introduction

Structures are graphical representations of the loops and case statements of text-based programming languages. Structures located on the block diagram window are used to repeat blocks of code and to execute code conditionally or in a specific order. LabVIEW includes the following structures – the While Loop, For Loop, Case structure, Stacked Sequence structure, Flat Sequence structure, Event structure, and Formula Node which are described in this section. This section also describes on arrays which are used to group data elements of the same type and clusters which group data of different type. In the later sections, the section focuses on waveform charts, waveform graphs, XY graphs, and intensity plots along with strings and files I/Os.

3.2 Control Structures

In any programming language the user probably requires repetition of a section of code. LabVIEW offers two loop structures namely, the *For Loop* and *While Loop* to control repetitive operation in a VI. A For Loop executes a specific number of times; a While Loop executes until a specified condition is no longer true. These structures are found under the **Structures** subpalette of the **Functions** palette as shown in Fig. 3.1.

3.2.1 The For Loop

A *For Loop* executes the code inside its borders, called its *subdiagram*, for total of *count* times, where the count equals the value contained in the *count terminal.* The count can be set by wiring a value from outside the loop of the count terminal. If '0' is wired to the count terminal, the loop does not execute. The For Loop is shown in Fig. 3.2.

The *iteration terminal* contains the current number of completed loop iterations; 0 during the first iteration, 1 during the second, and so on, up to N-1 (where N is the number of times the loop executes).

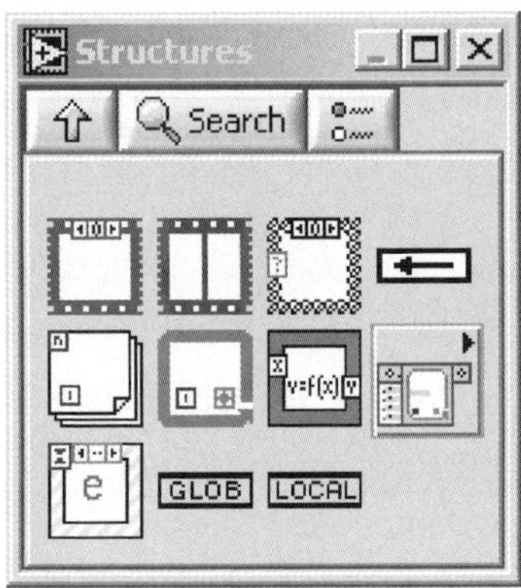

Fig. 3.1. Structures palette

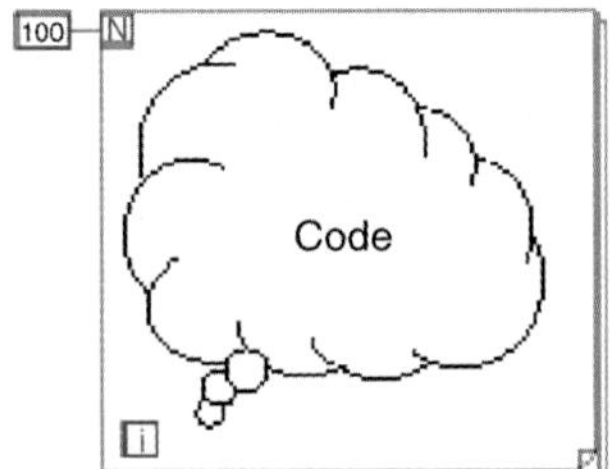

Fig. 3.2. LabVIEW For Loop

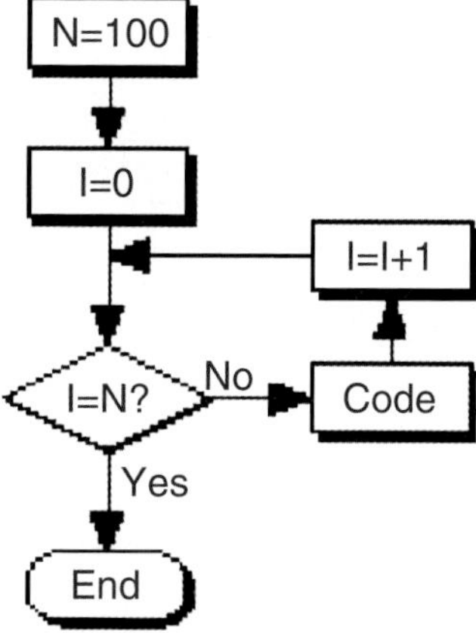

Fig. 3.3. Flow chart

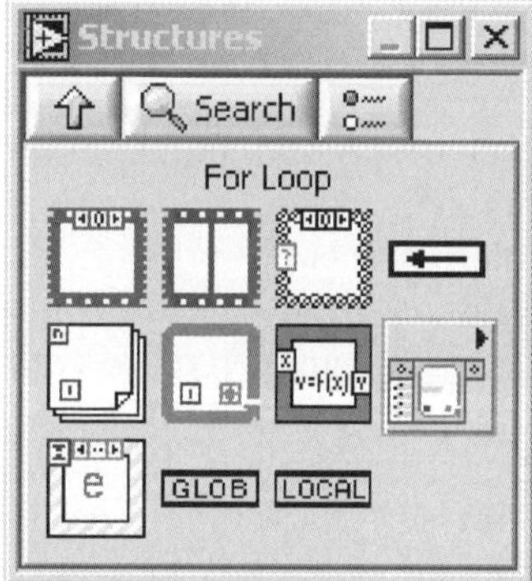

Fig. 3.4. Structures palette

The For Loop is equivalent to the following pseudocode:

For $I = 0$ *to* $N - 1$

Execute subdiagram

The flowchart of the For Loop to illustrate the pseudocode is shown in Fig. 3.3. The For Loop is located on the **Functions≫All Functions≫Structures** palette as shown in Fig. 3.4. A While Loop can also be placed on the block diagram and by right-clicking the border of the While Loop, and selecting **Replace with For Loop** from the shortcut menu, changes a While Loop to a For Loop.

Count Terminal

The value in the count terminal (an input terminal), shown in Fig. 3.5 indicates how many times to repeat the subdiagram.

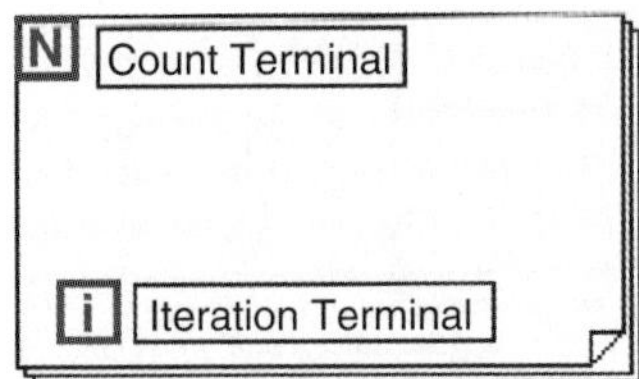

Fig. 3.5. For Loop terminals

Iteration Terminal

The iteration terminal (an output terminal), shown in Fig. 3.5 contains the number of iterations completed. The iteration count always starts at zero. During the first iteration, the iteration terminal returns 0.

Numeric Conversion

LabVIEW can represent numeric data types as signed or unsigned integers (8-bit, 16-bit, or 32-bit), floating-point numeric values (single-, double-, or extended-precision), or complex numeric values (single-, double-, or extended-precision). When two or more numeric inputs of different representations are wired to a function, the function usually returns output in the larger or wider format. The functions coerce the smaller representations to the widest representation before execution, and LabVIEW places a coercion dot on the terminal where the conversion takes place. For example, the For Loop count terminal is a 32-bit signed integer. If a double-precision, floating-point numeric is wired to the count terminal; LabVIEW converts the numeric to a 32-bit signed integer. A coercion dot appears on the count terminal of the first For Loop. If two different numeric data types are wired to a numeric function that expects the inputs to be the same data type, LabVIEW converts one of the terminals to the same representation as the other terminal. LabVIEW chooses the representation that uses more bits. If the number of bits is the same, LabVIEW chooses unsigned over signed.

Wait Functions

The *Wait Until Next ms Multiple function*, shown in Fig. 3.6, monitors a millisecond counter and waits until the millisecond counter reaches a multiple of the amount specified by the user. This function is used to synchronize activities. The function is placed within a loop to control the loop execution rate.

The Wait Until Next ms Multiple function waits until the internal computer timer is at the multiple specified. It is possible that the first loop period might be short as illustrated in Fig. 3.7.

Fig. 3.6. Wait Until Next ms Multiple function

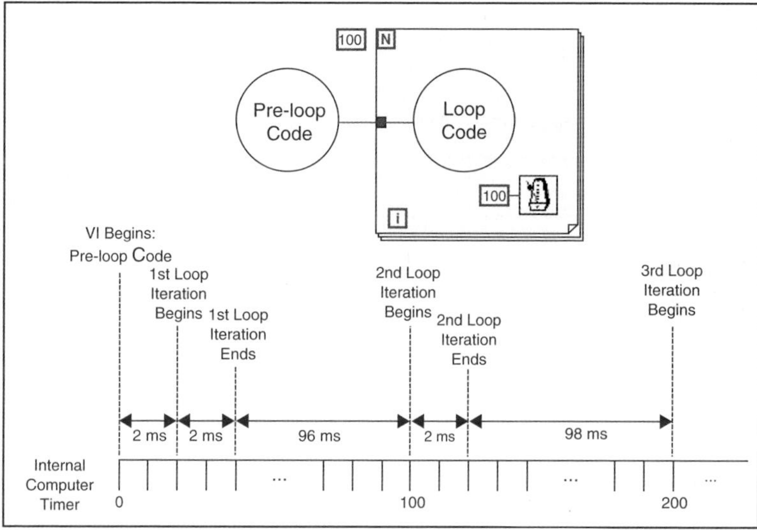

Fig. 3.7. Timing diagram for Wait Until Next ms Multiple function

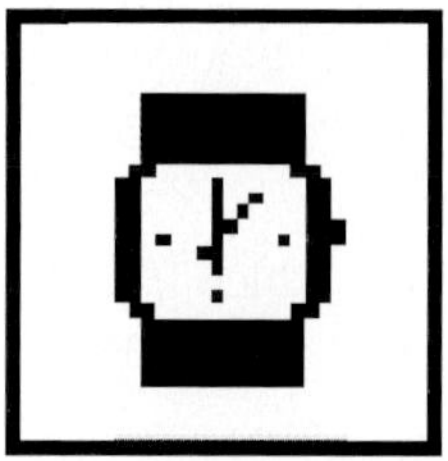

Fig. 3.8. Wait ms function

The Wait (ms) function, shown in Fig. 3.8, adds the wait time to the code execution time, as illustrated in Fig. 3.9. This can cause a problem if code execution time is variable.

The following example illustrates the application of For Loop.

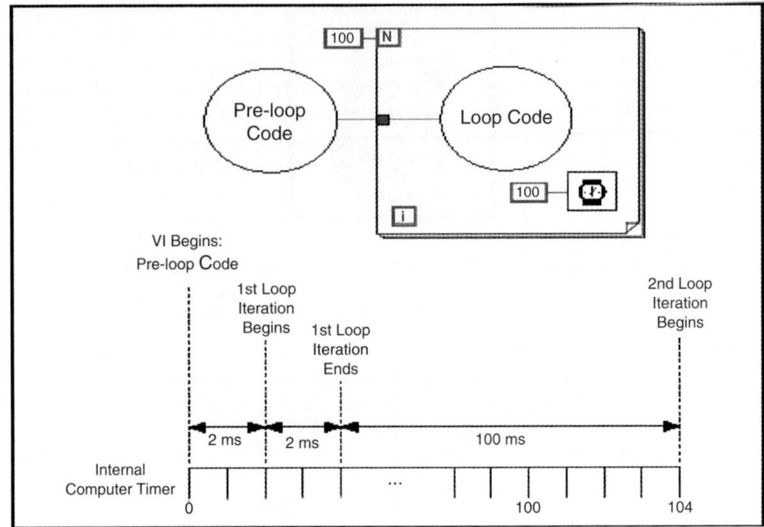

Fig. 3.9. Timing diagram for Wait ms function

Example 1

Problem statement. To find the sum of first 10 natural numbers using For Loop.

Block diagram construction:

- The For Loop is added from the structures subpalette on the functions palette located on the block diagram.
- The count terminal is wired with the value 10 for 10 natural numbers.
- The iteration terminal is wired the greater than or equal to node from the comparison subpalette from the Functions palette.
- The Boolean Output of the greater than or equal to is wired to the conditional terminal as in Fig. 3.10.

Front panel construction:

- The control and indicator are added from the controls palette of the front panel. Using the labeling tool the Owned label is changed as input and output, respectively.
- The front panel with the result is shown in Fig. 3.11.

Example 2

Problem statement. To find the factorial of a given number using For Loop.

Block diagram construction:

- The *For Loop* is added from the Functions palette on the block diagram.
- The code to compute the factorial of a given number is added inside the loop.

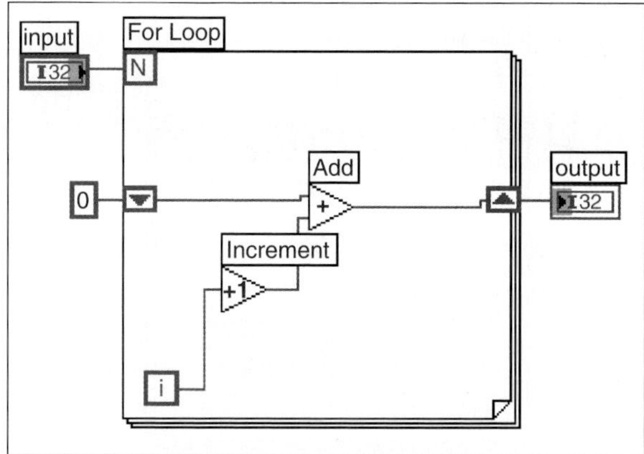

Fig. 3.10. Block diagram for Example 1

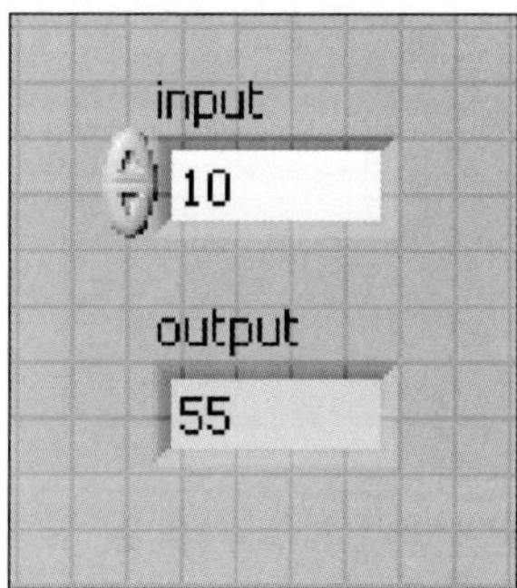

Fig. 3.11. Front panel for Example 1

- The increment and multiply nodes are added from the subpalette of the Functions palette.
- The *Wait ms* are added from the subpalette of the Functions palette.
- The number whose factorial is to be computed is wired to the count terminal of the For loop.
- The iteration terminal is wired to the increment node inside the loop.
- Shift registers are added by right clicking on the border of the loop.
- Input and output are wired as shown in Fig. 3.12.

Front panel construction:

- The control and indicator are added from the controls palette of the front panel. Using the labeling tool the Owned label is changed as input and output, respectively.
- The front panel with the result obtained is shown in Fig. 3.13.

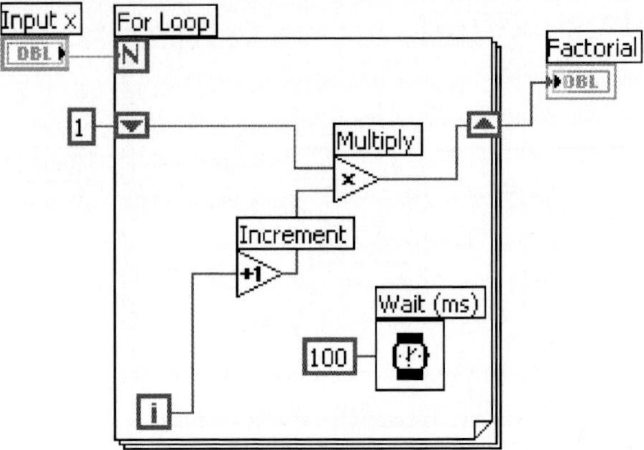

Fig. 3.12. Block diagram for Example 2

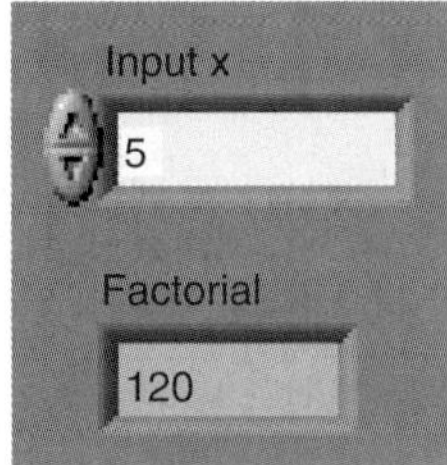

Fig. 3.13. Front panel for Example 2

3.2.2 The While Loop

The *While Loop* executes the subdiagram inside its borders until the Boolean value wired to its *conditional terminal* is FALSE. LabVIEW checks the conditional terminal value at the *end* of iteration. If the value is TRUE, the iteration repeats. The default value of the conditional terminal is FALSE, so if left unwired, the loop iterates only once.

The While Loop's *iteration terminal* shown in Fig. 3.14 behaves exactly like the one in the For Loop.

The While Loop is equivalent to the following pseudocode:

Do
Execute subdiagram
While condition is TRUE

We can also change the state that the conditional terminal of the While Loop checks, so that instead looping *while true*, we can have it loop *unless it's true*. To do this, we pop-up on the conditional terminal, and select **"Stop if True."**

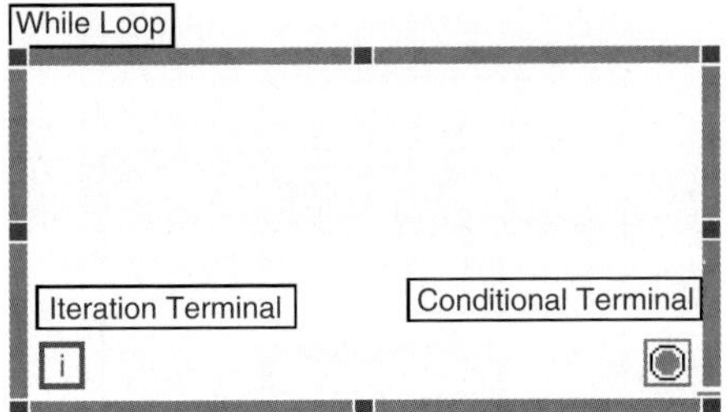

Fig. 3.14. While Loop terminals

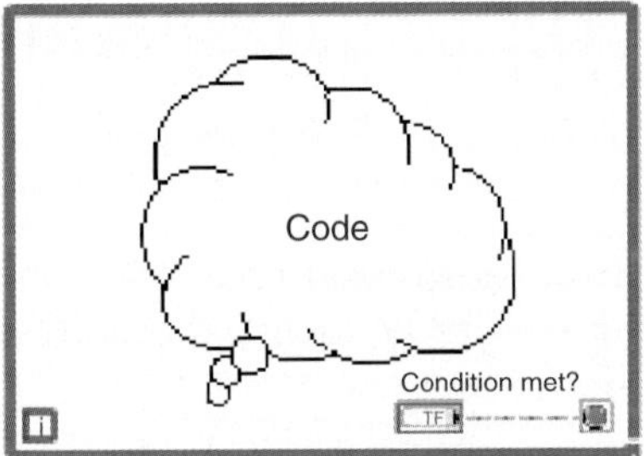

Fig. 3.15. LabVIEW While Loop

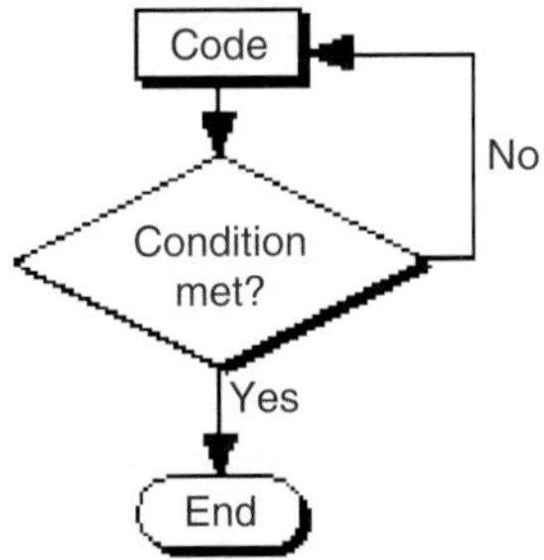

Fig. 3.16. Flow chart

The While Loop is equivalent to the following pseudocode:

Do

Execute subdiagram

While condition is NOT TRUE

The while loop is illustrated in Fig. 3.15. The flow chart for while loop is depicted in Fig. 3.16.

The While Loop is located on the **Functions≫Execution Control** palette shown in Fig. 3.17.

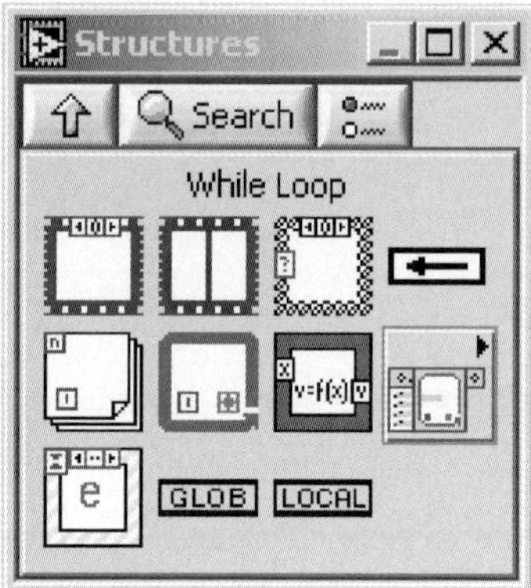

Fig. 3.17. Structures palette

The section of code to be added inside the While loop is dragged or while loop encloses the area of code to be executed conditionally.

Condition Terminal

The While Loop executes the subdiagram until the conditional terminal, an input terminal, receives a specific Boolean value. The default behavior and appearance of the conditional terminal is **Stop If True**. When a conditional terminal is **Stop If True**, the While Loop executes its subdiagram until the conditional terminal receives a True value.

Iteration Terminal

The iteration terminal, an output terminal, contains the number of completed iterations. The iteration count always starts at zero. During the first iteration, the iteration terminal returns 0.

Example 2

Problem statement. To find the sum of first 10 natural numbers using while loop

Block diagram construction:

- The While Loop is added from the structures subpalette on the functions palette located on the block diagram.
- The count terminal is wired with the value 10 for 10 natural numbers.
- The iteration terminal is wired the greater than or equal to node from the comparison subpalette from the Functions palette.
- The Boolean Output of the greater than or equal to is wired to the conditional terminal of the While Loop as in Fig. 3.18.

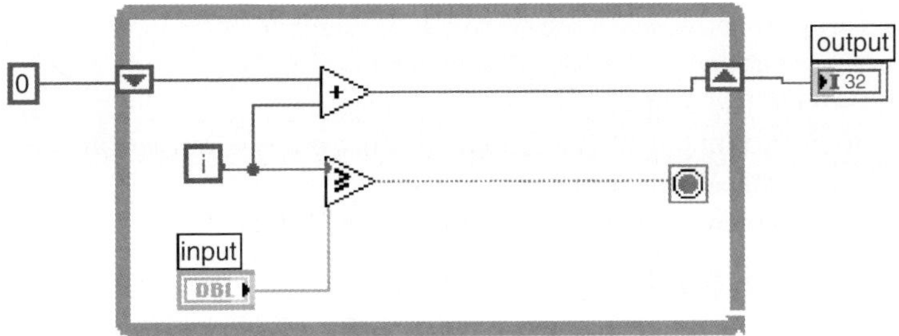

Fig. 3.18. Block diagram

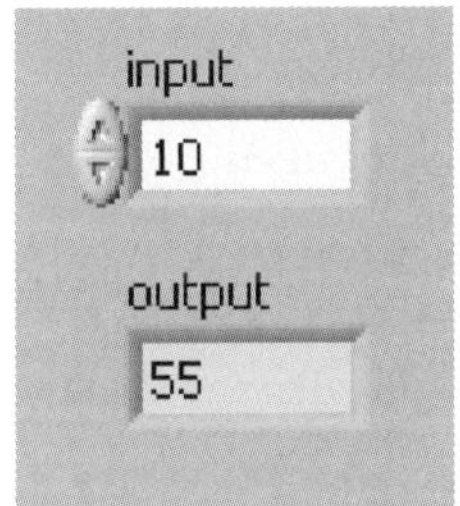

Fig. 3.19. Front panel

Front panel construction:

- The control and indicator are added from the controls palette of the front panel as in Fig. 3.19. Using the labeling tool the Owned label is changed as input and output, respectively.

Example 2

Problem statement. To find the factorial of a given number.
Block diagram construction:

- The While Loop is added from the Functions Palette on the Block Diagram.
- The code to compute the factorial of a given number is added inside the loop.
- The multiply function is added from the **Functions≫Arithmetic & Comparison≫Express Numeric** palette by 10.0.
- The increment and equal to functions are added from Express Comparison subpalette of the Arithmetic and Comparison palette in the Functions palette.
- The Wait ms is added from the Execution Control subpalette of the Functions palette.
- The number whose factorial is to be computed is wired to the count terminal of the For loop.

- The iteration terminal is wired to the increment node inside the loop.
- Tunnels are added automatically when the control and indicator are wired from outside the loop.
- The Boolean Output of the Equal to function is wired to the conditional terminal of the While loop.
- Input and Output are wired as shown in Fig. 3.20.

Front panel construction:

- The control and indicator are added from the controls palette of the front panel as in Fig. 3.21. Using the labeling tool the Owned label is changed as input and output, respectively.

Placing Objects Inside Structures

When a structure is first selected from the **Structures** subpalette of the **Functions** palette, the cursor appears as a miniature of the structure selected,

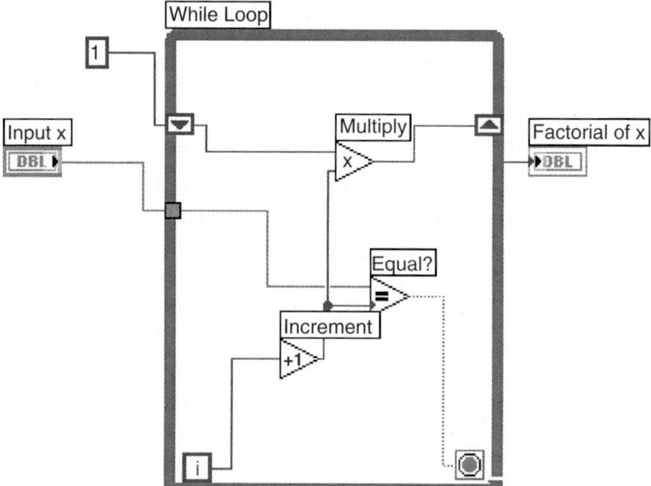

Fig. 3.20. Block diagram

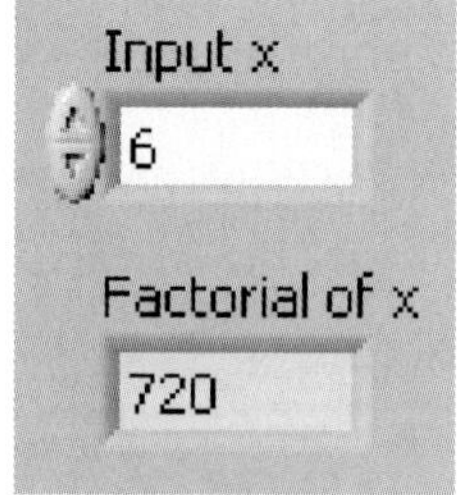

Fig. 3.21. Front panel

for example, the For Loop or the While Loop. The corner of the structure can be dragged to define the borders of the structure. When the mouse button is released the structure will appear containing all objects captured in the borders.

Once the structure appears on the block diagram, other objects can be placed inside, either by dragging them in or by placing them inside as they are selected from the **Functions** palette. When objects are being dragged *into* a structure, the structure's border will highlight as the object moves inside. When an object is dragged out of a structure, the *block diagram's* border (or that of an outer structure) will highlight as the object moves outside.

The existing structure can be resized by grabbing and dragging any corner with the Positioning tool. If an existing structure is moved so that it overlaps another object, the overlapped object will be visible above the edge of the structure. If an existing structure is dragged completely over another object, that object will display a thick shadow to warn that the object is *over* or *under* rather than *inside* the structure.

Terminals Inside Loops and Other Behavioral Issues

Data pass into and out of a loop through a little box on the loop border called a *tunnel*. Since LabVIEW operates according to dataflow principles, inputs a loop must pass their data in before the loop executes. Loop outputs pass data out only after tile loop completes all iterations.

Also, according to dataflow, a terminal should be placed inside a loop in order to check or update the terminal, during loop iteration. To remove a loop without deleting its contents, the user can pop-up on its border and select **Remove While Loop** or **Remove For Loop**, respectively. The user can also simply highlight the loop with the Positioning tool and delete it, all the objects inside will also be deleted.

The following section explains some of the terminals used inside the loop such as tunnels, shift registers, etc.

Structure Tunnels

Data can be passed out of or into a While Loop through a tunnel. Tunnels feed data into and out of structures. The tunnel appears as a solid block on the border of the While Loop. The block appears in the the color of the data type wired to the tunnel. Data pass out of a loop after the loop terminates. When a tunnel passes data into a loop, the loop executes only after data arrive at the tunnel.

In the block diagram shown in Fig. 3.22, the input is connected to a tunnel. The value in the tunnel does not get passed until the While Loop has finished execution.

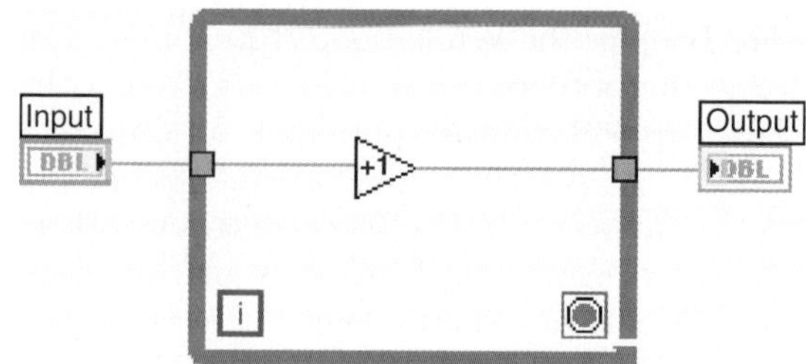

Fig. 3.22. Illustration for structure tunnel

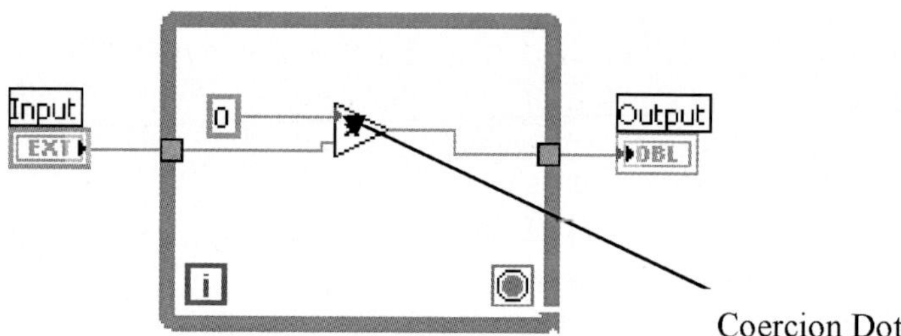

Fig. 3.23. Illustration of coercion dot

The Coercion Dot

The *coercion dot* is a little grey dot as in Fig. 3.23. It is named so, because LabVIEW is coercing one numeric representation to fit another. If two terminals of different numeric representations are wired together, LabVIEW converts one to the same representation as the other. If the count terminal has a 32-bit integer representation, while the Number of Iterations control is by default, LabVIEW converts the double-precision floating – point number to a long integer. In doing so, LabVIEW makes a new copy of the number in memory, in the proper representation. This copy takes up space. Although the extra space is negligible for scalar numbers (single-valued data types), it can add up quickly if arrays (which store multiple values) are used. The appearance of the coercion dot can be minimized on large arrays by changing the representation of the controls and indicators to exactly match the representation of the data they carry.

When a VI converts floating-point numbers to integers, it rounds to the nearest integer. A number with a decimal value of ".5" is rounded to the nearest even integer. An easy way to create a count terminal input with the correct data type and representation is to pop-up on the count terminal and select **Create constant** (for a block diagram constant) or **Create Control** (for a front panel control).

3.2.3 Shift Registers

Shift registers, available for while Loops and For Loops, are a special type of variable used to transfer values from one iteration of a loop to the next. They are unique to and necessary for LabVIEW's graphical structure. Popping-up on the left or right loop border and selecting **Add Shift Register** from the Pop-up menu can create a shift register as in Fig. 3.24.

A shift register comprises a pair of terminals directly opposite to each other on the vertical sides of the loop border. The right terminal stores the data upon the completion of iteration. These data are "shifted" at the end of the iteration and appear in the left terminal at the beginning of the next iteration. A shift register can hold any data type – numeric, Boolean, string, array, and so on. The shift register automatically adapts to the data type of the first object that is wired to it. It appears black when first created, but the shift register assumes the color of the data type wired to it.

Any number of shift registers can be used to store different variables on the same loop. This can be done by popping up on the loop border and adding shift registers until as many pairs as required is available as illustrated in Fig. 3.25. The left terminal will always stay parallel to its right terminal; if any one is moved, the other one also moves. If the user has a large number of shift registers in the loop, cannot predict exactly which ones are parallel, then select one and its partner will be automatically selected, or move one terminal a little and watch its mate follow. The result for the corresponding block diagram is shown in Fig. 3.26.

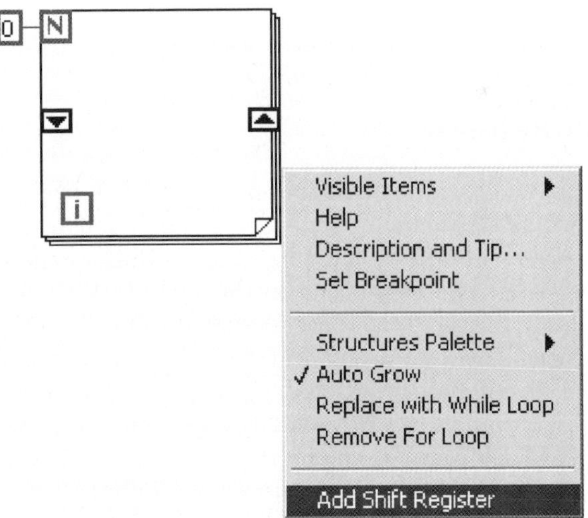

Fig. 3.24. Creating shift register from a For Loop

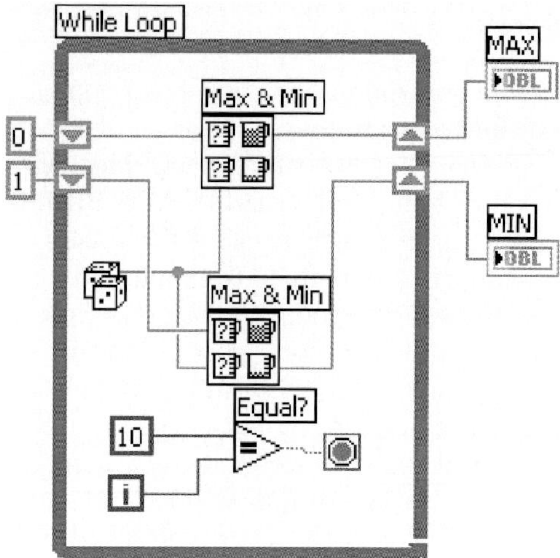

Fig. 3.25. Block diagram

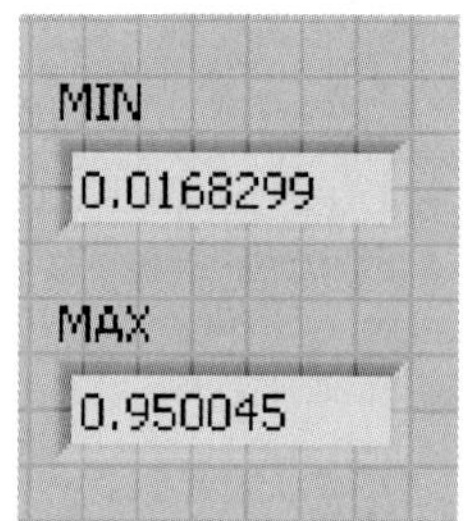

Fig. 3.26. Front panel

Stacked Shift Registers

The shift register can be configured to remember values from several previous iterations known as stacked shift registers. Stacked shift registers remember values from previous iterations and carry those values to the next iterations. Stacked shift registers can only occur on the left side of the loop because the right terminal only transfers the data generated from the current iteration to the next iteration. This is a useful feature when the user is averaging data values acquired in different iterations. To access values from previous iterations, additional terminals can be created by popping-up on the *left* terminal and choosing **Add Element** from the Pop-up menu as shown in Fig. 3.27.

If two more elements are added to the left terminal, values from the last three iterations carry over to the next iteration, with the most recent iteration value stored in the top shift register. The second terminal stores the data passed to it from the previous iteration, and the bottom terminal stores data from two iterations ago.

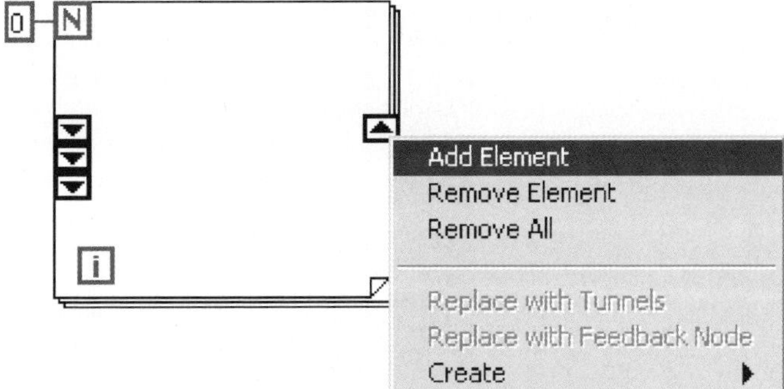

Fig. 3.27. Stacked Shift Registers

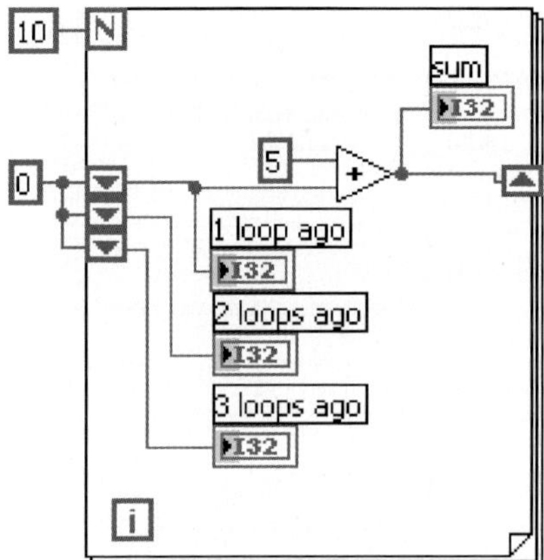

Fig. 3.28. Block diagram

For example if the user is creating a running sum of the iteration count. Each time through the loop, the new sum is saved in the shift register as in Fig. 3.28. At the end of the loop, the total sum is passed out to the numeric indicator as in Fig. 3.29. Similarly it is used in cases where it is required to average values from successive loop iterations.

Initializing Shift Registers

To avoid unforeseen and possibly nasty behavior, *shift registers* should always be *initialized*. To initialize the shift register with a specific value, the initial value is wired to the left terminal of the shift register from outside the loop. If shift registers are not initialized, the initial value will be the default value when the program runs for the first time. In subsequent runs, the shift register will contain whatever values are left over from previous runs.

For example, if the shift register data type is Boolean, the initial value will be FALSE for the first run. Similarly, if the shift register data type is

Fig. 3.29. Front panel

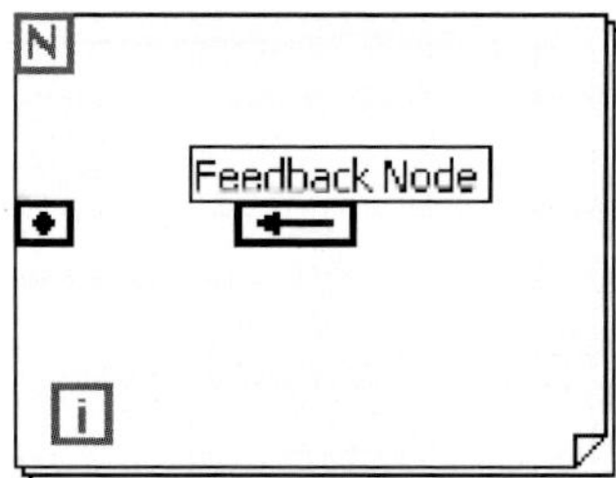

Fig. 3.30. For Loop with Feedback Node

numeric, the initial value will be zero. The second time the user runs the VI, an uninitialized shift register will contain values left over from the first run. LabVIEW does not discard values stored in the shift register until the VI is closed and removed from memory. In other words, if a VI containing uninitialized shift registers is available, the initial values for the subsequent run will be the ones left over from the previous run which leads to serious errors in problem solving.

3.2.4 Feedback Nodes

The Feedback Node, shown in Fig. 3.30, appears automatically in a For Loop or While Loop if the output of a subVI, function, or group of subVIs and functions are wired to the input of that same VI, function, or group. Like a shift register, the Feedback Node stores data when the loop completes iteration, sends that value to the next iteration of the loop, and transfers any data type. It is also used to avoid unnecessary long wires in the loops. The Feedback Node arrow indicates in the direction of dataflow along the wire.

The Feedback Node can be selected from the Structures palette as shown in Fig. 3.31 and placed inside a For Loop or While Loop.

If the Feedback Node is placed on the wire after the user branches the wire that connects data to the tunnel, the Feedback Node passes each value back

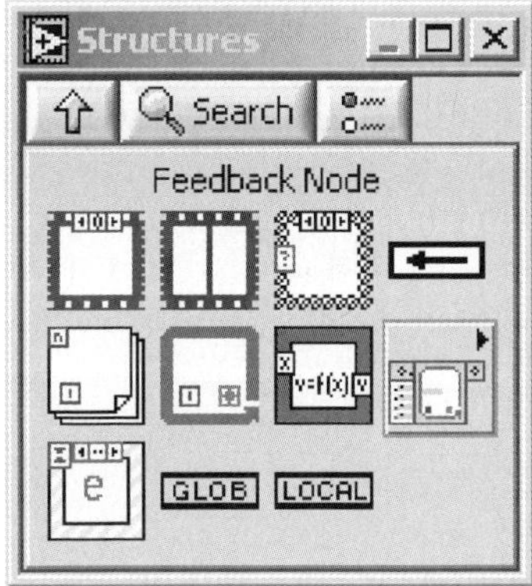

Fig. 3.31. Feedback Node in Structures palette

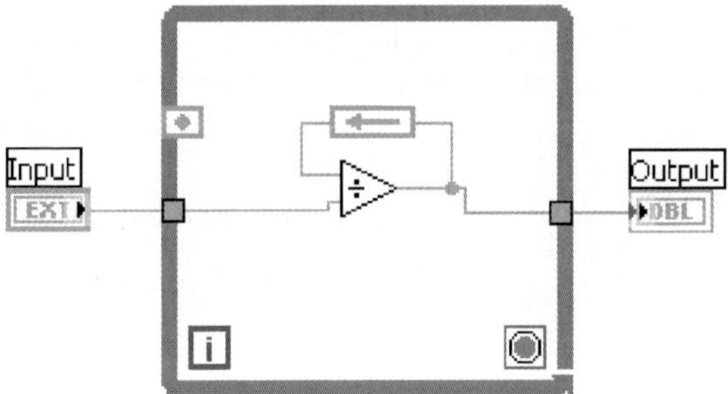

Fig. 3.32. Feedback Node after

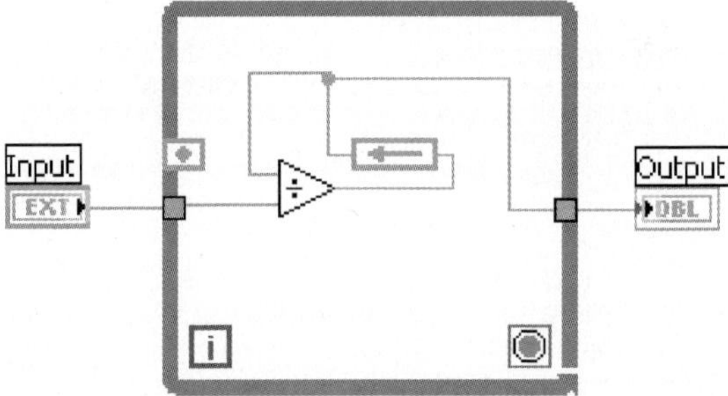

Fig. 3.33. Feedback Node before

to the input of the VI or function and then passes the last value to the tunnel as in Fig. 3.32. If the Feedback Node is placed on the wire before the user branches the wire that connects the data to the tunnel, the Feedback Node passes each value to the tunnel as in Fig. 3.33.

3.3 Selection Structures

Depending on the flow of data there are cases when a decision must be made in a program. For example, if *a* happens, do *b*; else if *c* happens, do *d*. In text-based programs, this can be accomplished with if–else statements, case statements, switch statements, and so on. LabVIEW includes many different ways of making decisions such as select function, case structure, and formula node. The simplest of these methods is the Select function.

Select Function

The Select function, located on the **Functions≫Express Comparison** palette, selects between two values dependent on a Boolean input as shown in Fig. 3.34. The select function is shown in Fig. 3.35.

If the Boolean input 's' is True, this function returns the value wired to the '**t**' input. If the Boolean input is False, this function returns the value wired to the '**f**' input. If the decision to be made is more complex than a Select function can execute, a Case structure may be required.

Example. To illustrate select function

The reciprocal of the given number 'x' is computed and if the result is greater than zero, select function returns the answer else it returns the number itself. This example is illustrated in Fig. 3.36.

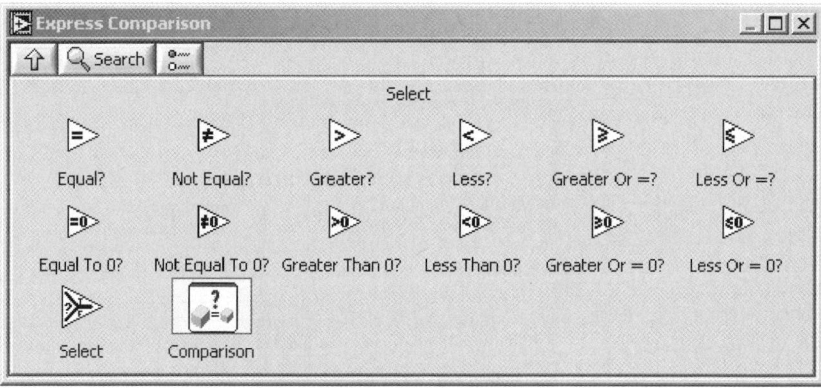

Fig. 3.34. Express Comparison palette

Fig. 3.35. Select function

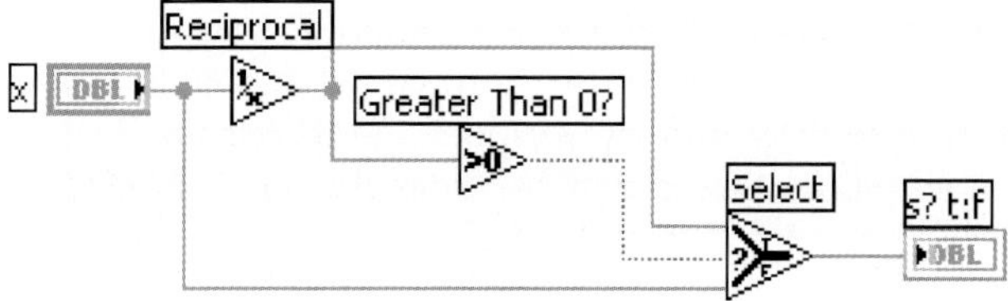

Fig. 3.36. Block diagram to illustrate the usage of select function

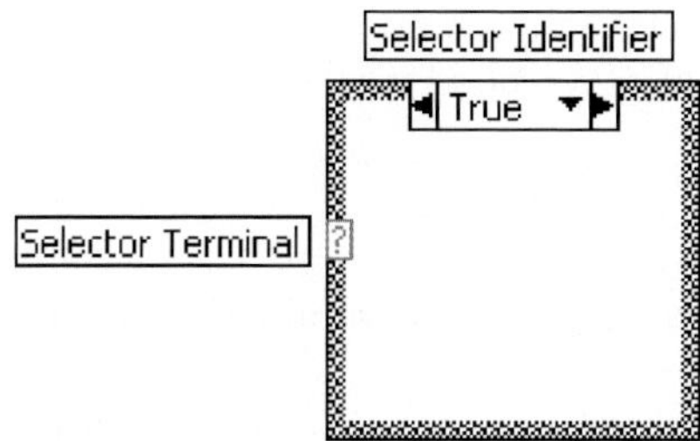

Fig. 3.37. Case Structure with terminals

3.3.1 Case Structures

A powerful structure is the *Case Structure* shown in Fig. 3.37. The case structure is LabVIEW's method of executing conditional text, sort of like an "if–then–else" statement. It is located in the **Structures** subpalette of the **Functions** palette. The Case Structure, has two or more subdiagrams, or cases; only one of them executes, depending on the value of the Boolean, numeric, or string value wired to the *selector terminal*.

If a Boolean value is wired to the selector terminal, the structure has two cases, FALSE and TRUE. If a numeric or string data type is wired to the selector, the structure can have from zero to almost unlimited cases. Initially only two cases are available, but number of cases can be easily added. More than one value can be specified for a case, separated by commas. In addition, the user can always select a "Default" case that will execute if the value wired to the selector terminal doesn't match any of the other cases. When a case structure is first placed on the panel, the Case Structure appears in its Boolean form; it assumes numeric values as soon as a numeric data type is wired to its selector terminal.

Case Structures can have multiple subdiagrams, but the user can see only one case at a time, sort of like a stacked deck of cards. Clicking on the decrement (left) or increment (right) arrow at the top of the structure displays the previous or next subdiagram, respectively. The user can also click on the display at the top of the structure for a pull-down menu listing all cases, and then pop-up on the structure border and select **Show Case**. If a floating-point number is wired to the selector, LabVIEW rounds that number to the nearest integer value. LabVIEW coerces negative numbers to 0 and reduces any value higher than the highest-numbered case to equal the number of that case.

The selector terminal can be positioned anywhere along the left border. If the data type wired to the selector is changed from a numeric to a Boolean, cases 0 to 1 change to FALSE and TRUE. If other cases exist (2 to n), LabVIEW does not discard them, in case the change in data type is accidental. However, these extra cases must be deleted before the structure can execute. For string data types wired to case selectors, the user should always specify the case values as strings between quotes. The only exception is the keyword **Default**, which should never be in quotes.

Wiring Inputs and Outputs

The data at all Case Structure input terminals (tunnels and selector terminal) are available to all cases. Cases are not required to use input data or to supply output data, but if any one case outputs a value, all must output a value. When an output from one case is wired, a little white tunnel appears in the same location on all cases. The run arrow will be broken until data to this output tunnel is wired from every case, at which time the tunnel will turn black and the run arrow will be whole again. If wiring is not done directly to the existing output tunnel, more tunnels might be created accidentally.

Adding Cases

If the user pops-up on the Case Structure border, the resulting menu gives options to **Add Case After** and **Add Case Before** the current case as in Fig. 3.38. The user can also choose to copy the currently shown case by selecting Duplicate Case. The user can also delete the current case (and everything in it) by selecting **Remove** Case.

Dialog Boxes

The **One button Dialog** and **Two button Dialog** functions bring up a dialog box containing a message according to the choice of the user. These functions are located in the **Time & Dialog** subpalette of the **Functions** palette. The **One button Dialog** shown in Fig. 3.39 stays open until the OK button is pressed, while the **Two button Dialog** shown in Fig. 3.40 box remains open until the OK or the CANCEL button is pressed. These buttons can also be renamed by inputting "button name" strings to the functions. These dialog boxes are modal; in other words, any other LabVIEW window cannot be activated while they are open. They are very useful for delivering messages to or soliciting input from the program's operator as shown in Fig. 3.39, where **message** is the text to display in the dialog box; **button name** is the name displayed in the dialog box button. The default is **OK**; **true**

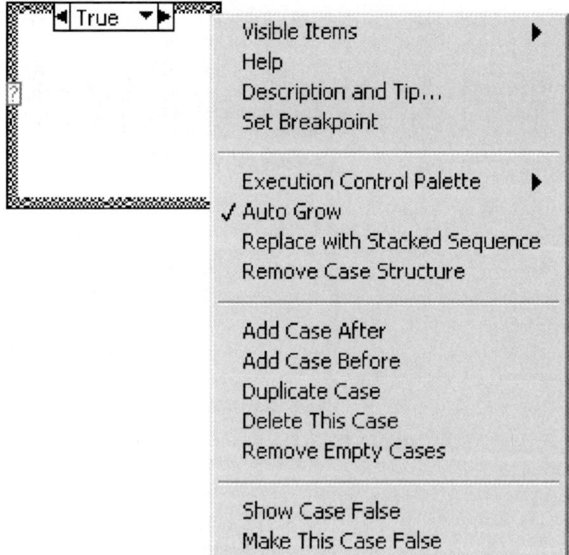

Fig. 3.38. Shortcut menu for case structure

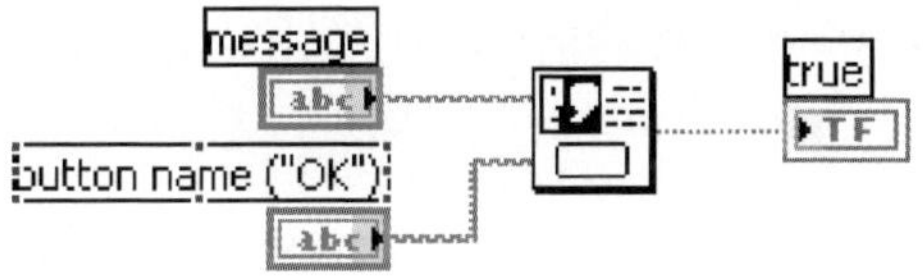

Fig. 3.39. One button Dialog

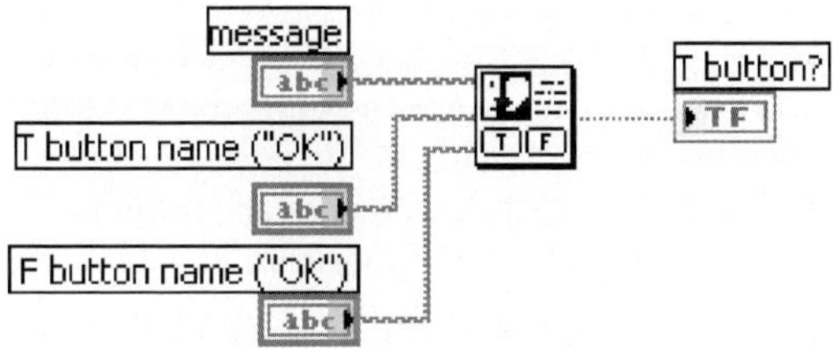

Fig. 3.40. Two button Dialog

contains a value of TRUE when one click the button as shown in Fig. 3.40, where **Message** is the text to display in the dialog box; **T button name** is the name displayed on one of the dialog box buttons. The default is **OK**; **F button name** is the name displayed on one of the dialog box buttons. The default is **Cancel**; **T button?** Returns a value of TRUE if one click the dialog box button named **T button name**. If one click the dialog box button named **F button name**, **T button?** Returns a value of FALSE. Based on

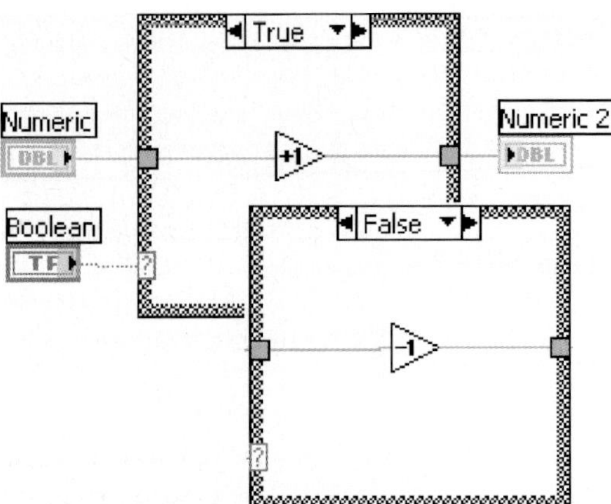

Fig. 3.41. Boolean Case Structure

the data type wired to the selector terminal of the case structure different case structures are available and are explained in the following sections:

Boolean Case Structure

The following example is a Boolean Case structure shown in Fig. 3.41. The cases are shown overlapped to simplify the illustration. If the Boolean control wired to the selector terminal is True, the VI increments the numeric value. Otherwise, the VI decrements the numeric value.

Integer Case Structure

The following example is an integer Case structure shown in Fig. 3.42. **Integer** is a text ring control located on the **Controls≫Text Controls** palette that associates numeric values with text items. If the text ring control wired to the selector terminal is 0 (add), the VI decrements the numeric values. If the value is 1 (subtract), the VI increments the numeric values. If the text ring control is any other value than 0 (add) or 1 (subtract), the VI adds the numeric values, because that is the default case.

String Case Structure

The following example is a string Case structure as shown in Fig. 3.43. If **String** is "Increment," the VI increments the numeric values. If **String** is "Decrement," the VI decrements the numeric values.

Enumerated Case Structure

The following example is an enumerated Case structure as shown in Fig. 3.44. An enumerated control gives users a list of items from which to select.

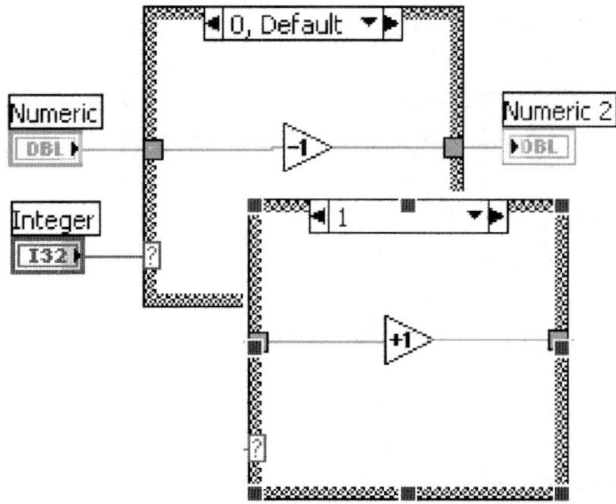

Fig. 3.42. Integer Case structure

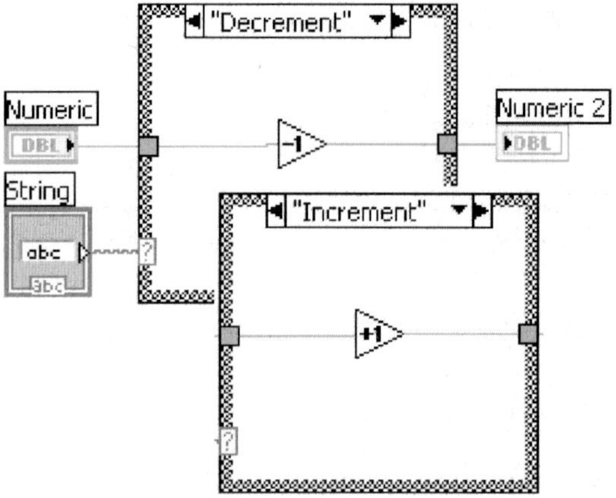

Fig. 3.43. String Case structure

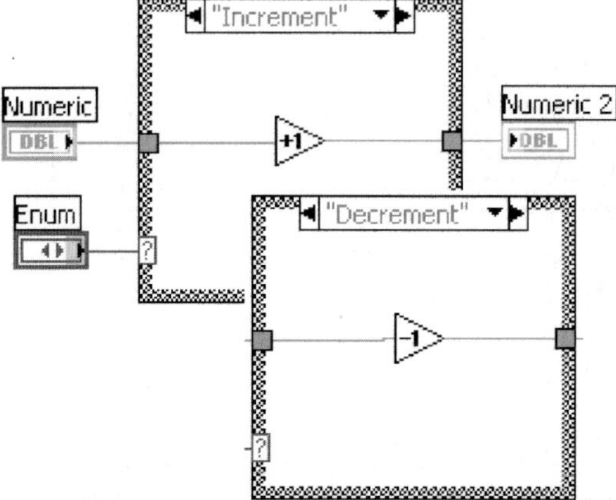

Fig. 3.44. Enumerated Case structure

The data type of an enumerated control includes information about the numeric values and string labels in the control. When an enumerated control is wired to the selector terminal of a Case structure, the case selector displays a case for each item in the enumerated control. The Case structure executes the appropriate case subdiagram based on the current item in the enumerated control. If **Enum** is "Increment," the VI increments the numeric values. If **Enum** is "Decrement," the VI decrements the numeric values.

Error Case Structure

The following example is an error cluster Case structure as in Fig. 3.45. When an error cluster is wired to the selector terminal of a Case structure, the case selector label displays two cases, Error and No Error, and the border of the Case structure changes color – red for Error and green for No Error.

The Case structure executes the appropriate case subdiagram based on the error state. When an error cluster is wired to the selection terminal, the Case structure recognizes only the **status** Boolean of the cluster.

Example

Problem statement. To find the square root of a given number
Block diagram construction:

- The Case Structure is added from the Structures subpalette of the Functions Palette on the Block Diagram.
- Based upon the terminal wired to the selector terminal, the case structure chooses one of the types.

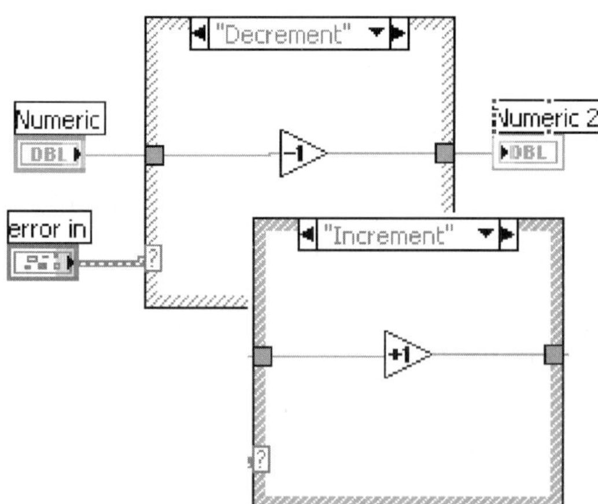

Fig. 3.45. Error Case structure

- The number whose square root is to be computed is compared to check if it is greater than zero. The Boolean output is wired to the selector terminal, thus the case structure is Boolean with two cases True and False. If the input value is greater than zero then the square root is computed else an error message is displayed along with the output number as −99999.
- The selector identifier is selected to the True case and the function to compute the square root of the given number is added. The square root icon is found on the Express Numeric subpalette of the Arithmetic & Comparison Palette located on the Functional Palette of the Block Diagram.
- Similarly the False case is selected and the string control is selected by right clicking the message terminal of the One button Dialog function. The user can type Error Negative Number in the constant function.
- Input and Output are wired as shown in Fig. 3.46.

Front panel construction:

- The control and indicator are added from the controls palette of the front panel as shown in Fig. 3.47. Using the labeling tool the Owned label is changed as input and output, respectively

3.3.2 Sequence Structures (Flat and Stacked Structures)

Determining the execution order of a program by arranging its elements in a certain sequence is called control flow. Visual Basic, C, and most other

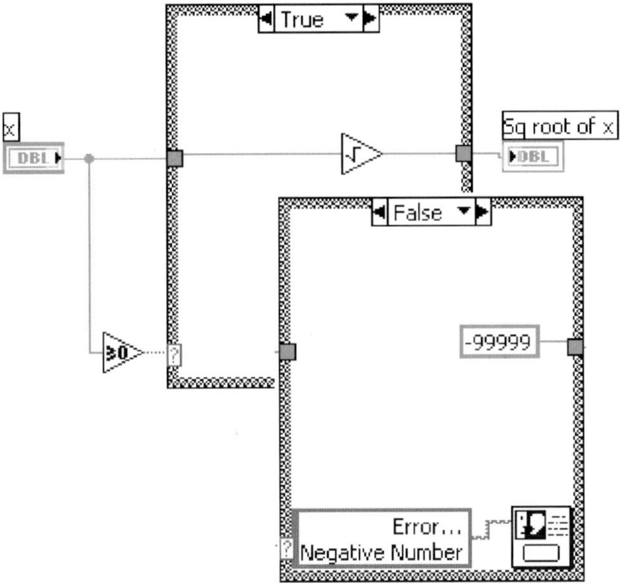

Fig. 3.46. Block diagram

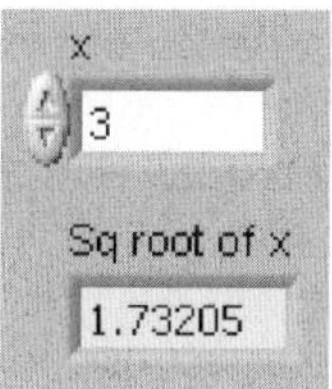

Fig. 3.47. Front panel

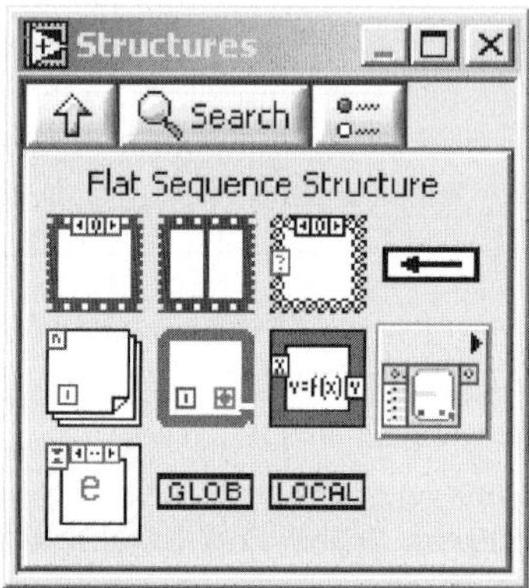

Fig. 3.48. Structures palette

procedural programming languages have inherent control flow because statements executes in the order in which they appear in the program. LabVIEW uses the *Sequence Structure* to obtain control flow within a dataflow framework. A Sequence Structure executes frame 0, followed by frame 1, then frame 2, until the last frame executes. Only when the last frame completes data leaves the structure.

The Sequence Structure as shown in Fig. 3.48, looks like a frame of film. It can be found in the **Structures** subpalette of the **Functions** palette.

Like the Case Structure, only one frame is visible at a time – the arrows at the top of the structure can be selected to see other frames; or the top display can be clicked to obtain a listing of existing frames, or the user can pop-up on the structure border and choose **Show Frame**. When a Sequence Structure is first dropped on the block diagram, it has only one frame; thus, it has no arrows or numbers at the top of the structure to designate which frame is showing. New frames can be created by popping up on the structure border and selecting **Add Frame After** or **Add Frame Before** as shown in Fig. 3.49.

The Sequence Structure is used to control the order of execution of nodes that are not data dependent on each other. Within each frame, as in the rest of the block diagram, data dependency determines the execution order of nodes.

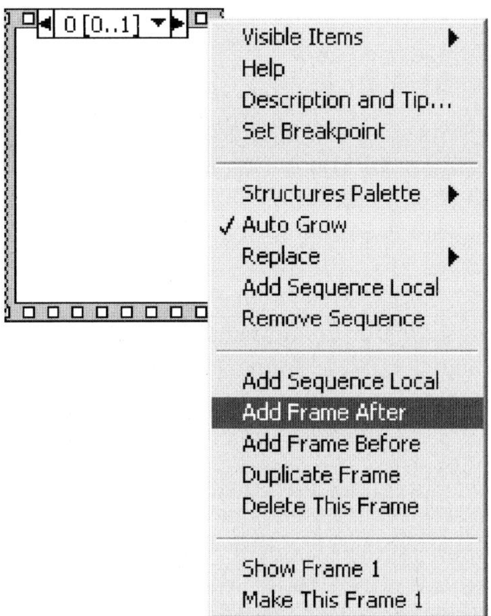

Fig. 3.49. Properties of Sequential Structure

Output tunnels of Sequence Structures can have only one data source, unlike Case Structures, whose outputs must have one data source per case. The output can originate from any frame, but that data are passed out of the structure only when the structure completes execution entirely, not when the individual frames finish. Data at input tunnels are available to all frames.

Sequence Locals

To pass data from one frame to any subsequent frame, the user must use a terminal called a *sequence local*. To obtain a sequence local, choose **Add Sequence Local** from the *structure border* Pop-up menu. This option is not available if the user pops up too close to another sequence local or over the subdiagram display window. The user can drag the sequence local terminal to any unoccupied location on the border. Use the **Remove** command from sequence local Pop-up menu to remove a terminal, or just select and delete it.

When it first arrives on the diagram, a sequence local terminal is just a small yellow box. When the source data is wired to the sequence local, an outward-pointing arrow appears in the terminal of the frame containing the data source. The terminals in subsequent frames contain an inward-pointing arrow, indicating that the terminal is a data source for that frame. In frames before the source frame, the user cannot use the sequence local (after all, it has not been assigned a value), and it appears as a dimmed rectangle.

Example

Block diagram construction:

- The Sequence Structure is added from the Structures subpalette of the Functions Palette on the Block Diagram.
- The inputs x and y are wired as shown in Fig. 3.50. Similarly the arithmetic functions are also wired as shown in the figure.

Front panel construction:

- The control and indicator are added from the controls palette of the front panel as shown in Fig. 3.51. Using the labeling tool the Owned label is changed as input and output, respectively

Timing

Sometimes it is found useful to control or monitor the timing of a VI. **Wait (ms)**, **Tick Count (ms)**, and **Wait until Next ms Multiple**, located in the **Time & Dialog** subpalette of the Functions palette, accomplish these tasks.

Wait (ms) causes VI to wait a specified number of milliseconds before it continues execution as explained in this section. **Wait Until Next ms Multiple** causes LabVIEW to wait until the internal clock equals or has

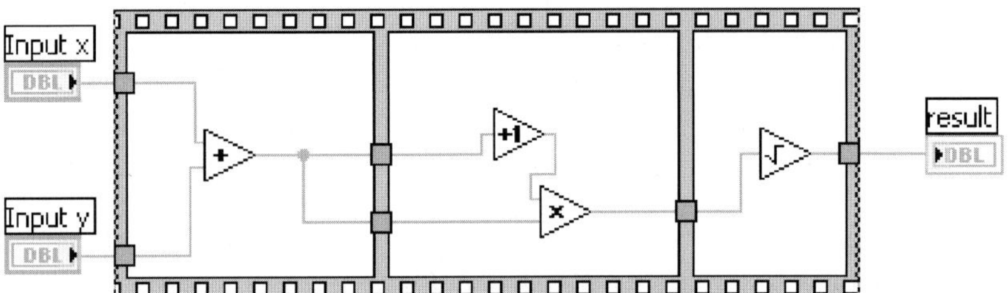

Fig. 3.50. Block diagram

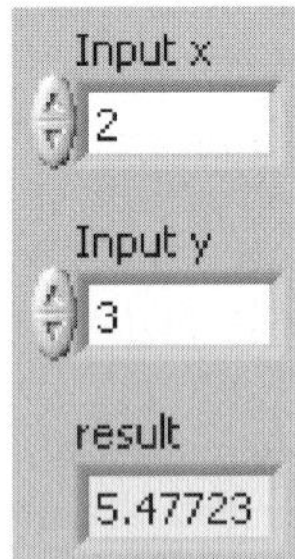

Fig. 3.51. Front panel

passed a multiple of the millisecond multiple input number before continuing VI execution; it is causing for making loops execute at specified intervals and synchronizing activities. These two functions are similar but not identical. For example, **Wait Until Next ms Multiple** will probably wait less than the specified number of milliseconds in the first loop iteration, depending on the value of the clock when it sorts (that is, how long it takes until the clock is at the next multiple and the VI proceeds). In addition, if the loop is still executing when the clock passes a millisecond multiple, the VI will wait until the clock reaches the *next* multiple, so the VI may become "out of synch" and slow down. The **Wait ms** and the **Wait Until Next ms Multiple** function are detailed under For loop section.

Tick Count (ms) returns the value of the operating system's internal clock in milliseconds; it is commonly used to calculate elapsed time. The internal clock does not always have great resolution – one tick of the clock can be up to 55 milliseconds (ms) on Windows 95/98, 10 ms on Windows 2000/NT, 17 ms on MacOS 9.x and earlier, and 1 ms on Linux, Solaris, and MacOS X machines; LabVIEW cannot work around this operating system limitations.

3.4 The Formula Node

The *Formula Node* is a resizable box that is used to enter algebraic formulas directly into the block diagram. The Formula Node is a convenient text-based node used to perform mathematical operations on the block diagram. Formula Nodes are useful for equations that have many variables or are otherwise complicated and for using existing text-based code. The existing text-based code can be copied and pasted into a Formula Node rather than recreating it graphically on the block diagram. This feature is found to be extremely useful when the user has a long formula to solve. For example, consider the fairly simple equation $y = x^{**}3 + sin(x)$. Even for this simple formula, when implemented equation using regular LabVIEW arithmetic functions, the block diagram is a little bit harder to follow than the text equations as shown in Fig. 3.52.

The same equation can be implemented using formula node. With the Formula Node, the user can directly enter a formula or formulas, in lieu of creating complex block diagram subsections. The task is very simple; the user can simply enter the formula inside the box. The input and output terminals of the Formula Node can be created by popping up on the border of the node and choosing **Add Input** or **Add Output** from the Pop-up menu as illustrate in Fig. 3.53. Then enter variable names into the input and output boxes. Names are case sensitive, and each formula statement must terminate with a semicolon (;). The equation is typed in the structure. Formula Nodes can also be used for decision-making similar to case structure and select function discussed in the previous sections.

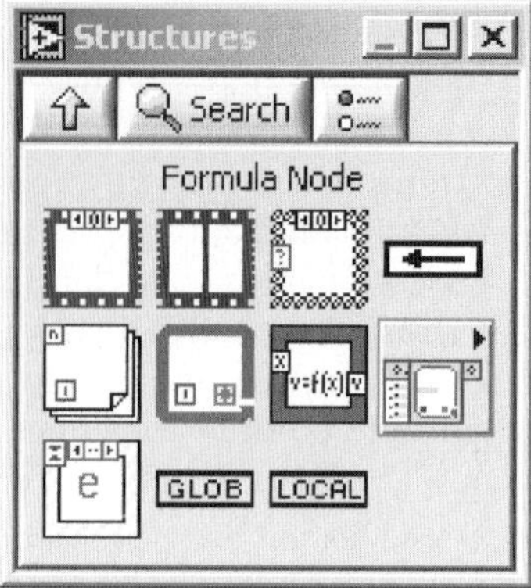

Fig. 3.52. Structures palette

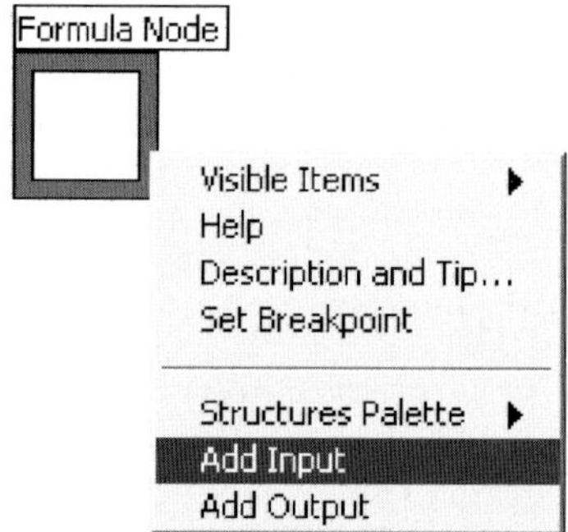

Fig. 3.53. Pop-up menu of Formula Node

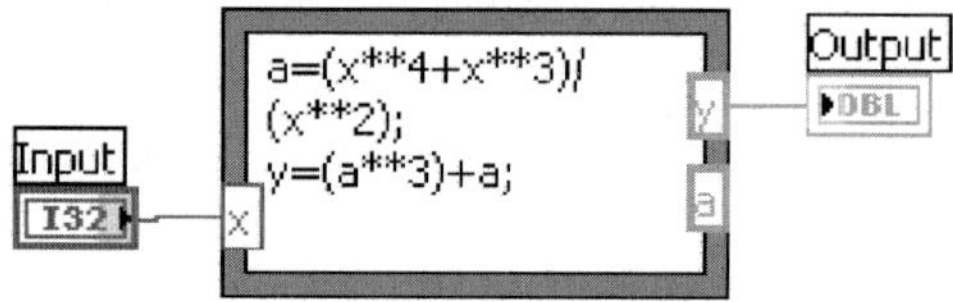

Fig. 3.54. Illustration of Formula node

The following example illustrates the usage of formula node. The input is 'x' which is given as input to the Formula Node. The node computes the given function and returns the output in terms of 'y'. Figure 3.54 shows the implementation of a formula in the node with input and output terminals.

3.5 Arrays

A LabVIEW array is a collection of data elements that are all the same type similar to traditional programming languages. An array can have one or more dimensions, and has 2^{31} elements per dimension (memory permitting, of course). An array element can have any type except another array, a chart, or a graph.

3.5.1 Single and Multidimensional Arrays

This section illustrates on creating single and multidimensional arrays.

Single Dimensional Array

Array elements are accessed by their indices; each element's *index* is in range 0 to $N-1$, where N is the total number of elements in the array. A one-dimensional (1D) array shown in Table 3.1 illustrates this structure.

Notice that the *first* element has index 0, the *second* element has index 1, and so on. Generally, waveforms are often stored in arrays, with each point in the waveform comprising an element of the array. Arrays are also useful for storing data generated in loops, where each loop iteration generates one element of the array.

Creating Array Controls and Indicators

Two steps are involved to make the controls and indicators for compound data types such as arrays and clusters.

Step 1. The array control or indicator known as the Array shell is created from the **Array** subpalette of the **Controls** palette available in the front panel

Step 2. The *array shell* is combined with a *data object*, which can be numeric, Boolean, or string (or cluster).

Using the above two steps the created structure resembles Fig. 3.55.

Table 3.1. Single dimensional array

Index	0	1	2	3	4	5	6	7	8	9
10-element array	12	32	82	8.0	4.8	5.1	6.0	1.0	2.5	10

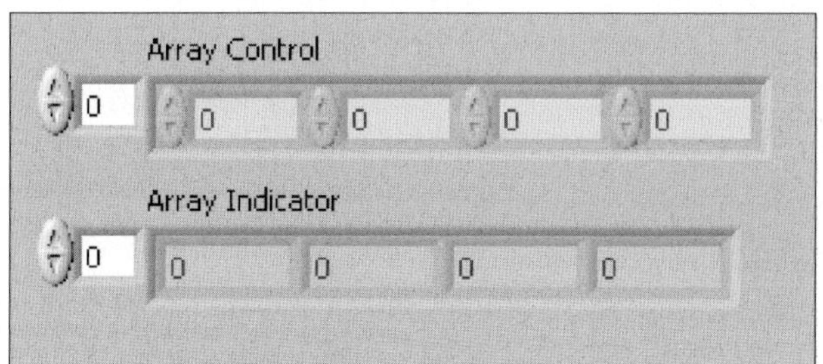

Fig. 3.55. Array Control and Indicator

The element display window resizes to accommodate its new data type, but remains grayed out until data is entered into it. Note that all elements of an array must be either controls or indicators, not a combination. When an array shell is first dropped on the front panel, its block diagram terminal is black, characteristic of an undefined data type. The terminal also contains brackets, which are LabVIEW's way of denoting an array structure. When a data type is assigned to the array, then the array's block diagram terminal assumes its new type color and lettering. Generally, the array wires are thicker than wires carrying a single value.

Data can be entered into the array by using the Labeling or Operating tool to type in a value, or if the data are numeric, the arrow of the index can be clicked to increment or decrement the value. To resize the object in the display window, the Positioning tool is used. To show more elements at the same time, move the Positioning tool around the window corner until the user finds the grid cursor, and then stretch either horizontally or vertically. Now multiple elements are visible. The element closest to the index display always corresponds to the element number displayed there.

The user can also create array constants on the block diagram just like creating numeric, Boolean, or string constants. **Array Constant** can be chosen from the **Array** subpalette of the **Functions** palette. Then the data is placed in an appropriate data type (usually another constant) similar to that created on the front panel. This feature is useful when the user needs to initialize shift registers or provide a data type to a file or network functions. To clear an array control, indicator, or constant of data, the user can pop-on the index display.

Two-Dimensional Arrays

A two-dimensional, or 2D, array stores elements in a gridline fashion. It requires two indices to locate an element: a column index and a row index, both of which are zero-based in LabVIEW. The six-column by four-row array that contains six times four elements is stored in the following manner as shown in Fig. 3.56.

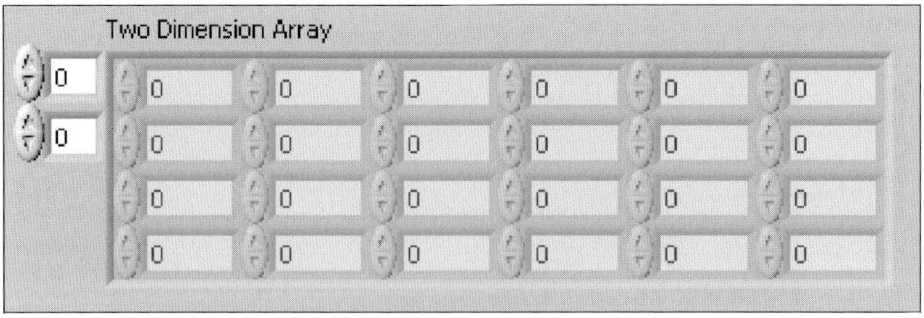

Fig. 3.56. 4×6 2D array

The user can add dimensions to an array control or indicator by popping up on its index display (not on the element display) and choosing **Add Dimension** from the Pop-up menu. In a 2D array of digital controls the user will have two indices to specify each element. The grid cursor of the Positioning tool can be used to expand the element display in two dimensions so that more elements can be seen. Unwanted dimensions can be removed by selecting **Remove Dimension** from the index display's Pop-up menu.

For example, if waveforms from several channels being read from a data acquisition (DAQ) board, they will be stored in a 2D array, with each column in the 2D array corresponding to one channel's data.

Creating Two-Dimensional Array

Two For Loops, one inside the other, can be used to create a 2D array on the front panel. The inner For Loop creates a row, and the outer For Loop "stacks" these rows to fill in the columns of the matrix. In two For Loops creating a 2D array of random numbers using autoindexing, notice that a 2D array wire is even thicker than a 1D array wire as shown in Fig. 3.57.

3.5.2 Autoindexing

The For Loop and the While Loop can index and accumulate arrays at their boundaries automatically, adding one new element for each loop iteration. This capability is called ***autoindexing***. One important thing to remember is that autoindexing enabled by default on For Loops but disabled by default on While Loops. The For Loop autoindexes, an array at its boundary. Each iteration creates the next array element. After the loop completes, the array passes out of the loop to the indicator; none of the array data are available until the Loop finishes. Notice that the wire becomes thicker as it changes to an array wire type at the Loop border. Figure 3.58 shows autoindex disabled in a For loop. Here only one value is passed out of the loop. Figure 3.59 shows

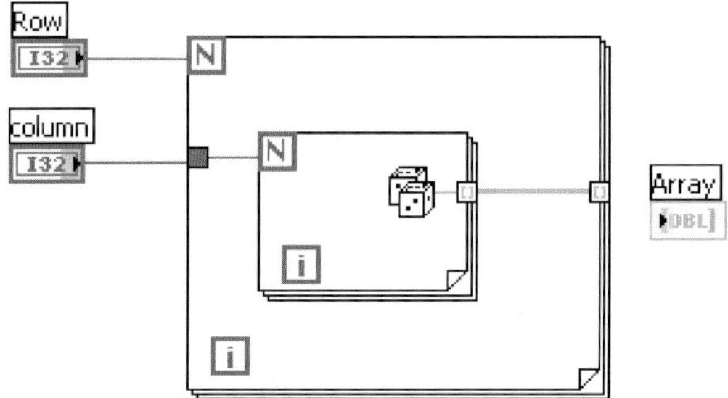

Fig. 3.57. For Loop for creating 2D array

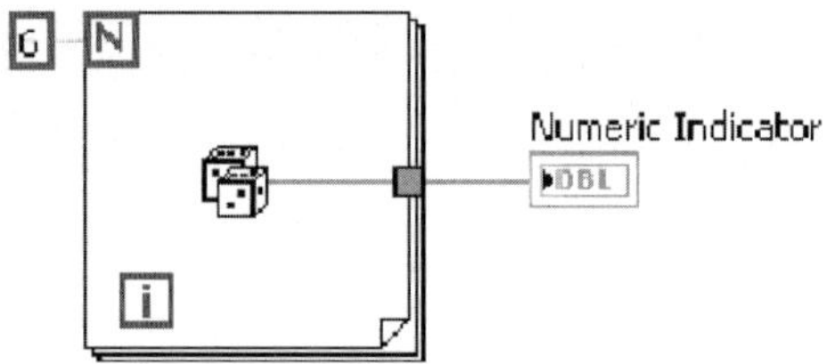

Fig. 3.58. Autoindexing disabled

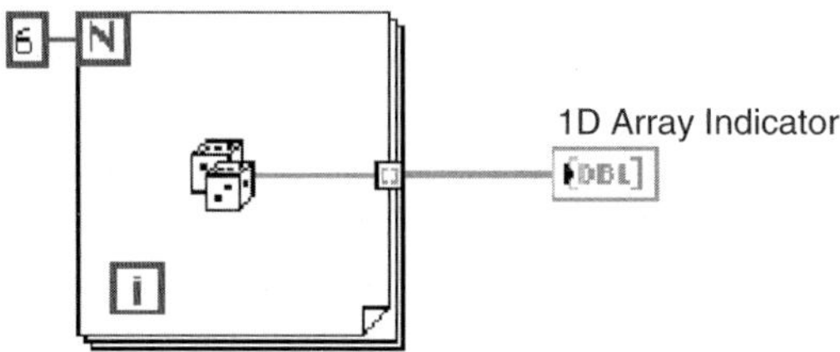

Fig. 3.59. Autoindexing enabled

autoindex enabled in a For loop. Here the wire appears thicker since the output is wired to a 1D array indicator. To wire a scalar value out of a For Loop without creating an array, autoindexing must be disabled. This can be done by popping up on the tunnel (the square with the [] symbol) and choosing **Disable Indexing** from the tunnel's Pop-up menu. Since autoindexing is disabled by default, to wire array data out of a While Loop, the user must pop-up on the tunnel and select **Enable Indexing**. When autoindexing is disabled, only the last value returned from the **Random Number (0–1)** function passes out of the loop. Notice that the wire remains the same size after it leaves the loop. Pay attention to this wire size, because autoindexing is a common source of problems among beginners.

Autoindexing also applies when the user is wiring arrays into loops. If indexing is enabled, the loop will index off one element from the array each time it iterates (note how the wire becomes thinner as it enters the loop). If indexing is disabled, the entire array passes in to the loop at once. Because For Loops are often used to process arrays, LabVIEW enables autoindexing by default when an array is wired into or out of them.

Using Autoindexing to Set the For Loop Count

When autoindexing is enabled on an array *entering* a For Loop, LabVIEW automatically sets the *count* to the array size, thus eliminating the need to wire a value to the count terminal. If a conflicting count is given for example, by setting the count explicitly and by autoindexing (or by autoindexing two different size arrays), LabVIEW sets the count to the smallest of the choices. The array size determines the number of For Loop iterations instead of the

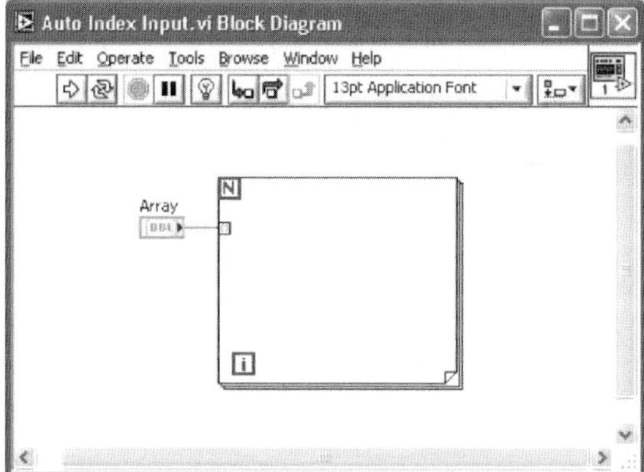

Fig. 3.60. Autoindex input

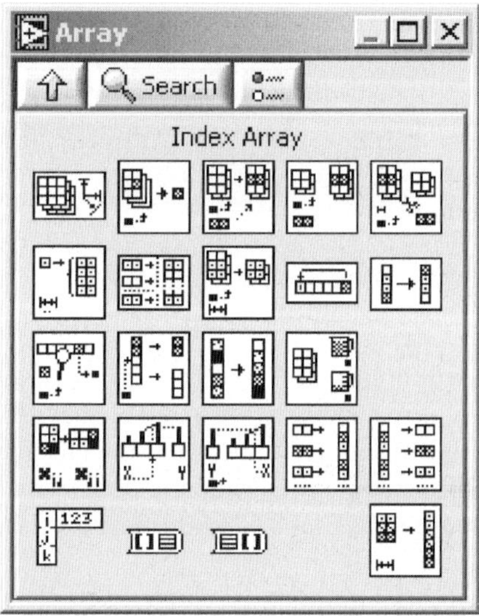

Fig. 3.61. Array subpalette in Functions palette

value wired to the count terminal, because the array size is the smaller of the two. In Fig. 3.60 it is found that the run arrow does not appear broken when the count terminal is wired as explained above.

3.5.3 Functions for Manipulating Arrays

LabVIEW has many functions for arrays in the **Array** subpalette of the **Functions** palette. Arrays (and all other Loop structures) are zero indexed – the first elements have an index of zero, second have an index of one, and so on. Some common functions are shown in Fig. 3.61.

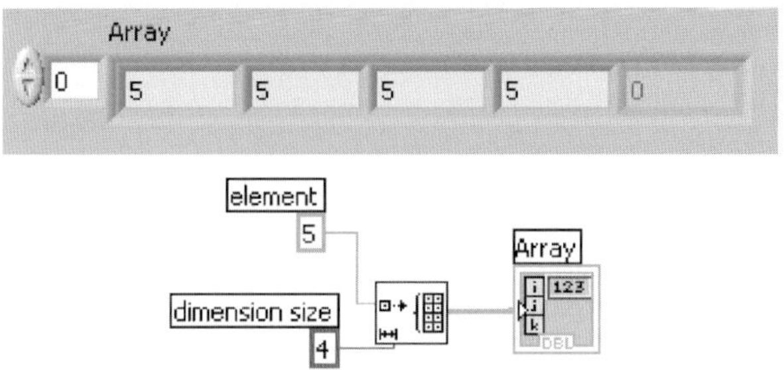

Fig. 3.62. Initialize Array

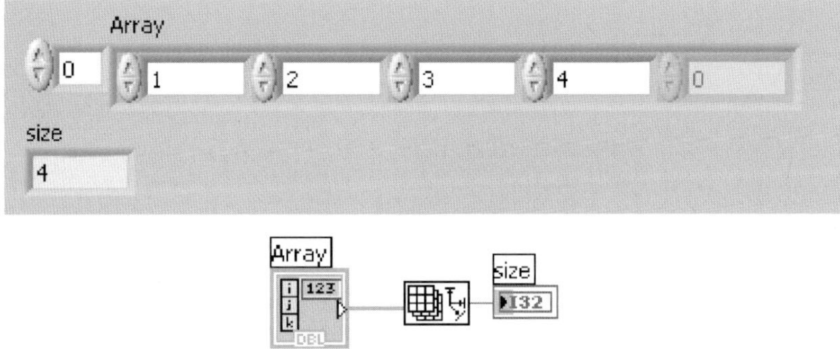

Fig. 3.63. Array Size

Initialize Array will create and initialize an n-dimensional array. It can be configured for larger dimensions by "growing" it with the Positioning tool to provide more **dimension** inputs. This function is useful for allocating memory for arrays of certain size or for initializing shift registers with array data type. The following example in Fig. 3.62 shows front panel and block diagram of Initialize Array function.

Array Size returns the number of elements in the input array. If the input array is n dimensional, **Array Size** returns an n-element, 1D array, with each element containing the size of one of the array's dimensions. The following example in Fig. 3.63 shows the size of the given array is 4.

Index Array. Returns the **element or subarray** of **n-dimension array** at **index**. Figure 3.64 illustrates the usage of Index array with an example. Index array can also be used to extract a row or column of a 2D array to create a subarray of the original. To do so, the 2D array is wired to the input of the function. Here two **index** terminals are available. The top **index** terminal indicates the row, and the second terminal indicates the column. The user can wire inputs to both **index** terminals to index a single element, or wire only one terminal to extract a row or column of data as illustrated in Fig. 3.65.

Build Array concatenates, or combines, two arrays into one, or adds extra elements to an array. The function is explained with an example in

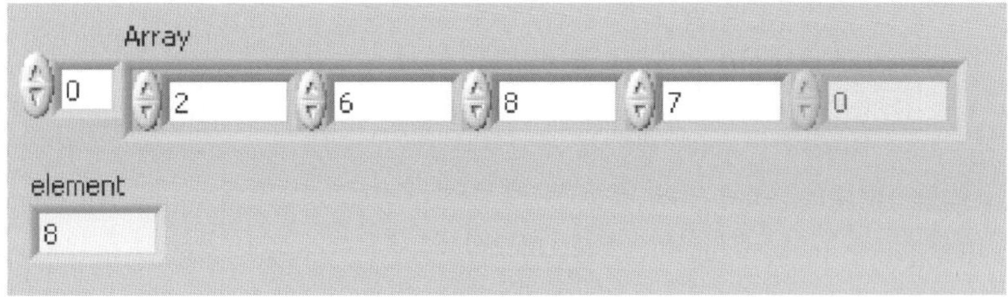

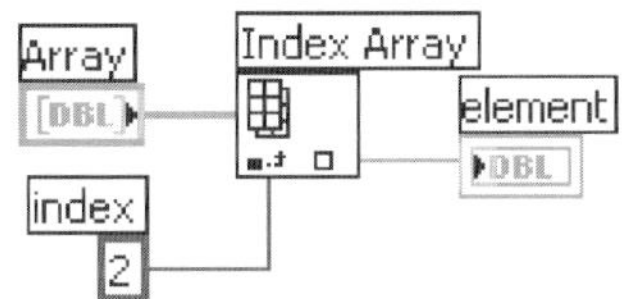

Fig. 3.64. Index Array for a 1D array

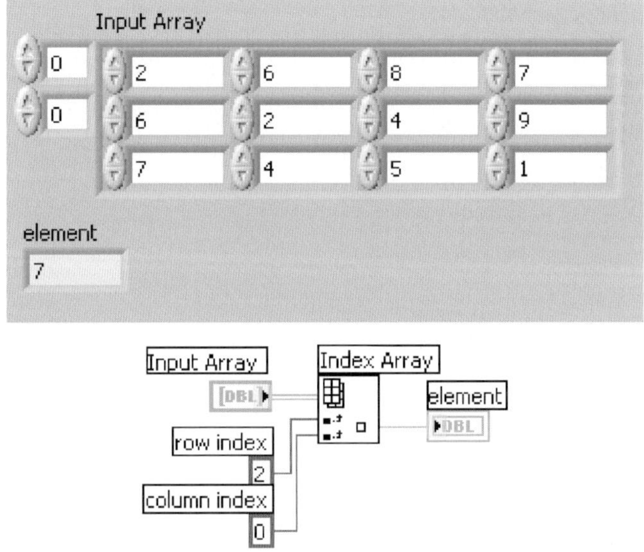

Fig. 3.65. Index Array for a 2D array

Fig. 3.66. The user can resize or "grow" this function to increase the number of inputs. **Build Array** has two types of inputs, *array* and *element*, so that it can assemble an array from both array and single-valued inputs. The **Build Array** function input will automatically adapt to an element or an array input, depending on the value wired to it. **Build Array** can also build or add elements to multidimensional arrays. To add an element to a multidimensional array, the element must be an array of smaller dimension. 2D arrays can also be built by using the **Build Array** function and wiring 1D array together,

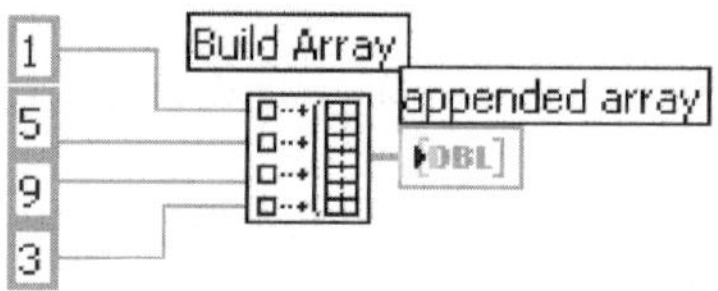

Fig. 3.66. Build Array function

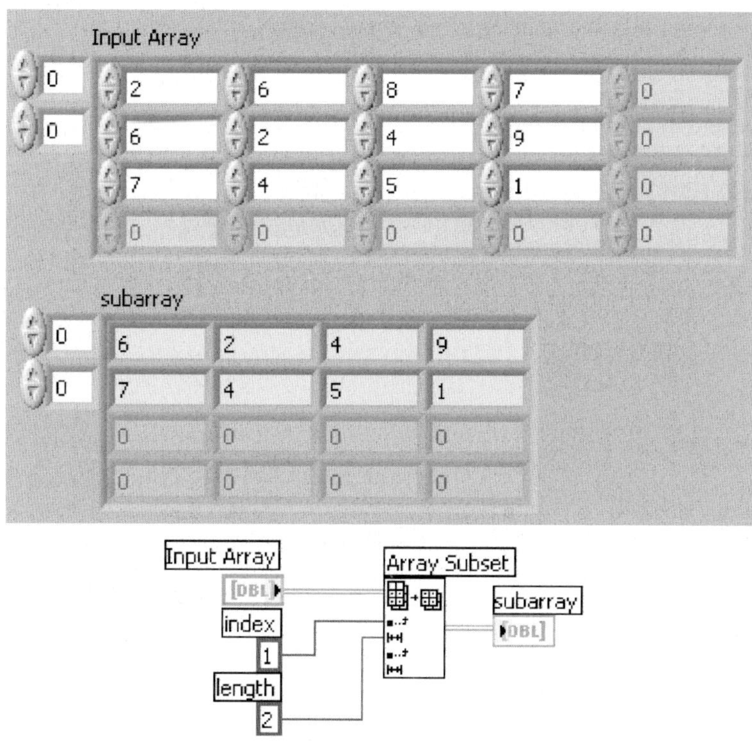

Fig. 3.67. Array subset function

instead of building a 2D array. In this case, the user can pop-up on the **Build Array** function, and choose **Concatenate Inputs**.

Array Subset returns a portion of an array starting at **index** and containing **length** elements. Notice that the third element's index is two because the index starts at zero; that is, the first element has an index of zero. Figure 3.67 shows the front panel and block diagram of the array subset function.

Insert into Array inserts an element or subarray into **n-dim array** at the point specified in **index** as shown in Fig. 3.68. When an array is wired to this function, the function resizes automatically to display **index** inputs for each dimension in the array. If the user does not wire any **index** inputs, the function appends the new element or subarray to the end of the **n-dim array**.

Rotate 1D Array rotates the elements of **array** by the number of places and in the direction indicated by **n**. The connector pane displays the default data types for this polymorphic function. Figure 3.69 illustrates the usage of this function.

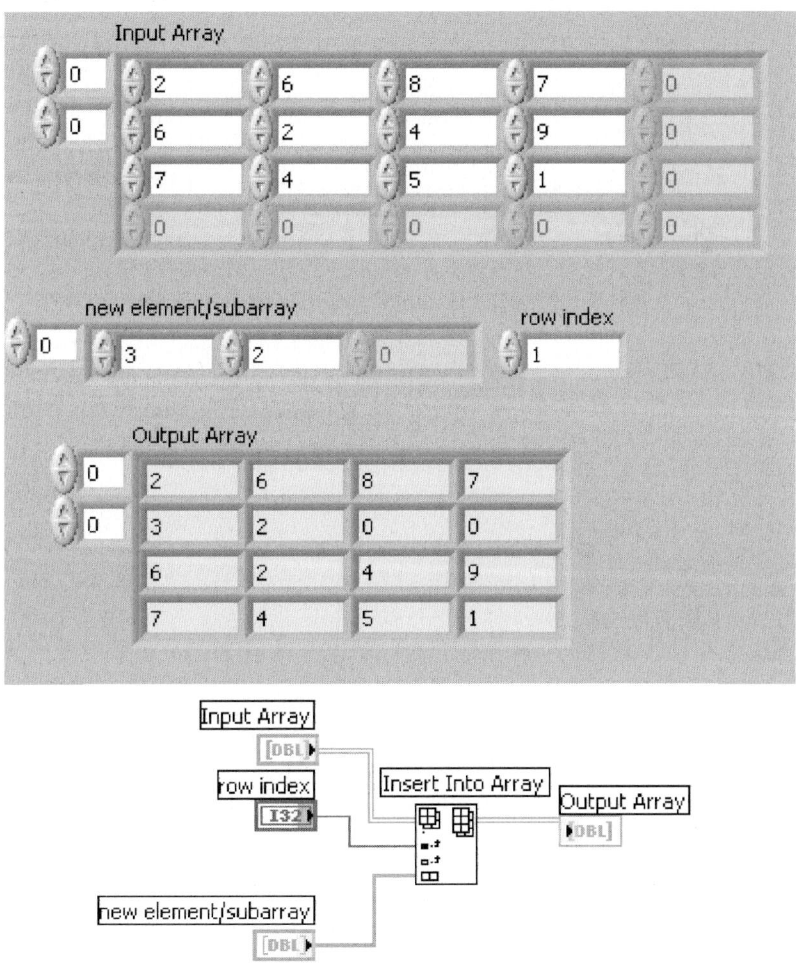

Fig. 3.68. Insert into Array function

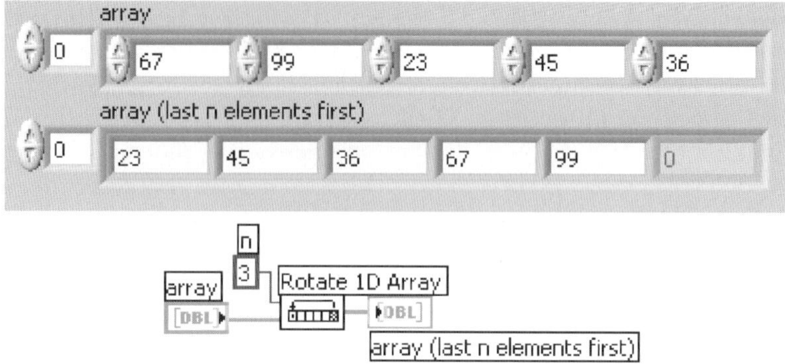

Fig. 3.69. Rotate 1D Array function

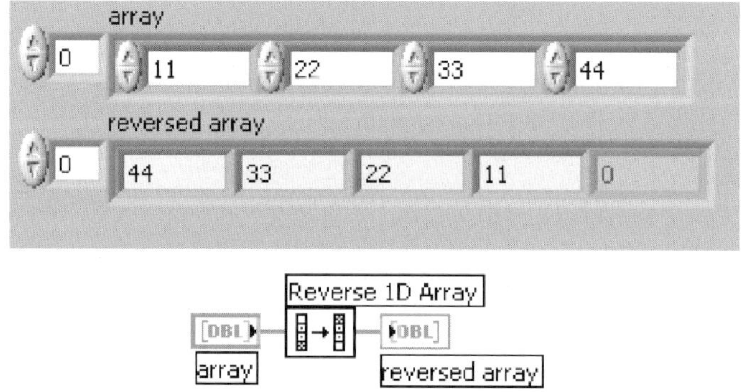

Fig. 3.70. Reverse 1D Array function

Reverse 1D Array reverses the order of the elements in **array** as shown in Fig. 3.70. The connector pane displays the default data types for this polymorphic function.

Search 1D array searches for an **element** in a **1D array** starting at **start index** as illustrated in Fig. 3.71. There is no need to sort the array before calling this function, because the search is linear. The connector pane displays the default data types for this polymorphic function

Split 1D Array divides **array** at **index** and returns the two portions as shown in Fig. 3.72. The connector pane displays the default data types for this polymorphic function.

Sort 1D array returns a sorted version of **array** with the elements arranged in ascending order. If **array** is an array of clusters, the function sorts the elements by comparing the first elements. If the first elements match, the function compares the second and subsequent elements. The connector pane displays the default data types for this polymorphic function. The illustration for this function is shown in Fig. 3.73.

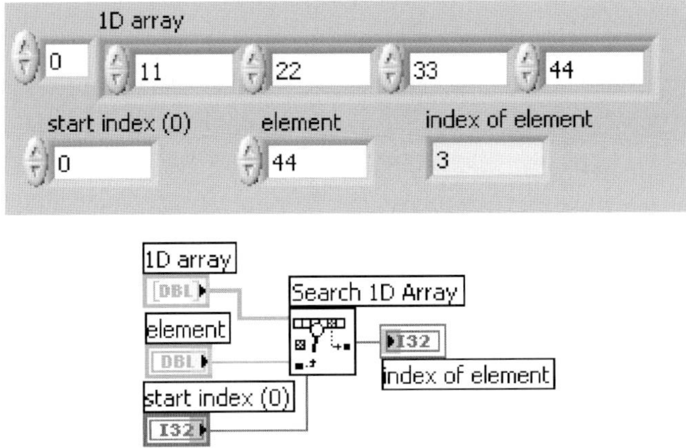

Fig. 3.71. Search 1D array function

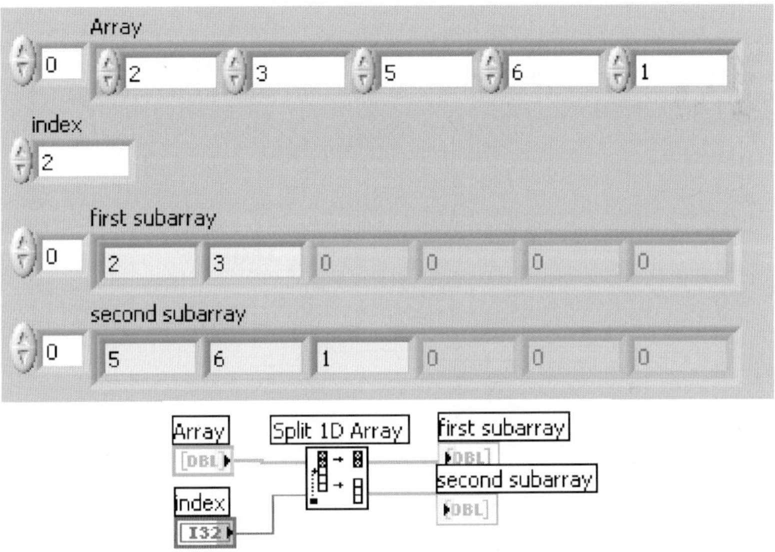

Fig. 3.72. Split 1D array function

Array max and min returns the maximum and minimum values found in **array**, along with the indexes for each value as shown in Fig. 3.74. The connector pane displays the default data types for this polymorphic function. If a numeric **array** has one dimension, the **max index** and **min index** outputs are scalar integers. If a numeric **array** has more than one dimension, these outputs are 1D arrays that contain the indexes of the maximum and minimum values.

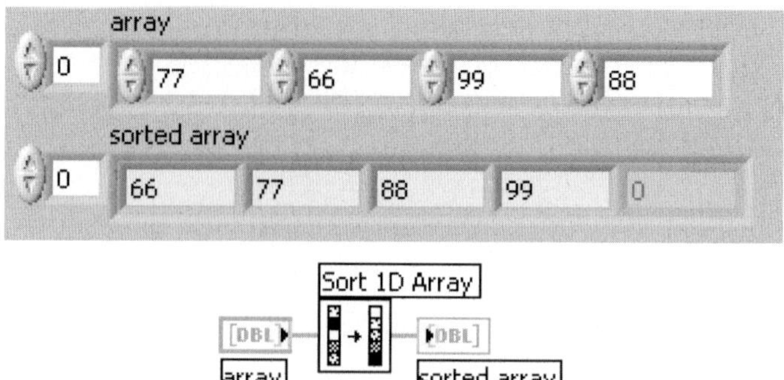

Fig. 3.73. Sort 1D array function

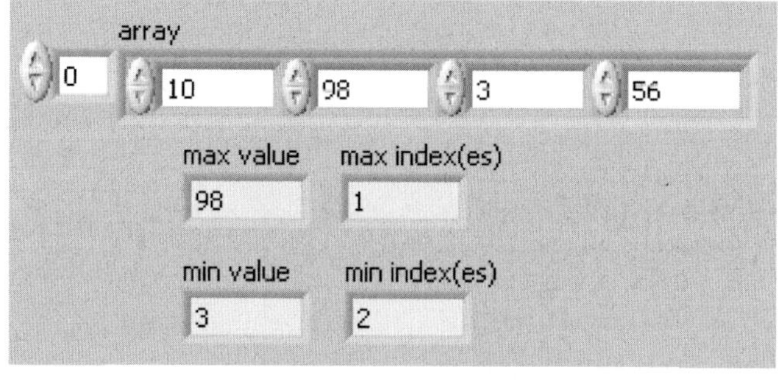

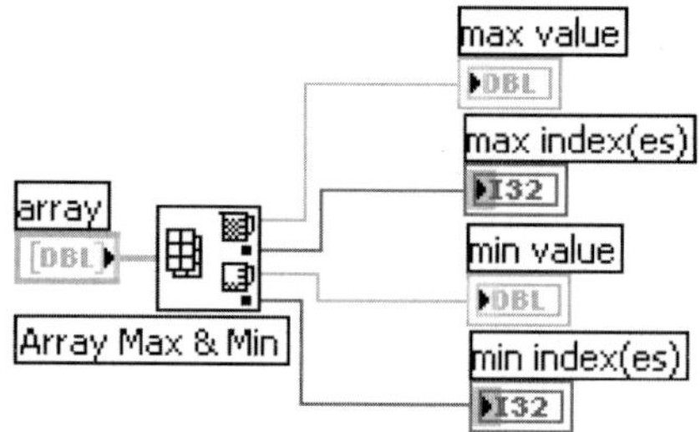

Fig. 3.74. Array min and max function

Example

Problem statement. To multiply two arrays and store the result in an array.
Block diagram construction:

- Multiply function is added from the Arithmetic and Comparison subpalette of the Functions palette and the wiring is done as shown in Fig. 3.75

Front panel construction:

- The arrays are created first.
- To create the arrays, the following steps are followed:

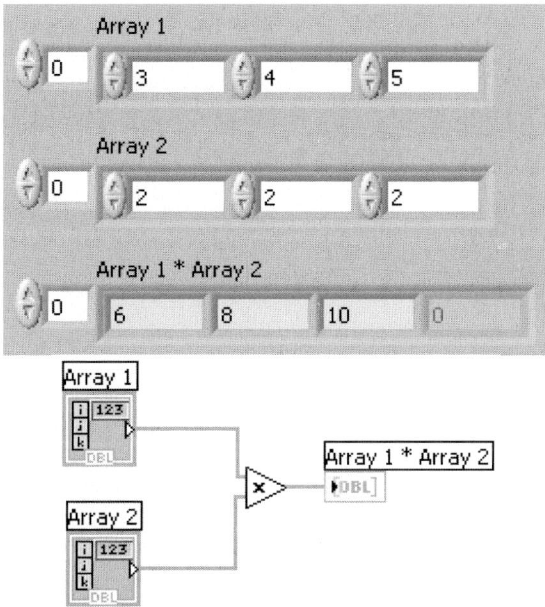

Fig. 3.75. Front panel and block diagram

Step 1. The array control or indicator known as the Array shell is created from the **Array** subpalette of the **Controls** palette available in the front panel.

Step 2. The *array shell* is combined with a *data object*, which can be numeric, Boolean, or string (or cluster).

- Three arrays are created of which two are controls and the third is an indicator.

3.5.4 Polymorphism

Another handy feature found most useful is the polymorphism of the LabVIEW arithmetic functions, **Add**, **Multiply**, **Divide**, and so on. Polymorphism is just a big word for a simple principle. The inputs to these functions can be of different size and representation. For example, a scalar can be added to an array or two arrays can be added together using the same function.

The example in Fig. 3.76 illustrates the addition of an array with a scalar constant, which gives an array as the result.

If two arrays are wired to the Add function, the function adds each element of one array to the corresponding element of the other array and returns the array as shown in Fig. 3.77.

If two arrays of different sizes are wired to the Add function, the function adds each element of one array to the corresponding element of the other

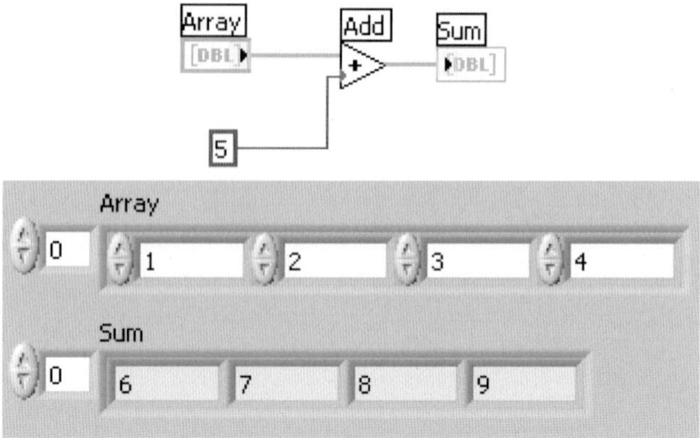

Fig. 3.76. Addition of scalar and an array

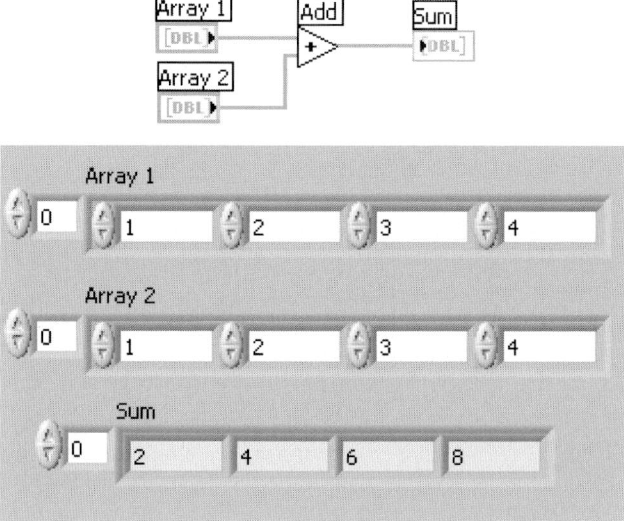

Fig. 3.77. Addition of two arrays

array and returns an array with size equal to the lowest dimension as shown in Fig. 3.78.

3.6 Clusters

Like an array, a *cluster* is a data structure that groups data. However, unlike an array, a cluster can group data of different types (i.e., numeric, Boolean, etc.); it is analogous to a *struct* in C or the data members of a class in C++ or Java. A cluster may be thought of as a bundle of wires, much like a telephone

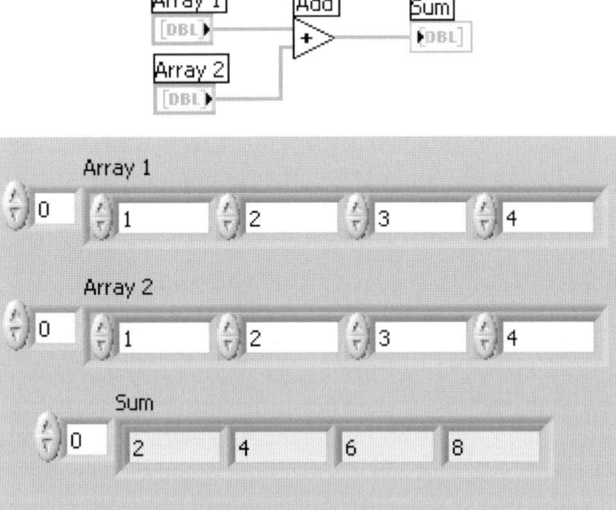

Fig. 3.78. Addition of two arrays of different size

cable. Each wire in the cable represents a different element of the cluster. Because a cluster has only one "wire" in the block diagram (even though it carries multiple values of different data types), clusters reduce wire clutter and the number of connector terminals that subVIs need. The cluster data type appears frequently when data are plotted on graphs and charts.

Cluster elements can be accessed by *unbundling* them all at once or by indexing one at a time, depending on the function chosen; each method has its place. Unbundling a cluster is similar as unwrapping a telephone cable and having access to the different-colored wires. Unlike arrays, which can change size dramatically, clusters have a fixed size, or a fixed number of wires in them. The Unbundled By Name function is used to access specific cluster elements.

Cluster terminals can be connected with a wire only if they have exactly the same type; in other words, both clusters must have the same number of elements, and corresponding elements must match in both data type and order. The principle of polymorphism applies to clusters as well as arrays, as long as the data types match.

Clusters are often used in error handling. The error clusters, **Error In.ctl** and **Error Out.ctl**, are used by LabVIEW to pass a record of errors between multiple VIs in a block diagram (for example, many of the DAQ and file I/O VIs have error clusters built into them). These error clusters are so frequently used that they appear in the **Array & Cluster** subpalette of the Controls palette for easy access.

Bundling several data elements into clusters eliminates wire clutter on the block diagram and reduces the number of connector pane terminals that sub-VIs need. The connector pane has, at most, 28 terminals. If a front panel contains more than 28 controls and indicators that are used programmatically, then the user can group some of them into a cluster and assign the cluster to a terminal on the connector pane. Like an array, a cluster is either a control or an indicator. A cluster cannot contain a mixture of controls and indicators.

3.6.1 Creating Cluster Controls and Indicators

A cluster is created by placing a **Cluster** shell (**Array & Cluster** subpalette of the **Controls** palette) on the front panel. Like arrays, objects can be added directly inside when they are pulled out from the **Controls** palette, or the user can drag an existing object into a cluster. **Objects inside a cluster must be all controls or all indicators**. A cluster cannot have a combination of both controls and indicators; this is based on the status of the first object one place inside it. The cluster can be resized with the Positioning tool if necessary. The cluster can conform exactly to the size of the objects inside it, by popping up on the *border* (not inside the cluster) and choosing an option in the **Auto sizing** menu.

The following example shown in Fig. 3.79 is a cluster of three controls.

Creating Cluster Constants

Create a cluster constant on the block diagram by selecting a cluster constant on the **Cluster** palette, placing it on the block diagram, and dragging a constant into the cluster shell. If a cluster control or indicator is available on the front panel and if the user wants to create a cluster constant containing the same elements on the block diagram, then the user can either drag that cluster from the front panel to the block diagram or right-click the cluster on the front panel and select **Create≫Constant** from the shortcut menu.

Fig. 3.79. Cluster with three controls

Cluster Order

Cluster elements have a logical order unrelated to their position within the shell. The first object placed in the cluster element is zero; the second is element one, and so on. If an element is deleted, the order adjusts automatically. The order and data types must be identical in order to connect a cluster with another cluster.

The order of the elements can be changed within the cluster by popping up on the cluster border and choosing **Reorder Controls in Cluster** from the Pop-up menu. A new set of buttons appears in the Toolbar and the cluster appearance changes. The while boxes on the elements as shown in Fig. 3.80 indicate current places in the cluster order. The black boxes show the new places. Clicking on an element with the cluster order cursor sets the element's place in the cluster order to the number displayed on the Toolbar. A new number can be typed into that field before an object is selected. The user can also revert to the old order by clicking on the Revert button.

Using Clusters to Pass Data to and from SubVIs

The connector pane of a VI can have a maximum of 28 terminals. To pass information to all 28 terminals when calling a subVI, the wiring can be very tedious and the user can easily make a mistake by bundling a number of

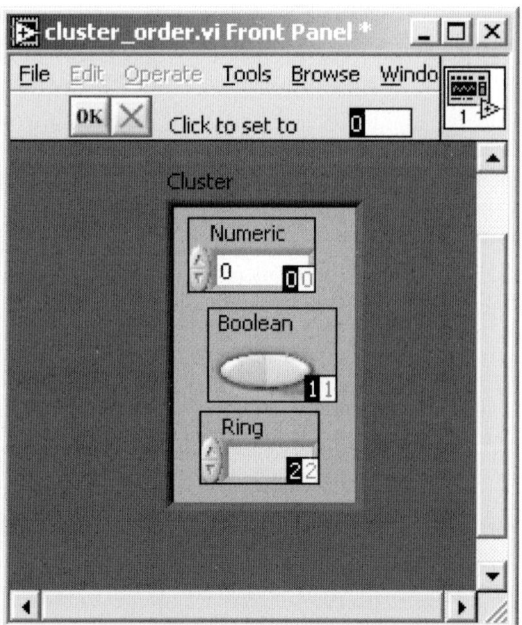

Fig. 3.80. Cluster order

controls or indicators into a cluster. To avoid this user can use a single terminal and wire to pass several values into or out of the subVI.

3.6.2 Cluster Functions

The Cluster functions located on the **Functions≫All Functions≫Cluster** palette is used to create and manipulate clusters. The Bundle and Bundle by Name functions are used to assemble and manipulate clusters. The Unbundled and Unbundled by Name functions are used to disassemble clusters. The Bundle, Bundle by Name, Unbundled, and Unbundled by Name functions can also be created by right-clicking a cluster terminal on the block diagram and selecting **Cluster Palette** from the shortcut menu as shown in Fig. 3.81. The Bundle and Unbundled functions automatically contain the correct number of terminals. The Bundle by Name and Unbundled by Name functions appear with the first element in the cluster. The Positioning tool can be used to resize the Bundle by Name and Unbundled by Name functions to show the other elements of the cluster.

Replacing a Cluster Element

To *replace* an element in a cluster, first the **Bundle** function is sized to contain the same number of input terminals as the elements in this cluster. Then the cluster is wired to the middle terminal of the **Bundle** function (symbols for data types inside the cluster will appear in the **Bundle** inputs) and wire the new value(s) of the element(s) to be replaced to the corresponding input.

Bundling Data

The **Bundle** function (**Cluster** palette) in Fig. 3.82 assembles individual components into a new cluster or allows us to replace elements in an existing cluster. The function appears as the icon at the left when one places it in the diagram window. Dragging a corner of the function with positioning tool can increase the number of inputs. When wired on each input terminal, a

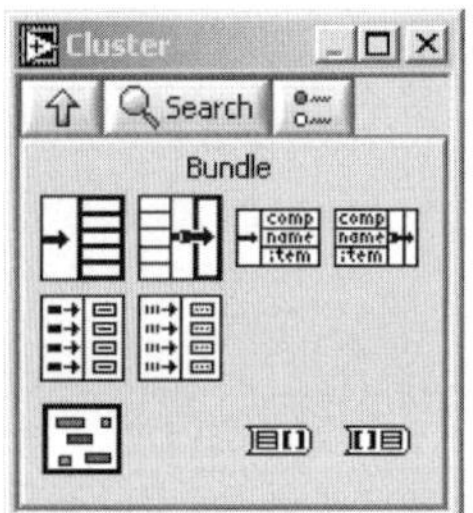

Fig. 3.81. Cluster palette

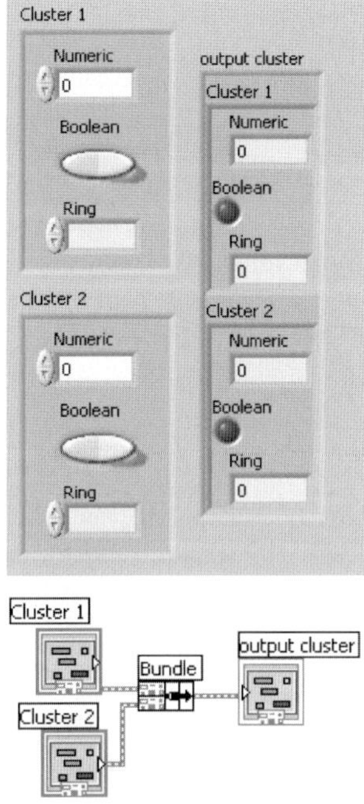

Fig. 3.82. Bundle Function

symbol representing the data type of the wired element appears on the empty terminal. The order of the resultant cluster will be the order of inputs to the **Bundle**.

To create a new cluster the user need not wire an input to the center **cluster** input of the **Bundle** function. This needs to be wired only if elements are replaced in the cluster.

Unbundling Clusters

The **unbundled** function (**Cluster** palette) in Fig. 3.83 splits a cluster into each of its individual components. *The output components are arranged from top to bottom in the same order they have in the cluster.* If they have the same data type, the elements' order in the cluster is the only way to distinguish among them. Dragging a corner in the function with the Positioning tool can increase the number of outputs. The **Unbundled** function must be sized to contain the same number of outputs as there are elements in the input cluster, or it will produce bad wires. When an input cluster is wired to the correctly sized **Unbundled**, the previously blank output terminals will assume the symbols of the data types in the cluster example has been shown in Fig. 3.83.

LabVIEW does have a way to bundle and unbundled clusters using element names.

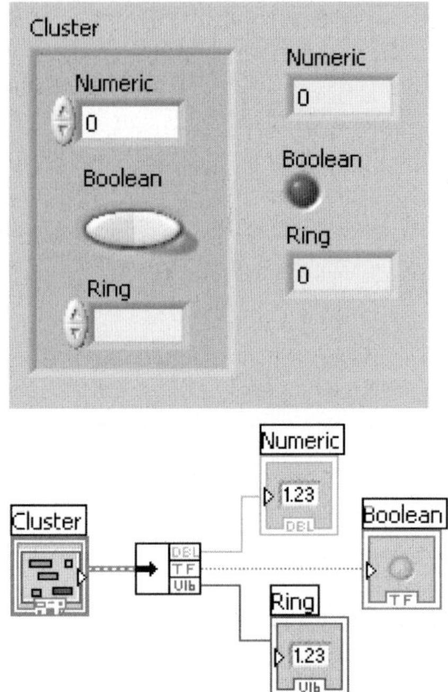

Fig. 3.83. Unbundle function

Bundling and Unbundling by Name

Sometimes there is no need to assemble or disassemble an entire cluster – the user just needs to operate on an element or two. This is accomplished using **Bundle By Name** and **Unbundle By Name** functions.

Unbundle By Name, also located in the **Cluster** palette, returns elements whose name(s) are specified. There is no need to consider cluster order to correct **Unbundle** function size. The unbundle function is illustrated in Fig. 3.84.

Bundle by Name, found in the **Cluster** palette, references elements by name instead of by position (as **Bundle** does). Unlike **Bundle**, we can access only the elements that are required. However, **Bundle by Name** cannot create new clusters; it can only replace an element in an existing cluster. Unlike **Bundle**, **Bundle by Name's** middle input terminal should be wired to allow the function know which element in the cluster has to be to replaced. This function is illustrated in Fig. 3.85.

All cluster elements should have owned labels when the **By Name** functions are used. As soon as the cluster input of **Bundle By Name** or **Unbundled By Name** is wired, the name of the first element in the cluster appears in the name input or output. To access another element, the user can click on the name input or output with the Operating or Labeling tool to see

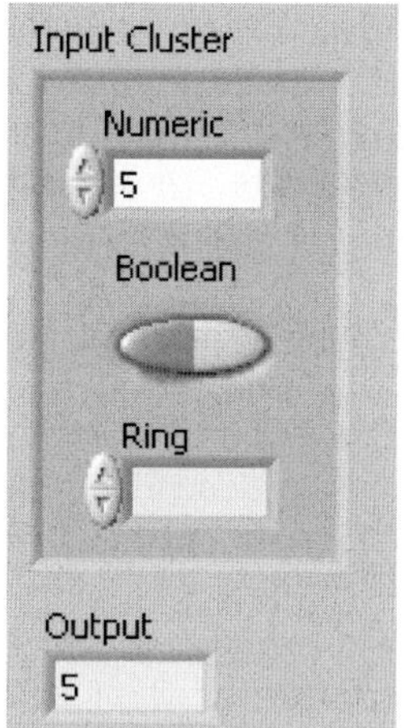

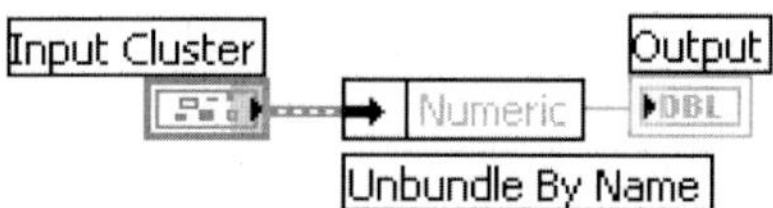

Fig. 3.84. Unbundled by Name function

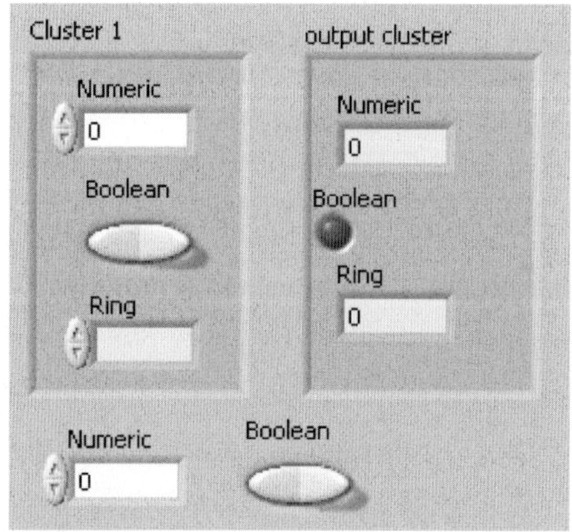

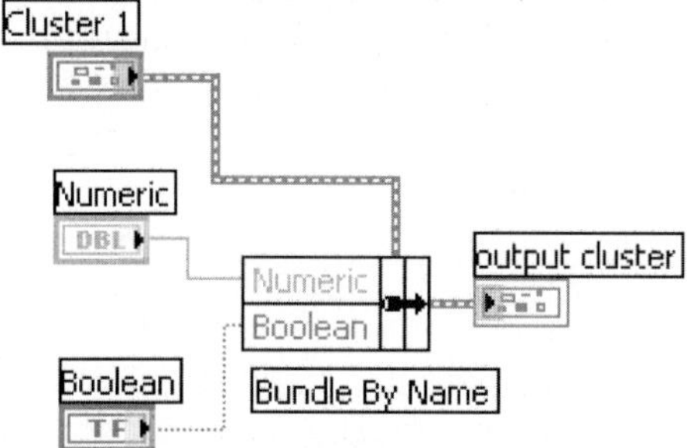

Fig. 3.85. Bundle by Name function

a list of the name of all labeled elements in the cluster. The element can be selected from this list, and the name will appear in the **name** terminal. This list can also be accessed by popping up on name and choosing **Select Item**.

Both functions can be resized to accommodate as many elements as need. The **By Name** functions do not break unless we remove an element they reference.

Build Cluster Array bundles each component input into a cluster and assembles all **component** clusters into an array of clusters as indicated in Fig. 3.86. Each cluster contains a single component. The connector pane displays the default data types for this polymorphic function. For example, if the user wires four arrays of single-precision, floating-point numbers to four component inputs, the function returns a four-element array of clusters, each cluster containing one array of single-precision, floating-point numbers. Element 0 of **array of clusters** has the value of the top array component, and so on.

Index and Bundle Cluster Array indexes a set of arrays and creates a cluster array in which the ith element contains the jth element of each input array. The connector pane displays the default data types for this polymorphic function. The front panel and block diagram for this function are shown in Fig. 3.87.

Interchangeable Arrays and Clusters

Sometimes the user may find it convenient to change arrays to clusters, and viceversa. This trick can be extremely useful, especially since LabVIEW includes many more functions that operate on arrays than clusters. For example,

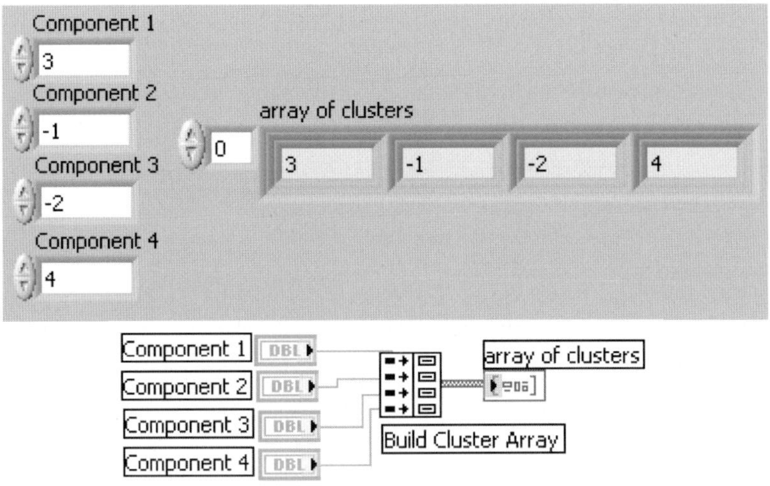

Fig. 3.86. Build Cluster Array

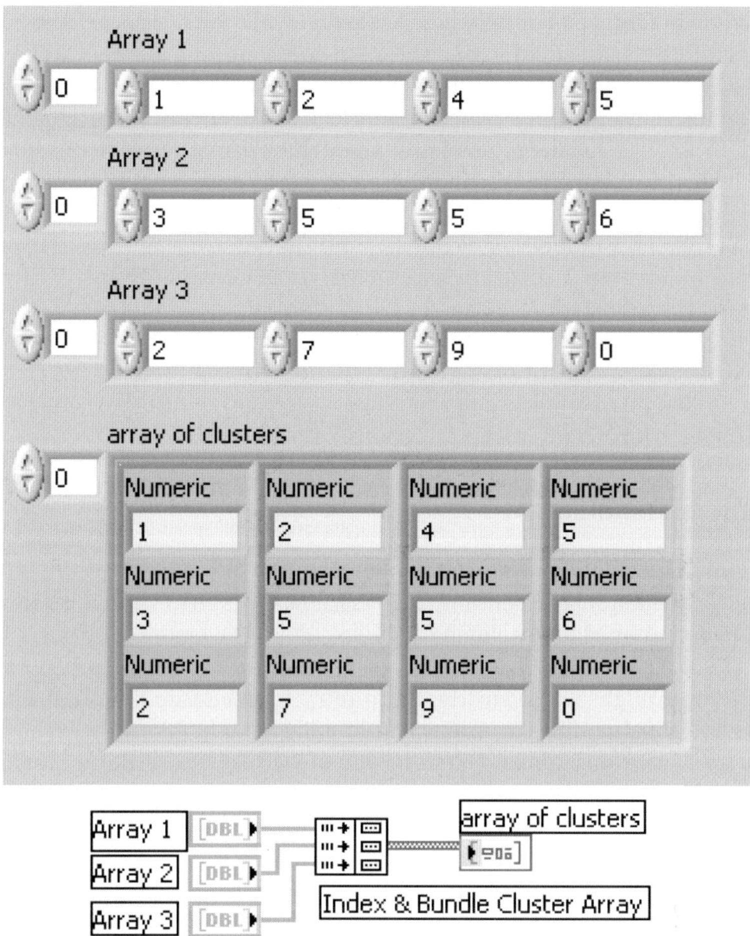

Fig. 3.87. Index and Bundle Cluster Array Function

if a cluster of buttons are available on the front panel and is required to reverse the order of the button's values. **Reverse 1D** Array would be perfect, but it only works on arrays. So the user can use the **Cluster to Array** function to change the cluster to an array, use **Reverse ID** Array to switch the values around, and then use **Array to Cluster** to change back to a cluster.

Cluster to Array converts a cluster of N elements of the same data type into an array of N elements of that type. Array index corresponds to cluster order (i.e., cluster element 0 becomes the value at array index 0). This function cannot be used on a cluster containing arrays as elements, because LabVIEW doesn't allow creation of an array of arrays. Note that, to use this function, all elements in the cluster must have the same data type.

Array to Cluster converts an N-element, 1D array into a cluster of N elements of the same data type; the user must pop-up on the **Array To Cluster** terminal and choose **Cluster Size** to specify the size of the output cluster since clusters don't size automatically like arrays. The cluster size defaults to nine; if the array has fewer than the number of elements specified in the cluster size, LabVIEW will automatically fill in the extra cluster values with the default value for the data type of the cluster. However, if the input array has a greater number of elements than the value specified in the cluster size window, the block diagram wire going to the output cluster will break until its size is adjusted.

Both functions are very handy if the user wants to display elements in a front panel cluster control or indicator but need to manipulate the elements by index value on the block diagram. They can be found in both the **Array** and **Cluster** subpalettes of the Functions palette.

3.6.3 Error Handling

By default, LabVIEW automatically handles any error that occurs when a VI runs by suspending execution, highlighting the subVI or function where the error occurred, and displaying a dialog box. For example, if an I/O VI on the block diagram times out, it is not feasible for the entire application to stop. The user might also want the VI to retry for a certain period of time. In LabVIEW, error handling decisions can be made on the block diagram of the VI. VIs and functions return errors in one of two ways – with numeric error codes or with an error cluster. Typically, functions use numeric error codes and VIs use an error cluster, usually with error inputs and outputs.

Error handling in LabVIEW follows the dataflow model similar to data flowing through a VI, the error information also flows. The error information is wired from the beginning of the VI to the end. An error handler VI can be included at the end of the VI to determine if the VI has completed execution without errors. The error in and error out clusters can be used in each VI created to pass error information through the VI.

As the VI runs, LabVIEW tests for errors at each execution node. If LabVIEW does not find any errors, the node executes normally. If LabVIEW detects an error, the node passes the error to the next node without executing. The next node does the same thing, and so on. The **Simple Error Handler VI**, is used to handle the error at the end of the execution flow. The **Simple Error Handler VI** is located on the **Functions≫All Functions≫Time & Dialog** palette.

Error Clusters

The error clusters located on the **Functions≫All Functions≫Array & Cluster** palette includes the components of information shown in Fig. 3.88:

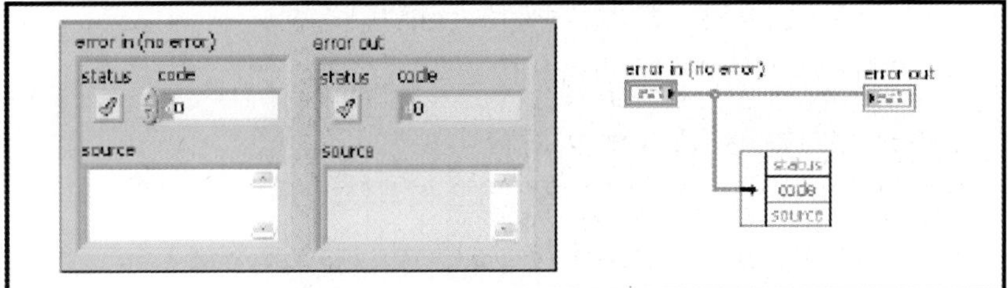

Fig. 3.88. Error cluster – front panel and block diagram

- **Status** is a Boolean value that reports true if an error occurred. Most VIs, functions, and structures that accept Boolean data also recognizes this parameter. For example, the user can wire an error cluster to the Boolean inputs of the Stop, Quit LabVIEW, or Select functions. If an error occurs, the error cluster passes a True value to the function.
- **Code** is a 32-bit signed integer that identifies the error numerically. A nonzero error code coupled with a **status** of "False" indicates a warning rather than a fatal error.
- **Source** is a string that identifies where the error occurred. The error cluster controls and indicators are used to create error inputs and outputs in subVIs.

Checking for errors in VIs helps the user identify the following problems:

- If communications are initialized incorrectly or wrote improper data to an external device.
- An external device lost power, is broken, or is not working properly.
- We upgraded the operating system software, which changed the path to a file or the functionality of a VI or library. There may be a problem in a VI or a system program.

Explain Error

When an error occurs, the user can right-click within the cluster border and select **Explain Error** from the shortcut menu to open the **Explain Error** dialog box. The **Explain Error** dialog box contains information about the error. The shortcut menu includes an **Explain Warning** option if the VI contains warnings but no errors. The **Explain Error** dialog box can also be accessed from the **Help≫Explain Error** menu.

Using While Loops for Error Handling

The user can wire an error cluster to the conditional terminal of a While Loop to stop the iteration of the While Loop as shown in Fig. 3.89. When the error

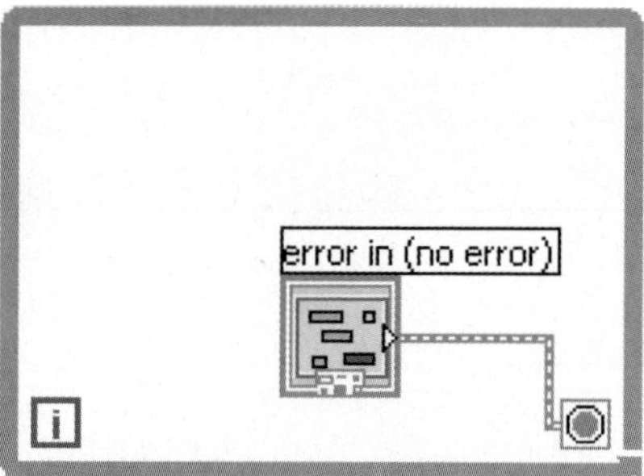

Fig. 3.89. While Loop for error handling

cluster is wired to the conditional terminal, only the True or False value of the status parameter of the error cluster is passed to the terminal. When an error occurs, the While Loop stops. When an error cluster is wired to the conditional terminal, the shortcut menu items **Stop if True** and **Continue if True** change to **Stop on Error** and **Continue while Error**.

3.7 Waveform Charts

A plot is simply a graphical display of X versus Y values. Often, Y values in a plot represent the data value, while X values represent time. The waveform chart, located in the **Graph** subpalette of the **Controls** palette, is a special numeric indicator that can display one or more plots of data. The waveform chart with its components is illustrated in Fig. 3.90. Most often used inside loops, charts retain and display previously acquired data, appending new data as they become available in a continuously updating display. In chart, the Y values represent the new data, and X values represent time (often, each Y value is generated in a loop iteration, and so the X value represents the time for one loop). LabVIEW has only one kind of chart, but the chart has three different update modes for interactive data display.

Chart Update Modes

The waveform chart has three update modes – *strip chart mode, scope chart mode*, and *sweep chart mode*. The update mode can be changed by popping up on the waveform chart and choosing one of the options from the **Advanced≫update Mode≫** menu. Modes can be changed while the VI is running by selecting **Update Mode** from the chart's runtime Pop-up menu. The default mode is **Strip Chart**.

The strip chart has a scrolling display similar to a paper strip chart. The scope chart and the sweep chart have retracing displays similar to that of an

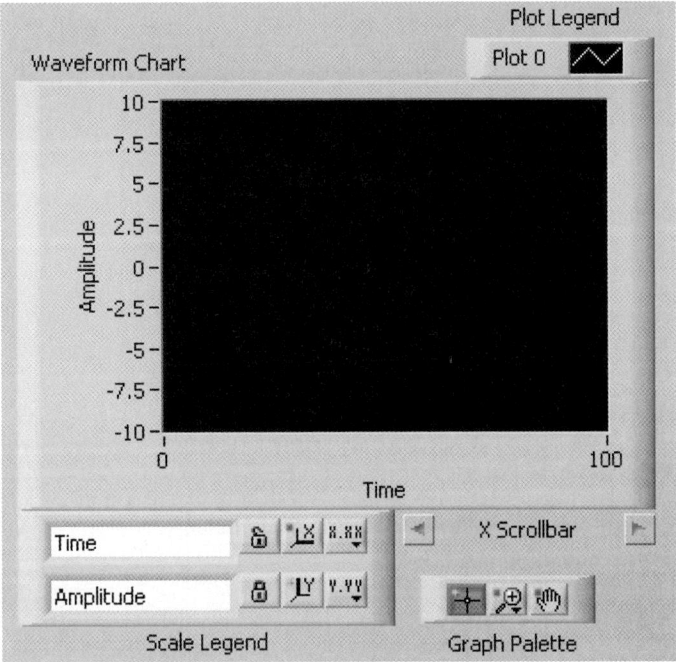

Fig. 3.90. Waveform chart with its components

oscilloscope. On the scope chart, when the plot reaches the right border of the plotting area, the plot erases, and plotting begins again from the left border. The sweep chart acts much like the scope chart, but the display does not go blank when the data reaches the right border. Instead a moving vertical line marks the beginning of new data and moves across the display as new data are added.

Because there is less overhead in retracing a plot, the scope chart and the sweep chart operate significantly faster than the strip chart.

Strip Chart

A strip chart shows running data continuously scrolling from left to right across the chart as shown in Fig. 3.91.

Scope Chart

A scope chart shows one item of data, such as a pulse or wave, scrolling partway across the chart from left to the right as shown in Fig. 3.92.

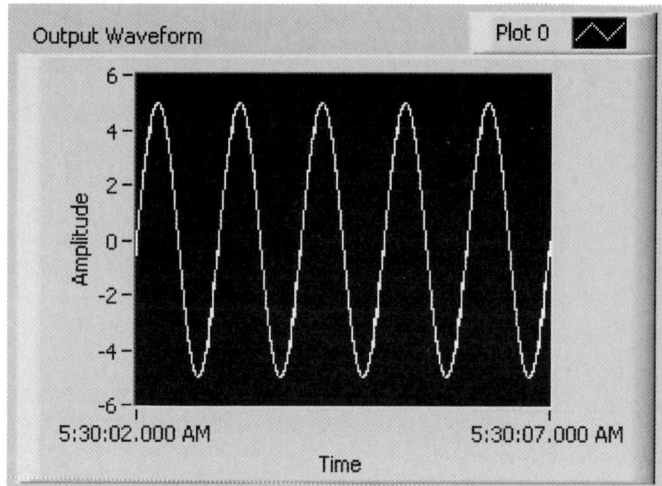

Fig. 3.91. Strip chart mode

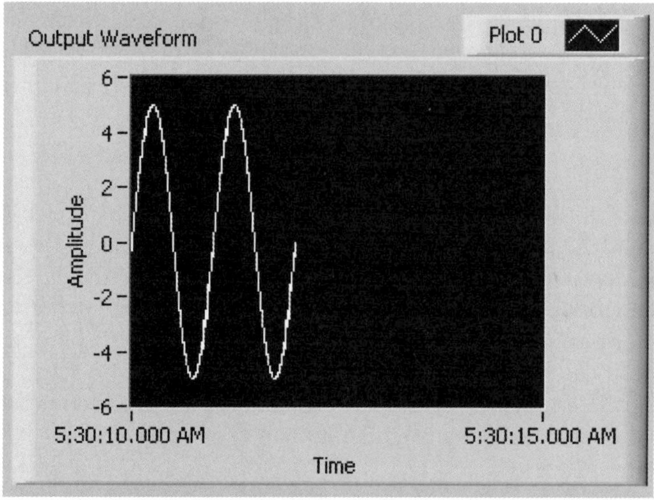

Fig. 3.92. Scope chart mode

Sweep Chart

A sweep chart is similar to an EKG display. A sweep chart works similarly to a scope except it shows the older data on the right and the newer data on the left separated by a vertical line as shown in Fig. 3.93.

The scope chart and sweep chart have retracing displays similar to an oscilloscope. Because there is less overhead in retracing a plot, the scope chart and the sweep chart display plots significantly faster than the strip chart.

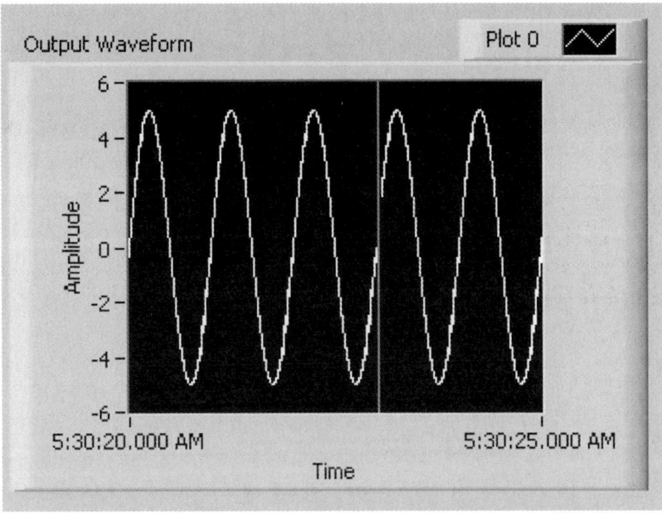

Fig. 3.93. Sweep chart

Wiring Charts

A scalar output can be wired directly to a waveform chart. The data type in the following waveform chart in Fig. 3.94 terminal matches the input data type.

Waveform charts can display multiple plots. Bundle multiple plots together using the Bundle function located on the **Cluster** palette. In the following block diagram illustrated in Fig. 3.95, the Bundle function bundles the outputs of the three VIs to plot on the waveform chart. The waveform chart terminal changes to match the output of the Bundle function. To add more plots, use the Positioning tool to resize the Bundle function.

Example

Block diagram construction:

- The While Loop is added into the block diagram form the Structures subpalette of the Functions palette
- The sine wave form VI is added from the subpalette
- Here Wait until next ms function is added and execution waits for 500 ms as specified
- The error in and error out are wired as shown in Fig. 3.96
- The conditional terminal is wired with the Boolean terminal.

Front panel construction:

- The waveform chart is added from the of the controls palette
- A Boolean control is added from the of the controls palette to control the operation of the waveform chart

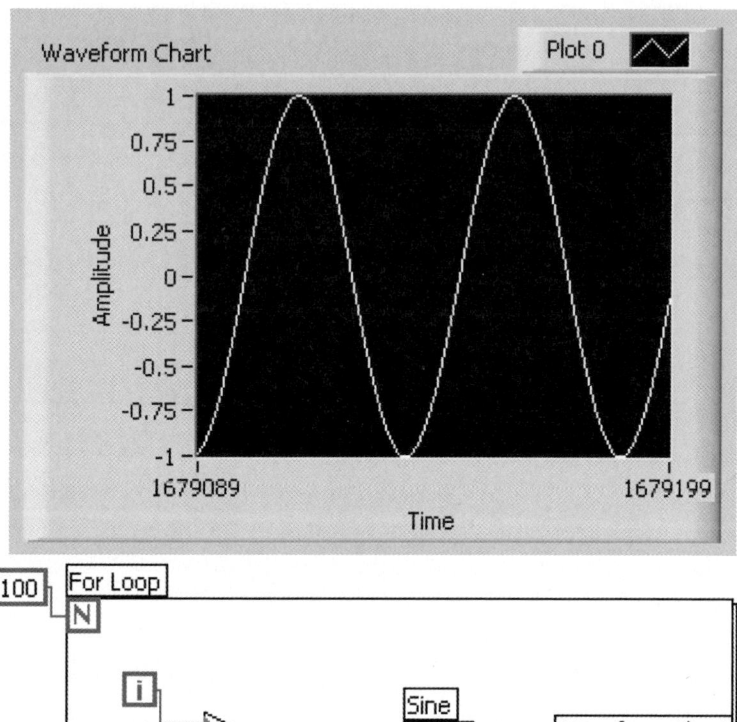

Fig. 3.94. Single plot chart

3.7.1 Chart Components

Show the Digital Display

Like many other numeric indicators, charts have the option to show or hide the digital display (popping up on the chart to get the **Visible Items≫ option**). The digital display shows the most recent value displayed by the chart.

The Scrollbar

Charts also have scrollbars that can be shown or hidden. The user can use the scrollbar to display older data that have scrolled of the chart.

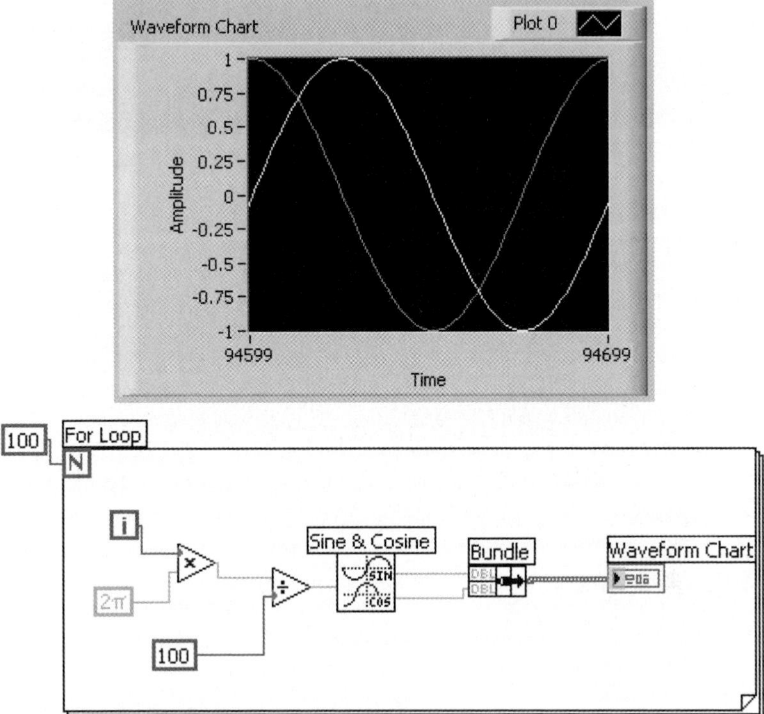

Fig. 3.95. Multiple plot chart

Clearing the Chart

Sometimes the user will find it useful to remove all previous data from the chart display. This operation can be performed by selecting **Data Operations≫Clear Chart** from the chart's Pop-up menu to clear a chart from edit mode. If the VI is in run mode, **Clear Chart** is a Pop-up menu option instead of being hidden under **Data Operations**.

Stacked and Overlaid Plots

If a multiple-plot chart is available, the user can choose whether to display all plots on the same Y axis, called an overlaid plot, or display each plot with its own Y scale, called a stacked plot. **Stack Plots** or **Overlay Plots** can be selected from the chart's Pop-up menu to toggle the type of display.

Multiple X and Y Scales

In a multi-plot chart, if the overlaid plot has different scales for each plot; for example, if one plot's Y range is from 0.0 to 1.0 and the other is $-1,000$

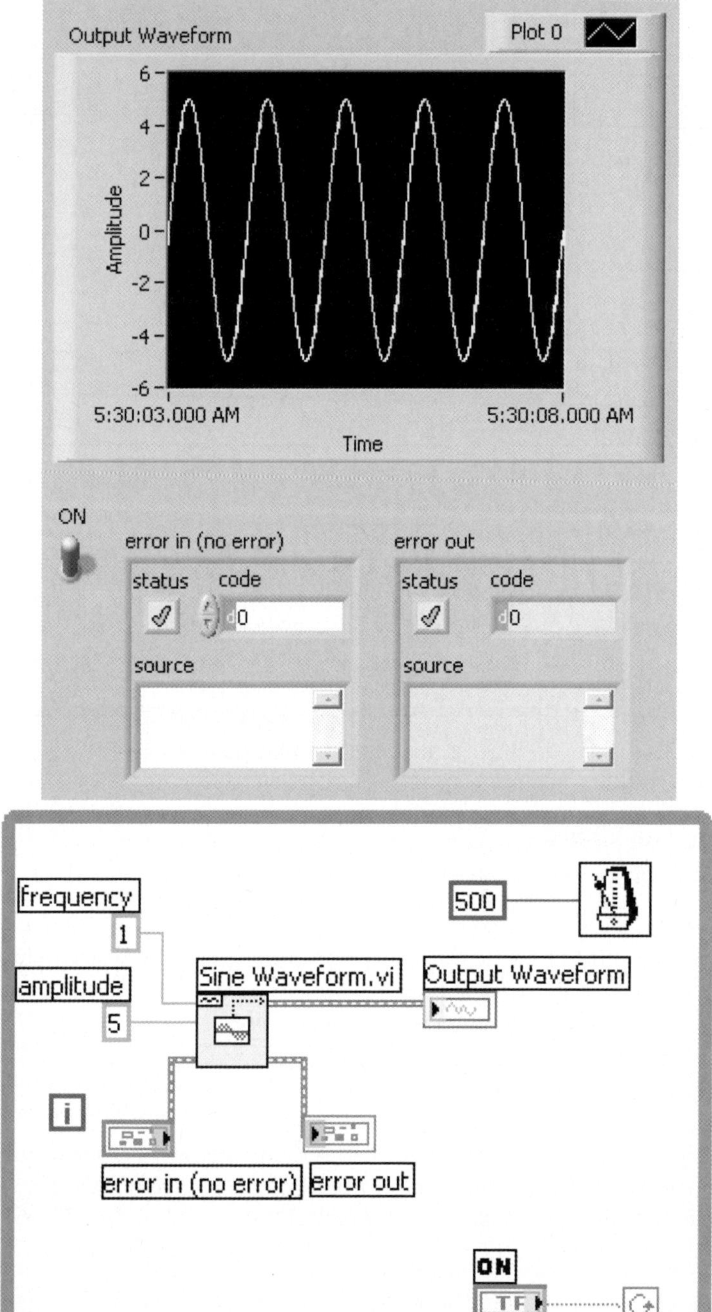

Fig. 3.96. Front panel and block diagram

to $+1,000$, it would be difficult to see both of them overlaid without a separate scale for each. Separate scales can be created for both the X and Y axes by popping up on the corresponding axis and selecting **Y/X Scale≫Duplicate Scale**.

Chart History Length

By default, a chart can store up to 1,024 data points. The user can store more data by selecting **Chart History Length** from the Pop-up menu and specifying a new value of up to 100,000 points. Changing the buffer size does not affect how much data are shown on the screen – resize the chart to show more or less data at a time. However, increasing the buffer size does increase the amount of data that can be scrolled back through.

3.7.2 Mechanical Action of Boolean Switches

Each time the Boolean switch VI is run, the user must first turn on the vertical Enable switch before clicking the run button, otherwise the loop will only execute once. The mechanical action of a Boolean control can be modified to change its behavior and circumvent this inconvenience. LabVIEW offers six possible choices for the mechanical action of a Boolean control.

Switch When Pressed action changes the control's value each time the user clicks on the control with the Operating tool. This action is the default for Booleans and is similar to that of a ceiling light switch.

Switch When Released action changes the control's value only when the mouse button is released within the graphical boundary of the control. The action is not affected by how often the VI reads the control. This mode is similar to the operation when the user clicks on a check mark in a dialog box; it becomes highlighted but does not change until the mouse button is released.

Switch until Released action changes the control's value when the user clicks on the control. It retains the new value until the mouse button is released, at which time the control reverts to its original value. The action is similar to that of a door buzzer and is not affected even if the VI reads the control very often.

Latch When Pressed action changes the control's value when the user clicks on the control. It retains the new value until the VI reads it once, at which point the control reverts to its default value. This action happens whether or not the user continues to press the mouse button.

Latch When Pressed is similar in functionality to a circuit breaker and is useful when the user wants the VI to do something only once for each time the control is set, such as to stop a While Loop when the STOP button is pressed.

Latch When Released action changes the control's value only after the mouse button is released. When the VI reads the value once, the control reverts to the old value. This action guarantees at least one new value. As with **Switch When Released**, this mode is similar to the behavior of buttons in a dialog box; the button becomes highlighted when clicked on it and latches a reading when the mouse button is released.

Latch until Released changes the control's value when the user clicks on the control. It retains the value until the VI reads the value once or until the mouse button is released, whichever occurs last.

3.8 Waveform Graphs

VIs with graphs usually collects the data in an array and then plot the data to the graph. The following illustration in Fig. 3.97 shows a graph with its elements. The graphs located on the **Controls≫Graph Indicators** palette include the waveform graph and XY graph. The waveform graph plots only single-valued functions, as in $y = \mathrm{f}(x)$, with points evenly distributed along the x-axis, such as acquired time-varying waveforms.

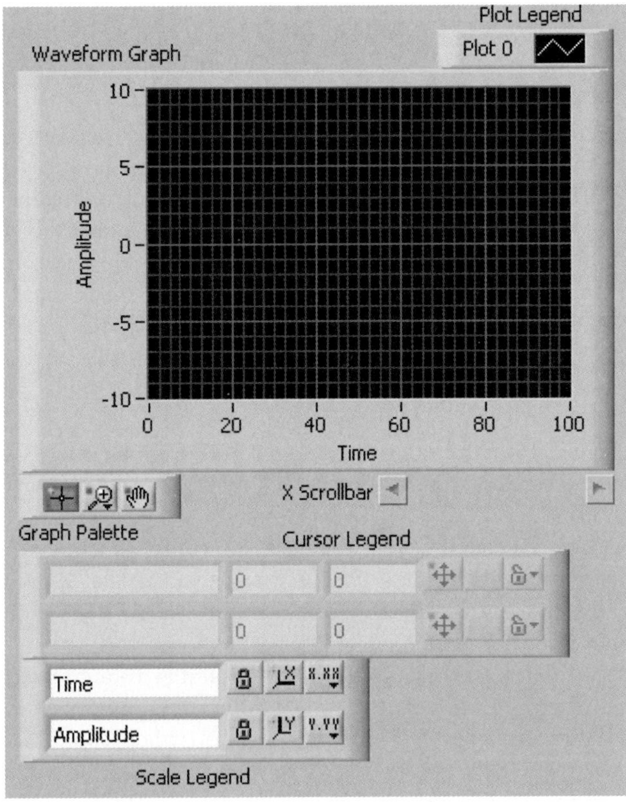

Fig. 3.97. Waveform graph with its components

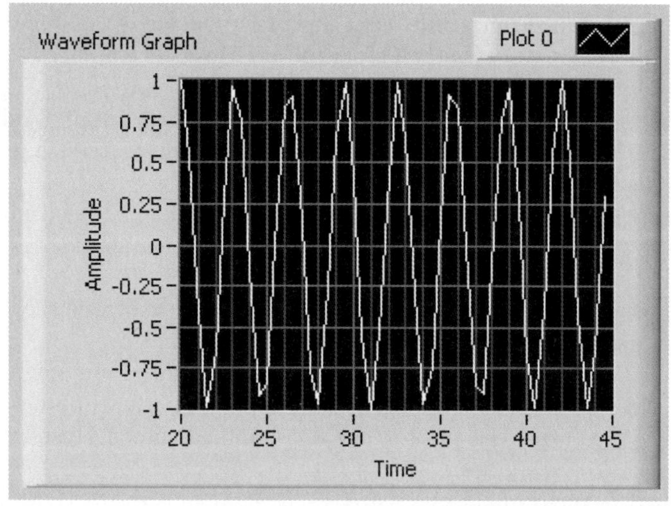

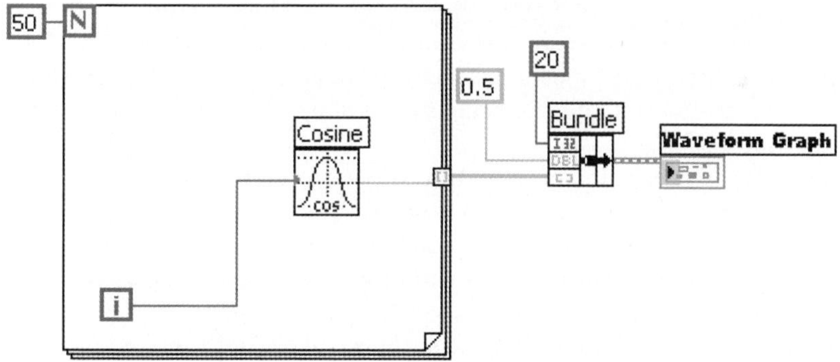

Fig. 3.98. Single plot graph

3.8.1 Single-Plot Waveform Graphs

The waveform graph accepts a single array of values and interprets the data as points on the graph and increments the x index by one starting at $x = 0$. The graph also accepts a cluster of an initial x value, and an array of y data. Figure 3.98 shows the front panel and block diagram of a waveform graph.

3.8.2 Multiple-Plot Waveform Graphs

A multi plot waveform graph accepts a 2D array of values, where each row of the array is a single plot. The graph interprets the data as points on the graph and increments the x index by one, starting at $x = 0$. Wire a 2D array data type to the graph, right-click the graph, and select **Transpose Array** from the shortcut menu to handle each column of the array as a plot.

A multiplot waveform graph also accepts a cluster of an x value, a Δx value, and a 2D array of y data. The graph interprets the y data as points on the graph and increments the x index, starting at $x = 0$.

A multi plot waveform graph accepts a cluster of an initial x value, a Δx value, and an array that contains clusters. Each cluster contains a point array that contains the y data. The Bundle function is used to bundle the arrays into clusters, and the Build Array function is used to build the resulting clusters into an array. The Build Cluster Array can be used, which creates arrays of clusters that contain inputs one specify.

Sometimes the user may want to change the flexibility of the time base for the graph. For example, if the user start sampling at a time other than "initial $X = 0$" (or $X_0 = 0$), or samples are spaced more than one unit apart, or "delta $X = 1$" (also written $\Delta X = 1$). To change the time base, the X_0 value, ΔX value, and the data array are bundled into a cluster; then the cluster is wired to the graph.

The **Build Array** function creates a 2D array out of two 1D arrays. This 2D array has two rows with 100 columns per row thus creating 2×100 array. By default, graphs plot each *row* of a 2D array as a separate waveform. If the data are organized by column, the user must make sure to transpose the array when it is plotted. Transposing means simply switching row values with column values; for example, if an array with three rows and ten columns is transposed, the user ends up with an array with ten rows and three columns. LabVIEW makes it easy to do this – simply pop-up on the graph and select **Transpose Array**. The **Transpose 2D Array** function found in the **Array** subpalette of the **Functions** menu can also be used to transpose arrays.

The X_0 value and ΔX (or delta X) value for each array are specified. These X parameters do not need to be the same for both sets of data. In this diagram, the **Build Array** function (**Array** palette) creates an array out of its two cluster inputs. Each input cluster consists of an array and two scalar numbers. The final result is an array of clusters, which the graph can accept and plot. In this case, the specified X_0 and ΔX are the same as the default values, but we can always use other values as well. The user can also create multiple-plot waveform graphs by wiring the data types normally passed to a single-plot graph, first to a **Build Array** function and then to the graph. **Build Array function (Array palette)** creates the proper data structure to plot two arrays on a waveform graph. Enlarge the **Build Array** function to include two inputs by dragging a corner with the Positioning tool.

3.9 XY Graphs

XY graphs display any set of points, evenly sampled or not. Resize the plot legend to display multiple plots. Use multiple plots to save space on the front panel and to make comparisons between plots. XY and waveform graphs automatically adapt to multiple plots. The waveform graphs we have been using are designed for plotting event sampled waveforms. However, if we sample at irregular intervals or are plotting a mathematical function that has multiple Y values for every value, we will need to specify points using their (X, Y) coordinates. X *graphs* plot this different type of data; they require input of a different data type than waveform graphs.

The XY graph expects an input of a bundled X array (the top input) and Y array (the bottom input). The **Bundle** function (**Cluster** palette) combines the X and Y arrays into a cluster wired to the XY graph. The XY graph terminal now appears as a cluster indicator.

Single Plot XY Graphs

The single-plot XY graph accepts a cluster that contains an x array and a y array. The XY graph also accepts an array of points, where a point is a cluster that contains an x value and a y value.

Multiplot XY Graphs

The multiplot XY graph accepts an array of plots, where a plot is a cluster that contains an x array and a y array. The multi plot XY graph also accepts an array of clusters of plots, where a plot is an array of points. Figure 3.99 illustrates the block diagram and front panel of an XY graph.

For a multipleplot XY graph, simply build an array of the clusters have X and Y values used for single plots.

X and Y Scale Menus

The X and Y scales each have a submenu of options. Use **Auto Scale** to turn the autoscaling option on or off. Normally, the scales are set to the exact range of the data when an autoscale operation is performed. The **Loose Fit** option can be used if the user wants LabVIEW to round the scale to numbers. With a loose fit, the numbers are rounded to a multiple of the increment used for the scale. For example, if the markers increment by five, then the minimum and maximum values are set to a multiple of five instead of the exact range of the data:

- The **Formatting** option brings up a dialog box that allows us to configure the following:
- The **Select Scale** is a pull-down menu allowing us to select the scale by name that is being formatted.
- The **Scale Style** menu allows the user select major and minor tick marks for the scale, or none at all. Clicking on this icon can see choices. A major tick mark corresponds to a scale label, while a minor tick mark denotes an interior point between labels. This menu also lets the user select the markers for a given axis either visible or invisible.

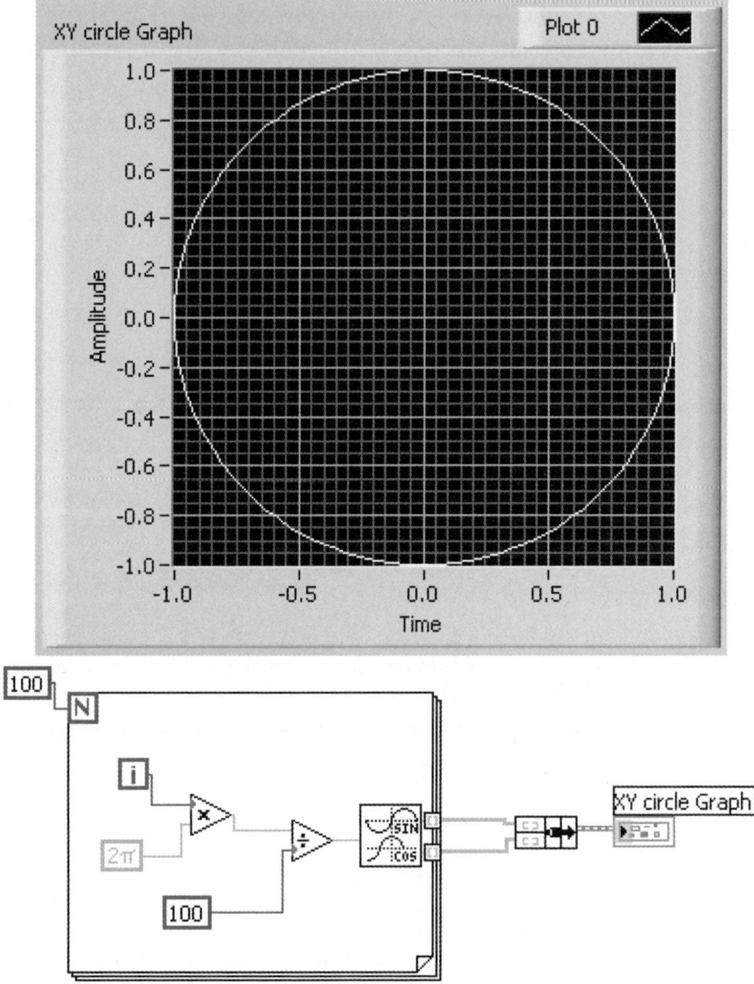

Fig. 3.99. XY graph illustration

- The **Mapping Mode** allows the user to select either a linear or logarithmic scale for the data display.
- The **Grid Options** allows the user to choose between no gridlines, gridlines only at major tick mark locations, or gridlines at both major and minor tick marks.

The Scale Legend

The scale legend allows the user to create labels for the X and Y scales (or for multi-XY scales, if we have more than one) and have easy pop-up access to the configuration. The scale legend gives one buttons for scaling the X or Y

as changing the display format, and autoscaling. The user can click on the buttons to configure the same options as in the **X Scale Formatting** pop-up window. It is just a more convenient place to access this information for some people. The Operating tool can be used to click on the Scale Lock button to toggle and to perform scaling for each scale, the visibility of scales, etc.

The Plot Legend

Charts and graphs use a default style for each plot unless the user has created a custom plot style for it. The plot legend allows the user to label, color, select line style, and choose point style for each plot. The legend can be shown or hidden using the **Visible Items≫Plot Legend** option of the chart or graph Pop-up menu. The user can also specify a name for each plot in the legend.

When the user selects **Plot Legend**, only one plot is displayed in the box that appears. The user can show more plots by dragging down a corner of the legend with the Positioning tool. After set plot characteristics in the **Plot Legend** is chosen, the plot retains those settings, regardless of whether the legend is visible. If the chart or graph receives more plots than are defined in the legend, LabVIEW draws the extra plots in default style.

When the chart or graph body is moved, the legend also moves along with it. The user can change the position of the legend relative to the graph by dragging only the legend to a new location. The legend can be resized on the left to give labels more room in the window or on the right to give plot samples more room.

By default, each plot is labeled with a number, beginning with zero. This can be modified similar to other LabVIEW labels with the Labeling tool. Each plot sample has its own menu to change the plot, line, color, and point styles of the plot. The user can also access the menu by clicking on the legend with the Operating tool:

- The **Common Plots** option allows the user to easily configure any plot, including a scatter plot, a bar plot, and a fill to zero plot. Options in this subpalette configure the point, line, and fill styles of the graph in one step (instead of setting these options individually, as listed next).
- The **Color** option displays the Color palette so that the user can select the plot color. The user can also color the plots on the legend with the Color tool and change the plot colors while running the VI.
- The **Line Style** and **Line Width** offer different types of patterns and widths for the lines to plot.
- The **Anti-Aliased** option makes line plots appear smoother and nicer overall, but be aware that anti-aliased line drawing can be computationally intensive and slow performance.
- The **Bar Plots** option allows the user to create bar plots of 100%, 75%, or 1% width, either horizontally or vertically. The **Fill Baseline** option

controls the baseline of the bars; plots can have no fill or they can fill to zero, negative infinity, or infinity.

- The interpolation option determines the method in which LabVIEW draws lines between data points. The first option does not draw any lines, making it suitable for a scatter plot. The option at the bottom left draws a straight line between points. The four stepped options link points with a right-angled elbow, useful for histograms.
- The **Point Style** options display different point styles (none, round, square, hollow, filled, etc.) from which the user can choose.
- The X Scale and Y Scale are useful for specifying which plot corresponds to which X and Y scale, when there are multiple scales on the user's chart or graph.

Using the Graph Palette

The *graph palette* is the little box that shows up at the right bottom of the graph or chart when it is first dropped on the front panel. With the graph palette, the user can pan (i.e., scroll the display area) and focus in on a specific area using palette buttons (called zooming). The user can also move cursor around.

Normally, the users are in standard mode, meaning that they can click on the graph cursors to move them around. If the user presses the pan button, then the user can switch to a mode where the user can scroll the visible data by dragging sections of the graph with the pan cursor. If the zoom button is selected, a Pop-up menu arises that allows the user to choose from several methods of zooming (focusing on specific sections of the graph by magnifying a particular area). Some of the operations of the zoom function are explained as follows:

Zoom by Rectangle. Drag the cursor to draw a rectangle around the area to be zoomed in. When the mouse button is released, the axes will rescale to show only the selected area.

Zoom by Rectangle in X provides zooming unrestricted to X data (the Y scale remains unchanged).

Zoom by Rectangle in Y provides zooming unrestricted to Y data (the X scale remains unchanged).

Undo Last Zoom is pretty obvious. It also resets the scales to their previous setting.

Zoom In about a Point. If the user holds down the mouse on a specific point, the graph will continuously zoom in until the mouse button is released.

Zoom Out about a Point. If the user holds down the mouse on a specific point, the graph will continuously zoom out until the mouse button is released.

Graph Cursors

LabVIEW graphs have *cursors* to mark data points on plots to further animate the existing work. On a graph, the cursors and the **Cursor Legend** are visible. Cursors cannot be found on charts

The Cursor palette can be viewed by selecting **Visible Items≫Cursor Legend** from the graph's Pop-up menu. When the palette first appears, it is grayed out. By clicking on the Active Cursor button the cursor can be made active. The cursors can be moved around manually, or programmatically using property node.

To manually move a cursor, the cursor is dragged around the graph with the Operating tool. If the user drags the intersection point, then the user can move in all directions. If the user drags the horizontal or vertical lines of the cursor, then the user can move only horizontally or vertically.

The Cursor Movement Control can also be used to move a cursor up, down, left, and right. A graph can have as many cursors as required. The **Cursor** legend helps the user to keep track of them, and the user can stretch it to display multiple cursors.

Popping up on the Cursor legend, and choosing Data Operations≫Delete Element can delete a cursor.

Intensity Charts and Graphs – Color as a Third Dimension

Intense charts and graphs display three dimensions of data on a 2D plot by using color to display values of the third dimension of data (the "Z" values). The intensity chart features a scrolling display, where as the intensity graph display is fixed. Intensity plots are extremely useful for displaying patterned data such as terrain, where color represents altitude over a two-dimensional area, or tempera patterns, where color represents temperature distribution. The intensity graphs and charts are also used to display patterned data, such as temperature patterns and topographical terrain, where the magnitude represents altitude.

Intensity plots function much like two-dimensional charts and graph most ways, with the addition of color to represent the third variable. A Color Scale is available so that the user can set and display the color-mapping scheme. Intensity graph cursor displays also include a Z value.

Intensity charts and graphs accept 2D arrays of numbers, where each number in the array is a color value. The indices of each array element represent the plot location for that color. The user defines the mapping of numbers to colors using the color scale. Instead of doing this the user can also do it programmatically using property nodes. On a 3×4 array plotted on an intensity graph, the colors mapped are red (1.0), brown (2.0), and green (3.0).

To assign a color to a value on the Color Scale, the user can pop-up on the corresponding marker and select **Marker Color≫**. The Z scale of intensity charts and graphs has Arbitrary Marker Spacing by default, so the user can alter the "gradient" of colors by dragging a marker with the Operating tool. New markers can be created by selecting **Add Marker** from the Color Scale Pop-up menu; then dragged to a desired location and a new color is assigned.

Like the waveform graph and chart, the intensity graph features a fixed display while the intensity chart features a scrolling display. The intensity graph and chart accept a 2D array of numbers. Each number in the array represents a specific color. The indexes of the elements in the 2D array set the plot locations for the colors. The intensity graph or chart can display up to 256 discrete colors.

Intensity Graph and Chart Options

The intensity graphs and charts share many of the optional parts of the waveform graphs and charts, which can be shown or hidden by right-clicking the graph or chart and selecting **Visible Items** from the shortcut menu. In addition, because the intensity graphs and charts include color as a third dimension, a scale similar to a color ramp control defines the range and mappings of values to colors. The following illustration in Fig. 3.100 shows the elements of

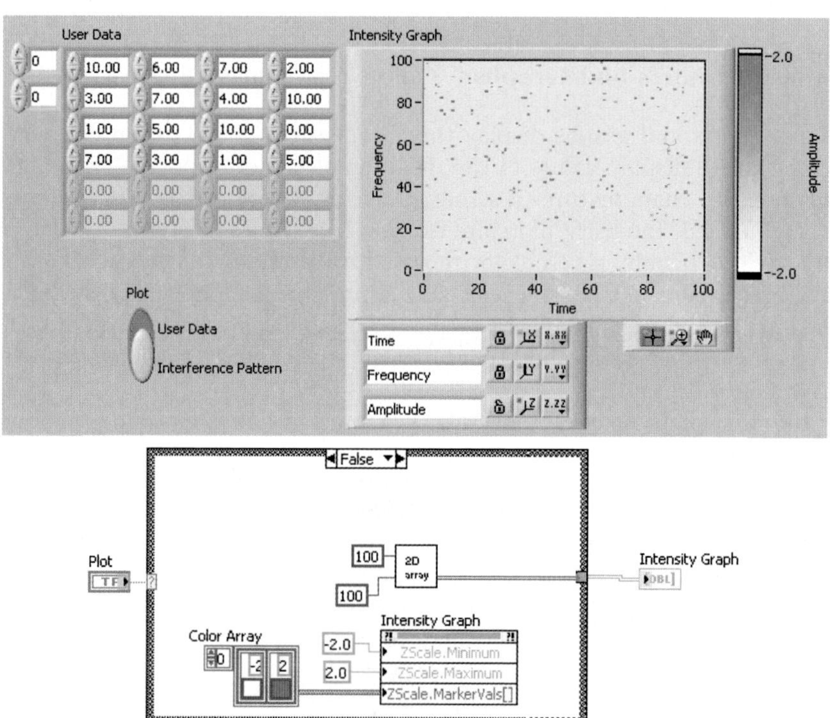

Fig. 3.100. Intensity plot illustration

an intensity graph. The Operating or Positioning tools can be used to right-click the marker next to the color ramp. By selecting **Marker Color** from the shortcut menu the color associated with a marker can be changed. Now the user can select the color required from the color picker that appears. Markers can be added to a color ramp by right-clicking the color ramp and select **Add Marker** from the shortcut menu. Using the Operating tool to drag the marker to the value required can change the value of an arbitrary marker on a color ramp. The Labeling tool can also be used to highlight the text of the marker and enter a new value.

3.10 Strings

A string is a sequential collection of displayable or nondisplayable ASCII characters. Strings provide a platform-independent format for information and data. Often, strings may be used for displaying simple text message. For example, in instrument control, numeric data can be passed as character strings. Then these strings can be converted to numbers to process the data. Storing numeric data to disk can also use strings; in many of the file I/O VI. LabVIEW first converts numeric values to string data before it saves them in a file.

Some of the more common applications of strings include the following:

- Creating simple text messages.
- Passing numeric data as character strings to instruments and then converting the strings to numeric values.
- Storing numeric data to disk. To store numeric values in an ASCII file, the numeric values must be first converted to strings before writing the numeric values to a disk file.
- Instructing or prompting the user with dialog boxes. On the front panel, strings appear as tables, text entry boxes, and labels.

3.10.1 Creating String Controls and Indicators

The string control and indicator located on the **Controls≫Text Controls** and **Controls≫Text Indicators** palettes are used to simulate text entry boxes and labels. Using the Operating tool or labeling tool text data can be typed or edited in a string control. The Positioning tool is used to resize a front panel string object. The space occupied by a string object can be minimized by right-clicking the object and selecting the **Visible Items≫Scrollbar** option from the shortcut menu. The display types can be selected by right-clicking a string control or indicator on the front panel. The display type along with their examples are illustrated in Table 3.2.

Table 3.2. Display types with examples

Display type	Description	Message
Normal Display	Displays printable characters using the font of the control. Nonprintable characters generally appear as boxes	`There are four display types.` `\ is a backslash.`
'\' codes display	Displays backslash codes for all non-displayable characters	`There\sare\sfour\sdisplay\` `stypes.\n\\\sis\sa\sbackslash.`
Password display	Displays an asterisk (*) for each character including spaces	*********************** ************************

Hex display	Displays the ASCII value of each character in hex instead of the character itself	`5468 6572 6520 6172 6520 666F`
		`7572 2064 6973 706C 6179 2074`
		`7970 6573 2E0A 5C20 6973 2061`
		`2062 6163 6B73 6C61 7368 2E`

Choosing Display Type

String controls and indicators have several options, which are found useful. For example, they can display and accept characters that are normal displayable, such as backspaces, carriage returns, and tabs. If '\' **Codes Display** (instead of **Normal Display**) is selected from a string's Pop-up menu, displayable characters appear as a backslash(\) followed by the appropriate code.

Uppercase letters are used for hexadecimal characters and lowercase letters for the special characters, such as form feed and backspace are used. LabVIEW interprets the sequence \BFare as hex BF followed by the word "are," whereas LabVIEW interprets \bFare and \bfare as a backspace followed by the words "Fare" and "fare". In the sequence \Bfare, \B is not the back space code, and \Bf is not a valid hex code. In a case like this, when a back slash is followed by only part of a valid hex character, LabVIEW assumes a zero follows the backslash, and so LabVIEW interprets \B as hex 0B. Any time a backslash is not followed by a valid character code, LabVIEW ignores the backslash character.

The data in the string does not change when the display mode is toggled; only the display of certain characters changes. '\'**Codes Display** mode is very useful for debugging programs and for specifying nondisplayable characters required by instruments, the serial port, and other devices.

Strings also have a **Password Display** option, which sets the string control or indicator to display a "*" for every character entered into it, so that the types are not visible. While the front panel shows only a stream of "* * **," the block diagram reads the actual data in the string. Obviously, this display can be useful if the user needs to programmatically password-protect all or part of the VIs.

The Hex Display option can be used to view the string as hexadecimal characters instead of alphanumeric characters.

Single-Line Strings

If **Limit to Single Line** is chosen from a string's Pop-up menu, the string cannot exceed one line of text; that is, no carriage returns are allowed in the string. If the user hits <center> or <return>, text entry will be automatically terminated. If strings are not limited to a single line, hitting <return> causes the cursor to jump to a new line to let the user type more.

Updating While One Type

Normally, string controls do not change their value on their terminal in the block diagram until the user finishes typing and hits <center>. By clicking outside the string box, or clicking on the "√" button on the VI toolbar indicates the string entry is complete.

Popping-up on the string control and selecting **Update Value While Typing** updates the values as the user types.

The Scrollbar

If the **Visible Items≫Scrollbar** option is chosen from the string Pop-up **Visible Items** submenu, a vertical scrollbar appears on the string control or indicator. This option can be used to minimize the space taken up on the front panel by string controls that contain a large amount of text. This option will be grayed out unless the size of string is increased enough for a scrollbar to fit.

3.10.2 String Functions

Like arrays, strings can be much more useful while using many of the built-in functions provided by LabVIEW. This section examines a few of the functions from the **String** subpalette of the **Functions** palette.

String Length returns the number of characters in a given string as shown in Fig. 3.101.

Concatenate Strings concatenates all input strings into a single output string as shown in Fig. 3.102.

The function appears as the icon shown in block diagram in Fig. 3.102. The function can be resized with the Positioning tool to increase the number of inputs. In addition to simple strings, the user can also wire a 1D array of strings as input; the output will be a single string containing a concatenation of strings in the array.

String Subset accesses a particular section of a string. It returns the substring beginning at **offset** and containing **length** which indicates the number

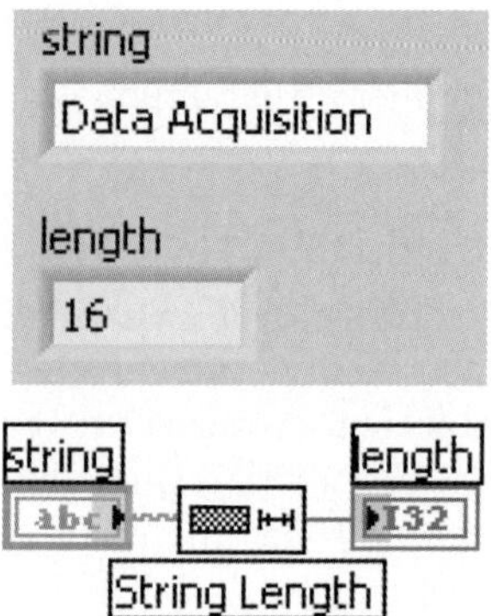

Fig. 3.101. String Length function

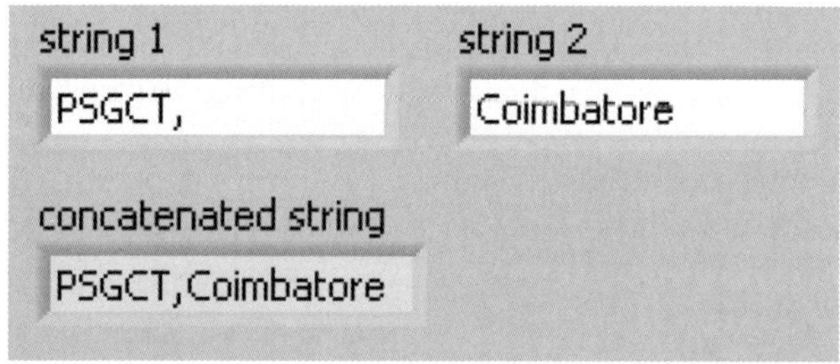

Fig. 3.102. Concatenate Strings function

of characters. The **offset** of the first character in **string** is 0. For example, in Fig. 3.103 if the string "Data Acquisition" is used as the input, the String Subset function returns the following **substring** for an **offset** of 2 and a **length** of 4: 'ta A'.

To upper case converts all alphabetic characters in **string** to uppercase characters as shown in Fig. 3.104. The function evaluates all numbers in **string** as ASCII codes for characters. This function does not affect nonalphabetic characters. The connector pane displays the default data types for this polymorphic function. If **string** is a numeric value or an array of numeric values, each number is evaluated as an ASCII value. The To Upper Case function translates all values in the range of 97–122 into values over the range of 65–90. All other values remain unchanged.

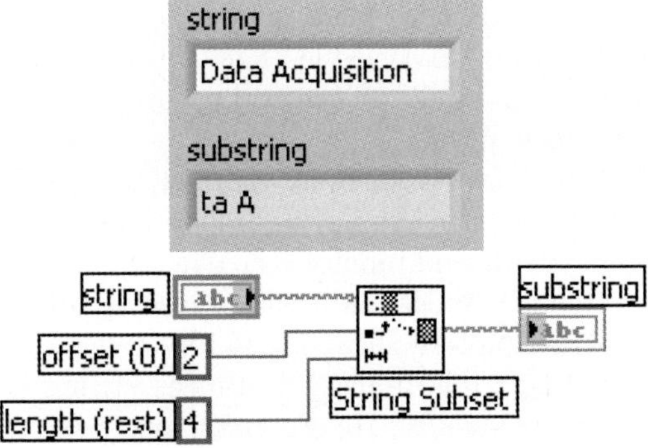

Fig. 3.103. String Subset function

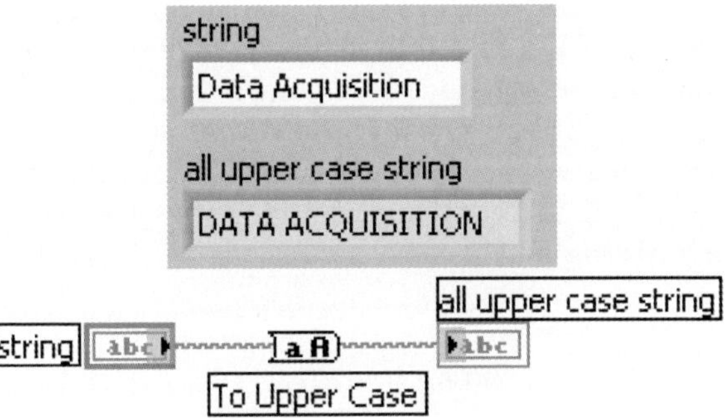

Fig. 3.104. To upper case function

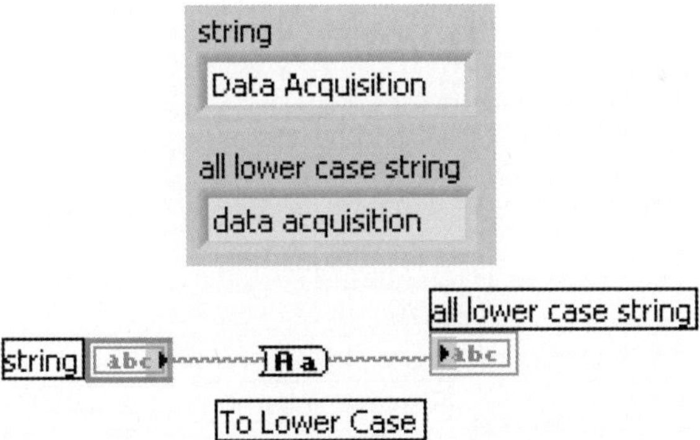

Fig. 3.105. To Lower Case function

To Lower case converts all alphabetic characters in **string** to lowercase characters as shown in Fig. 3.105. Evaluates all numbers in **string** as ASCII codes for characters. This function does not affect nonalphabetic characters. The connector pane displays the default data types for this polymorphic function. If **string** is a numeric value or an array of numeric values, each number is evaluated as an ASCII value. The To Lower Case function translates all values in the range of 65–90 into values over the range of 97–122. All other values remain unchanged.

Replace Substring inserts, deletes, or replaces a substring at the offset specified in **string**. The Replace Substring function deletes **length** characters in **string** starting at **offset**, and replaces the deleted portion with **substring**. If **length** is 0, the Replace Substring function inserts **substring** at **offset**. If **substring** is empty, the Replace Substring function deletes **length** characters at **offset** as shown in Fig. 3.106. In many instances, strings are to be converted to numbers or numbers to strings. The **Format Into String** and **Scan From String** functions have these capabilities.

Format into String formats string, numeric, path, or Boolean data as text which can be written into a file. For example, the function converts the floating-point number 1.28 to the 6-byte string "1.2800". **Format into String** formats the input argument (which is in numeric for mat) as a string, according to the format specifications in **format string**. The function appends the newly converted string to the input wired to **initial string**, if there is one, and outputs the results in **resulting string**. Table 3.3 gives some examples of **Format into String's** behavior.

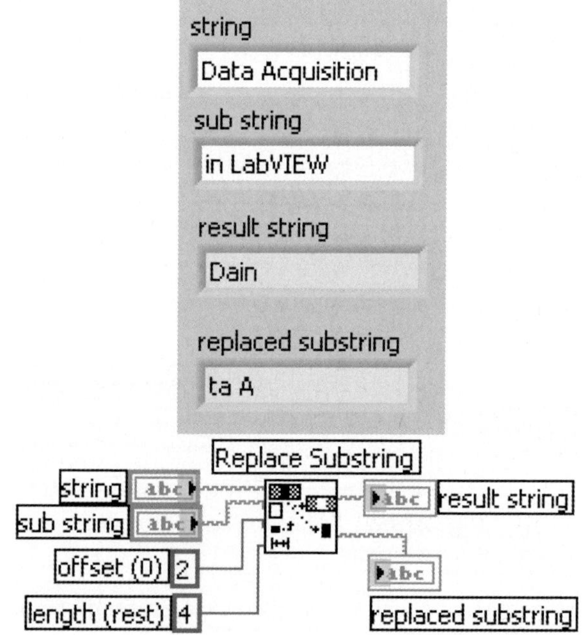

Fig. 3.106. Replace Substring function

Table 3.3. Format into String function

Initial String	Format String	Number	Resulting String
(empty)	score = %2 d%%	87	score = 87%
score =	%2 d%%	87	score = 87%
(empty)	leveL = %7.2 eV	0.03632	level = 3.64E-2 V
(empty)	%5.3f	5.67 N	5.670 N

The "%" character begins the formatting specification. Given *"%number1.number2," number 1* specifies field width of the resulting string and *number 2* specifies the precision (i.e., number of digits after the decimal point). An "f" formats the input number as a floating-point number with fractional format, "d" formats it as a decimal integer, and "e" formats it as a floating-point number with scientific notation:

Format into String can be resized to convert multiple values to a single string simultaneously.

Scan from String is the "opposite" of **Format into String**, which converts a string containing valid numeric characters (0 to 9, +, −, e, E, and period) to numeric data. This function starts scanning the **input string at initial search location** and converts the data according to the specifications in format string. **Scan from String** can be resized to convert multiple values simultaneously.

Both **Format into String** and **Scan from String** have an **Edit Scan String** interface that is used to create the format string. In this dialog box, the user can specify format, precision, data type, and width of the converted value. Double-click on the function or pop-up on it and select **Edit Format String** to access the **Edit Scan String** or **Edit Format String** dialog box.

Get Date/Time String (found in the **Time & Dialog** palette) outputs the date string, which contains the current date, and the time string, which contains the current time. This function is useful for time stamping the data.

Match Pattern is used to look for a given pattern of characters in a string. It searches for and returns a **matched substring**. **Match Pattern** looks for the **regular expression** in a string, beginning at **offset**; if it finds a match, it splits the string into three substrings. If no match is found, the match substring empty and **offset past match** is set to −1. This function is explained as shown in Fig. 3.107.

3.11 Tables

Tables are found in **Lists & Table** subpalette of the Controls palette. Tables have row and column headings that can be shown or hidden; the headings are separated from the data space by a thin open border space. Text headings

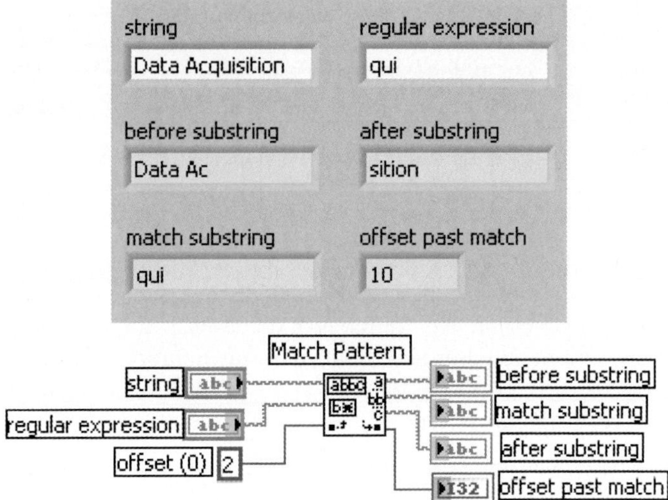

Fig. 3.107. Match Pattern function

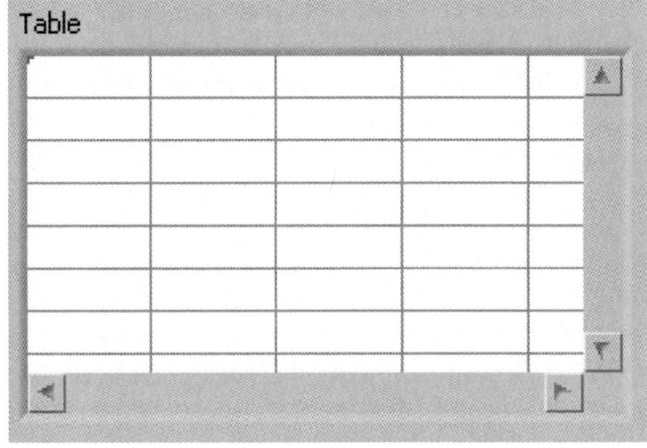

Fig. 3.108. Table

can be entered using the Labeling tool or operating tool. The user can also update or read headings using property nodes. Like an array index display, a table index display indicates which cell is visible at the upper-left corner of the table.

Each cell in a table is a string, and each cell resides in a column and a row. Therefore, a table is a display for a 2D array of strings. The illustration in Fig. 3.108 shows a table. Define cells in the table by using the Operating tool or the Labeling tool to select a cell and typing text in the selected cell. The table displays a 2D array of strings, so the user must convert 2D numeric arrays to 2D string arrays before they are displayed in a table indicator. The row and column headers are not automatically displayed as in a spreadsheet.

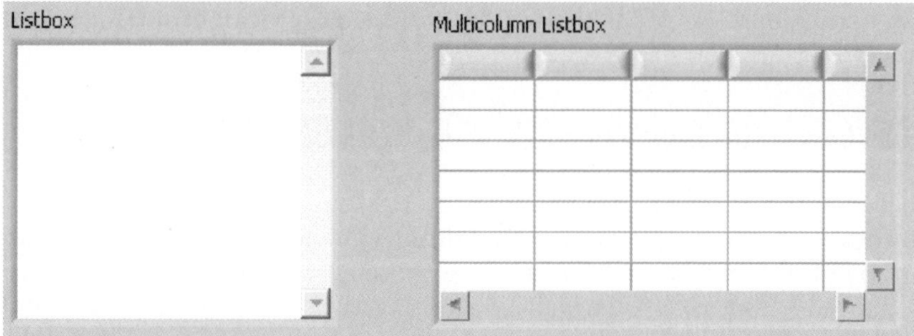

Fig. 3.109. List box

3.12 List Boxes

There are two kinds of list boxes in LabVIEW, **List box** and **Multi-co List box**, as shown in Fig. 3.109. These boxes are located in the **List & Table** subpalette of the controls palette.

A list box in LabVIEW is similar to a table, but it behaves very different at runtime. During edit mode, any text can be typed into a list box similar to that of a table. When the VI is run, the list box acts as a "choice" menu where the user can click on any row to highlight it and select.

List boxes are useful to present data (either in columns or in multiple columns) to a user.

3.13 File Input/Output

File input and output (I/O) operations retrieve information from and store information in a disk file. LabVIEW has a number of very versatile file I/O functions, as well as some simple functions that take care of almost all aspects of file I/O in one shot. Some of the simple file functions are discussed in this section. All functions are located in the **File I/O** subpalette of the **Functions** palette.

3.13.1 File I/O VIs and Functions

File I/O operations pass data to and from files. The File I/O VIs and functions are located on the **Functions≫All Functions≫File I/O** palette to handle all aspects of file I/O, including the following:

- Opening and closing data files
- Reading data from and writing data to files
- Reading from and writing to spreadsheet-formatted files
- Moving and renaming files and directories
- Changing file characteristics
- Creating, modifying, and reading configuration files

The File I/O palette is divided into four types of operations: high-level, low-level, advanced, and express.

High-Level File I/O VIs

The high-level File I/O VIs are located on the top row of the **Functions≫All Functions≫File I/O** palette to perform common I/O operations. The user can save time and programming effort by using the high-level VIs to write to and read from files. The high-level VIs perform read or write Operations in addition to opening and closing the file. If an error occurs, the high-level VIs displays a dialog box that describes the error. However, because high-level VIs encapsulate the entire file operation into one VI, they are difficult to customize to any use other than the one intended.

Low-Level and Advanced File I/O VIs and Functions

Low-level VIs is used for more specific tasks. The low-level File I/O VIs and functions are located on the middle row of the **Functions≫All Functions≫File I/O** palette and the Advanced File I/O functions located on the **Functions≫All Functions≫File I/O≫Advanced File Functions** palette to control each file I/O operation individually. The principal low-level functions are used to create or open a file, write data to or read data from the file, and close the file. The low-level VIs and functions can handle most file I/O needs.

Low-Level File I/O VI and Functions

The following low-level File I/O VI and functions are used to perform basic file I/O operations:

- **Open/Create/Replace File**. Opens an existing file, creates a new file, or replaces an existing file, programmatically or interactively using a file dialog box. The user can optionally specify a dialog **prompt**, default file name, **start path**, or filter **pattern**. If **file path** is empty, the VI displays a dialog box from which we can select a file.
- **Read File**. Reads data from an open file specified by **refnum** and returns it in **data**. Reading begins at the current file mark or a location specified by **pos mode** and **pos offset**. The data is read according to the format of the specified file.
- **Write File**. Writes data to an open file specified by **refnum**. Writing begins at a location specified by **pos mode** and **pos offset** for byte stream files and at the end of the file for data log files. **Data**, **header**, and the format of the specified file determine the amount of data written.

– **Close File**. Closes an open file specified by **refnum** and returns the path to the file associated with the refnum. Error I/O operates uniquely in this function, which closes regardless of whether an error occurred in a preceding operation. This ensures that files are closed correctly.

3.13.2 File I/O Express VIs

The Express VIs on the **File I/O** palette includes the Read LabVIEW Measurement File Express VI and the Write LabVIEW Measurement File Express VI. The LabVIEW measurement data file (.lvm) is a tab-delimited text file that can be opened with a spreadsheet application or a text-editing application. In addition to the data an Express VI generates, the .lvm file includes information about the data, such as the date and time the data was generated.

The File functions expect a file path input, which looks like a kind of string. A path is a specific data type that provides a platform-specific way to enter path of a file. If a file path is not wired, the File functions will pop-up a dialog box asking the user to select or enter a filename. When called, File functions open or create a file, read or write the data, and then close file. The files created with the VIs are just ordinary files. Once data has been written to a file, the user can open the file using word processing program to see the data.

One very common application for saving data to file is to format the file so that we can open it in a spreadsheet program. In most spreadsheet tabs separate columns and EOL (End of Line) characters separate **Write to Spreadsheet File** and **Read from Spreadsheet File** deal with in spreadsheet format:

Write Characters to File writes a character string to a new file or applies the string to an existing file. The VI opens or creates the file before writing to it and closes it afterwards.
Read Characters from File reads a specified number of characters from file beginning at a specified character offset. The VI opens the file before reading from it and closes it afterwards.
Read Lines from File reads a specific number of lines from a file beginning at a specified character offset. The VI opens the file before reading from it and closes it afterwards.
Write to Spreadsheet File converts 2D or 1D array of single-precision numbers to a text string and then writes the string to a new file or appends the string to an existing file. The user can optionally transpose the data. The text files created by this VI are readable by most spreadsheet applications.
Read From Spreadsheet File reads a specified number of lines or rows from a numeric text file, beginning at a specified character offset, and converts the data to a 2D single-precision array of numbers. The user can

optionally transpose the array. This VI will read spreadsheet saved in text format.
Binary File VIs are read from and write to binary files. Data can be integers or single-precision numbers.

These file functions are very high level and easy to use. All are found in the **File I/O** palette. LabVIEW contains other file functions that are much more versatile but more complicated.

LabVIEW Data Directory

The default LabVIEW Data directory is used to store the data files LabVIEW generates, such as .lvm or .txt files. LabVIEW installs the LabVIEW Data directory in the default file directory for the operating system to help the user organize and locate the data files LabVIEW generates. By default, the Write LabVIEW Measurement File Express VI stores the .lvm files it generates in this directory, and the Read LabVIEW Measurement File Express VI reads from this directory. The Default Data Directory constant, shown at left and the Default Data Directory property also return the LabVIEW Data directory by default.

By selecting **Tools≫Options** and selecting **Paths** from the top pull-down menu the user can specify a different default data directory. The default data directory differs from the default directory, which is the directory specified for new VIs, custom controls, VI templates, or other LabVIEW documents that are created.

Error Handling

The low-level File I/O VIs and functions return error information. The error information can be wired from the beginning of the VI to the end. An error handler VI is included, such as the Simple Error Handler VI located on the **Time & Dialog** palette, at the end of the VI to determine if the VI ran without errors. The **error in** and **error out** clusters in each VI are used to build or pass the error information through the VI.

Saving Data in a New or Existing File

Any data type can be written to a file that is opened or created with the File I/O VIs and functions. If other users or applications need to access the file, string data can be written in ASCII format to the file. Files can either be accessed programmatically or interactively through a file dialog box. To access a file through a dialog box, there is no need to wire **file path** in the Open/Create/Replace File VI. However, the user can save time by programmatically wiring the default filename and path to the VI. The Open/Create/Replace File VI opens the file test1.dat. The VI also generates a **refnum** and an error cluster. When a file is opened, device, or network connection, LabVIEW creates a refnum associated with that file, device, or network connection.

All operations that are performed on open files, devices, or network connections use refnums to identify each object. The error cluster and **refnum** pass in sequence from one node to the next. Because a node cannot execute until it receives all its inputs, passing these two parameters forces the nodes to run in order and creates a data dependency. The Open/Create/Replace File VI passes the refnum and error cluster to the Write File function, which writes the data to disk. When the Write File function finishes execution, it passes the refnum and error cluster to the Close File function, which closes the file. The Simple Error Handler VI examines the error cluster and displays a dialog box if an error occurred. If an error occurs in one node, subsequent nodes do not execute, and the VI passes the error cluster to the Simple Error Handler VI.

Formatting Spreadsheet Strings

To write data to a spreadsheet file, the user must format the string as a spreadsheet string, which is a string that includes delimiters, such as tabs. In many spreadsheet applications, the tab character separates columns, and the end of line character separates rows. The end of line constant located on the **Functions≫All Functions≫String** palette is used to ensure portability of VIs among platforms. **(Windows)** The constant inserts a carriage return and a linefeed. **(Mac OS)** The constant inserts a carriage return. **(UNIX)** The constant inserts a linefeed. The Write to Spreadsheet File VI or the Array to Spreadsheet String function can be used to convert a set of numbers from a graph, a chart, or an acquisition into a spreadsheet string. To write numbers and text to a spreadsheet or word processing application, the String functions and the Array functions are used to format the data and to combine the strings. Then the data is written to a file.

Format into File

The Format into File function is used to format string, numeric, path, and Boolean data as text and write the text to a file. Often the user can use this function instead of separately formatting the string with the Format into String function or Build Text Express VI and writing the resulting string with the Write Characters to File VI or Write File function. The Format into File function can also be used to determine the order in which the data appears in the text file. However, the user cannot use this function to append data to a file or overwrite existing data in a file. For these operations, use the Format into String function with the Write File function. The user can wire a refnum or path to the **input file** terminal of the Format into File function, or leave this input unwired for a dialog box to prompt for the filename.

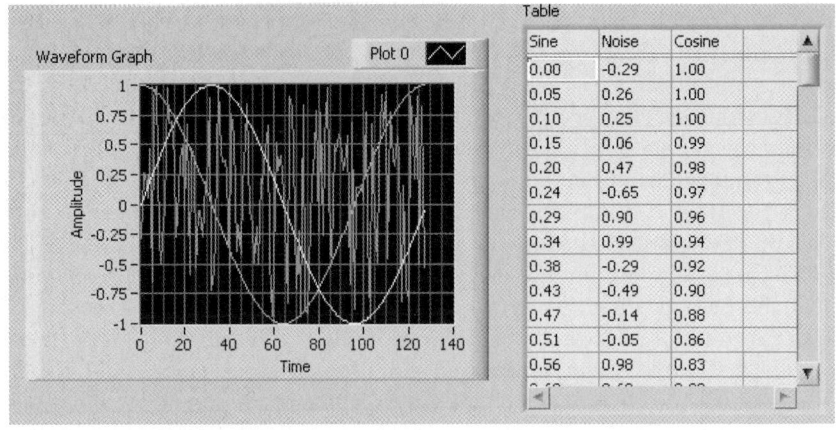

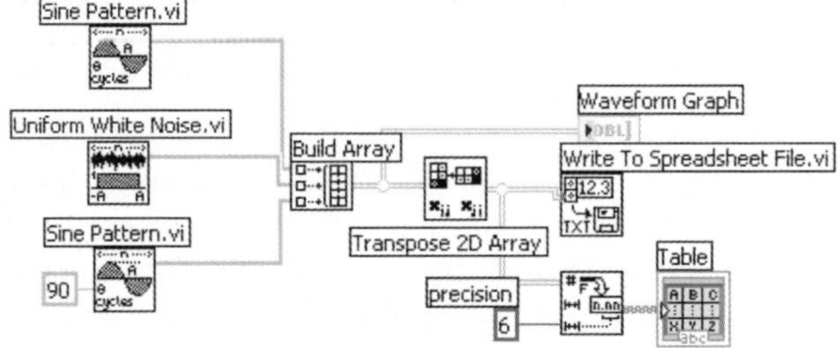

Fig. 3.110. File IO illustration

Write LabVIEW Measurement File

The Write LabVIEW Measurement File Express VI includes the open, write, close, and error handling functions. It also handles formatting the string with either a tab or comma delimiter. The Merge Signals function combines the iteration count and the random number into the dynamic data type. The dialog box displays the configuration for the Write LabVIEW Measurement File Express VI. This VI creates a .lvm file, which can be opened in a spreadsheet application. Figure 3.110 shows the front panel and block diagram of an example of File I/O.

Summary

- The structures on the block diagram is used to repeat blocks of code and to execute code conditionally or in a specific order.
- The While Loop executes the subdiagram until the conditional terminal receives a specific Boolean value. By default, the While Loop executes its subdiagram until the conditional terminal receives a TRUE value.
- The For Loop executes a subdiagram a set number of times.

- Loops are created by using the cursor and dragging a selection rectangle around the section of the block diagram
- The Wait Until Next ms Multiple function makes sure that each iteration occurs at certain intervals. This function is used to add timing to loops.
- The Wait (ms) function waits a set amount of time.
- Coercion dots appear where LabVIEW coerces a numeric representation of one terminal to match the numeric representation of another terminal.
- Shift registers on For Loops and While Loops are used to transfer values from one loop iteration to the next.
- Shift registers are created by right-clicking the left or right border of a loop and selecting **Add Shift Register** from the shortcut menu.
- To configure a shift register to carry over values to the next iteration, right-click the left terminal and select **Add Element** from the shortcut menu.
- The Feedback Node stores data when the loop completes an iteration, sends that value to the next iteration of the loop, and transfers any data type.
- The Feedback Node is used to avoid unnecessarily long wires.
- Arrays group data elements of the same type. The user can build arrays of numeric, Boolean, path, string, waveform, and cluster data types.
- The array index is zero-based, which means it is in the range 0 to $n - 1$, where n is the number of elements in the array.
- To create an array control or indicator, the user can select an array on the **Controls≫ Array & Cluster** palette, place it on the front panel, and drag a control or indicator into the array shell.
- If an array is wired to a For Loop or While Loop input tunnel, the user can read and process every element in that array by enabling autoindexing.
- The Array functions located on the **Functions≫All Functions≫Array** palette are used to create and manipulate arrays.
- By default, LabVIEW enables autoindexing in For Loops and disables autoindexing in While Loops.
- Polymorphism is the ability of a function to adjust to input data of different data structures.
- Clusters group data elements of mixed types. A cluster cannot contain a mixture of controls and indicators.
- To create a cluster control or indicator, select a cluster on the **Functions≫All Functions≫Array & Cluster** palette, place it on the front panel, and drag controls or indicators into the cluster shell.
- Error checking informs the user why and where errors occur.
- The error cluster reports the **status**, **code**, and **source** of the error.
- The error cluster controls and indicators are used to create error inputs and outputs in subVIs.
- The waveform chart is a special numeric indicator that displays one or more plots.
- The waveform chart has the following three update modes:

- Waveform graphs and XY graphs display data from arrays.
- The intensity charts and graphs are used to plot three-dimensional data. The third dimension is represented by different colors corresponding to a color mapping that is defined by the user. Intensity charts and graphs are commonly used in conjunction with spectrum analysis, temperature display, and image processing.
- The Select function selects between two inputs dependent on a third Boolean input.
- A case structure has two or more subdiagrams, or cases. Only one subdiagram is visible at a time, and the structure executes only one case at a time.
- If the case selector terminal is a Boolean value, the structure has a TRUE case and a FALSE case. If the selector terminal is an integer, string, or enumerated type value, the structure can have up to $2^{31} - 1$ cases.
- Inputs are available to all subdiagrams of a Case structure, but subdiagrams do not need to use each input. If at least one output tunnel is not defined, all output tunnels on the structure appear as white squares.
- When creating a subVI from a Case structure, wire the error input to the selector terminal, and place all subVI code within the No Error case to prevent the subVI from executing if it receives an error.
- Formula Nodes are useful for equations that have many variables or are otherwise complicated and for using existing text-based code. Each equation statement must terminate with a semicolon.
- Strings group sequences of ASCII characters.
- To minimize the space that a string object occupies, the user can right-click the object and select **Show Scrollbar** from the shortcut menu.
- The String functions located on the **Functions≫All Functions≫String** palette are used to edit and manipulate strings on the block diagram.
- The File I/O VIs and functions are used to handle all aspects of file I/O.
- The high-level File I/O VIs are used to perform common I/O operations.
- The low-level File I/O VI and functions and the Advanced File I/O functions are used to control each file I/O operation individually.
- The Express File I/O VIs are used for simple datalogging operations.
- When writing to a file, the user can open, create, or replace a file, write the data, and close the file. Similarly, when reading from a file, the user can open an existing file, read the data, and close the file.
- To access a file through a dialog box, leave **file path** unwired in the Open/Create/Replace File VI.

Review Questions

1. How code could be conditionally operated in specific order?
2. Explain the loop structures of LabVIEW with examples.
3. Describe the For Loop structure with an example.

4. Define FOR, WHILE with their subdiagram.
5. Describe the While loop structure with an example.
6. Give the block diagram construction steps to find the sum of first 10 natural numbers using For Loop.
7. Give the block diagram construction steps to find the factorial of a given number using For Loop.
8. Explain with an example how Wait function works?
9. Give the block diagram construction steps to find the sum of first 10 natural numbers using While loop.
10. Give the block diagram construction steps to find the factorial of a given number using While loop.
11. State the purpose of Wait until Next ms Multiple function
12. With suitable diagram explain the Use shift registers in For Loops and While Loops.
13. What does the Feedback Node mean in VI?
14. Explain about selection structures in LabVIEW with neat sketches.
15. State arrays and its activity.
16. Explain in detail about the concept of an array.
17. What is meant by autoindexing?
18. Describe the array operations with suitable examples.
19. Explain the enumerated and error case structures with examples.
20. Write the block diagram construction steps to find the square root of a given number.
21. How can we insert an object in the array shell before we use the array in the block diagram?
22. Explain the concept of Polymorphism with an example.
23. How to create an array control or indicator?
24. Define the terms: Polymorphism and Clusters.
25. Explain the various methods of clustering with sketches.
26. What is the use the error cluster controls and indicators.
27. Explain the purpose of waveform chart.
28. What are the various modes of waveform chart? Explain.
29. Explain about the single and multiple plot waveform graphs with an example.
30. State the use of a subVI from a Case structure and the case structure itself in VI.
31. Define Formula Node.
32. Discuss about Tables and List boxes.
33. Explain about the single and multi plot XY graphs.
34. Explain the string functions with examples.
35. What are the various file input and output operations?
36. What is Express VI and what for is it used?
37. What are various file operations available?
38. How file input and output are related with express VI?
39. Chart out the various advantages of express VI.

4

Inputs and Outputs

Learning Objectives. On completion of this chapter the reader will have a knowledge on:

- Inputs and Outputs
- Components of Measurement System
- Origin of Signals
- Transducers and types of transducers
- Sensors and Biosensors – Definitions
- General Signal Conditioning Functions
- Analog to Digital Control
- Digital to Analog Control

4.1 Introduction

Inputs and outputs of virtual instruments are representations of the loops and case statements of text-based programming languages. So the study of these inputs and outputs along with their characteristic feature is essential in all kinds of applications. LabVIEW inputs and outputs include transducer, data acquisition, signal conditioning, and PC peripheral devices. Sensors are used to measure process conditions and valves to influence process operations are essential for process control. Since successful process control requires appropriate instrumentation, engineers should understand the principles of common sensors and transducers introduced in this section. This section also describes the various components of measuring systems and concentrates on the origin of signals and related devices to acquire signals. A brief explanation is given on analog-to-digital (A/D) and digital-to-analog (D/A) converters. This section introduces the fundamentals of using signal conditioning hardware with PC-based DAQ systems. First, the signal conditioning requirements of the most common transducers are discussed. This section also describes some general signal conditioning functions and briefly discusses the role of signal conditioning products such as the National Instruments Signal Conditioning extensions for Instrumentation (SCXI) product line.

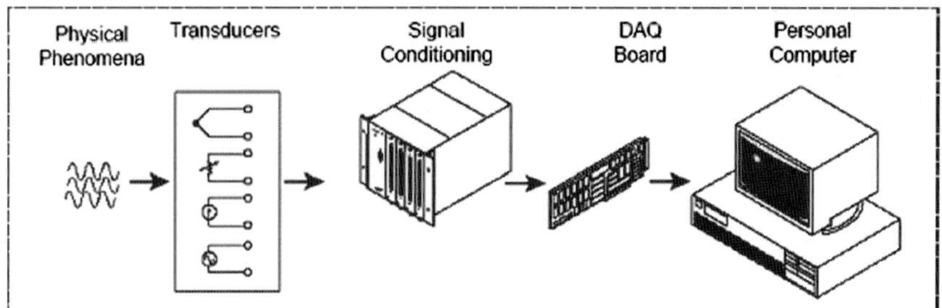

Fig. 4.1. A general PC-based DAQ system with signal conditioning

4.2 Components of Measuring System

PC-based data acquisition (DAQ) systems and plug-in boards are used in a wide range of applications in the laboratory, in the field, and on the manufacturing plant floor. Typically, DAQ plug-in boards are general-purpose data acquisition instruments that are well suited for measuring voltage signals. However, many real-world sensors and transducers output signals that must be conditioned before a DAQ board can effectively and accurately acquire the signal.

This front-end preprocessing, which is generally referred to as *signal conditioning*, includes functions such as signal amplification, filtering, electrical isolation, and multiplexing. In addition, many transducers require excitation currents or voltages, bridge completion, linearization, or high amplification for proper and accurate operation. Therefore, most PC-based DAQ systems include some form of signal conditioning in addition to the plug-in DAQ board and personal computer, as shown in Fig. 4.1.

An instrument may be defined as a device or a system, which is designed to maintain a functional relationship between prescribed properties of physical variables and must include ways and means of communication to a human observer. The functional relationship remains valid only as long as the static calibration of system remains constant. On the other hand, the performance of a measurement system can be described in terms of static and dynamic characteristics.

It is possible and desirable to describe the operation of a measuring instrument or a system in a generalized manner without resorting to intricate details of the physical aspects of a specific instrument or a system. The whole operation can be described in terms of functional elements. Most of the measurement systems contain three main functional elements such as the primary sensing element, variable conversion element, and data presentation element. Each functional element is made up of a distinct component or groups of components, which perform the required and definite steps in the measurement. These may be taken as basic elements, whose scope is determined by their functioning rather than their construction.

Primary sensing element. The quantity under measurement makes its first contact with the primary sensing element of a measurement system. In other words the measurand is first detected by primary sensor. This act is then immediately followed by the conversion of measurand into an analogous electrical signal performed by a transducer. A transducer in general, is defined as a device which converts energy from one form to another. But in Electrical measurement systems, this definition is limited in scope. A transducer is defined as a device which converts a physical quantity into an electrical quantity. The physical quantity to be measured, in the first place is sensed and detected by an element which gives the output in a different analogous form. This output is then converted into an electrical signal by a transducer. This is true of most of the cases but is not true for all. In many cases the physical quantity is directly converted into an electrical quantity by a transducer. The first stage of a measurement system is known as a detector transducer stage.

Variable conversion element. The output of the primary sensing element may be electrical signal such as a voltage, a frequency or some other electrical parameter. Sometimes this output is not suited to the system. For the instrument to perform the desired function, it may be necessary to convert this output to some other suitable form while preserving the information content of the original signal. For example, suppose output is in analog form and the next stage of the system accepts input signals only in digital form and therefore, an A/D converter will have to be used for converting signals from A/D form for them to be acceptable for the next stage of the system.

Variable manipulation element. The function of this element is to manipulate the signal presented to it preserving the original nature of the signal. Manipulation here means only a change in numerical value of the signal. For example, an electronic amplifier accepts a small voltage signal as input and produces an output signal which is also voltage but of greater magnitude. Thus voltage amplifier acts as a variable manipulation element. It is not necessary that a variable manipulation element should follow the variable conversion element as shown in Fig. 4.2. It may precede the variable conversion element in many cases. In case, the voltage is too high, attenuators are used which lower the voltage or power for the subsequent stages of the system.

As discussed earlier, the output of transducer contains information needed for further processing by the system and the output signal is usually a voltage or some other form of electrical signal. The two most important properties of voltage are its magnitude and frequency though polarity may be a consideration in some cases. Many transducers develop low voltages of the order

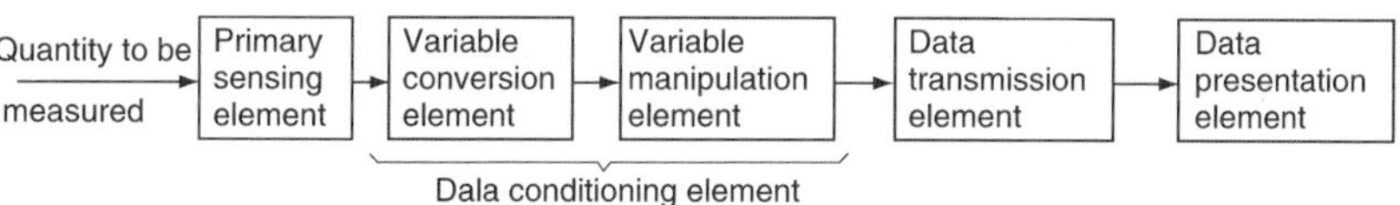

Fig. 4.2. Functional elements of an instrumentation system

of mV and some even μV. A fundamental problem is to prevent this signal being contaminated by unwanted signals like noise due to an extraneous source which may interfere with the original output signal. Mother problem is that a weak signal may be distorted by processing equipment. The signal after being sensed cannot transmitted to the next stage without removing the interfering sources, as otherwise highly distorted results may be obtained which ale far from true. Many a times it becomes necessary to perform certain operations on the signal before it is transmitted further. These processes may be linear like amplification, attenuation, integration, differentiation, addition and subtraction. Some nonlinear processes like modulation, detection, sampling, filtering, chopping and clipping, etc. are also performed on the signal to bring it to the desired form to be accepted by the next stage of measurement system. This process of conversion is called Signal Conditioning. The term signal conditioning includes many other functions in addition to variable conversion and variable manipulation. In fact the element that follows the primary sensing element in any instrument or measurement system is called Signal Conditioning Element.

When the elements of an instrument are actually physically separated, it becomes necessary to transmit data from one to another. The element that performs this function is called a Data Transmission Element. For example, space-crafts are physically separated from the earth where the control stations guiding their movements are located. Therefore control signals are sent from these stations to space-crafts by complicated telemetry systems using radio signals. The signal conditioning and transmission stage is commonly known as Intermediate Stage.

In the signal conditioning stage, the signals are passed through a slew of steps like:

- Amplification
- Rectification
- Filtering
- Scaling

Data presentation element. The information about the quantity under measurement has to be conveyed to the personnel handling the instrument or the system for monitoring, control, or analysis purposes. The information conveyed must be in a form intelligible to the personnel or to the intelligent instrumentation system. This function is done by data presentation element. In case data is to be monitored, visual display devices are needed. These devices may be analog or digital indicating instruments like ammeters, voltmeters, etc. In case the data is to be recorded, recorders like magnetic tapes, high speed camera and TV equipment, storage type C.R.T, printers, analog and digital computers, or microprocessors may be used. For control and analysis purpose microprocessors or computers may be used.

The final stage in a measurement system is known as terminating stage. As an example of a measurement system, consider the simple bourdon tube

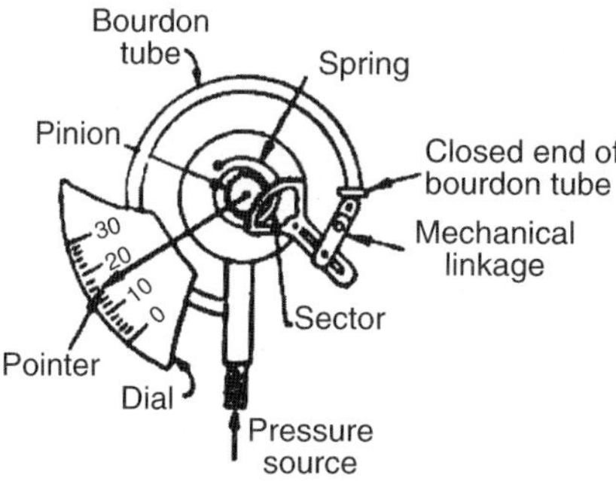

Fig. 4.3. Bourdon tube pressure gauge

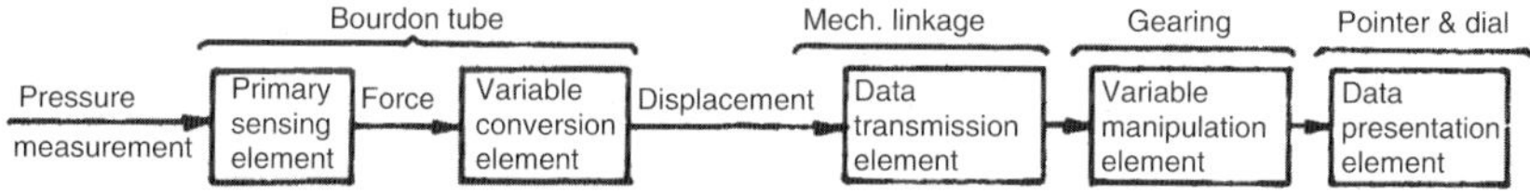

Fig. 4.4. Schematic diagram of a Bourdon tube pressure gauge

pressure gauge as shown in Fig. 4.3. This gauge offers a good example of a measurement system. In this case the bourdon tube acts as the primary sensing element and a variable conversion element. It senses the input quantity (pressure in this case). On account of the pressure the closed end of the bourdon tube is displaced. Thus the pressure is converted into a small displacement. The closed end of the bourdon tube is connected through mechanical linkage to a gearing arrangement. The gearing arrangement amplifies the small displacement and makes the pointer to rotate through a large angle. The mechanical linkage thus acts as a data transmission element while the gearing arrangement acts as a data manipulation element.

The final data presentation stage consists of the pointer and dial arrangement; which when calibrated with known pressure inputs, gives an indication of the pressure signal applied to the bourdon tube. The schematic diagram of this measurement system is given in Fig. 4.4.

When a control device is used for the final measurement stage, it is necessary to apply some feedback to the input signal to accomplish the control objectives. The control stage compares the signal representing the measured variable with a reference signal of the same form. This reference signal has a value the measured signal should have and is presented to a controller. If the measured signal agrees with the reference value, the controller does nothing. However, if there is a difference between the measured value and the reference value, an error signal is generated. Thus the controller sends a signal to a device which acts to alter the value of the measured signal. Suppose the measured variable is flow of a liquid, then the control device is a motorized valve placed in the flow system. In case the measured flow rate is too low than the preset flow rate, then the controller would cause the valve to open, thereby increasing the flow rate. If on the other hand, the flow rate were too high, the valves are closed. The operation of closing or opening of valve will cease when the output flow rate is equal to preset value of flow rate. The physical phenomenon, which is the signal from external world faces signal abnormalities, so these have to be rectified in order to make the signal more effective.

4.3 Origin of Signals

Data acquisition deals with the elements shown in Fig. 4.5. The physical phenomenon may be electrical, optical, mechanical, or any other phenomena that is required to measure. The sensor changes that phenomenon into a signal that is easier to transmit, record, and analyze-usually a voltage or current. Signal conditioning amplifies and filters the raw signal to prepare it for A/D conversion, which transforms the signal into a digital pattern such that it is suitable for the computer.

4.3.1 Transducers and Sensors

The transducer converts one physical phenomenon into another; but more commonly the resulting phenomenon is required to be an electrical signal. For instance, a thermocouple produces a voltage that is related to temperature. An example of a transducer with a non-electrical output is a liquid-in-glass thermometer. It converts temperature changes to visible changes in the volume of fluid. A transducer can also be called a sensor. Sensor starts with a transducer as the front end, but then add signal conditioning (such as amplifiers), computation (such as linearization), and a means of transmitting the signal over some distance without degradation. All these processes involved resulted in increased cost adds weight and appears bulky.

Figure 4.6 is a general model of all the world's sensors. If the instrument has only the first couple of blocks indicated with a dotted box, then it is probably a transducer. Table 4.1 contains examples of some sensors. The first example, a temperature transmitter, uses a thermocouple with some built-in signal conditioning. Many times the user can use thermocouples without a transmitter, but always some form of signal conditioning is required. The

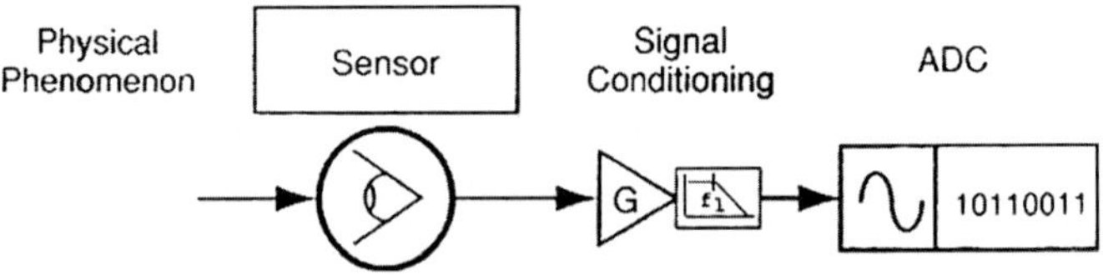

Fig. 4.5. Elements of the data acquisition process

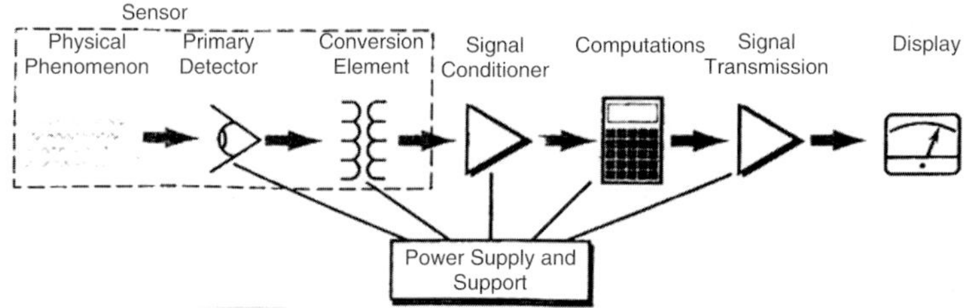

Fig. 4.6. A complete general sensor model

Table 4.1. Three practical examples of sensor systems

Block	Example A	Example B	Example C
Phenomenon	Temperature	Pressure	Magnetic field
Detector	–	Diaphragm displacement	Faraday rotation
Transducer	Thermocouple	LVDT	Laser and photodiode
Signal conditioner	Cold junction; amplifier	Demodulator	ADC
Computations	Linearize	Linearize; scale	Ratio; log; scale
Transmission	0–10 VDC	4–20 mA	RS-232 serial
Display	Analog meter	Analog meter	Computer system
Support	DC	DC (2-wire current loop)	DC; cooling

second example, a pressure transmitter, is a slightly more complex instrument. The third example, a magnetic field sensor uses an optical principle called Faraday rotation in which the polarization of light is affected by magnetic fields. If the data acquisition system interprets light intensity as an input signal, this technology is well suitable for such an application.

4.3.2 Acquiring the Signal

Until now, there was a discussion on the real (analog) world of signals. Now the section focuses on digitizing those signals for use in LabVIEW. By definition, analog signals are continuous-time, continuous-valued functions that is they can take on any possible value and are defined over all possible time resolutions.

An ADC samples the analog signals on a regular basis and converts the amplitude at each sample time to a digital value with finite resolution. These are termed discrete-time, discrete-value functions. Unlike their analog counterparts, discrete functions are defined only at times specified by the sample rate and may have values determined by the resolution of the ADC. In other words, when an analog signal is digitized, an approximation has to be made based on the signal and the specifications used for data analysis. Resolution

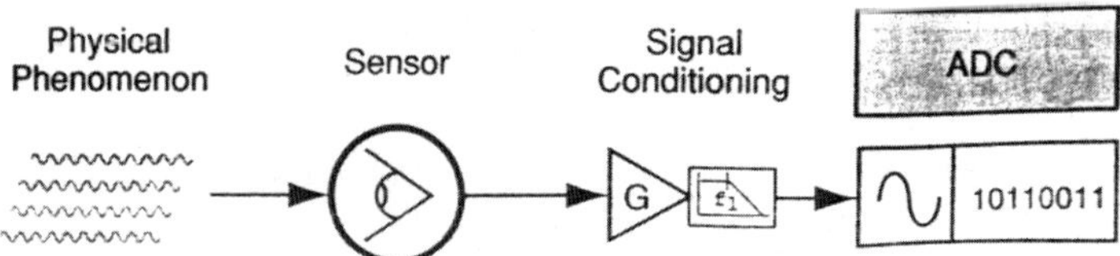

Fig. 4.7. Acquiring and digitizing a signal

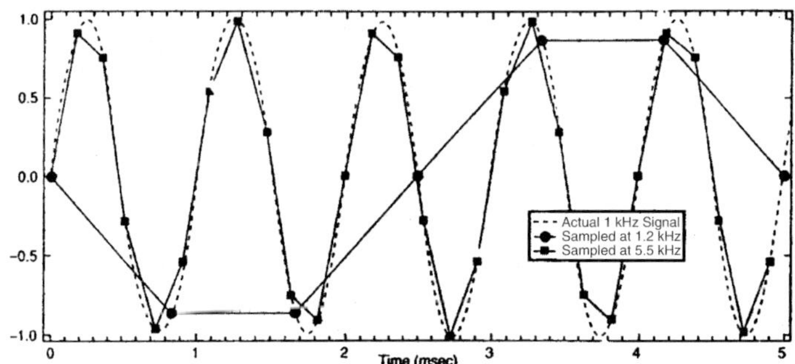

Fig. 4.8. Effects of sampling rates

also plays a very important role here. The additional amplitude and temporal resolution are the two important factors that are to be considered, but they may be expensive. To overcome this, there should be a clear idea on sampling. An example of acquiring and digitizing a signal is illustrated in Fig. 4.7. The process of sampling is based on Nyquist Shannon's Sampling Theorem, which is discussed in the following section.

Figure 4.8 shows a graphical display of the effects of sampling rates. When the original 1 kHz sine wave is sampled at 1.2 kHz, it is totally unrecognizable in the data samples. In the figure, the dotted notation indicates the process of sampling at 5.5 kHz which produces better representation.

4.3.3 Sampling Theorem

A fundamental rule of sampled data systems is that the input signal must be sampled at a rate greater than twice the highest frequency component in the signal. This is known as the *Nyquist* criterion. Stated as a formula, it requires that $f_s/2 > f_a$, where f_s is the sampling frequency and f_a is the signal being sampled. Violating the Nyquist criteria is called under sampling, and results in aliasing.

4.3.4 Filtering and Averaging

To get rid of aliasing, a type of low-pass filter has to be used, known as an anti-aliasing filter. Analog filters are absolutely mandatory, regardless of

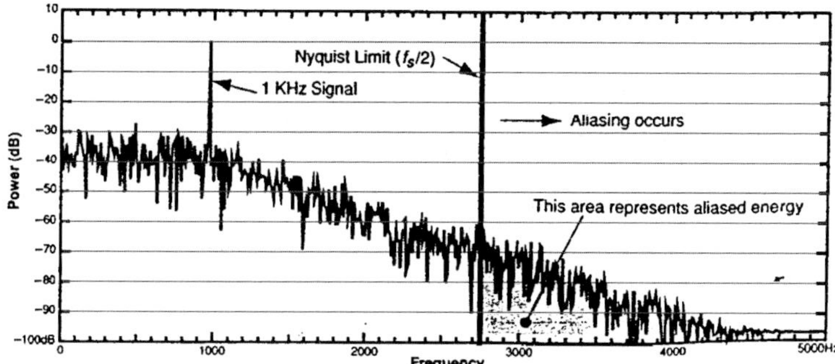

Fig. 4.9. Power spectrum of a 1 kHz sine wave with low pass-filtered noise added

the sampling rate, unless the signal's frequency characteristics are known. In order to limit the bandwidth of the raw signal to $f_s/2$ an analog filter is used. The analog filter can be in the transducer, the signal conditioner, on the A/D board, or in all three sections. The filter is usually made up of resistors, capacitors, and sometimes, operational amplifiers, in which case it is called an active filter. One problem with analog filters is that they are very complex and expensive. If the desired signal is fairly close to the Nyquist limit, the filter needs to cut off very quickly, which requires lots of stages, which depends on the order of the filter.

Digital filters can augment, but cannot replace, analog filters. Digital filter VIs is included with the LabVIEW analysis VIs, and they are functionally equivalent to analog filters. The simplest type of digital filter is a moving averager, which has the advantage of being usable in real-time on a sample-by-sample basis. One way to simplify the anti-aliasing filter problem is to over sample the input. If the A/D hardware is fast enough, the sampling rate is first increased and then a digital filter is used to eliminate the higher frequencies that are of no interest. This makes the analog filtering design problem simpler because the Nyquist frequency has been raised much higher, so the analog filter need not be sharp. The signal has to be sampled at a rate high enough to avoid significant aliasing with a modest analog filter, but sampling at too high rate may not be practical because the hardware is too expensive and extra data may overload the CPU. Figure 4.9 illustrates the occurrence of aliasing in a 1 kHz sine wave with low pass filtered noise in it.

A potential problem with averaging comes up while handling nonlinear data. The process of averaging is defined to be the summation of several values, divided by the number.

If the main concern is rejecting 60 Hz line frequency interference, an old trick is to grab an array of samples over one line period (16.66 ms). This should be done for every channel. Using plug-in boards with LabVIEW's data acquisition drivers, the user can adjust the sampling interval with high precision, making this a reasonable option. First, a simple experiment is set to

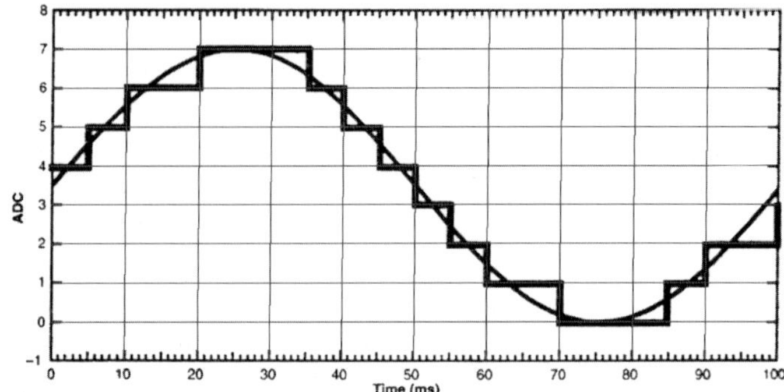

Fig. 4.10. A sine wave and its representation by a 3-bit ADC sampling every 5 ms

acquire and average data from a noisy input. The sampling period is varied and checked for a null noise at each 16.66 ms multiple.

If there is an attempt to average recurrent waveforms to reduce noise, the arrays of data that are acquired must be perfectly in-phase. If a phase shift occurs during acquisition, then the waveforms will partially cancel each other or distort in some other way. Triggered data acquisition is the normal solution because it helps to guarantee that each buffer of data is acquired at the same part of the signal's cycle.

Some other aspects of filtering that may be important to some of the applications are impulse response and phase response. For ordinary data logging, these factors are generally ignored. But if the user is doing dynamic testing like vibration analysis, acoustics, or seismology, impulse and phase response can be very important. As a rule of thumb, when filters become very complex (high-order), they cutoff more sharply, have more radical phase shifts around the cutoff frequency, and (depending on the filter type) exhibit ringing on transients.

The best way to analyze filtering needs is to use a spectrum analyzer since we are able to know exactly what signals are present and what has to be filtered out. Some common methods used for filtering are:

- A dedicated spectrum analyzer instrument (very expensive)
- A digital oscilloscope with FFT capability (LabVIEW computes the power spectrum)
- Even a multifunction I/O board running as fast as possible with Lab VIEW doing the power spectrum.

Looking at the power spectrum in Fig. 4.10, some of the noise power is above the floor for the ADC and is also at frequencies above 2.75 kHz. This little triangle represents aliased energy and gives the user a qualitative feel for how much contamination the user can expect. It depends on the nature of the out-of-band noise.

4.3.5 Triggering

Triggering refers to any method by which an ADC or DAC can be synchronized to some event. If there is a regular event that causes each individual A/D conversion, it is called the time base, and is usually generated by a crystal-controlled clock oscillator. For this discussion, triggering is defined as an event that starts or stops a series of conversions that are individually paced by a time base.

One situation when there is a need for triggering is when the user is waiting for a transient event to occur, which is a single pulse. Another use of triggering is to force the data acquisition to be in-phase with the signal. Signal analysis may be simplified if the waveform always starts with the same polarity and voltage. Or, the user may need to acquire many buffers of data from a recurrent waveform (like a sine wave) to average them, thus reducing the noise.

Trigger sources come in three varieties:

- External
- Internal
- Software-generated

External triggers are digital pulses, usually produced by specialized hardware or a signal coming from the equipment that is been interfaced with. For example, a function generator has a connector on it called sync that produces a TTL pulse every time the output waveform crosses 0 V in the positive direction. While dealing with pulsed light sources (like some lasers), an optical detector such as a photodiode can be used to trigger a short and a high-speed burst of data acquisition. Signal conditioning is generally required for external triggers because most data acquisition hardware demands a clean pulse with a limited amplitude range.

Internal triggering is built into many data acquisition devices, including oscilloscopes. Transient recorders, and multifunction boards. It is basically an analog function where a device called a comparator detects the signal's crossing of a specified level. The slope of the signal may also be part of the triggering criteria. Really sophisticated instruments permit triggering on specified patterns, especially useful in digital logic and communications signal analysis. The on-board triggering features of the National Instruments boards are easy to use in LabVIEW, due to the data acquisition VIs.

Software-generated triggers require a program that evaluates an incoming signal or some other status information and decides when to begin saving data or generating an output. A trivial example is the Run button in LabVIEW that starts up a simple data acquisition program. On-the-fly signal analysis is a bit more complex and quickly runs into performance problems when fast signals are used. For instance, the user may want to save data from a special analyzer only when the process temperature gets above a certain limit. Since the temperature probably changes very slowly, the difficult problem would be to evaluate the distortion of an incoming audio signal and save only the waveforms that are defective.

4.3.6 Throughput

A final consideration in of converters is the throughput of the system, a yardstick for overall performance, usually measured in samples per second. Major factors determining throughput are:

- A/D or D/A conversion speed
- Use of multiplexers and amplifiers, which may add delays between channels
- Disk system performance, if streaming data to or from disk
- Use of DMA, which speeds data transfer
- CPU speed, especially if a lot of data processing is required
- Operating system overhead

The nature of electrical output from the transducer depends on the basic principle involved in the design. The output may be analog, digital or frequency modulated.

Basically, there are two types of transducers, electrical, and mechanical.

4.4 Transducer

Analytical methods in chemistry have mainly been based on photometric transducers, as in spectroscopic and colorimetric methods. However, most sensors have been developed around electrochemical transducers, because of simplicity of construction and cost. While electrons drive microprocessors, the directness of an electrical device will tend to have maximum appeal. However, with the rapid development of photon-driven devices through the use of optical fibers, it could well be that electrical appliances will soon become obsolete – starting with the telephone. In addition, the use of micro-mass-controlled devices, based mainly on piezoelectric crystals, may become competitive in the near future.

Transducers can be subdivided into the following four main types.

Electrochemical transducers. Electrochemical transducers are classified as Potentiometric, Voltmetric, Conductometric, and FET-based transducers:

(a) *Potentiometric transducers.* These involve the measurement of the emf (potential) of a cell at zero current. The emf is proportional to the logarithm of the concentration of the substance being determined.

(b) *Volt metric transducers.* An increasing (decreasing) potential is applied to the cell until oxidation (reduction) of the substance to be analyzed occurs and there is a sharp rise (fall) in the current to give a peak current. The height of the peak current is directly proportional to the concentration of the electroactive material. If the appropriate oxidation (reduction) potential is known, one may step the potential directly to that value and observe the current. This mode is known as amperometric.

(c) *Conductometric transducers.* Most reactions involve a change in the composition of the solution. This will normally result in a change in the electrical conductivity of the solution, which can be measured electrically.
(d) *FET-based transducer.* Miniaturization can sometimes be achieved by constructing one of the above types of electrochemical transducers on a silicon chip-based field-effect transistor. This method has mainly been used with Potentiometer sensors, but could also be used with voltmetric or conductometric sensors.

Optical transducers. These have taken a new lease of life with the development of fibre optics, thus allowing greater flexibility and miniaturization. The techniques used include absorption spectroscopy, fluorescence spectroscopy, luminescence spectroscopy, internal reflection spectroscopy, surface plasmon spectroscopy, and light scattering.

Piezoelectric devices. These devices involve the generation of electric currents from a vibrating crystal. The frequency of vibration is affected by the mass of material adsorbed on its surface, which could be related to changes in a reaction. Surface acoustic wave devices are a related system.

Thermal sensors. All chemical and biochemical processes involve the production or absorption of heat. This heat can be measured by sensitive thermistors and hence be related to the amount of substance to be analyzed.

Table 4.2 illustrates transducer characteristics which define many of the signal conditioning requirements of a DAQ system.

4.4.1 Selecting a Transducer

The transducer or sensor has to be physically compatible with its intended application. The following properties should be considered while selecting a transducer:

1. *Operating range.* Chosen to maintain range requirements and good resolution.
2. *Sensitivity.* Chosen to allow sufficient output.
3. *Frequency response and resonant frequency.* Flat over the entire desired range.
4. *Environmental compatibility.* Temperature range, corrosive fluids, pressure, shocks, interaction, size, and mounting restrictions.
5. *Minimum sensitivity.* To expected stimulus, other than the measurand.
6. *Accuracy.* Repeatability and calibration errors as well as errors expected due to sensitivity to other stimuli.
7. *Usage and ruggedness.* Ruggedness, both of mechanical and electric intensities versus size and weight.
8. *Electrical parameters.* Length and type of cable required, signal to noise ratio when combined with amplifiers, and frequency response limitations.

Table 4.2. Transducer characteristics

Sensor	Electrical characteristics	Signal conditioning requirements
Thermocouple	Low-voltage output	Reference temperature sensor (for cold-junction compensation)
	Low sensitivity	
	Nonlinear output	High amplification
		Linearization
RTD	Low resistance (100 3/4 typical)	Current excitation
	Low sensitivity	Four-wire/three-wire configuration
	Nonlinear output	Linearization
Strain gauge	Low-resistance device	Voltage or current excitation
	Low sensitivity	Bridge completion
	Nonlinear output	Linearization
Current-output device	Current loop output (4–20 mA typical)	Precision resistor
Thermistor	Resistive device	Current excitation or voltage excitation with reference resistor
	High resistance and sensitivity	
	Very nonlinear output	Linearization
Integrated circuit (IC) temperature sensor	High-level voltage or current output	Power source
	Linear output	Moderate gain

4.4.2 Electrical Transducer

An electrical transducer is a sensing device by which the physical, mechanical or optical quantity to be measured is transformed directly by a suitable mechanism into an electrical voltage/current proportional to the input measurand.

An electrical transducer must have the following parameters:

1. *Linearity.* The relationship between a physical parameter and the resulting electrical signal must be linear.
2. *Sensitivity.* This is defined as the electrical output per unit change in the physical parameter (for example V/°C for a temperature sensor). High sensitivity is generally desirable for a transducer.
3. *Dynamic range.* The operating range of the transducer should be wide, to permit its use under a wide range of measurement conditions.
4. *Repeatability.* The input/output relationship for a transducer should be predictable over a long period of time. This ensures reliability of operation.
5. *Physical size.* The transducer must have minimal weight and volume, so that its presence in the measurement system does not disturb the existing conditions.

Advantages of electrical transducers. The main advantages of electrical transducers (conversion of physical quantity into electrical quantities) are as follows:

1. Electrical amplification and attenuation can be easily done.
2. Mass inertia effects are minimized.
3. Effects of friction are minimized.
4. The output can be indicated and recorded remotely at a distance from the sensing medium.
5. The output can be modified to meet the requirements of the indicating or controlling units. The signal magnitude can be related in terms of the voltage current (The analog signal information can be converted in to pulse or frequency information. Since output can be modified, modulated or amplified at will, the output signal can be easily used for recording on any suitable multi channel recording device.)
6. The signal can be conditioned or mixed to obtain any combination with outputs of similar transducers or control signals.
7. The electrical or electronic system can be controlled with a very small power level.
8. The electrical output can be easily used, transmitted and processed for the purpose of measurement.

Electrical transducers can be broadly classified into two major categories: (a) active and (b) passive.

An *active transducer* generates an electrical signal directly in response to the physical parameter and does not require an external power source for its operation. Active transducers are self-generating devices, which operate under energy conversion principle and generate an equivalent output signal (for example, from pressure to charge or temperature to electrical potential).

Typical example of active transducers are piezoelectric sensors (for generation of charge corresponding to pressure) and photovoltaic cells (for generation of voltage in response to illumination).

Passive transducer operates under energy controlling principles, which makes it necessary to use an external electrical source with them. They depend upon the change in an electrical parameter (R, L, and C).

Typical examples are strain gauges (for resistance change in response to pressure), and thermistors (for resistance change corresponding to temperature variations).

Electrical transducers are used mostly to measure nonelectrical quantities. For this purpose a detector or sensing element is used, which converts the physical quantity into a displacement. This displacement actuates an electric transducer, which acts as a secondary transducer and produces an output that is electrical in nature. This electrical quantity is measured by the standard method used for electrical measurement. The electrical signals may be current, voltage, or frequency; their production is based on R, L, and C effects.

A transducer which converts a nonelectrical quantity into an analog electrical signal may be considered as consisting of two parts, the sensing element, and the transduction element. The sensing or detector element is that part of a transducer which responds to a physical phenomenon or to a change in a physical phenomenon. The response of the sensing element must be closely related to the physical phenomenon.

The transduction element transforms the output of a sensing element to an electrical output. This, in a way, acts as a secondary transducer.

Transducers may be further classified into different categories depending upon the principle employed by their transduction elements to convert physical phenomena into output electrical signals. The different electrical phenomena employed in the transduction elements of transducers are as follows:

1. Resistive
2. Inductive
3. Capacitive
4. Electromagnetic
5. Piezoelectric
6. Photoemissive
7. Photoresistive
8. Potentiometric
9. Thermoelectric
10. Frequency generating

Resistive Transducer

Resistive transducers are those in which the resistance changes due to a change in some physical phenomenon. The change in the value of the resistance with a change in the length of the conductor can be used to measure displacement.

Strain gauges work on the principle that the resistance of a conductor or semiconductor changes when strained. This can be used for the measurement of displacement, force, and pressure. The resistivity of materials changes with changes in temperature. This property can be used for the measurement of temperature.

Potentiometer

A resistive potentiometer (pot) consists of a resistance element provided with a sliding contact, called a wiper. The motion of the sliding contact may be translatory or rotational.

Translatory resistive elements, as shown in Fig. 4.11a, are linear (straight) devices. Rotational resistive devices are circular and are used for the measurement of angular displacement, as shown in Fig. 4.11b. Some have a combination of both, with resistive elements in the form of a helix, as shown in Fig. 4.11c. They are known as helipots.

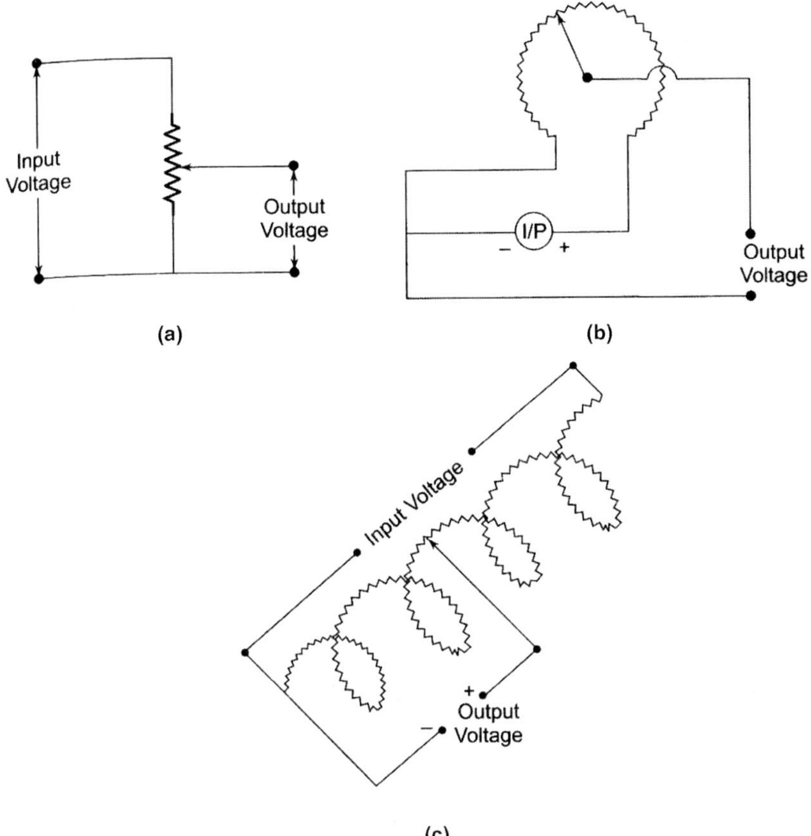

Fig. 4.11. (**a**) Translatory type; (**b**) rotational type and (**c**) helipot (rotational)

Helical resistive elements are multi turn rotational devices which can be used for the measurement of either translatory or rotational motion. A potentiometer is a passive transducer since it requires an external power source for its operation.

Advantages

1. They are inexpensive.
2. Simple to operate and are very useful for applications where the requirements are not particularly severe.
3. They are useful for the measurement of large amplitudes of displacement.
4. Electrical efficiency is very high, and they provide sufficient output to allow control operations.

Disadvantages

1. When using a linear potentiometer, a large force is required to move the sliding contacts.
2. The sliding contacts can wear out, become misaligned and generate noise.

Resistance Pressure Transducer

Measurement in the resistive type of transducer is based on the fact that a change in pressure results in a resistance change in the sensing elements. Resistance pressure transducers are of two main types. First, the electromechanical resistance transducer, in which a change of pressure, stress, position, displacement, or other mechanical variation is applied to a variable resistor. The other resistance transducer is the strain gauge, where the stress acts directly on the resistance. It is very commonly used for stress and displacement measurement in instrumentation.

In the general case of pressure measurement, the sensitive resistance element may take other forms, depending on the mechanical arrangement on which the pressure is caused to act.

Figure 4.12a and b show two ways by which the pressure acts to influence the sensitive resistance element, i.e., by which pressure varies the resistance element. They are the bellow type and the diaphragm type.

In each of these cases, the element moved by the pressure change is made to cause a change in resistance. This resistance change can be made part of a bridge circuit and then taken as either ac or dc output signal to determine the pressure indication.

Inductive Transducer

Inductive transducers may be either of the self-generating or the passive type. The self-generating type utilizes the basic electrical generator principles i.e., a motion between a conductor and magnetic field induces a voltage in the conductor (generator action). This relative motion between the field and the conductor is supplied by changes in the measurand.

An inductive electromechanical transducer is a device that converts physical motion (position change) into a change in inductance.

Transducers of the variable inductance type work upon one of the following principles:

1. Variation of self-inductance
2. Variation of mutual inductance

Inductive transducers are mainly used for the measurement of displacement. The displacement to be measured is arranged to cause variation in any of three variables:

1. Number of turns
2. Geometric configuration
3. Permeability of the magnetic material or magnetic circuits

For example, let us consider the case of a general inductive transducer. The inductive transducer has N turns and a reluctance R. When a current "i" is passed through it, the flux is

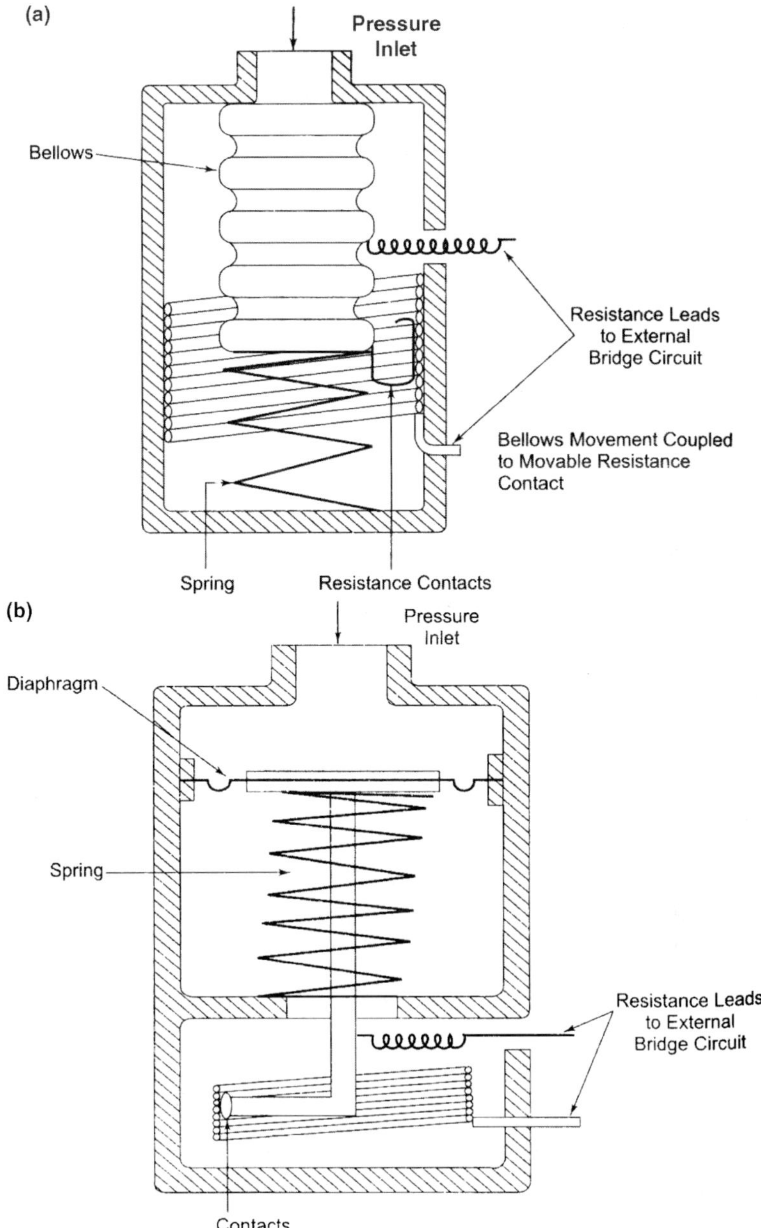

Fig. 4.12. (**a**) Resistance pressure transducer and (**b**) sensitive diaphragm moves the resistance contact

$$\text{Ø} = \frac{\text{Ni}}{\text{R}}$$

Therefore

$$\frac{\text{dØ}}{\text{dt}} = \frac{\text{N}}{2} \times \frac{\text{di}}{\text{dt}} - \frac{\text{Ni}}{\text{R}} \times \frac{\text{dR}}{\text{dt}}$$

If the current varies very rapidly,

$$\frac{\text{dØ}}{\text{dt}} = \frac{\text{N}}{2} \times \frac{\text{di}}{\text{dt}}$$

But emf induced in the coil is given by $\text{e} = \text{N} \times \text{dØ}/\text{dt}$. Therefore

$$\text{e} = \text{N} \times \frac{\text{N}}{2} \times \frac{\text{di}}{\text{dt}} = \frac{\text{N}^2}{\text{R}} \times \frac{\text{di}}{\text{dt}}$$

Also the self-inductance is given by

$$\text{L} = \frac{\text{e}}{\text{di}/\text{dt}} = \frac{\text{Ni}}{\text{R}}$$

Therefore, the output from an inductive transducer can be in the form of either a change in voltage or a change in inductance.

Change in Self-Inductance with Number of Turns

The output may be caused by a change in the number of turns. Figure 4.13a, b is transducers used for the measurement of displacement of linear and angular movement, respectively.

Figure 4.13a is an air cored transducer for measurement of linear displacement. Figure 4.13b is an iron cored coil used for the measurement of angular displacement. In both cases, as the number of turns are changed, the self inductance and the output also changes.

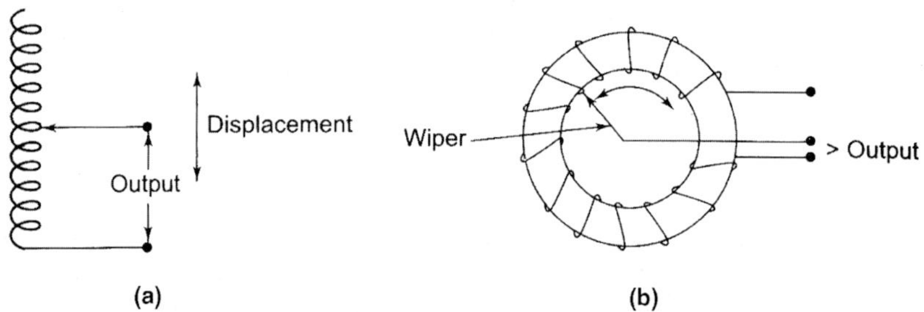

Fig. 4.13. (**a**) Linear inductive transducer and (**b**) angular inductive transducer

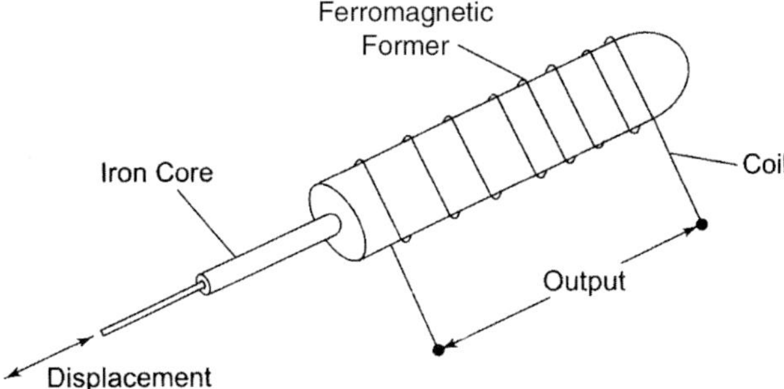

Fig. 4.14. Inductive transducer working on the principle of variation of permeability

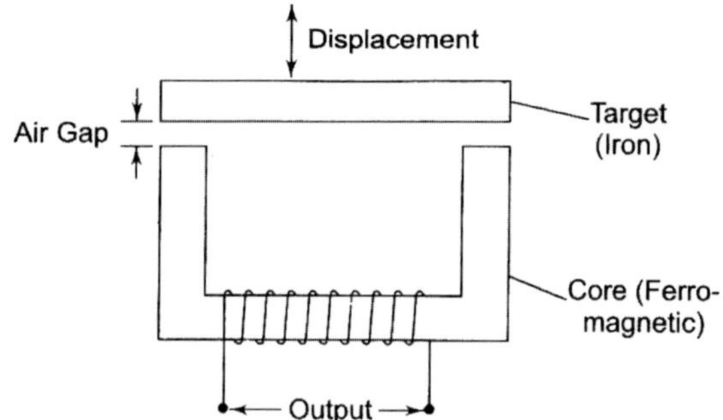

Fig. 4.15. Variable reluctance transducer

Transducer Working on the Principle of Change in Self-Inductance with Change in Permeability

Figure 4.14 shows an inductive transducer which works on the principle of the variation of permeability causing a change in self inductance. The iron core is surrounded by a winding. If the iron core is inside the winding, its permeability is increased, and so is the inductance. When the iron core is moved out of the winding the permeability decreases resulting in a reduction of the self-inductance of the coil. This transducer can be used for measuring displacement.

Variable Reluctance Type Transducer

A transducer of the variable type consists of a coil wound on a ferromagnetic core. The displacement which is to be measured is applied to a ferromagnetic target does not have any physical contact with the core on which it is mounted. The core and the target are separated by an air gap, as shown in Fig. 4.15.

The reluctance of the magnetic path is determined by the air gap. The inductance of the coil depends upon the reluctance of the magnetic circuits. The self-inductance of the coil is given by

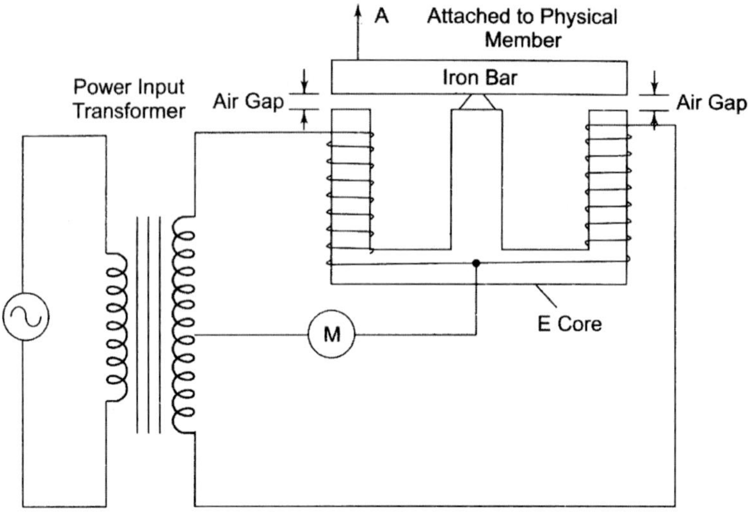

Fig. 4.16. Variable reluctance bridge circuit

$$L = \frac{N^2}{R_i + R_g}$$

where N = number of turns; R_i = reluctance of iron parts and R_g = reluctance of air gap. The reluctance of the iron part is negligible compared to that of the air gap. Therefore $L = N^2/R_g$. But reluctance of the air gap is given by

$$Rg = \frac{lg}{\mu_0 \times Ag}$$

where 1 = length of the air gap; Ag = area of the flux path through air; μ_0 = permeability
R_g is proportional to l_g, as μ_0 and A_g are constants. Hence L is proportional to l/μ_0, i.e., the self-inductance of the coil is inversely proportional to the length of the air gap.

Figure 4.16 shows the variable reluctance bridge circuit. When the target is near the core, the length is small and therefore the self inductance large. But when the target is away from the core the reluctance is large, resulting in a smaller self-inductance value. Hence the inductance of the coil is a function of the distance of the target from the core, i.e., the length of the air gap.

Capacitive Transducer

A linear change in capacitance with changes in the physical position of the moving element may be used to provide an electrical indication of the element's position. The capacitance is given by C = KA/d where K = the

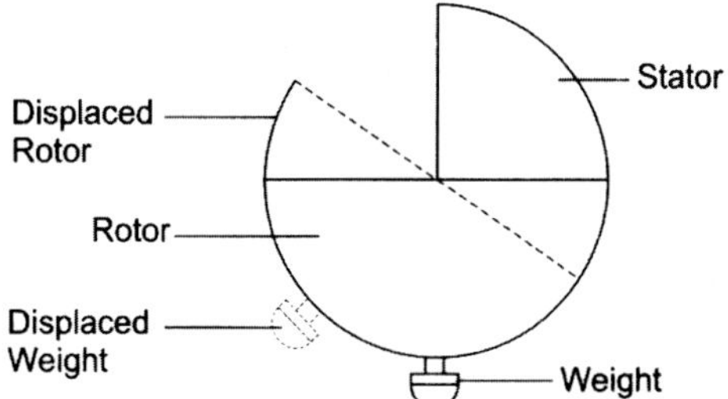

Fig. 4.17. Capacitive transducer

dielectric constant; A = the total area of the capacitor surfaces; d = distance between two capacitive surfaces and C = the resultant capacitance.

From the above equation, it is seen that capacitance increases (a) if the effective area of the plate is increased, and (b) if the material has a high dielectric constant.

The capacitance is reduced if the spacing between the plates is increased. Transducers which make use of these three methods of varying capacitance have been developed. With proper calibration, each type yields a high degree of accuracy. Stray magnetic and capacitive effects may cause errors in the measurement produced, which can be avoided by proper shielding. Some capacitive dielectrics are temperature sensitive, so temperature variations should be minimized for accurate measurements.

A variable plate area transducer is made up of a fixed plate called Stator and a movable plate called the *rotor*. The rotor is mechanically coupled to the member under test. As the member moves, the rotor changes its position relative to the stator, thereby changing the effective area between the plates. A transducer of this type is shown in Fig. 4.17. Such a device is used to detect the amount of roll in an aircraft. As the aircraft rolls to the left, the plates moves to the relative position shown by dashed lines in Fig. 4.17 and the capacitance decreases by an amount proportional to the degree of roll. In this case the stator, securely attached to the aircraft is the moving element. The weight on the rotor keeps its position fixed with reference to the surface of the earth, but the relative position of the plates changes and this is the factor that determines the capacitance of the unit.

Figure 4.18 shows a transducer that makes use of the variation in capacitance resulting from a change in spacing between the plates This particular transducer is designed to measure pressure (in vacuum).

Enclosed in an airtight container is a metallic diaphragm which moves to the left when pressure is applied to the chamber and to the right when vacuum is applied. This diaphragm is used as one plate of a variable capacitor. Its distance from the stationary plate to its left, as determined by the pressure applied to the unit, determines the capacitance between the two plates.

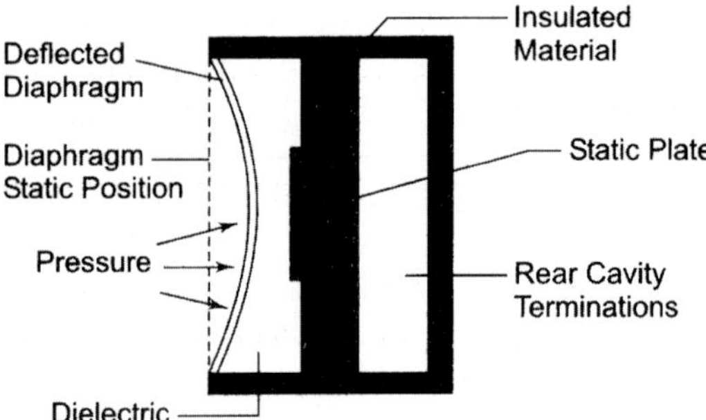

Fig. 4.18. Capacitive pressure transducer

The monitor indicates the pressure equivalent of the unit's capacitance by measuring the capacitor's reactance to the ac source voltage.

The portion of the chamber to the left of the moving plate is isolated from the side into which the pressurized gas or vapor is introduced. Hence the dielectric constant of the unit does not change for different types of pressurized gas or vapor. The capacity is purely a function of the diaphragm position and is a linear device. Changes in pressure may be easily detected by the variation of capacity between a fixed plate and another plate free to move as the pressure changes. The resulting variation follows the basic capacity formula.

$$C = 0.885 \frac{K(n-1)A}{t} \text{pf}$$

where A = area of one side of one plate in cm; n = number of plates; t = thickness of dielectric in cm; and K = dielectric constant. The capacitive transducer, as in the capacitive microphone, is simple to construct and inexpensive to produce. It is particularly effective for HF variations.

However, when the varying capacitance is made part of an AC bridge to produce an AC output signal, the conditions for resistive and reactive balance generally require much care to be taken against unwanted signal pickup in the high impedance circuit, and also compensation for temperature changes. As a result, the receiving instrument for the capacitive sensor usually calls for more advanced and complex design than is needed for other transducers.

4.5 Sensors

Sensors are provided for every controlled variable and many additional variables used to monitor the process. As a rough guideline, the total number of measurements is about three times the number of controlled variables. Some of the general factors considered while selecting the proper sensor for each application are:

- *Accuracy* is always a factor; however, very high accuracy is not always required. Since very accurate sensors are usually more expensive, the engineer should ensure that the required accuracy is achieved with reasonable cost.
- *Good reproducibility* is required for sensors used for control, even when accuracy is not needed.
- *Fast dynamic* response is desired, especially when the instrumentation is located in feedback control loop because slower loop dynamics leads to poorer control performance.
- *Reliability* the instrumentation should be reliable, i.e., it should perform its function with a low probability of failure. Matching the instrument to the process conditions and performing periodic maintenance contributes to reliability.

Another common decision made for all sensors is the *range* over which the sensor will measure the variable. The range should be set to:

- Include the typical values experienced in normal operation
- Include expected disturbances
- Provide good accuracy.

For most sensors, a larger range reduces the accuracy and reproducibility; thus, a balance is required. In some cases, all three goals might not be attainable with a single sensor; then, two sensors are required to provide the range and accuracy. The sensors discussed in this section measure the most common process variables: temperature, flow, pressure, and level. They are often referred to as "direct" sensors because the principles of measurement are much simpler than a composition analyzer. However, no sensor truly measures the process variable "directly." Each sensor measures the effect of the process variable as indicated by a physical position, force, voltage, or other more easily measured property. The relationship between the process variable and the quantity actually measured is not exact, and the engineer should understand the accuracy, range, and process limits, so that the most appropriate sensor is selected.

Human beings have at least five sensors among the available sensing devices. The main types of sensors are nose, tongue, ears, eyes, and fingers. In the laboratory, one of the best known types of sensor is the litmus paper test for acids and alkalis, which gives a qualitative indication, by means of a color reaction, indicating the presence or absence of an acid. A more precise method of indicating the degree of acidity is the measurement of pH, either by the more extended use of color reactions in special indicator solutions, or even by simple pH papers.

However, the best method of measuring acidity is the use of the pH meter, which is an electrochemical device giving an electrical response which can be read by a needle moving on a scale or on a digital read-out device or input to a microprocessor. In such methods, the sensor that responds to the degree of

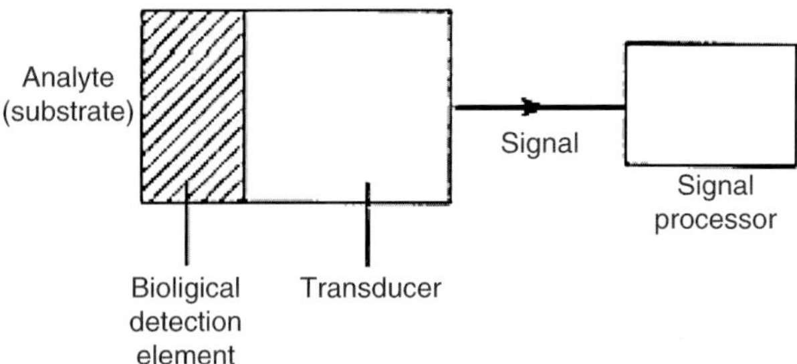

Fig. 4.19. Schematic layout of a sensor

acidity is either a chemical – the dye litmus, or a more complex mixture of chemical dyes in pH indicator solutions – or the glass membrane electrode in the pH meter.

The chemical or electrical response has to be converted into a signal that can be observed visually. With litmus as in the above example, the process of conversion is simpler. In the case of the pH meter, the electrical response (a voltage change) has to be converted into an observable response – movement of a meter needle or a digital display.

Sensors can be divided basically into three types, namely:

- Chemical sensors which measure chemical substances by chemical or physical responses
- Physical sensors for measuring distance, mass, temperature, pressure
- Biosensors which measure chemical substances by using a biological sensing element

All of these devices have to be connected to a transducer of some kind, so that a visibly observable response occurs. Chemical sensors and biosensors are generally concerned with sensing and measuring particular chemicals which may or may not be biological themselves. Figure 4.19 shows the general schematic arrangement of a sensor.

4.5.1 The Nose as a Sensor

One might consider the ears, eyes, and fingers to be physical sensors as they detect physical sensations of sound, light and heat, etc., respectively. What we humans detect with the nose – smells – are in fact small quantities of chemicals. The nose is an extremely sensitive and selective instrument which is very difficult to emulate artificially. It can distinguish between many different chemical substances qualitatively and can give a general idea of "quantity" down to very low detection limits. The chemicals to be detected pass through the olfactory membrane to the olfactory bulbs, which contain biological receptors that sense the substrate. The response is an electrical signal which is transmitted to the brain via the olfactory nerves. The brain then transducers

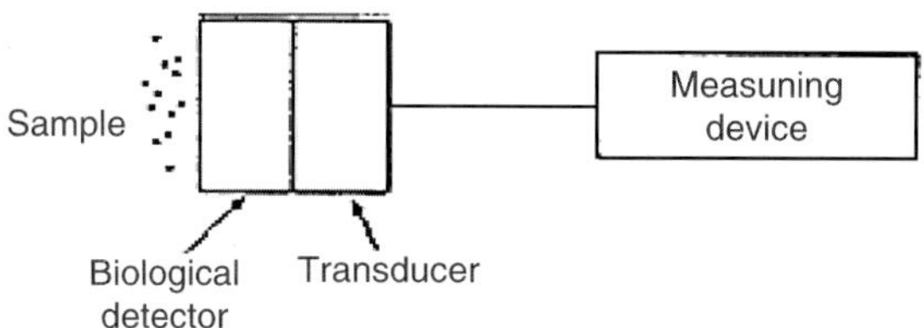

Fig. 4.20. Schematic of a sensor

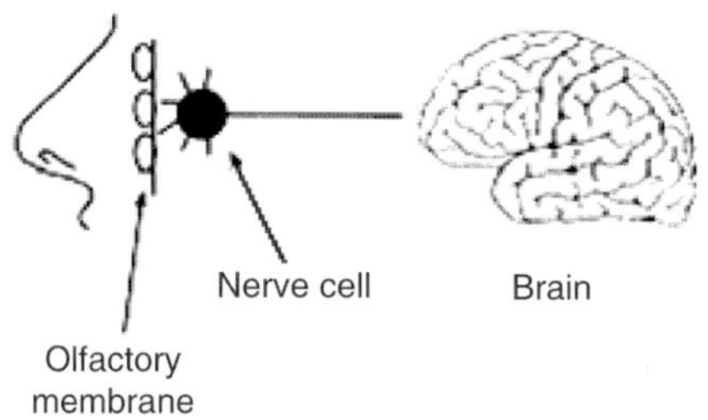

Fig. 4.21. Analogy with the nose as a sensor

this response into the sensation we know as smell. The tongue operates in a similar way.

Schematic of a sensor is shown in Fig. 4.20. Figure 4.21 shows a schematic diagram of the nasal olfactory system, illustrating the comparison with our generalized sensor. The nostrils collect the "smell sample," which is then sensed by the olfactory membrane, i.e., the sensing element. The responses of the olfactory receptors are then converted by the olfactory nerve cell, which is the equivalent of the transducer, into electrical signals which pass along the nerve fiber to the brain for interpretation. Thus, the brain acts as a microprocessor, turning the signal into a sensation which can be smelled.

4.5.2 Sensors and Biosensors: Definitions

There are sometimes differences of usage for the terms sensors, transducers, biosensors, and actuators, so it is necessary to define them:

- The term sensor is used to describe the whole device. A sensor is a device that detects or measures a physical property and records indicates or otherwise responds to it.
- A transducer is defined as a device that converts an observed change (physical or chemical) into a measurable signal. In chemical sensors, the latter is usually an electronic signal whose magnitude is proportional to the concentration of a specific chemical or set of chemicals.
- The term actuator is defined as put into action. This is the part of the device, which produces the display.

4.5.3 Differences Between Chemical Sensors, Physical Sensors, and Biosensors

A chemical sensor is defined as a device which responds to a particular analyze in a selective way through a chemical reaction and can be used for the qualitative or quantitative determination of the analyze. Such a sensor is concerned with detecting and measuring a specific chemical substance or set of chemicals.

Physical sensors are concerned with measuring physical quantities such as length, weight, temperature, pressure, and electricity – for their own sakes.

Biosensors are really a subset of chemical sensors, but are often treated as a topic in their own right. A biosensor can be defined as a device incorporating a biological sensing element connected to a transducer. The data that this sensor detects and measures may be purely chemical (even inorganic), although biological components may be the target analyze. The key difference is that the recognition element is biological in nature.

Recognition Elements

Recognition elements are the key component of any sensor device. They impart the selectivity that enables the sensor to respond selectively to a particular analyze or group of analyses, thus avoiding interferences from other substances. Methods of analysis for specification ions have been available for a long time using ion-selective electrodes, which usually contains a membrane selective for analyze of choice. In biosensors, the most common recognition element is an enzyme. Others include antibodies, nucleic acids, and receptors.

Performance Factors

- *Selectivity*. This is the most important characteristic of sensors – the ability to discriminate between different substances. Such behavior is principally a function of the selective component, although sometimes the operation of the transducer contributes to the selectivity.
- *Sensitivity range*. This usually needs to be sub-mill molar, but in special cases can go down to the femtomolar (10–15 M) range.
- *Accuracy*. This needs to be better than ±5%.
- *Nature of solution*. Conditions such as pH, temperature and ionic strength must be considered.
- *Response time*. This is usually much longer (30 s or more) with biosensors than with chemical sensors.
- *Recovery time*. This is the time that elapses before the sensor is ready to analyze the next sample – it must not be more than a few minutes.
- *Life time*. The working lifetime is usually determined by the stability of the selective material. For biological materials this can be a short as a few days, although it is often several months or more.

- *Noise.* The limit to the signal-to-noise ratio of capacitive sensors is the ratio of excitation voltage to amplifier voltage noise. This provides a signal-to-noise ratio of 2 × 109 to 5 × 1011 in a 1 Hz bandwidth. This theoretic limit will be degraded if amplifier current noise becomes a factor, as with very high impedance sensors, or if amplifier input capacitance is larger than sensor capacitance. The lowest noise is achieved with a JFET amplifier (or an op amp with JFET input stages) and with the JFET input capacitance equal to the sensor capacitance.
- *Stability.* With stable materials and careful construction, sensors with a long-term stability of better than one part in 10–9 per day are possible.
- *Leakage.* Small sensor plates can have an impedance of many megohms, and amplifier input impedance and leakage should be many hundreds of megohms for good accuracy. Leakage paths on the printed circuit board are usually on the surface and can be caused by:
  - Rosin residue or lubricant collecting airborne impurities
  - Conductive residue from water-base cleaning fluids
  - Paint, pigmented with carbon
  - High humidity
- *Static charge.* Static charge can build up on the insulators near capacitive sense plates due to turboelectric charging, causing a tiny spark in serious cases. In less serious cases, a charge buildup of 50–100 V will not arc over but can cause an unwanted sensitivity to mechanical vibration. This sensitivity results from the voltage V = Q/C caused by constant capacitor charge and a capacitance which varies with spacing change due to vibration and can drive the amplifier input over its rails.

To Reduce Static Charge

- *Raise the humidity.* The surface leakage resistance of almost all insulators is a strong function of humidity. With RH 20% and higher, a little surface leakage drains off the charge.
- *Redesign the circuit.* With a low-input-impedance AC amplifier, the over-the-rails problem can be handled, and the signal processing electronics can be designed to ignore a single large transient.
- *Usage of Bare metal plates.* Bare metal will not have tribo electric-effect static discharge problems. Corrosion of bare copper is minimized by use of corrosion-resistant metal plating such as nickel or gold over nickel.
- *Usage of Higher frequency carrier.* Mechanical vibrations and resonance's are usually in a low frequency range, below 10 kHz. A 100 kHz carrier and a high pass filter will minimize mechanical resonance and static charge effects.

4.5.4 Thermocouples

The most popular transducer for measuring temperature is the *thermocouple.* Because the thermocouple is inexpensive, rugged, and can operate over a very

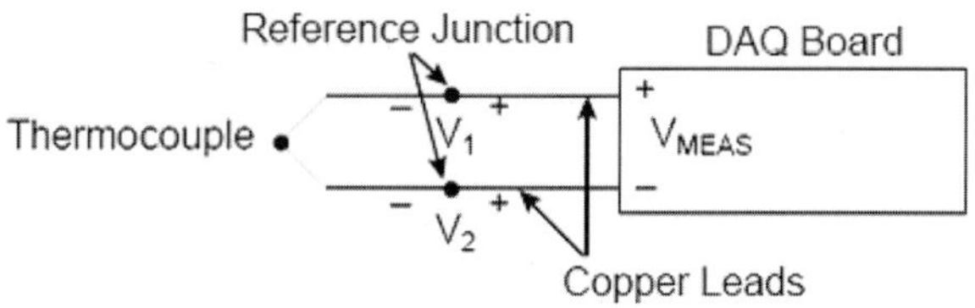

Fig. 4.22. A thermocouple connected to a DAQ board

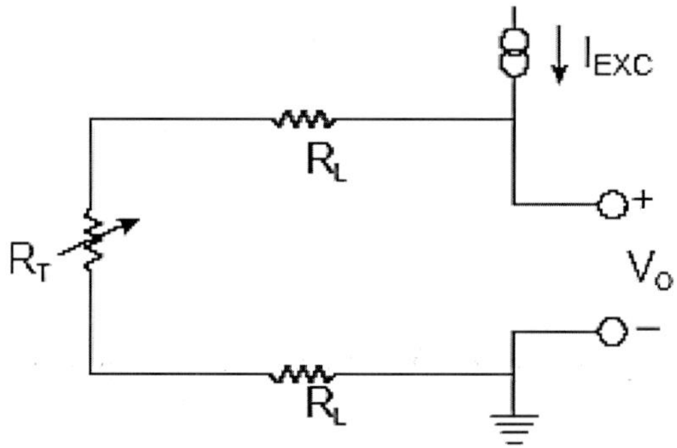

Fig. 4.23. A two-wire RTD

wide range of temperatures, the thermocouple is a very versatile and useful sensor. However, the thermocouple has some unique signal conditioning requirements. A thermocouple operates on the principle that the junction of two dissimilar metals generates a voltage that varies with temperature. However, measuring this voltage is difficult because connecting the thermocouple to DAQ board measurement wires creates the reference junction or cold junction, as shown in Fig. 4.22. These junctions act as thermocouples and produce their own voltages. Thus, the final measured voltage, V_{MEAS}, includes both the thermocouple and reference-junction voltages. The method of compensating for these unwanted reference-junction voltages is called *cold-junction compensation* as shown in Fig. 4.23.

There are two general approaches to cold-junction compensation – hardware and software compensation. Hardware compensation uses a special circuit that applies the appropriate voltage to cancel the cold-junction voltage. Although there is no need for software to implement hardware compensation, each thermocouple type must have its own compensation circuit that works at all ambient temperatures, which can be expensive.

Software cold-junction compensation, on the other hand is very flexible and requires only knowing the ambient temperature. If an additional sensor is used to measure the ambient temperature at the cold junction, software can compute the appropriate compensation for the unwanted thermoelectric voltages. This is the reason for many signal conditioning accessories equipped with direct-reading temperature sensors, such as thermistors or semiconductor sensors, installed at the terminals. Software cold-junction compensation follows this process:

1. Measure the temperature of the reference junction and compute the equivalent thermocouple voltage for this Junction using standard thermocouple tables or polynomials.
2. Measure the output voltage (V_{MEAS}) and add – *not* subtract – the reference-junction voltage computed in step 1.
3. Convert the resulting voltage to temperature using standard thermocouple polynomials or look-up tables.

Many software packages, such as LabVIEW, LabWindows, and NI-DAQ, include routines that perform these temperature-to-voltage and voltage-to-temperature conversions for different types of thermocouples according to National Institute of Standards and Technology (NIST) standard reference tables. Sensitivity is another characteristic to consider with thermocouple measurements. Thermocouple outputs are very low level and change only 7–50 μV for every 1°C change in temperature. The sensitivity of the system can be increased with a low-noise, high-gain amplification of the signal. For example, a plug-in DAQ board with an analog input range of ±5 V, an amplifier gain of 100, and a 12-bit ADC has the following resolution:

$$\frac{10\,V}{(2^{12}) \times 100} = 24.4\,\mu V/bit$$

However, the same DAQ board with a signal conditioner amplifier gain of 1,000 has a resolution of 2.4 μV/bit, which corresponds to a fraction of a degree celsius. More importantly, an external signal conditioner can amplify the low-level thermocouple signal close to the source, which minimizes noise corruption. A high-level amplified signal suffers much less corruption from radiated noise in the environment and in the PC than a low-level unamplified signal.

4.5.5 RTD: Resistance Temperature Detector

Another popular temperature-sensing device is the *RTD*, which is known for its stability and accuracy over a wide temperature range. An RTD consists of a wire coil or deposited film of pure metal whose resistance increases with temperature. Although RTDs constructed with different metals and resistance are available, the most popular type of RTD is made of platinum and has a nominal resistance of 100 3/4 at 0°C.

Because an RTD is a passive resistive device, a current is passed through the RTD to produce a voltage that the DAQ board can measure. RTD have relatively low resistance (100 3/4) that changes only slightly with temperature (less than 0.43/4/°C), so special configurations are used to minimize errors

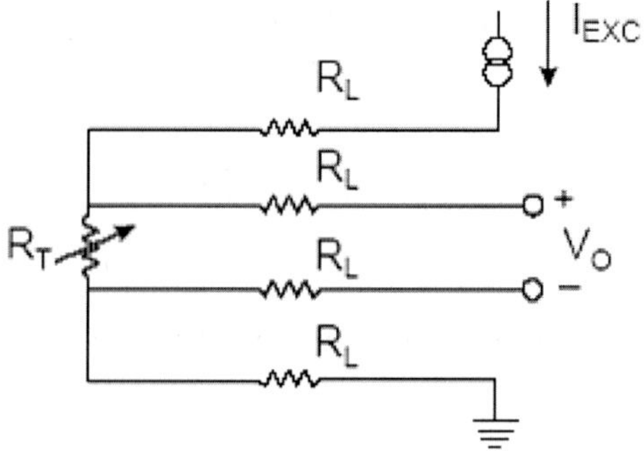

Fig. 4.24. A four-wire RTD

from lead wire resistance. For example, consider the measurement of a two-wire RTD as shown in Fig. 4.23. With this RTD, labeled R_T, the voltage drops caused by the excitation current, I_{EXC}, passing through the lead resistances, R_L, add to the measured voltage, V_O. For longer lead lengths, therefore, the four-wire RTD shown in Fig. 4.24 is a better choice. With a four-wire RTD, one pair of wires carries the excitation current through the RTD; the other pair senses the voltage across the RTD. Because only negligible current flows through the sensing wires, the lead resistance error is very small.

To keep costs down, RTDs are also available in three-wire configurations. The three-wire RTD is most effective in a Wheatstone bridge configuration. In this configuration, the lead resistances are located in opposite arms of the bridge, so their errors cancel each other.

4.5.6 Strain Gauges

The *strain gauge* is the most common device used in mechanical testing and measurements. The most common type is the bonded-resistance strain gauge, which consists of a grid of very fine foil or wire. The electrical resistance of the grid varies linearly with the strain applied to the device. When using a strain gauge, we bond the strain gauge to the device under test apply force and measure the strain by detecting changes in resistance. Strain gauges are also used in sensors that detect force or other derived quantities, such as acceleration, pressure, and vibration.

These sensors generally contain a pressure-sensitive diaphragm with strain gauges mounted to the diaphragm. Because strain measurement requires detecting relatively small changes in resistance, the Wheatstone bridge circuit is almost always used. The Wheatstone bridge circuit consists of four resistive elements with a voltage excitation supply applied to the ends of the bridge. Strain gauges can occupy one, two, or four arms of the bridge, with any remaining positions filled with fixed resistors. Figure 4.25 shows a configuration with a half-bridge strain gauge consisting of two strain elements, R_{G1} and R_{G2}, combined with two fixed resistors, R_1 and R_2.

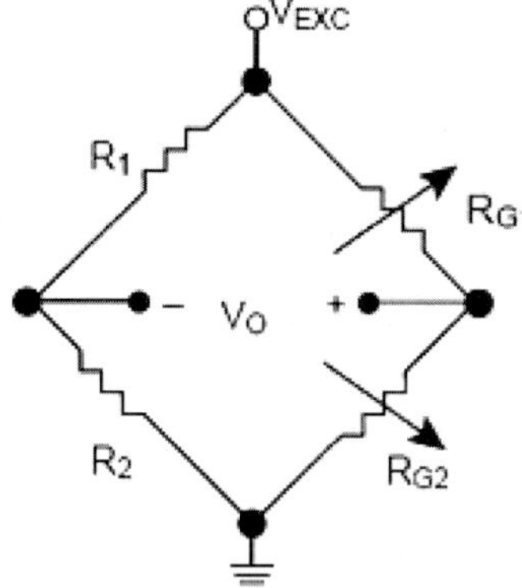

Fig. 4.25. Half-bridge strain gauge configuration

When the ratio of R_{G1} to R_{G2} equals the ratio of R_1 to R_2, the measured voltage V_O is 0 V. This condition is referred to as a *balanced bridge*. As strain is applied to the gauges, their resistance values change, causing a change in the voltage at V_O. Full-bridge and half-bridge strain gauges are designed for maximum sensitivity by arranging the strain gauge elements in opposing directions. For example, the half-bridge strain gauge includes an element R_{G1}, which is installed so that its resistance increases with positive strain, and an element R_{G2}, whose resistance decreases with positive strain.

The resulting V_O responds with sensitivity that is twice that of a quarter-bridge configuration. Some signal conditioning products have voltage excitation sources, as well as provisions for bridge-completion resistors. Bridge completion resistors should be very precise and stable. Because strain-gauge bridges are rarely perfectly balanced, some signal conditioners also null offsets, a process in which the resistance ratio of the unstrained bridge is adjusted to balance the bridge and remove any initial DC offset voltage. Alternatively, the user can measure this initial offset voltage and use this measurement in the conversion routines to compensate for the unbalanced initial condition.

Current Signals

Many sensors that are used in process control and monitoring applications output a current signal, usually 4–20 or 0–20 mA. Current signals are sometimes used because they are less sensitive to errors such as radiated noise and lead resistance voltage drops. Signal conditioners must convert this current signal to a voltage signal. To do this easily, the current signal is passed through a resistor, as shown in Fig. 4.26.

A DAQ system can be used to measure the voltage $V_O = I_S R$ that will be generated across the resistor, where I_S is the current and R is the resistance. Select the resistor value that has a usable range of voltages, and use a high-precision resistor with a low temperature coefficient. For example, the SCXI

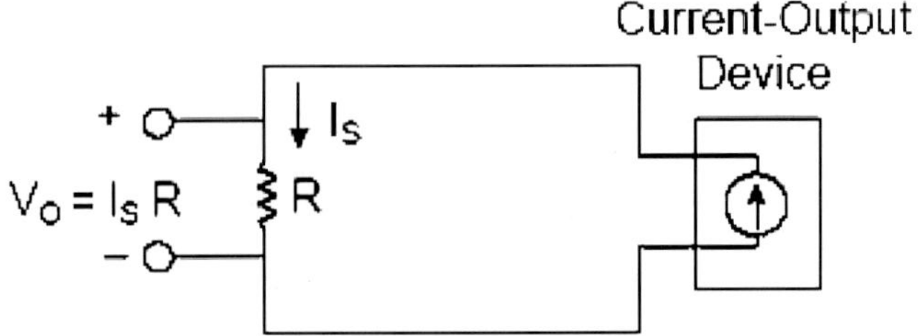

Fig. 4.26. Current signals and signal conditioning

Process-Current Resistor Kit consists of 249 3/4, 0.1%, 5 ppm/°C resistors. These resistor values will convert a 4–20 mA current loop into a voltage signal that varies from 0.996 to 4.98 V.

4.6 General Signal Conditioning Functions

Regardless of the types of sensors or transducers that are used, the proper signal conditioning equipment can improve the quality and performance of the system. Signal conditioning functions are useful for all types of signals, including amplification, filtering, and isolation.

4.6.1 Amplification

Unwanted noise can play havoc with the measurement accuracy of a PC-based DAQ system. Signal conditioning circuitry with amplification, which applies gain outside of the PC chassis and near the signal source, can increase measurement resolution and effectively reduce the effects of noise. An amplifier, whether located directly on the DAQ board or in external signal conditioners, can apply gain to the small signal before the ADC converts the signal to a digital value. Boosting the input signal uses as much of the ADC input range as possible. However, many transducers produce voltage output signals on the order of mill volts or even microvolt. Amplifying these low-level analog signals directly on the DAQ board also amplifies any noise picked up from the signal lead wires or from within the computer chassis. When the input signal is as small as microvolt, this noise can drown out the signal itself, leading to meaningless data.

A simple method for reducing the effects of system noise on the signal is to amplify the signal as close to the source as possible, which boosts the analog signal above the noise level before noise in the lead wires or computer chassis can corrupt the signal. For example, a J-type thermocouple outputs a very low-level voltage signal that varies by 50 μV/°C.

4.6.2 Filtering and Averaging

Filters are used to reject unwanted noise within a certain frequency range. Many systems will exhibit 60 Hz periodic noise components from sources such as power supplies or machinery. Lowpass filters on our signal conditioning circuitry can eliminate unwanted high-frequency components. However, be sure to select the filter bandwidth carefully so that the time response of the signals is not affected. Although many signal conditioners include lowpass noise filters to remove unwanted noise, an extra precaution is to use software averaging to remove additional noise.

Software averaging is a simple and effective technique of digitally filtering acquired readings; for every data point that is required, the DAQ system acquires and averages many voltage readings. For example, a common approach is to acquire 100 points and average those points for each measurement needed.

4.6.3 Isolation

Improper grounding of the DAQ system is the most common cause of measurement problems and damaged DAQ boards. Isolated signal conditioners can prevent most of these problems by passing the signal from its source to the measurement device without a galvanic or physical connection. Isolation breaks ground loops, rejects high common-mode voltages, and protects expensive DAQ instrumentation. Common methods for circuit isolation include the usage of optical, magnetic, or capacitive isolators. Magnetic and capacitive isolators modulate the signal to convert it from a voltage to a frequency.

The frequency can then be transmitted across a transformer or capacitor without a direct physical connection before being converted back to a voltage value. This voltage is referred to as *common-mode voltage*. If a single-ended measurement system is used, as shown in Fig. 4.27, the measured voltage includes the voltage from the desired signal, V_S, as well as this common-mode voltage from the additional ground currents in the system, V_G.

If a DAQ board with differential inputs is used, some of the common-mode voltage, typically up to 12 V can be rejected. However, larger ground potential differences, or ground loops, will damage unprotected DAQ devices.

4.6.4 Multiplexing

Signal conditioners equipped with signal multiplexers can effectively expand the input/output (I/O) capabilities of the plug-in DAQ board. The typical plug-in DAQ board has 8–16 analog inputs and 8–24 digital I/O lines. External multiplexers can increase the I/O capacity of a plug-in board to hundreds and even thousands of channels.

Analog input multiplexers use solid-state or relay switches to sequentially switch, or *scan*, multiple analog input signals onto a single channel of the DAQ board.

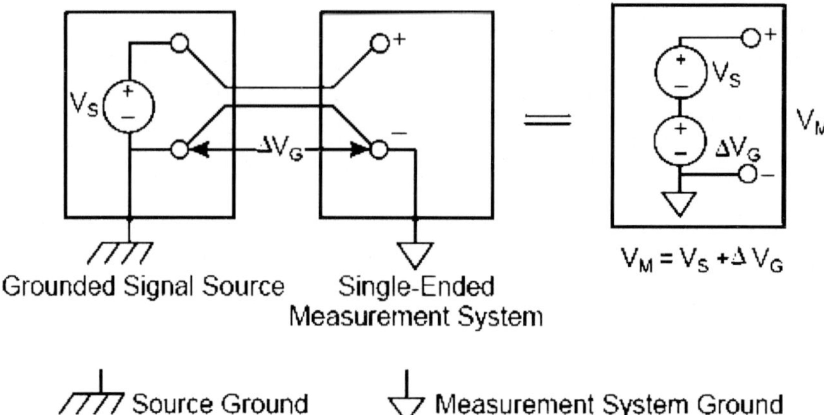

Fig. 4.27. Single-ended DAQ measurement system

4.6.5 Digital Signal Conditioning

Digital signals require signal conditioning peripherals to directly connect the signals with the DAQ board. But it is not a good practice to directly connect digital signals used in research and industrial environments to a DAQ board without some type of isolation because of the possibility of large voltage spikes or large common-mode voltages. Some signal conditioning modules and boards optically isolate the digital I/O signals to remove these problems. Digital I/O signals can control electromechanical or solid-state relays to switch loads such as solenoids, lights, motors, and so on. Solid-state relays can also be used to sense high-voltage field signals and convert them to digital signals.

4.6.6 Pulse Operation

A single pulse can be used to sample a variable capacitor, like a microcomputer read pulse, or a train of pulses can be used. This method can result in simpler electronics but will have higher noise.

4.6.7 Signal Conditioning Systems for PC-Based DAQ Systems

The signal conditioning functions discussed in this section is implemented in different types of signal conditioning products. These products cover a very wide range of price and capability. For example, the National Instruments SC-207X Series termination boards have a temperature sensor for use with thermocouples and a breadboard area with silk-screened component locations for easy addition of current measurement resistors, simple resistance–capacitance (RC) filters, and other signal conditioning circuitry.

The 5B Series of signal conditioning I/O modules is for specific types of transducers or signals. Up to 16 I/O modules can be installed in a backplane and directly connect the modules to a plug-in DAQ board. For external signal multiplexing, the AMUX-64T analog multiplexer board expands the analog input capability of the I/O multifunction board up to 256 channels. The AMUX-64T also includes a temperature sensor and silk screened component locations. The SCXI product line is a signal conditioning system that combines the expandability of multiplexing with the flexibility of modular signal conditioning.

4.6.8 Signal Conditioning with SCXI

Signal conditioning extension for instrumentation (SCXI) is a signal conditioning and instrumentation front end for plug-in DAQ boards in LabVIEW. An SCXI system consists of an SCXI chassis that houses one or more signal conditioning modules that multiplex, amplify, isolate, and condition both analog and digital signals. The SCXI system then passes the conditioned signals to a single plug-in DAQ board for acquisition directly into the PC. A variety of SCXI modules are available with different signal conditioning capabilities.

For example, the SCXI-1120 module is an eight-channel isolation amplifier module. Each of the input channels includes an isolation amplifier with gains of up to 2,000 and a low pass filter configurable for gains of 4 Hz or 10 kHz. The SCXI-1121 module is a four-channel isolation amplifier module that also has four channels of excitation. The user can configure each excitation channel for voltage or current. The module also includes half-bridge completion circuitry for strain-gauge measurements. Terminal blocks for the SCXI modules include temperature sensors for cold-junction compensation with thermocouples.

The signal conditioning connector (SCC) series is a modular portable signal conditioning system. SCC consists of single and dual-channel signal conditioning modules with built-in signal connectors. For example, the SCC-TC02 provides amplification, filtering, cold-junction compensation, and a convenient mini plug connector for one thermocouple input. Any combination of SCC modules can be installed onto an SCC carrier or backplane, such as the SC-2345. The SC-2345 holds up to 18 SCC modules and cables directly to an E Series DAQ board or module.

Signal conditioning is an important component of a complete PC-based DAQ system. Signal conditioning has many features that are used to connect sensors such as thermocouples, RTDs, strain gauges, and current-output devices to PC-based DAQ boards. Signal conditioning improves the accuracy, effectiveness and safety of the measurements, for any type of sensor used in the measurement because of capabilities such as amplification, isolation, and filtering. The National Instruments SCXI product line can supply the signal conditioning and instrumentation front end required for the PC-based DAQ systems.

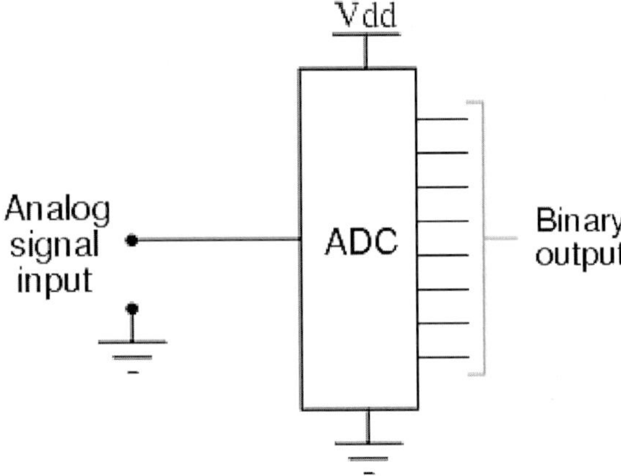

Fig. 4.28. Analog-to-digital converter

4.7 Analog-to-Digital Control

Connecting digital circuitry to sensor devices is simple if the sensor devices are inherently digital in nature. Switches, relays, and encoders are easily interfaced with gate circuits due to the on/off nature of their signals. However, when analog devices are involved, interfacing becomes much more complex. Some mechanism is needed to electronically translate analog signals into digital (binary) quantities, and visa-versa. An analog-to-digital converter, or ADC, performs the former task while a digital-to-analog converter, or DAC, performs the latter. An ADC inputs an analog electrical signal such as voltage or current and outputs a binary number. In block diagram form, it can be represented as shown in Fig. 4.28.

4.7.1 Understanding Integrating ADCs

Integrating ADCs provide high resolution A/D conversions, with good noise rejection. These ADCs are ideal for digitizing low bandwidth signals, and are used in applications such as digital multimeters and panel meters. They often include LCD or LED drivers and can be used stand alone without a microcontroller host. The following section explains how integrating ADCs work. Discussions include single-, dual- and multislope conversions. Also, an in-depth analysis of the integrating architecture will be discussed.

Finally a comparison against other ADC architectures will aid in the understanding and selection of integrating ADCs. Integrating analog-to-digital converters (ADCs) provide high resolution and can provide good line frequency and noise rejection. Having started with the ubiquitous 7106, these converters have been around for quite some time. The integrating architecture provides a novel and straightforward approach to converting a low bandwidth analog signal into its digital representation. These types of converters often include built-in drivers for LCD or LED displays and are found in many portable instrument applications, including digital panel meters and digital multimeters.

Single-Slope ADC Architecture

The simplest form of an integrating ADC uses single-slope architecture (Figs. 4.31a, b). Here, an unknown input voltage is integrated and the value is compared against a known reference value. The time it takes for the integrator to trip the comparator is proportional to the unknown voltage (V_{INT}/V_{IN}). In this case, the known reference voltage must be stable and accurate to guarantee the accuracy of the measurement.

One drawback to this approach is that the accuracy is also dependent on the tolerances of the integrator's R and C values. Thus in a production environment, slight differences in each component's value change the conversion result and make measurement repeatability quite difficult to attain. To overcome this sensitivity to the component values, the dual-slope integrating architecture is used.

Dual-Slope ADC Architecture

A dual-slope ADC (DS-ADC) integrates an unknown input voltage (V_{IN}) for a fixed amount of time (T_{INT}), then "disintegrates" (T_{DEINT}) using a known reference voltage (V_{REF}) for a variable amount of time (Fig. 4.29).

The key advantage of this architecture over the single-slope is that the final conversion result is insensitive to errors in the component values. That is, any error introduced by a component value during the integrate cycle will be canceled out during the de-integrate phase. In equation form:

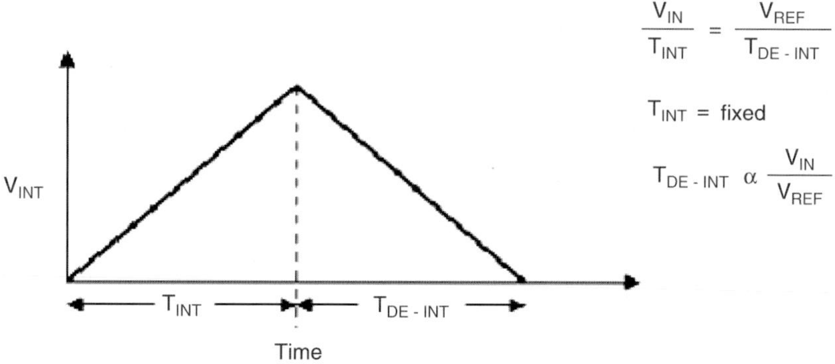

Fig. 4.29. Dual slope integration

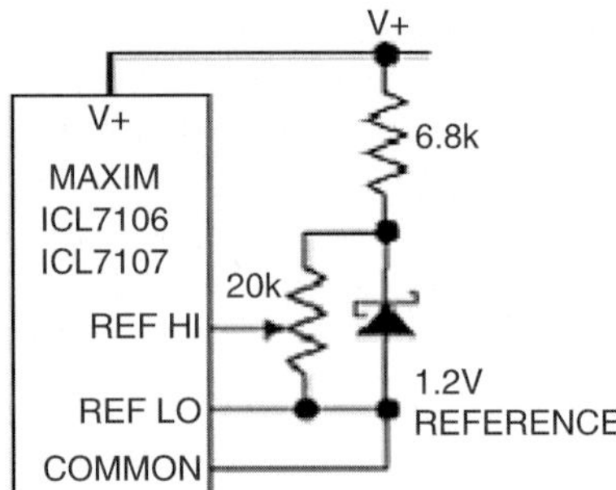

Fig. 4.30. Dual Slope Converter

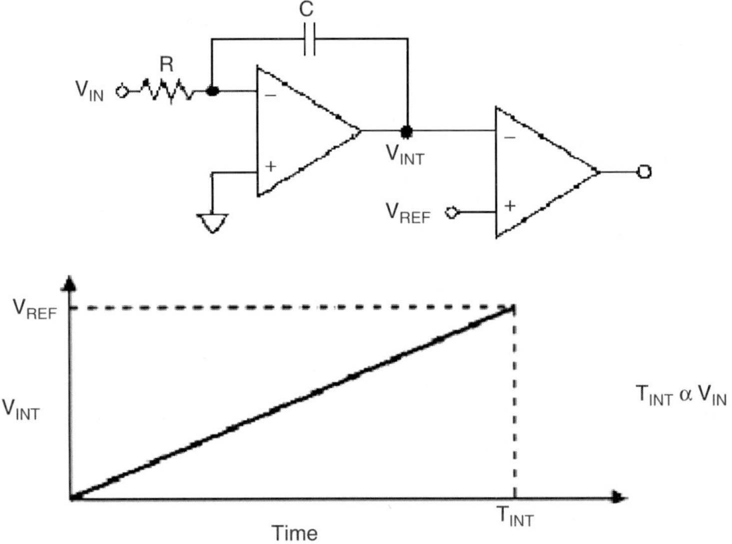

Fig. 4.31. (a) Single-slope ADC circuit (b) Response of single slope ADC

$$V_{IN}{}^{*}T_{INT} = V_{REF}{}^{*}T_{DEINT} \text{ or}$$
$$T_{DEINT} = T_{INT}(V_{IN}/V_{REF})$$

From this equation, it is found that the de-integrate time is proportional to the ratio of V_{IN}/V_{REF}. A complete circuit diagram of a dual-slope converter is shown in Fig. 4.30.

As an example, to obtain 10-bit resolution, integration is performed for 1024 (210) clock cycles, then disintegrated for 1024 clock cycles (giving a maximum conversion of two 210 cycles). For more resolution, increase the number of clock cycles. This tradeoff between conversion time and resolution is inherent in this implementation. It is possible to speed up the conversion time for a given resolution with moderate circuit changes. Unfortunately, all improvements shift some of the accuracy to matching, external components, charge injection, etc. In other words, all speed-up techniques have larger error budgets. Even in the simple converter in Fig. 4.1, there are many potential error sources to consider such as power supply rejection (PSR), common-mode rejection (CMR), finite gain, over-voltage concerns, integrator saturation, comparator speed, comparator oscillation, "rollover," dielectric absorption, capacitor leakage current, parasitic capacitance, charge injection, etc.

Multislope Integrating ADC

The normal limit for resolution of the dual-slope architecture is based on the speed of the error comparator. For a 20-bit converter (approximately 1 part in a million) and a 1 MHz clock, the conversion time would be about 2 s. The ramp rate seen by the error comparator is about 2 V/106 divided by 1 µs. This is about $2 \mu V\,\mu s^{-1}$. With such a small slew rate, the error comparator would allow the integrator to go well beyond its trip point by a considerable amount. This overshoot (measured at the integrator output) is called the "residue." This brute force technique is not likely to achieve a 20-bit converter.

Versus Other ADC Architectures

This section focuses on the integrating ADC versus an successive approximate register (SAR) and sigma–delta ADC. The flash and pipeline ADC architectures will be ignored since they rarely (if ever) compete against the slower speed integrating architecture.

Versus Successive Approximate Register ADC

Both the SAR and integrating architectures work well with low bandwidth signals. The SAR ADC has a much wider bandwidth range, as they can easily convert signals at speeds in the low MHz range, while the integrating architecture is limited, about 100 samples s^{-1}. Both architectures have low power consumption. Since SAR ADCs can be shut down between conversions, the effective power consumption is similar to the integrating ADC (to the first order). The biggest difference between the two converters is the common mode rejection and the number of external components required. Because the user sets the integration time, unwanted frequencies, such as 50 Hz or 60 Hz can effectively be notched out.

In addition, in SAR ADC since integration basically is a method of averaging, the integrating ADC typically will have better noise performance. A SAR ADC has code-edge noise and spurious noise that is converted will have a more adverse affect with the SAR ADC than with the integrating ADC. The integrating ADC easily converts low-level signals. Since the integrator ramp is set by the value of the integrating resistor, it is fairly easy to match the input signal range to the ADC. Most SARs expect a large signal at the ADC input.

Thus for small (i.e., mV) signals, front-end signal conditioning circuitry is required. The integrating ADC needs more external components than the SAR. A SAR typically needs a couple bypass capacitors. The integrating ADC requires a good integrating and reference capacitors and also a low-drift integrating resistor. In addition, the reference voltage is often a non-standard value (like 100 or 409.6 mV) so a reference voltage divider circuit is often used.

Versus Sigma–Delta ADC

The sigma–delta ADC uses over sampling to obtain very high resolution. It also allows input bandwidths in the low MHz range. Like the integrating ADC, this architecture can have excellent line rejection. It also provides a very low-power solution and it allows low level signals to be converted. Unlike the integrating ADC, the sigma–delta does not require any external components. In addition, it requires no trimming or calibration due to its digital architecture. Due to the over sampling nature and the fact that the sigma delta includes a digital filter, an anti-aliasing filter is not often required on the front end.

Sigma–delta converters typically are available in 16-bit to 24-bit resolutions while integrating ADCs target the 12-bit to 16-bit range. Due to its straightforward architecture and its maturity, integrating ADCs are fairly inexpensive especially at the 12-bit level. However, at 16-bits, the sigma–delta also provides a low cost solution.

4.7.2 Understanding SAR ADC

Successive-approximation-register (SAR) analog-to-digital converters (ADCs) represent the majority of the ADC market for medium to high resolution ADCs. SAR ADCs provide up to 5 Msps (mega samples per second) sampling rate with resolution from 8 to 18 bits. The SAR architecture allows for high performance, low power ADCs to be packaged in small form factors for today's demanding applications. It also provides an explanation for the heart of the SAR ADC, the capacitive DAC and also the high-speed comparator.

Finally, the article will contrast the SAR architecture against pipeline, flash ADCs. SAR ADCs are frequently the architecture of choice for medium-to-high-resolution applications with sample rates under 5 Msps. SAR ADCs most commonly range in resolution from 8 to 16 bits and provide low power consumption as well as a small form factor. This combination makes them ideal for a wide variety of applications, such as portable/battery-powered instruments, pen digitizers, industrial controls, and data/signal acquisition.

As the name implies, the SAR ADC basically implements a binary search algorithm. Therefore, while the internal circuitry may be running at several

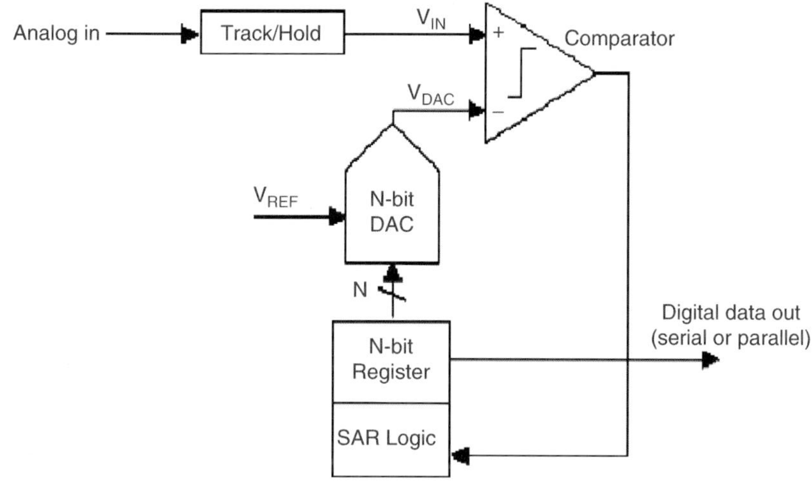

Fig. 4.32. Simplified N-bit SAR ADC architecture

megahertz (MHz), the ADC sample rate is a fraction of that number due to the successive-approximation algorithm.

Architecture

Although there are many variations in the implementation of an SAR ADC, the basic architecture is quite simple (Fig. 4.32). The analog input voltage (V_{IN}) is held on a track/hold. To implement the binary search algorithm, the N-bit register is first set to mid scale (that is, 100. . . . 00, where the MSB is set to "1"). This forces the DAC output (V_{DAC}) to be $V_{REF}/2$, where V_{REF} is the reference voltage provided to the ADC. A comparison is then performed to determine if V_{IN} is less than or greater than V_{DAC}.

If V_{IN} is greater than V_{DAC}, the comparator output is at logic high or "1" and the MSB of the N-bit register remains at "1." Conversely, if V_{IN} is less than V_{DAC}, the comparator output is logic low and the MSB of the register is cleared to logic "0." The SAR control logic then moves to the next bit down, forces that bit high, and does another comparison. The sequence continues all the way down to the LSB. Once this is done, the conversion is complete, and the N-bit digital word is available in the register.

The y-axis (and the bold line in Fig. 4.33) represents the DAC output voltage. In the example, the first comparison shows that $V_{IN} < V_{DAC}$. Thus, bit 3 is set to "0." The DAC is then set to 01002 and the second comparison is performed. As $V_{IN} > V_{DAC}$, bit 2 remains at "1." The DAC is then set to 01102, and the third comparison is performed. Bit 1 is set to "0," and the DAC is then set to 01012 for the final comparison. Finally, bit 0 remains at "1" because $V_{IN} > V_{DAC}$.

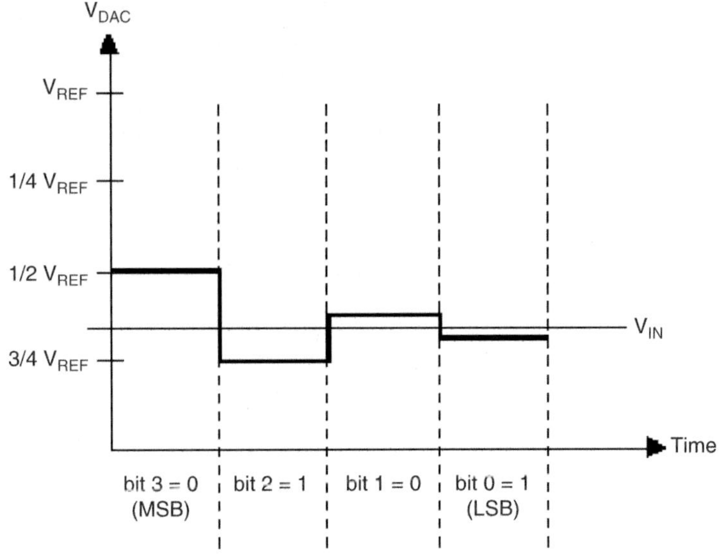

Fig. 4.33. SAR operation

Four comparison periods are required for a 4-bit ADC. Generally speaking, an N-bit SAR ADC will require N comparison periods and will not be ready for the next conversion until the current one is complete. Thus, these types of ADCs are power and space-efficient. Some of the smallest ADCs available on the market are based on the SAR architecture.

The MAX1115-MAX1118 series of 8-bit ADCs as well as their higher-resolution counterparts, the MAX1086 and the MAX1286 (10 and 12 bits, respectively), fit in tiny SOT23 packages measuring 3 mm by 3 mm. Another feature of SAR ADCs is that power dissipation scales with the sample rate, unlike flash or pipelined ADCs, which usually have constant power dissipation versus sample rate. This is especially useful in low-power applications or applications where the data acquisition is not continuous as in the case of PDA Digitizers.

SAR ADC Versus Other ADC Architectures

This section compares the SAR ADC with pipelined ADC, Flash ADC and sigma–delta ADC.

SAR ADC Versus Pipelined Architecture

A pipelined ADC employs a parallel structure in which each stage works on one to a few bits (of successive samples) concurrently. The inherent parallelism increases throughput, but at the expense of power consumption and latency.

Latency in this case is defined as the difference between the time an analog sample is acquired by the ADC and the time when the digital data is available at the output. For instance, a five-stage pipelined ADC will have at least five clock cycles of latency, whereas an SAR has only one clock cycle of latency. Note that the latency definition applies only to the throughput of the ADC, not the internal clock of an SAR, which runs at many times the frequency of the throughput.

Pipelined ADCs frequently have digital error correction logic to reduce the accuracy requirement of the flash ADCs (that is, comparators) in each pipeline stage. On the other hand, SAR ADC requires the comparator to be as accurate as the overall system. A pipelined ADC generally takes up significantly more silicon area than an equivalent SAR. Like an SAR, a pipelined ADC with more than 12 bits of accuracy usually requires some form of trimming or calibration.

SAR ADC Versus Flash

A flash ADC is made up of a large bank of comparators, each consisting of wideband, low-gain preamp(s) followed by a latch. The preamps only have to provide gain but do not need to be linear or accurate; meaning, only the comparators' trip points have to be accurate. As a result, a flash ADC is the fastest architecture available. The primary trade-off for speed is the significantly lower power consumption and the smaller form factor. Although extremely fast 8-bit flash ADCs (or their folding/interpolation variants) exist with sampling rates as high as $1.5\,\mathrm{GS\,s^{-1}}$ (for example, the MAX104/MAX106/MAX108), it is much harder to find a 10-bit flash, while 12-bit (and above) flash ADCs are not commercially viable products.

This is simply because in a flash the number of comparators goes up by a factor of 2 for every extra bit of resolution, and, at the same time, each comparator has to be twice as accurate. In an SAR ADC, however, the increased resolution requires more accurate components, yet the complexity does not increase exponentially. Of course, SAR ADCs are not capable of speeds anywhere close to those of flash ADCs.

SAR ADC Versus Sigma–Delta Converter

Traditional over sampling/sigma–delta-type converters commonly used in digital audio applications have limited bandwidths of about 22 kHz. Recently, some high-bandwidth sigma–delta-type converters have reached bandwidths of 1–2 MHz with 12–16 bits of resolution. These are usually very-high-order sigma–delta modulators (for example, fourth-order or higher) incorporating a multibit ADC and multibit feedback DAC. Sigma–delta converters have the innate advantage of requiring no special trimming or calibration, even to attain 16–18 bits of resolution. They also do not require anti-alias filters with steep roll-offs at the analog inputs, because the sampling rate is much higher than the effective bandwidth; the backend digital filters take care of this.

The over sampling nature of the sigma–delta converter may also tend to "average out" any system noise at the analog inputs. However, sigma–delta converters trade speed for resolution. The need to sample many times (at least 16 times and often more) to produce one final sample indicates that the internal analog components in the sigma–delta modulator operate much faster than the final data rate. The digital decimation filter is also a challenge to design and consumes a lot of silicon area. The fastest high-resolution sigma–delta converters are not expected to have significantly higher bandwidth than a few MHz in the near future.

In summary, the primary advantages of SAR ADCs are low power consumption, high resolution and accuracy, and a small form factor. Because of these benefits, SAR ADCs can often be integrated with other larger functions. The main limitations of the SAR architecture are the lower sampling rates and the requirements for the building blocks (such as the DAC and the comparator) to be as accurate as the overall system.

4.7.3 Understanding Flash ADCs

Flash ADCs, also known as parallel ADCs, are the fastest way to convert an analog signal to a digital signal. Flash ADCs are ideal for applications requiring very large band width, however, they typically consume more power than other ADC architectures and are generally limited to 8-bits resolution.

Flash ADCs are made by cascading the high-speed comparators. Each comparator represents one least significant bit (LSB), and the output code can be determined in one compare cycle. Flash analog-to-digital converters, also known as parallel ADCS, are the fastest way to convert an analog signal to a digital signal. They are suitable for applications requiring very large bandwidths. However, flash converters consume a lot of power, have relatively low resolution, and can be quite expensive. This limits them to high frequency applications. Examples include data acquisition, satellite communication, radar processing, sampling oscilloscopes, and high-density disk drives.

Architecture

Figure 4.34 shows a typical flash ADC block diagram. For an "N" bit converter, the circuit employs 2N-1 comparators. A resistive divider with 2N resistors provides the reference voltage. The reference voltage for each comparator is one LSB greater than the reference voltage for the comparator immediately below it. Each comparator produces a "1" when its analog input voltage is higher than the reference voltage applied to it. Otherwise, the comparator output is "0." Thus, if the analog input is between V_{x4} and V_{x5}, comparators x_1 to x_4 produce "1"s and the remaining comparators produce "0"s.

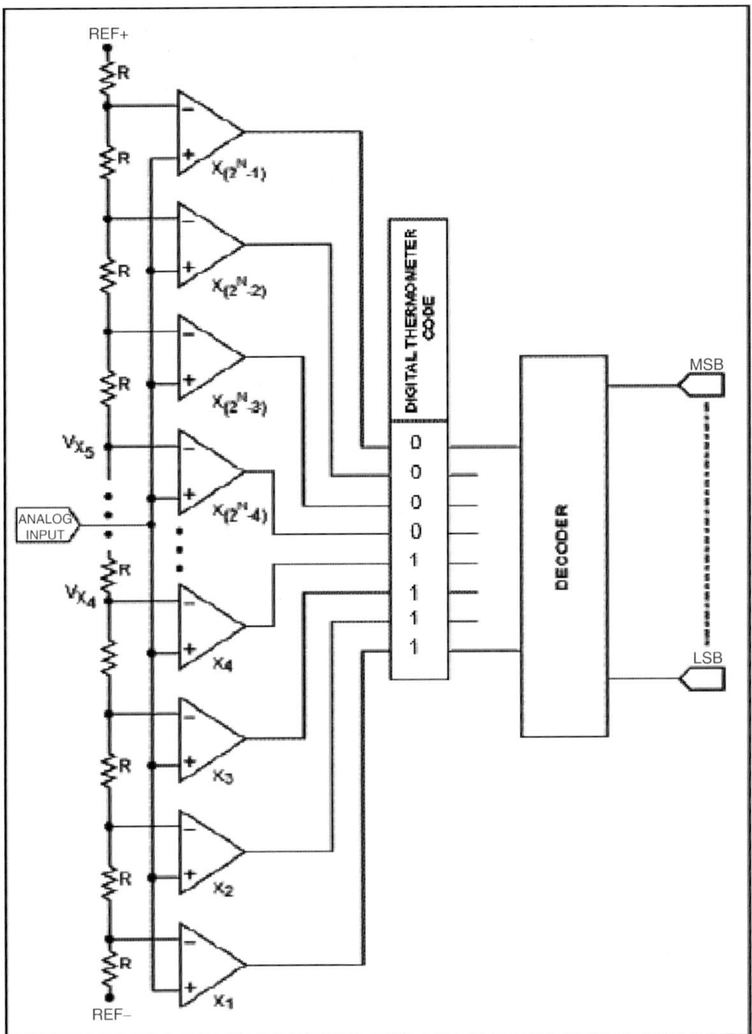

Fig. 4.34. Flash ADC architecture

The point where the code changes from ones to zeros is the point where the input signal becomes smaller than the respective comparator reference voltage levels. This is known as thermometer code encoding, so named because it is similar to a mercury thermometer, where the mercury column always rises to the appropriate temperature and no mercury is present above that temperature. The thermometer code is then decoded to the appropriate digital output code. The comparators are typically a cascade of wideband low gain stages. They are low gain because at high frequencies it is difficult to obtain both wide bandwidth and high gain. They are designed for low voltage offset, such that the input offset of each comparator is smaller than an LSB of the

ADC. Otherwise, the comparator's offset could falsely trip the comparator, resulting in a digital output code that does not represent a thermometer code. A regenerative latch at each comparator output stores the result. The latch has positive feedback, so that the end state is forced to either a "1" or a "0."

Sparkle Codes

Normally, the comparator outputs will be a thermometer code, such as 00011111. Errors may cause an output like 00010111 (i.e., there is a spurious zero in the result). This out of sequence "0" is called a sparkle. This may be caused by imperfect input settling or comparator timing mismatch. The magnitude of the error can be quite large. Modern converters like the MAX104 employ an input track-and-hold in front of the ADC along with an encoding technique that suppresses sparkle codes.

Meta Stability

When a digital output of a comparator is ambiguous (neither a one nor a zero), the output is defined as meta stable. Meta stability can be reduced by allowing more time for regeneration. Gray-code encoding can also greatly improve meta stability. Gray-code encoding allows only one bit in the output to change at a time. The comparator outputs are first converted to gray-code encoding and then later decoded to binary if desired. Another problem occurs when a meta stable output drives two distinct circuits. It is possible for one circuit to declare the input a "1" while the other circuit thinks it's a "0." This can create major errors. To avoid this, only one circuit should sense a potential output.

Input Signal Frequency Dependence

When the input signal changes before all the comparators have completed their decision, the ADC performance is adversely impacted. The most serious impact is a drop-off in signal-to-noise ratio plus distortion (SINAD) as the frequency of the analog input frequency increases. Measuring spurious free dynamic range (SFDR) is another good way to observe converter performance. The "effective bits" achieved is a function of input frequency. This can be improved by adding a track-and-hold (T/H) circuit in front of the ADC. This allows dramatic improvement, especially when input frequencies approach the Nyquist frequency, as shown in Fig. 4.35 (taken from the MAX104 data sheet). Parts without the track-and-hold show a significant drop-off in SFDR.

Clock Jitter

Signal-to-noise ratio (SNR) is degraded when there is jitter in the sampling clock. This becomes noticeable for high analog input frequencies. To achieve

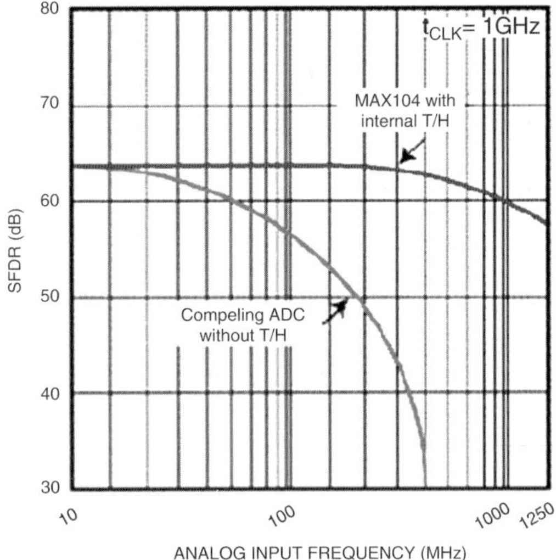

Fig. 4.35. Spurious free dynamic range as a function of input frequency

accurate results, it is critical to provide the ADC with a low jitter, sampling clock source.

Architecture Tradeoffs

ADCs can be implemented by employing a variety of architectures. The principal tradeoffs between these alternatives are:

- *The time it takes to complete a conversion (conversion time).* For flash converters, the conversion time does not change materially with increased resolution. The conversion time for SAR or pipelined converters increases approximately linearly with an increase in resolution (Fig. 4.34a). For integrating ADCs, the conversion time doubles with every bit increase in resolution.
- *Components matching the requirements in the circuit.* Flash ADC component matching typically limits resolution to around 8-bits. Calibration and trimming are sometimes used to improve the matching available on chip. Component matching requirements double with every bit increase in resolution. This applies to flash, successive approximation or pipelined converters, but not integrating converters. For integrating converters, component matching does not materially increase with an increase in resolution (Fig. 4.34b).

Die size, cost, and power. For flash converters, every bit increase in resolution almost doubles the size of the ADC core circuitry. The power also doubles. In contrast, an SAR, Pipelined, or sigma delta ADC die size will increase linearly with an increase in resolution, and an integrating converter core die size will

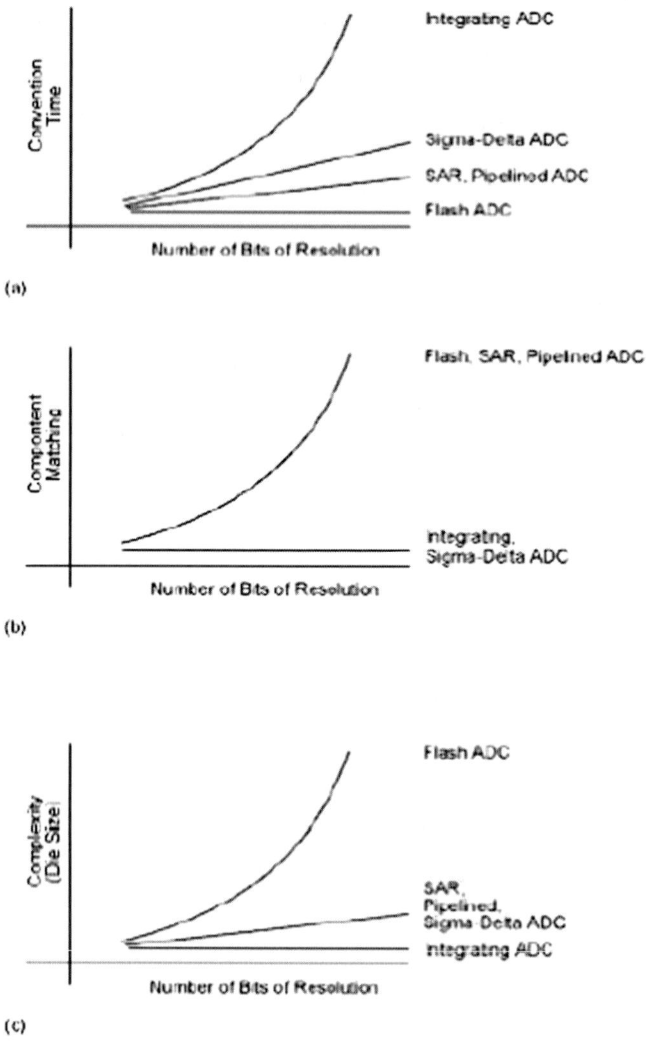

Fig. 4.36. Architecture tradeoffs

not materially change with an increase in resolution (Fig. 4.36). An increase in die size increases cost.

Flash Versus Successive Approximation Register ADCs

In an SAR converter, the bits are decided by a single high-speed, high-accuracy comparator one bit at a time (from the MSB down to the LSB), by comparing the analog input with a DAC whose output is updated by previously decided bits and thus successively approximates the analog input. This serial nature of the SAR limits its speed to no more than a few Msps, while flash ADCs exceed gigasample per second (Gsps) conversion rates. SAR converters are available in resolutions up to 16-bits. An example of such a device is the MAX1132. Flash ADCs are typically limited to around 8-bits. The slower speed also allows the SAR ADC to be much lower in power.

For example, the MAX1106, an 8-bit SAR converter, uses 100 μA at 3.3 V with a conversion rate of 25 ksps. The MAX104 dissipates 5.25 W. This is about 16,000 times higher in terms of power consumption compared to the MAX1106, but also 40,000 times faster in terms of its maximum sampling rate. The SAR architecture is also less expensive. The MAX1106 at 1 k volumes sells at approximately $1.51, while the MAX104 sells at roughly $398. Package sizes are larger for flash converters. In addition to a larger die size requiring a larger package, the package needs to dissipate a lot of power and needs many pins for power and ground signal integrity. The package size of the MAX104 is more than 50 times larger than the MAX1106.

Flash Versus Pipelined ADCs

A pipelined ADC employs a parallel structure in which each stage works on one to a few bits of successive samples concurrently. This improves speed at the expense of power and latency. However, each pipelined stage is much slower than a flash section. The pipelined ADC requires accurate amplification in the DACs and inter-stage amplifiers, and these stages have to settle to the desired linearity level. By contrast, in a flash ADC, the comparator only needs to be low offset and be able to resolve its inputs to a digital level (i.e., there is no linear settling time involved). However, some flash converters require pre amplifiers to drive the comparators. Gain linearity needs to be carefully specified. Pipelined converters are capable of conversion speeds of around 100 Msps at 8–14-bit resolutions.

An example of a pipelined converter is the MAX1449, a 105 MHz, 10-bit ADC. For a given resolution, pipelined ADCs are around 10 times slower compared to flash converters of similar resolution. Pipelined converters are possibly the optimal architecture for ADCs that need to sample at rates up to around 100 Msps with resolution at 10-bits and above. At resolutions of up to 10-bits, and conversion rates above a few hundred Msps, flash ADCs dominate. Interestingly, there are some situations where flash ADCs are hidden inside a converter employing architecture to increase its speed. This is the case, for example, in the MAX1200; a 16-bit pipelined ADC that includes an internal 5-bit flash ADC.

Flash Versus Integrating ADCs

Single, dual, and multislope ADCs can achieve high resolutions of 16-bits or more are relatively inexpensive and dissipate materially less power. These devices support very low conversion rates, typically less than a few hundred samples per second. Most applications are for monitoring DC signals in the instrumentation and industrial markets. This architecture competes with sigma–delta converters.

Flash Versus Sigma–Delta ADCs

Flash ADCs do not compete with this architecture because currently the achievable conversion rates differ by up to two orders of magnitude. The sigma–delta architecture is suitable for applications with much lower bandwidth, typically less than 1 MHz, with resolutions in the 12–16-bit range. These converters are capable of the highest resolution possible in ADCs. They require simpler anti-alias filters (if needed) to band limit the signal prior to conversion. They trade speed for resolution by over sampling, followed by filtering to reduce noise. However, these devices are not always efficient for multi-channel applications.

This architecture can be implemented by using sampled data filters (also known as modulators) or continuous time filters. For higher frequency conversion rates the continuous time architecture is potentially capable of reaching conversion rates in the hundreds of Msps range with low resolution of 6–8-bits. This approach is still in the early research and development stage and offers competition to flash alternatives in the lower conversion rate range. Another interesting use of a flash ADC is the presence of a building block inside a sigma–delta circuit to increase the conversion rate of the ADC.

Subranging ADCs

When higher resolution converters or smaller die size and power for a given resolution are needed, multistage conversion is employed. This architecture is known as a subranging converter. This is also sometimes referred to as a multistep or half-flash converter. This approach combines ideas from successive approximation and flash architectures. Subranging ADCs reduce the number of bits to be converted into smaller groups, which are then run through a lower resolution flash converter. This approach reduces the number of comparators and reduces the logic complexity, compared to a flash converter (Fig. 4.37). The tradeoff results in slower conversion speed compared to flash.

The MAX153 is an 8-bit, 1 Msps ADC implemented with a sub-ranging architecture. This circuit employs a two-step technique. Here a first conversion is completed with a 4-bit converter. A residue is created, where the result of the 4-bit conversion is converted back to an analog signal (with an 8-bit accurate DAC) and subtracted from the input signal. This residue is again converted by the 4-bit ADC and the results of the first and second pass are combined to provide the 8-bit digital output.

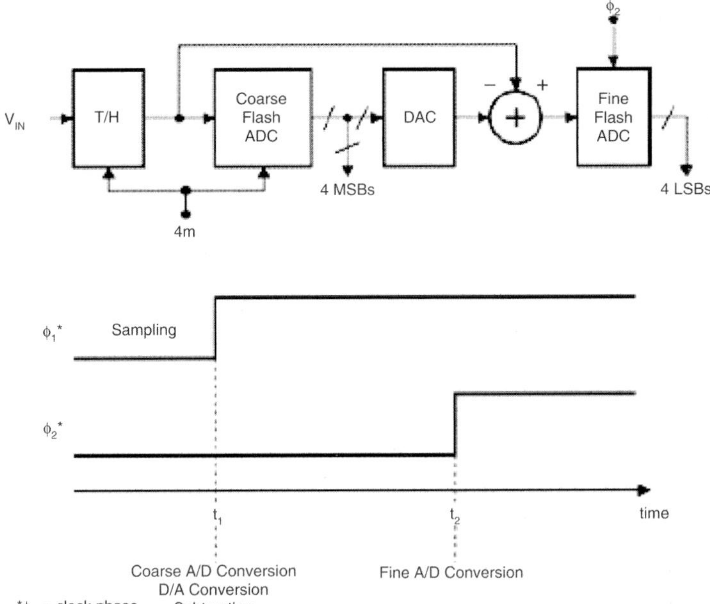

Fig. 4.37. Subranging ADC architecture

4.7.4 Understanding Pipelined ADCs

This section explains the architecture and operation of pipelined ADCs. It discusses key performance characteristics of pipeline ADCs such as architecture, latency, digital error correction, component accuracy, and digital calibration. The pipelined ADC has become the most popular ADC architecture for sampling rates from a few mega samples per second (Msps) up to 100 Msps, with resolutions from 8 bits at the faster sample rates up to 16 bits at the lower rates.

These kind of resolutions and sampling rates cover a wide range of applications, including CCD imaging, ultrasonic medical imaging, digital receiver, base station, digital video (for example, HDTV), xDSL, cable modem, and fast Ethernet. Lower-sampling-rate applications are still the domain of the successive approximation register (SAR) and integrating architectures (and more recently over sampling/sigma–delta ADCs), whereas the highest sampling rates (a few hundred $\mathrm{MS\,s^{-1}}$ or higher) are still obtained using flash ADCs and their variants. However, it is safe to say that pipelined ADCs of various forms have improved greatly in speed, resolution, dynamic performance, and low power in recent years.

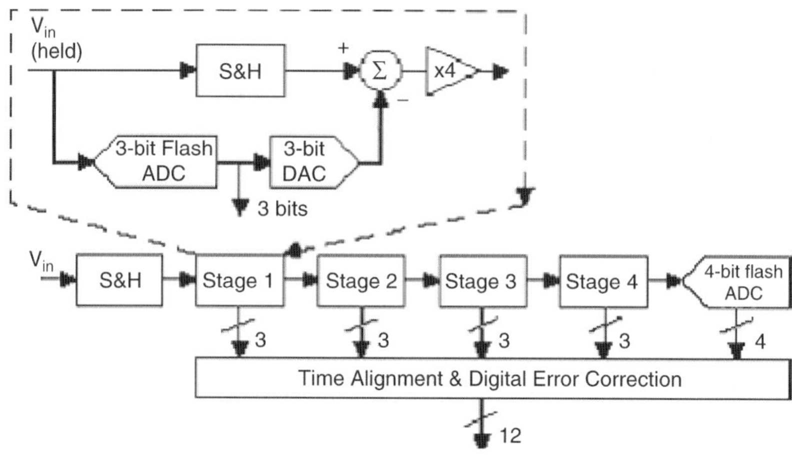

Fig. 4.38. Pipelined ADC architecture with four 3-bit stages

Pipelined ADC Architecture

Figure 4.38 shows a possible block diagram of a 12-bit pipelined ADC. Here, the analog input V_{IN} is first sampled and held steady by a sample-and-hold (S&H), while the flash ADC in stage one quantizes it to 3 bits. The 3-bit output is then fed to a 3-bit DAC (accurate to about 12 bits), and the analog output is subtracted from the input. This "residue" is then gained up by a factor of 4 and fed to the next stage (stage two). This gained-up residue continues through the pipeline, providing 3 bits per stage until it reaches the 4-bit flash ADC, which resolves the last 4LSB bits. Because the bits from each stage are determined at different points in time, all the bits corresponding to the same sample are time-aligned with shift registers before being fed to the digital-error-correction logic. Note that as soon as a certain stage finishes processing a sample, determining the bits and passing the residue to the next stage, it can start processing the next sample due to the sample-and-hold embedded within each stage. This pipelining action accounts for the high throughput.

Latency

Because each sample has to propagate through the entire pipeline before all its associated bits are available for combining in the digital-error-correction logic, data latency is associated with pipelined ADCs. In the example in Fig. 4.39, this latency is about three cycles.

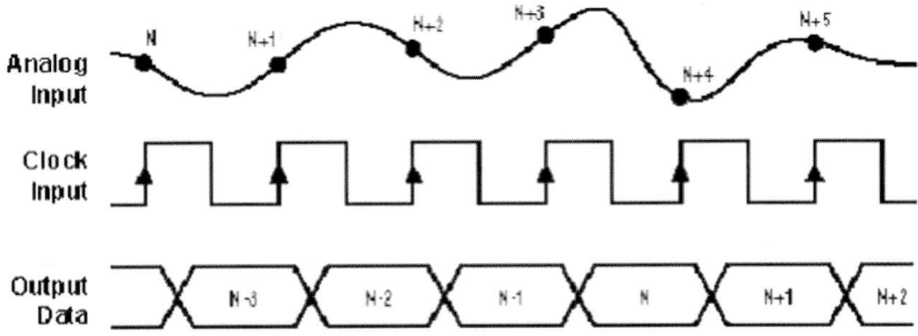

Fig. 4.39. Data latency in a pipelined ADC

Digital Error Correction

Most modern pipelined ADCs employ a technique called "digital error correction" to greatly reduce the accuracy requirement of the flash ADCs (and thus the individual comparators). In Fig. 4.39, the 3-bit residue at the summation-node output has a dynamic range one-eighth that of the original stage-one input (VIN), yet the subsequent gain is only 4. Therefore, the input to stage two occupies only half the range of the 3-bit ADC in stage two (that is, when there is no error in the first 3-bit conversion in stage one).

If one of the comparators in the first 3-bit flash ADC has a significant offset, when an analog input close to the trip point of this comparator is applied, an incorrect 3-bit code and thus an incorrect 3-bit DAC output would result, producing a different residue. However, it can be proven that, as long as this gained-up residue doesn't over-range the subsequent 3-bit ADC, the LSB code generated by the remaining pipeline when added to the incorrect 3-bit MSB code will give the correct ADC output code. The implication is that none of the flash ADCs in Fig. 4.39 has to be as accurate as the entire ADC. In fact, the 3-bit flash ADCs in stages one through four require only about 4 bits of accuracy. The digital error correction will not correct for errors made in the final 4-bit flash conversion. However, any error made here is suppressed by the large cumulative gain preceding the 4-bit flash, requiring the final stage to be only more than 4-bits accurate. In Fig. 4.39, although each stage generates 3 raw bits, because the inter-stage gain is only 4, each stage (stages one to four) effectively resolves only 2 bits. The extra bit is simply to reduce the size of the residue by one half, allowing extra range in the next 3-bit ADC for digital error correction, as mentioned above. This is called "1-bit overlap" between adjacent stages. The effective number of bits of the entire ADC is therefore $2 + 2 + 2 + 2 + 4 = 12$ bits.

Component Accuracy

Digital error correction does not correct gain or linearity errors in the individual DAC and gain amplifiers. In particular, the front-end Sample and Holds and DAC need about 12-bit accuracy, whereas the components in subsequent stages require less accuracy (for example, 10-bit for stage two, 8-bit for stage three, and so forth) because their error terms are divided down by the preceding interstage gain(s). This fact is often exploited to further save power by making the pipelined stages progressively smaller.

In most pipelined ADCs designed with CMOS or BiCMOS technology, the S&H, the DAC, the summation node, and the gain amplifier are usually implemented as a single switched-capacitor circuit block called a multiplying DAC (MDAC). The major factor limiting MDAC accuracy is the inherent capacitor mismatch. A purely bipolar implementation would be more complicated and would suffer mainly from resistor mismatch in the current source DAC and the interstage gain amplifier. In general, for about 12 bits of accuracy or higher, some form of capacitor/resistor trimming or digital calibration is required, especially for the first couple of stages.

Digital Calibration

The MAX1200, MAX1201, and the MAX1205 have ADCs with sampling rates of 16-bit, 14-bit, and 2-bit Msps, respectively. The MAX family employs digital calibration to ensure its excellent accuracy and dynamic performance. The MAX1200 family is a CMOS pipelined ADC with four 4-bit stages (with 1-bit overlap) and a 5-bit flash ADC at the end, giving a total of 3+3+3+3+5 = 17 raw bits as shown in Fig. 4.40. The extra 1–3 bits are required by the digital calibration to quantize the error terms to greater accuracy than the ADC itself and are discarded to give either 14 bits or 16 bits overall.

Calibration starts from the multiplying digital-to-analog converter (MDAC) in the third stage; beyond the third stage the MDAC error terms are

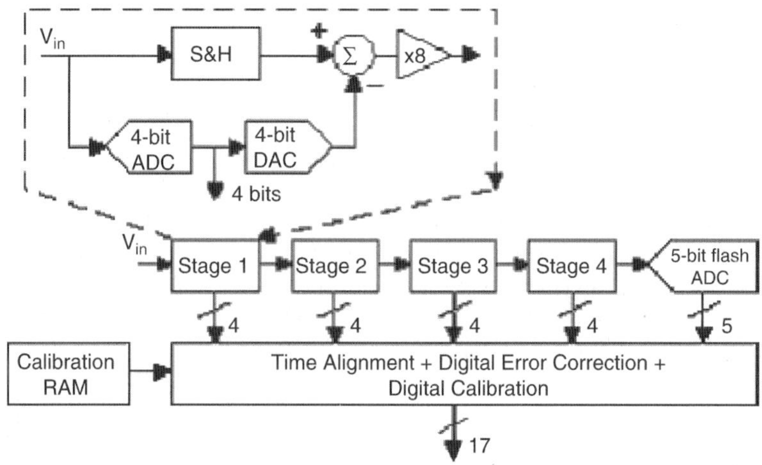

Fig. 4.40. MAX1200 pipelined ADC architecture

small enough that calibration is not needed. The third-stage output is digitized by the remaining pipelined ADC, and the error terms are stored in on-chip RAM. Once the third MDAC is calibrated, it can be used to calibrate the second MDAC in a similar fashion. Likewise, once the second and third MDAC are calibrated, they are used to calibrate the first MDAC. Averaging is used (especially in the first and second MDAC) to ensure that the calibration is noise-free. During normal conversions, those error terms are recalled from the RAM and are used to adjust the outputs from the digital error correction logic.

Pipelined ADC Versus SAR ADC

In a successive approximation register (SAR) ADC, the bits are decided by a single high-speed, high-accuracy comparator bit by bit, from the MSB down to the LSB, by comparing the analog input with a DAC whose output is updated by previously decided bits and successively approximates the analog input. This serial nature of SAR limits its operating speed to no more than a few MS/s, and still slower for very high resolutions (14–16 bits). A pipelined ADC, however, employs a parallel structure in which each stage works on 1 to a few bits (of successive samples) concurrently. Although there is only one comparator in an SAR, this comparator has to be fast (clocked at approximately the number of bits x the sample rate) and as accurate as the ADC itself. In contrast, none of the comparators inside a pipelined ADC needs this kind of speed or accuracy.

However, a pipelined ADC generally takes up significantly more silicon area than an equivalent SAR. SAR also displays latency of only one cycle (one cycle = 1/Fsample), versus about three or more cycles in a typical pipeline. Like a pipeline, an SAR with more than 12 bits of accuracy usually requires some form of trimming or calibration.

Pipelined ADC Versus Flash ADC

Despite the inherent parallelism, a pipelined ADC still requires accurate analog amplification in DACs and interstage gain amplifiers, and thus significant linear settling time. A purely flash ADC, on the other hand, has a large bank of comparators, each consisting of wideband, low-gain preamps followed by a latch. The preamps, unlike those amplifiers in a pipelined ADC, need to provide gains that do not even have to be linear or accurate—meaning, only the comparators' trip points have to be accurate. As a result, a pipelined ADC cannot match the speed of a well-designed flash ADC.

Although extremely fast 8-bit flash ADCs (or their folding/interpolation variants) exist with sampling rates as high as 1.5 Gsps (for example, the MAX104/MAX106/MAX108), it is much harder to find a 10-bit flash, while 12-bit (or above) flash ADCs are not commercially viable products. This is simply because in a flash the number of comparators goes up by a factor of 2 for every extra bit of resolution, and at the same time each comparator has to be twice as accurate. In a pipeline, however, to a first order the complexity only increases linearly with the resolution, not exponentially. At sampling rates obtainable by both a pipeline and a flash, a pipelined ADC tends to have much lower power consumption than a flash. A pipeline also tends to be less susceptible to comparator meta stability.

Comparator meta stability in a flash can lead to sparkle-code errors (a condition in which the ADC provides unpredictable, erratic conversion results)

Pipelined ADC Versus the Sigma–Delta Converter ADC

Traditionally, over sampling/sigma–delta-type converters commonly used in digital audio have a limited bandwidth of about 22 kHz or so. But recently some high-bandwidth sigma–delta-type converters have reached a bandwidth of 1–2 MHz with 12–16 bits of resolution. These are usually very-high-order sigma–delta modulators (for example, fourth or even higher) incorporating a multi-bit ADC and multibit feedback DAC, and their main applications are in ADSL. Sigma–delta converters have the innate nature of requiring no special trimming/calibration, even for 16–18 bits of resolution.

They also require no steep rolling-off anti-alias filter at the analog inputs, because the sampling rate is much higher than the effective bandwidth; the backend digital filters take care of it. The oversampling nature of the sigma-delta converter also tends to "average out" any system noise at the analog inputs. However, sigma–delta converters trade speed for resolution. The need to sample many times (for example, at least 16 times, but often much higher) to produce one final sample cause the internal analog components in the sigma–delta modulator to operate much faster than the final data rate.

The digital decimation filter is also nontrivial to design and takes up a lot of silicon area. The fastest, high-resolution sigma–delta-type converters are not expected to have more than a few MHz of bandwidth in the near future. Like pipelined ADCs, sigma–delta converters also have latency.

Pipelined ADC Versus Half (Two-Step) Flash ADC

A two-step flash can be generalized as a two-stage pipeline. However, as the number of bits goes up (for example, 12 bits or higher) with digital error correction, each stage would need to incorporate a 6- to 7-bit flash ADC. The inter stage gain amplifier would also need very high gain. Therefore, for higher resolution, it is wiser to use more than two stages.

The pipelined ADC is the architecture of choice for sampling rates from a few $\mathrm{MS\,s^{-1}}$ up to $100\,\mathrm{MS\,s^{-1}}$+. Complexity goes up only linearly (not exponentially) with the number of bits, providing converters with high speed, high resolution, and low power at the same time. They are very useful for a wide range of applications, most notably in the digital communication area, where a converter's dynamic performance is often more important than traditional DC specifications like differential nonlinearity (DNL) and integral nonlinearity (INL). Their data latency is of little concern in most applications.

4.8 Digital-to-Analog Control

A DAC, on the other hand, inputs a binary number and outputs an analog voltage or current signal. In block diagram form, it appears ads shown in Fig. 4.41.

Despite the popularity of digital devices, real world signals are typically represented by analog signals. Digital control systems process real world analog signals by using an ADC to convert analog signals to digital. Conversions back to analog signals are accomplished using a DAC. Maxim offers a complete line of precision DACs from 8 to 16 bits. It is important to find a DAC that meets to requirements of the application. A design engineer needs to look at parameters such as linearity, resolution, speed, and accuracy when selecting a DAC.

Despite the widespread use of digital electronics, the real world is still analog. As a result, DACs are needed at the point of contact between digital and analog sections of a system. The following discussion covers important parameters that should be taken into account when selecting a DAC, and it also highlights some interesting features available in new devices. Most electronic devices now include digital control circuitry, but analog values are still needed to control valves, loudspeakers, and other actuators.

The conversion from digital to analog is usually accomplished within a dedicated D/A converter. This DAC can be selected from the vast array of available standard circuits, in accordance with the conversion requirements and system settings.

In selecting a DAC, the first step is to determine the necessary resolution N, where 2N LSBs equal the maximum analog output. The market offers greatest product variety for DACs that specify a resolution in the range 8–16 bits (256–65,536 steps). Resolution gives no indication of accuracy, however, because other sources of error must be taken into account. The next parameter of interest is INL, which describes the deviation of a DAC's transfer function from a straight line. For DACs, this deviation is measured at every step (Fig. 4.42a).

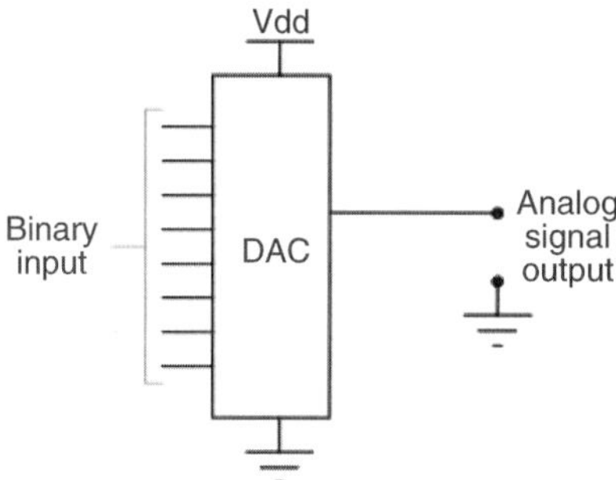

Fig. 4.41. Digital-to-analog converter

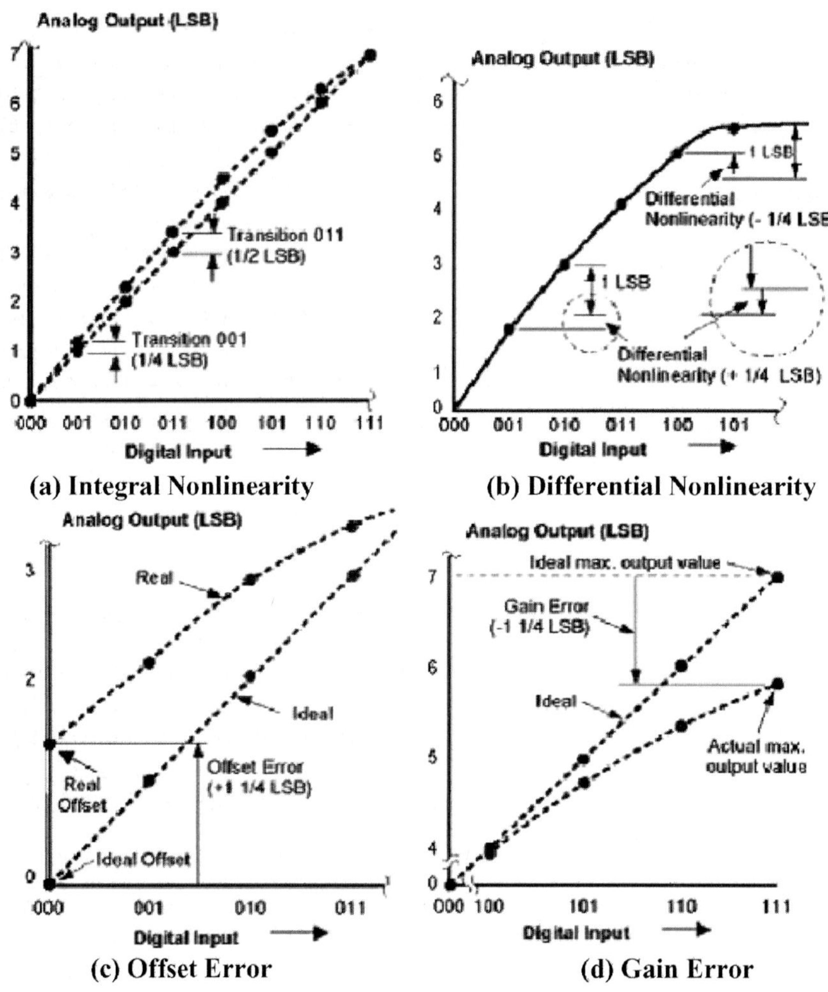

Fig. 4.42. Major performance parameters for a DAC

The straight line can be a best approximation to the actual transfer function, or a line drawn between the transfer function's end points (after subtracting the gain and offset errors). Though low-cost devices specify this parameter as high as ±16 LSBs, it can often be improved through the use of correction coefficients in the operation software. For high-end DACs, INL values are better than ±1 LSB. DNL is the difference between an actual step height and the ideal value of one LSB.

The target value for DNL (< 1 LSB) ensures that the DAC is monotonic. This means no data is lost because the output always changes in accord with the digital input; it increases in response to a digital increment, and decreases in response to a digital decrement. Figure 4.42b defines DNL in terms of the transfer function. The deviation of the actual output from the ideal value is graphed in Fig. 4.42c. For DACs, offset error equals the output voltage when the digital input is zero. This error remains constant for all input values, and normally can be compensated by calibrating the circuit. Offset error is often specified as an absolute value in millivolts rather than LSBs. (Offset error in LSBs depends on the output step height, which in turn depends on the level of reference voltage.) An acceptable offset error is usually less than $\pm 10\,\text{mV}$.

Gain error is defined as the difference between the ideal maximum output voltage and the actual maximum value of the transfer function after subtracting the offset error (Fig. 4.42d). Because gain error changes the slope of the transfer function, it delivers the same percentage error for each step. Gain error is expressed in LSB or millivolts, as a percent of the maximum value.

Dynamic Output Characteristics

An ideal DAC would deliver its analog output immediately after a digital value is applied to the input. Actual DACs, however, impose a settling time that consists of the internal propagation delay and a limited slew rate in the output driver. Settling time begins with the start of a conversion, ends when the DAC output becomes stable, and includes any static error that couples to the output. Though caused by the DAC itself, this effect is made *worse by improper grounding and board layout.*

Data Interface

Formerly, the most widely used data interface was the parallel type. It still offers the advantage of fast transmission and a simple data protocol. The lower pin count of a serial interface, on the other hand, requires less board space and allows a smaller package. The SPI (Serial Peripheral Interface) and related types, which require simpler handling by the IC manufacturer and the user, have a much larger market share than does the 2-wire I2C-compatible interface.

SPI is a 3-wire interface (data input, data output, and clock) that also requires a chip-select wire for device addressing. Because galvanic isolation of the signal path is often required as well, the unidirectional data lines of a 3-wire interface offer yet another advantage (isolation of unidirectional lines is easily implemented with opto-couplers). New DACs like the 12-bit MAX5539 and the MAX5543 integrate this isolation within the package, thereby easing the design of analog data outputs in industrial equipment.

Reference Voltage

To a large extent, the characteristics of a DAC are defined by its reference voltage, whether generated within the DAC or applied externally.

Table 4.3. Allowable temperature range for maximum drift of ±1 LSB in 2.5 V references

Resolution (bit)	Step count	1 LSB at 2.5 V (mV)	Max. temperature drift (ppm °C^{-1}) at 0–70°C	Max. temperature drift (ppm °C^{-1}) at −40–85°C
8	256	9.766	111.61	62.50
10	1024	2.441	27.90	15.63
12	4096	0.610	6.98	3.91
13	8192	0.305	3.49	1.95
14	16384	0.153	1.74	0.98
16	65536	0.038	0.44	0.24

First, the reference voltage (V_{REF}) sets the DAC's maximum output voltage if the output signal is not amplified by an additional output stage. V_{REF} also defines the voltage step by which the output changes in response to a 1-LSB transition at the input. One step equals $V_{REF}/2$ N, where N is the DAC resolution. At a constant temperature, the output voltage of a reference varies within a range specified by its initial accuracy. For a changing temperature, the output voltage drift has a direct bearing on the DAC's quality.

Table 4.3 makes obvious that a DAC requires minimum drift in its reference. Integrated references are typically 100 ppm/°C, and are therefore qualified only for a limited temperature range. An exception is the 12-/13-bit MAX5122/MAX5132, whose integrated precision reference specifies a maximum drift of 10 ppm/°C (3 ppm/°C typically). When connecting an external reference, we should consider not only the current required and the voltage range of the DAC's reference input, but also any dynamic effects produced by the DAC's inner structure. With variation of the applied digital value, the reference input resistance can also change.

Thus, the reference selected must be capable of following each load step within the required time, or a capacitor or an op-amp buffer is to be added. DACs with an external reference input (like the MAX5170) can also be operated in the multiplying mode. A variable voltage (rather than a constant one) is applied to the reference input (Fig. 4.43). The variable voltage is multiplied with the adjusted digital input value and transferred to the output, producing the effect of an accurate digital potentiometer.

For this operating mode, we should consider the DAC's bandwidth and voltage range, as well as dynamic characteristics of the reference input; such as voltage feed through from the reference input to the output at a digital value of zero.

Output Stages

The output stage of a DAC can be designed to provide a voltage output or a current output, but the simpler voltage output has a much larger market share.

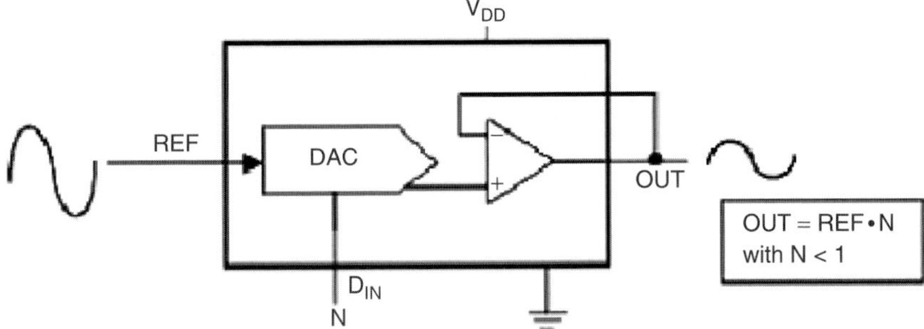

Fig. 4.43. Multiplying DAC

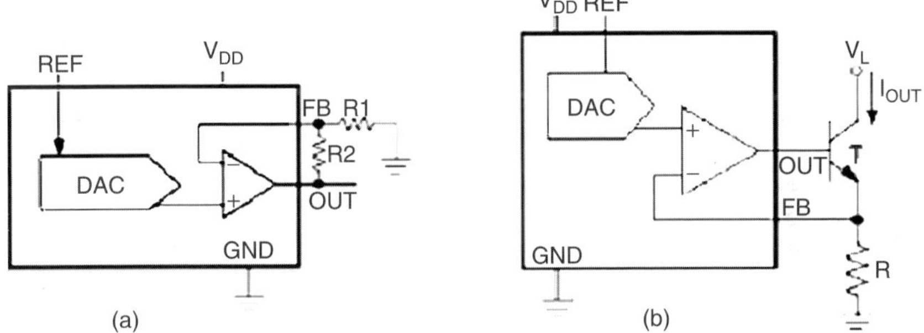

Fig. 4.44. (**a**) Force-sense output and (**b**) current output

Some Maxim devices offer a voltage output with fixed gain or an uncommitted amplifier as options, the so-called "force-sense" outputs. That arrangement aids to set an individual gain with two external resistors (Fig. 4.44a). The force-sense voltage output also aids to realize a current output (Fig. 4.44b).

The MAX5120/MAX5170 family provides a special power-up feature called glitch prevention. Without this feature, the DAC output simply follows the supply voltage at power-up until the integrated circuits start to work. That action corrupts the output with an impulse up to 3 V high, which can lead to a malfunction in the following circuitry. MAX5120/MAX5170 devices suppress this impulse. They also deliver a power-up reset that clears all DAC registers. This reset can be adjusted for power-up at 0 V or at half of the maximum output voltage (midrange), as is required for bipolar output stages. Most new devices operate on a unipolar power supply, but they can deliver bipolar output signals if we add an external bipolar amplifier and define midrange of the maximum output voltage as zero.

The 12-bit MAX530 operates on bipolar supplies and delivers bipolar output voltages directly. As another consideration for the DAC power supply, latch-up can occur if a digital-input voltage is 0.3 V higher then the supply voltage. In particular, data signals should not be applied to the DAC input

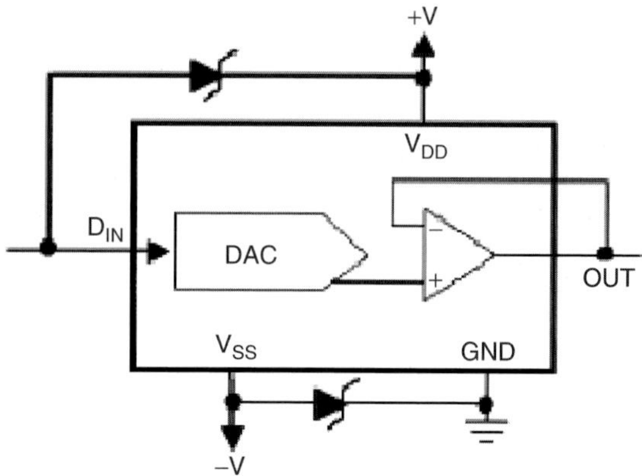

Fig. 4.45. Schottky diodes for protection against latch-up effects

at power-up or power-down. Schottky diodes shown in Fig. 4.45 provide protection against this problem.

Summary

- Data acquisition in virtual instruments contain three main stages of devices – data acquisition, data conditioning, and data manipulation.
- Data acquisition is done using various equipments such as DAQ, transducers, and so on. The characteristic nature and key feature of these instruments were discussed in detail.
- Data conditioning is done using various techniques such as amplification, filtering, isolation, and so forth. The characteristic nature and key feature of these instruments were also delineated in detail.
- Special care has been taken with utmost solemn to explain about various kinds of DAQ and ADC available in the market. Few comparisons were also made between various devices to prove the supremacy of its characters.

Review Questions

1. Give the components of measuring systems with a neat block diagram and explain the functions of each block.
2. Draw the schematic of a sensor.
3. Write the differences between chemical, physical, and biosensors.
4. What are the characteristic features of sensors?
5. Explain in detail about capacitive sensors and its uses.
6. Define transducer.

7. What are the types of transducers?
8. Give the characteristics of transducer.
9. Explain the following: RTD and strain gauge.
10. List out the general signal conditioning functions.
11. With sketches explain the ADC architecture types.
12. With sketches explain the DAC architecture types.
13. Discuss about the flash ADC architecture.
14. Explain about various types of sensors.
15. Draw the human sensor architecture.
16. How to use transducers for signal sensing?
17. What are the various stages of signal conditioning?
18. How signal conditioning with SCXI is used?

5

Common Instrument Interfaces

Learning Objectives. On completion of this chapter the reader will have a knowledge on:

- Common Instrument Interfaces
- 4–20 MA Current Loop
- 60 MA Current Loop
- RS 232, RS222, and RS485
- Concepts of GPIB, Advantages of GPIB
- VISA Programming, Attributes of VISA, Advantages of VISA

5.1 Introduction

Instrument interfaces play pivot role in all kind of digital equipments. It is the job of these equipments to make a symbiotic liaison between the external environment data and the process data. The information that received by these instrument aids in conditioning so the data available could be relied more efficiently. Some of the main considerations for PC based instrument interfaces depends on the type of connector (pinouts) on the instrument, type of cables needed, electrical properties involved, signal levels, grounding, cable length, communication protocols used, and the software drivers available. This section describes the operation and characteristic features of serial communication devices such as 4–20, 60 mA current loops along with the RS232C standard. The IEEE standard GPIB is also detailed in later part of this section. VISA, which is a high level API capable of controlling VXI, GPIB, or Serial instruments. It also makes the appropriate driver calls depending on the type of instrument interfaced with PC.

5.2 4–20 mA Current Loop

Current-mode data transmission is the preferred technique in many environments, particularly in industrial applications. This section describes the design of a 4–20 mA current loop using the MAX1459 sensor signal conditioner.

Current-mode data transmission is the preferred technique in many environments, particularly in industrial applications. Most systems employ the familiar 2-wire, 4–20 mA current loop, in which a single twisted-pair cable supplies power to the module and carries the output signal as well.

The 3-wire interface is less common but allows the delivery of more power to the module electronics. A 2-wire system provides only 4 mA at the line voltage (the remaining 16 mA carries the signal). Current loops offer several advantages over voltage-mode output transducers:

- They do not require a precise or stable supply voltage.
- Their insensitivity to IR drops makes them suitable for long distances.
- A 2-wire twisted-pair cable offers very good noise immunity.
- The 4 mA of current required for information transfer serves two purposes: it can furnish power to a remote module, and it provides a distinction between zero (4 mA) and no information (no current flow).

In a 2-wire, 4–20 mA current loop, supply current for the sensor electronics must not exceed the maximum available, which is 4 mA (the remaining 16 mA carries the signal). Because a 3-wire current loop is easily derived from the 2-wire version, the following discussion focuses on the 2-wire version.

Need for a Current Loop

The 4–20 mA current loop shown in Fig. 5.1 is a common method of transmitting sensor information in many industrial process-monitoring applications. A sensor is a device used to measure physical parameters such as temperature, pressure, speed, liquid flow rates, etc. Transmitting sensory information through a current loop is particularly useful when the information has to be sent to a remote location over long distances (1,000 ft, or more). The loop's operation is straightforward: a sensor's output voltage is first converted to a proportional current, with 4 mA normally representing the sensor's zero-level output, and 20 mA representing the sensor's full-scale output.

Then, a receiver at the remote end converts the 4–20 mA current back into a voltage which in turn can be further processed by a computer or display module. However, transmitting a sensor's output as a voltage over long distances has several drawbacks. Unless very high input-impedance devices are used, transmitting voltages over long distances produces correspondingly lower voltages at the receiving end due wiring and interconnect resistances. However, high-impedance instruments can be sensitive to noise pickup since the lengthy signal-carrying wires often run in close proximity to other electrically noisy system wiring. Shielded wires can be used to minimize noise pickup, but their high cost may be prohibitive when long distances are involved.

Sending a current over long distances produces voltage losses proportional to the wiring's length. However, these voltage losses also known as "loop

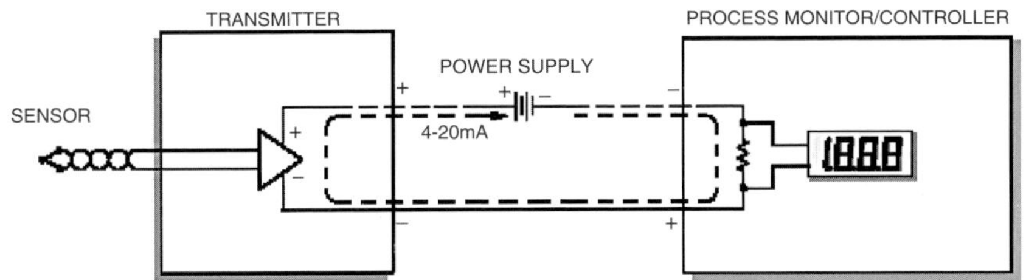

Fig. 5.1. Typical components in a Loop Powered Application

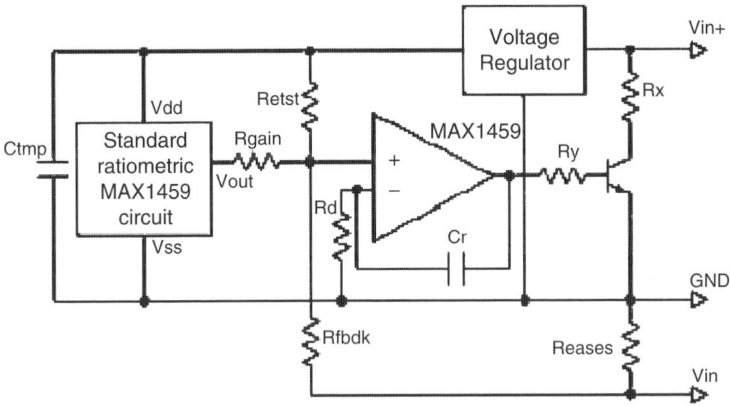

Fig. 5.2. 4–20 mA transmitter

drops" do not reduce the 4–20 mA current as long as the transmitter and loop supply can compensate for these drops. The magnitude of the current in the loop is not affected by voltage drops in the system wiring since all of the current (i.e., electrons) originating at the negative (−) terminal of the loop power supply has to return back to its positive (+) terminal.

5.2.1 Basic 2-wire Circuit

A voltage output should be converted to current when configuring a ratio metric 4–20 mA circuit, because current-mode applications require a 4 mA offset and 16 mA span. This section presents the circuit details and results obtained from a current-loop configuration based on the MAX1459 sensor-signal conditioner. In principle, a voltage regulator is to be added, which converts the 10–32 V loop voltage to a fixed 5 V for operating the MAX1459. Figure 5.2 shows the circuitry required for implementing a standard ratio metric version of the MAX1459 circuit. The voltage regulator can be any low-cost device whose quiescent current is sufficiently low for the 4 mA budget.

Rx is more strongly recommended, because it protects the transistor from large transient currents during power-up and during voltage transients. Rx also ensures reliability by helping to protect the transistor during high-temperature operation. By sharing load power it maintains the transistor junction temperature in a safe region. The value of Rx is determined by the minimum operating voltage (Vin), the maximum operating current (20 mA), the voltage drop across R sense, and the transistor's Vce saturation voltage. For applications in which the minimum Vin is low, it may be necessary to eliminate Rx altogether. Capacitor Cf is recommended for compensating the amplifier.

5.2.2 4–20 mA Equation

From Fig. 5.2, it is clear that all system currents must flow through R_{sense} and R_{fdbk}. The ratio of these resistors should be chosen such that most of the current flows through Rsense, though the voltage drop across it is small. On this basis, the user has to select convenient values for R_{sense} and R_{fdbk}. The MAX1459 output (Vout) is calibrated for 0.5 V at minimum pressure and 4.5 V at maximum pressure. Thus, Vout = 4 V. The transmitter offset current is 4 mA, the FSO current is 16 mA, and Vdd = 5 V. Because the Rd current is negligible, current through Rfdbk is the sum of currents flowing through Rofst and Rgain. The current 'I' range between 4 and 20 mA. Therefore

$$\frac{Vdd}{Rofst} = \frac{Rsense^{*}I}{Rfdbk}$$

Solving for Rofst,

$$Rofst = \frac{Rfdbk^{*}Vdd}{Rsense^{*}I}$$

The value for Rgain can be determined by noting that a 4 V FSO across Rgain (Vout) must add 16 mA across Rsense. Therefore,

$$\frac{\Delta Vout}{Rgain} = \frac{Rsense * 16\,mA}{Rfdbk}$$

Solving for Rgain,

$$Rgain = \frac{\Delta Vout * Rfdbk}{Rsense * 16\,mA}$$

The values determined for Rofst and Rgain are ideal values, and they provide very little extra range to accommodate variations in the component values. In practice, it is preferable to obtain a larger adjustment range for FSO by lowering the value of Rgain. The value of Rofst should also be changed by a small amount. This approach allows use of low accuracy resistors, and their TC is of little concern because the MAX1459 compensates for the collective effect of temperature on the resistors. Because line- and load-regulation errors cannot be compensated by the MAX1459, the basic regulator requirements are low line-regulation error, and (to a lesser extent) low load-regulation error as shown in Fig. 5.3 and the parameters are shown in Table 5.1.

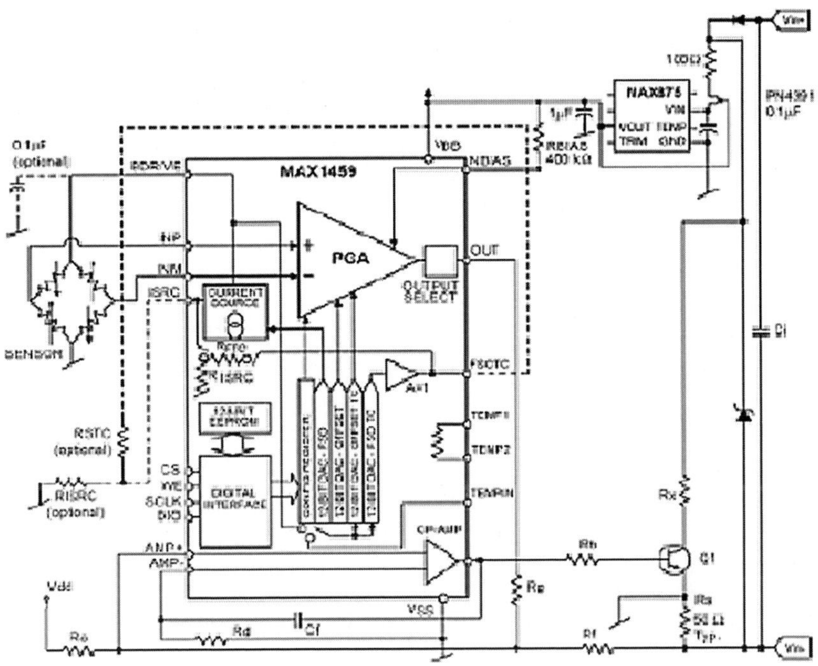

Fig. 5.3. Architecture diagram of 4–20 mA

Table 5.1. Component values for 4–20 mA Applications

Parameter	Value	Comments
Supply (min)	10 V	Determined by regulator dropout and Rx
Supply (max)	32 V	Limited by regulator and transistor
1 Minimum	4 mA	Zero pressure reference output
1 Mid-scale	12 mA	Mid-scale current output
1 Maximum	20 mA	Full-scale pressure output
Rsense	50 Ω	1 volt drop at 20 mA
Rfdbk	100 kΩ	Should be > than Rsense
Rofst	5 MΩ	Determines current loop offset value
Rgain	500 kΩ	Determines current loop gain value
Rx	390 Ω	Protects the output transistor
Ry	5 k Ω	Optional component
Cbyp	2.2 μF	Reduces digital noise in loop output
Cc	0.1 μF	Recommended for stability over temp
Rd	100 kΩ	Recommended for stability over temp

To meet the 4 mA budget, the quiescent operating current should be less than 500 μA. Voltage-calibration accuracy and temperature coefficient are of secondary importance because the MAX1459 compensates these parameters.

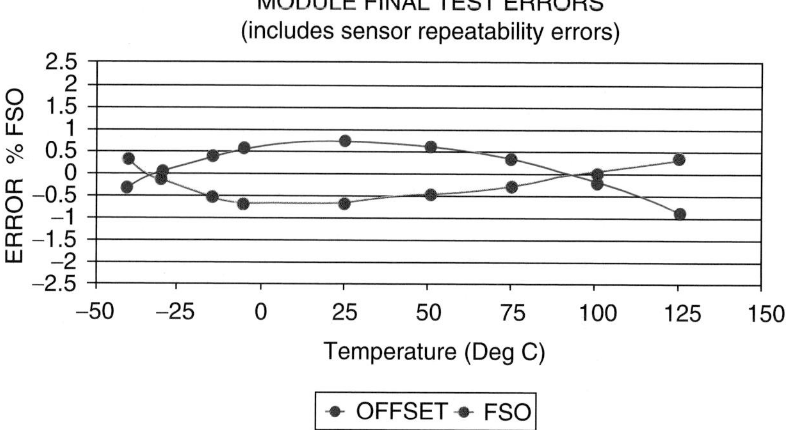

Fig. 5.4. Compensation results

The regulator's maximum input voltage should be greater than the maximum loop-supply voltage (Vin).

A power-supply bypass capacitor Cbyp (in addition to the one required within the standard ratio metric circuit) has been added to prevent the noise of digital activity in the MAX1459 EEPROM from appearing on the output-current loop. This capacitor value depends on the circuit layout and the regulator characteristics. The EEPROM supply can be separated from the MAX1459 with an RC network consisting of a large value capacitor and a small value resistor.

The test system ground return should be at the junction of Rsense and the emitter of the output transistor. This node is equivalent to the MAX1459's Vss node. Because the ground return is floating, it must not be connected in parallel with other devices in a test system with multiple transducers. Nor should loop current be measured in the presence of digital activity between the MAX1459 and the test system. DIP relays are recommended for selecting and deselecting both the ground return and the digital lines required for programming individual modules. Results obtained after compensating the sensor are shown in Fig. 5.4. Large raw sensor errors are temperature compensated to within ±1%. The compensation procedure assumes an output voltage from the pressure transducer that is ratio metric to the supply voltage (the MAX1459 current source).

This procedure requires a minimum of two test pressures (zero and full scale) at two arbitrary test temperatures. Ideally, the two temperatures are those for which we wish to perform a best linear-fit compensation.

5.2.3 3 V/5 V DACs Support Intelligent Current Loop

This section explains 4–20 mA current loops and intelligent transmitters and explains their need for high-resolution low-power DACs. An optimum power-supply chip is not the only critical choice to be made in designing a mixed 3 V/5 V microprocessor-controlled system. Because more and more systems require low-power components capable of low-voltage startup, state-of-the-art A/D and D/A converters with near-ideal specifications, including 3 V/5 V operation, are also available. Today's popular industrial-control applications, such as programmable-logic controllers, factory process control, computer numeric control (CNC), and intelligent transmitters, all demand low-power semiconductors. Consistent with this trend is the use of 4–20 mA current loops, which have become a well-established part of the analog communications between the host computer and smart transmitters in a factory environment.

5.2.4 Basic Requirements for 4–20 mA Transducers

For transmitting low-amplitude low-frequency signals over several hundred yards in a noisy industrial control environment, current is preferred over voltage, because the current at any instant is constant over the entire length of the cable. Voltage transmission is not recommended, because the voltage at any point depends on line resistance and capacitance, which change with the cable's length. Current transmission also allows a single 2-wire cable to carry power and signals at the same time. At the end of the transmission line, a precise termination resistor converts the loop current to an accurate voltage.

This resistor (typically 50–750) establishes the current-loop receiver's input impedance. High signal-source impedance minimizes voltage fluctuations across the termination resistor caused by variations in line resistance, but it also picks up more EMI and other industrial interference. Large-valued bypass capacitors reduce EMI pickup by helping to lower the signal-source impedance.

Advantages of 4–20 mA Current Loop

Current loops offer four major advantages such as:

- Long-distance transmission without amplitude loss
- Detection of offline sensors, broken transmission lines, and other failures
- Inexpensive 2-wire cables
- Lower EMI sensitivity

5.2.5 Digitally Controlled 4–20 mA Current Loops

Smart transmitters incorporate a processor or a controller that linearizes the sensor data and communicates it to the host system. As shown in Fig. 5.5,

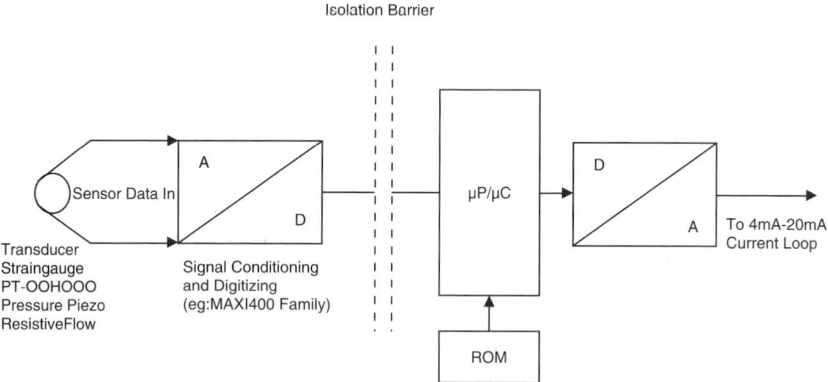

Fig. 5.5. Smart transmitter using 4–20 ma

these systems employ five general building blocks: an A/D converter, a microcontroller (µC), RAM, a D/A converter with an optional integrated amplifier, and a sensor or a transducer (thermocouple, strain gauge, PT100 RTD1, etc.).

The ADC in a single-channel transmitter often includes compensation circuitry, and the ADC in a multi channel system often includes one or more op-amps and multiplexers. Sophisticated ADC/DAC combinations can compensate the sensor for offset, offset temperature coefficient, full-span output, full span output temperature coefficient, and nonlinearity. A new line of intelligent signal conditioners with 0.1% accuracy were designed by Maxim Corporation for these applications. The first products in this four-member family (MAX1450 and MAX1457) are available now. Future members include the MAX1458, a signal conditioner equipped for internal calibration and the temperature compensation of piezoresistive sensors (via electronic trim using an on-board EEPROM). Another future product optimized for the same purpose (MAX1460) is a highly integrated, signal-processor-based, digitally compensated signal conditioner.

If the sensor must reside in an explosive atmosphere, safety measures require not only an isolation barrier that prevents ground loops but operation that is "intrinsically safe" as well; this is a rule that limits the level of transmitter energy below that capable of generating an electrical discharge. The isolation barrier for such a system typically resides on the power-supply side. For smart transmitter systems that are not intrinsically safe, the isolation barrier can be placed between the microcontroller (µC) and the conditioned and digitized sensor data. Data can be transferred across this barrier by general-purpose opt couplers such as the 6N136, the 4N26, or the IL300.

The transducer voltage must be digitized with precision by an A/D converter, preferably one with high resolution and an on-board calibration capability that removes system- and component-drift errors before the signal reaches a controlling processor. The processor then reads the data, processes

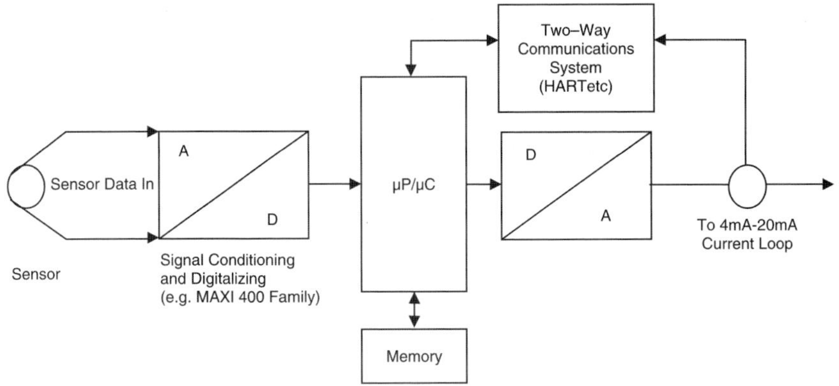

Fig. 5.6. Smart transmitter using 4–20 mA

it, and transmits it through a low-power high-resolution D/A converter to the 4–20 mA current loop. The next step up is an even smarter circuit called an intelligent transmitter (Fig. 5.6). An intelligent transmitter combines a transducer signal and memory for storing transmitter information with the two-way communication skills of a microcomputer. With an additional A/D converter, the system can produce data on the current loop's condition that enables adjustment and calibration by the μC.

New low-power low-voltage D/A converters from Maxim meet both requirements for a digitally adjustable 4–20 mA current loop: a 3 V/5 V supply-voltage capability, and internal amplifiers able to control the gate voltage of an external MOSFET. The only drawback in this configuration is the need to drive an external n-channel MOSFET, which requires a much higher supply voltage. If not provided on-board, this voltage must be realized with an external power-boost circuit. Fortunately, most industrial-control applications provide both high and low voltages for support of 3 V/5 V programmable-logic control as well as the sensors (piezoelectric, pressure, temperature, and flow) that demand voltages as high as 36 V (24 V typical).

As component suppliers adapt to the industry's need for lower power, the use of common-threshold MOSFETs (controllable by a single +5 V power supply) is spreading rapidly. We can create a reliable, digitally adjustable 4–20 mA current loop by combining an op-amp and an MOSFET with any kind of the D/A converters. Most of these D/A converters have Schmitt-trigger inputs, which enable a direct interface to the opto couplers in an isolated system.

5.3 60 mA Current Loop

With the HS RS-232.60-mA Current-Loop converter, the user can connect an asynchronous RS-232 device to a 60-mA current loop device. The converter

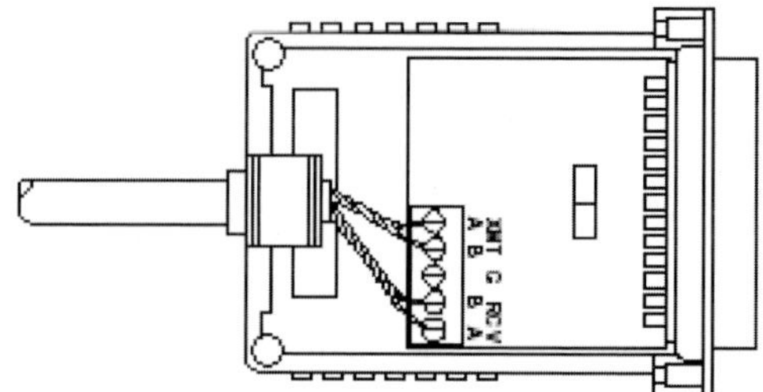

Fig. 5.7. Current-loop connection of the strain-relief assembly

requires no AC power or batteries to operate since it derives ultra-low power from the interface, and it supports data rates up to 115.2 kbps at short distances. When operating in full-duplex mode, the converter supports communication distances up to 4 miles (at a speed of about 150–300 bps) over two unconditioned twisted-pair wires. The general architecture of 60 mA current loop is shown in Fig. 5.7.

To guard against data loss caused by ground loops, the converter has 2,500 V RMS optical isolators on the line side. Models are available with male or female DB25 connectors for the RS-232 side, and terminal block, RJ-11, or RJ-45 connectors for the current-loop side. The converter connects directly to the RS-232 interface. Two-pair cable running to the 60-mA current-loop device attaches to the converter via an RJ-11 jack, RJ-45 jack, or terminal blocks with built-in strain relief. The converter is easy to use. On all models, there are no internal jumpers or DIP switches to set. The only configuration required is setting an external DCE/DTE switch on the converter, which eliminates the need for a crossover cable on the RS-232 interface.

Key Features

- Only one converter is needed for each 60-mA conversion to RS-232.
- Connect to the current loop via two twisted pairs.
- External switch to set DTE/DCE configuration.
- Operate in full-duplex mode only.
- Support data rates up to 115.2 kbps.
- Models available with male or female DB25 RS-232 connectors and terminal block, RJ-11, or RJ-45 current loop connectors.

Connecting PC to a Current-Loop Device with the HS Current-Loop Converter

The only configuration necessary for operation of the converter is setting the external DCE/DTE switch. If the RS-232 device connected to the converter is a modem or multiplexer, the switch is set to DTE. This setting causes the converter to behave like Data Terminal Equipment and transmit data on pin 2. If the RS-232 device connected to the converter is a PC, terminal, or host computer, the switch is set to DCE. This setting causes the converter to behave like Data Communications Equipment and transmits data. Only one converter is needed for each RS-232 to 60-mA current loop conversion. The converter connects to the current loop device via two twisted pairs.

5.4 RS232

The RS-232/485 port sequentially sends and receives bytes of information one bit at a time. Although this method is slower than parallel communication, which allows the transmission of an entire byte at once, it is simpler and can be used over longer distances because of lower power consumption.

For example, the IEEE 488 standard for parallel communication states that the cable length between the equipments should not exceed 20 m total, with a maximum distance of 2 m between any two devices. RS-232/485 cabling, however, can extend 1,200 m or greater. Typically, RS-232/485 is used to transmit American Standard Code for Information Interchange (ASCII) data.

Although National Instruments serial hardware can transmit 7-bit as well as 8-bit data, most applications use only 7-bit data. Seven-bit ASCII can represent the English alphabet, decimal numbers, and common punctuation marks. It is a standard protocol that virtually all hardware and software understand. Serial communication is mainly using the three transmission lines: (1) ground, (2) transmit, and (3) receive. Because RS-232/485 communication is asynchronous, the serial port can transmit data on one line while receiving data on another. Other lines such as the handshaking lines are not required. The important serial characteristics are baud rate, data bits, stop bits, and parity. To communicate between a serial instrument and a serial port on computer, these parameters must match.

The RS-232 port, or ANSI/EIA-232 port, is the serial connection found on IBM-compatible PCs. It is used for many purposes, such as connecting a mouse, printer, or modem, as well as industrial instrumentation. The RS-232 protocol can have only one device connected to each port. The RS-485 (EIA-485 Standard) protocol can have 32 devices connected to each port. With this enhanced multidrop capability the user can create networks of devices connected to a single RS-485 serial port. Noise immunity and multi drop capability make RS-485, the choice in industrial applications requiring many distributed devices networked to a PC or other controller for data collection. USB was designed primarily to connect peripheral devices to PCs, including keyboards, scanners, and disk drives.

RS-232 (Recommended standard-232) is a standard interface approved by the Electronic Industries Association (EIA) for connecting serial devices. In other words, RS-232 is a long established standard that describes the physical interface and protocol for relatively low-speed serial data communication

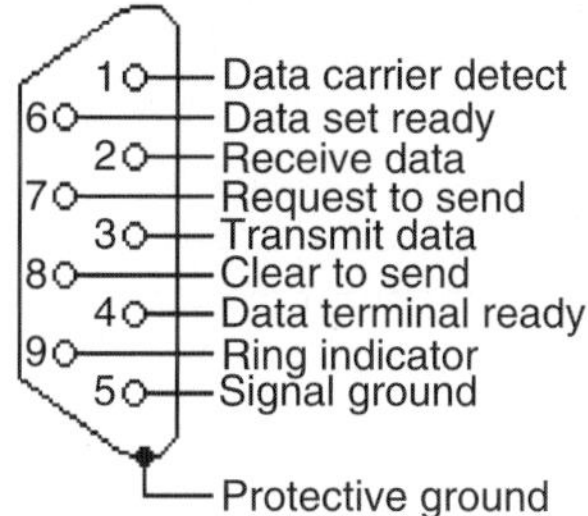

Fig. 5.8. RS232C DB9 pinout

Table 5.2. 9-pin D-type pin assignment

Pin Number	Signal	Description
1	DCD	Data carrier detect
2	RxD	Receive data
3	TxD	Transmit data
4	DTR	Data terminal ready
5	GND	Signal ground
6	DSR	Data set ready
7	RTS	Ready to send
8	CTS	Clear to send
9	RI	Ring indicator

between computers and related devices. An industry trade group, the Electronic Industries Association (EIA), defined it originally for teletypewriter devices. In 1987, the EIA released a new version of the standard and changed the name to EIA-232-D. Many people, however, still refer to the standard as RS-232C, or just RS-232. RS-232 is the interface that the computer uses to exchange data with the modem and other serial devices. The serial ports on most computers use a subset of the RS-232C standard.

RS232 on DB9 (9-Pin D-Type Connector)

There is a standardized pin out for RS-232 on a DB9 connector as indicated in Fig. 5.8. Table 5.2 gives a description of each of the pins.

RS232 on DB25 (25-Pin D-Type Connector)

In DB-25 connector most of the pins are not needed for normal PC communications, and indeed, most new PCs are equipped with male D type connectors having only 9 pins. Using a 25-pin DB-25 or 9-pin DB-9 connector, its normal cable limitation of 50 feet can be extended to several hundred feet with high-quality cable. The standardized pin out for RS-232 on a DB-25 connector, as shown in Fig. 5.9. Table 5.3 gives a description of each of the pins.

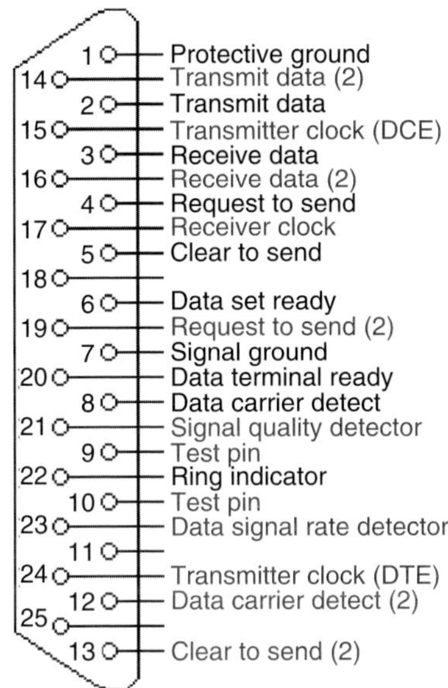

Fig. 5.9. RS232C DB25 pinout

Signal Descriptions

- TxD – This pin carries data from the computer to the serial device
- RXD – This pin carries data from the serial device to the computer
- DTR signals – DTR is used by the computer to signal that it is ready to communicate with the serial device like modem. In other words, DTR indicates the modem that the DTE (computer) is ON.
- DSR – Similarly to DTR, Data set ready (DSR) is an indication from the modem that it is ON.
- DCD – Data carrier detect (DCD) indicates that carrier for the transmit data is ON.
- RTS – This pin is used to request clearance to send data to a modem.
- CTS – This pin is used by the serial device to acknowledge the computer's RTS signal. In most situations, RTS and CTS are constantly on throughout the communication session.
- Clock signals (TC, RC, and XTC) – The clock signals are only used for synchronous communications. The modem or DSU extracts the clock from the data stream and provides a steady clock signal to the DTE. The transmit and receive clock signals need not have to be the same, or even at the same baud rate.
- *CD – CD stands for Carrier detect.* Carrier detect is used by a modem to signal that it has a made a connection with another modem, or has detected a carrier tone. In other words, this is used by the modem to signal that a carrier signal has been received from a remote modem.

Table 5.3. 25-pin D-type pin assignment

Pin Number	Signal	Description
1	PG	Protective ground
2	TD	Transmitted data
3	RD	Received data
4	RTS	Request to send
5	CTS	Clear to send
6	DSR	Data set ready
7	SG	Signal Ground
8	CD	Carrier detect
9	+	Voltage (testing)
10	−	Voltage (testing)
11		
12	SCD	Secondary CD
13	SCS	Secondary CTS
14	STD	Secondary TD
15	TC	Transmit clock
16	SRD	Secondary RD
17	RS	Receiver clock
18		Ready to send
19	SRS	Secondary RTS
20	DTR	Data terminal ready
21	SQD	Signal quality detector
22	RI	Ring indicator
23	DRS	Data rate select
24	XTC	External clock
25		

– *RI – RI stands for ring indicator.* A modem toggles (keystroke) the state of this line when an incoming call rings in the phone. In other words, this is used by an autoanswer modem to signal the receipt of a telephone ring signal. The carrier detect (CD) and the ring indicator (RI) lines are only available in connections to a modem. Because most modems transmit status information to a PC when either a carrier signal is detected

(i.e., when a connection is made to another modem) or when the line is ringing, these two lines are rarely used.

Limitations of RS-232

RS-232 has some serious shortcomings as discussed later.

First, the interface presupposes a common ground between the DTE and DCE. This is a reasonable assumption where a short cable connects a DTE and DCE in the same room, but with longer lines and connections between devices that may be on different electrical busses, this may not be true.

Second, a signal on a single line is impossible to screen effectively for noise. By screening the entire cable we can reduce the influence of outside noise, but internally generated noise remains a problem. As the baud rate and line length increase, the effect of capacitance between the cables introduces serious crosstalk until a point is reached where the data itself is unreadable. Using low capacitance cable can reduce crosstalk. Also, as it is the higher frequencies, control of slew rate in the signal (i.e., making the signal more rounded, rather than square) decreases the crosstalk.

The original specifications for RS-232 had no specification for maximum slew rate. Voltage levels with respect to ground represent the RS-232 signals. There is a wire for each signal, together with the ground signal. This interface is useful for point-to-point communication at slow speeds. For example, port COM1 in a PC can be used for a mouse, port COM2 for a modem, etc. This is an example of point-to-point communication: one port, one device. Due to the way the signals are connected, a common ground is required. This implies limited cable length – about 30–60 m maximum. Shortly, RS 232 was designed for communication of local devices, and supports one transmitter and one receiver.

5.5 RS422 and RS485

When communicating at high data rates, or over long distances in real world environments, single-ended methods are often inadequate. Differential data transmission (balanced differential signal) offers superior performance in most applications. EIA has recently released new serial interface, RS-422 and RS-485. These standards were designed for high speed communication.

RS422 Serial Communication

RS422 is a Standard interface approved by the EIA, and designed for greater distances and higher baud rates than RS232. In its simplest form, a pair of converters from RS232 to RS422 (and back again) can be used to form an “RS232 extension card.” Data rates of up to 100 Kbits s^{-1} and distances up to 4,000 ft can be accommodated with RS422. RS422 is also specified for multidrop or party-line applications where only one driver is connected to, and transmits on, a “bus” of up to 10 receivers. RS422 devices cannot be used to construct a truly multipoint network. A true multipoint network consists of multiple drivers and receivers connected on a single bus, where any node can transmit or receive data.

RS485 Serial Communication

RS485 is an EIA standard for multipoint communications. It supports several types of connectors, including DB-9 and DB-37. RS-485 is similar to RS-422 but can support more nodes per line. RS485 meets the requirements for a truly multipoint communications network, and the standard specifies up to 32 drivers and 32 receivers on a single (2-wire) bus. With the introduction of "automatic" repeaters and high-impedance drivers/receivers this "limitation" can be extended to hundreds or even thousands of nodes on a network. The RS-485 and RS-422 standards have much in common, and are often confused for that reason. RS-485, which specifies bi-directional, half-duplex data transmission, is the only EIA standard that allows multiple receivers and drivers in "bus" configurations. RS-422, on the other hand, specifies a single, unidirectional driver with multiple receivers.

Converters

Converters in general can be used to change the electrical characteristic of one communications standard into another, to take advantage of the best properties of the alternate standard selected. For example, an Automatic RS232<=>RS485 converter, could be connected to a computer's RS232, full-duplex port, and transform it into an RS485 half-duplex, multi-drop network at distances up to 4,000 ft. Converters in most instances, pass data through the interface without changing the timing and/or protocol. While the conversion is "transparent" the software must be able to communicate with the expanded network features.

An "Automatic Converter" (RS232<=>RS485) will turn on the RS485 transmitter when data is detected on the RS232 port, and revert back into the receive mode after a character has been sent. This avoids timing problems that are difficult to deal with in typical systems. When full duplex is converted into half-duplex only one device at a time can transmit data. Automatic Converters take care of the timing problems and allow fast communications without software intervention. The summary of serial communication standards is given in Table 5.4.

5.6 GPIB

The purpose of this section is to provide guidance and understanding of the General Purpose Interface Bus (GPIB) bus to new GPIB bus users or to provide more information on using the GPIB bus's features.

GPIB Data Acquisition and Control Module provides analog and digital signals for controlling virtually any kind of a device and the capability to read back analog voltages, digital signals, and temperatures. The 4867 Data Acquisition and Control module is an IEEE-488.2 compatible device and has a Standard Commands for Programmable Instruments (SCPI) command parser that accepts SCPI and short form commands for ease of programming. Applications include device control, GPIB interfacing and data logging. The 4867 is housed in a small 7 in. × 7 in. Mini box that can either rest on a bench or be

Table 5.4. Summary of serial communication standards

Specifications	RS232	RS423	RS422	RS485
Mode of operation	Single-ended	Single-ended	Differential	Differential
Allowed no. of Tx and Rx	1 Tx, 1 Rx	1 Tx, 10 Rx	1 Tx, 10 Rx	32 Tx, 32 Rx
Maximum cable length	50 ft.	4,000 ft.	4,000 ft.	4,000 ft.
Maximum data rate	20 kbps	100 kbps/10 mbps	100 kbps/10 mbps	100 kbps/ 10 mbps
Minimum driver output range	±5 V to ±15 V	±3.6 V	±2 V	±1.5 V
Maximum driver output range	±25 V	±6 V	±6 V	±6 V
Tx load impedance (ohms)	3–7 k	>= 450	100	54
Rx input sensitivity	±3 V	±200 mV	±200 mV	±200 mV
Rx input voltage range	±15 V	±12 V	±7 V	−7 V to +12 V
Maximum Rx input resistance (ohms)	3–7 k	4 k min	4 k min	>= 12 k

rack mounted in a test system. Analog, digital, and thermocouple connections are made via a 62-pin metal-shell connector on the 4867. GPIB signals are on a standard GPIB connector. The 4867's GPIB address and its configuration settings are set with SCPI commands and saved internally in flash memory. The SCPI commands can also be used to query the current configuration.

5.6.1 History and Concept

The GPIB bus was invented by Hewlett-Packard Corporation in 1974 to simplify the interconnection of test instruments with computers. At that time, computers were bulky devices and did not have standard interface ports. Instruments were lucky to have a connector with parallel BCD output lines that could be connected to a 10–20 column printer. A special computer interface had to be designed and built for each instrument that the engineer wanted to add to the test system. Even the simplest automated test systems were several month long projects. As conceived by HP, the new Hewlett-Packard

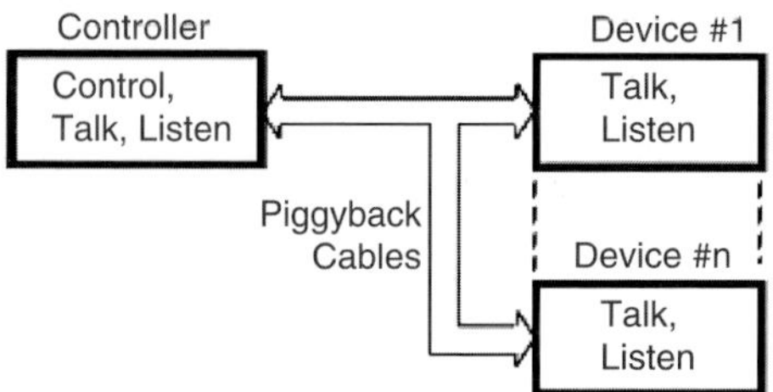

Fig. 5.10. IEEE 488 bus concept

Instrument Bus (HP-IB) uses a standard cable to interconnect multiple instruments to the computer.

Each instrument had its own interface electronics and a standard set of responses to commands. The system is easily expandable so multi-instrument test systems could be put together by piggy backing cables from one instrument to another. There were restrictions on the number of instruments that a driver could drive and the length of the bus cable. The number of devices was limited to 14 and the cable length to 20 m. Hewlett-Packard proposed the concept to US and International Standards bodies in 1974. It was adopted by the IEC committee in Europe in 1975. In the United States, other instrument companies objected to the HPIB name and so a new name, the General Purpose Instrument Bus (GPIB) was created.

The GPIB bus was formally adopted as the IEEE STD 488 in 1978. The IEEE-488 concept of Controllers and Devices is shown in Fig. 5.10. Controllers have the ability to send commands, to talk data onto the bus and to listen to data from devices. Devices can have talk and listen capability. Control can be passed from the active controller (Controller-in-charge) to any device with controller capability. One controller in the system is defined as the System Controller and it is the initial controller-in-charge (CIC).

Devices are normally addressable and have a way to set their address. Each device has a primary address between 0 and 30. Address 31 is the unlisten or untalk address. Devices can also have secondary addresses that can be used to address device functions. An example is ICS's 4896 GPIB to Quad Serial Interface which uses secondary addresses to address each channel. Although there are 31 primary addresses, IEEE 488 drivers can only drive 14 physical devices. Some devices can be set to talk only or to listen only. This lets two devices communicate without the need for a controller in the system. An example is a DVM that outputs readings and a printer that prints the data.

The IEEE-488 Standard defined an instrument with interface and device partitions as shown in Fig. 5.11. Interface messages and addresses are sent from the controller-in-charge to the device's interface function. Instrument particular commands such as range, mode, etc., are data messages that are passed through the Interface to the device.

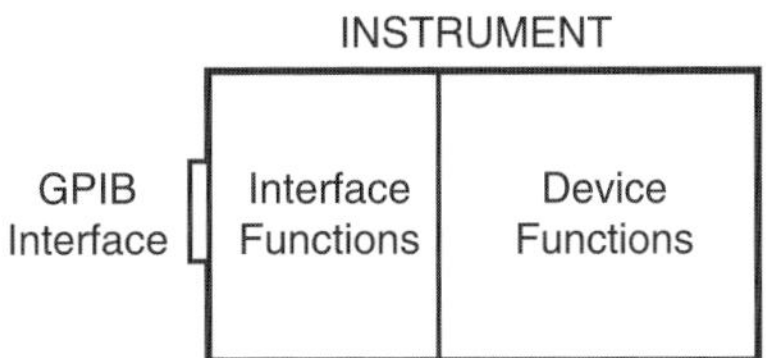

Fig. 5.11. IEEE 488 instrument

5.6.2 Types of GPIB Messages

GPIB devices communicate with other GPIB devices by sending device-dependent messages and interface messages through the interface system.

Device-dependent messages, often called data or data messages, contain device-specific information, such as programming instructions, measurement results, machine status, and data files.

Interface messages manage the bus. Usually called commands or command messages, interface messages perform such functions as initializing the bus, addressing and unaddressing devices, and setting device modes for remote or local programming.

The term "command" as used here should not be confused with some device instructions that are also called commands. Such device-specific commands are actually data messages as far as the GPIB interface system itself is concerned.

5.6.3 Physical Bus Structure

GPIB is a 24 conductor shown in Fig. 5.12. Physically, the GPIB interface system consists of 16 low-true signal lines and eight ground-return or shield-drain lines. The 16 signal lines, discussed later, are grouped into data lines (eight), handshake lines (three), and interface management lines (five).

Data Lines

The eight data lines, DIO1 through DIO8, carry both data and command messages. The state of the Attention (ATN) line determines whether the information is data or commands. All commands and most data use the 7-bit ASCII or ISO code set, in which case the eighth bit, DIO8, is either unused or used for parity.

Handshake Lines

Three lines asynchronously control the transfer of message bytes between devices. The process is called a 3-wire interlocked handshake. It guarantees that message bytes on the data lines are sent and received without transmission error.

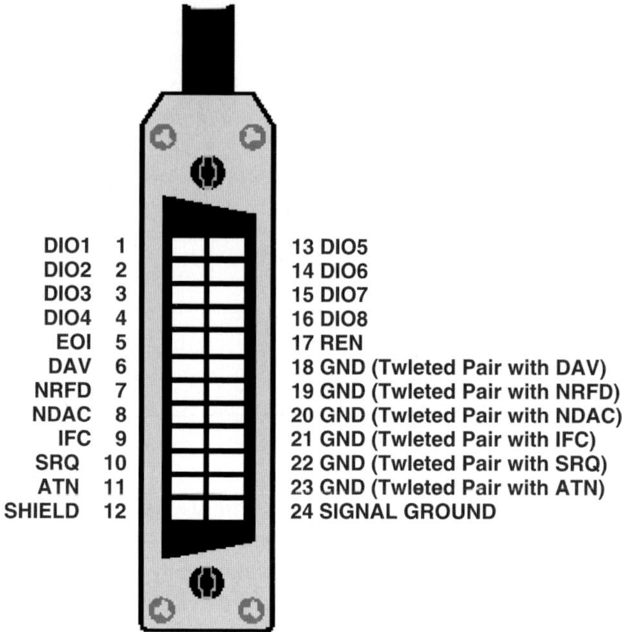

Fig. 5.12. Physical bus structure of GPIB

- *NRFD (not ready for data)* – Indicates when a device is ready or not ready to receive a message byte. The line is driven by all devices when receiving commands by listeners when receiving data messages, and by the talker when enabling the HS488 protocol.
- *NDAC (not data accepted)* – Indicates when a device has or has not accepted a message byte. The line is driven by all devices when receiving commands, and by Listeners when receiving data messages.
- *DAV (data valid)* – Tells when the signals on the data lines are stable (valid) and can be accepted safely by devices. The Controller drives DAV when sending commands, and the Talker drives DAV when sending data messages.

Three of the lines are handshake lines, NRFD, NDAC, and DAV, which transfer data from the talker to all devices who are addressed to listen. The talker drives the DAV line; the listeners drive the NDAC and NRFD lines. The remaining five lines are used to control the bus's operation.

- *ATN (attention)* is set true by the controller-in-charge while it is sending interface messages or device addresses. ATN is false when the bus is transmitting data.
- *EOI (end or identify)* can be asserted to mark the last character of a message or asserted with the ATN signal to conduct a parallel poll.

- *IFC (interface clear)* is sent by the system controller to unaddressed all devices and places the interface function in a known quiescent state.
- *REN (remote enable)* is sent by the system controller and used with other interface messages or device addresses to select either local or remote control of each device.
- *SRQ (service request)* is sent by any device on the bus that wants service.

Physical and Electrical Characteristics

Devices are usually connected with a shielded 24-conductor cable with both a plug and receptacle connector at each end. The user can link devices in either a linear configuration as shown in Fig. 5.13 or a star configuration as shown in Fig. 5.14, or a combination of the two.

Interface Messages

Table 5.5 lists the GPIB Interface Messages and Addresses with their common mnemonics. MLA, LAD, and UNL are listen addresses with hex values of 20–3 F. MTA, TAD, and UNT are talk addresses with hex values of 40–5 F. A device normally responds to both talk and listen addresses with the same value. i.e., LAD 4 and TAD 4. Secondary addresses have hex values of 60–7 F. Devices are designed with different IEEE 488 capabilities so not all devices respond to all of the interface messages. Universal messages are recognized by all devices regardless of their address state. Addressed commands are only recognized by devices that are active listeners.

The Standard also defined a Status Byte in the instrument that could be read with a Serial Poll to determine the device's status. Bit 6 of the Status

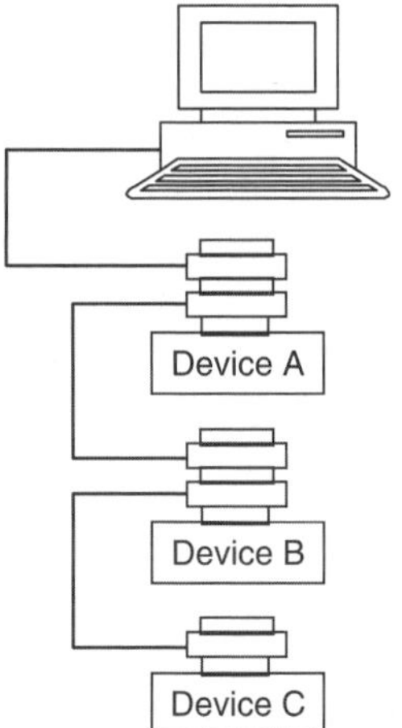

Fig. 5.13. Linear configuration

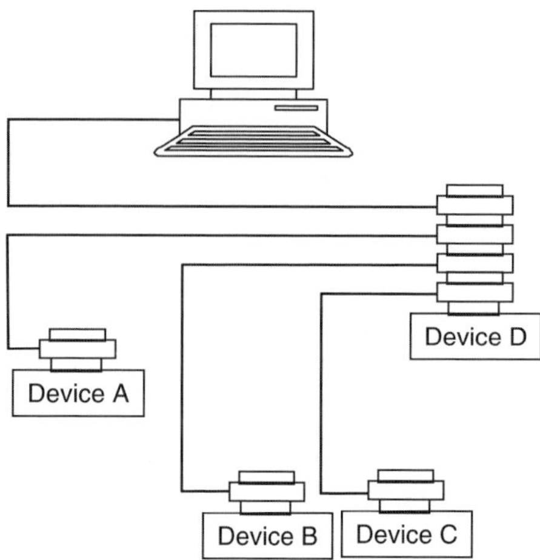

Fig. 5.14. Star configuration

Table 5.5. 488.1 Interface messages and addresses

Command	Function
Address commands	
MLA	My listen address (controller to self)
MTA	My talk address (controller to self)
LAD	Device listen address (0-30)
TAD	Device talk address (0-30)
SAD	Secondary device address (device optional address of 0-31)
UNL	Unlisten (LAD 31)
UNT	Listen (TAD 31)
Universal commands (to all devices)	
LLO	Local lockout
DCL	Device clear
PPU	Parallel poll unconfigure
SPE	Serial poll enable
SPD	Serial poll disable
Addressed commands (to addressed listeners only)	
SDC	Selected device clear
GTL	Go to local
GET	Device trigger
PPC	Parallel poll configure
TCT	Take control

Byte was defined as the Service Request bit that could be set when other bits in the Status Byte are set. The other bits were user defined. The Service Request pulls the SRQ line low to interrupt the controller. The Service

Request is reset when the device is Serial Polled or when the service request cause is satisfied.

5.6.4 Physical Standards

The physical standards available on the 4867 Data Acquisition and Control Module are:

- *Temperature measurements.* The 4867 have four thermocouple inputs for reading temperatures. The 4867's on card compensation circuit accepts J type thermocouples and provides calibrated readings from −100 to +400 C with a resolution of 0.1 C. Temperature readings are digitally filtered to reduce 60 Hz noise.
- *Analog inputs.* The 4867 have six single ended analog inputs with programmable unipolar and bipolar input ranges. The ranges are 0 to +5, 0 to +10, −5 to +5 and −10 to +10 Vdc. Resolution is 1 part in 12 bits. All analog inputs are continually read and digitally filtered to reduce measurement errors and signal noise. Each channel can be individually scaled so it is reading matches the measured parameter.
- *Analog outputs.* The 4867 has four analog outputs which provide 0 to +5 V-DC signals. Resolution is 1 part in 10 bits. Outputs can be scaled and offset to match the controlled parameter.
- *Digital I/O.* The 4867 have 32 digital I/O lines that can be configured as inputs or outputs in 8 bit byte increments. When used as outputs the lines are latched and can sink 48 mA or source 24 mA. As inputs, the lines have pull up resistors, it can sense TTL, CMOS, or contact closures.

5.6.5 IEEE 488.2 STANDARD

The GPIB concept expressed in IEEE-STD 488 made it easy to physically interconnect instruments but it did not make it easy for a programmer to talk to each instrument. Some companies terminated their instrument responses with a carriage return; others used a carriage return-linefeed sequence, or just a linefeed. Number systems, command names and coding depended upon the instrument manufacturer. In an attempt to standardize the instrument formats, Tektronix proposed a set of standard formats in 1985.

This is the basis for the IEEE-STD 488.2 standard that was adopted in 1987. At the same time, the original IEEE-488 Standard was renumbered to 488.1. The new IEEE-488.2 Standard established standard instrument message formats, a set of common commands, a standard Status Reporting Structure and controller protocols that would unify the control of instruments made by hundreds of manufacturers. The standard instrument message format terminates a message with a linefeed and or by asserting EOI on the last character. Multiple commands in the same message are separated by semicolons. Fixed point data are the default format for numeric responses.

Table 5.6. IEEE 488.2 commands

Command	Function
Required common commands are:	
⋆**CLS**	Clear Status Command
⋆**ESE**	Standard Event Status Enable Command
⋆**ESE?**	Standard Event Status Enable Query
⋆**ESR?**	Standard Event Status Register Query (0.255)
⋆**IDN?**	Identification Query (Company, model, serial number and revision)
⋆**OPC**	Operation Complete Command
⋆**OPC?**	Operation Complete Query
⋆**RST**	Reset Command
⋆**SRE**	Service Request Enable Command
⋆**SRE?**	Service Request Enable Query (0-255)
⋆**STB?**	Status Byte Query Z (0-255)
⋆**TST?**	Self-Test Query
⋆**WAI**	Wait-to-Continue Command
Devices that support parallel polls must support the following three commands:	
⋆**IST?**	Individual Status Query?
⋆**PRE**	Parallel Poll Register Enable Command
⋆**PRE?**	Parallel Poll Register Enable Query
Devices that support Device Trigger must support the following commands:	
⋆**TRG**	Trigger Command
Controllers must support the following command:	
⋆**PCB**	Pass Control Back Command
Devices that save and restore settings support the following commands:	
⋆**RCL**	Recall configuration
⋆**SAV**	Save configuration

The common set defined a subset of ten commands that each IEEE-488.2 compatible instrument must respond to plus additional optional commands for instruments with expanded capabilities. The required common commands simplified instrument programming by giving the programmer a minimal set of commands that one can count on being recognized by each 488.2 instrument. Table 5.6 lists the 488.2 Common Commands and their functions.

Except the *IDN, the remaining commands are used with the Status Reporting Structure. The IEEE-488.2 Standard Status Reporting Structure is shown in Fig. 5.15. The new Status Reporting Structure expanded on the 488.1 Status Byte by adding a Standard Event Status Register (ESR Register) and an Output Queue. Enable registers and summation logic was added to the Status Registers so that a user could enable selected bits in any status register. The ESR Register reports standardized device status and command errors. Bit 6 in the ESR Register is not used and can be assigned for any use by the device designer. The Standard Event Status Enable Register is used to select the specific event bits.

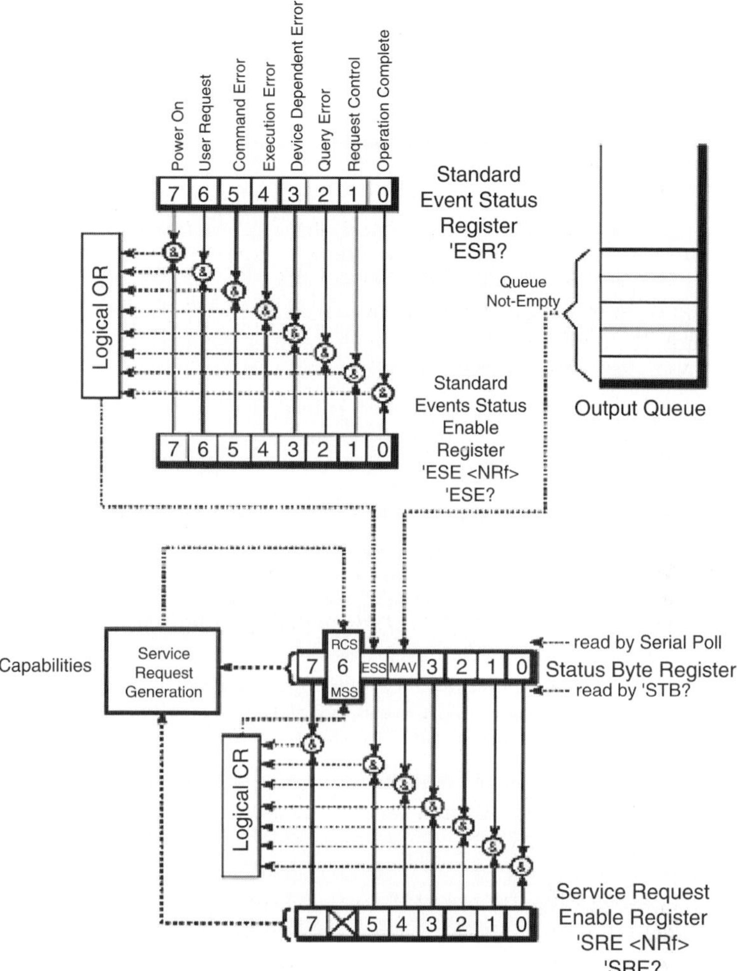

Fig. 5.15. Status reporting structure

When an enabled bit in the Event Status Register becomes true, the summary output sets the ESB bit (bit 5) in the Status Byte Register. Bits in the ESR Register are set until the register is read by the ESR query or cleared by the CLS command.

The Output Queue contains responses from the 488.2 queries. Its status is reported in the MAV bit (bit 4) of the Status Byte. Typically this bit is not enabled because the user normally follows a query by reading the response. The 488.2 Status Byte contains the ESB and MAV bits plus five user definable bits. Bit 6 is still the RQS bit but it now has a dual personality. When the status Byte is read by a Serial Poll, the RQS bit is reset. The Service Request generation is a two step process.

When an enabled bit in the ESR Register is set, the summary output sets the ESB bit in the Status Byte Register. If the ESB bit is enabled, then the RQS bit is set and an SRQ is generated. Reading or clearing the ESR Register drops the summary output which in turn, resets the ESB bit in the Status Byte. If no other enabled bits in the Status Byte are true, bit 6 and the SRQ line will be reset

Saving the Enable Register Settings

The enable register settings cannot be saved with the SAV command. The 488.2 Standard defined a PSC flag which enables clearing the ESE and SRE registers at power turn-on. The enable registers are restored to a 0 value at power turn-on when the PSC flag is set on. The PSC command disables the PSC flag and saves the enable register values.

5.6.6 Advantages of GPIB

The advantages of GPIB are summarized as follows:

1. Combines analog I/O, digital I/O, relay drivers, and temperature readings in a ICS Minibox.
2. Versatile GPIB module controls almost any device.
3. 32-line Digital Interface configurable as gated inputs or latched outputs.
4. User configurable to match the application.
5. Six analog inputs with programmable ranges measures up to 10 V dc.
6. Four 0–5 V dc analog outputs.
7. Controls analog devices.
8. High current Relay.
9. Drivers sink 300 mA.

5.7 VISA

Virtual Instrumentation Software Architecture (VISA) is a standard I/O language for instrumentation programming. VISA by itself does not provide instrumentation programming capability. VISA is a high-level API that calls into lower level drivers. The hierarchy of NI-VISA is shown in Fig. 5.16. VISA is capable of controlling VXI, GPIB, or Serial instruments and makes the appropriate driver calls depending on the type of instrument being used. When debugging VISA problems, it is important to keep in mind that this hierarchy exists.

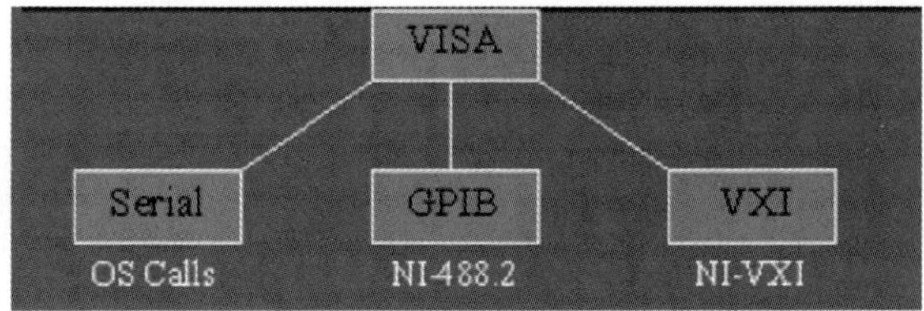

Fig. 5.16. VISA architecture

5.7.1 Supported Platforms and Environments

The platforms and environments that are supported by NI-VISA are shown in Table 5.7.

Because VISA is the language used in writing instrument drivers, most instrument drivers currently written by National Instruments supports all of these environments.

5.7.2 VISA Programming

The terminology related with VISA are explained as follows:

Resources. The most important objects in the VISA language are known as resources.

Operations. In object-oriented terminology, the functions that can be used with an object are known as operations.

Attributes. In addition to the operations that can be used with an object, the object has variables associated with it that contain information related to the object. In VISA, these variables are known as attributes.

A simplified outline of the internal structure of the VISA language is shown in Fig. 5.17.

There is a default resource manager at the top of the VISA hierarchy that can search for available resources or open sessions to them. Resources can be GPIB, serial message based VXI, or register-based VXI. The most common operations for message based instruments are read and write. The most common operations for register based Instruments are In's and Out's. In

Table 5.7. VISA platform and environment

Platform	**Environments**
Macintosh	LabVIEW, C
Windows 3.1	MSVC, Borland C + +, CVI, LabVIEW, VB
Windows 95/NT	C, CVI, LabVIEW, VB
Solaris 1	CVI, LabVIEW
Solaris 2	CVI, LabVIEW
Hp-Ux	CC, CVI, LabVIEW

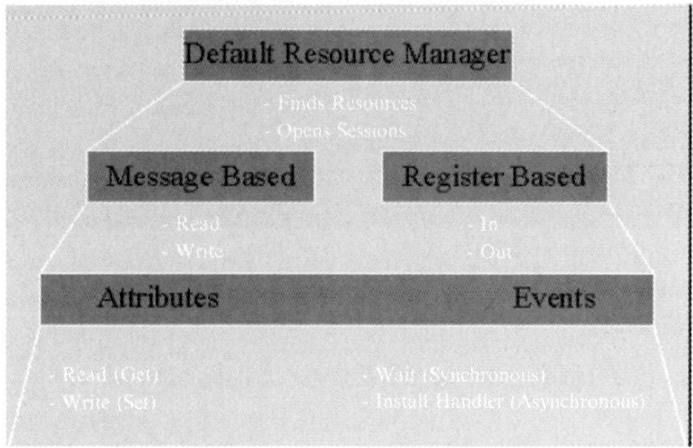

Fig. 5.17. Internal structure of VISA

addition, resources have a variety of properties associated with them known as attributes. These can be read or modified with Attribute Nodes.

5.7.3 DEFAULT Resource Manager, Session, and Instrument Descriptors

The Default Resource Manager is at the highest level of VISA operations. Communication must be established with the Resource Manager at the beginning of any VISA program. This immediately brings up two terms that need to be defined: Resource and Session.

- *Resource.* An instrument or controller.
- *Session.* A connection, or link, to the VISA Default Resource Manager or a resource.

The reason the VISA Default Resource Manager is so important is because one of its operations are to open sessions to other resources. Therefore communication with the Default Resource Manager must be established first within an application. The VISA Default Resource Manager also has another operation that it can carry out. That operation is to search for available resources in the system. Sessions can then be opened to any of these resources.

The VISA Find Resources VI that carries out the operation of searching for available resources in the system is shown in Fig. 5.18. This VI is a common starting point for a VISA Program. It can be used to determine if all of the necessary resources for the application to run are available. The only necessary input to the VISA Find Resources VI is a string called the search expression. This string determines the types of resources returned by the Find Resources VI. Possible strings for the search expression are shown in Table 5.8.

The important return values of the VI are the return count which simply reports the Number of resources that were found and the find list. The find list is an array of strings. Each string contains the description of one of the resources that was found. These strings are known as instrument descriptors.

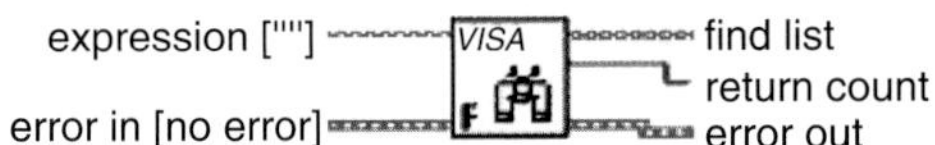

Fig. 5.18. VISA Find Resource

Table 5.8. Strings for search expression

Interface	Expression
GPIB	GPIB[0-9]*::?*INSTR
GPIB-VXI	GPIB-VXI?*INSTR
GPIB or GPIB-VXI	GPIB?*INSTR
VXI	VXI?*INSTR
All VXI	?*VXI[0-9]*::?*INSTR
Serial	ASRL[0-9]*::?*INSTR
All	?*INSTR

- *INSTRUMENT Descriptor*. The exact name and location of a VISA resource, this string has the format Interface Type [Board Index]::Address ::VISA Class The instrument descriptors are simply specific instruments found by the search query. The board index only needs to be used if more than one interface type is present in the system. For example, if the system contains two GPIB plug-in boards which could be referred to as GPIB0 and one as GPIB1. In this case, the board index needs to be used in Instrument descriptors. For VXI instruments, the "Address" parameter is the Logical Address of the instrument. For GPIB instruments, it is the GPIB primary address. For Serial instruments the address is not used. An example is "ASRL1::INSTR" as the Descriptor for the COM1 serial port on a personal computer.
- *Class*. The VISA Class is a grouping that encapsulates some or all of the VISA operations. INSTR is the most general class that encompasses all VISA operations. In the future, other classes may be added to the VISA specification. Currently the VISA Class does not need to be included as part of the instrument descriptor. Nevertheless, this should be done to ensure future compatibility. Currently, most applications simply use the INSTR.

LabVIEW supplies another way to set the class for a VISA session that can be used now. This is done by popping up on the front panel VISA Session control and selecting the VISA Class as shown in the Fig. 5.19.

The instrument descriptors are used to open sessions to the resources in the system using the Default Resource Managers ability to open sessions. The VISA Open VI shown in Fig. 5.20 carries out this operation.

The resource name input is the VISA instrument descriptor for the resource to which a session will be opened. In order to open sessions to resources in LabVEIW, a VISA Session front panel control is needed. The VISA session

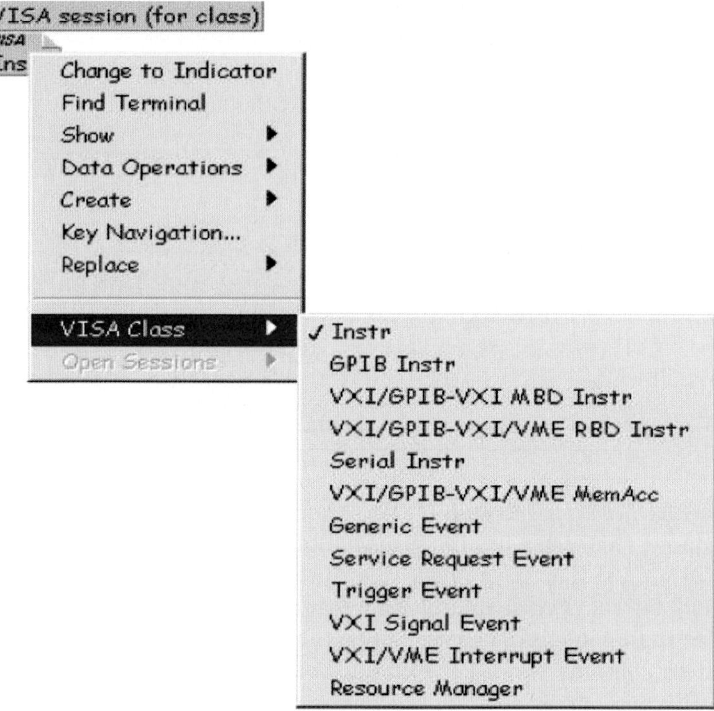

Fig. 5.19. VISA session and control

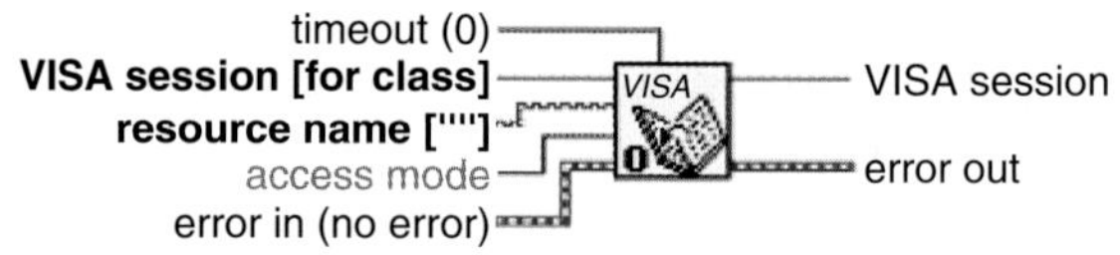

Fig. 5.20. VISA Open VI

front panel control shown in Fig. 5.21 can be found in the Controls Palette in the Path and Refnum subpalette.

To properly end a VISA program all of the sessions opened to VISA resources should be closed. To do this, a VISA Close VI is available that is shown in Fig. 5.22.

The important input to the VISA Close VI is the session to be closed. This session originally comes from the output session terminal of the VISA Open VI. If a session is not closed when a VI is run, that session will remain open. It is a good idea to always close any sessions that are opened in an application so open sessions do not build up and cause problems with system resources. If a VI runs and leaves a session open without closing it – this session can still be used in the later VIs. An already open session can be accessed by popping up on a VISA Session Front panel control and selecting the Open sessions.

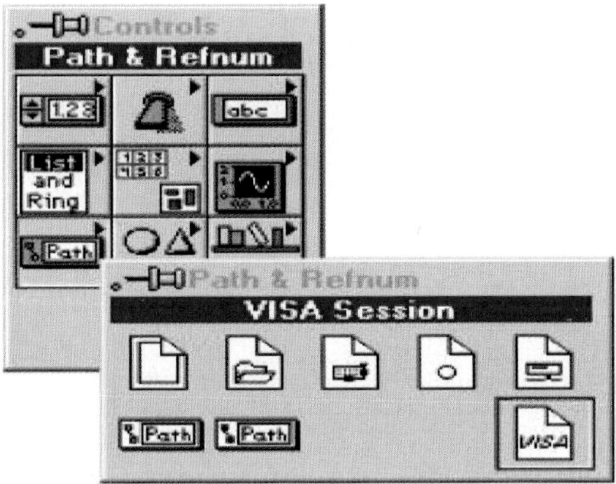

Fig. 5.21. VISA session front panel

Fig. 5.22. VISA Close VI

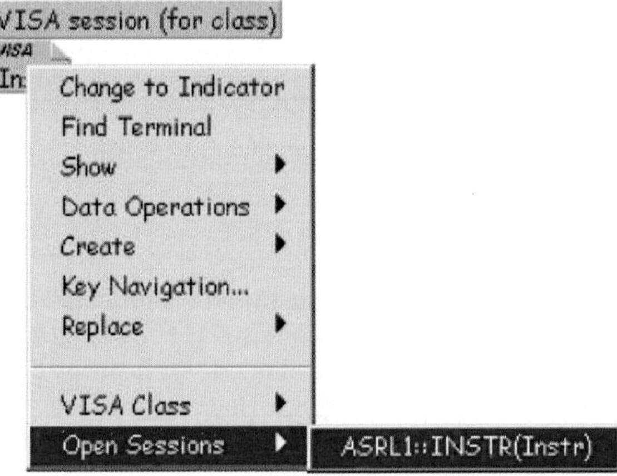

Fig. 5.23. Open session

The output of the VISA Session front panel control will then be the already open session that is selected. An example of selecting an Already open session is shown in the Fig. 5.23.

The open session's choice of the front panel VISA Session control can be used to check if any sessions are currently open. Accessing open sessions by popping up on front panel VISA Session Controls also provides a convenient way to interactively run parts of an Instrument driver.

Fig. 5.24. Error cluster

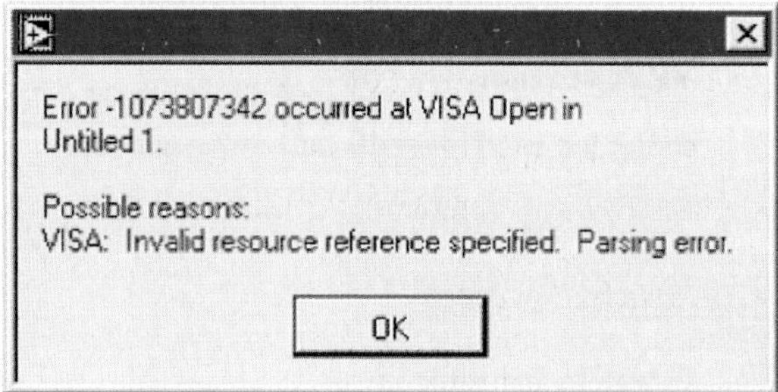

Fig. 5.25. Front Panel error cluster

Error Handling with VISA

Error handling with VISA VIs is similar to error handling with other I/O VIs in LabVIEW. Each of the VISA VIs contains Error Input and Error Output terminals that are used to pass error clusters from one VI to another. The error cluster shown in Fig. 5.24 contains a Boolean flag indicating whether an error has occurred, a numeric VISA error code, and a string containing the location of the VI where the error occurred.

If an error occurs, subsequent VIs will not try to execute and will simply pass on the error cluster. A front Panel error cluster indicator showing the output from the error out terminal of a VISA function is shown in Fig. 5.25.

VISA error codes are 32-bit integers that are usually referred to in hexadecimal format. It is not always convenient to handle errors with pop-up dialogue boxes through the LabVIEW error Handling VIs. VISA also provides an operation that will take a VISA error code and produce the error message string corresponding to the code as an output. This VI is shown in Fig. 5.26.

The inputs to this VI are a VISA session and a VISA error cluster. The VI will check the VISA code in the error cluster that was passed to it and output the text description of the code from the status description terminal. Even if

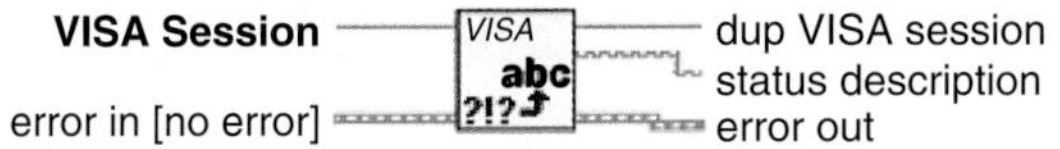

Fig. 5.26. Error message

the input error cluster has the Error value set, a status string description indicating success or a warning will be produced. This VI allows a text description of the status after a VISA VI executes to be used in a program. A LabVIEW string indicator displaying the error string returned from the VISA Status Description VI is also obtained. The exact method used for implementing error handling will depend on the nature of the program. However, some sort of error handling mechanism should be implemented in any program involving VISA.

5.7.4 VISAIC and Message-Based Combination

VISA comes with a utility called VISA Interactive Control (VISAIC) on all platforms that Support VISA and LabVIEW (except the Macintosh). This utility provides access to all of VISA's functionality interactively, in an easy-to-use graphical environment. It is a convenient starting point for program development and learning about VISA. When VISAIC runs it automatically finds all of the available resources in the system and lists the instrument descriptors for each of these resources under the appropriate resource type. The VISAIC opening window is shown in Fig. 5.27.

The Soft Front Panels tab of the main VISAIC panel will give the option to launch the soft front panels of any VXI plug and play instrument drivers that have been installed on the system. The NI I/O tab will give the option to launch the NI-VXI interactive utility or the NI-488 interactive utility. This gives convenient links into the interactive utilities for the drivers VISA calls in case would like to try debugging at this level. Double-clicking on any of the instrument descriptors shown in the VISA IC window will open a session to that instrument.

Opening a session to the instrument produces a window with a series of tabs for interactively running VISA commands. The exact appearance of these tabs depends on which compatibility mode VISAIC is in. The compatibility mode and other VISAIC preferences are accessed by selecting Edit ≫ Preferences. The window shown in Fig. 5.25 appears.

For VXI instruments, VISAIC also attempts to access one of the configuration registers on each instrument to verify its existence. In upcoming lessons VISAIC will be used as a learning and development tool. It is often easier to develop an application interactively before doing the actual programming. VISAIC can also serve as a very useful debugging tool when problems arise in writing a program or in using a pre-existent application using VISA VIs.

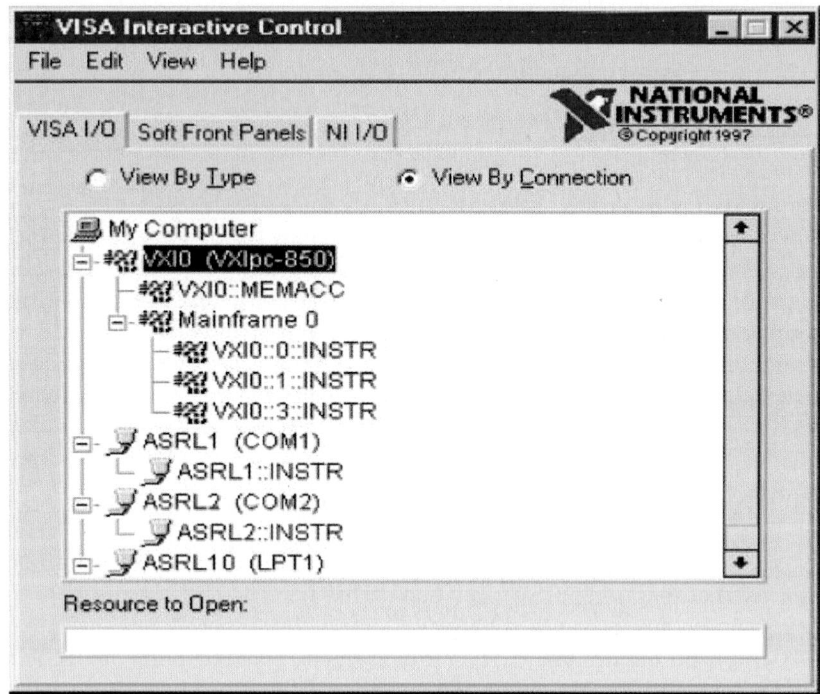

Fig. 5.27. VISAIC opening window

5.7.5 Message-Based Communication

Serial, GPIB, and many VXI devices recognize a variety of message-based command strings. At the VISA level, the actual protocol used to send a command string to an instrument is transparent. The user can only write a message or read a message from a message-based device. The VIs that are used to perform these operations are VISA Write and VISA Read.

The VISA Write VI is shown in Fig. 5.29. This function writes data to the device synchronously or asynchronously depending on the platform it is being used.

The VISA Read VI is equally easy to use. It is shown in Fig. 5.30. The VISA Read VI must be given a byte count input equal to the maximum number of bytes that should be read from the instrument. The VI will terminate the read when this number of bytes has been read or when the end of the transfer is indicated.

The actual message-based commands that the instrument recognizes vary from manufacturer to manufacturer. The GPIB IEEE 488.2 protocol and SCPI (Standard Communication Programming Interface) standards made some attempts at standardizing the commands for message-based instruments.

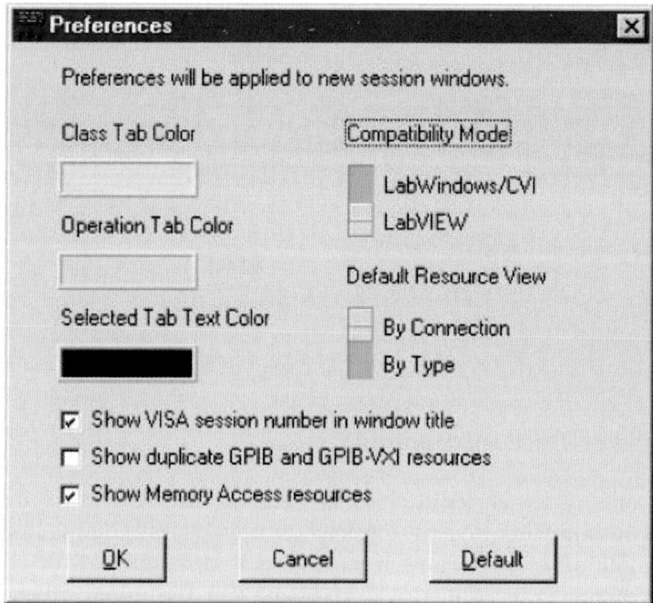

Fig. 5.28. VISAIC preferences

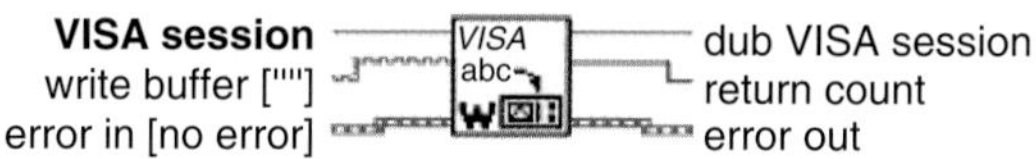

Fig. 5.29. VISA Write

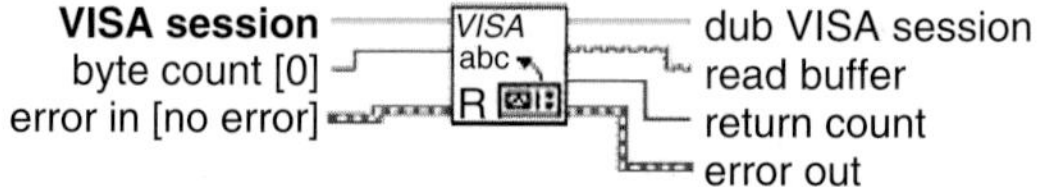

Fig. 5.30. VISA Read

However, pre-written instrument drivers exist for many message-based devices. These instrument drivers contain functions that put together the appropriate ASCII command strings and send them to the instrument. A simple example that writes the *IDN string to a message-based instrument and reads the response is shown in Fig. 5.31.

This program could be used successfully with any device that recognizes the *IDN command. The device could be Serial, GPIB, or VXI. This device selection can be made in the instrument descriptor.

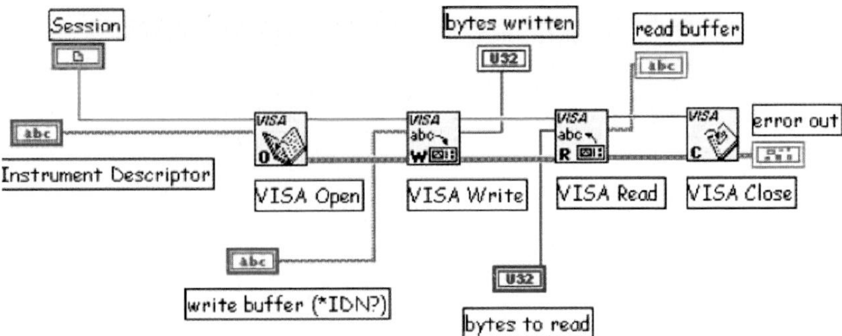

Fig. 5.31. Message-based instrument

5.7.6 Register-Based Communication

VISA contains a set of register access VIs for use with VXI instruments. Some VXI instruments do not support message-based commands. The only way to communicate with these instruments is through register accesses. All VXI instruments have configuration registers in the upper 16 kbytes of the memory space. Therefore, register access functions can also be used to read from and write to the configuration registers for message-based devices. The basic VISA operation used to read a value from a register is VISA In, which is shown in Fig. 5.32. There are actually three different versions of this operation for reading a 8-, 16-, or 32-bit value. This function reads in a 16-bit word from a specified address space assigned with a memory base and offset. The valid address space values are:

- A16 Space: 1
- A24 Space: 2
- A32 Space: 3

This VI and other basic register access VIs can be found in the High Level Register Access subpalette under the main VISA function palette as shown in Fig. 5.33.

The address space input simply indicates which VXI address space is to be used. The offset input sometimes causes confusion. But VISA keeps track of the base memory address that a device requests in each address space. The offset input is relative to this base address.

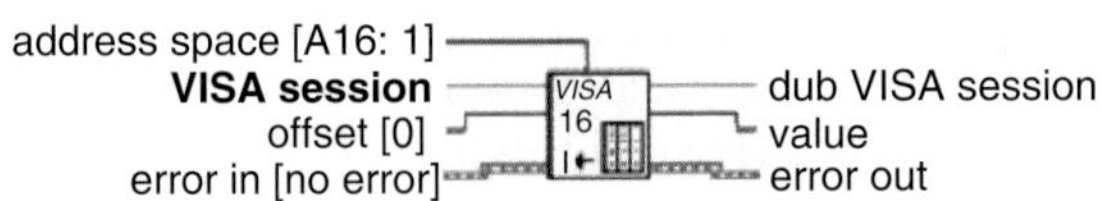

Fig. 5.32. VISA In 16 VI

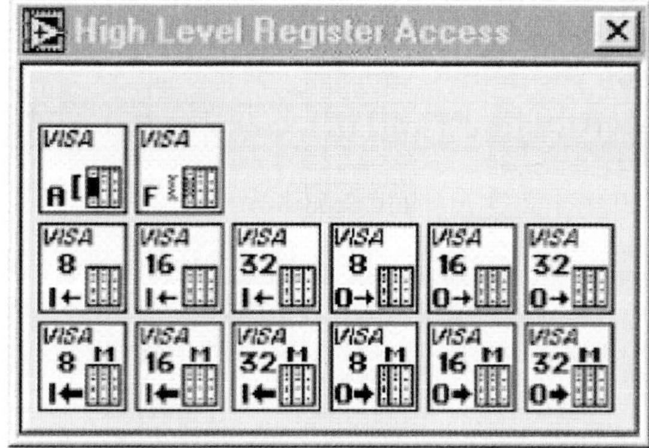

Fig. 5.33. High level register

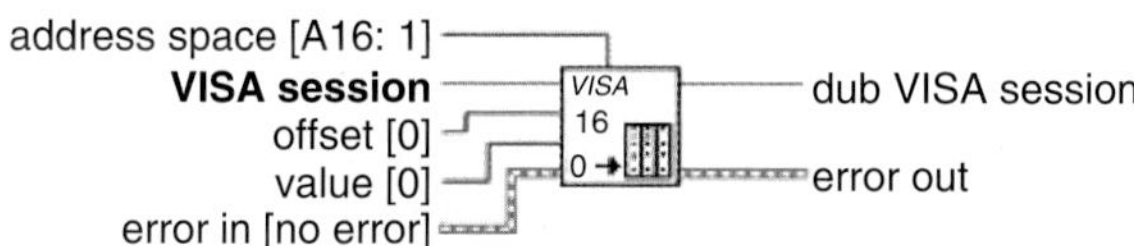

Fig. 5.34. VISA out VI

There is another set of high level register access operations that parallel the VISA In operations, but are used for writing to registers. These operations are the VISA Out operations shown in Fig. 5.34.

This VI is similar to the VISA In 16 VI except the value to write must be provided to the value terminal.

An example of using the high level VISA access functions in a VI is shown in the simple program in Fig. 5.35. The program asks the user for the instrument descriptor and the offset of the configuration register to be read and returns the value from the register. If an error occurs in the sequence of VISA VIs that execute in the program, the Simple Error Handler will pop up a dialogue box informing the user about the status of the error and outputs a text message associated with the VISA Error Code.

5.7.7 VISA Attributes

The basic operations that are associated with message and register-based resources in VISA operations allow register access, and message based communication. In addition to the basic communication operations, VISA resources have a variety of attributes whose values can be read or set in a program. In a LabVIEW program, these attributes are handled programmatically similar to the properties of front panel controls and indicators. Attribute nodes are used to read or set the values of VISA attributes. A VISA attribute node is shown in Fig. 5.36. This node grows and evaluates from the top proceeding down the node until the final attribute occurs.

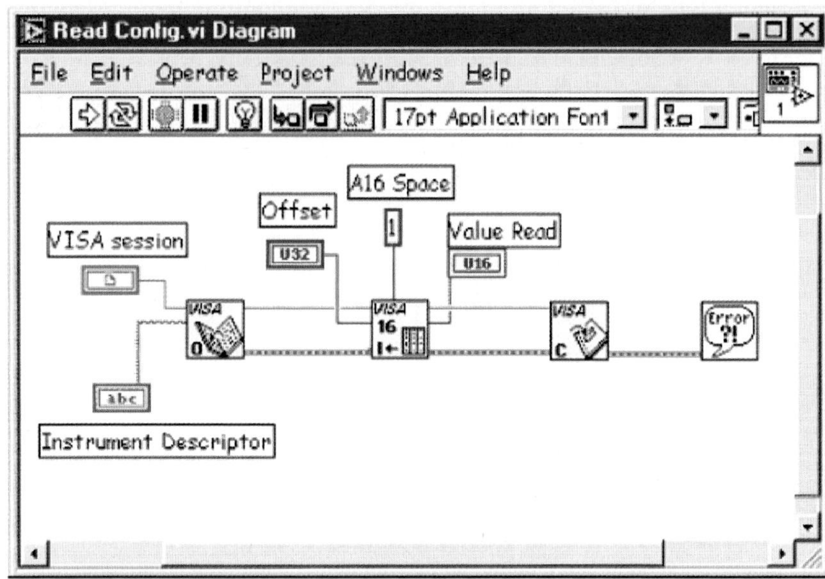

Block Diagram

Fig. 5.35. Illustration of VISA Access Functions

The VISA attribute node contains a single attribute terminal when it is placed initially on the block diagram. However, it can be resized to contain as many different terminals as necessary. The initial terminal on the VISA attribute node is a read terminal. This means that the value of the attribute selected in that terminal will be read. This is indicated by the small arrow pointing to the right at the right edge of the terminal. Terminals can

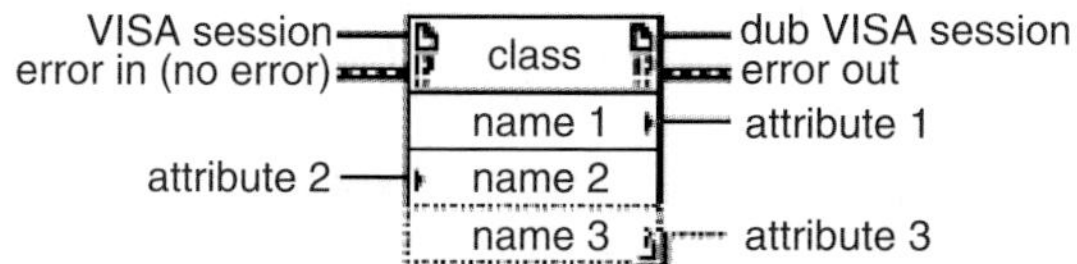

Fig. 5.36. VISA attribute node

individually be changed from a read terminal to a write terminal by popping-up on the attribute.

The attribute in each terminal of the attribute node can be selected by popping up on the attribute node terminal and choosing "Select Item." This will provide a list of all the possible attributes that can be set in the program. This will produce a long list of attributes. The number of different attributes that are shown under the Select Item choice of the VISA Attribute Node can be limited.

Several different classes can be selected by popping up the VISA Attribute Node and selecting class. The default INSTR class encompasses all the VISA attributes. These classes simply limit the attributes displayed to those related to that class instead of all the VISA attributes. Once a session is connected to the Session input terminal of the attribute, the VISA Class will be set to the class associated with that session. Initially, the VISA attributes will be somewhat unfamiliar and their exact nature may not be clear from the name alone.

Some of the common attributes for each resource type are shown in the list.

Serial

Serial baud rate. The baud rate for the serial port.
Serial data bits. The number of data bits used for serial transmissions.
Serial parity. The parity used for serial transmissions.
Serial stop bits. The number of stop bits used for serial transmissions.

GPIB

GPIB Readdressing. Specifies if the device should be readdressed before every write operation.
GPIB Unaddressing. Specifies if the device should be unaddressed after read and write Operations.

VXI

Mainframe logical address. The lowest logical address of a device in the same chassis With the resource.

Manufacturer identification. The manufacturer ID number from the device's Configuration registers.

Model code. The model code of the device from the device's configuration registers.

Slot. The slot in the chassis that the resource resides in.

VXI logical address. The logical address of the device.

VXI memory address space. The VXI address space used by the resource.

VXI memory address base. The base address of the memory region used by the resource.

VXI memory address size. The size of memory region used by the resource.

There are many other attributes besides those given in this list. There are also attributes that are not specific to a certain interface type. The timeout attribute which is the timeout used in message-based I/O operations is a good example of such an attribute.

5.7.8 Advantages of VISA

One of VISA's advantages is that it uses many of the same operations to communicate with instruments regardless of the interface type. For example, the VISA command to write an ASCII string to a message-based instrument is the same whether the instrument is Serial, GPIB, or VXI. Thus, VISA provides interface independence. This facility can make it easy to switch between interfaces.

VISA is also designed so that programs written using VISA function calls are easily portable from one platform to another. To ensure this, VISA strictly defines its own data types such that, issues like the size of an integer variable from one platform to another should not affect a VISA program.

The VISA function calls and their associated Parameters are uniform across all platforms; software can be ported to other platforms and then recompiled. In other words, a C program using VISA can be ported to other Platforms supporting C. A LabVIEW program can be ported to other platforms supporting LabVIEW. At first, the abundance of new VISA data types might seem like a Burden, but LabVIEW makes them virtually transparent when using VISA. Another advantage of VISA is that it is an object-oriented language which will easily adapt to new instrumentation interfaces as it is developed in the future. The interface independence that is built into the language will also make it even easier for programmers to move to new interfaces with VISA in the future.

VISA's greatest advantage, perhaps, is that it is an easy language to learn and use. Its Object-oriented structure makes the language and its operations intuitive to learn. This is due to the fact that VISA provides interface independence by using the same operations for different interfaces and also by the fact VISA presents a very simple and Easy-to-use API.

VISA provides the most frequently used functionality for Instrumentation programming in a very compact command set.

Summary

Various interface instruments available were explained along with their configuration details. Characteristic feature of serial ports like RS232, 4–20 mA current loop, 60 mA current loop were discussed and compared. GPIB architecture and its advantage also were delineated in detail for the user's choice. VISA architecture and its configuration utility were detailed along with some of the common attributes for each resource type such as GPIB, VXI, and serial ports.

Review Questions

1. Write the need for a current loop.
2. Explain in detail about 20 mA current loop interface.
3. Compare and contrast RS 232C and current loop interfaces.
4. Write the basic requirements for 4–20 mA transducers.
5. List the advantages of 4–20 mA current loop.
6. Discuss about the smart transmitter using 20 mA current loop.
7. Give the key features of 60 mA current loop.
8. Write the features of RS 232C interface.
9. What are the limitations of RS 232C.
10. List the RS 232C signals and explain its functions.
11. Write notes on RS 422 and 485.
12. Compare all serial communication standards.
13. Explain in detail about functional features and signals of IEEE 488 with neat sketches.
14. List the commands of IEEE 488.2 bus.
15. Give the advantages of GPIB and VISA.
16. Explain the internal structure of VISA with a diagram.
17. How 40 and 60 mA current loop can be differentiated?
18. What are the rules for data transmission in RS 232?
19. What is the necessity for GPIB?
20. Explain the stages involved in data reception process using GPIB with a neat schematic diagram.
21. What is meant by VISA.
22. State the principles architecture of VISA.
23. Explain the performance of VISA.
24. How does the VISA methodology works?

6

Interface Buses

Learning Objectives. On completion of this chapter the reader will have a knowledge on:

- Interface bus standards and types
- Architecture of USB, need for USB, Data Formats
- USB Functions, USB descriptor, advantages of USB
- PCI, Architecture of PCI
- Features of PCI
- Types of PCI Bus
- PCI Express
- Need for PCI Express
- Types of PCI Express Architecture
- Performance of PCI Express
- PXI Architecture
- Features of PCMCIA
- Types of PC Cards
- Features of PC Card Technology
- SCXI Hardware and Software
- Analog Input Signal Connections
- SCXI Software-Configurable Settings
- Need for VXI
- Features of VXI
- Message-Based Communication and Serial Protocol
- LXI Modular Switching Chassis
- LXI/PXI Module Selection

6.1 Introduction

In computer architecture, a bus is a subsystem that transfers data or power between computer components inside a computer or between computers. Unlike a point-to-point connection, a bus can logically connect several peripherals over the same set of wires. Each bus defines its set of connectors to physically plug devices, cards or cables together. Early computer buses were literally parallel electrical buses with multiple connections, but the term is now used for any physical arrangement that provides the same logical functionality as a parallel electrical bus. Modern computer buses can use both parallel and bit-serial connections, and can be wired in either a multidrop or daisy chain topology, or connected by switched hubs, as in the case of Universal Serial Bus (USB). The users are not limited to the type of instrument that is required to be controlled. This section describes some of the most common and latest PC interface buses such as USB, PCI, PXI, PCMCIA, VXI, and LXI. The chapter includes the motivation and scope for USB with its standard specifications. The section also focuses on the goals of PCI and its higher versions such as the PXI and PCMCIA which are Plug'n'Play and simplicity to the end user. It introduces the readers with the advanced interface buses like VXI and LXI.

6.2 USB

In the latest PCs, many of the peripherals such as the keyboard, mouse, scanners, printers, modem and even external CD-ROM drives and CD-writers are connected to the computer through a USB cable. This section gives a brief introduction to the Universal Serial Bus or USB. As personal computers became more and more powerful and began to handle photographic images, audio, video and other bulky data, it became evident that existing communications buses were not good enough to carry the data back-and-forth between the computers and peripherals. The users wanted the bulky data to be moved around faster, rather than more slowly. So a group of leading computer and telecom firms including IBM, Intel, Microsoft, Compaq, Digital Equipment, NEC, and Northern Telecom got together and developed USB.

6.2.1 Architecture of USB

The USB is a medium-speed serial data bus designed to carry relatively large amounts of data over relatively short cables: up to about 5 m long. It can support data rates of up to $12\,\mathrm{Mb\,s^{-1}}$ (megabits per second), which is fast enough for most PC peripherals such as scanners, printers, keyboards, mice, joysticks, graphics tablets, low-res digital cameras, modems, digital speakers, low speed CD-ROM and CD-writer drives, external Zip disk drives, and so on. The USB is an addressable bus system, with a 7-bit address code so it can support up to 127 different devices or nodes. However, it can have only one host. So a PC with its peripherals connected through the USB forms a star Local Area Network (LAN).

On the other hand any device connected to the USB can have a number of other nodes connected to it in daisy-chain fashion, so it can also form the hub for a mini-star subnetwork. Similarly, it can have a device which purely

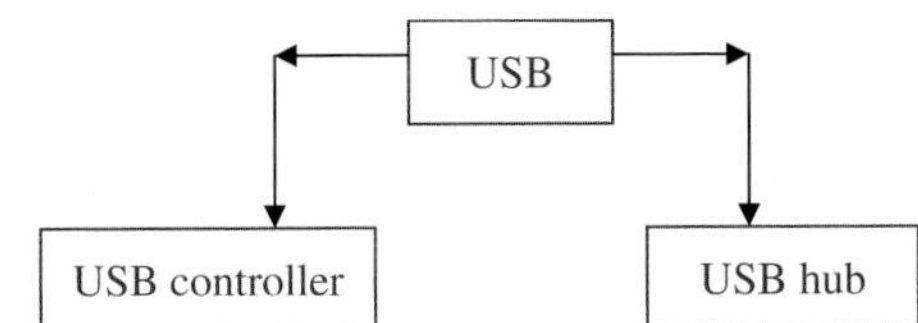

Fig. 6.1. USB structure – star topology

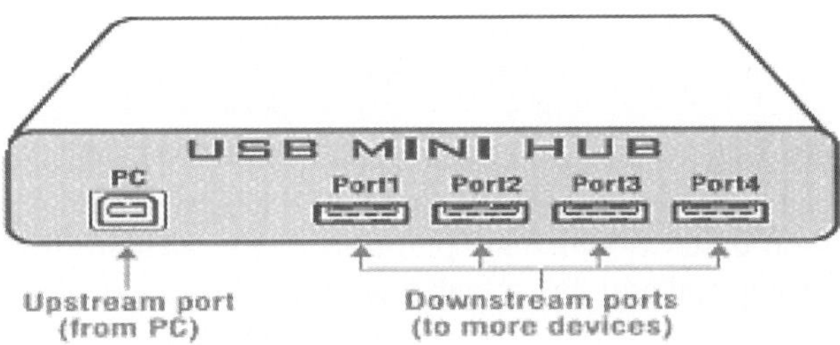

Fig. 6.2. USB ports

functions as a hub for other node devices, with no separate function of its own. This expansion through hubs is available because the USB supports a tiered star topology, as shown in Fig. 6.1.

Each USB hub acts as a traffic cop for its network, routing data from the host to its correct address and preventing bus contention clashes between devices trying to send data at the same time. On a USB hub device, the single port is used to connect to the host PC either directly or via another hub known as the upstream port, while the ports used for connecting other devices to the USB are known as the downstream ports as illustrated in Fig. 6.2.

Most hubs provide either four or seven downstream ports, or less. Another important feature of the USB is that it is designed to allow hot swapping. Devices can be plugged into and unplugged from the bus without having to turn the power off and on again, re-boot the PC or even, manually start a driver program. A new device can simply be connected to the USB, and the PCs operating system should recognize it and automatically set up the necessary driver to service it.

6.2.2 Need for USB

The USB is host controlled. There can only be one host per bus. The specification in itself does not support any form of multi master arrangement. This is aimed at and limited to single point to point connections such as a mobile phone and personal organizer and not multiple hub, multiple device desktop configurations. The USB host is responsible for undertaking all transactions and scheduling bandwidth. Data can be sent by various transaction methods using a token-based protocol. One of the original intentions of USB was to reduce the amount of cabling at the back of PC. The idea came from the Apple Desktop Bus, where the keyboard, mouse and some other peripherals could be connected together (daisy chained) using the one cable.

However, USB uses a tiered star topology, similar to that of 10BaseT Ethernet. This imposes the use of a hub somewhere, which adds to greater expense, more boxes on desktop and more cables. However it is not as bad as it may seem. Many devices have USB hubs integrated into them. For example, keyboard may contain a hub which is connected to computer. Mouse and other devices such as digital camera can be plugged easily into the back of keyboard. Monitors are just another peripheral on a long list which commonly has in-built hubs.

This tiered star topology, rather than simply daisy chaining devices together has some benefits. First power to each device can be monitored and even switched-off if an over current condition occurs without disrupting other USB devices. High, full and low speed devices can be supported, with the hub filtering out high speed and full speed transactions so lower speed devices do not receive them. Up to 127 devices can be connected to any one USB bus at any one given time. To extent devices simply add another port/host. While earlier USB hosts had two ports, most manufacturers have seen this as limiting and are starting to introduce 4 and 5 port host cards with an internal port for hard disks etc. The early hosts had one USB controller and thus both ports shared the same available USB bandwidth. As bandwidth requirements have increased, multiport cards with two or more controllers allowing individual channels are used.

6.2.3 Power Cables

USB cables consist of two twisted pairs of wires, one pair used to carry the bidirectional serial data and the other pair 5 V DC power. This makes it possible for low-powered peripherals such as a mouse, joystick or modem to be powered directly from the USB or strictly from the host PC (or the nearest hub) upstream, through the USB. Most modern PCs have two USB ports, and each can provide up to 500 mA of 5 V DC power for bus powered peripherals. Individual peripheral devices (including hubs) can draw a maximum of 100 mA from their upstream USB port, so if it require less than this figure for operation it can be bus Powered. If it needs more, it has to use their own power supply such as a plug-pack adaptor. Hubs should be able to supply up to 500 mA at 5 V from each downstream port, if it is not bus powered. Serial data is sent along the USB in differential or push-pull mode, with opposite polarities on the two signal lines.

This improves the signal-to-noise ratio (SNR), by doubling the effective signal amplitude and also allowing the cancellation of any common-mode noise

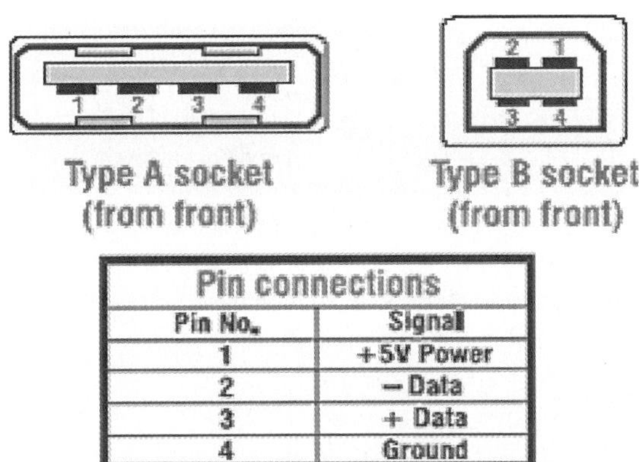

Pin connections	
Pin No.	Signal
1	+5V Power
2	− Data
3	+ Data
4	Ground

Fig. 6.3. USB pin and signals

induced into the cable. The data is sent in nonreturn-to-zero (NRZ) format, with signal levels of 3.3 V peak (i.e., 6.6 V peak differential).

USB cables use two different types of connectors:

- *Type A*. Plugs for the upstream end.
- *Type B*. Plugs for the downstream end.

Hence the USB ports of PCs are provided with matching Type A sockets, are the downstream ports of hubs, while the upstream ports of USB devices (including hubs) have Type B sockets. Type A plugs and sockets are flat in shape and have the four connections in line, while Type B plugs and sockets are much squarer in shape and have two connections on either side of the center spigot.

Both types of connector are polarized so it cannot be inserted the wrong way around. Figure 6.3 shows the pin connections for both type of connector, with sockets shown and viewed from the front. Though USB cables having a Type A plug at each end are available, it should never be used to connect two PCs together, through their USB ports. This is because a USB network can only have one host, and both would try to claim that role. In any case, the cable would also short their 5 V power rails together, which could cause a damaging current to flow.

USB is not designed for direct data transfer between PCs. All normal USB connections are made using cables with a Type A plug at one end and a Type B plug at the other; although extension cables with a Type A plug at one end and a Type A socket at the other can also be used, providing the total extended length of a cable is within 5 m. Moreover, USB cables are usually easy to identify as the plugs have a distinctive symbol molded into them. (Fig. 6.4)

Fig. 6.4. USB socket

6.2.4 Data Formats

USB data transfer is essentially in the form of packets of data, sent back-and-forth between the hosts and peripheral devices. However, USB is designed to handle different types of data. The two main formats are the asynchronous mode and synchronous mode.

Asynchronous Mode. This mode is used for transferring data that is not time critical. This format is used to transmit data from a scanner or digital scanner for example to a printer. The packets can be interleaved on the USB with others being sent to or from other devices.

Synchronous Mode. The other main format is synchronous mode, used to transfer data that is time critical, such as audio data to digital speakers, or to/from a modem. These packets must not be delayed by those from other devices.

The two other data formats are interrupt format, used by devices to request servicing from the PC/host, and control format, used by the PC/host to send token packets to control bus operation, and by all devices to send handshake packets to indicate whether the data it has just received was acknowledged (ACK) or had errors (NAK).

Data Signaling Rate

The speeds of the USB along with the tolerance of the USB clocks as per the specification are:

- High speed data is clocked at 480 $\mathrm{Mb\,s^{-1}}$ with a data signaling tolerance of $\pm$500 ppm.
- Full speed data is clocked at 12 $\mathrm{Mb\,s^{-1}}$ with a data signaling tolerance of $\pm$0.25% or 2,500 ppm.
- Low speed data is clocked at 1.50 $\mathrm{Mb\,s^{-1}}$ with a data signaling tolerance of $\pm$1.5% or 15,000 ppm.

This allows resonators to be used for low cost low speed devices, but rules them out for full or high speed devices.

Data Packets

Data on the USB bus is transmitted with the LSB first. USB packets consist of the following fields:

- *Sync.* All packets must start with a sync field. The sync field is 8 bits long, which is used to synchronize the clock of the receiver with the transmitter. The last two bits indicate the location of the PID field.
- *PID.* PID stands for Packet ID. This field is used to identify the type of packet that is being sent. Table 6.1 shows the possible values with 4 bits for the PID. To ensure that the data is received correctly, the 4 bits are complemented and repeated, making an 8 bit PID in total.
- *ADDR.* The address field specifies the device that is designated by the packets. Since the packet size is 7 bits, USB can address 127 devices. Address 0 is not valid, as any device which is not assigned an address must respond to packets sent to address zero.
- *ENDP.* The endpoint field is made up of 4 bits, allowing 16 possible endpoints. Low speed devices, can only have 2 endpoint additional addresses on top of the default pipe.
- *CRC.* Cyclic Redundancy Checks are performed on the data within the packet payload. All token packets have a 5-bit CRC while data packets have a 16-bit CRC.

Packet Types

USB has four different packet types. Token packets indicate the type of transaction to follow, data packets contain the payload, handshake packets are used

Table 6.1. PID values

Group	PID Value	Packet Identifier
Token	0001	OUT Token
	1001	IN Token
	0101	SOF Token
	1101	SETUP Token
Data	0011	DATA0
	1011	DATA1
	0111	DATA2
	1111	MDATA
Handshake	0010	ACK Handshake
	1010	NAK Handshake
	1110	STALL Handshake
	0110	NYET (No Response Yet)
Special	1100	PREamble
	1100	ERR
	1000	Split
	0100	Ping

for acknowledging data or reporting errors and start of frame packets indicate the start of a new frame:

1. *Token Packets.* There are three types of token packets:
   - In – Informs the USB device that the host is ready to read information.
   - Out – Informs the USB device that the host is ready to send information.
   - Setup – Used to begin control transfers.

   Token Packets must conform to the following format:

Sync	PID	ADDR	ENDP	CRC5	EOP

2. *Data Packets.* There are two types of data packets each capable of transmitting 0–1,023 bytes of data:
   - Data0
   - Data1

   Data packets have the following format:

Sync	PID	Data	CRC16	EOP

- *Handshake Packets.* There are three type of handshake packets which consist simply of the PID
  - ACK – Acknowledges that the packet has been successfully received.
  - NAK – Reports that the device cannot send nor receive data temporary. It is also used during Interrupt transaction to inform the host that there is no data to send.
  - STALL – The device finds that it is in a state that it requires intervention from the host.

  Handshake Packets have the following format:

Sync	PID	EOP

3. *Start of Frame Packets.* The SOF packet consists of a 11-bit frame number is sent by the host every 1–500 ns. SOF Packets have the following format:

Sync	PID	Frame Number	CRC5	EOP

6.2.5 Speed

USB supports two different speeds, a full speed mode of 12 Mb s^{-1} and a low speed mode of 1.5 Mb s^{-1}. The 1.5 Mb s^{-1} mode is slower and less susceptible to EMI, thus reducing the cost of ferrite beads and quality components. For example, crystals can be replaced by low cost resonators. The latest USB supports 480 Mb s^{-1} and is known as High Speed mode. This High speed mode competes with the Fire wire Serial Bus.

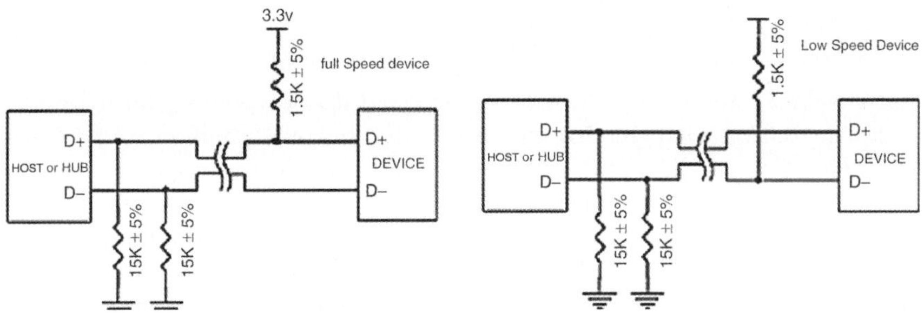

Fig. 6.5. Full speed and low speed with pull up resistor

A USB device must indicate its speed by pulling either the D+ or D– line high to 3.3 V. A full speed device, pictured later will use a pull up resistor attached to D+ to specify itself as a full speed device. These pull up resistors at the device end will also be used by the host or hub to detect the presence of a device connected to its port. Without a pull up resistor, USB assumes that there is no device connected to the bus. Some devices have this resistor built into its silicon, which can be turned on and off under firmware control, others require an external resistor.

High speed devices will start by connecting as a full speed device. Once it has been attached, it will do a high speed chirp during reset and establish a high speed connection if the hub supports it. If the device operates in high speed mode, then the pull up resistor is removed to balance the line as shown in Fig. 6.5. To summarize, USB Speeds are grouped as:

- High Speed – 480 $\mathrm{Mb\,s^{-1}}$
- Full Speed – 12 $\mathrm{Mb\,s^{-1}}$
- Low Speed – 1.5 $\mathrm{Mb\,s^{-1}}$

6.2.6 Electrical Specifications

USB uses a differential transmission pair for data. This is encoded using non-return to zero (NRZ) and is bit stuffed to ensure adequate transitions in the data stream. On low and full speed devices, a differential '1' is transmitted by pulling D+ over 2.8 V with a 15 kΩ resistor pulled to ground and D– under 0.3 V with a 1.5 kΩ resistor pulled to 3.6 V. A differential '0' on the other hand is a D– greater than 2.8 V and a D+ less than 0.3 V with the same appropriate pull down/up resistors. The receiver defines a differential '1' as D+ 200 mV greater than D– and a differential '0' as D+ 200 mV less than D–.

The polarity of the signal is inverted depending on the speed of the bus. Therefore the terms 'J' and 'K' states are used in signifying the logic levels. In low speed a 'J' state is a differential 0. In high speed a 'J' state is a differential 1. USB transceivers will have both differential and single ended outputs. Certain bus states are indicated by single ended signals on D+, D− or both. For example a Single Ended Zero (SE0) can be used to signify a device reset if held for more than 10 mS. SE0 is generated by holding both D− and D+ low (<0.3 V). Single ended and differential outputs are important to note while using a transceiver and FPGA as USB device. The low speed/full speed bus has a characteristic impedance of 90 Ω +/ − 15%.

High speed (480 Mb s^{-1}) mode uses a 17.78 mA constant current for signaling to reduce noise. One of the benefits of USB is bus-powered devices – devices which obtain its power from the bus and requires no external plug packs or additional cables. There are three classes of USB functions,

- Low-powered bus functions
- High-powered bus functions
- Self-powered functions

Low powered bus functions draw all its power from the VBUS and cannot draw any more than one unit load. The USB specification defines a unit load as 100 mA. Low powered bus functions must also be designed to work down to a Bus Voltage (VBUS) of 4.40 V and up to a maximum voltage of 5.25 V measured at the upstream plug of the device.

High powered bus functions will draw all its power from the bus and cannot draw more than one unit load until it has been configured, after which it can then drain 5 unit loads (500 mA Max) provided it asked for this in its descriptor. High power bus functions must be able to be detected and enumerated at a minimum 4.40 V. When operating at a full unit load, a minimum VBUS of 4.75 V is specified with a maximum of 5.25 V. Once again, these measurements are taken at the upstream plug. Self-power functions may draw up to 1 unit load from the bus and derive the rest of its power from an external source. If this external source fails, it must have provisions in place to draw no more than 1 unit load from the bus.

Self-powered functions are easier to design according to specifications as it does not involve power consumption. A 1 unit bus powered load allows the detection and enumeration of devices without mains/secondary power applied. No USB device, whether bus powered or self-powered can drive the VBUS on its upstream facing port.

Other VBUS considerations are the in rush current which must be limited. Inrush current is contributed to the amount of capacitance on device between VBUS and ground. The specification therefore specifies that the maximum decoupling capacitance it can have on device is 10 μF. When the device is disconnected after current flow through the inductive USB cable, a large fly back voltage occurs on the open end of the cable. To prevent this, a 1 μF decoupling capacitance is used. For the typical bus powered device, it cannot drain any more than 500 mA.

6.2.7 Suspend Mode

A USB device will enter suspend mode when there is no activity on the bus for greater than 3.0 ms. It then has a further 7 ms to shutdown the device and draws no more than the designated suspend current. Thus the device must draw the rated suspend current from the bus 10 ms after the bus activity has stopped. In order to maintain connected to a suspended hub or host, the device must still provide power to its pull up speed selection resistors during suspend. USB has a start of frame packet sent initially to prevent an idle bus from entering suspend mode in the absence of data.

- A high speed bus will have microframes sent every 125.0 μs ± 62.5 ns.
- A full speed bus will have a frame sent down every 1.000 ms ±500 ns.
- A low speed bus will have an End of Packet (EOP) every 1 ms only in the absence of any low speed data.

The term "Global Suspend" is used when the entire USB bus enters suspend mode collectively. However, selected devices can be suspended by sending a command to the hub that the device is connected too. This is referred to as a "Selective Suspend." The device will resume operation when it receives any non idle signaling. If a device has remote wakeup enabled then it may signal to the host to resume from suspend.

6.2.8 Cables or Pipes

While the device sends and receives data on a series of endpoints, the client software transfers data through pipes. A pipe is a logical connection between the host and endpoint(s). Pipes will also have a set of parameters associated with them such as the amount of bandwidth that is allocated to it, transfer type (Control, Bulk, ISO, or Interrupt) it uses, a direction of dataflow and maximum packet/buffer size. For example the default pipe is a bi-directional pipe made up of endpoint zero in and endpoint zero out with a control transfer type.

USB defines two types of pipes:

- *Stream pipes* have no defined USB format that can send any type of data down a stream pipe and can retrieve the data out of the other end. Data flows sequentially and has a pre-defined direction, either in or out. Stream pipes will support bulk, isochronous and interrupt transfer types. Stream pipes can either be controlled by the host or device.
- *Message pipes* have a defined USB format. It is host controlled, which are initiated by a request sent from the host. Data is then transferred in the desired direction, dictated by the request. Therefore message pipes allow data to flow in both directions but will only support control transfers.

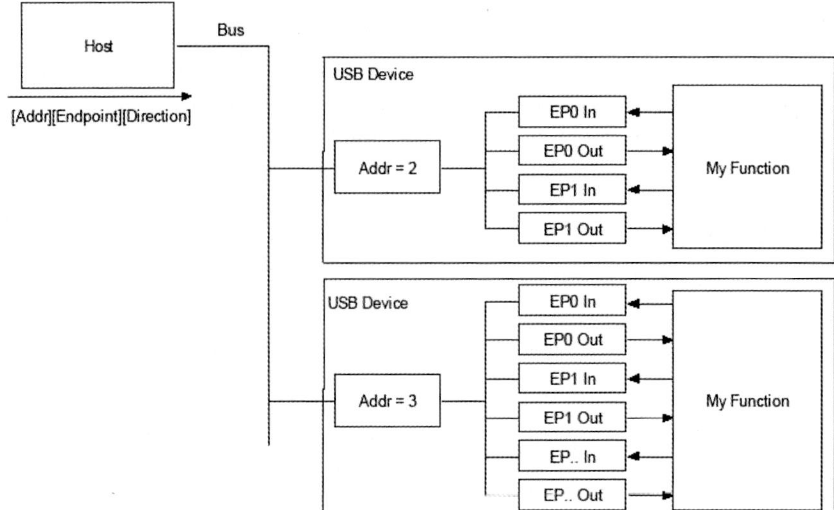

Fig. 6.6. USB functions

6.2.9 USB Functions

Most functions have a series of buffers, typically 8 bytes long. Each buffer belongs to an endpoint – EP0 IN, EP0 OUT, etc. For example, the host sends a device descriptor request. The function hardware will read the setup packet and determine from the address field whether the packet is for itself, and if so will copy the payload of the following data packet to the appropriate endpoint buffer dictated by the value in the endpoint field of the setup token. It will then send a handshake packet to acknowledge the reception of the byte and generate an internal interrupt within the semiconductor/microcontroller for the appropriate endpoint signifying it has received a packet as shown in Fig. 6.6. This is typically done in hardware. The software now gets an interrupt, and reads the contents of the endpoint buffer and parse the device descriptor request.

Endpoints

Endpoints can be described as sources or sinks of data. As the bus is host centric, endpoints occur at the end of the communications channel at the USB function. At the software layer, device driver may send a packet to devices EP1 for example. As the data is flowing out from the host, it will end up in the EP1 OUT buffer. Endpoints can also be seen as the interface between the hardware of the function device and the firmware running on the function device.

All devices must support endpoint zero. This is the endpoint which receives all of the devices control and status requests during enumeration and throughout the duration while the device is operational on the bus.

The USB specification defines four transfer/endpoint types:

- Control transfers
- Interrupt transfers
- Isochronous transfers
- Bulk transfers

Bandwidth Management

The host is responsible in managing the bandwidth of the bus. This is done at enumeration when configuring Isochronous and Interrupt Endpoints throughout the operation of the bus. The specification places limits on the bus, allowing no more than 90% of any frame to be allocated for periodic transfers (Interrupt and Isochronous) on a full speed bus. On high speed buses this limitation gets reduced to more than 80% of a microframe can be allocated for periodic transfers.

6.2.10 USB Descriptor

All USB devices have a hierarchy of descriptors which describe the device type, manufacturer of the device, version of USB supported, number of configuration methods, the number of endpoints and their types etc as shown in Fig. 6.7.

The more common USB descriptors are:

- Device descriptors
- Configuration descriptors

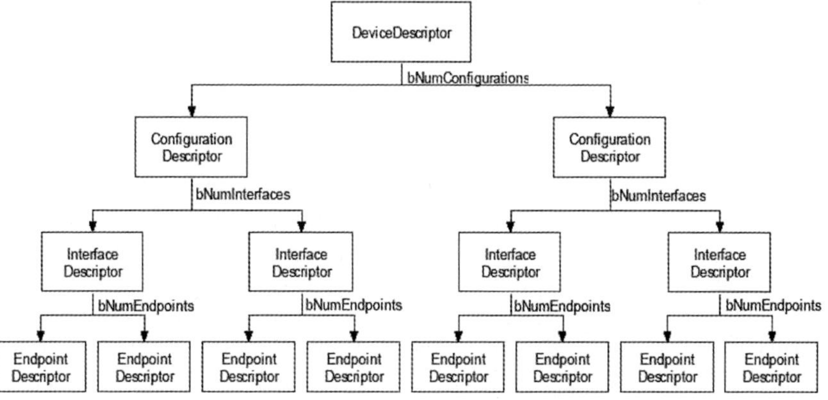

Fig. 6.7. USB descriptor

- Interface descriptors
- Endpoint descriptors
- String descriptors

USB devices can only have one device descriptor. The device descriptor includes information such as the USB version with which the device complies with, the Product and Vendor IDs used to load the appropriate drivers and the number of possible configurations the device can have. The configuration descriptor specifies values such as the amount of power this particular configuration uses, if the device is self- or bus-powered and the number of interfaces it possesses. When a device is enumerated, the host reads the device descriptors and can make a decision of which configuration to enable. It can only enable one configuration at a time.

For example, it is possible to have a high power bus-powered configuration and a self-powered configuration. If the device is plugged into a host with a mains power supply, the device driver may choose to enable the high bus-powered configuration enabling the device to be powered without a connection to the mains, yet if it is connected to a laptop or personal organizer it could enable the second configuration (self-powered) requiring the user to plug device into the power point.

The configuration settings are not limited to power differences. Each configuration could be powered in the same way and draw the same current yet have different interface or endpoint combinations. However, it should be noted that changing the configuration requires all activity on each endpoint to stop. While USB offers this flexibility, very few devices have more than 1 configuration.

The interface descriptor could be seen as a header or grouping of the endpoints into a functional group performing a single feature of the device. Unlike the configuration descriptor, there is no limitation as to having only one interface enabled at a time. A device could have 1 or many interface descriptors enabled at once.

Interface descriptors have a Interface Number field specifying the Interface number and a Alternate Setting which allows an interface to change settings on the fly. For example it could have a device with two interfaces, interface one and interface two. Interface one has Interface Number set to zero indicating it is the first interface descriptor and an Alternative Setting of zero. Interface two would have an Interface Number set to one indicating it is the second interface and a Alternative Setting of zero (default). It could then throw in another descriptor, also with an Interface Number set to one indicating that it is the second interface, but setting the Alternative Setting to one, indicates that the interface descriptor can be an alternative setting to that of the other interface descriptor two. When this configuration is enabled, the first two interface descriptors are used with Alternative Settings set to zero.

However during operation the host can send a Set Interface request directed to that of Interface one with an alternative setting of one to enable the other interface descriptor. This gives an advantage over having two configurations, in that it can be transmitting data over interface zero while it change the endpoint settings associated with interface one without effecting interface zero. Each endpoint descriptor is used to specify the type of transfer, direction, polling interval, and maximum packet size for each endpoint. Endpoint zero, the default control endpoint is always assumed to be a control endpoint and as such never has a descriptor.

6.2.11 Advantages of USB

- Always ON, always connected and always ready to go
- Capable of downloading up to 100 times faster than Traditional devices
- Supports upstream data rates up to 30 Mbps
- Enables access to real-time online games
- Downloads megafiles in seconds
- Stand-by switch disconnects the modem from the outside world when not in use
- No telephone line needed
- USB or Ethernet connectivity simplifies installation

6.3 PCI

First released in 1992, the Peripheral Component Interface (PCI) has rapidly evolved into a viable replacement for the ISA bus, and is the most commonly used method for adding boards to a PC. It solves many of the problems with older architectures, while at the same time delivering a substantial increase in processing speed. PCI provides a new way of connecting peripherals to both the system memory and the CPU, with the goal of alleviating many problems encountered when installing new cards in an ISA based system (IRQ conflicts, address conflicts, etc.). To ensure the longevity of the bus, PCI bus is backward compatible with those designed to older ones. PCI boards may also be used in a system that also uses ISA bus cards. However, this is becoming increasingly usable as ISA is rapidly becoming obsolete.

The PCI is the high speed bus made by Intel. It is used today in all PCs and other computers for connecting adapters, such as network-controllers, graphics cards, sound cards, etc. The widely adopted PCI standard has defeated a large number of competing standards in the marketplace.

One of the most widely adopted standards in desktop and server computing is Peripheral Component Interconnect, or PCI. The original IBM PC architecture had an elaborate series of related bus architectures, rooted in the original Industry Standard Architecture (ISA) bus.

6.3.1 A 32-Bit Bus

The PCI is actually 32-bit wide, but in practice it functions like a 64-bit bus. Running at 33 MHz, it has a maximum transmission capacity of 132 MBps. According to the specifications, it can have up to 8 units with a speed of up to 200 MHz. The bus is processor independent. Therefore, it can be used with all 32- or 64-bit processors, and it is also found in other computers than personal computers. The PCI bus is compatible with the ISA bus in that it can react on ISA bus signals, and perform the same functions like creating the same IRQs.

Buffering and PNP

The PCI bus is buffered in relation to the CPU and the peripheral components. This means, that the CPU can deliver its data to the buffer, and then proceed with other tasks. The bus handles the further transmission in its own method. Conversely, the PCI adapters can also transmit data to the buffer, regardless of whether the CPU is free to process them. They are placed in a queue, until the system bus can forward them to the CPU. Under optimal conditions, the PCI bus transmits 32 bits per clock tick. Sometimes, it requires two clock ticks. Because of this, the peripheral PCI units operate asynchronous. Therefore, the PCI (contrary to the VL bus) is not a local bus in a strict sense. Finally, the PCI bus is intelligent relative to the peripheral components, in that Plug-and-Play (PnP) is included in the PCI specifications. All adapter cards for the PCI configure themselves.

6.3.2 History

The original ISA bus specification allowed for a 16-bit bus, but early PCs only supported the 8-bit subset. The original specification of the bus was called the "AT bus," but this name was trademarked. The "ISA" name was adopted to get around this trademark issue and to allow everyone to get down to the business of setting jumpers. The ISA bus had a number of problems. In short, it was slow, it was hard to configure, and the standard for it was not quite complete enough. As a result, not all implementations were fully compatible, although they were mostly pretty good.

Several other bus architectures were developed during this time period, some technically far superior to ISA. However, most of them were at least partially proprietary. IBM had the Micro Channel Architecture (MCA) bus, Apple Corporation had NuBus, Sun Corporation had SBus, Amiga used Zorro II (16-bit), Zorro III (32-bit), and Video Electronics Standards Association (VESA) had the VESA Local bus (VLB), which was higher-performance than ISA.

PCI started out as a way of interconnecting things on a motherboard, but Intel introduced systems that used it for cards, too. The PCI bus had a number of advantages over competing bus architectures. PCI devices can have direct access to system memory without waiting on the CPU. A single PCI bus can have up to five devices connected to it, and more importantly, PCI supports bridges, allowing one of those connections to be used for a connection to even more devices. Some individual cards may actually contain an entire PCI bridge chip and separate bus; for instance, dual and quad Ethernet cards often follow this procedure. PCI also allows for a great deal of autoconfiguration; this was a feature most other buses had offered, which was lacking from ISA.

PCI was proposed as a standard in 1991. But, in 1995, Windows 95 showed up, and with it there was a good support for PCI. This made PCI a mainstream alternative to users who did not want to mess around with jumpers. Suddenly, the large market of end-user PC computers was a large market for PCI expansion cards. Once this market was established, everyone else found on board to be compatible. The PCI bus is used in everything from tiny embedded systems to enterprise servers. It is fast, cheap, and it solves a large number of problems. It is not quite universal anymore: PCI-X and PCI Express are starting to show up on the scene. For instance, recently POWER4 servers use PCI-X for expansion cards.

6.3.3 Architecture of PCI with Two Faces

On modern system boards, the PCI bus shown in Fig. 6.8 has two "faces":

- Internal PCI bus, which connects to Enhanced Integrated Development Environment (EIDE) channels on the motherboard.
- The PCI expansion bus, which typically has 3–4 slots for PCI adapters.

The PCI bus is continuously being developed further by a Special Interest Group, consisting of the most significant companies (Intel, IBM, Apple, and others), which coordinate and standardize the development.

6.3.4 Features of PCI

Some of the features of PCI such as Plug and Play capability, Speed, Bandwidth, Voltage requirements, and bus mastering are discussed in the following sections.

Plug and Play

Implementing PCI control registers is vitally important to ensure that Plug and Play, one of PCIs most attractive features, works properly. Setting jumpers and switches to configure address and IRQ is not required. The system configures itself by having the PCI BIOS access configuration registers on each add-in board at boot-up time. These configuration registers indicate the system about the resources required, (I/O space, memory space,

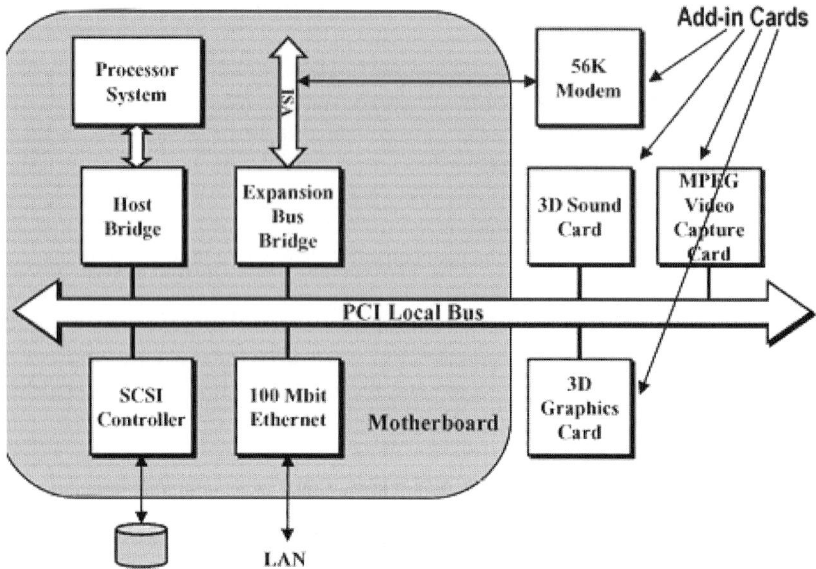

Fig. 6.8. PCI architecture

interrupts, etc.). The system can allocate its resources accordingly, making sure that no two devices conflict. Another implication of the PCI implementation is that a board's I/O addresses and interrupt is not fixed, meaning that they can change every time the system boots.

High Speed and Wide Bandwidth

More than any other bus, PCI can take full advantage of today's high-power microprocessors to deliver extremely high speed data transfers. The original PCI bus was designed to operate with a 33 MHz clock, to provide data transfer speeds up to 132 Mbytes s^{-1}. These 32-bit adapters use multiplexing to achieve 64-bit data transfers. Later versions of PCI enable true 64-bit data transfers using up to a 133 MHz clock to enable transfer speeds of up to 1,066 Mbytes s^{-1}. These boards use a longer connector that adds additional 32-bits of data signals. This is done by using the same set of pins to address and send data, the former implemented on the first clock cycle and the latter on the second. PCIs burst mode facilitates this operation as it allows a single address cycle to be followed by multiple data cycles. A special bus signal called a Cycle Frame is used to signal the beginning and end of a transfer cycle. Parity signals are used to ensure signal integrity, which is particularly vulnerable in such a complex transfer system.

The high speed data transfers across the PCI bus limits the number of PCI expansion slots that can be built into a single bus to 3 or 4, as opposed to the 6 or 7 available on the ISA bus. To expand the number of available expansion slots, PCI-to-PCI bridges are used. These bridges create a primary and a secondary PCI bus, each of which is electrically isolated from the other. Multiple bridges can be cascaded, theoretically allowing unlimited numbers of PCI slots to be configured in a single system. Practically, it is important not to overload the CPU, as adding too many devices through expansion slots could considerably compromise system bandwidth. The bridge enables bus transfers to be forwarded upstream or downstream, from one bus to another until the target or destination is reached.

Flexible Bus Mastering

Several pins on the PCI bus are reserved for implementing bus mastering. Bus Mastering means that any PCI device can take control of the bus at any time, even allowing it to shut out the CPU. Devices use bandwidth as available, and can potentially use all bandwidth in the system if no other demands are made for it. Bus mastering works by sending Request signals to the Central Resource (circuitry on the motherboard shared by all bus devices) when a device wants control of the bus, and when that control is ceded a Grant signal is received by the device. This flexible approach, which separates the arbitration and control signals allow greater control over the arbitration process.

Interrupt sharing on the PCI bus is also implemented to provide maximum flexibility. Four level-sensitive interrupts are located on the bus at pins A6, B7, A7, and B8, each of which can be assigned to from one to 16 separate devices. These interrupts can be activated at any time because they are not synchronized with the other signals on the bus. The PCI specification does not define how interrupts are to be shared. The process is implemented on a case-by-case basis by the motherboard manufacturer. For instance, Quatech multiport serial PCI boards require only one slot to provide up to eight serial ports. These eight ports share a single interrupt, and our boards provide an interrupt status register that will indicate which of the eight ports triggered an interrupt.

Voltage Requirements

The original PCI standard required that plug-in boards use 5 V power provided by the PCs motherboard. As the PCI standard evolved, the option was added for a 3.3 V power source. Now, with the latest PCI 2.3 release the 3.3 V power supply is required and the 5 V power supply is obsolete.

In order to make sure that a plug-in board receives the correct voltage (as getting the wrong voltage would cause the PC to be badly damaged) a different set of keys was designed for each voltage type. The key is the arrangement of gold fingers at the bottom of the board. When a board is inserted into a PCI slot, these gold fingers fit into the appropriate socket. Motherboards supplying

5.5 V slots can only accept PCI cards keyed for 5 V. Motherboards supplying 3.3 V slots can only accept PCI cards keyed for 3.3 V.

Both PC manufacturers and PCI board developers realize that users would optimally like a combination of both options. So, some motherboards provide universal connectors that can accept both 3.3 and 5 V keyed PCI cards. Likewise, some PCI cards specifically designed to function with either a 3.3 or 5 V power supply use a special key that will fit into either type of motherboard connector. These boards are called Universal PCI boards.

6.3.5 Low Profile PCI

Clearly, traditional PCI add-in boards are not practical for portable systems or for systems that use small-size cases. Two new standards have been incorporated into the PCI 2.3 specification that addresses these issues: Low Profile PCI and Mini-PCI. Low Profile PCI was designed to provide greater flexibility in desktop and server environments. The card is mechanically similar to a standard PCI card, but uses a shorter card and a different mounting bracket.

Low Profile PCI cards are designed to fit into systems as low as 3.350″ with out using riser cards. There are two types of Low Profile PCI boards: MD1 and MD2. Both are built on 32-bit addressing, and differ only in length, MD1 being shorter than MD2. Systems can be designed to support one or both configurations. Existing PCI backplanes used for standard cards can also accept the Low Profile cards. However, the Low Profile specification also includes a new bracket design that cannot be used with standard PCI boards.

Mini-PCI, on the other hand, is not backward compatible with any older devices. It is designed to be used by system integrators to add additional functionality for mobile computers. It requires a completely new interface, and is typically used to add communication peripherals such as modems and network interface cards to notebook computers, docking stations, or sealed case PCs.

Mini-PCI boards are even smaller than Low Profile boards, with a minimum size specification of 2.75 in. × 1.81 in. × 0.22 in. They are functionally equivalent to standard and Low Profile boards, using the same protocols, PC signals and software drivers. However, because of the small size they require higher density, more compact, and thus more expensive components. So, while Mini-PCI is extremely useful for the mobile applications for which it was intended, it is not the most economical or flexible choice for desktop expansion.

6.3.6 PCI-X

The PCI-X specification is a high-performance enhancement to the conventional PCI bus specification. The first release of PCI-X doubled the maximum clock frequency that can be used by PCI devices from 66 to 133 MHz, thus enabling communication at speeds over 1 Gbyte s^{-1}. It also improved the efficiency of the PCI bus itself and the devices attached to it, by providing new features such as split transactions and transaction byte counts. PCI-X was developed for applications such as Gigabit Ethernet, Fiber Channel and other Enterprise server applications.

PCI-X 2.0, the current revision, increases speed and performance even further. Maximum clock frequencies of 266 and 533 MHz are now supported to permit lightning fast transfers of up to 4.3 Gbyte s^{-1} of bandwidth that is 32 times faster than the original conventional PCI 2.1 specification. PCI-X 2.0 also adds additional features for systems reliability that will minimize errors at high speeds and keep up with other advances in high-bandwidth architecture.

6.3.7 PCI for Data Communication

PCI is the standard for board-level expansion slots in PC-based systems. However, for serial and parallel communication, the potential of PCI goes largely untapped. Limitations of these communication protocols considerably slow down the system. The PCI specification shown in Table 6.2 anticipates this situation, and the bus is designed to slow down when dealing with low speed devices. So, though running serial and parallel devices through a PCI bus will not cause system problems, neither will it significantly improve system performance over ISA. But, as ISA slots are now essentially obsolete, there is typically no choice but to use PCI add-in boards in recent systems. For higher speed devices such as audio, streaming video, interactive gaming, high-speed modems, etc., PCI provides a clear advantage over older bus architectures.

In addition, PCs are becoming smaller and smaller in size, and many will not accommodate a standard PCI board. For these systems, most notably Thin Clients and embedded systems, the Low Profile PCI form factor provides robust functionality on a very small size board. PCI competes with USB for high-bandwidth applications. Both provide the advantage of Plug and Play installation, which may be important to some users. Currently, only USB supports hot swapping, but support for it is planned in a future PCI specification. USB has been positioned as a low cost solution for a variety

Table 6.2. PCI Bus specifications

PCI specifications	
Bus Clock Signal	133 MHz
Bus Width	64-bit
Theoretical Maximum Transfer Rate	1 GByte s^{-1} (8 Gb s^{-1}, gigabit per sample)
Advantages	Very High Speed, PNP, Dominant Board level bus
Disadvantages	Incompatible with older systems, can cost more

of desktop applications. However, PCI still has a speed advantage even over the newer USB 2.0 when using true 64-bit data transfers. In fact, in many systems' USB ports are added to through the PCI Bus.

6.3.8 PCI IDE Bus Mastering

The PCI bus also allows us to set up compatible IDE/ATA hard disk drives to be bus masters. Under the correct conditions this can increase performance over the use of PIO modes, which are the default way that IDE/ATA hard disks transfer data to and from the system. When PCI bus mastering is used, IDE/ATA devices use DMA modes to transfer data instead of PIO; IDE/ATA DMA modes are described in detail here.

Since this capability was made available to the latest developed machines, it has been one of the most important functions of the modern PC. In particular, there are a lot of misconceptions about its performance advantages. In addition, there have been a lot of problems with compatibility in getting this new technology to work.

IDE bus mastering requires all of the following in order to function at all:

- *Bus Mastering Capable System Hardware.* This includes the motherboard, chipset, bus and BIOS. Most recent motherboards using the Intel 430 Pentium chipset family (FX, HX, VX, TX) or the Intel 440FX Pentium Pro chipset, will support bus mastering IDE.
- *Bus Mastering Hard Disk.* Normally this means that the drive must be capable of at least multiword DMA mode 2 transfers. All Ultra ATA hard disks also support bus mastering.
- *32-Bit Multitasking Operating System.* This means usually Windows NT, Windows 95 or Linux.
- *Bus Mastering Drivers.* A special driver must be provided to the operating system to enable bus mastering to work.

In particular, the following are common problems encountered when trying to set up bus mastering:

- Driver bugs and incompatibility issues, especially with older drivers that are "generic" and not tuned to a particular motherboard.
- Older hard disk drives not working properly.
- Problems with dissimilar drives connected to the same IDE channel as master and slave.
- Problems when using a CD-ROM drive alone on an IDE channel without a hard drive.
- Bus mastering drivers that don't work on certain motherboards; also, some motherboards or systems just will not work with bus mastering at all.

Bus mastering over the IDE/ATA interface will be improved and these problems will be just a distant memory. With the creation of Ultra ATA and the DMA-33 high-speed transfer mode, it appears that the future lies in the use of PCI bus mastering with the IDE/ATA interface. There is just some work to do until this support is both universal and well-implemented.

6.3.9 PCI Internal Interrupts

The PCI bus uses its own internal interrupt system for dealing with requests from the cards on the bus. These interrupts are often called "#A," "#B," "#C," and "#D" to avoid confusion with the normal numbered system IRQs, though they are sometimes called "#1"–"#4" as well. These interrupt levels are not generally seen by the user except in the BIOS setup screen for PCI, where they can be used to control how PCI cards operate.

These interrupts, if needed by cards in the slots, are mapped to regular interrupts, normally IRQ9 through IRQ12. The PCI slots in most systems can be mapped to at most four regular IRQs. In systems that have more than four PCI slots, or that have four slots and a USB controller (which uses PCI), two or more of the PCI devices share an IRQ.

In Windows 95 OEM SR2, additional entries appear under the Device Manager for the PCI devices. Each device may have an additional entry entitled "IRQ Holder for PCI Steering." PCI steering is in fact a feature that is part of the Plug and Play portions of the system, and enables the IRQ used for PCI devices to be controlled by the operating system to avoid resource problems.

6.3.10 PCI Bus Performance

The PCI bus provides superior performance to the VESA local bus; in fact, PCI is the highest performance general I/O bus currently used on PCs. This is due to several factors:

- *Burst mode.* The PCI bus can transfer information in a burst mode, where after an initial address is provided multiple sets of data can be transmitted in a row. This works in a way similar to cache bursting.
- *Bus mastering.* PCI supports full bus mastering, which leads to improved performance.
- *High bandwidth options.* The PCI bus specification version 2.1 calls for expandability to 64 bits and 66 MHz speed; if implemented this would quadruple bandwidth over the current design. In practice the 64-bit PCI bus has yet to be implemented on the PC (it does exist in non-PC platforms such as Digital Equipment's Alpha and is also found now on servers) and the speed is currently limited to 33 MHz in most PC designs, most likely for compatibility reasons. For mainstream PCI, there may be a limitation to 32 bits and 33 MHz. However, it appears that the higher-performance PCI options are going to live on, albeit in modified form, through the new Accelerated Graphics Port.

The speed of the PCI bus can be set synchronously or asynchronously, depending on the chipset and motherboard. In a synchronized setup (used by most PCs), the PCI bus runs at half the memory bus speed; since the memory bus is usually 50, 60, or 66 MHz, the PCI bus would run at 25, 30, or 33 MHz respectively. In an asynchronous setup the speed of the PCI bus can be set independent of the memory bus speed. This is normally controlled through jumpers on the motherboard, or BIOS settings. Over clocking the system bus on a PC that uses synchronous PCI will cause PCI peripherals to be over clocked as well, often leading to system stability problems.

6.3.11 PCI Expansion Slots

The PCI bus offers more expansion slots than most VESA Local Bus (VLB) implementations, without the electrical problems that plagued the VESA bus. Most PCI systems support 3 or 4 PCI slots, with some going significantly higher.

The PCI bus offers a great variety of expansion cards compared to VLB. The most commonly found cards are video cards (of course), SCSI host adapters, and high-speed networking cards. Hard disk drives are also on the PCI bus but are normally connected directly to the motherboard on a PCI system. However, it should be noted that certain functions cannot be provided on the PCI bus. For example, serial and parallel ports must remain on the ISA bus.

Variants

Although PCI is "fast", as stated by most standards, modern hardware has pushed its limits. The original PCI bus runs at 33 MHz, and is 32 bits wide with a straightforward theoretical capacity of 133 Mbytes s^{-1}. Unfortunately, that is not nearly fast enough for some users. For instance, a single gigabit Ethernet card would consume nearly the entire bus, even if there were no overhead at all for communication.

One alternative is to go with faster or wider variants. Sixty-six megahertz and 64-bit wide PCI buses exist, and so do 66 MHz, 64-bit wide ones. These buses provide a capacity of 533 Mbytes s^{-1}, which is pretty good.

Another direction of variance is the development of MiniPCI, which is a fairly complete subset of PCI, missing only a few features, such as 64-bit support, and built with a much smaller form factor. MiniPCI cards fit into a specialized slot designed to be a good fit for laptops or embedded systems.

The PCI-SIG has introduced the new PCI-X 2.0 standard, which allows for dramatically faster transfers with new devices, while retaining backwards compatibility with older devices. These are probably going to be seen as transitional specs, but will offer hardware owners an upgrade path.

Finally, there are latest specs, such as Hyper-Transport and PCI Express, which are trying to push the performance envelope a good deal further, not just for one or two cards, but for whole systems. But even if PCI is due to be replaced or upgraded, it has had a good long run.

6.3.12 Standardization

The PCI specification was not originally developed by a standards group, but simply by Intel engineers, working to solve a problem that was keeping them from selling as many chips as possible. When Intel started pitching PCI as a standard, though, other people got involved with development. The PCI-SIG group was formed in 1992 presently with more than 900 members.

The PCI standard is currently at Revision 2.3. Devices like to specify which version of the PCI spec they require for compatibility; nearly all PC hardware says it requires PCI cards compatible with Version 2.1. Backwards compatibility is one of the highest priorities of the PCI-SIG, which is not available on the running older hardware.

6.3.13 Using PCI

PCI is of the greatest interest to hardware developers and OS developers. End users do not really need to know much about it, except that PCI devices mostly work in PCI computers. For OS developers, PCI provides a clean interface to getting information about a piece of hardware and communicating with it. Developers for systems like Linux™ or NetBSD, who are likely to need to write drivers for other peoples' hardware, benefit immensely from the reasonably stable basic interface to expansion devices.

Hardware developers get a lot of benefits, too. Someone developing a computer, and wants expansion options for it, will probably be best off using PCI: it provides more options than nearly anything else. If the form factor is too large, there's always MiniPCI – although the variety of hardware is a lot smaller. The smaller hardware variety probably ties into economies of scale. No one would use MiniPCI for something that could be done just as easily with full-sized PCI. The majority of MiniPCI cards appear to be wireless networking cards, used in various laptops. Some are combo cards, providing an interface to a wireless network and a modem. MiniPCI cards have to be fairly tiny, so anything that requires a lot of chips or wiring is unlikely to be available in that form factor.

Unfortunately, boot ROMs are not quite so portable. Some vendors have produced cards with boot ROMs that work in multiple computers. The biggest systems to target are the x86 family and Open Firmware. For now, that covers a variety of major vendors. In fact, some non-x86 systems have had boot firmware that can emulate enough x86 instructions to handle some boot ROMs. So, the PCI spec is not quite the universal solution to making all devices work on all hardware. In practice, vendors still need to worry about at least a little bit of software support.

6.4 PCI Express

PCI Express architecture is an industry standard high-performance, general-purpose serial I/O interconnect designed for use in enterprise, desktop, mobile, communications and embedded platforms. It has many state-of-the art attributes, including:

- PCI compatibility using the established PCI software programming models, thus facilitating a smooth transition to new hardware while allowing software to evolve to take advantage of PCI Express features.
- A cost-effective, low-pin count interface offering maximum bandwidth per pin, reducing cost and design complexity and enabling smaller form factors.
- Scalable bandwidth of 16 Gbytes s^{-1} at its initial signaling rate of 2.5 GHz with much higher transfer rates in the future using higher frequency signaling technologies.
- Support for multiple interconnect widths through 1, 2, 4, 8, 12, 16, and 32 LAN configurations aggregated to match application bandwidth needs.
- Support for new and innovative, hot-plug/hot-swap add-in card and module devices Unique, advanced features such as Power Management, Quality of Service and other native functions not available in other I/O architectures.

The Express Module Specification is a new PCI Express modular form factor defined for Servers & Workstations. Its ease-of-use packaging and integrated features provide benefits to many customers and system integrators (OEM, IHV, VAR, and End-users). The Express Module enables and promotes fundamental improvements in the way Servers & Workstations are built and serviced.

6.4.1 Need for PCI Express

The interconnect demands of emerging communications platforms exceed the capabilities of traditional parallel busses such as PCI. Innovations such as 10 GHz and higher CPU speeds, faster memories, high end graphics, gigabit networking and storage interfaces, and high-performance consumer devices have driven the need for a system interconnect with much greater bandwidth scalability and lower cost of implementation. PCI Express architecture uniquely meets these varying market segments requirements.

6.4.2 Types of PCI Express Architecture

The PCI Express architecture suite of specifications can be described as follows:

1. *PCI Express Base Specification.* The core specification for the multilayered PCI Express architecture.

2. *PCI Express Card Electromechanical Specification.* The ATX-based form factors.
3. *PCI Express Mini Card Electromechanical Specification.* The BTO/CTO mobile platform form factor.
4. *PCI Express to PCI/PCI-X Bridge Specification.* The bridging architecture to enable PCI Express components to function with legacy PCI devices. It is currently at revision 1.0.
5. *PCI Express x16 Graphics 150W-ATX.* The considerations for power and thermal issues in the next generation of platforms with leading-edge graphics and multimedia applications to deliver higher performance capabilities. It is currently at version 1.0.
6. *PCI Express Express Module.* A hot-pluggable modular form factor defined for servers and workstations.

6.4.3 Performance

The Express Module Specification is a new PCI Express modular form factor defined for Servers & Workstations. Its ease-of-use packaging and integrated features provide benefits to many customers and system integrators. The Express Module enables and promotes fundamental improvements in the way Servers & Workstations are built and serviced.

6.4.4 Express Card

The Express Card standard was created by a broad coalition of PCMCIA member companies, including Dell, Hewlett Packard, IBM, Intel, Lexar Media, Microsoft, SCM Microsystems and Texas Instruments. PCMCIA developed the new standard with assistance from the USB Implementers Forum (USB-IF) and PCI-SIG. PCMCIA is a nonprofit trade association founded in 1989 to establish technical standards for PC Card technology and to promote interchangeability among computer systems. Express Card technology is the name of a new standard introduced by PCMCIA in 2003. The Express Card standard promises to deliver thinner, faster and lighter modular expansion to desktop and notebook computer users. Consumers will be able to add hardware capabilities such as memory, wired and wireless communications and security features by simply inserting a module into their systems. All Express Card slots accommodate modules designed to use either USB 2.0, or the emerging PCI Express standards. Over time, express Card technology is expected to become the preferred solution for hot-pluggable internal I/O expansion for desktop and mobile computers.

6.5 PXI

PCI Extension for Instrumentation (PXI) was created in response to the needs of a variety of instrumentation and automation for users who require ever increasing performance, functionality, and reliability. Existing industry standards are leveraged by PXI to benefit from high component availability at lower costs. Most importantly, by maintaining software compatibility with industry-standard personal computers, PXI allows industrial customers to use the same software tools and environments.

PXI leverages the electrical features defined by the widely adopted Peripheral Component Interconnect (PCI) specification. It also leverages the Compact PCI form factor, which combines the PCI electrical specification. This combination allows Compact PCI and PXI systems to have up to seven peripheral slots versus four in a desktop PCI system.

Systems with more expansion slots can be built by using multiple bus segments with industry-standard PCI–PCI bridges. The PXI Hardware Specification adds electrical features that meet the high-performance requirements of instrumentation applications by providing triggering, local buses, and system clock capabilities. PXI also offers two-way interoperability with Compact PCI products. By implementing desktop PCI in a rugged form factor, PXI systems can leverage the large base of existing industry-standard software. Desktop PC users have access to different levels of software, from operating systems to low-level device drivers to high-level instrument drivers to complete graphical APIs.

6.5.1 PXI Architecture

The PXI Architecture is shown in Fig. 6.9. PXI supports the two module form factors that are depicted in Fig. 6.10. The 3U module form factor defines modules that are 100 by 160 mm (3.94 by 6.3 in.) and have two interface connectors. J1 carries the signals required for the 32-bit PCI local bus and J2 carries the signals for 64-bit PCI transfers and the signals for implementing PXI electrical features. The 6U module form factor defines modules that are 233.35 mm by 160 mm (9.19 in. × 6.3 in.) and in the future may carry one additional connector for expansion of the PXI Hardware Specification.

A PXI system is composed of a chassis that supports the PXI backplane and provides the means for supporting the system controller module and the peripheral modules. The chassis must have one system slot and may have one or more peripheral slots. Any number of controller expansion slots may be available to the left of the system slot. The optional star trigger controller, when used, must reside next to the system controller module. If a star trigger controller is not used in a system, a peripheral module can be installed in the slot next to the system controller module. The backplane carries the interface connectors (P1, P2,) and provides the interconnection between the controller and peripheral modules. A maximum of seven peripheral modules can be used

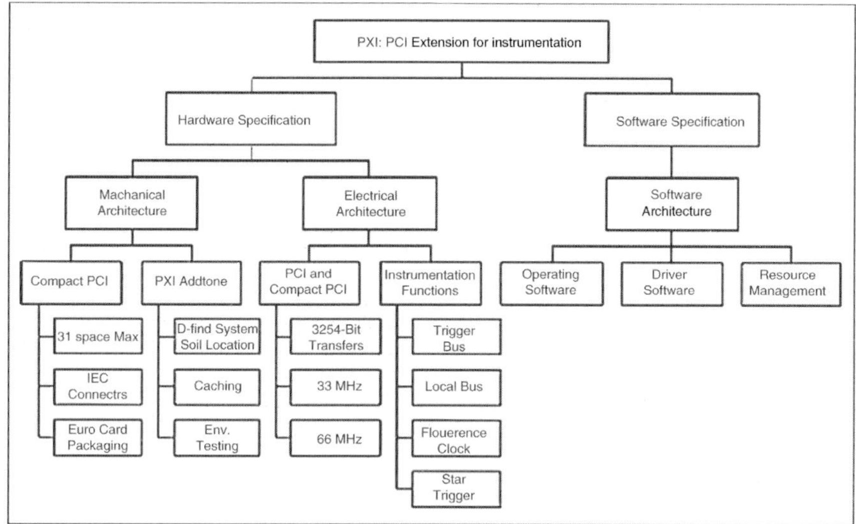

Fig. 6.9. PXI architecture

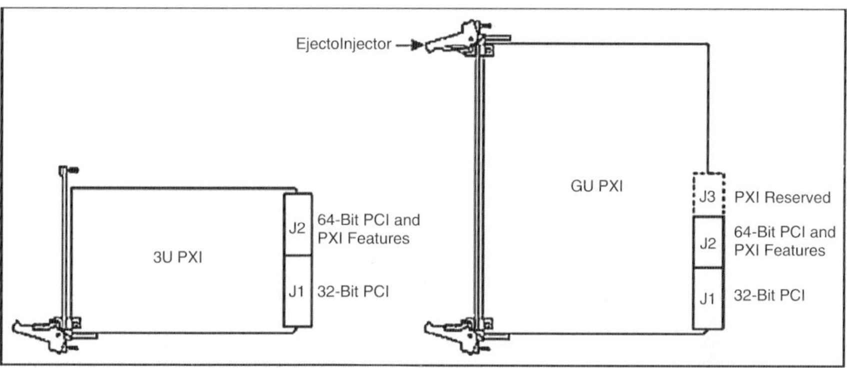

Fig. 6.10. PXI peripheral form factors and connectors

in a single 33 MHz PXI bus segment, and a maximum of four peripheral modules can be used in a single 66 MHz PXI bus segment. PCI–PCI bridges can be used to add bus segments for additional peripheral slots. Figure 6.11 presents an example of a typical PXI system to illustrate the following keywords.

Chassis Supporting Stacking 3U Modules in a 6U Slot

For efficient use of 3U modules in a 6U chassis, 6U chassis can support stacking two 3U modules in a single 6U slot. This allows one 3U module to be plugged into the P1/P2 position of a 6U slot and another 3U module to be plugged into the P4//P5 position of the same 6U slot simultaneously. A 6U PXI chassis

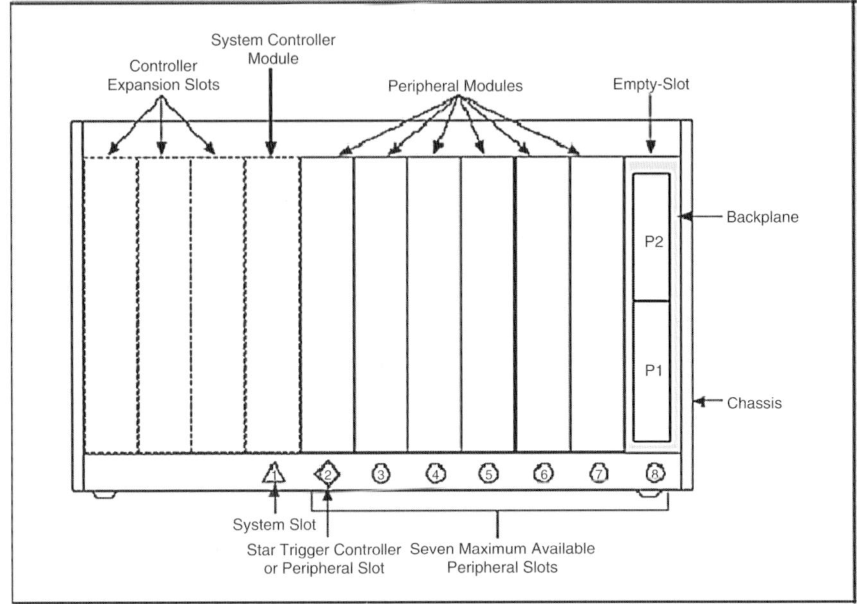

Fig. 6.11. PXI system

may have any number of 6U slots that support this feature. Electrically this configuration is accomplished by populating connectors in the P4 and P5 positions of the 6U backplane and routing them as would P1/P2. PCI bridging is required on the 6U backplane between the PCI bus residing on the P1 and P2 connectors and the PCI bus residing on the P4 and P5 connectors if PCI bus length and loading requirements cannot be met.

PCI bridging between the PCI bus residing on the P1 and P2 connectors and the PCI bus residing on the P4 and P5 connectors is not allowed on the system controller module. Mechanically this configuration can be accomplished by making use of center extrusions fixed within the chassis to physically support the insertion, extraction, and mounting of the lower and upper 3U modules residing in a 6U slot. Alternatively, this may be accomplished mechanically by a stacking adapter attached to the two 3U modules prior to insertion into the 6U slot. Figure 6.12 shows an example of a 6U chassis that supports stacking 3U modules.

System Slot Location

PXI defines the system slot location to be on the left end of the PCI bus segment in a basic PXI system. This defined arrangement is a subset of the numerous possible configurations allowed by Compact PCI. Defining a single location for the system slot simplifies integration and increases the degree of compatibility between PXI controllers and Chassis. Furthermore, the PXI

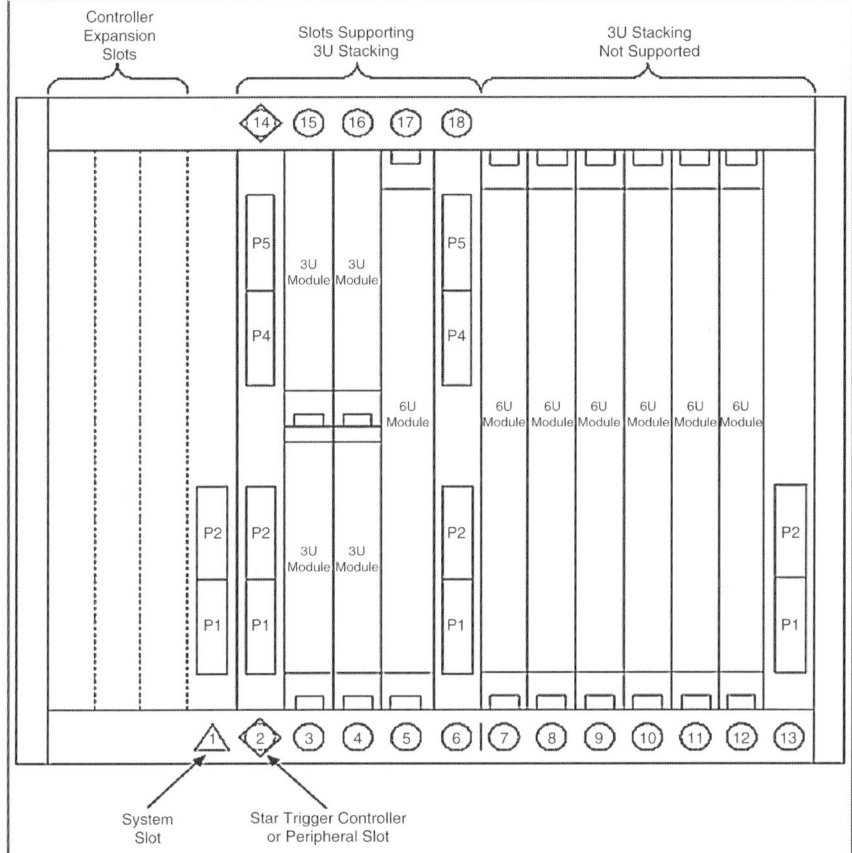

Fig. 6.12. 6U chassis

Hardware Specification requires that if necessary the system controller module should expand to the left into the controller expansion slots. Expanding to the left prevents system controllers from using valuable peripheral slots.

PXI also defines the location of a star trigger controller slot that can accept a peripheral module or a special star trigger module that can provide individual triggers to all other peripheral modules. The star trigger signals are routed from the star trigger slot to each peripheral slot on all PXI backplanes.

6.5.2 Interoperability with Compact PCI

Interoperability among PXI compatible products and standard Compact PCI products is a very important feature provided by this specification. Many PXI-compatible systems will require components that do not implement PXI-specific features. For example, a user may want to use a standard Compact PCI network interface module in a PXI chassis. Likewise, some users may

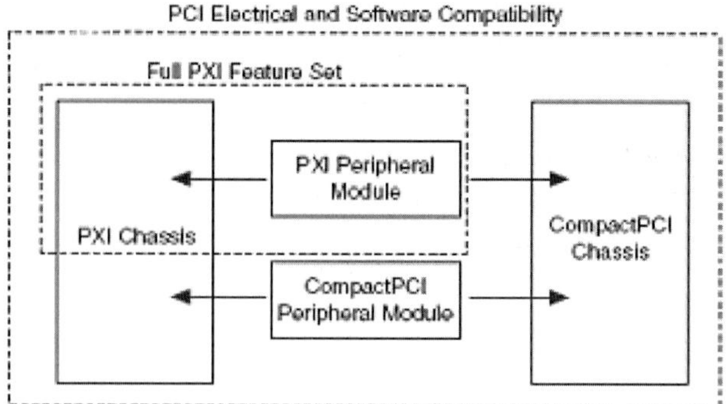

Fig. 6.13. Compact PCI

choose to use a PXI compatible module in a Standard Compact PCI chassis. The compact PCI chassis is shown in Fig. 6.13. In these cases, the user will not be able to use PXI-specific functions but will still be able to use the basic functions of the module. The interoperability between PXI-compatible products that uses the J2 connector for PXI-defined signals and other application-specific implementations of Compact PCI chassis which may define other signal definitions for sub-buses on the P2 backplane connector is not guaranteed. Of course, both Compact PCI and PXI leverage the PCI local bus, thereby ensuring software and electrical compatibility.

6.5.3 Electrical Architecture Overview

Many instrumentation applications require system timing capabilities that cannot be directly implemented across standard desktop computer backplanes such as ISA, PCI, or PCMCIA. PXI uses the standard PCI bus and adds specific signals for instrumentation including bused trigger lines, slot-specific triggers, a dedicated system clock, and slot-to-slot local buses to address the need for advanced timing, synchronization, and sideband communication.

PXI Features

PXI offers the same performance features defined by the desktop PCI specification, with one notable exception. A PXI system can have up to eight slots per 33 MHz segment that is one system slot and seven peripheral slots, whereas a desktop PCI system can have only five per 33 MHz segment that is one motherboard or system slot and four peripheral slots. Similarly, a PXI system can have up to five slots per 66 MHz segment that is one system slot and four peripheral slots, whereas a desktop PCI system can have only three per 66 MHz segment that is one motherboard or system slot and two peripheral slots. The capability to have three additional 33 MHz peripheral slots and two additional 66 MHz peripheral slots is defined in the Compact PCI specification upon which PXI draws. Otherwise, all the features of PCI transfer into PXI:

- 33/66 MHz performance
- 32-bit and 64-bit data transfers
- 132 Mbytes s^{-1} (32-bit, 33 MHz) to 528 Mbytes s^{-1} (64-bit, 66 MHz) peak data rates
- System expansion via PCI–PCI bridges
- 3.3 V migration
- Plug and Play capability

Similar to the PCI specification, it is recommended that PXI modules do not have their internal registers mapped to PCI I/O Space. I/O Space is limited, fragmented, and increasingly more difficult to guarantee allocation.

Local Bus

The PXI local bus is a daisy-chained bus that connects each peripheral slot with its adjacent peripheral slots to the left and right. Thus, the right local bus of a given peripheral slot connects to the left local bus of the adjacent slot, and so on. Each local bus is 13 lines wide and can be used to pass analog signals between modules or to provide a high-speed sideband digital communication path that does not affect the PXI bandwidth. Local bus signals can range from high-speed TTL signals to analog signals as high as 42 V. Keying of adjacent modules is implemented by initialization software that prohibits the use of incompatible modules. This software uses the configuration information specific to each peripheral module to evaluate compatibility before enabling local bus circuitry. This method provides a flexible means for defining local bus functionality that is not limited by hardware keying. The local bus lines for the leftmost peripheral slot of a PXI backplane are used for the star trigger. Figure 6.14 schematically shows a complete PXI system demonstrating the local buses.

System Reference Clock

The PXI 10 MHz system clock (PXI_CLK10) is distributed to all peripheral modules in a system. This common reference clock can be used for synchronization of multiple modules in a measurement or control system. The PXI backplane specification defines implementation guidelines for PXI_CLK10. As a result, the low skew qualities afforded by this reference clock make it ideal for precise multimodal synchronization by using trigger bus protocols to qualify individual clock edges.

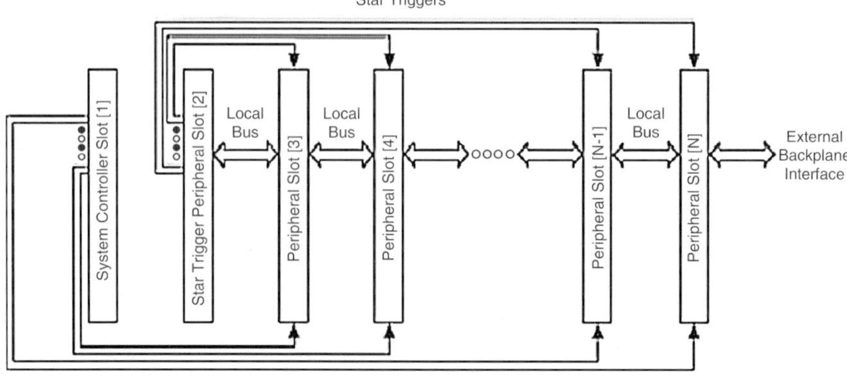

Fig. 6.14. PCI Arbitration and clock signals

Trigger Bus

The PXI bused trigger lines are eight in number and are highly flexible that can be used in a variety of ways. For example, triggers can be used to synchronize the operation of several different PXI peripheral modules. In other applications, one module can control timed sequences of operations performed on other modules in the system. Triggers may be passed from one module to another, allowing precisely timed responses to asynchronous external events that are being monitored or controlled. The number of triggers that a particular application requires varies with the complexity and number of events involved.

Star Trigger

The PXI star trigger bus offers ultra-high performance synchronization features to users of PXI systems. The star trigger bus implements a dedicated trigger line between the first peripheral slot (adjacent to the system slot) and the other peripheral slots. A star trigger controller can be installed in this slot and can be used to provide very precise trigger signals to other peripheral modules. Systems that do not require this advanced trigger can install any standard peripheral module in this slot. Through the required use of line-length equalization techniques for routing the star triggers, PXI systems can meet demanding triggering requirements for which bused triggers are not appropriate. The star trigger can be used to communicate information back to the star trigger controller, as in the case of reporting a slot's status, as well as responding to information provided by the star trigger controller.

This trigger architecture for PXI gives two unique advantages in augmenting the bused trigger lines. The first advantage is a guarantee of a unique trigger line for each module in the system. For large systems, this eliminates the need to combine multiple module functions on a single trigger line or to artificially limit the number of trigger times available. The second advantage is the low-skew connection from a single trigger point. The PXI backplane defines specific layout requirements such that the star trigger lines provide matched propagation time from the star trigger slot to each module for very precise trigger relationships between each module.

System Expansion with PCI–PCI Bridge Technology

A PXI system can be built with more than one bus segment by using standard PCI–PCI bridge technology. The bridge device takes up one PCI load on each of the bus segments that it links together. Thus, a 33 MHz system with two bus segments offers 13 expansion slots for PXI peripheral modules where two bus segments +8 slots per segment +1 system controller slot +2 slots for PCI-PCI Bridge = 13 available expansion slots.

Similarly, a three-bus segment 33 MHz system would offer 19 expansion slots for PXI peripheral modules. The trigger architecture defined by PXI has implications for systems with multiple bus segments. The PXI trigger bus provides connectivity within a single bus segment and does not allow physical connection to an adjacent bus segment.

This maintains the high performance characteristics of the trigger bus and allows multi segment systems to partition instruments into logical groups. Multiple segments may be logically linked by providing buffers between physical segments. The star trigger provides the means to independently access all 13 peripheral slots in a two-segment system for applications in which a high number of instruments require synchronization and controlled timing. In PXI systems where there are more than two segments, it is recommended that the star triggers are only routed to the slots in the first two segments, but other routings are allowed. Figure 6.15 shows the PXI trigger architecture for a PXI system with two bus segments.

6.5.4 Software Architecture Overview

Revision 2.0 and earlier of the PXI Specification included software requirements for PXI modules, chassis and systems. From PXI Hardware Specification revision 2.1 and higher, the PXI Systems Alliance maintains a separate software specification for PXI modules, chassis and systems. PXI modules, chassis, and systems developed to comply with this PXI Hardware Specification must also comply with the PXI Software Specification.

6.6 PCMCIA

Personal Computer Memory Card International Association (PCMCIA) is an international standards body and trade association founded in 1989 developed a standard for small, credit card-sized devices, called PC Cards. Originally designed for adding memory to portable computers, the PCMCIA

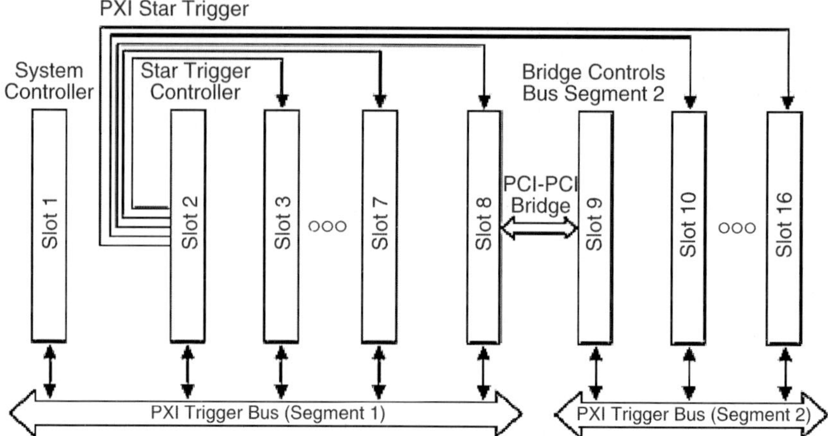

Fig. 6.15. Triggering architecture

standard has been expanded several times and is now suitable for many types of devices. The inclusion of PCMCIA technology in PCs delivers a variety of benefits. Besides providing an industry standard interface for third-party cards (PC Cards), PCMCIA allows users to easily swap cards in and out of a PC as needed, without having to deal with the allocation of system resources for those devices. These useful features hot swapping and automatic configuration, as well as card slot power management and other PCMCIA capabilities – are supported by a variety of software components on the PCMCIA-based PC. In most cases, the software aspect of PCMCIA remains relatively transparent to the user. As the demand for notebook and laptop computers began skyrocketing in the late 1980s, users realized that their expansion options were fairly limited. Mobile machines were not designed to accept the wide array of available expansion cards that their desktop counterparts could enjoy.

6.6.1 Features of PCMCIA

- One rear slot, access from rear of PC
- Accept Type I/II/III Cards
- Comply with PCI Local Bus Specification Rev.2.2
- Comply with 1995 PC Card Standard
- Extra compatible registers are mapped in memory
- Use T/I 1410 PCMCIA controller
- Support PC Card with Hot Insertion and Removal
- Support 5 V or 5/3.3 V 16-bit PC Cards
- Support Burst Transfers to Maximize Data Throughput on both PCI Buses
- Supports Distributed DMA and PC/PCI DMA

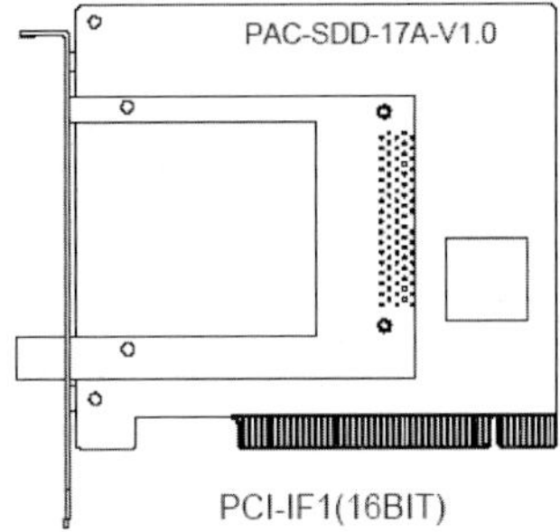

Fig. 6.16. Board lay out

6.6.2 Specifications

- Power Supply Voltage: 5 V
- Operation Temperature: 32–158° F (0–70° C)
- Storage Temperature: −4–185° F (−20 – 85° C)
- Relative Humidity: Max 90%
- Dimensions 128(H) × 134(W) × 20(D) mm

6.6.3 Board Layout and Jumper Settings

The board layout of the PC card is shown in Fig. 6.16. This model does not require any jumper settings.

6.6.4 Types of PC Cards

The PC Card Standard provides physical specifications for three types of PC Cards, with provisions for extended cards. Each card type has features that fit the needs of different applications.

1. Type I PC Cards are typically used for memory devices such as RAM, Flash, OTP, and SRAM cards.
2. Type II PC Cards are typically used for I/O devices such as data/fax modems, LANs, and mass storage devices.
3. Type III PC Cards are used for devices whose components are thicker, such a rotating mass storage devices.

These Extended cards allow the addition of components that must remain outside the system for proper operation, such as antennas for wireless applications. All three-card types measure the same in length and width, and use the same 68-pin connector along the edge for connecting to a computer.

The only difference between the card types is thickness, which vary as 3.3, 5.0, and 10.5 mm for Type I, Type II, and Type III cards, respectively. Because it differs only in thickness, a thinner card can be used in a thicker slot, but a thicker card cannot be used in a thinner slot.

6.6.5 Features of PC Card Technology

The features of PC card technology include:

1. *Card Information Structure.* The CIS describes the characteristics and capabilities of the card so the host system can automatically configure it. CIS is defined in the Meta format Specification.
2. *Card Bus.* Card Bus allows PC Cards and hosts to use 32-bit bus mastering and to operate at speeds up to 33 MHz.
3. *DMA* The Standard allows cards to utilize Direct Memory Access technology directly in the hardware when matched with a corresponding host system.
4. *Execute In Place (XIP).* Allows operating system and application software to run directly from the PC Card, eliminating the need for large amounts of system RAM.
5. Low Voltage Operation The Standard enables 3.3 and 5-V operation. A physical keying mechanism for 3.3-V cards protects them from being damaged in a 5-V slot.
6. Multiple Function PC Cards. The Standard enables truly standardized multiple function PC-Cards.
7. *Plug and Play.* PC Cards can be inserted or removed while the system is on because power and ground contact pins are the longest contacts, ensuring that disconnect signals disengage first, preserving data integrity.
8. *Power Management.* The Standard provides a means to interface to Advanced Power Management (APM) through the Card Services Specification.
9. *Zoomed Video (ZV).* ZV is a connection between a PC Card and host system that allows the card to write video data directly to the VGA controller. The data is transferred with no buffering requirements because it is transferred over the ZV bus and not the system bus.

6.6.6 Utilities of PCMCIA Card in the Networking Category

Under the networking category, typically PCMCIA slot supports 4 types of cards such as:

- LAN card
- Wireless LAN card
- Modem card
- ATA flash disk card

Wireless networking is one of the most recent major advances in the world of telecommunications. The ability to communicate and send data without a physical connection is becoming more popular at a very quick rate. In other words, a wireless LAN is one in which a mobile user can connect to a local area network (LAN) through a wireless (radio) connection. A standard, IEEE 802.11, specifies the technologies for wireless LANs. The standard includes an encryption method, the Wired Equivalent Privacy algorithm.

Features and Benefits

- One, two, or four independent serial ports
- Speeds up to 921.6 kbps
- Auto enable/disable feature
- Removable cable standard
- Permanently attached rugged cable option for one and two port RS 232 cards
- 3.3 and 5 V compatible
- Plug and Play and Hot Swap capability
- Low power requirements
- Linux support
- Windows CE and Pocket PC support
- Windows 9x/Me/NT/2000/XP and OS/2 support

6.7 SCXI

The Signal Conditioning Extension for Instrumentation (SCXI) modules are for signal conditioning of thermocouples, low-bandwidth volt and millivolt sources, 4–20 mA current sources, and 0–20 mA process-current sources. When used with the SCXI precision current source module, the user can use the SCXI to measure RTDs and thermistors. The SCXI-1102/B/C has 32 differential analog input channels and one cold-junction sensor channel. Each channel also has an amplifier with a selectable gain of 1 or 100 can multiplex the SCXI inputs to a single output, which drives a single DAQ device channel.

On each channel the modules have the following:

- SCXI-1102 – a two-pole low pass filter with a 2 Hz cutoff frequency to reject 60 Hz noise.
- SCXI-1102B – a three-pole low pass filter with a 200 Hz cutoff frequency.
- SCXI-1102C – a three-pole low pass filter with a 10 kHz cutoff frequency.

Can multiplex several SCXI modules and other SCXI modules into a single channel on the DAQ device, greatly increasing the number of analog input signals that can digitize.

6.7.1 SCXI Hardware and Software

Hardware includes:

1. At least one of the following modules:
   - SCXI-1102
   - SCXI-1102B
   - SCXI-1102C
2. At least one of the following terminal blocks:

  - SCXI-1300
  - SCXI-1303
  - SCXI-1308
  - SCXI-1310
  - TC-2095
  - TBX-1303
  - TBX-96
3. An SCXI chassis or PXI/SCXI combination chassis has one of the following:
  - E/M Series DAQ device
  - SCXI-1600

Software includes

1. NI-DAQ with one of the following software packages
  - LabVIEW
  - Measurement Studio
  - LabWindows/CVI

6.7.2 Analog Input Signal Connections

The signal terminals for the positive input channel are located in column B of the connector. The signal terminal for each corresponding negative input channel is located in column C of the connector. Each input goes to a separate filter and amplifier that is multiplexed to the module output buffer. If the terminal block has a temperature sensor, the sensor output—connected to pins A3 and A4 (CJ SENSOR) – is also filtered and multiplexed to the module output buffer. The differential input signal range of an SCXI-1102/B/C module input channel is ±10 V when using a gain of 1 or ±0.1 V when using a gain of 100. This differential input range is the maximum measurable voltage difference between the positive and negative channel inputs. The common-mode input signal range of an SCXI-1102/B/C module input channel is ±10 V. This common-mode input range for either positive or negative channel input is the maximum input voltage that results in a valid measurement. Each channel includes input protection circuitry to withstand the accidental application of voltages up to ±42 VAC peak or 60 VDC.

Ground-Referencing the Signals

The input signals can be ground-referenced, as shown in Fig. 6.17, or floating, as shown in Fig. 6.18. Before connecting the thermocouple or any other signal, it is necessary to determine whether it is floating or ground-referenced. If it is a floating signal, it must be ground-referenced in one of two ways:

- Can connect the negative channel input to chassis ground through a bias resistor

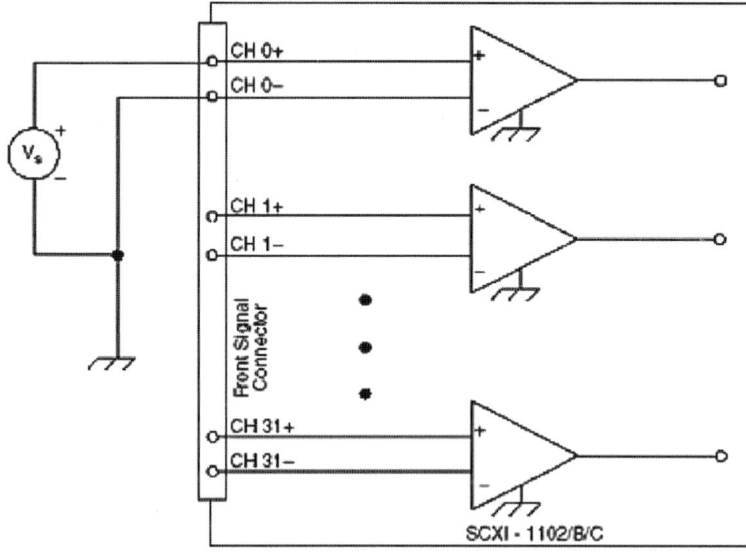

Fig. 6.17. Ground reference signal

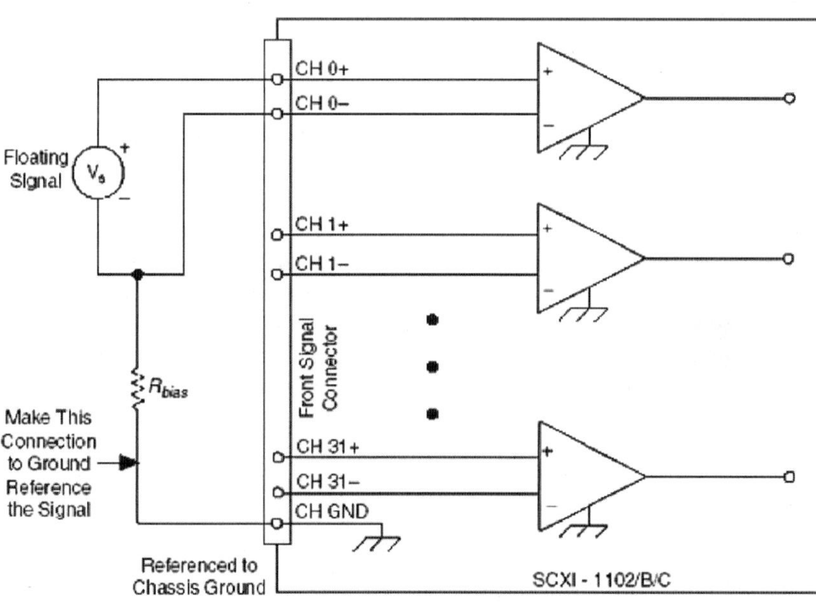

Fig. 6.18. Floating point signal

– Can use the bias resistors that are included with some terminal blocks such as the SCXI-1303 module

The SCXI-1303 also has a resistor pack for pulling up the positive inputs for open-thermocouple detection. The ground signals that are already ground-referenced need not be grounded; doing so results in a ground loop, which adversely affects the measurement accuracy. Directly grounding floating signals to the chassis ground without using a bias resistor is not recommended as this can result in noisy readings.

6.7.3 SCXI Software-Configurable Settings

This section describes the method to set the gain/input signal range and to configure software for compatible sensor types. It also describes the method to perform configuration of these settings for the SCXI-1102/B/C in NI-DAQmx and Traditional NI-DAQ. This section describes the most frequently used software-configurable settings for the SCXI-1102/B/C.

Gain/Input Range

Gain/input range is a software-configurable setting that allows choosing the appropriate amplification to fully utilize the range of the E/M Series DAQ device. In most applications NI-DAQ chooses and sets the gain for determined by the input range. Otherwise, it determines the appropriate gain using the input signal voltage range and the full-scale limits of the SCXI-1102/B/C output. A gain of 1 or 100 on a per channel basis is allowable.

CJC Source/Value

When using a terminal block that has a CJ sensor for thermocouple measurements, the user can set the CJC source as internal, which scans the sensor at the beginning of each measurement and scales the readings accordingly. This section describes where users can access each software-configurable setting for modification in MAX. The location of the settings varies depending on the version of NI-DAQ use. The DAQ Assistant can also be used to graphically configure common measurement tasks, channels, or scales which will be explained in detail in Chap. 7.

NI-DAQmx

Using NI-DAQmx, the user can configure software settings such as sensor type and gain/input signal range. The SCXI-1102/B/C module is used to make the following types of measurements:

- Voltage input
- Thermocouple
- RTD
- Thermistors
- Current input

Creating a Global Channel or Task

To create a new voltage, temperature, or current input using NI-DAQmx global task or channel, the following steps are to be performed:

1. Double-click Measurement & Automation on the desktop.
2. Right-click Data Neighborhood and select Create New.
3. Select NI-DAQmx Task or NI-DAQmx Global Channel, and click next.
4. Select Analog Input.
5. Select one of the following:
   - Voltage
   - Temperature and then select one of the following:
     - Thermistor
     - RTD
     - Thermocouple
     - Current
6. If the user is creating a task, a range of channels can be selected by holding down the <<Shift>> key while selecting the channels. The user can also select multiple individual channels by holding down the <<Ctrl>> key while selecting channels.
7. Name the task or channel and click Finish.
8. Select the channel(s) want to configure. We can select a range of channels by holding down the <<Shift>> key while selecting the channels. The user can select multiple individual channels by holding down the <<Ctrl>> key while selecting channels.
9. Enter the specific values for application in the Settings tab. Context help information for each setting is provided on the right side of the screen. Configure the input signal range using either NI-DAQmx Task or NI-DAQmx Global Channel. When set the minimum and maximum range of NI-DAQmx Task or NI-DAQmx Global Channel, the driver selects the best gain for the measurement.
10. If are creating a task and want to set timing or triggering controls, Enter the values in the Task Timing and Task Triggering tabs.

Traditional NI-DAQ (Legacy)

Using Traditional NI-DAQ (Legacy), the user can configure software settings, such as gain/input signal range in the following three ways:

- Module property pages in MAX
- Virtual channels properties in MAX
- Functions in ADE

Most of these settings are available in module properties and/or using virtual channels:

- Gain/input signal range – configure gain using module properties.

Creating a Virtual Channel

To create a virtual channel, the following steps are to be followed:

1. Right-click Data Neighborhood and select Create New.
2. Select Traditional NI-DAQ Virtual Channel and click Finish.
3. Select Analog Input from the drop-down menu and click Next.
4. Enter the Channel Name and Channel Description, and click Next.
5. Select one of the following measurement types from the drop-down menu:
   - Voltage
   - X Thermocouple
   - X RTD
   - Current
6. Click Next.
7. The next windows ask for information that is dependent upon the selection made in step 5. Supply the needed information and click next as needed.
8. Click Finish.

Verifying the Signal

This section describes how to take measurements using test panels in order to verify signal, and configuring and installing a system in NI-DAQmx and Traditional NI-DAQ (Legacy). The user can verify the signals on the SCXI-1102/B/C using NI-DAQmx by completing the following steps:

1. Click + next to Data Neighborhood.
2. Click + next to NI-DAQmx Tasks.
3. Click the task.
4. Select the channel(s) want to verify. The user can select a block of channels by holding down the <<Shift>> key or multiple channels by holding down the <<Ctrl>> key. Click OK.
5. Enter the appropriate information on the Settings tab.
6. Click the Test button.
7. Click the Start button.
8. After have completed verifying the channels, click the Stop button.

The user has now verified the SCXI-1102/B/C configuration and signal connection.

Verifying the Signal Using Virtual Channel

If the user has already created a virtual channel, the following steps are to be followed to verify the signal:

1. Right-click the virtual channel to verify and select Test.
2. In Channel Names, select the channel to verify.
3. When completed verifying the channel, click close.

Verifying the Signal Using Channel Strings

The following steps are followed to use channel strings in verifying the signal:

1. Click + next to Devices and Interfaces.
2. Click + next to Traditional NI-DAQ Devices.
3. Right-click the appropriate E Series DAQ device.
4. Click Test Panels.
5. Enter the channel string.
6. Enter the input limits.
7. Select the Data Mode.
8. Select the Y Scale Mode.

6.7.4 Theory of Operation

This section gives a brief overview and a detailed discussion of the circuit features of the SCXI Architecture module as shown in Fig. 6.19. The major components of the SCXI modules are as follows:

- Rear signal connector
- SCXIbus connector
- SCXIbus interface
- Digital control circuitry
- Analog circuitry

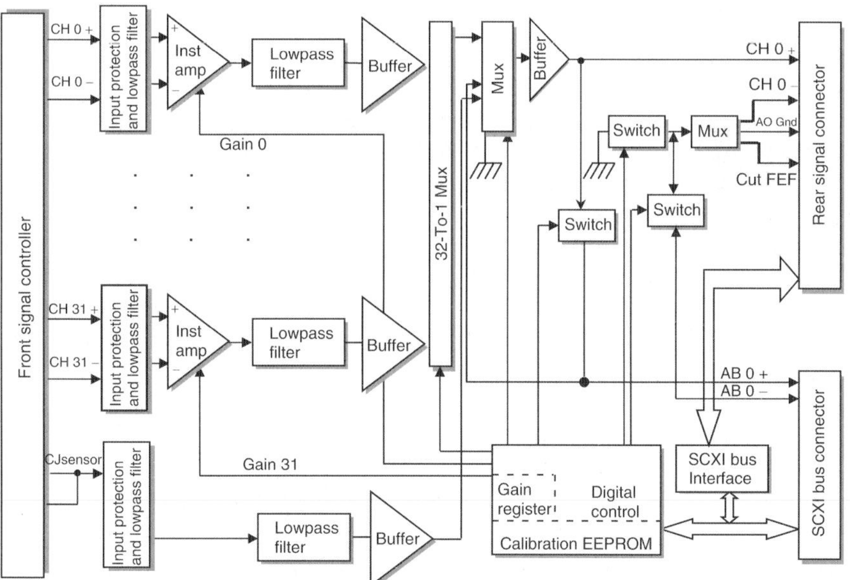

Fig. 6.19. SCXI architecture

The SCXI modules consist of 32 multiplexed input channels, each with a software-programmable gain of 1 or 100. Each input channel has its own lowpass filter. The SCXI modules also have a digital section for automatic control of channel scanning, temperature sensor selection, and gain selection.

Rear Signal Connector, SCXI bus Connector, and SCXI bus Interface

The SCXIbus controls the SCXI-1102/B/C module. The SCXIbus interface connects the rear signal connector to the SCXIbus, allowing a DAQ device to control the SCXI-1102/B/C module and the rest of the chassis.

Digital Control Circuitry

The digital control circuitry consists of the Address Handler and registers that are necessary for identifying the module, starting calibration information, setting the gain, and selecting the appropriate channel.

Analog Circuitry

The analog circuitry per channel consists of a low pass filter and an amplifier with a software selectable gain of 1 or 100. The CJ SENSOR channel also has a buffered low pass filter but has no amplifier. The channels and CJ SENSOR are multiplexed to a single output buffer.

Analog Input Channels

Each of the 32 analog input channels feeds to a separate amplifier with a programmable gain of 1 or 100. Then the signal passes through a fixed low pass The CJ SENSOR input channel is used to read the sensor temperature from the terminal blocks that have one, such as the SCXI-1300 and SCXI-1303. The temperature sensor is for cold-junction compensation thermocouple measurements. The CJ SENSOR channel also passes through a 2 Hz low pass filter to reject unwanted noise on the SCXI-1102/B/C. Along with the other 32 input channels, the CJ SENSOR is multiplexed to the output buffer, where it can be read by the DAQ device.

Theory of Multiplexed Operation

The SCXI-1102/B/C operates in multiplexed mode, which means that all input channels of the SCXI module are multiplexed into a single analog input channel of the E/M Series DAQ device. Multiplexed mode operation is ideal for high channel count systems. The power of SCXI multiplexed mode scanning is its ability to route many input channels to a single channel of the E/M

Series DAQ device. The multiplexing operation of the analog input signals is performed entirely by multiplexers in the SCXI modules, not inside the E/M Series DAQ device or SCXI chassis.

The SCXI-1102/B/C scanned channels are kept by the NI-DAQ driver in a scan list. Immediately prior to a multiplexed scanning operation, the SCXI chassis is programmed with a module scan list that controls which module sends its output to the SCXIbus during a scan through the cabled SCXI module. The list can contain channels in any physical order, but the SCXI-1102/B/C multiplexer can only sequence the channels in the order 0, 1... 31. The SCXI-1102/B/C cannot skip channels or scan channels in random order. The ordering of scanned channels must be sequential. The scan list is limited to a total of 512 channels per E/M Series DAQ device. When configure a module for multiplexed mode operation, the routing of multiplexed signals to the E/M Series DAQ device depends on which module in the SCXI system is cabled to the E/M Series DAQ device.

There are several possible scenarios for routing signals from the multiplexed modules to the E/M Series DAQ device. If the scanned SCXI-1102/B/C module is not directly cabled to the E/M Series DAQ device, the module sends its signals through the SCXIbus to the cabled module. The cabled module routing controlled by the SCXI chassis, routes the SCXIbus signals to the E/M Series DAQ device through the CH 0 pin on its rear signal connector. If the E/M Series DAQ device scans the cabled module, the module routes its input signals through the CH 0 pin on its rear signal connector to the E/M Series DAQ device CH 0. Multiplexed mode scanning acquisition rates have limitations that are determined based on the hardware in the system, and the mode of operation.

The maximum multiplexing rate of SCXI is 333 kHz. If the E/M Series DAQ device can sample more quickly than 333 kHz, and then the maximum multiplexing rate of SCXI is the limiting factor. If the E/M Series DAQ device cannot sample at 333 kS s^{-1}, the sample rate of the E/M Series DAQ device is the limiting factor on the maximum acquisition rate of the system. Since the user must scan the SCXI-1102/B/C sequentially, the driver automatically scans channels not included in the scan list if a sequential order is not maintained. When this happens, the maximum sample rate is also factored into the channels that are scanned and discarded for the purpose of completing the scan list.

For measurement accuracy of 0.012% of full scale, the minimum scan interval is 3 μs, which is the smallest interval in which we can switch between analog channels on the module and still measure accurate voltages. The 3 μs scan interval gives a maximum sampling rate of 333 kHz. For better accuracy, we must increase the scan interval in accordance with the specifications, which will reduce the maximum aggregate sample rate.

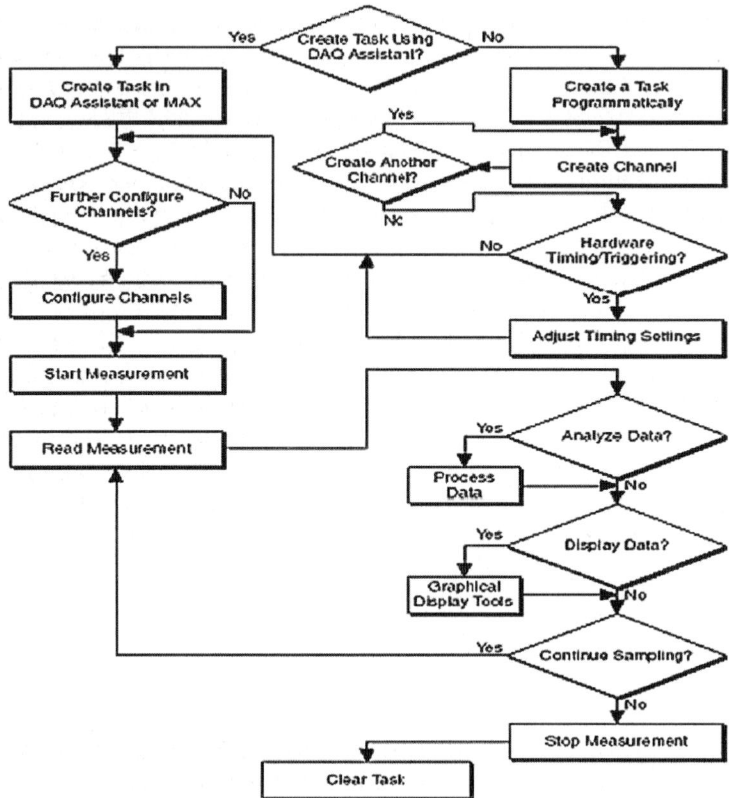

Fig. 6.20. Voltage measurement flowchart

6.7.5 Typical Program Flowchart

Figure 6.20 shows a typical program voltage measurement flowchart for creating a task to configure channels, take a measurement, analyze the data, present the data, stop the measurement, and clear the task.

General Discussion of Typical Flowchart

The following sections briefly discuss some considerations for a few of the steps in the flowchart. These sections are meant to give an overview of some of the options and features available when programming with NI-DAQmx.

Creating a Task Using DAQ Assistant or Programmatically

When creating an application, the user must first decide whether to create the appropriate task using the DAQ Assistant or programmatically in the ADE. Developing application using DAQ Assistant gives us the ability to configure most settings such as measurement type, selection of channels, excitation voltage, signal input limits, task timing, and task triggering can access the DAQ Assistant through MAX or NI ADE. Choosing to use the DAQ Assistant can simplify the development of application.

NI recommends creating tasks using the DAQ Assistant for ease of use, when using a sensor that requires complex scaling, or when many properties differ between channels in the same task. If are using an ADE other than an NI ADE, or if want to explicitly create and configure a task for a certain type of acquisition, can programmatically create the task from ADE using functions or VIs. If create a task using the DAQ Assistant, can still further configure the individual properties of the task programmatically with functions or property nodes in ADE. NI recommends creating a task programmatically if need explicit control of programmatically adjustable properties of the DAQ system.

Programmatically adjusting properties for a task created in the DAQ Assistant overrides the original, or default, settings only for that session. The changes are not saved to the task configuration. The next time load the task, the task uses the settings originally configured in the DAQ Assistant.

Adjusting Timing and Triggering

There are several timing properties that can configure through the DAQ Assistant or programmatically using function calls or property nodes. To create a task in the DAQ Assistant, the user can still modify the timing properties of the task programmatically in the application. When programmatically adjusting timing settings, can set the task to acquire continuously, acquire a buffer of samples, or acquire one point at a time.

For continuous acquisition, the must use a while loop around the acquisition components even if the task is configured for continuous acquisition using MAX or the DAQ Assistant. For continuous and buffered acquisitions, the user can set the acquisition rate and the number of samples to read in the DAQ Assistant or programmatically in the application. By default, the clock settings are automatically set by an internal clock based on the requested sample rate. Also the user can select advanced features such as clock settings that specify an external clock source, internal routing of the clock source, or select the active edge of the clock signal.

Configuring Channel Properties

All ADEs used to configure the SCXI-1102/B/C access an underlying set of NI-DAQmx properties. Table 6.3 shows some of these properties. Using Table 6.3 the user can determine the kind of properties required to set to configure the module for application.

Table 6.3. Properties of SCXI

Property	Short Name	Description
Analog Input≫Maximum Value	AI.Max	Specifies the maximum value you expect to measure. The SCXI-1102/B/C gain and E/M Series DAQ device range are computed automatically from this value.
Analog Input≫Minimum Value	AI.Min	Specifies the minimum value you expect to measure. The SCXI-1102/B/C gain and E/M Series DAQ device range are computed automatically from this value.
Analog Input≫General Properties ≫Advanced≫Range≫High	AI.RNG.High	Specifies the upper limit of the E/M Series DAQ device input range in voltage.
Analog Input≫General Properties ≫Advanced≫Range≫Low	AI.RNG.Low	Specifies the lower limit of the E/M Series DAQ device input range in voltage.
Analog Input≫General Properties ≫Advanced≫Gain and Offset ≫Gain Value	AI.Gain	Specifies a gain factor to apply to the signal conditioning portion of the channel
Analog Input≫Temperature ≫Thermocouple≫Type	AI.Thermepl.Type	Specifies the type of thermocouple connected to the channel
Analog Input≫Temperature ≫Thermocouple ≫ CJC Source (read only)	AI.Thermcpl.CJCSrc	Indicates the source of cold-junction compensation
Analog Input≫Temperature ≫Thermocouple≫CJC Value	AI.Thermcpl.CJCVal	Specifies the temperature of the cold-junction if the CJC source is constant value
Analog Input≫Temperature ≫Thermocouple≫CJC Channel (read only)	AI.Thermcpl.CJCChan	Indicates the channel that acquires the temperature of the cold junction if CJC is channel
Analog Input≫Temperature ≫Advanced≫Force≫Read From Channel	AI.ForceReadFromChan	Specifies whether to return the reading from the CJC channel during a read operation

Table 6.3. *Continued*

Property	Short Name	Description
Analog Input≫Temperature ≫RTD≫Type	AI.RTD.Type	Specifies the type of RTD connected to the channel
Analog Input≫Temperature ≫RTD≫R0	AI.RTD.R0	Specifies the resistance in ohms of the sensor at 0° C

Property	Short Name	Description
Analog Input≫Temperature≫ RTD≫Custom≫A, B, C	AI.RTD.A AI.RTD.B AI.RTD.C	Specifies the A, B, or C constant of the Callendar-Van Dusen equation when using a custom RTD type
Analog Input≫General Properties≫Signal Conditioning≫ Resistance Configuration	AI.Resistance.Cfg	Specifies the resistance configuration for the channel, such as 2-wire, 3-wire, or 4-wire

Property	Short Name	Description
Analog Input≫Temperature ≫Thermistor≫R1	AI.Thrmistr.R1	Specifies the resistance in ohms of the sensor at 0° C.
Analog Input≫Temperature≫	AI.Thrmistr.A	Specifies the A, B, or C constant of the Callendar-Van Dusen equation when using a custom thermistor type.
Thermistor≫Custom≫A, B, C	AI.Thrmistr.B AI.Thrmistr.C	

Property	Short Name	Description
Analog Input≫General Properties≫Signal Conditioning ≫Current Shunt Resistors≫Location	AI.CurrentShunt.Loc	Specifies the shunt resistance location
Analog Input≫General Properties≫Signal Conditioning≫Value ≫Current Shunt Resistor	AI.CurrentShunt. Resistance	Specifies the resistance, in ohms, of the external shunt resistance

Acquiring, Analyzing, and Presenting

After configuring the task and channels, the user can start the acquisition, read measurements, analyze the data returned, and display it according to the needs of application. Typical methods of analysis include digital filtering, averaging data, performing harmonic analysis, applying a custom scale, or adjusting measurements mathematically. NI provides powerful analysis toolsets for each NI ADE to help us perform advanced analysis on the data without requiring having a programming background. After acquiring the data and performing the required analysis, the data is displayed in a graphical form or logged to a file.

NI ADEs provide easy-to-use tools for graphical display, such as charts, graphs, slide controls, and gauge indicators. NI ADEs have tools that allow the user to easily save the data to files such as spread sheets for easy viewing, ASCII files for universality, or binary files for smaller file sizes.

Completing the Application

After completing the measurement, analysis, and presentation of the data, it is important to stop and clear the task. This releases any memory used by the task and frees up the DAQ hardware for use in another task.

6.8 VXI

VXI bus is an exciting and fast-growing platform for instrumentation systems. The IEEE officially adopted the VXI specification, IEEE 1155, 1993. The VXI plug and play Systems Alliance, founded in 1993, sought a higher level of system standardization to cover all VXI system components. By focusing on software standardization, the alliance defined standards to make VXI systems easy to integrate and use while maintaining multivendor software interpretability. VXI is used in many different applications ranging from test and measurement and ATE, to data acquisition and analysis in both research and industrial automation. Many users are migrating to VXI by integrating it into existing systems consisting of GPIB instruments, VME cards, or plug-in data acquisition (DAQ) boards. The user can control a VXI system with a remote general-purpose computer using the high-speed Multisystem extension Interface (MXI) bus interface or GPIB. The user can also embed a computer into a VXI chassis and control the system directly. Despite the system configurations, VXI offers the flexibility and performance to take on today's most challenging applications.

6.8.1 Need for VXI

The demand for an industry-standard instrument-on-a-card architecture has been driven by the need for physical size reduction of rack-and-stack instrumentation systems, tighter timing and synchronization between multiple instruments, and faster transfer rates than the 1 Mbytes s^{-1} rate of the 8-bit GPIB. The modular form factor, high bandwidth, and commercial success of the VME bus made it particularly attractive as an instrumentation platform. The tremendous popularity of GPIB also made it attractive as a model for device communication and instrument control protocols. The VXI bus specification adds the standards necessary to combine the VME bus with GPIB to create a new, modular instrumentation platform that can meet the needs of future instrumentation applications.

6.8.2 Features of VXI

VXI brings the following benefits to instrumentation users:

- Open, multivendor standards maximize flexibility and minimize obsolescence
- Increased system throughput reduces test time and/or increases capabilities
- Smaller size and higher density reduce floor space, enhance mobility or portability, and give close proximity to devices(s) being tested or controlled
- More precise timing and synchronization improve measurement capability
- Standardized VXI plug and play software eases system configuration, programming, and integration
- Modular, rugged design improves reliability, increases Mean-Time Between Failure (MTBF), and decreases Mean-Time To Repair (MTTR)

Members of the VXI bus Consortium and the VXI plug & play Systems Alliance have combined their expertise to develop technically sound standards for both hardware and software, bringing the entire industry into a new generation of easy-to-use instrumentation.

6.8.3 VXI Bus Mechanical Configuration

Physically, a VXI bus system consists of a mainframe chassis that has the physical mounting and backplane connections for plug-in modules, as shown in Fig. 6.21. The VXI bus uses the industry-standard IEEE-1014 VME bus as a base architecture to build upon. As shown in Fig. 6.22, VXI uses the full 32-bit VME architecture, but adds two board sizes and one connector.

The P1 connector and the center row of the P2 connector are retained exactly as defined by the VME specification. The VME user-definable pins on the P2 connector and the additional pins on P3, the third VXI connector, implement instrumentation signals between plug-in modules directly on the backplane.

The VXI bus specification includes packaging requirements, electromagnetic compatibility, power distribution, cooling and air flow for VXI bus mainframes and plug-in modules. The modules are installed in the mainframe slots. LED s, switches, test points, and I/O connections are accessible from the module front panel.

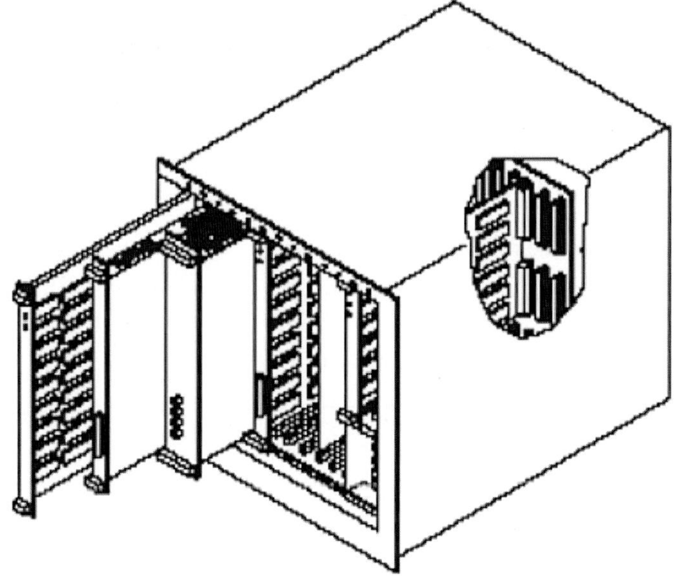

Fig. 6.21. VXI bus

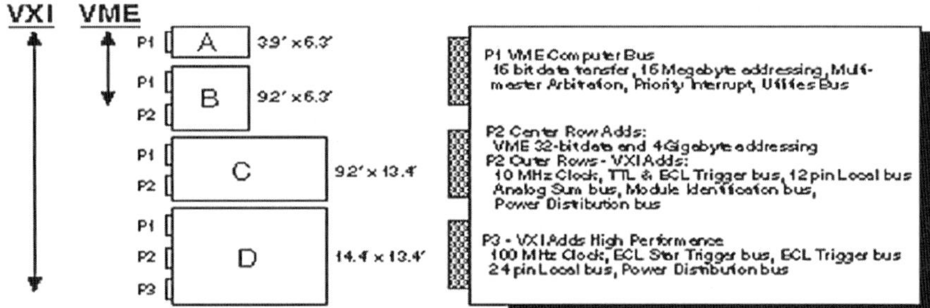

Fig. 6.22. VXI modules

6.8.4 Noise Incurred in VXI

The addition of a new module to a VXI bus system must not degrade the performance of any other module. The VXI bus specification includes near-field radiation and susceptibility requirements, which prevent one module from interfering with the operation of other modules. To meet these requirements, the VXI bus module width was increased from the 0.8 in. VME requirement to 1.2in., so that there is enough room for the module to be completely enclosed in a metal case for shielding. The metal cases connect to backplane grounds. Thus, the user can use existing VME boards in a VXI bus chassis, but not vice versa.

The VXI bus specification also has conducted-emissions and susceptibility requirements, which prevent any power supply noise from affecting the performance of a module. For far-field radiated emissions such as FCC and VDE, each module must not contribute more than its share of the total. For example, in a mainframe that holds 13 modules, each module must not contribute more than 1/13 of the allowed total. Because of the desire for extremely precise time coupling between modules using the backplane, it is necessary to minimize the noise and crosstalk on the backplane clock and trigger signal lines. The backplane is required to be a single, monolithic board across any one slot.

6.8.5 Hardware Registers

VXI modules must have a specific set of registers located at specific addresses, as shown in Fig. 6.23. The upper 16 kB of the 64 kB A16 address space are reserved for VXI bus devices. Each VXI device has an 8-bit logical address that specifies where its registers are located in this address space. A single VXI system can have up to 256 VXI devices. The logical address of a VXI device, which can be manually set or automatically configured by the system at startup, is analogous to the GPIB address of a GPIB device.

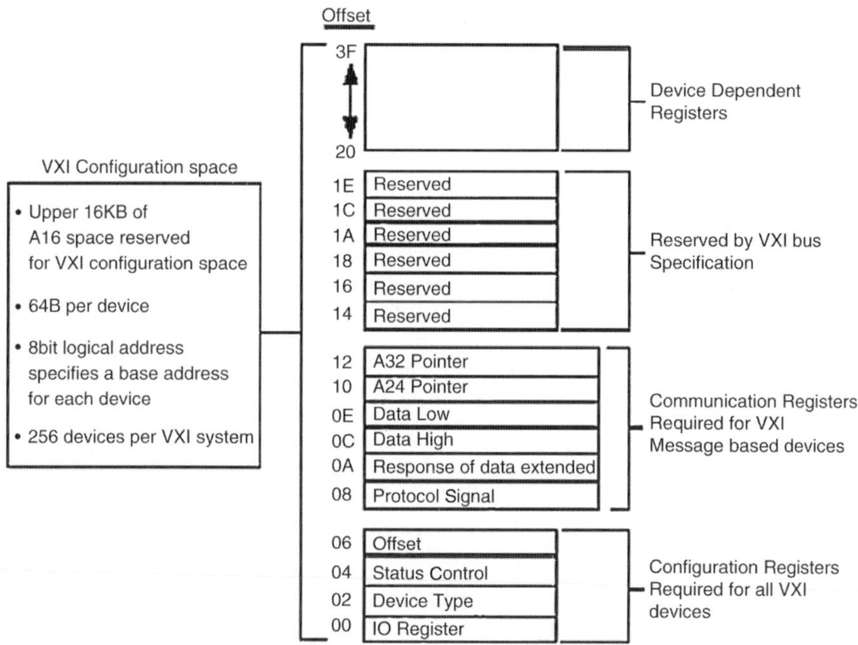

Fig. 6.23. VXI registers

6.8.6 Register-Based Devices

Because of the VXI configuration registers, which are required for all VXI devices, the system can identify each VXI device, its type, model and manufacturer, address space, and memory requirements. VXI bus devices with only this minimum level of capability are called Register-Based devices. With this common set of configuration registers, the Centralized Resource Manager (RM), essentially a software module, can perform automatic system and memory configuration when the system is initialized.

6.8.7 Message-Based Communication and Serial Protocol

In addition to Register-Based devices, the VXI bus specification also defines Message-Based devices, which are required to have communication registers and configuration registers. All Message-Based VXI bus devices, regardless of the manufacturer, can communicate at a minimum level using the VXI-specified.

When minimum communication is possible, higher-performance communication channels, such as shared-memory channels, can be established to take advantage of the VXI bus bandwidth capabilities. The VXI bus, Word Serial Protocol is functionally very similar to the IEEE-488 protocol, which transfers data messages to and from devices 1 byte (or word) at a time. Thus, VXI Message-Based devices communicate in a fashion very similar to IEEE-488 instruments. In general, Message-Based devices typically contain some level of local intelligence that uses or requires a high level of communication. All VXI Message-Based devices are required to use the Word Serial Protocol to communicate in a standard way.

The protocol is called word serial, because if want to communicate with a Message-Based device, do so by writing and reading 16-bit words one at a time to and from the Data In (write Data Low) and Data Out (read Data Low) hardware registers located on the device itself. Word Serial communication is paced by the bits in the response register of the device, indicating whether the Data In registers is empty and whether the Data Out registers is full. This operation is very similar to Universal Asynchronous Receiver Transmitter (UART) on a serial port.

6.8.8 Commander/Servant Hierarchies

The VXI bus defines a Commander/Servant communication protocol so that the user can construct hierarchical systems using conceptual layers of VXI devices. This structure is like an inverted tree. A Commander is any device in the hierarchy with one or more associated lower-level devices, or Servants. A Servant is any device in the sub-tree of a Commander. A device can be both a Commander and a Servant in a multiple-level hierarchy. A Commander has exclusive control of the communication and configuration registers of its immediate Servants (one or more). Any VXI module has one and only one Commander.

Commanders communicate with Servants through the communication registers of the Servants using the Word Serial Protocol if the Servant is a Message-Based device or by device-specific register manipulation if the Servant is a Register-Based device. Servants communicate with their Commander by responding to the Word Serial commands and queries from their Commander through the Word Serial protocol if it is Message-Based, or by device-specific register status if it is Register-Based.

Interrupts and Asynchronous Events

Servants can communicate asynchronous status and events to their Commander through hardware interrupts or by writing specific messages (signals) directly to their Commander's hardware Signal Register. Non bus master devices always transmit such information through interrupts, whereas devices that have bus master capability can either use Interrupts or send signals. Some Commanders can receive signals only, whereas others might be only interrupting handlers.

The VXI bus specification contains defined Word Serial commands so that a Commander can understand the capabilities of its Message-Based Servants and configure them to generate interrupts or signals in a particular way. For example, a Commander can instruct its Servants to use a particular interrupt line, to send signals rather than generate interrupts, or configure the reporting of only certain status or error conditions. Although the Word Serial Protocol is reserved for Commander/Servant Communications, peer-to-peer communication between two VXI devices can be established through a specified shared-memory protocol or by simply writing specific messages directly to the signal register of the device.

Slot 0 and the Resource Manager

The leftmost slot of a VXI chassis has special system resources such as backplane clocks, configuration signals, and synchronization (trigger) signals and therefore must be occupied by a device with VXI "Slot 0" capabilities. The VXI Resource Manager (RM) function, essentially a software module, can reside on any VXI module or even on an external computer.

The RM, in combination with the Slot 0 device, identifies each device in the system, assigns logical addresses, memory configurations, and establishes Commander/Servant hierarchies using the Word Serial Protocol to grant Servants to the Commanders in the system. After establishing the Commander/Servant hierarchy, the RM issues the Begin Normal Operation Word Serial command to all top-level Commanders. During normal system operation, the RM may also halt the system and/or remap the hierarchy if necessary.

6.8.9 Three Ways to Control a VXI System

System configuration is divided into three categories as follows:

The first type of VXI system consists of a VXI mainframe linked to an external controller through the GPIB. The controller talks across the GPIB to a GPIB-VXI interface module installed inside the VXI mainframe. The GPIB-VXI interface transparently translates the GPIB protocol to and from the VXI Word Serial protocol.

The second configuration involves a VXI-based embedded computer. The embedded computer is a VXI module that resides inside the VXI mainframe and connects directly to the VXI backplane. This configuration offers the smallest physical size for a VXI system as well as performance benefits due to direct connection to the VXI backplane.

The third configuration uses a high-speed MXI bus link from an external computer to control the VXI backplane. The external computer operates as though it is embedded directly inside the VXI mainframe. This configuration is functionally equivalent to the embedded method, except that it has the flexibility for use with a wide variety of computers and workstations.

6.8.10 Software Standards

As a step toward industry-wide software compatibility, the VXI plug and play alliance developed one specification for I/O software – the Virtual Instrument System Architecture, or VISA. The VISA specification, VPP-4.1, defines a next-generation I/O software standard not only for VXI, but also for GPIB and serial interfaces. The VXI software bus is shown in Fig. 6.24.

With the VISA standard endorsed by over 50 of the largest instrumentation companies in the industry including Tektronix, Hewlett-Packard, and National Instruments, VISA unifies the industry by facilitating the development of interoperable and reusable software components able to stand the test of time. Before VISA, there were many different commercial implementations of I/O software for VXI, GPIB, and serial interfaces; however, none of these I/O software products were standardized or interoperable.

6.9 LXI

LXI is an acronym for LAN eXtensions for Instrumentation, a standard that has been created by the leading test and measurement companies to provide control of instrumentation products. The standard has been created by the members of the LXI Consortium and is published as a popularly available standard accessible Membership of the LXI consortium is open to all companies that wish to support the aims of the consortium.

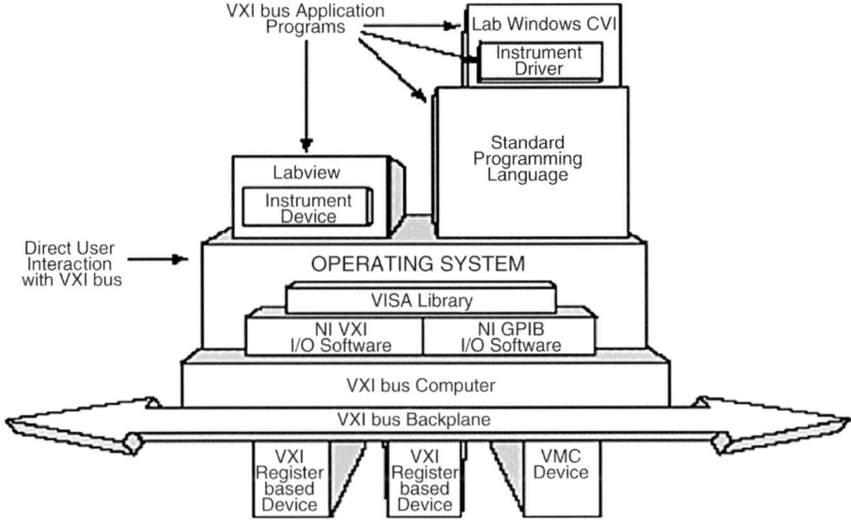

Fig. 6.24. VXI software bus

6.9.1 LXI Modular Switching Chassis

The features of LXI Modular Switching Chassis are

- Fully Compliant LXI Interface
- Ethernet 100base t Interface
- Accepts any Pickering Interfaces PXI 3u Switching module
- More than 500 switching module Configurations available
- Applications from Simple Switching to RF and optical
- Supports 7 or 13 user Slots
- Support for Selected Instrument modules

The 60–100 and 60–101 chassis are fully compliant with Functional Class C of the LXI standard. It will allow PXI switching modules to be supported in a LXI compliant environment. The 60–100 is suitable for modules occupying seven or less slots and the 60–101 can support up to one slot. The chassis allows any of the large number of PXI modules from Pickering Interfaces to be installed and controlled through a standardized Ethernet interface. The chassis are supplied with a generic IVI driver to control the PXI modules in accordance with the LXI specification, but can also be controlled through a kernel driver.

Modules available for installation in the 60–100/101 chassis cover a large variety of switching functions. It includes simple switches, matrices, and multiplexers based on reed relay and electromechanical technology which can switch from microamps to 40 A. A range of RF, microwave, and optical switching functions as well as some instrument functions such as variable resistor cards are supported. This makes the 60–100 and 60–101 chassis into universal switching platforms.

LXI Modular Switching Chassis

The LXI compliant Ethernet interface enables the chassis to be controlled through standard LAN interfaces found on all controllers. It allows the remote operation of switching products with simple cabling over long distances, even permitting a chassis to be controlled over world wide networks. The separation of the chassis from the controller's PCI bus simplifies power-on sequencing of systems and provides a high degree of independence from the migration of the Windows and VISA environment. Availability of the 60–100/101 chassis permit users to choose Pickering PXI switching products regardless of whether the test system is working in a PXI or LXI environment. The 60–100 and 60–101 are ideal for operation in environments where high acoustic noise levels cannot be accepted through the use of low noise temperature controlled fans. The architecture for conversion of PXI to LXI is shown in Fig. 6.25.

6.9.2 LXI/PXI Module Selection

The 60–100 and 60–101 chassis are capable of supporting a large variety of PXI modules from Pickering Interfaces. The modules are the same design as those used in a PXI chassis. The following information provides a convenient way of selecting from the large number of modules available the product most suited for the application. From Table 6.4 the user can select the short form data sheet that best describes requirements.

PXI Module Compatibility

Support of PXI modules requires availability of the correct software driver. Chassis is supplied with drivers for all Pickering system 40 switching modules.

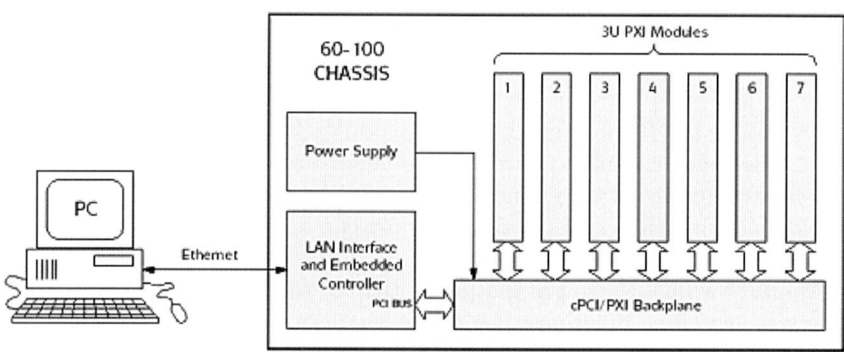

Fig. 6.25. PXI into LXI

Table 6.4. Data sheet

Short Form Data Sheet	Configurations Covered	Max Relay Count (up to)	Max Current or Voltage
Low Density General Purpose Relay Modules	– Uncommitted	32	1.25 A
High Density General Purpose Relay Modules	– Uncommitted	100	1.25 A
General Purpose Power Relay Modules	– Uncommitted	39	3 A
General Purpose High Power Relay Modules	– Uncommitted	12	40 A
High Voltage Switch Modules	– Uncommitted – Multiplexer	24	500 V d.c.
Low Density Matrix Modules	– Matrix	128	2 A/250 V
Medium Density Matrix Modules	– Matrix	184	1 A/150 V
High Density Matrix Modules	– Matrix	440	1 A/150 V
Power Matrix Modules	– Matrix	80	5 A/250 V
Large Scale Matrix Modules	– Matrix	4,416	2 A
Fault Insertion Modules	– Fault Insertion Matrix	2,384	10 A/250 V
Low Density MUX Modules	– Multiplexer – Low Thermal Switch MUX	48–1 MUX	1.2 A/150 V
High Density MUX Modules	– Multiplexer	198–1 MUX	1.2 A/150 V
Power MUX Modules	– Multiplexer	64–1 MUX	5 A/250 V
RF Switching Modules	– Uncommitted – Multiplexer – Matrix	96	–
Microwave Switching Modules	– Uncommitted – Multiplexer – Matrix	36 i/p MUX	–
Optical Switching Modules	– Multiplexer – Insert/ bypass Switch	8–1 MUX	–

Table 6.4. *Continued*

Short Form Data Sheet	Configurations Covered	Max Relay Count (up to)	Max Current or Voltage
Telecoms Switching Modules	– Daisy chain tributary switch – Datacoms MUX	16 channel 2 pole switch or 36–1 MUX	1A/100V
Programmable Resistor Modules	– Selectable resistor – Programmable Resistor	48 Channels or 24 bit resolution	–
Digital Input/Output Cards and Prototyping Modules	– TTL or Open Collector – Optically Isolated	32 bit in, 32 bit out	–

Switching Support

Any Pickering Interface's PXI switching modules includes all 4 slot BRIC matrices, featuring up to 08 crosspoints. Additionally, the 60–101 can support any eight-slot BRIC matrix with up to 4,416 crosspoints.

Power Supply

AC Input. 90–64 V, 50/60 Hz universal input, rear panel with on/off switch.
Cooling. Front and side selectable air intake, rear exhaust, variable speed low noise fans.

Monitoring

Front panel monitoring LEDs:

- Power
- Ready
- Error
- LAN
- Active

LAN Interface

Connector: RJ45 Connector.
Connection Speed: 100baseT interface.
Designed to comply with the LXI Standard Version 1.0 Class C

Summary

The usage of interface buses for reception and transmission of data were discussed in detail. The purpose of USB along with its architecture was discussed, which says the advantage in using USB for interfacing. The configuration and the required properties for interfacing, its slot allotment were also delineated. The chapter also focused on the advanced and enhanced version of PCI bus, that is PCI express along with the essential and the various factors that govern the interfacing process of PXI, PCMCIA, SCXI, LXI, and VXI.

Review Questions

1. Explain in detail about architecture of USB.
2. Write the need for USB.
3. Give the electrical specifications of USB.
4. What are the three classes of USB functions? Explain in detail.
5. Discuss in detail about the packet types of USB.
6. Explain in detail about the USB functions with neat sketches.
7. Discuss in detail about the USB descriptor with suitable diagrams.
8. List the advantages of USB.
9. Explain in detail the architecture of PCI.
10. Write the features of PCI.
11. Give the specifications of PCI bus.
12. Explain the concept of PCI IDE bus mastering.
13. Write notes on PCI internal interrupts and PCI bus performance.
14. Write the need for PCI express.
15. What are the types of PCI express architecture.
16. Explain the architecture of PXI with neat sketches.
17. Give the features of PXI and PCMCIA.
18. Write the benefits of PCMCIA.
19. Explain the functional features of SCXI architecture.
20. List out the properties of SCXI.
21. Write the need and features of VXI bus.
22. Write the need and features of LXI bus.

7

Hardware Aspects

Learning Objectives. On completion of this chapter the reader will have a knowledge on:

- Hardware Aspects
- Signal Grounding concepts
- Digital I/O Techniques
- Data Acquisition in LabVIEW
- Hardware Installation and Configuration of Data Acquisition
- Components of DAQ
- DAQ Signal Accessory
- DAQ Assistant
- Steps to Create a MAX-Based Task
- Steps to Create a Project-Based Task
- DAQ Hardware
- Hardware Triggering
- DAQ Software

7.1 Introduction

Today, most scientists and engineers are using personal computers with ISA, EISA, PCI, PCMCIA, Macintosh NuBus, or parallel or serial ports for data acquisition in laboratory research, test and measurement, and industrial automation. Many applications use plug-in boards to acquire data and transfer it directly to computer memory. Others use DAQ hardware remote from the PC that is coupled via parallel or serial port. Obtaining proper results from a PC-based DAQ system depends each of the components such as personal computer, transducers, signal conditioning, DAQ hardware, and software. This chapter defines the terminology common to each of the elements of a PC-based DAQ system. The hardware aspects of the system such as the signal grounding concepts have also been detailed. The digital IO techniques required for the DAQ assistant operation along with the installation and configuration details of DAQ are elaborated in this chapter.

7.2 Signal Grounding

This section explains the methodology to create analog input connections to achieve the optimum performance from the user's board. Prior to jumping into actual connection schemes, the user should have a basic understanding of single-ended/differential inputs and system grounding/isolation. The board provides 16 differential input channels or eight single-ended input channels. The theory of single ended and differential inputs is described in the following sections.

7.2.1 Single-Ended Inputs

A single-ended input measures the voltage between the input signal and ground. In single-ended mode the board measures the voltage between the input channel and low-level ground (LLGND – Fig. 7.1). The single-ended input configuration requires only one physical connection (wire) per channel and allows the board to monitor more channels than the 2-wire differential configuration using the same connector and onboard multiplexer. For clarity, the input multiplexers are not shown in the following diagrams.

Because the board is measuring the input voltage relative to its own LLGND, single-ended inputs are more susceptible to both electromagnetic interference (EMI) and any ground noise at the signal source. Figure 7.2 shows an

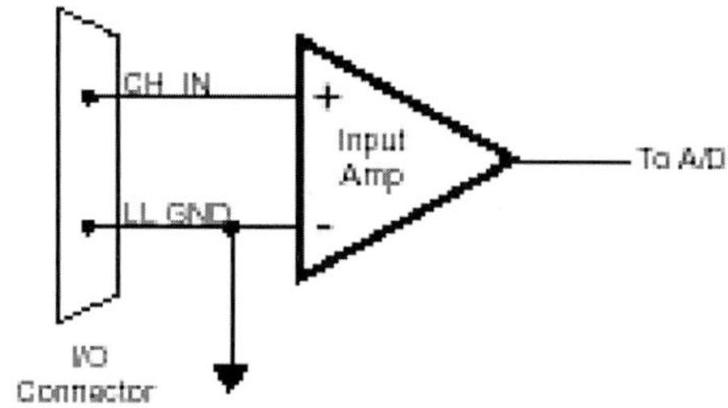

Fig. 7.1. Single-ended input

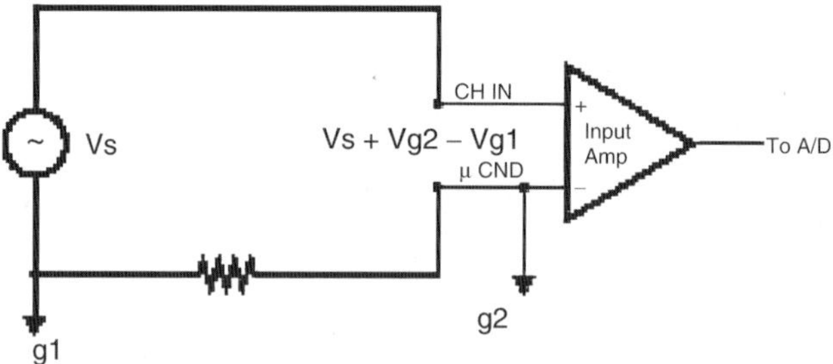

Fig. 7.2. Single-ended input with common mode voltage

input having a common mode voltage. In the figure, any voltage differential between grounds G1 and G2 shows up an error signal at the input amplifier.

7.2.2 Differential Inputs

Differential inputs measure the voltage between two distinct input signals. Within a certain range – called the *common mode range* – the measurement is almost independent of signal source to board ground variations. A differential input is more immune to EMI than a single-ended one. Most EMI noise induced in one lead is also induced in the other. The input measures only the difference between the two leads, and the EMI common to both is ignored. This effect is a major reason for twisted pair wiring because the twisting ensures that both wires are subject to virtually identical external influence. Figure 7.3 shows the basic differential input configuration.

Before describing grounding and isolation, it is important to explain the concepts of common mode voltage, and common mode range. Common mode voltage is shown in Fig. 7.4 as *V cm*. This common mode voltage is ignored by differential input configuration. However, V cm + Vs should remain within the amplifier's common mode range of ±10 V.

Although differential inputs measure the voltage between two signals – almost without respect to the either signal's voltages relative to ground – there is a voltage limit on the signal. Although the board has differential

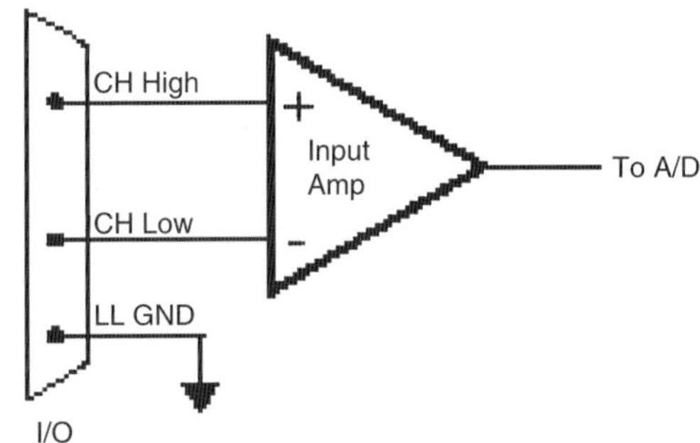

Fig. 7.3. Differential input configuration

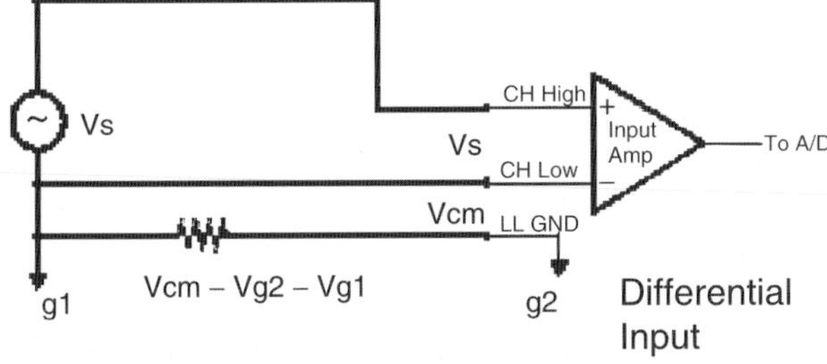

Fig. 7.4. Differential input – voltages defined

inputs, it cannot measure the difference between 100 and 101 V as 1 V. Even in differential mode, an input signal cannot be measured if it is more than 10 V from the board's LLGND.

7.2.3 System Ground and Isolation

There are three ways to connect a signal source to the board: The board's input and the signal source may have the same (or common) ground. This signal source can be connected directly to the board. The board input and the signal source may have an offset voltage between their grounds (AC and/or DC). This offset is the common mode voltage. Depending on the magnitude of this voltage, the user may be able to connect the board directly to the signal source. The board and the signal source may already have isolated grounds. This signal source can be directly connected to the board.

Systems with Common Grounds

In the simplest (but perhaps the least likely) case, the signal source will have the same ground potential as the board. This would typically occur when providing power or excitation to the signal source directly from the board. It is important to note that any voltage between the board's ground and the signal ground is a potential error voltage if the system is set assuming there is a common ground.

Systems with Common Mode (Ground Offset) Voltages

The most frequently encountered grounding scenario involves grounds that are somehow connected, but have offset voltages between the board and signal source grounds. This offset voltage may be AC, DC, or both. The offset can be caused by a wide array of phenomena, such as EMI pickup, and resistive voltage drops in ground wiring and connections. Ground offset voltage is a more descriptive term for this type of system, but the term *common mode* is more frequently used.

Small Common Mode Voltages

If the voltage between the signal source ground and board ground is small, the combination of the ground voltage and input signal does not exceed the allowable ±10 V common mode range. Specifically, when the voltage between grounds is added to the maximum input voltage, the result is within ±10 V. This input is compatible with the board, and the user can connect the system without additional signal conditioning. Fortunately, most systems fall into this category and have small voltage differentials between grounds.

Large Common Mode Voltages

If the ground differential is large enough, the allowable ±10 V does exceed the common mode range. Specifically, when the voltage between board and signal source grounds are added to the maximum input voltage, the result exceeds ±10 V. In this case, the board should not be directly connected to the signal source. The system's grounding configuration should be changed or isolation signal conditioning circuitry has to be added.

7.2.4 Wiring Configurations

All grounding and input type combinations are summarized in Table 7.1.

The following sections contain recommended input wiring schemes for each of the acceptable input configuration/grounding combinations.

Common Ground/Single-Ended Inputs

Single-ended is the recommended configuration for common ground connections. However, if some inputs are common ground and some are not, the differential mode should be used for all inputs. There is no performance penalty (other than loss of channels) for using a differential input rather than a single-ended input to measure a common ground signal source. However, the reverse is not true. Figure 7.5 shows the recommended connections for a common ground/single-ended input system.

Common Ground/Differential Inputs

Using differential inputs to monitor a signal source with a common ground is an acceptable configuration, though it requires more wiring and offers fewer channels than selecting a single-ended configuration illustrated in Fig. 7.6.

Table 7.1. Input configuration recommendations

Ground category	Input configuration	Our recommendation
Common ground	Single-ended inputs	Recommended
Common ground	Differential inputs	Acceptable
Common mode Voltage > ±0 V, < ±10 V	Single-ended inputs	Not Recommended
Common mode Voltage < ±10 V	Differential inputs	Recommended
Common mode Voltage > ±10 V	Single-ended inputs	Unacceptable without adding Isolation
Common mode Voltage > ±10 V	Differential inputs	Unacceptable without adding Isolation
Already isolated grounds	Single-ended inputs	Acceptable
Already isolated grounds	Differential inputs	Recommended

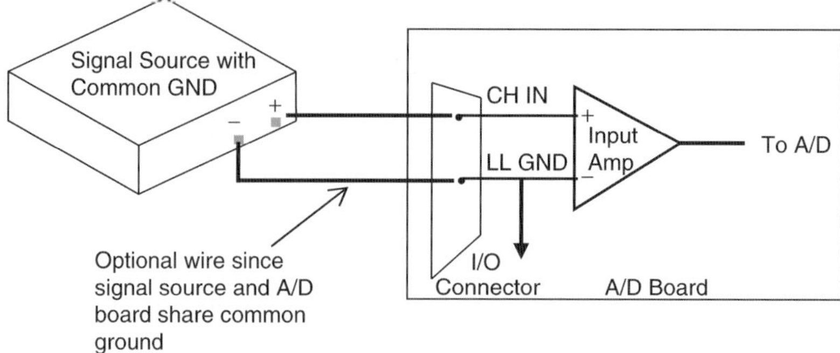

Fig. 7.5. Common ground/single-ended input system

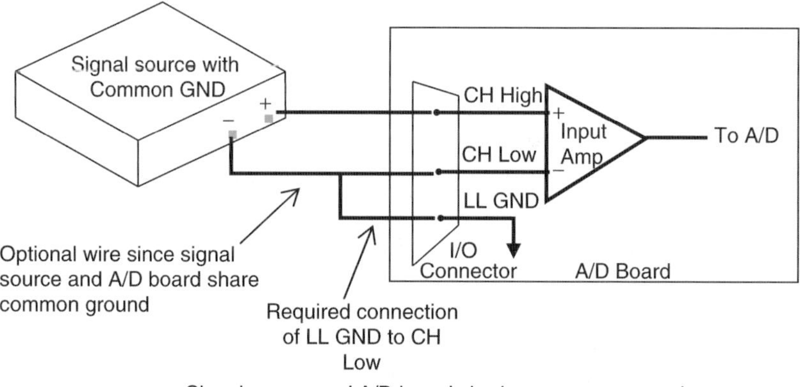

Fig. 7.6. Common ground/differential inputs

Common Mode Voltage < ±10 V/Single-Ended Inputs

The phrase *common mode* has no meaning in a single-ended system, and this is not a recommended configuration. This case can be described as a system with offset grounds. Depending on the overall accuracy that is required, the user may receive acceptable results. Figure 7.7 illustrates the operation of this mode.

Common Mode Voltage < ±10 V/Differential Inputs

Systems with varying ground potentials should always be monitored in the differential mode. The sum of the input signal and the ground differential (referred to as the common mode voltage) should not exceed the common mode range of the A/D board (generally ±10 V). Figure 7.8 shows recommended connections in this configuration.

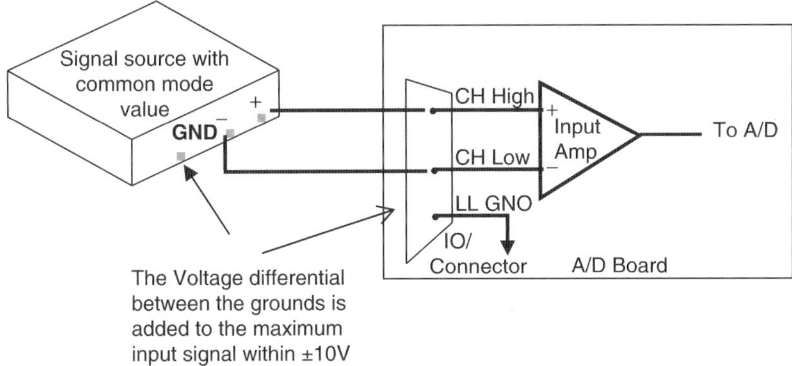

Fig. 7.7. Common mode/differential inputs (< ±10 V) Common Mode Voltage > ±10 V

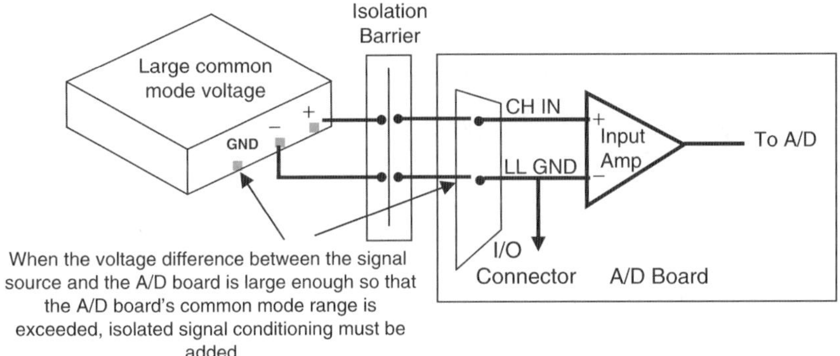

Fig. 7.8. Common mode voltage < ±10 V/differential inputs

The board cannot directly monitor signals with common mode voltages greater than ±10 V. Hence the system ground configuration is to be modified to reduce the overall common mode voltage, or an isolated signal conditioning circuit can be added between the source and the board.

Isolated Grounds/Single-Ended Inputs

Single-ended inputs are used to monitor isolated inputs, whereas differential mode increases the system's noise immunity. Figure 7.9 shows the recommended connections in this configuration.

Isolated Grounds/Differential Inputs

To ensure optimum performance with isolated signal sources the differential input setting is used. Figure 7.10 shows the recommend connections for this configuration.

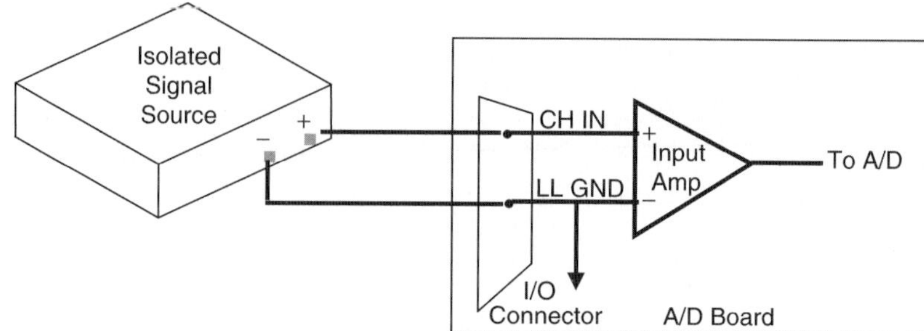

Fig. 7.9. Isolated grounds/single-ended inputs

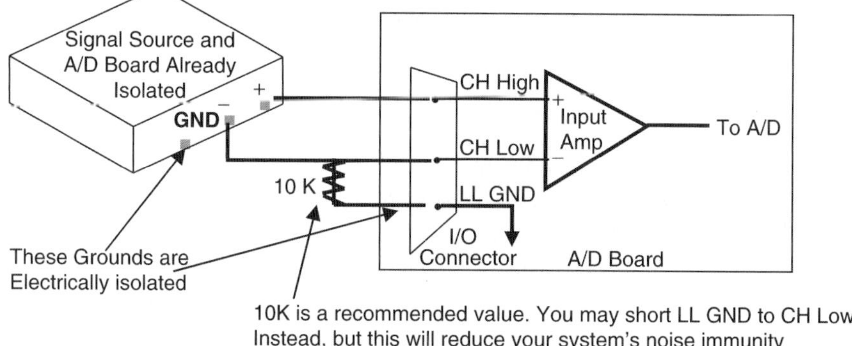

Fig. 7.10. Isolated grounds/differential inputs

7.3 Digital I/O Techniques

This short introduction describes a few topics often needed by board users. It covers a few key application techniques such as pull-up an pull-down resistors, transistor-to-transistor logic (TTL) to solid-state relays and voltage dividers used with digital I/O.

7.3.1 Pull-Up and Pull-Down Resistors

Whenever the board is powered on or reset, the digital I/O control registers are set to a known state. If pull-up/pull-down resistors were not used, the input bits would typically float high when it is in the input mode. There may also be enough drive current available from the inputs to turn on connected devices. If the inputs of the device that is being controlled are left to float, they may float up or down. The way they float depends on the characteristics of the circuit and the electrical environment and may be unpredictable. The result is that the controlled device may be turned on. This is why pull-up or pull-down resistors are needed.

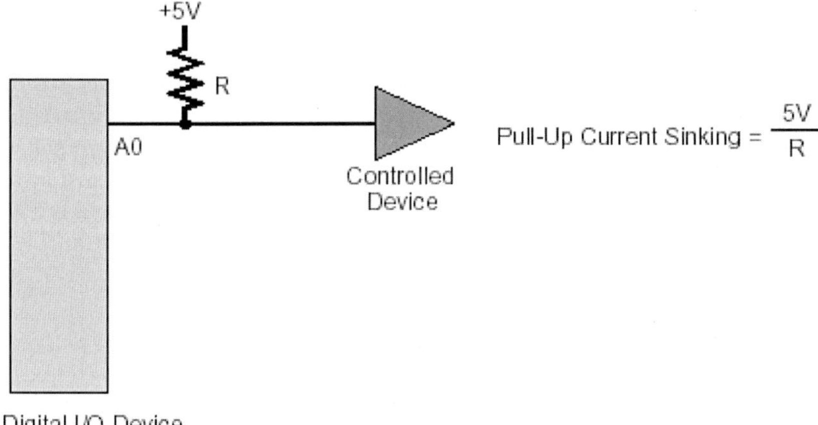

Fig. 7.11. Pull-up resistor configuration

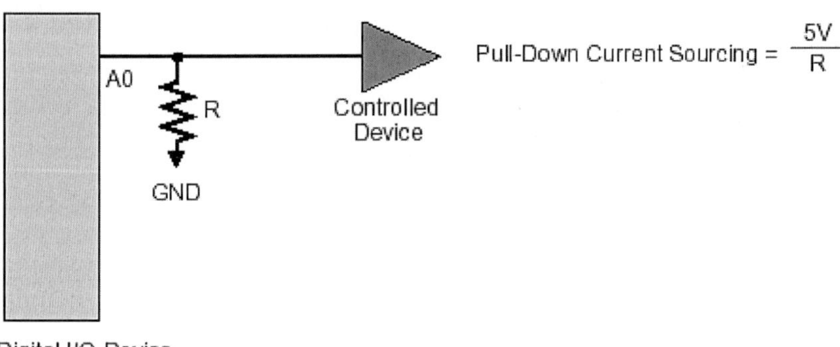

Fig. 7.12. Pull-down resistor configuration

Using a pull-up scheme as indicated in Fig. 7.11, I/O lines are connected to logic power through resistors. The high resistance values require very small drive currents. If the board is reset and enters high impedance input mode, the I/O lines are pulled high. At that point, both the board and the attached device will sense a high signal. If the board is in output mode, it has enough power to override the pull-up resistor's high signal and drive the line to 0 V.

Pull-down resistors accomplish similar tasks except that the I/O lines are connected to logic ground through resistors. When the board is reset, and enters high impedance input mode, the lines are pulled low. If the board is in output mode, it has enough power to override the pull-down resistor's low signal and drive the lines high. The pull-down scheme is shown in Fig. 7.12.

7.3.2 TTL to Solid-State Relays

Many applications require digital outputs to switch high AC and DC voltages on and off and to monitor high AC and DC voltages. Obviously, these high voltages cannot be controlled or read directly by the TTL digital lines of a board. Solid-state relays, such as those available from Measurement Computing Corp. (MCC) allow to control and monitor AC and high DC voltages and provide 750 V isolation. Solid-state relays (SSRs) are the recommended method of interfacing to AC and high DC signals. The most convenient way to use SSRs is to purchase an SSR rack. An SSR rack is a circuit board having sockets for SSRs and buffer amplifiers powerful enough to switch the SSRs. SSR racks are available from MCC and most manufacturers of SSRs.

7.3.3 Voltage Dividers

If the user requires detecting a signal that varies over a range greater than the maximum input specification of a digital input, a voltage divider is used to reduce the voltage of the input signal to a safe level.

The voltage divider works on the principle of Ohm's law, which states

$$Voltage = Current \times Resistance$$

In a voltage divider, the voltage across one of the resistors in a circuit is proportional to the resistance to the total resistance in the circuit. The consideration in using a voltage divider is to choose two resistors with the proper proportions relative to the full value of the input voltage to the desired output voltage to the board input.

Dropping the voltage proportionally is called attenuation. Digital inputs often require voltage dividers. For example, to detect a field signal that is at 0 V when OFF and 24 V when ON, the user cannot connect the signal directly to the board's digital inputs. The voltage must be dropped to 5 V maximum when ON. The Attenuation required is 24:5 or 4.8. Using the equation in Fig. 7.13 an appropriate value of R1 can be determined if R2 is 1K. It should be noted that a TTL input is "ON" when the input voltage is greater than 2.5 V.

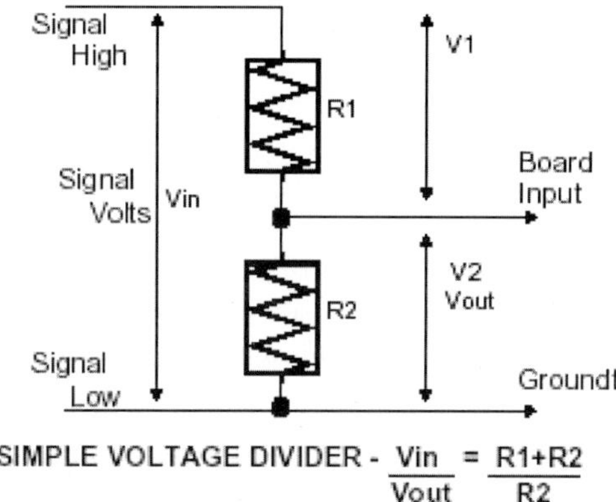

Fig. 7.13. Voltage divider schematic

7.4 Data Acquisition in LabVIEW

The LabVIEW Data Acquisition VIs are located on the Data Acquisition palette and the DAQmx – Data Acqusition palette. The Data Acquisition palette contains the traditional NI-DAQ VIs. The DAQmx – Data Acquisition palette contains the VIs for NI-DAQmx. The DAQmx – Data Acquisition palette contains all of the VIs necessary to perform analog I/O, digital I/O, and counter/timer operations. The VIs are organized so that the most common operations can be performed using the VIs. The task can be configured to perform a very specific function by using the Property Nodes in the palette. Many applications that do not require advanced timing and using the DAQ Assistant Express VI can perform synchronization. The DAQ Assistant Express VI is used to easily configure the data acquisition device. When the DAQ Assistant Express VI is placed on the block diagram, a dialog box appears where a local task can be configured to perform a specific measurement function. The local task allows the user to specify the exact type of measurement. After a task is created, the information for the local task is stored in the DAQ Assistant Express VI. Double-clicking the VI and creating a new task can reconfigure the DAQ Assistant Express VI.

7.5 Hardware Installation and Configuration

The VIs in Data Acquisition and Instrument I/O palettes are at a level beyond what any other function in LabVIEW does – they let the user to communicate with, read from, write to, measure, control, turn or and off, or blow up stuff in the external world via DAQ and GPIB boards. The following sections will take the user a whirlwind tour of the analog, digital, GPIB, serial, and VISA functions available in LabVIEW. To do this effectively, the user first needs to understand a little bit about the interface between LabVIEW and the boards.

As shown in Fig. 7.14, initiating a DAQ operation involves LabVIEW calling *NI-DAQ* (which is the catch-all driver for the board in use), which in turn signals the hardware to initiate the I/O operation. DAQ boards use onboard *buffers* (called FIFOs, for "First-In, First-Out") and RAM buffers as

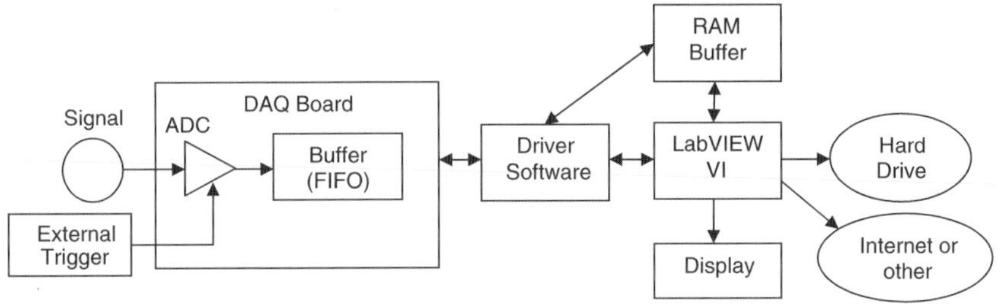

Fig. 7.14. DAQ operation

an intermediate place to store the data they acquire. The software is not the only place an initiation of an I/O operation takes place – an external piece of hardware can also trigger the operation.

The two important characteristics which classify the type of DAQ operation performed are:

- Whether a buffer is used
- Whether an external trigger is used to start, stop, or synchronize an operation

7.5.1 Buffers

A *buffer*, as used in this context, is an area of memory in the PC reserved for data to reside temporarily before it reaches the destination. For example, it may be required to acquire a few thousand data samples in one second. It would be difficult to display or graph all these data in the one same second. But by acquiring the data into a buffer, the data can be stored there first, and then later retrieve them for display or analysis. Buffers are related to the speed and volume of the DAQ operation (generally analog I/O). If the board has DMA capability, analog input operations have fast hardware path to the RAM in computer, meaning the data can be acquired directly into the computer's memory. When a buffer is not used, each data point has to be acquired one at a time before the samples are handled.

Buffered I/O is used when:

- Many samples are to be acquired or generated at a rate faster than it is practical to display, store on a hard drive, or analyze in real time.
- Analog data ($>$ 10 samples s^{-1}) are to be acquired or generated continuously and be able to do analysis or display some of the data on the computer
- The sampling period must be precise and uniform throughout the data samples

Non buffered I/O is used when:

- The data set is small and short (for example, acquiring one data point from each of 2 channels-every second)

7.5.2 Triggering

Triggering refers to the method by which the user can initiate, terminate, or synchronize a DAQ event. A trigger is usually a digital or analog signal whose condition is analyzed to determine the course of action. Software triggering is the easiest and most intuitive. The user can control the trigger directly from the software, such as by using a Boolean front panel control to start or stop data acquisition. Hardware triggering lets the circuitry in the board take care of the triggers, adding much more precision and control over the timing of the DAQ events. Hardware triggering can be further subdivided into external and internal triggering. An example of an internal trigger is the programming of the board to output a digital pulse when an analog-in channel reaches a certain voltage level. All National Instrument's DAQ boards have an external trigger pin, which is a digital input used for triggering. Many instruments provide a digital output (often called "trigger out") used specifically to trigger some other device or instrument, in this case, the DAQ board uses software triggering when:

- The user needs to have explicit control over all DAQ operations
- The timing of the event (such as when an analog input operation be gins) need not be very precise

The DAQ board uses hardware triggering when

- Timing a DAQ event needs to be very precise
- The user wants to reduce software overhead (for example, a While Loop that watches for a certain occurrence can be eliminated)
- The DAQ events must be synchronized with an external device.

7.6 Components of DAQ

LabVIEW includes a set of VIs that can acquire data from, and send data to DAQ devices. Often, one device can perform a variety of functions – analog-to-digital (A/D) conversion, digital-to-analog (D/A) conversion, digital I/O, and counter/timer operations. Each device supports different DAQ and signal generation speeds. Also, each DAQ device is designed for specific hardware platforms and operating systems.

7.6.1 System Components

Before a computer-based measurement system can measure a physical signal, such as temperature, a sensor or transducer must convert the physical signal into an electrical one, such as voltage or current. The user might consider the plug-in DAQ device to be the entire measurement system, but it is actually only one system component. It is not possible to directly connect signals to a plug-in DAQ device. In these cases, signal conditioning accessories are required to condition the signals before the plug-in DAQ device converts them to digital information. The software controls the DAQ system by acquiring the raw data, analyzing, and presenting the results. Consider the following options for a DAQ system:

- The plug-in DAQ device that resides in the computer can be plugged into the PCI slot of a desktop computer or the PCMCIA slot of a laptop computer for a portable DAQ measurement system.

– The DAQ device is external and connects to the computer through an existing port, such as the serial port or Ethernet port, which means the user can quickly and easily place measurement nodes near sensors.

The computer receives raw data through the DAQ device. The application that the user develops presents and manipulates the raw data in a form which can be easily understood. The software also controls the DAQ system by commanding the DAQ device when and from which channels to acquire data. Typically, DAQ software includes drivers and application software. Drivers are unique to the device or type of device and include the set of commands the device accepts. Application software, such as LabVIEW, sends the drivers commands, such as acquire and return a thermocouple reading. The application software also displays and analyzes the acquired data. NI measurement devices include NI-DAQ driver software, a collection of VIs used to configure, acquire data from, and send data to the measurement devices.

7.6.2 NI-DAQ

NI-DAQ 7.0 contains two NI-DAQ drivers – Traditional NI-DAQ and NI-DAQmx – each with its own application programming interface (API), hardware configuration, and software configuration.

– Traditional NI-DAQ is an upgrade to NI-DAQ 6.9.x, the earlier version of NI-DAQ. Traditional NI-DAQ has the same VIs and functions and works the same way as NI-DAQ 6.9.x. The Traditional NI-DAQ can be used on the same computer as NI-DAQmx, which cannot be done with NI-DAQ 6.9.x.
– NI-DAQmx is the latest NI-DAQ driver with new VIs, functions, and development tools for controlling measurement devices. The advantages of NI-DAQmx over previous versions of NI-DAQ include the DAQ Assistant for configuring channels and measurement tasks for a device; increased performance, including faster single-point analog I/O and multithreading; and a simpler API for creating DAQ applications using fewer functions and VIs than earlier versions of NI-DAQ. Traditional NI-DAQ and NI-DAQmx support different sets of devices.

The following illustration in Fig. 7.15 shows the measurement software framework.

When programming an NI measurement device, NI application software such as LabVIEW, LabWindows/CVI, and Measurement Studio, or open ADEs that support calling dynamic link libraries (DLLs) through ANSI C interfaces can be used. Using NI application software greatly reduces development time for data acquisition and control applications regardless of which programming environment that is used:

– LabVIEW supports data acquisition with the LabVIEW DAQ VIs, a series of VIs for programming with NI measurement devices.

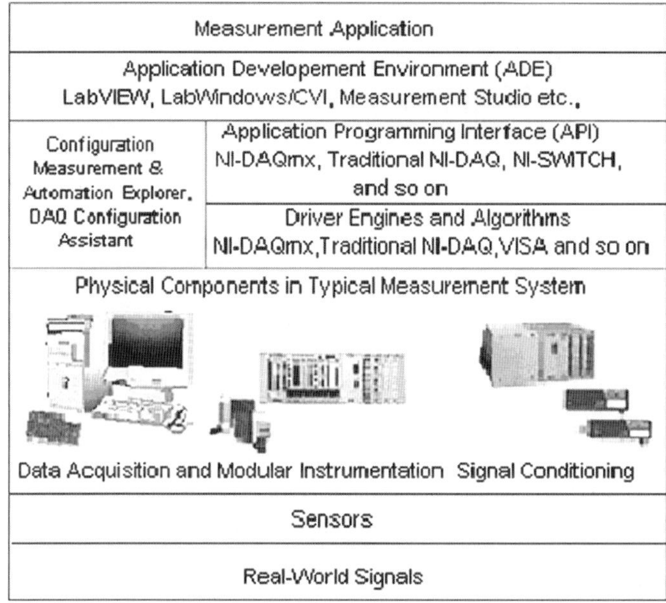

Fig. 7.15. Framework of measurement software

- For C developers, LabWindows/CVI is a fully integrated ANSI C environment that provides the LabWindows/CVI Data Acquisition library for programming NI measurement devices.
- Measurement Studio development tools are used for designing test and measurement software in Microsoft Visual Studio .NET. Measurement Studio includes tools for Visual C#, Visual Basic .NET, and Visual C++ .NET.

7.7 DAQ Signal Accessory

The DAQ signal accessory tests and demonstrates the use of National Instruments multifunction I/O (MIO) and Lab/1200 Series data acquisition (DAQ) devices. The DAQ signal accessory as in Fig. 7.16 has a function generator, a microphone jack, four LEDs, a solid-state relay, a thermocouple jack, an integrated circuit (IC) temperature sensor, a noise generator, a digital trigger, access to two counters, and a 24-pulse per revolution quadrature encoder. The DAQ signal accessory supports the following DAQ devices:

- 68-pin MIO E Series devices
- 68-pin AI E Series devices
- Am9513-based MIO Series devices
- Lab/1200 Series devices including the SCXI-1200, Lab-PC-1200, DAQPad-1200, and DAQCard-1200

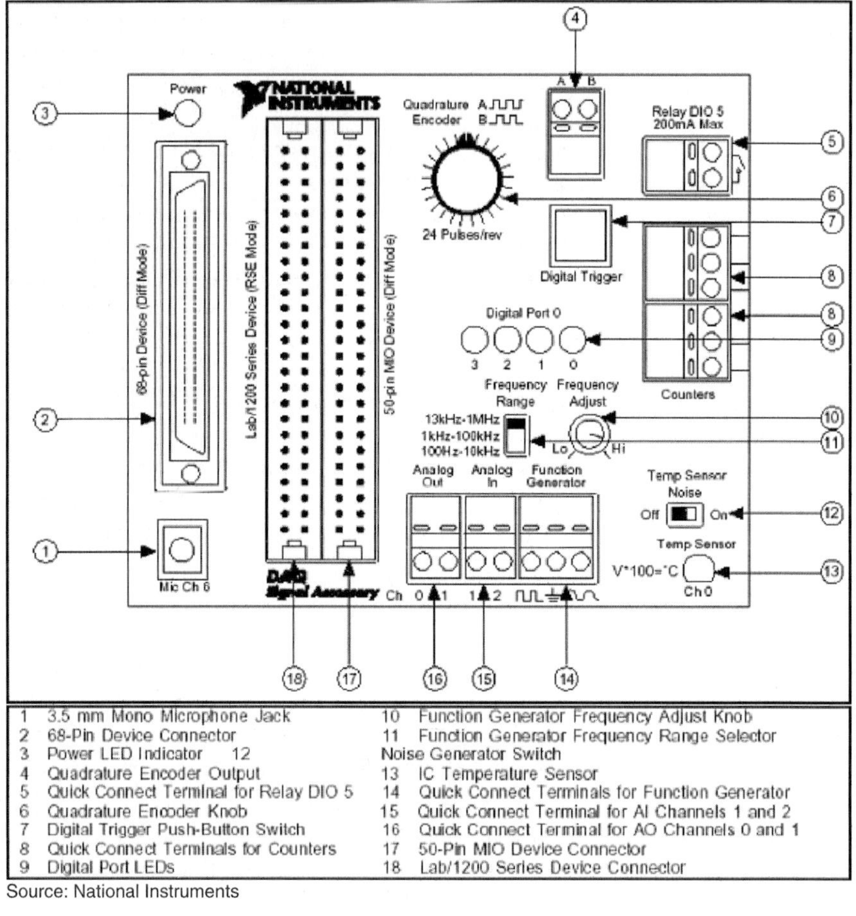

Fig. 7.16. DAQ signal accessory and its components

- The DAQ signal accessory does *NOT* support the following devices:
  - PC-LPM-16 and DAQCard-700/500 devices
  - DSP and A21*xx* Series devices
  - SCXI devices, except the SCXI-1200

The DAQ Signal Accessory includes the following functional blocks whose functions are explained as:

1. Analog input
2. Analog output
3. Power supply
4. Function generator
5. Microphone
6. LEDs

7. Relay
8. Thermocouple and IC temperature sensor
9. Noise generator
10. Digital trigger
11. Counter/timers
12. Quadrature encoder

Analog Input

Analog input channels 1 and 2 are available at the quick connect terminal located in the lower central portion of the top panel. The DAQ device must be configured in bipolar mode to use the AI signal connections on the DAQ signal accessory. If MIO Series device is used, then the device has to be configured for differential (DIFF) mode. If Lab/1200 Series device is used, then the device has to be configured for nonreferenced single-ended (NRSE) mode.

Analog Output

Analog output channels 0 and 1 are available at the quick connect terminal located at the lower central portion of the top panel.

Power Supply

The 5 V line from the DAQ device supplies power to the DAQ signal accessory. Laptops do not have any problem in supplying this power. If the power LED on the DAQ signal accessory is not lit, the fuse on DAQ device has to be checked. If the power LED is not lit and if the DAQ device does not have a fuse, it should be ensured that the computer is powered up and the SH6868 or NB1 cable is connected. Table 7.2 presents the DAQ signal accessory's power consumption relative to the operations that are used.

7.7.1 Function Generator

The function generator produces a 2 Vpp (peak-to-peak voltage) sine wave and a TTL square wave. Using the function generator Frequency Range selector switch, the frequency range can be selected. The choices are 100 Hz to 10 kHz, 1–100 kHz, or 13 kHz to 1 MHz. The Frequency Range knob can be used to further adjust the frequency.

Table 7.2. Power consumption

DAQ signal accessory operations	Power usage (mA)
Circuitry	34.2
4 LEDs + circuitry	82.2
4 LEDs + relay + circuitry	95.5

7.7.2 Microphone

The 3.5 mm microphone jack is wired internally to AI channel 6. The jack is wired as a mono jack and hence the user should use mono microphones.

LEDs

Four LEDs using inverted logic are wired internally to digital port 0. For example, driving digital input/output (DIO) line 0 low will illuminate LED 0.

Relay

Turn on a solid-state relay by driving DIO line 5 low. DIO line 5 is port B, line 1 of the Am9513-based MIO devices. This relay has no moving parts and is not subject to the limited lifetimes of electromechanical relays. Figure 7.17 shows both the internal and external relay connections.

In the closed state, the relay has a maximum resistance of 8 W (6 W typical) and carries up to 200 mA of current. In the open state, the relay blocks up to 60 VDC or 30 Vrms with 42.4 Vpeak with a maximum leakage current of 1 mA. Isolation prevents damage to the DAQ device. On power up of the DAQ device, the relay default setting is open because the digital lines of the DAQ devices are in a high-impedance or input state. If the DAQ signal accessory is unplugged, the relay returns to the open state. When inductive loads are connected to the solid-state relay, a large counter-electromotive force occurs at relay switching time because of the energy stored in the inductive load. These flyback voltages can severely damage the relay. The effect of flyback voltages are minimized in inductive load by installing a flyback diode across the inductive load for DC loads or a Metal Oxide Varistor (MOV) for AC loads. This type of installation is shown in Fig. 7.18.

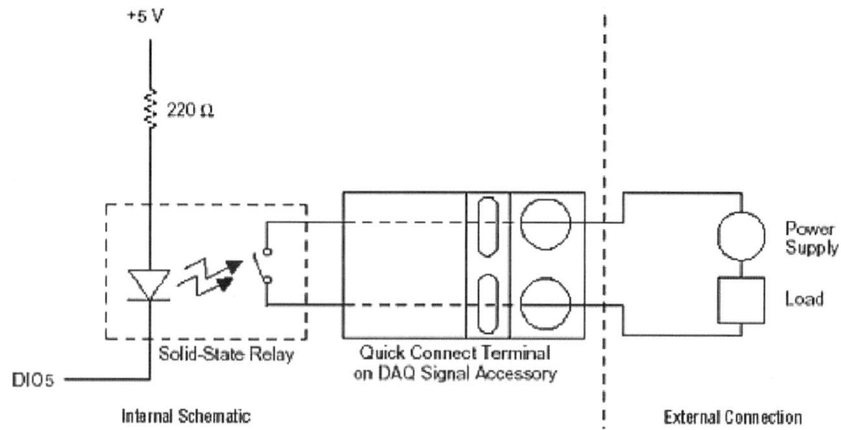

Fig. 7.17. Relay connection

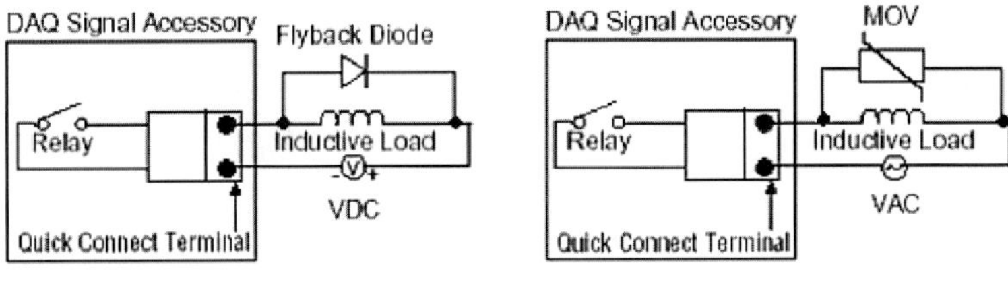

Fig. 7.18. Relay protection when switching inductive loads

7.7.3 Thermocouple and IC Temperature Sensor

The standard slotted thermocouple jack on the side panel is connected internally to AI channel 4. Mating connectors (type SMB) are available from Omega and can use any type of thermocouple wire. The IC temperature sensor, wired to AI channels 0 and 5, provides cold-junction compensation through software. The IC sensor voltage is linearly proportional to the sensor temperature where °C = Volts ∗ 100. The sensor is accurate to 1.5°C. For convenience, AI channel 3 is tied to analog ground so the user can measure and compensate for the amplifier offset.

7.7.4 Noise Generator

The noise generator switch on the DAQ signal accessory can be used to induce noise on IC temperature sensor signals. Using this switch, software techniques, such as averaging, to reduce the effects of noise can be demonstrated. The noise is equivalent to about 2°C and is produced by the sine wave of the function generator. Adjusting the function generator frequency to about 10 kHz produces a typical noise signal on the temperature sensor output.

7.7.5 Digital Trigger

The digital trigger switch is used to trigger acquisitions. Pushing the switch creates a debounced falling edge on the signal line.

7.7.6 Counter/Timers

The DAQ signal accessory has connections for two counters from each DAQ device as shown in Table 7.3.

7.7.7 Quadrature Encoder

A 24-pulse per revolution mechanical quadrature encoder measures the position of a shaft as it rotates. The DAQ signal accessory demonstrates the use

Table 7.3. Counter connections

Counter Line	MIO/AI E Series	Am9513-based MIO Series	Lab/1200 Series
SOURCE/CLOCK	Counter 0	Counter 5	Counter 2
GATE	Counter 0	Counter 5	Counter 2
OUT	Counter 0	Counter 5	Counter 0
SOURCE/CLOCK	Counter 1	Counter 1	Counter 1
GATE	Counter 1	Counter 1	Counter 1
OUT	Counter 1	Counter 1	Counter 1

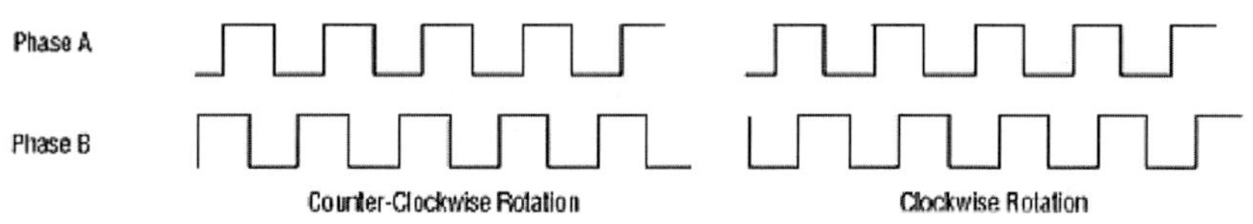

Fig. 7.19. Quadrature encoder pulse trains

of the quadrature encoder using a knob located in the upper central portion of the top panel. The quadrature encoder produces two pulse train outputs corresponding to the shaft position as the knob is rotated. Depending on the direction of rotation, phase A leads phase B by 90° or phase B leads phase A by 90°. This relationship is illustrated in Fig. 7.19.

7.8 DAQ Assistant

The DAQ Assistant is a graphical interface used to configure measurement tasks and channels, and to customize timing, triggering, and scales without programming. Using the DAQ Assistant, the user can interactively build a measurement channel or task for use in LabVIEW, LabWindows/CVI, or Measurement Studio application developments of version 7.x or higher and VI Logger 2.0 or higher. The user can then incorporate the tasks into the application. The DAQ Assistant can be launched from LabVIEW 7.x or higher, LabWindows/CVI 7.x of higher, Visual Studio .NET with Measurement Studio 7.x or higher; from Measurement & Automation Explorer (MAX) 3.0 or higher; from VI Logger 2.0 or higher. To launch the DAQ Assistant in MAX 3.0, right-click Data Neighborhood, then select Create New. Select NI-DAQmx Task. The DAQ Assistant opens. The DAQ Assistant Express VI allows the user to easily configure the data acquisition device. When the DAQ Assistant Express VI is placed on the block diagram, a dialog box as shown in Fig. 7.20 appears, where the user can configure a local task to perform a specific measurement function. Creating a local task allows the user to specify the exact type of measurement to take. The DAQ Assistant Express VI allows easy configuration of the data acquisition device.

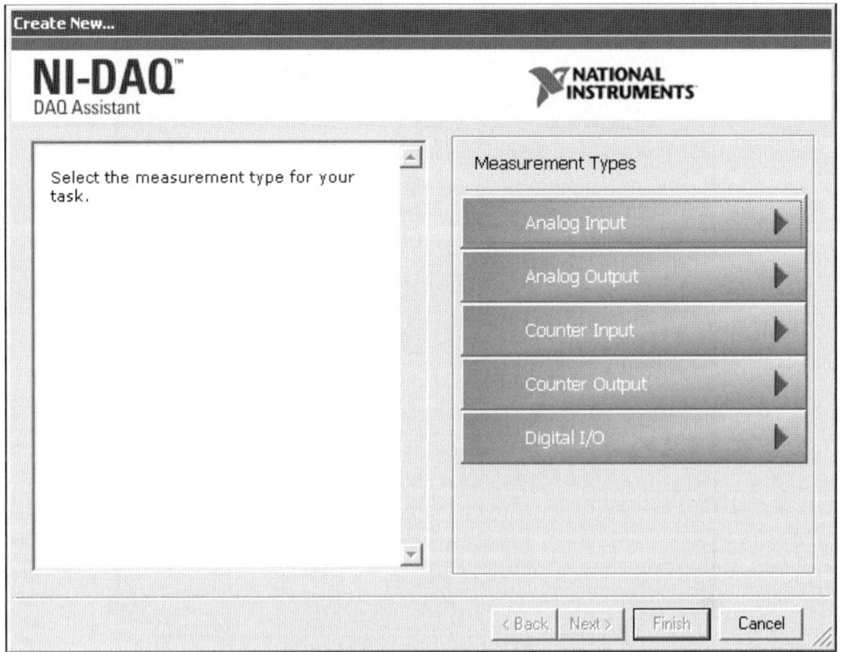

Fig. 7.20. Dialog box of DAQ Assistant

This section explains the use of DAQ Assistant, to quickly develop and deploy a data acquisition (DAQ) application in LabWindows/CVI. The DAQ Assistant is an easy-to-use graphical interface for configuring measurement tasks and channels and for customizing timing, triggering, and scales without programming. Using the DAQ Assistant, the user can configure a measurement task for all of the DAQ applications and then generate code to configure and use the task in the application program. The user can use the DAQ Assistant to quickly create DAQ tasks. A *task* is a collection of one or more virtual channels with timing, triggering, and other properties. Conceptually, a task represents a measurement or generation to be performed. The user can set up and save all of the configuration information in a task and use the task in an application. The DAQ Assistant is launched from LabWindows/CVI by selecting Tools≫Create/Edit DAQmx Tasks. The dialog box shown in Fig. 7.21 appears.

Two options are available while creating tasks for use in a LabWindows/CVI application.

- *MAX-based tasks* are global to the machine on which they are created and can be used by other programs on the same machine. The user can view, edit, and test this type of task directly from within MAX. MAX-based tasks can also be used in multiple projects, even those that are not in created in LabWindows/CVI.

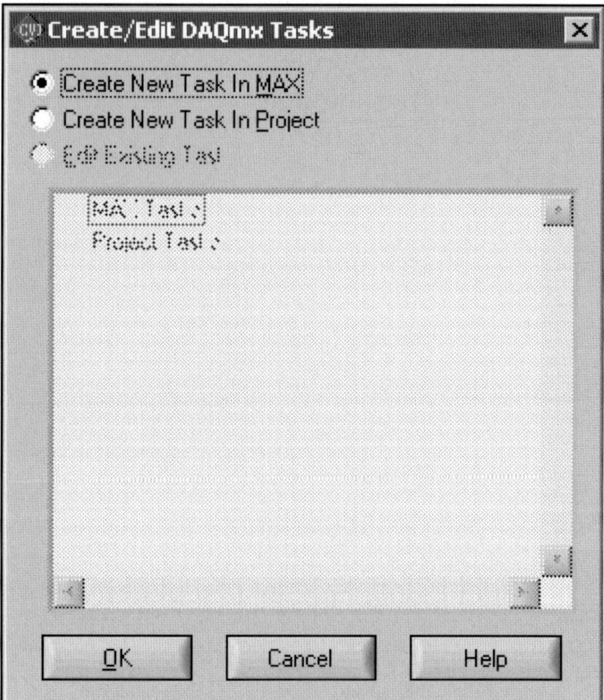

Fig. 7.21. Create/edit DAQmx Dialog Box

- *Project-based tasks* are stored in a project. Tasks stored in the project are local to that project. Project-based tasks can be more convenient if the user needs to share a task definition among multiple developers or store the task definition in a source code control system.

7.8.1 MAX-Based Tasks

To use the MAX-based task, DAQmxLoadTask can be used, as shown in Fig. 7.22. The function panel for DAQmxLoadTask contains a list of all MAX-based DAQmx tasks on the computer. Press <Enter> on the **Task Name** control in this function panel to view the list of tasks.

7.8.2 Steps to Create a MAX-Based Task

1. Select **Create New Task in MAX** from the Create/Edit DAQmx Tasks dialog box.
2. Click **OK** to launch the DAQ Assistant.
3. In the next dialog box, select the measurement type for the task and click **Next**.

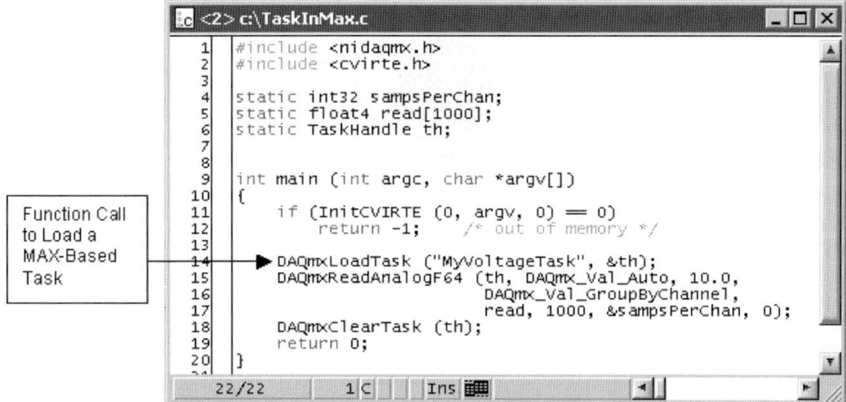

Fig. 7.22. DAQmx Load Task

4. Select the virtual channels and click **Next**.
5. Enter a name for the task and click **Finish**. The DAQ Assistant creates the task.
6. The DAQ Assistant panel shown in Fig. 7.23 appears. In this panel, the user can configure channel-specific settings such as scaling, input limits, and terminal configuration. The task-specific settings such as timing and triggering can also be configured.
7. When the user specifies the various settings, the user can test the task. The **Test** button can be used to verify that the task works properly.
8. The DAQ Assistant saves the task to MAX.

7.8.3 Project-Based Tasks

To use the project-based task, the task creation function (CreateDAQTask-InProject in this example) declared in the generated header file is used. The header file should also be included into the main project file.

7.8.4 Steps to Create a Project-Based Task

1. Select **Create New Task in Project** from the Create/Edit DAQmx Tasks dialog box.
2. Click **OK** to launch the DAQ Assistant.
3. In the next dialog box, select the measurement type for the task and click **Next**.
4. Select the virtual channels and click **Next**.
5. After creating the task, the DAQ Assistant panel shown in Fig. 7.24 appears. In this panel, the user can configure channel-specific settings such as scaling, input limits, and terminal configuration. The user can also configure task-specific settings such as timing and triggering.

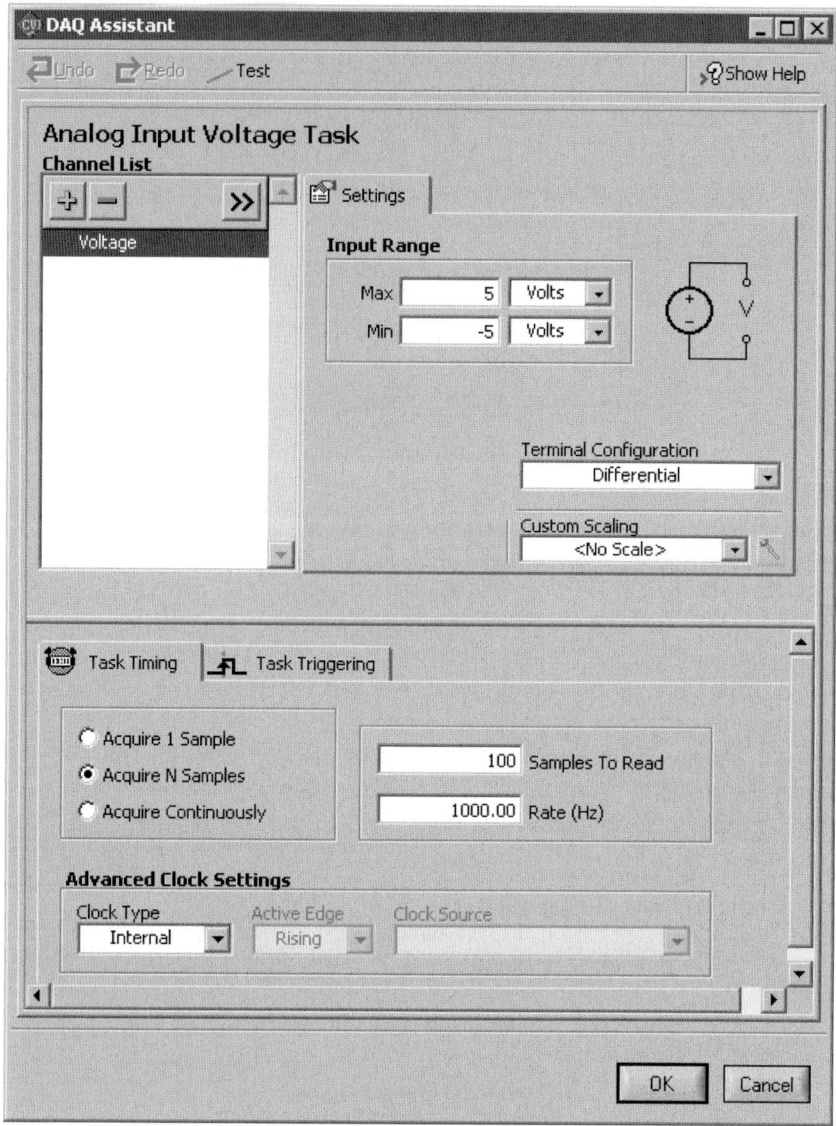

Fig. 7.23. DAQ Assistant panel

6. When the user specifies the various settings, the task can be tested. Click the **Test** button to verify that the task works properly.
7. Specify a task name, create task function, and a target directory. Enter this information in the area of the DAQ Assistant panel shown in Fig. 7.25.
   - **Task Name**. The name associated with the task as well as the prefix to the .c, .h, and .mxb files that will be created. For example, use DAQ TaskInProject as the task name.

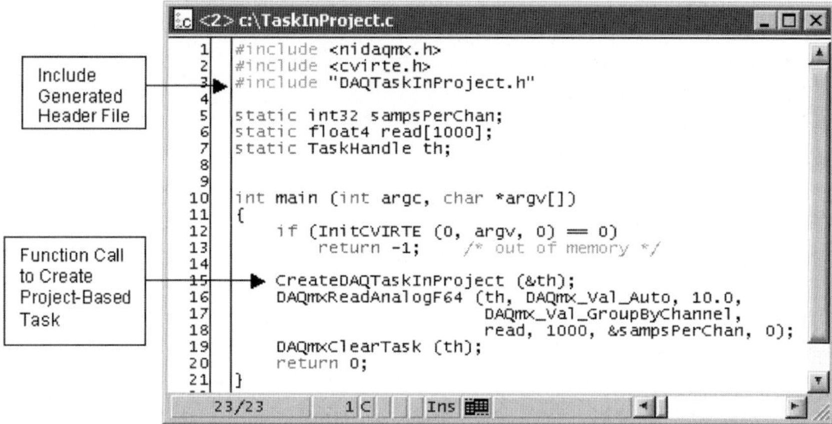

Fig. 7.24. Create DAQ – function call

- **Create Task Function**. The name of the function that is used to create and configure the task programmatically. For example, use CreateDAQTaskInProject as the create task function.
- **Target Directory**. The directory where the three generated files will be stored.

8. Click OK to generate the task.

The DAQ Assistant creates three files to describe the task in LabWindows/CVI: a source (.c), a header (.h), and a task description (.mxb) file. LabWindows/CVI automatically adds these files to the project. The generated source and header files define the task creation function, which contains the code necessary to create and configure the specified task programmatically. The generated task description file (.mxb) contains a binary description of the task that is used when the user edits the task in the DAQ Assistant. Opening the .mxb file opens the DAQ Assistant for that task. These files are all linked to the task, so any changes made to the task from the DAQ Assistant are automatically reflected in the generated code.

7.8.5 Project-Based and MAX-Based Tasks

After the DAQmxLoadTask or the task creation function is called, other functions necessary to perform various actions can be added. For example, if data is read from analog input channels, the user might want to use the DAQmxReadAnalogF64 function. So the user can call DAQmxClearTask. This function stops the task and clears the task from memory. The user also can use the DAQ Assistant to generate the code needed to run the task.

- **Generate example code for a MAX-based task**. In the Source window, right-click the call to DAQmxLoadTask and select **Generate DAQ Example Code**.

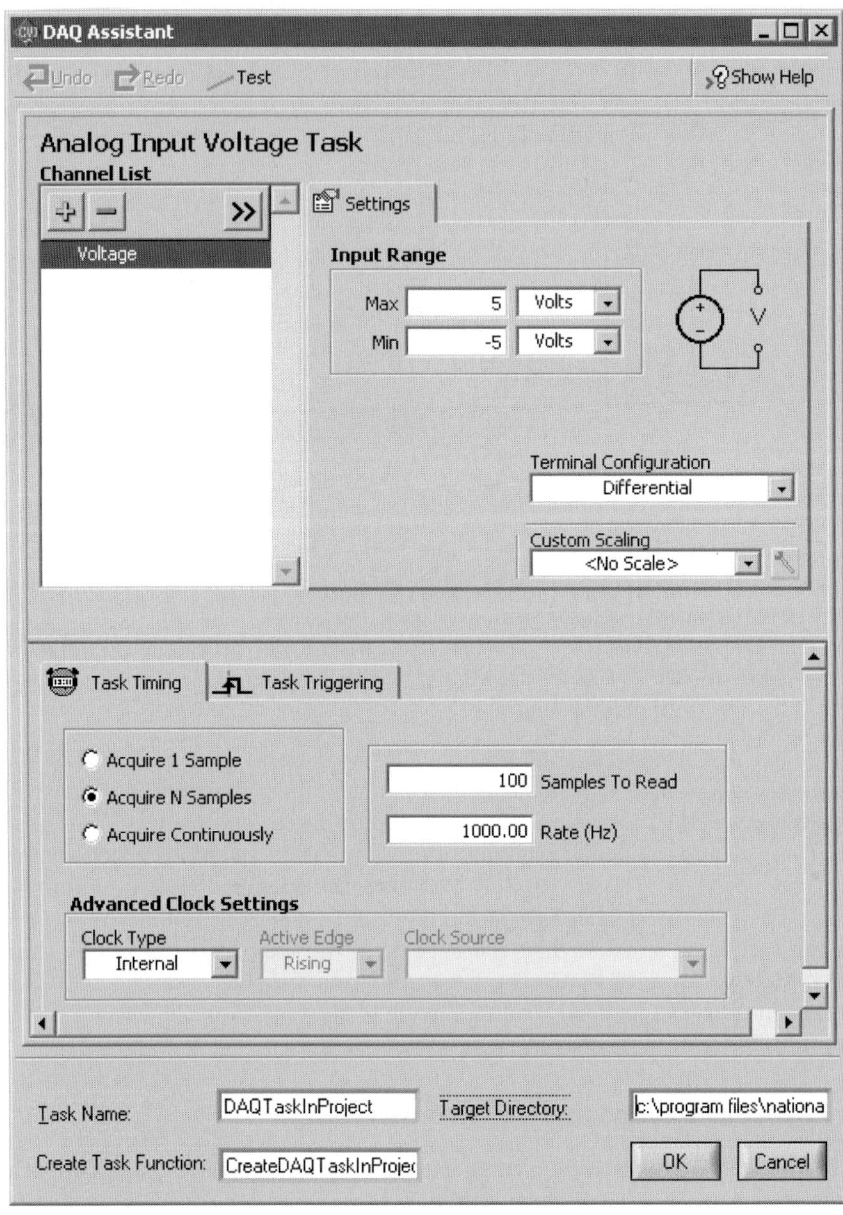

Fig. 7.25. DAQ Assistant panel

– **Generate example code for a project-based task**. In the Source window, right-click the call to the task creation function (for example, CreateDAQTaskInProject) and select **Generate DAQ Example Code**.

When the **Generate DAQ Example Code** is selected for either a project- or MAX-based task, the Generate DAQ Example Code dialog box, shown in Fig. 7.26, appears.

In the dialog box, specify the name of the function to run the task and the file to contain the generated example code, then click **OK**. The generated code includes source and header files that define the run task function. LabWindows/CVI automatically adds these files to the project and modifies the current source file to call the run task function. When called, the run task function programmatically creates a user interface that allows the user to start the task and display the data.

Figure 7.27 shows an example source file after selecting Generate DAQ Example Code for a project-based task.

7.8.6 Edit a Task

There are several ways to launch the DAQ Assistant to edit a task after creating a task. Any type of task can be edited by selecting **Tools≫Create/Edit DAQmx Tasks**. This command launches the Create/Edit DAQmx Tasks dialog box, in which the user can select the task to edit.

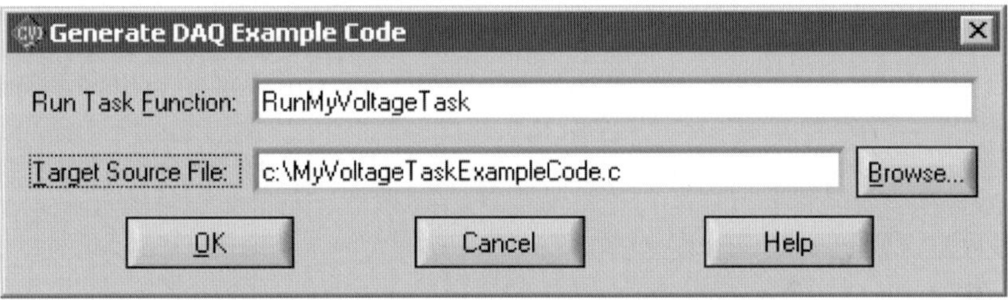

Fig. 7.26. Generate DAQ Example Code

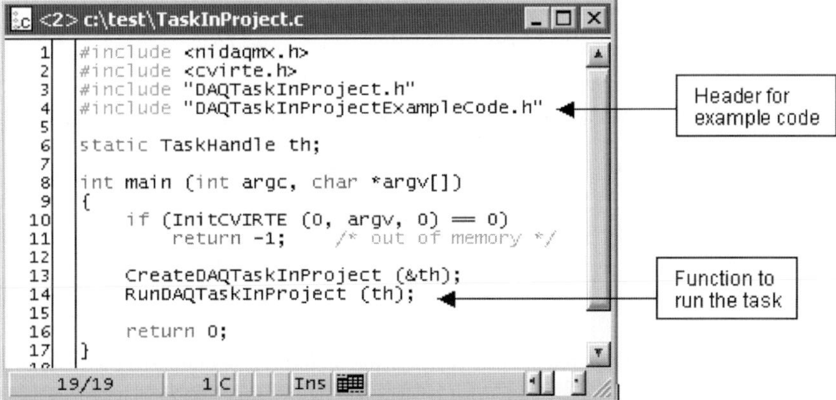

Fig. 7.27. Header and function call

The user can edit MAX-based tasks directly from within MAX by selecting **My System≫Data Neighborhood≫** NI-DAQmx Tasks. Project-based tasks can also be edited by double-clicking the .mxb file in the Project Tree in LabWindows/CVI. After completion of editing a task, LabWindows/CVI updates the MAX task or the configuration files in the project, as appropriate, with any changes made. However, example code that is generated is not automatically updated. Because the example code is not linked directly to the task, if any modifications are made to the task, the example code must be regenerated for the task to reflect the changes made.

7.8.7 Copy a MAX Task to Project

The user can create a new project-based task that is a copy of a MAX-based task. In the Source window, the user can right-click the call to DAQmxLoadTask and select **Copy DAQ Task to Project**. In the dialog box that appears, the user must specify a task name, task creation function, and target directory, similar to that specified in the DAQ Assistant for project-based tasks. After completion, LabWindows/CVI generates the source (.c), header (.h), and task description file (.mxb) and adds these files to the project; replaces the call to DAQmxLoadTask with a call to the new task creation function; and adds an include statement for the generated header file.

There is no link between the MAX task and the new project task that is created. Any changes made to the MAX task will not be automatically reflected in the new project task.

7.9 DAQ Hardware

This section discusses the configuration manager used to configure the DAQ devices and the hardware concepts such as channel configuration and triggering concepts. DAQ hardware and software are also dealt in the later part of this section along with the DAQ Assistant.

7.9.1 Windows Configuration Manager

The Windows Configuration Manager keeps track of all the hardware installed in the computer, including National Instruments DAQ devices. If a Plug & Play (PnP) device, such as an E Series MIO device is available, the Windows Configuration Manager automatically detects and configures the device. If the device is a non-PnP device, or legacy device, the user must configure the device manually using the Add New Hardware option in the Control Panel.

The user can verify the Windows Configuration by accessing the Device Manager. The **Data Acquisition Devices** can be seen, which lists all DAQ devices installed in the computer. The **General** tab on the dialog box displays overall information regarding the device. The **Resources** tab specifies the system resources to the device such as interrupt levels, DMA, and base address for software-configurable devices. The **NI-DAQ Information** tab specifies the bus type of the DAQ device. The **Driver** tab specifies the driver version and location for the DAQ device.

LabVIEW installs Measurement & Automation Explorer (MAX), which establishes all device and channel configuration parameters. After installing a DAQ device in the computer, the user must run this configuration utility. MAX reads the information the Device Manager records in the Windows Registry and assigns a logical device number to each DAQ device. Use the device number to refer to the device in LabVIEW. Access MAX either by double-clicking the icon on the desktop or selecting **Tools≫Measurement & Automation Explorer** in LabVIEW. MAX is also the means for SCXI and SCC configuration. MAX detects all the National Instruments hardware including the GPIB interface. MAX saves the logical device number and the configuration parameters in the Windows Registry. The plug and play capability of Windows automatically detects and configures switchless DAQ devices, such as the PCI-6024E. When a device is installed in the computer, the device is automatically detected.

7.9.2 Channel and Task Configuration

Using Traditional NI-DAQ a set of *virtual channels* can be configured. A collection of property settings that include a physical channel, the type of measurement or generation specified in the channel name, and scaling information are also configured using this utility. In Traditional NI-DAQ and earlier versions, virtual channels are a simple method to remember which channels are used for different measurements. NI-DAQmx *channels* are similar to the virtual channels of Traditional NI-DAQ. NI-DAQmx also includes *tasks* that are integral to the API. A *task* is a collection of one or more channels and the timing, triggering, and other properties that apply to the task itself. A task represents a measurement or generation the user want to perform.

Channels are created only inside tasks that are local. Channels defined outside a task are global and can be used separately. Configuring virtual channels is optional in Traditional NI-DAQ and earlier versions but is integral to every measurement the user take in NI-DAQmx. In Traditional NI-DAQ, virtual channels are configured using MAX. In NI-DAQmx, virtual channels are configured either in MAX or in a program, and channels are configured as part of a task or separately.

7.9.3 Hardware Triggering

There are two ways to begin a DAQ operation:

1. Through a software trigger.
2. Through a hardware trigger.

With software triggering, the DAQ operation begins when the software function that initiates the acquisition executes. For example, the "Easy" DAC VIs use software triggering. As soon as Lab VIEW executes the VI, the acquisition or generation starts. All DAQ boards support software triggering.

Another common method to begin a DAQ operation is to wait for a particular external event to occur. The user usually begins the acquisition depending on the characteristics of a digital or analog signal such as the signal state, level, or amplitude. Hardware circuitry on the plug-in board uses this analog or digital event to start the board docks that control the acquisition. Most DAQ boards support digital triggering to initiate an acquisition. Other boards also support analog triggering to initiate an acquisition. The input pin on the DAQ board that accepts the triggering signal is either EXTRIG or START. All National Instruments E-series boards support digital triggers. An external analog trigger can also be used to initiate the read of the data to and from the buffer in LabVIEW, rather than just initiate the DAQ Operation. This sort of triggering is called a conditional retrieval.

When performing a conditional retrieval, the DAQ board acquires data and stores it in the acquisition buffer using software triggering. However, the board does not receive data until it acquires a sample that meets certain level and slope conditions. The difference between conditional retrieval and hardware analog triggering is a very important factor to be considered. Systems using analog triggering do not store data in the acquisition buffer until the trigger condition occurs; systems using conditional retrieval do store the data in a buffer, but these data are not read into LabVIEW until the trigger occurs.

7.9.4 Analog Input

Analog input is used to perform A/D conversions. The available analog input measurement types for a task are voltage, temperature, strain, current, resistance, or frequency. Each measurement type has its own characteristics, such as resistor values for current measurements or strain gauge parameters for strain measurements. When performing analog input, the task can be timed to Acquire Single Sample, Acquire n Samples, or Acquire Continuously.

Acquire Single Sample

Acquiring a single sample is an on-demand operation. In other words, NI-DAQmx acquires one value from an input channel and immediately returns the value. This operation does not require any buffering or hardware timing.

Acquire n Samples

One way to acquire multiple samples for one or more channels is to acquire single samples in a repetitive manner. However, acquiring a single data sample on one or more channels over and over is inefficient and time consuming. Moreover, the user does not have accurate control over the time between each sample or channel.

Instead the user can use hardware timing, which uses a buffer in computer memory, to acquire data more efficiently. Programmatically, the user should include the timing function and specify the **sample rate** and the **sample mode (finite)**. As with other functions, the user can acquire multiple samples for a single channel or multiple channels. With NI-DAQmx also data can be gathered from multiple channels.

Acquire Continuously

In order to view a process, or log a subset of the samples as they are acquired, the user has to acquire samples continuously. For these types of applications, the sample mode is set to **continuous**.

Task Triggering

When a device controlled by NI-DAQmx does something, it performs an action. Two very common actions are producing a sample and starting a waveform acquisition. Every NI-DAQmx action needs a stimulus or cause. When the stimulus occurs, the action is performed. Causes for actions are called triggers. The start trigger starts the acquisition. The reference trigger establishes the reference point in a set of input samples. Data acquired up to the reference point is pretrigger data. Data acquired after the reference point is posttrigger data.

7.9.5 Analog Output

Analog output is used to perform D/A conversions. The available analog output types for a task are voltage and current. To perform a voltage or current task, a compatible device must be installed that can generate that form of signal. When performing analog output, the task can be timed to Generate Single Sample, Generate n Samples, or Generate Continuously.

Generate Single Sample

Single updates are used if the signal level is more important than the generation rate. For example, one sample at a time is generated to generate a constant, or DC, signal. A software timing can be used to control the generation of a signal. This operation does not require any buffering or hardware timing. For example, to generate a known voltage to stimulate a device, a single update would be an appropriate task.

Generate n Samples

One way to generate multiple samples for one or more channels is to generate single samples in a repetitive manner. However, generating a single data sample on one or more channels over and over is inefficient and time consuming. Moreover, the user does not have accurate control over the time between each sample or channel. Instead, the user can use a hardware timing, which uses a buffer in computer memory to generate samples more efficiently.

Software timing or hardware timing can be used to control whenever a signal is generated. With software timing, the rate at which the samples are generated is determined by the software and operating system instead of the measurement device. With hardware timing, a TTL signal, such as a clock on the device, controls the rate of generation. A hardware clock can run much faster than a software loop. A hardware clock is also more accurate than a software loop. Generate n Samples is most commonly used to generate a finite time-varying signal, such as an AC sine wave.

Generate Continuously

Continuous generation is similar to Generate n Samples, except that an event must occur to stop the generation. To generate signals continuously, such as generating a nonfinite AC sine wave, the timing mode is set to **continuous**.

Task Triggering

When a device controlled by NI-DAQmx does something, it performs an action. Two very common actions are producing a sample and starting a generation. Every NI-DAQmx action needs a stimulus or cause. When the stimulus occurs, the action is performed. Causes for actions are called triggers. The start trigger starts the generation. The reference trigger is not supported for analog output tasks.

7.9.6 Digital Output

Measuring and generating digital values is used in a variety of applications, including controlling relays and monitoring alarm states. Generally, measuring and generating digital values is used in laboratory testing, production testing, and industrial process monitoring and control. Digital I/O can read from or write to a line or an entire digital port, which is a collection of lines. The digital lines in a DAQ device are used to acquire a digital value. This acquisition is based on software timing. On some devices, the lines are configured individually to either measure or generate digital samples. Each line corresponds to a channel in the task.

The digital port(s) in a DAQ device are used to acquire a digital value from a collection of digital lines as shown in Fig. 7.28. This acquisition is

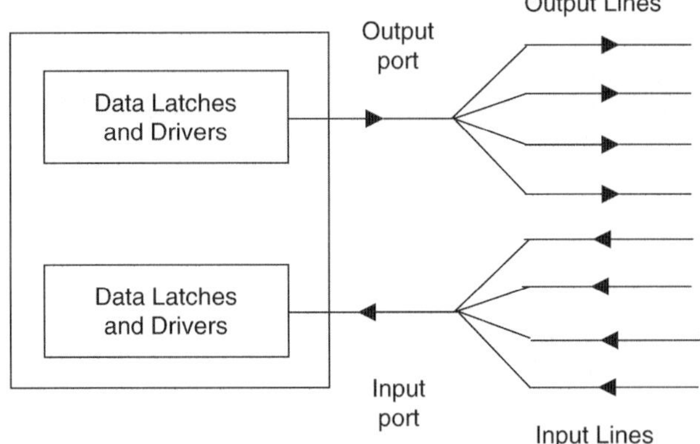

Fig. 7.28. Digital Input and Output

based on software timing. The user can configure the ports individually to either measure or generate digital samples. Each port corresponds to a channel in the task.

A digital line is the equivalent of an analog channel, a path where a single digital signal is set or retrieved. Digital lines are usually either input lines or output lines, but sometimes they can be bi-directional. On most boards, digital lines must be configured as input or output; they cannot act as both at the same time.

A port is a collection of digital lines that are configured in the same direction and can be used at the same time. The number of digital lines in each port depends on the board, but most ports consist of four or eight lines. For example, a multifunction DAQ board could have eight digital lines, configurable as one 8-line port, two 4-line ports, or even eight 1-line ports. Ports are specified as digital channels, just like analog channels.

Port width is the number of lines in a port.

State refers to the one of two possible cases for a digital line: a Boolean TRUE (same as logical 1 or "on") or a Boolean FALSE (same as logical 0 or "off").

A pattern is a sequence of digital states, often expressed as a binary number, which denotes the states of each of the lines on a port. For example, a four line port might be set with the pattern "1101," meaning the first, third and fourth lines are TRUE, and the second line is FALSE. The first bit, or least-significant bit (LSB), is the rightmost bit on the pattern. The last (fourth bit in this example), or most-significant bit (MSB), is the leftmost bit in the pattern.

Incidentally, National Instrument DAQ boards use TTL positive logic, which means a logical low is somewhere in the 0–0.8 V range; a logical high is between 2.2 and 5.5 V.

7.9.7 Counters and Timers

A counter is a digital timing device. Typical applications of counters are event counting, frequency measurement, period measurement, position measurement, and pulse generation. A counter contains the following four main components:

- **Count Register**. Stores the current count of the counter. The user can query the count register with software.
- **Source**. An input signal that can change the current count stored in the count register. The counter looks for rising or falling edges on the source signal. Whether a rising or falling edge changes the count is software selectable. The type of edge selected is referred to as the active edge of the signal. When an active edge is received on the source signal, the count changes. Whether an active edge increments or decrements the current count is also software selectable.
- **Gate**. An input signal that determines if an active edge on the source will change the count. Counting can occur when the gate is high, low, or between various combinations of rising and falling edges. Gate settings are made in software.
- **Output**. An output signal that generates pulses or a series of pulses, otherwise known as a pulse train. When a counter is configured for simple event counting, the counter increments when an active edge is received on the source. In order for the counter to increment on an active edge, the counter must be started. A counter has a fixed number it can count to as determined by the resolution of the counter. For example, a 24-bit counter can count to:

  $$2^{(\text{Counter Resolution})} - 1 = 2^{24} - 1 = 16,777,215$$

  When a 24-bit counter reaches the value of 16,777,215, it has reached the terminal count. The next active edge will force the counter to roll over and start at 0.

7.10 DAQ Software

DAQ software is specialized software that has been developed specifically for use with data acquisition boards, cards and systems. This software can range from simple software used for one application such as data logging, or it can be extremely robust, offering a wide range of uses for measurement and control and laboratory applications. DAQ software in conjunction with a personal computer is used to control the data acquisition board's functions.

Simple Functions

The software is used to control basic tasks such as sending a convert command to an A/D converter, starting or stopping a counter/clock, clear or reset a bank of digital I/O lines, and more.

High-Level Functions

More elaborate functions provided by DAQ software includes acquiring, converting to engineering units, and displaying data graphically on a PC monitor; analyzing a complex analog input signal, providing linearization and cold-junction compensation calculations for various types of thermocouple signals; or simultaneously sampling multiple channels of analog input data, comparing them, and plotting them in a three-dimensional presentation.

High-Level Mathematical Functions

FFT and frequency analysis, signal generation, mathematics, curve fitting and interpolation, along with time and frequency domain analysis, allow users to derive meaningful information from data.

Application Development

Whether the user is taking temperature measurements with an A/D board, analyzing signals using a stand-alone oscilloscope board, or measuring pressure with a sophisticated signal conditioner; using specialized application development DAQ software with the appropriate driver is ideal for the application. From data acquisition to instrument control, and image acquisition to motion control, application development software provides the tools to rapidly develop the acquisition system.

Summary

MAX is the primary configuration and testing utility that is available for the DAQ device. The DAQ Assistant is used to configure the DAQ device and perform data acquisition. Most programs can use the DAQ Assistant. For programs that require advanced timing and synchronization, use the VIs that come with NI-DAQmx. The DAQ Assistant can perform analog input, analog output, counter, and digital I/O and operations. The DAQ software including high level mathematical functions and the application development environment are also detailed in this chapter.

Review Questions

1. Explain in detail the concept of signal grounding with neat sketches.
2. Discuss the concept of digital I/O techniques with its application techniques.
3. Describe the application techniques used with digital I/O.
4. Explain how data acquisition is done in LabVIEW.
5. With a block diagram explain the DAQ operation.
6. Write the characteristics of DAQ operation.
7. Explain in detail about the components of DAQ.
8. Explain the process of measurement software framework with neat sketches.
9. What are the functional blocks of DAQ signal accessory? Explain the functions of each.
10. Write notes on DAQ Assistant.
11. List the steps to create project-based tasks.
12. List the steps to create MAX-based task.
13. Explain in detail about DAQ hardware.
14. Discuss about hardware and software triggering.
15. Explain in detail about DAQ software.
16. Give some of the data acquisition methods used at present.

8

Data Transmission Concepts

Learning Objectives. On completion of this chapter the reader will have a knowledge on:

- Data Transmission Concepts
- Pulse Codes and their types
- Analog and Digital Modulation Techniques
- Wireless Communication
- Trends in Wireless Communication
- Software Defined Radio
- RF Network Analyzer
- Distributed Automation and Control Systems
- CORBA-Based Automation Systems
- SCADA – Architecture

8.1 Introduction

Data transmission is the conveyance of any kind of information from one space to another. Historically this could be done by courier, a chain of bonfires or semaphores, and later by Morse code over copper wires. In recent computer terms, it means sending a stream of bits or bytes from one location to another using any number of technologies, such as copper wire, optical fibre, laser, radio, or infrared (IR) light. Practical examples include moving data from one storage device to another and accessing a website, which involves data transfer from web servers to a user's browser. A related concept to data transmission is the data transmission protocol used to make the data transfer legible. Data transmission plays a very important role in all kind of digital equipments as it is the responsibility of these devices to transmit the data without being lost. It is the role of these equipments to make sure that external environment data are transmitted in the same manner without being changed for processing. The information that received by these instrument aids in conditioning so the data available could be relied for more efficiency of the output. This section encompasses the operation and characteristic feature of pulse codes, analog and digital modulation, wireless communication, RF analyzer, distributed network, and the SCADA.

8.2 Pulse Codes

There are different code formats of pulse waveforms. The classification is based on three criteria such as:

Form of information-transmission
Relation to zero level
Direction based on the form of information-transmission

The format used can be any of the following:

(a) Full binary transmission, where both the "0" and "1" bits are part of the formats.
(b) Half binary transmission, where only the "1" are transmitted, recognizing the "0" by the absence of a pulse at the time of clock transition.
(c) Multiple binary transmission, where ternary and quadratic codes are used for each transmitted pulses.

Based on the relationship to the zero level, the transmission format can be either "return-zero" (RZ) in which there is a return to the zero level after the transmission of each bit of information.

Nonreturn to zero (NRZ) where there is a no voltage level change if consecutive bits are transmitted although there is a level change when there is an information variation from "0" to "1" or "1" to "0." In the third criterion of classification namely that of direction, the code format can be either unipolar, where the pulses are in a single direction, (or) bipolar, where the pulses are in both directions. Some of the codes are shown above for a binary word. Non Return to zero code bipolar format is a three-level code also known as "duo binary code."

8.2.1 RZ and RB Recording

Recording consists of passing the magnetizable material beneath the *recording head* and at the same time applying a recording current to the coil. The direction of the current which is applied to the coil determines the polarity of the magnet which is recorded on the medium. Turning the current on, letting it remain on, and then turning it off again creates a magnetic field through the medium which produces a complete magnet.

Return to Zero

The return to zero or RZ recording scheme is also known as bidirectional recording. The method of recording and playback is shown in Fig. 8.1. The

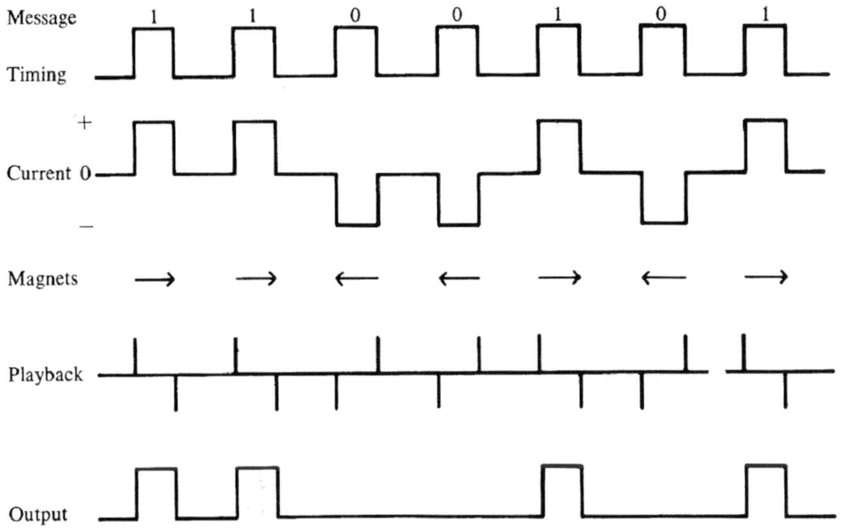

Fig. 8.1. Return to zero (RZ) recording

message used is: 1100101. It assumes that the medium is previously unrecorded. Otherwise an *erase head* must demagnitize the medium where the new recording is to take place by scrambling the magnetization of the particles. For each bit in the message, there is a timing pulse that is produced by the recording system to indicate when recording should begin, shown on the second line. This timing pulse causes current to be applied to the *recording head.* The polarity of the current is determined by the bit being recorded.

- For 1, a positive current is shown in the figure.
- For 0, a negative current is provided.

As this current is applied to the coil, a field is induced in the core; it appears at the gap and passes through the magnetic medium that is traveling at a constant speed beneath the head. The result of this current is to align the small magnetic particles on the medium as indicated by arrows in the next line of the figure.

Playback

To recover the information recorded on the medium, the medium is passed at a uniform velocity so that it is in contact with the *head.* The magnetic field in the medium is induced into the core and as the field changes a voltage is induced into the coil. The signal produced by the voltage is a pair of spikes for each magnet, as shown at the next to last line of the figure, Each pair of

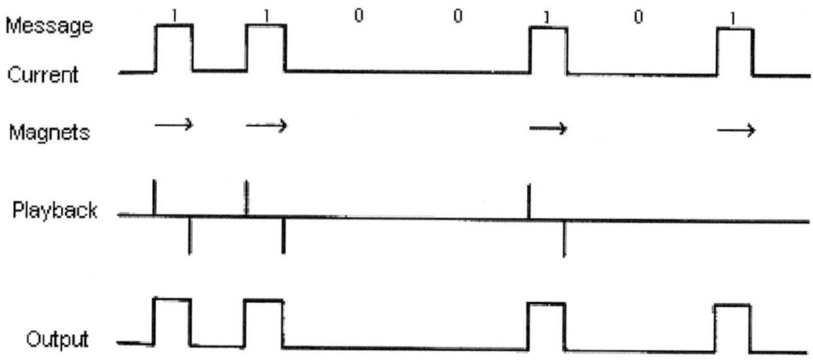

Fig. 8.2. Return to bias (RB) recording

spikes is interpreted electronically to produce the message – a positive spike followed by a negative spike becomes a 1 – a negative followed by a positive spike becomes a 0.

Return to Bias

The return to bias (RB) recording method is illustrated in Fig. 8.2. RB assumes an unrecorded or erased medium. It differs from the return to zero method because 0s are not recorded and only one direction of magnetization is used. Thus recording current flows in only one direction. There is a magnet recorded for each 1 in the message; there is no magnet recorded for 0s in the message. All magnets are polarized in the same direction.

There are two difficulties that arise from the RB recording method. The playback is not self-clocking since no pulses are produced by 0. Information previously recorded on the tape is not recorded over or obliterated for 0 in the message.

Clocking

The RB system can be used in the *magnetic tape unit* when all the bits of a character are recorded in paralleled by one set of heads and odd parity is used. It is the odd parity that provides the clocking. This guarantees that a 1 is recorded in some position of each character.

To produce the clocking, the 1s from each track are combined, using an OR. The first positive pulse received from some track turns on a clocking pulse. The negative signal turns off the clocking pulse.

Old Recording

To obliterate old recording, it may be erased with an erase head that precedes the recording head. An alternative is to provide an alternating bias current that also serves to randomize or erase the old information.

8.2.2 NRZ Recording

Both the RZ and the RB systems by their nature produce superfluous pulses. For the RZ method, two pulses per bit are recorded; for the RE method, two pulses are produced for each 1, although no pulses are produced for a 0. The more pulses that are required for each bit, the more this interferes with packing information close together. If one bit per pulse is sufficient, more magnets can be pushed together in the same area and thus the user can get more bits per running inch. This section describes the following two methods for achieving this goal.

Nonreturn to Zero

Figure 8.3 illustrates the action for recording information using the NRZ method. This method reverses the current flow only when there is a change in signal information. Thus, if there are a series of 1s in the message, only when the user goes from 1 to 0, information is recorded on the medium in the form of reversal of the direction of the magnetic field. Similarly, for a series of 0s, a single long magnet is recorded; the direction of recording changes only when a 1 appears after the 0s. Thus for NRZ, current is always flowing in one direction or another and no erase is required. The medium may be prerecorded.

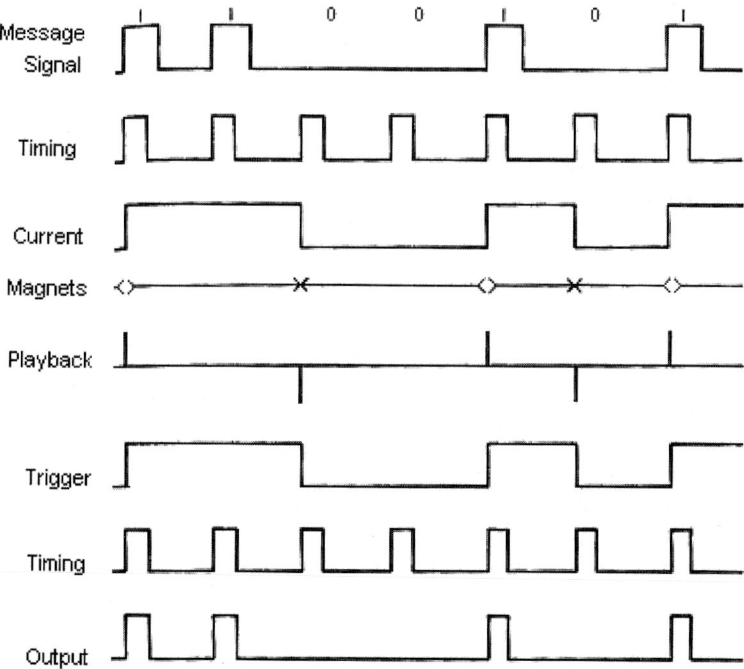

Fig. 8.3. Nonreturn to zero (NRZ) recording

Write Rule

For writing, current flows in one fixed direction for a 1 and in the opposite direction for a 0. This creates a magnet in one direction for a series of 1s and in the other direction for a series of 0s.

Recording

In Fig. 8.3, a signal occurs that corresponds to the message shown: pulses appear for 1s, none for 0s. When a signal and a timing pulse coincide, positive current flows in the recording; when a timing pulse occurs without corresponding signal pulse, negative current is passed to the recording head.

Corresponding to the duration of the positive current, one magnet is produced; corresponding to the negative current, a magnet is also produced, but its polarity is opposite.

Playback

When the material passes by the read head, a positive pulse is produced corresponding to the onset of positive current; during recording, a negative pulse occurs at the read head corresponding to the onset of negative C during recording. It is clear from the figure that a playback pulse occurs only for each change in signal polarity.

Read Rule

For each pulse that is read, a flip flop can be triggered. This flip flop records a 1 until the next pulse comes along, where it is triggered to 0, and so forth. Now the trigger flip flop information is sampled with a timing pulse, thus yielding the proper output information.

Difficulties

NRZ is not self-timing in the same sense as RZ; it is usable for magnetic tape recording where multitrack information is provided with an odd parity check. Unlike RB, all information is recorded positively so old information is destroyed, thus maintaining the integrity of the new data. If the trigger fails to respond to a playback pulse, however, the entire remainder of the message may be mixed up because the trigger is exactly reversed from its original form. This can be controlled to some degree by observing the polarity of the playback pulse. The polarity of a signal can be used to set the polarity of the flip flop instead of triggering it. But further failure to trigger during writing will cause improper recording that cannot be compensated during reading.

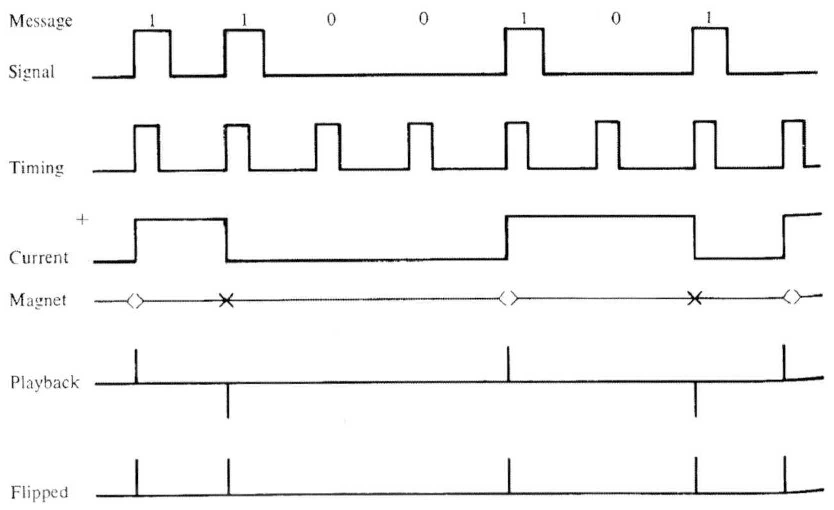

Fig. 8.4. Nonreturn to zero for 1s (NRZI)

NRZI

This method is called the NRZ for 1s method or NRZI. It works on a principle similar but not identical to NRZ.

Every 1 in the message causes recording current to change polarity.
Every 0 in the message causes no change in polarity.

Figure 8.4 shows the action for this type of recording.

Write Rule

Assume a write flip flop that monitors the direction of the current: a 1 in the message triggers this flip flop; a 0 in the message does not change the state of the flip flop.

Write Method

The medium records a magnetic force in a different direction for each 1 that occurs in the message. This magnet continues to be recorded in the same direction for all 0s that follow in the message. Polarity transitions occur only where there are 1s in the message, as the playback signal is produced. Thus a playback pulse, regardless of its polarity, corresponds to a 1 in the message.

Read Rule

A pulse in the playback corresponds to a "1" in the message.

Method

The negative playback pulses are simply inverted and combined with positive pulses to get the message.

Properties

There is no need to worry about the triggering getting out of step since 1s in the playback signal correspond to 1s in the message and 0s to 0s. Current is always flowing in the head during recording so that old information is erased. Only the timing problem remains, which is solved by odd parity when this method is used for magnetic tape recording.

8.2.3 Phase Encoding

The NRZ and NRZI methods have a further disadvantage that is not immediately apparent. The information that is recorded is different according to the nature of the data. In either system, a long sequence of 0s or 1s will cause a recording current that does not change direction (no pulses). This means that one very long magnet is recorded on the medium. This presents a technical pro in designing an amplifier to reproduce the playback signal. The amplifier must be able to respond properly to this long period of inactivity. A very low-frequency response is required of the amplifier; it must still have a good high-frequency response to respond properly to a recorded sequence of spikes.

The other problem that arises for single track recording is that of clocking. Whereas for multitrack recording the user can rely on the presence of 1s in some of the other tracks, this is not possible for the disk and the drum, where a single track must supply both the data and the timing. Phase encoding (PE) overcomes this problem.

Principle

For each bit to be written, a time period is described during which recording takes place. This time period is divided into two parts:

Start time is the beginning of this period
Center time, defines the center of the period

For each system of recording there are one or two current changes for each bit to be recorded. The number of changes depends on whether the bit is 0 or I. There is only one current change for a 0; there are two current changes when a 1 is recorded.

A further advantage of those described earlier is achieved because there are only two frequencies recorded. There are only two magnet lengths:

A small magnet proportional to half the signal time
A large magnet proportional to the full signal time

These magnets may be polarized in either direction. This makes for a simpler engineering design. It also permits packing information much closer there is less cancellation of signals. Detection is simpler because of the periodic nature of the recording.

The Diphase System

The diphase system is used by IBM, sometimes referred to as double-frequency NRZI. Here a 1 appears at center time for a 1 bit and no pulse appears at center time for a 0 bit.

To enforce this arrangement, the following write rule is necessary:

Current to a recording head changes direction at start time regarding of the bit to be recorded.

Current changes direction at center time only when the bit being recorded is 1.

This is illustrated in Fig. 8.5. Notice that the rule contains no statement about the direction of the current – only a change in current direction is required. Figure 8.5 shows a message to be recorded. A clock is present in the equipment that issues pulses at start time. Another clock issues pulses at change time but is gated by the signal so that these pulses occur only when there is a 1 in the message for this time period. The figure shows the timing pulses and the data pulses that occur for 1s only.

Assume that these pulses cause a change in current direction at the recording head. The current direction is then illustrated on the next line of the figure.

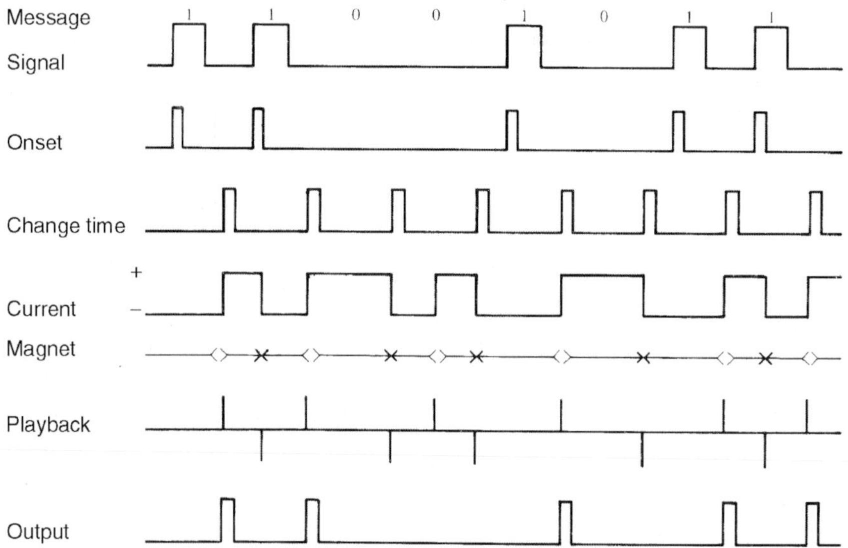

Fig. 8.5. Manchester phase encoding (PE) scheme

The magncts produced on the magnetic material depend on the orientation of the head.

Playback

As the magnetic material is played back under a similar set of read write heads, the signal produced for the message recorded is shown on the next line. Notice that pulses always occur at start time. A pulse may appear at center time but only when there is a 1 in the message.

Detection

To detect the information, a clock must be synchronized on the timing signals. As we see in the next chapter, some means must be provided to effect this synchronization. A clock running at the proper speed now activates a gate that allows only the data signals to pass through as shown in the figure. Since the data signals may be of either polarity, negative 1s must be flipped to get a proper sequence of 1s.

Manchester

The **Manchester Phase Encoding** system is used by IBM for all its PE magnetic tape units. For this system:

There is always a change of recording current at center time
The direction of change of this current is different according to whether there is a 1 or a 0 in the message
There is a change in current at start time only when this is necessary to get the proper current change at center time.

8.3 Analog and Digital Modulation Techniques

Different types of signals that are generally encountered in communication systems were discussed in Chap. 5. Many of these signals have frequency spectra that are not suitable for direct transmission especially when atmosphere is used as the transmission channel. In such a case, the frequency spectra of the signal may be translated by modulating high frequency carrier wave with the signal. Consider, for example, picture signal of a TV camera. It has frequency spectra of DC to 5.5 MHz. Such a wide band of frequencies cannot be propagated through ionosphere. However, if this signal is modulated with a carrier in VHF of UHF range, the percentage bandwidth becomes very small and the signal becomes suitable for transmission through atmosphere.

Apart from this primary requirement for modulation of signals, there are additional objectives which are met by modulation discussed in the following:

(a) *Ease of radiation.* As the signals are translated to higher frequencies, it becomes relatively easier to design amplifier circuits as well as antennae systems at these increased frequencies.
(b) *Adjustment of bandwidth.* Bandwidth of a modulated signal may be made smaller or larger than the original signal. Signal to noise ratio in the receiver which is a function of the signal bandwidth can thus be improved by proper control of bandwidth at the modulating stage.
(c) *Shifting signal frequency to an assigned value.* The modulation process permits changing the signal frequency to a reassigned band. This frequency may be changed many times by successive modulations.

Modulation may be defined as the process by which some parameter of a high frequency signal termed as carrier, is varied in accordance with the signal to be transmitted. Various modulation methods have been developed for transmission of signals as effectively as possible, with minimum possible distortion. The comparison of the effectiveness of these modulation methods may be based upon the signal power to noise power measured at the output of a receiver. Accordingly, a wide range of modulation techniques have been developed. These techniques may be broadly grouped into analogue techniques and pulse techniques. The analog methods of modulation are simpler and cheaper than pulse modulation techniques. The former employ the sinusoidal signals as carrier while the latter circuits use trains of pulses as the carrier signal.

Analog modulation may be divided into amplitude modulation (AM) and angle modulation. AM may be categorized as AM with carrier at sidebands (AM/DSB), vestigial sideband (VSB), double sideband suppressed carrier (DSB/SC), single sideband suppressed carrier (SSB/SC), and independent sideband suppressed carrier (ISB/SC).

AM/DSB is very popular for radiobroadcast and radiotelephony. For TV transmission with a large bandwidth, VSB is preferred because of reduced bandwidth of this modulation system. DSB/SC or SSB/SC provides a further reduction in power and bandwidth requirement. SSB/SC finds an extensive use in multiplexed coaxial system and can carry several messages simultaneously.

All AM systems are, however, prone to noise which directly affect the signal amplitude.

In angle modulation, the instantaneous angle of a sinusoidal carrier is varied as per the instantaneous amplitude of the modulating signal. The system leads to *phase modulation* (PM) and *frequency modulation* (FM). FM and PM waves require a much larger bandwidth than AM, but are capable of giving a sufficiently improved signal to noise ratio than the latter. It also leads to considerable saving in power.

Pulse modulation methods employ a pulse train as the carrier. The simplest type of pulse modulation is *pulse amplitude modulation* (PAM) which is similar to AM. Other pulse modulation techniques include *pulse duration modulation* (PDM), *pulse position modulation* (PPM) and *pulse code modulation* (PCM). While PAM may be compared with AM, PDM and PPM with angle modulation. The PCM has no analog equivalent. In PCM there have been certain methods developed such as *delta modulation* (DM) and *adaptive delta modulation* (ADM).

A third form of modulation consists of modulating a sinusoidal signal with pulse signals and may be termed as *digital modulation.* Digital modulation may be divided into *amplitude shift keying* (ASK), *frequency shift keying* (FSK) and *phase shift keying* (PSK). They are especially useful for data transmission systems.

8.3.1 Amplitude Modulation

The process of AM consists of varying the peak amplitude of a sinusoidal carrier wave in proportion to the instantaneous amplitude of the modulation signal. Though the modulating signal is usually an audio wave of speech or music which is complex in nature, the analysis of the AM system as well as others is restricted to a modulating signal that is sinusoidal and has single frequency.

Assume the modulating signal and the carrier be represented by $e_m = E_m \sin \omega_m t$ and $e_c = E_{cm} \sin \omega_c\, t$ where ω_m and ω_c are the signal and the carrier angular velocities, respectively. These components along with the resulting modulated waves are shown in Fig. 8.6.

The process of modulation increases the peak amplitude of the carrier so that the modulated wave peak amplitude is given as $E_{mod} = E_{cm} + E_m \sin \omega_m\, t$ and its instantaneous value is given as

$$\begin{aligned} e_{mod} &= E_{mod} \sin \omega_c\, t \\ &= (E_{cm} + E_m \sin \omega_m\, t) \sin \omega_c\, t \\ &= E_{cm} \sin \omega_m\, t + E_{cm} \sin \omega_m\, t \times \sin \omega_c\, t \end{aligned} \tag{8.1}$$

Expansion of Eq. (8.1) gives

$$e_{mod} = E_{cm} \sin \omega_c\, t + \frac{m_a}{2} E_{cm} \cos(\omega_c - \omega_m)t - \frac{m_a}{2} E_{cm} \cos(\omega_c + \omega_m)t \tag{8.2}$$

where $m_a = \dfrac{E_m}{E_{cm}}$ is termed as degree of modulation. Equation 8.2 shows that the AM wave consists of three components – $E_{cm} \sin \omega_c\, t$ is the original carrier

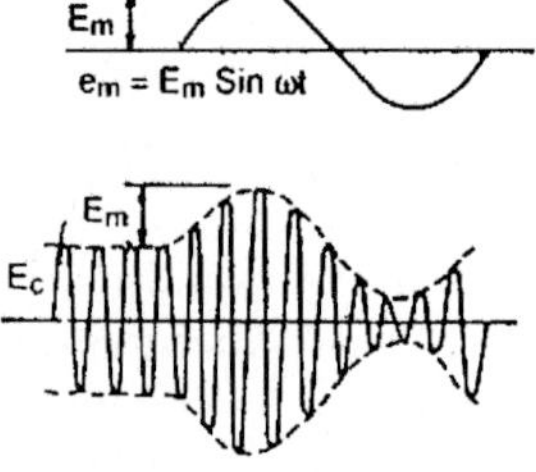

Fig. 8.6. Modulating signal and the AM wave

that is undisturbed, $\frac{m_a E_{cm}}{2}\cos(\omega_c - \omega_m)t$ is the component that has peak amplitude $m_a E_c/2$ and frequency that is the difference, between the carrier and the modulating signal and lastly the component $-\,m_a E_{cm}/2\cos(\omega_c - \omega_m)t$ is the component that has an amplitude equal to the second component but opposite in phase and a frequency that is, the sum of the carrier and signal frequencies. Consequently, these two components are termed as lower sideband (LSB) and upper sideband (USB). Each of the side bands is equally spaced from the carrier frequency and has a magnitude that $0.5\,m_a$ the carrier amplitude E_c. Frequency spectrum of the AM wave so obtained is shown in Fig. 8.7a for a single modulating frequency and for a band of frequencies in Fig. 8.7b.

Power in an AM Wave

Consider the wave as represented by Eq. (8.2) being fed to a load resistor of $1\,\Omega$. The average power in each of the component of the wave is given by the square of RMS values of the components concerned. Thus carrier power $(P_c) = \left(\frac{E_{cm}}{\sqrt{2}}\right)^2 = \frac{{E_{cm}}^2}{2}$ $\because$ E_c is the peak value of the carrier voltage.

Power in side bands $(P_c) = \left(\frac{m_a E_{cm}}{2\sqrt{2}}\right)^2 + \left(\frac{m_a E_{cm}}{2\sqrt{2}}\right)^2 = \frac{m_a^2 E_{cm}^2}{4}$

Total power: If the carrier voltage has an RMS value E_c; then total power

$$(P_t) = {E_c}^2\left(1 + \frac{m_a}{2}\right)^2 = P_c\left(1 + \frac{m_a}{2}\right)^2$$

$$\therefore \frac{P_{sb}}{P_t} = \frac{{m_a}^2{E_c}^2}{4} \div \frac{{E_c}^2}{2}\left(1 + \frac{m_a}{2}\right)^2 = \left(\frac{{m_a}^2}{2 + {m_a}^2}\right)^2 \tag{8.3}$$

As can be seen for $m_a = 1$, power in side bands is only 1/3 of the total power in the AM wave. The remaining two-third power is in the carrier. Since the carrier does not contain any intelligence, it follows that an AM wave has only 1/3 of its power as useful for $m_a = 1$. In practice, however, the value of m_a lies between 0.3 and 0.5 and useful power varies from 4.3% to 11% of the total power.

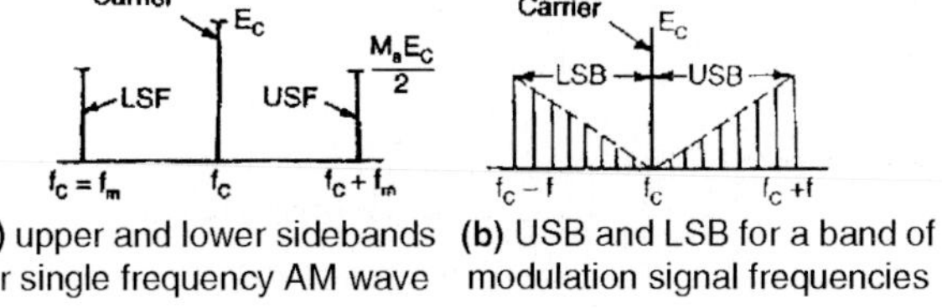

Fig. 8.7. Frequency spectrum of AM

8.3.2 Frequency Modulation (FM)

FM is the process of varying the frequency of a carrier wave in proportion to the instantaneous amplitude of the modulating signal without any variation in the amplitude of the carrier wave. Because the amplitude of the wave remains unchanged, the power associated with an FM wave is constant. Fig. 8.8 depicts an FM wave.

As can be seen from the figure, when the modulating signal is zero, the output frequency equals f_c (center frequency). When the modulating signal reaches its positive peak, the frequency of the modulated signal is maximum and equals (f_c+f_m). At negative peaks of the modulating signal, the frequency of the FM wave becomes minimum and equal to (f_c-f_m). Thus, the process of FM makes the frequency of the FM wave to deviate from its center frequency (f_c) by an amount $(\pm\ \Delta f)$ where Δf is termed as the *frequency deviation* of the system. During this process, the total power in the wave does not change but a part of the carrier power is transferred to the sidebands.

Assume the modulating signal to be represented by

$$e_m = E_m \cos \omega_m\ t$$

and the carrier wave being represented by

$$e_c = E_c \sin(\omega_m\ t + \theta) \tag{8.4}$$

$(\omega_m\ t + \theta)$ represents the total phase angle ϕ at a time t and θ represents the initial phase angle.

Thus $\phi = (\omega_c\ t + \theta)$. The angular velocity may be determined by finding the rate of change of this phase angle.

$$i.e., \text{angular velocity} = \frac{d\phi}{dt} = \omega_c \tag{8.5}$$

After FM takes place, angular velocity of the carrier wave varies in proportion to the instantaneous amplitude of the modulating signal. The instantaneous angular velocity ω_i is given by

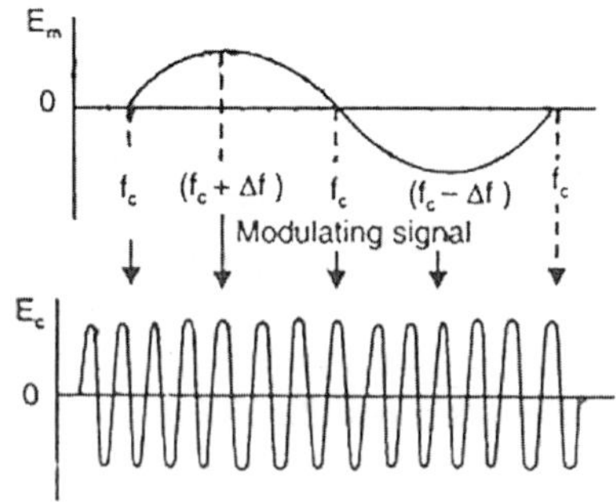

Fig. 8.8. Frequency modulating wave

$$\begin{aligned}\omega_{\mathrm{i}} &= \omega_c + K \cdot e_m \\ &= \omega_c + K \cdot E_m \cos \omega_m\, t \qquad (8.6)\end{aligned}$$

where K is a constant of proportionality.

Maximum frequency shift or deviation occurs when the cosine terms in Eq. (8.6) has a value ± 1. Under this condition, the instantaneous angular velocity is given by

$$\omega_{\mathrm{i}} = \omega_c + K \cdot e_m$$

so that the maximum frequency deviation Δf is given by

$$\Delta f = \frac{KE_m}{2\pi}$$

Equation (8.6) can be rewritten as

$$\omega_i = \omega_c + 2\pi\, \Delta f \cos \omega_m\, t \qquad (8.8)$$

Integration of Eq. (8.8) gives the instantaneous phase angle of the frequency modulated wave.

$$\begin{aligned}\therefore \phi &= \int \omega_i \cdot dt \\ &= \int (\omega_i + 2\pi\, \Delta f \cos \omega_m\, t) \cdot dt \\ &= \omega_c\, t + \frac{2\pi \Delta f}{\omega_m} \sin \omega_m\, t + \theta_1\end{aligned}$$

where θ_1 is a constant of integration representing a constant phase angle and may be neglected in the following analysis.

The instantaneous amplitude of the modulated waves is given by

$$\begin{aligned}e_{mod} &= E_c \sin \phi_i \\ &= E_c \sin \left(\omega_c + \frac{\Delta f}{f_m} \sin \omega_m t \right) \qquad (8.8a)\end{aligned}$$

The ratio $\frac{\Delta f}{f_m}$ is termed as the modulation index of the frequency modulated wave and is denoted by m_f. It should be noted that for a given frequency deviation Δf the modulation index varies with the modulating frequency f_m. A comparison of the modulation index m_a for the AM and m_f for the frequency modulated wave shows that while m_a is given as the ratio of the change in the carrier amplitude due to AM to the carrier amplitude, m_f is given as the ratio of the change in the carrier frequency during FM to the carrier frequency. Substituting m_f in Eq. (8.8a), we obtain equation for the frequency modulated wave.

$$e_{mod} = E_c \sin(\omega_c t + m_f \sin \omega_m t) \qquad (8.9)$$

8.3.3 Phase Modulation

As already pointed, FM and PM are very closely related and may be termed as Angle modulation. However, there exist some important differences between the two that warrant this discussion.

The process of PM consists in varying the phase of the carrier linearly in proportion to the modulating signal such that maximum phase shift occurs during positive and negative peaks of the modulating signal. If e_c is taken to represent the instantaneous value of the carrier, then

$$e_c = E_c \sin(\omega_c\, t + \theta)$$

where θ is the initial phase of the wave.

Let the modulating signal be $e_m = E_m \sin \omega_m\, t$. If the carrier phase varies sinusoidally with the modulating signal, then the phase modulated wave is represented by

$$e_c = E_c \sin(\omega_c\, t + \theta \pm \Delta\theta \sin \omega_m\, t) \tag{8.10}$$

Assuming that initially, $\theta = 0$, the

$$e_c = E_c \sin(\omega_c t + \Delta\theta \sin \omega_m t) \tag{8.11}$$

Equation (8.10) representing the phase modulated wave may be expanded in a way similar to FM. This expansion gives

$$\begin{aligned} e_o = E[&j_o(m_o) \sin \omega_c t + j_2(m_p) \sin(\omega_c + \omega_m)t - \sin(\omega_c - \omega_m)t \\ &+ j_2(m_p) \sin(\omega_c t + 2\omega_m)t + \sin(\omega_c + 2\omega_m)t \ldots\ldots] \end{aligned} \tag{8.12}$$

where (m_p) is termed as the modulation index for PM and equals $\Delta\theta$. The PM waves like FM waves have identical frequency spectrum. In PM, $\Delta\theta$ is given a fixed maximum value so that as the modulating frequency f_m varies, frequency deviation Δf also varies and $\Delta\theta = m_p = \dfrac{\Delta f}{f_m}$ remains constant. This is different from FM where Δf is constant and can be given a large value. As signal to noise ratio at the receiver output depends upon frequency deviation and not upon $\Delta\theta$, FM is preferred to PM. However, this gives an indirect method of producing frequency modulated waves.

It should be remembered that FM and PM are closely related and may be termed as *angle modulation*, because in both the cases, the changes in phase angle as well as frequency of the modulated carrier take place. Thus, consider a phase modulated wave given below.

Phase modulated wave $e_{PM} = E_c \cos[\omega_c\, t + \theta_o + m_p f(t)]$ where f(t) is the modulating signal. The term under the bracket represents the instantaneous phase angle ϕ_i

$$\therefore \phi_i = [\omega_c\, t + \theta_o + m_p f(t)]$$

Instantaneous angular velocity $\omega_i = \dfrac{d\phi_i}{dt}$

$$\therefore \omega_i = \omega_c + m_p \frac{df}{dt} \tag{8.13}$$

Equation (8.13) shows that in a PM wave, the frequency varies in proportion to the derivative of the modulating signal.

Now consider a frequency modulated wave given as

$$e_{PM} = E_c \cos[\omega_c\, t + m_f \cdot f(t)]$$

Instantaneous angular frequency $\omega_i = \omega_c + m_f \cdot f(t)$. The instantaneous phase change ϕ_i, in the FM wave can be computed by integrating the angular velocity ω_i.

$$\therefore \phi_i = \int \omega_i\, dt = \int [\omega_c + m_f \cdot f(t)]\, dt$$

$$= \omega_c\, t + m_f \int f(t)dt + \theta_0$$

where θ_0 is a constant of integration. The instantaneous phase angle in a frequency modulated wave varies directly as the integral of the modulating signal.

An important conclusion can be derived from the preceding discussion. If the modulating signal $f(t)$ is integrated before using it to phase modulate a carrier, the result is a FM wave Similarly, if the modulating signal is differentiated before allowing it to frequency modulate a carrier, the result is a PM wave. An important characteristic of PM is that it does not require additional pre-emphasis and de-emphasis like FM. Hence it may be considered as an FM system with perfect pre-emphasis and de-emphasis.

8.3.4 Need for Digital Modulation

The move to digital modulation provides more information capacity, compatibility with digital data services, higher data security, better quality communications, and quicker system availability. Developers of communications systems face these constraints:

- Available bandwidth
- Permissible power
- Inherent noise level of the system

Digital modulation schemes have greater capacity to convey large amounts of information than analog modulation schemes.

8.3.5 Digital Modulation and their Types

This section covers the main digital modulation formats, their main applications, relative spectral efficiencies and some variations of the main modulation types as used in practical systems. Fortunately, there are a limited number of modulation types which form the building blocks of any system. The three basic types of digital modulation are ASK, FSK, and PSK.

Phase Shift Keying

One of the simplest forms of digital modulation is binary or bi-phase shift keying (BPSK) as shown in Fig. 8.9. One application where this is used is for deep space telemetry. The phase of a constant amplitude carrier signal moves between zero and 180°. On an I and Q diagram, the I state has two different values. There are two possible locations in the state diagram, so a binary one or zero can be sent. The symbol rate is one bit per symbol.

A more common type of PM is quadrature phase shift keying (QPSK). It is used extensively in applications including code division multiple access (CDMA) cellular service, wireless local loop, Iridium (a voice/data satellite system) and digital video broadcasting-satellite (DVB-S). Quadrature means that the signal shifts between phase states which are separated by 90°. The signal shifts in increments of 90° from 45° to 135°, −45°, or −135°. These points are chosen as they can be easily implemented using an I/Q modulator. Only two I values and two Q values are needed and this gives two bits per symbol. There are four states because $2 \times 2 = 4$. It is therefore a more bandwidth-efficient type of modulation than BPSK, potentially twice as efficient.

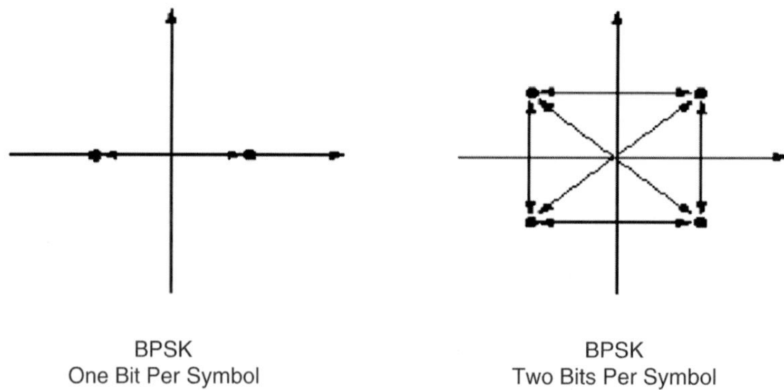

Fig. 8.9. Phase shift keying

Frequency Shift Keying

FM and PM are closely related. A static frequency shift of +1 Hz means that the phase is constantly advancing at the rate of 360° per second ($2\,\Pi\,\mathrm{rad\,s^{-1}}$), relative to the phase of the unshifted signal.

FSK shown in Fig. 8.10 is used in many applications including cordless and paging systems. Some of the cordless systems include Digital Enhanced Cordless Telephone (DECT) and Cordless Telephone 2 (CT2). In FSK, the frequency of the carrier is changed as a function of the modulating signal (data) being transmitted. Amplitude remains unchanged. In binary FSK (BFSK or 2FSK), a "1" is represented by one frequency and a "0" is represented by another frequency.

Minimum Shift Keying

Since a frequency shift produces an advancing or retarding phase, frequency shifts can be detected by sampling phase at each symbol period. Phase shifts of $(2N+1)\pi/2$ radians are easily detected with an I/Q demodulator. At even numbered symbols, the polarity of the I channel conveys the transmitted data, while at odd numbered symbols the polarity of the Q channel conveys the data. This orthogonality between I and Q simplifies detection algorithms and hence reduces power consumption in a mobile receiver.

The minimum frequency shift which yields orthogonality of I and Q is that which results in a phase shift of $\pm\,\Pi/2$ radians per symbol (90° per symbol). FSK with this deviation is called minimum shift keying (MSK). The deviation must be accurate in order to generate repeatable 90° phase shifts. MSK is used in the global system for mobile (GSM) communications cellular standard. A phase shift of +90° represents a data bit equal to "1," while −90 degrees represents a "0." The peak-to-peak frequency shift of an MSK signal is equal to one-half of the bit rate.

FSK and MSK produce constant envelope carrier signals, which have no amplitude variations. This is a desirable characteristic for improving the power

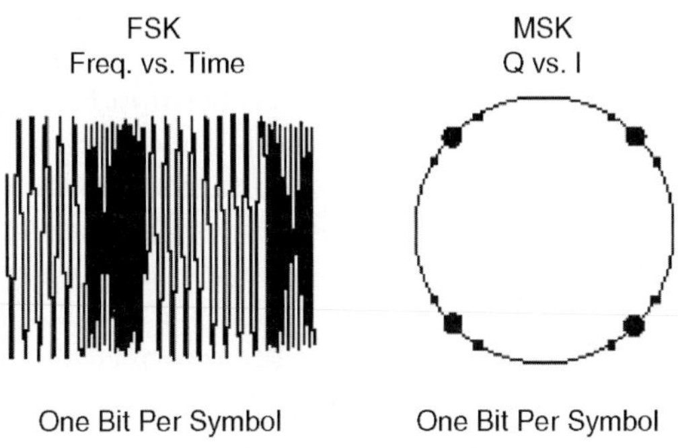

Fig. 8.10. Frequency shift keying

efficiency of transmitters. Amplitude variations can exercise nonlinearities in an amplifier's amplitude-transfer function, generating spectral regrowth, a component of adjacent channel power. Therefore, more efficient amplifiers (which tend to be less linear) can be used with constant-envelope signals, reducing power consumption. MSK has a narrower spectrum than wider deviation forms of FSK. The width of the spectrum is also influenced by the waveforms causing the frequency shift.

If those waveforms have fast transitions or a high slew rate, then the spectrum of the transmitter will be broad. In practice, the waveforms are filtered with a Gaussian filter, resulting in a narrow spectrum. In addition, the Gaussian filter has no time-domain overshoot, which would broaden the spectrum by increasing the peak deviation. MSK with a Gaussian filter is termed Gaussian MSK (GMSK).

Quadrature Amplitude Modulation

Another member of the digital modulation family is quadrature amplitude modulation (QAM). QAM is used in applications including microwave digital radio, Digital Video Broadcasting – Cable (DVB-C) and modems. In 16-state QAM (16QAM), there are four I values and four Q values. This results in a total of 16 possible states for the signal. It can have a transition from any state to any other state at every symbol time. Since $16 = 2^4$, four bits per symbol can be sent. This consists of two bits for I and two bits for Q. The symbol rate is one fourth of the bit rate. So this modulation format produces a more spectrally efficient transmission. It is more efficient than BPSK or QPSK.

Another variation is 32QAM. In this case there are six I values and six Q values resulting in a total of 36 possible states ($6 \times 6 = 36$). This is too many states for a power of two (the closest power of two is 32). So the four corner symbol states, which take the most power to transmit, are omitted. This reduces the amount of peak power the transmitter has to generate. Since $2^5 = 32$, there are five bits per symbol and the symbol rate is one fifth of the bit rate. The current practical limits are approximately 256QAM, though work is underway to extend the limits to 512 or 1024QAM. A 256QAM system uses 16 I-values and 16 Q-values giving 256 possible states. Since $2^8 = 256$, each symbol can represent eight bits. A 256QAM signal that can send eight bits per symbol is very spectrally efficient as shown in Fig. 8.11.

However, the symbols are very close together and are thus more subject to errors due to noise and distortion. Such a signal may have to be transmitted with extra power (to effectively spread the symbols out more) and this reduces power efficiency as compared to simpler schemes. In any digital modulation system, if the input signal is distorted or severely attenuated the receiver will eventually lose symbol lock completely. If the receiver can no longer recover the symbol clock, it cannot demodulate the signal or recover any information. With less degradation, the symbol clock can be recovered, but it is noisy, and the symbol locations themselves are noisy.

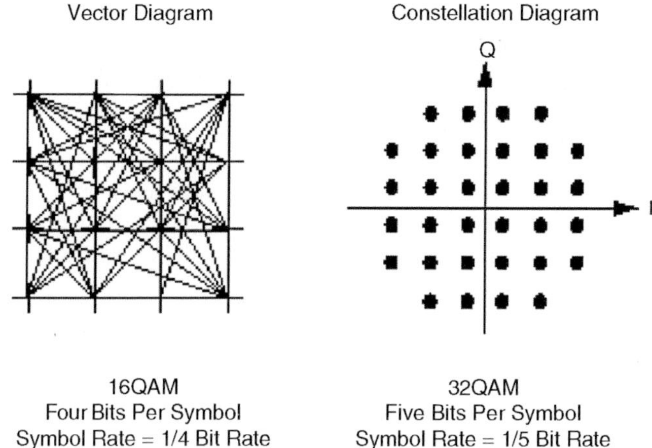

Fig. 8.11. Quadrature amplitude modulation

In some cases, a symbol will fall far enough away from its intended position that it will cross over to an adjacent position. The I and Q level detectors used in the demodulator would misinterpret such a symbol as being in the wrong location, causing bit errors. QPSK is not as efficient, but the states are much farther apart and the system can tolerate a lot more noise before suffering symbol errors. QPSK has no intermediate states between the four corner-symbol locations so there is less opportunity for the demodulator to misinterpret symbols. QPSK requires less transmitter power than QAM to achieve the same bit error rate.

8.3.6 Applications of Digital Modulation

Some of the applications of Digital Modulation are tabulated in Table 8.1.

8.4 Wireless Communication

The growing demand for wireless communication services has resulted in a convergence of industry, government, and public interests. Industry, under considerable pressure to establish reliable service, must identify and acquire sites suitable for communication facilities. Local governments must determine if selected sites conform with land-use zoning regulations. Some citizens voice concerns about neighborhood aesthetics and property values, and have questions about health effects of exposure to radiofrequency (RF) radiation emitted from wireless facilities. This section provides a brief summary of what is currently known about RF radiation exposure and health, exposure levels produced by wireless communication facilities, and federal RF standards.

Table 8.1. Applications of Digital Modulation

Modulation format	Application
MSK, GMSK	GSM, CDPD
BPSK	Deep space telemetry, cable modems
QPSK, 1/4 DQPSK	Satellite, CDMA, NADC, TETRA, PHS, PDC, LMDS, DVB-S, cable (return path), cable modems, TFTS
OQPSK	CDMA, satellite
FSK, GFSK	DECT, paging, RAM mobile data, AMPS, CT2, ERMES, land mobile, public safety
8, 16 VSB	North American digital TV (ATV), broadcast, cable
8PSK	Satellite, aircraft, telemetry pilots for monitoring broadband video systems
16 QAM	Microwave digital radio, modems, DVB-C, DVB-T
32 QAM	Terrestrial microwave, DVB-T
64 QAM	DVB-C, modems, broadband set top boxes, MMDS
256 QAM	Modems, DVB-C (Europe), Digital Video (US)

8.4.1 Background

Electric power, radiowaves, microwaves, infrared, visible and ultraviolet light, X-rays and gamma rays are all sources or examples of electromagnetic radiation. Ultraviolet light (UV-B & UV-C), X-rays and gamma rays, at the upper end of the electromagnetic spectrum, have sufficient energy to cause direct ionization of atoms. Exposure to ionizing radiation has been linked to cancer and genetic mutations in biological tissue. Microwaves, radiowaves (including those used by wireless communications), and electric power use frequencies that lack the energy to ionize atoms and, thus, are included as part of the nonionizing electromagnetic spectrum.

Nonionizing radiation interacts with atoms and molecules in the body through mechanisms other than ionization. Interactions include those that result in tissue heating (thermal effects), interactions that cannot induce tissue heating (nonthermal or athermal effects), and interactions where both thermal and athermal effects occur simultaneously. Although a majority of known effects of exposure to non-ionizing radiation are attributable to thermal exposures, athermal effects have been demonstrated in both strong and weak RF fields. Some evidence suggests that athermal effects may include changes in the immune system and neurological or behavioral effects.

However, contradictory experimental results have been reported, and further research is needed to determine whether these athermal effects are harmful to human health. The interest in effects of RF radiation has increased, in part, because growth in technologies utilizing RF energy has resulted in more of the population being potentially exposed. Familiar applications include AM and FM radio, television, microwave ovens, and citizens' band radio. Less familiar are newer technologies that have become common in the past decade such as satellite communications, paging services, and wireless communications services.

Wireless communication uses radiofrequencies in the 800–2,200 MHz range. Originally called cellular communication because of the way geographic areas are divided into service areas known as "cells," wireless systems rely on fixed facilities, or base stations, where antennas are built. As a person travels from one cell to another, the communication from the telephone is "handed off" from the cell being left to the cell being entered. Depending on the type of service available, cellular telephones transmit either analog or digitized voice messages. In analog communication systems, messages are transmitted by varying either the amplitude (height) or the frequency of the radiowave. Digital communication systems transmit messages as a series of digits in rapid bursts or pulses.

Both analog and digital systems, commonly known as "cellular systems," operate between 824 and 894 MHz. The latest generation of wireless communications, called Personal Communications Services (PCS), is similar to cellular telephone service. In this system, digital technology delivers voice, data, and video images. PCS operates in a higher frequency band (1,850–2,200 MHz) than cellular systems and utilizes a very localized wireless network requiring smaller cells and more antennas than cellular service has in the past.

Locations chosen by wireless companies to site antennas depend on a variety of factors, such as the proximity of adjacent cell sites, engineering and topographical considerations, community response and the existence of a willing property owner. A typical site consists of a small structure to house electronic equipment, radiotransmitters, and receivers. Some facilities are placed on existing structures such as rooftops or buildings. Other sites are created by placing antennas on towers or monopoles. Towers, buildings, and transmitters can be shared by multiple users.

8.4.2 Wireless Data

The original wireless networks were meant for voice traffic and, as such, are not particularly suitable for data traffic. As an example, delays of less than 100 ms are required for voice traffic, in order to avoid undesirable echoing effects, but larger delays may be tolerated for most if not all data. On the other hand, packetized speech can tolerate some packet loss, and bit-error rates (BERs) of 0.001. This may result in a slight quality loss, but no major aftermath. A BER of <0.00001 is required for data transmission, and no packet loss is allowed. Finally, telephone conversations last, on the average, between three and 20 minutes, so a setup time of a few seconds is acceptable.

Data transmissions can vary from a few seconds for a short e-mail to minutes for a large data transfer, so the setup time should be very small. These differences greatly affect wireless local-area networks (LANs) and PANs, as they are designed to accommodate both data and voice traffic.

WLANs

Wireless LANs use high-frequency electromagnetic waves, either IR or RF, to transmit information from one point to another. It is generally agreed that RF will be more practical than IR in home and office networking, since it can propagate through solid obstacles. Traffic from multiple users is modulated onto the radiowaves at the transmitter, and extracted at the receiver. Multiple radiocarriers can coexist in the same physical space, and at the same time, without interfering with each other by transmitting at different frequencies (frequency-division multiple access or FDMA), in different time slots, (time-division multiple access or TDMA), or using specific codes for each message (code-division multiple access or CDMA).

Wireless RF networks can feature an independent, peer-to-peer network, or an ad hoc network that connects communications devices with wireless adapters operating within a given frequency range (such as the ISM bands). Wireless LANs have been standardized by the IEEE 802.11 standards subgroup.

WPANs

WPANs use RF technologies similar to those of WLANs, but are meant for smaller communication coverage areas (10 s of meters versus 100 s). In 1998, the WPAN group published the original functionality requirement. Also in 1998, the same group invited participation from several organizations, such as Bluetooth, and others. Only the HomeRF and Bluetooth groups responded. In March, 1998, the Home RF group was formed. *In* May, 1998, the Bluetooth development was announced, and a Bluetooth special interest group (SIG) was formed within the WPAN group. Bluetooth has since been selected as the base specification for IEEE 802.15. In March, 1999, IEEE 802.15 was approved as a separate subgroup within the IEEE 802 group, to handle WPAN standardization. The IEEE 802.15 WPAN group focuses on developing standards for short-distance wireless networks. The resulting standard is intended to coexist with other wireless and wired networks within the ISM band.

8.4.3 Trends in Wireless Communication

Market demands for higher cellular density in urban areas, broadband internet wireless, and better data security, while using a minimum amount of frequency spectrum is driving wireless developments forward at an amazing speed. Forward leaps in computing power have resulted in GIGAFLOPS (billion floating point operations per seconds) or TERAMACS (trillion multiply-accumulate per second) of processing power that allows telecom engineers to do more and more processing in the digital domain. This offers obvious advantages in flexibility for the implementation of complex algorithms that permit a better utilization of the airwave spectrum, improved transmission quality, and better data security.

8.4.4 Software Defined Radio

The newest generation of radios are justly called software defined radio (SDR), since the entire signal processing chain, from intermediate frequency (say 50–125 MHz) down to base band frequency (like a voice channel at 8 kHz) is done digitally using reconfigurable firmware and software, replacing analog and ASIC digital processing. A typical SDR shown in Fig. 8.12 has the following components:

- Analog interface to the RF stage and antenna, through high-speed converters (ADC and DACs) able to digitize a wide portion of the spectrum
- High-speed front-end signal processing including digital down- or up-conversion
- Protocol-specific processing such as CDMA, TDMA, and satellite communications involve spreading/de-spreading, chip rate and frequency-hop rate recovery
- Code/decode functions, including modulation/demodulation, carrier and symbol rate recovery, channel interleaving/de-interleaving
- Data communications – Interfacing with carrier networks and backbone for data I/O and command-and-control processing, usually handled by general purpose ARM or PowerPC processors and Real Time Operating System (RTOS).

Recent trends in wireless communications boards implement the radio primarily in software through the integration of high performance computing in programmable logic and digital signal processors (DSPs) with the analog front-end. Designers are able to partition the high speed front-end signal processing from the lower speed base band and protocol functions for an optimal mix of performance and flexibility.

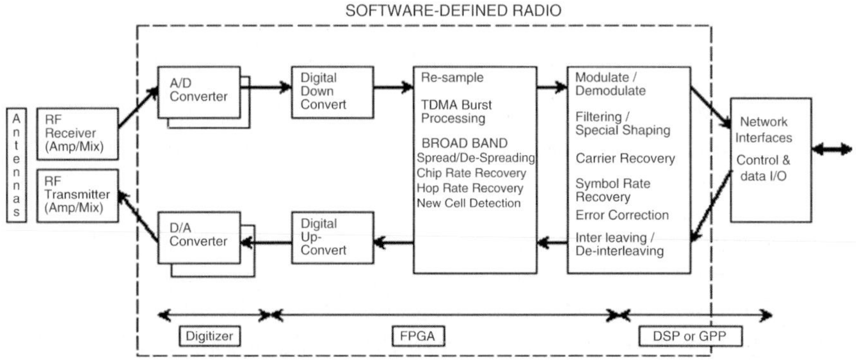

Fig. 8.12. Software defined radio

FPGAs for Ultra High-Speed Processing

The front-end signal processing is usually a very high speed process that is best suited for FPGA processing. The front end signal processing such as cascaded integrated comb filters and decimators used in up–down-conversion, fit well within most FPGA architectures and are rather simple arithmetic, but intensive in computation cycles because they run at the digitizing rate. These functions are ideal for FPGAs because they scale-up easily for parallel processing and usually do not involve highly complex algorithms. Other functions that require high speed processing in the FPGA typically include spreading/ de-spreading, code identification, and error correction.

When there is tight integration of the analog and signal processing functions, the designer has complete flexibility for application-specific front-end signal processing for improved signal quality, code identification and carrier acquisition Programmable logic density and speed are a driving force behind new SDR architectures. FPGA device sizes now approach the 10 million gate mark, offer speeds 200–400 MHz, and allow very complex interfaces to be mapped in logic.

There are huge offerings of IP cores that allow firmware engineers to rapidly integrate interfaces such as PCI, Ethernet, T1/E1, Rapid IO, and Hyper Transport as well as communication-specific functions like Digital Down Conversion, FIR and CIC filters, Viterbi and turbo decoder, FFTs, G.809. FEC and POS–PHY interface.

DSPs for Smarter, Complex Processing

The more complex algorithms found in the protocol-specific algorithms are more suited to programmable DSPs. The number-crunching capability of DSPs, lend themselves perfectly to the base band processing found in SDR. Algorithms such as demodulation, error correction, data packetizing, and radio control loops are well supported by DSP hardware and software. Many DSPs incorporate hardware acceleration for common functions such as Viterbi decoding, and a large body of software is available for most wireless applications. DSP software development tools now feature high performance RTOS that speed development with standardized plug-in software functions, thereby greatly reducing software development time.

DSP architectures and chip speeds are not only delivering GIGAFLOPS of performance, but also include built-in interfaces like PCI or Utopia (a flexible test and operations PHY interfaces to telephony backbone). Most DSP architectures allow a “glueless” interface between multiple DSPs for efficient clustering of DSPs, and efficient partitioning of the SDR functions. Application specific DSPs are also emerging and offer the optimum set of peripherals for very specific uses like direct audio I/O, video compression or multimedia for PDAs. Finally dual-core devices, combining a traditional DSP architecture with an ARM (or RISC) processor core with shared resources continue to blur the lines between DSP and system-level functions.

Multi-Boards for High-Channel Count and Redundancy

At the system level, SDR systems used for servicing high channel counts and requiring redundancy for high system availability continue to compartmentalize the above processing tasks on separate boards, usually hot-swappable 6U compact PCI. New switched fabric architectures within Compact PCI systems provide an efficient means for high bandwidth, low latency data transfer between boards. Now, high performance serial data buses such as Rapid IO or Channel link are beginning to replace PCI bus systems, reducing system complexity while delivering data transfer rates of 250 MBytes s^{-1} between boards.

Integrated Architecture for Development and Maximum Bandwidth

Development and test platforms for SDR systems go as far as integrating all the processing stages into a single compact PCI board: analog converters FPGA and DSP. The advantage of such integrated designs is that it provides radio firmware and software engineers a complete hardware platform that they can be used in the field to develop and test new IP algorithms (intellectual property) or as a powerful, reconfigurable station to test hardware. The advantage here is that all functions of the radio system, excluding network interfaces, can be fully configured and controlled from within a single development environment.

There is no need to procure multiple boards and understand the intricacies of efficient synchronization or to learn multiple tool sets. Wireless board technology continues to be driven by the emergence of newer, more powerful logic and DSP devices, software standards and system architectures. Tighter integration of analog with enormous signal processing power seems to be the name of the game and is leading wireless communications boards to new levels of performance.

8.5 RF Network Analyser

Root-mean-square (RMS) power detection is required to measure and control transmitted power in a multicarrier wireless infrastructure. Traditional power detectors using diode detection or log amps do not accurately measure power when the peak-to-average ratio of the transmitted signal is not fixed. Temperature stability of measurement circuitry is critical as is the linearity of the detector's transfer function. This section presents techniques to improve the temperature stability of an RMS power detector and the linearity of its transfer function to less than ± 0.3 Db over a dynamic range of more than 50 dB.

Modern wireless transmitters generally require strict control of transmitted RF power. In wireless cellular networks, strict power control allows the

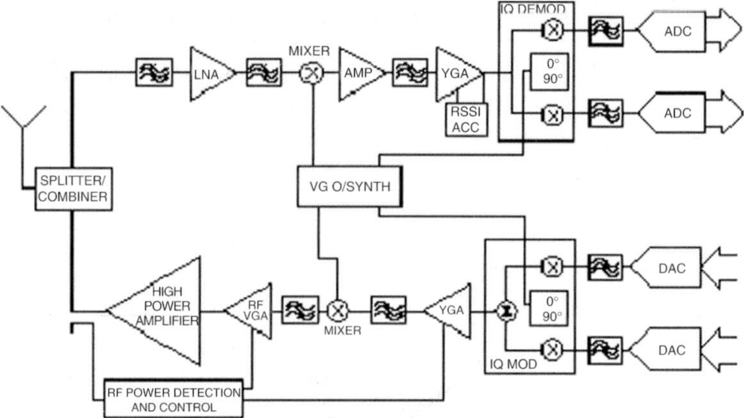

Fig. 8.13. RF network

size of cells to be precisely set to enhance coverage. Precise power control also avoids the need for excessive thermal dimensioning of the RF power amplifier (PA), which is required when there is uncertainty about the actual transmitted power. The RF network is shown in Fig. 8.13. For example, if a 50 W power amplifier has a transmit power uncertainty of just 1 dB, the PA must be dimensioned so that it can safely (i.e., without overheating) transmit 63 W. Power measurement and control is also used in receivers, usually at intermediate frequencies (IFs).

Here the objective is to measure and control the gain of the received signal so that IF amplifiers and analog-to-digital converters (ADCs) are not overdriven. While precision in the measurement of the received signal (commonly referred to as the received signal strength indicator or RSSI) is useful for maximizing the signal-to-noise ratio, it is less important than on the transmit side; the goal is to merely keep the received signal under a certain limit. RMS RF power detectors can measure RF power independently of signal peak-to-average ratio or crest factor. This capability is critical when the peak-to-average ratio of a measured signal is changing. This is common in wireless cellular networks due to the ever changing number of calls being carried by a cellular base station. The changing peak-to-average ratio results both from the transmission of multiple carriers at varying power levels and from the variations in code-domain power in a single code division multiple access (CDMA) carrier.

High Dynamic Range RMS-to-DC Converter

The AD8362 is an RMS-to-DC converter that can measure RMS voltages over a range of 60 dB or more, spanning from very low frequencies up to approximately 2.7 GHz. Figure 8.14 shows the transfer function of the AD8362

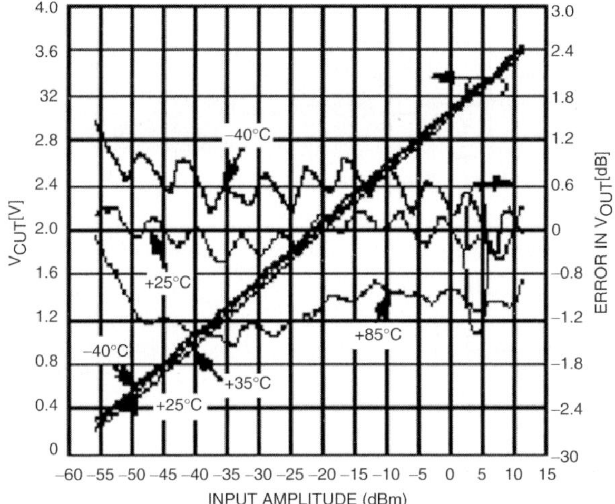

Fig. 8.14. Error rate

at 2.2 GHz in terms of volts out versus input signal strength in dBm relative to 50. Figure 8.14 also shows the deviation of this transfer function from a best fit line. This line has a slope and intercept that are calculated using a linear regression of the measured data. Once the slope and intercept of this line are calculated, an error plot, scaled in dB, can be plotted. In Fig. 8.14, this line is scaled on the right.

This plot shows a repetitious ripple with peak-to-peak amplitude as large as 0.75 dB. This ripple results in an equally large measurement uncertainty. The plot also shows that the transfer function changes over temperature. The temperature drift of the transfer is, in this case, dominated by the change in the intercept (i.e., the slope stays relatively constant).

Operation of the AD8362 Logarithmic RMS-to-DC Converter

Figure 8.15 shows a block diagram of the AD8362. The main components of the AD8362 are a linear-in-dB variable gain amplifier (VGA) that comprises a voltage controlled attenuator, a fixed gain amplifier, a low dynamic range RMS-to-DC converter, and an error amplifier. The input signal is applied to the VGA. The output of the VGA is applied to the low range RMS-to-DC converter. The output of this detector is proportional to the RMS voltage of the VGA output signal. A fixed reference voltage, also referred to as the target voltage, is applied to an identical low dynamic range RMS-to-DC converter.

The outputs of the two detectors are applied to an error amplifier/ integrator that produces an error signal. The output of the error amplifier is applied to the gain control input of the VGA. The gain control transfer function of the VGA is negative, that is, an increasing voltage decreases the gain.

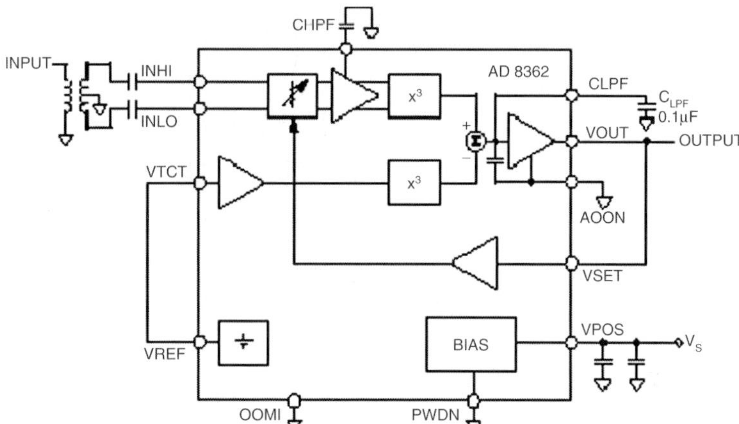

Fig. 8.15. Capacitor connected circuit

When a small input signal is applied to the circuit, the voltage from the signal path detector will be small, producing a decreasing error signal that drives the VGA. This error signal will continue to decrease, increasing the VGA gain, until the output of the signal chain detector is equal to the output of the reference detector.

Likewise, a large input signal will produce an increasing error signal that decreases the gain of the VGA until the voltage from the signal path detector is equal to the reference detector. In all cases, when the system reaches equilibrium, the input voltage to the RMS-to-DC converter settles to the same value. As a result, the low range RMS-to-DC converter needs very little operating range for the circuit to work.

The transfer function of the VGA is linear-in-dB, that is, the gain in dB is proportional to the control voltage, scaled in volts. In this case, the slope of the VGA's gain control is approximately $-50\,\mathrm{mV\,dB^{-1}}$. The result is a logarithmic transfer function for the overall circuit (i.e., the relationship between the input to the VGA and the output of the error amplifier), that is, the output voltage is proportional to the log or the RMS input voltage. The temperature stability of this gain control function is critical to the overall temperature stability of the RMS measurement.

Gaussian Interpolator

The source of the periodic ripple in the conformance curves in Fig. 8.14 is the Gaussian interpolator. The Gaussian interpolator determines the nodes from which the signal is taken from the variable attenuator. This is then applied to the fixed gain amplifier that comprises the output stage of the AD8362's VGA. The input ladder attenuator is composed of a number of sections, each of which attenuates the input signal by 6.33 dB. The signal is tapped off these sections via variable transconductance stages.

The Gaussian interpolator determines the transconductance stages that are active, based upon the control signal that is applied to the control port of the variable attenuator, thereby determining the amount of attenuation applied to the input signal. Attenuation levels that fall between tap points require that adjacent transconductance stages be active simultaneously to produce a weighted average of these tap points according to the transconductance cell that is directed to conduct harder. The manner in which the conductance of adjacent stages changes to slide the tap point along the attenuator is responsible for the ripple observed in the conformance curves.

Filtering of the Error Signal

The squaring cell in the low range RMS-to-DC converter produces a component at dc along with a component at twice the input frequency. This follows from the trigonometry identity If this signal is a single-tone sine wave, the output of the squaring cell will be a dc component and a sine wave tone at twice the input frequency. The dominant pole of the error amplifier/integrator will filter out the double frequency component, leaving just the dc component. If the input signal is a broadband signal, such as a CDMA or wideband CDMA (WCDMA) signal, the component that appears at dc spans the range from dc to half the original signal bandwidth.

Therefore, once the double frequency has been filtered out, the output of the circuit that is fed back to the VGA still contains significant ripple, which appears as a noise-like signal superimposed on a dc level. Normal practice is to increase the filtering in the error amplifier so that the noise on the signal at the output of the error amplifier is significantly reduced. This results in a noise-free output from the overall circuit.

Removal of Transfer Function Ripple

Figure 8.16 shows an alternate configuration for the circuit that takes advantage of this baseband noise. In contrast to the circuit in Fig. 8.15, the size of the integrator's external filter capacitor has been significantly reduced but is still kept large enough to effect valid RMS averaging. When a broadband signal is applied as an input to the circuit, the output of the error amplifier contains significant noise but is still centered on the correct RMS output level. The noise level at the output of the error amplifier is set to a level of at least 300 mV p–p, 300 mV being the distance in dB between adjacent taps on the R-2R ladder of the VGA.

The gain control slope of the VGA is 50 mV dB^{-1} i.e., 6 dB. As long as this output noise level is at least 300 mV p–p, its actual value is not critical. This lightly filtered signal is fed back to the VGA control input. The noise in this signal causes the gain of the VGA to fluctuate around a central point. The gain control slope of the VGA is 50 mV dB^{-1}. As a result, the noise will cause the instantaneous gain of the VGA to change by approximately 6 dB. The wiper of the Gaussian interpolator is moved back-and-forth across approximately one tap of the R-2R ladder.

Because the gain control voltage is constantly moving across at least one tap of the Gaussian interpolator, the relationship between the RMS signal strength of the VGA output and the VGA control voltage becomes independent of the VGA's gain control ripple. The signal being applied to the squaring cell is now lightly AM modulated. However, this modulation does not change the peak-to-average ratio of the signal. Because of the reduced filter capacitor, the RMS voltage appearing at the output of the error amplifier now contains significant peak-to-peak noise.

While it is critical to feed this signal back to the VGA gain control input with the noise intact, the RMS voltage going to the external measurement node can be filtered using a simple filter to yield a largely noise-free RMS voltage. Figure 8.16 shows the resulting reduction in transfer function ripple of the RMS-to-DC converter. The 600 mV peak-to peak noise fed back to the VGA gain control terminal may seem needlessly large because only enough noise to exercise the gain control voltage over 6 dB (one tap on the R-2R ladder) is required. However, as the call loading in a spread spectrum CDMA signal decreases, the peak-to-average ratio of the signal also decreases.

This results in less noise appearing at the detector output. As a result, the peak-to-peak noise is set so that it always spans at least one tap on the R-2R ladder. The peak in the error function at around −57 dB m is due to measurement. Figure 8.17 shows the response of the transfer function of the modified circuit when an un-modulated sine wave is applied. No reduction in transfer function ripples results. As already noted, when a sine wave is

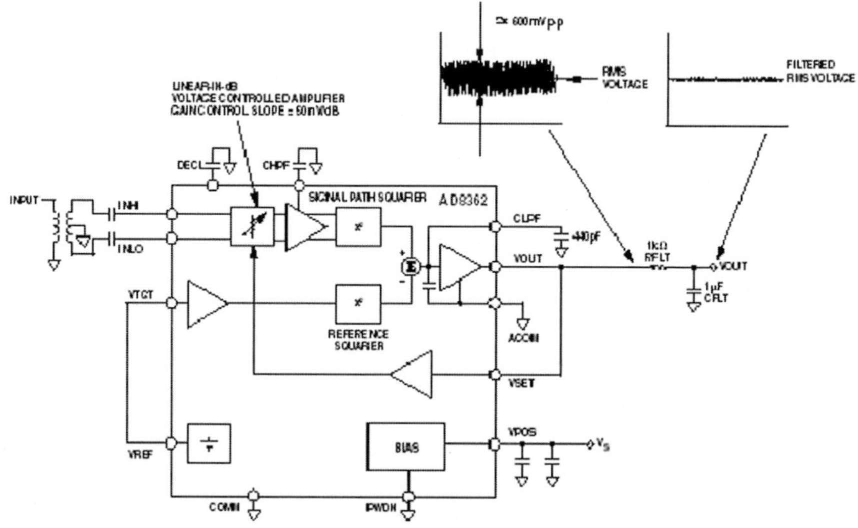

Fig. 8.16. Feedback circuit

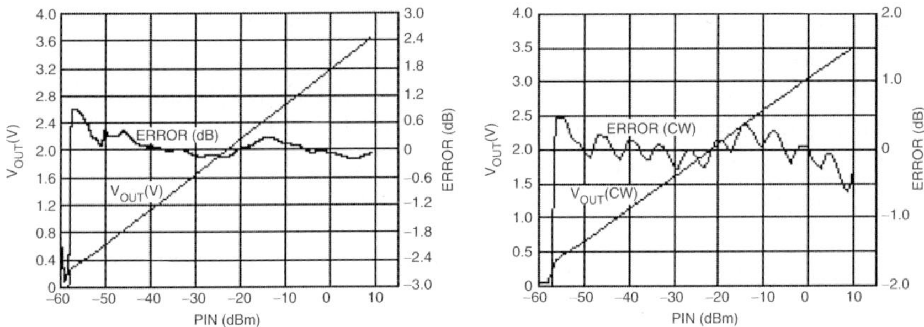

Fig. 8.17. Noise control

applied to the squaring cell, the output products are the double frequency and a dc voltage level. Because the sine wave is narrow band, no noise-like voltages appear close to dc. Once the double frequency has been removed, no ac component is available to exercise the VGA's gain control input over any range.

Although the AD8362, a 60 dB logarithmic detector, has excellent baseline performance, its measurement accuracy can be improved even further. The techniques used are simple, involve resistors, capacitors, and a temperature sensor, and can be mass produced due to the part-to-part repeatability of the temperature drift.

8.6 Distributed Automation and Control Systems

Nowadays, the level of integration determines the effectiveness of modern technical systems. With respect to technology, integration has three principal constituents: standardization, automation and rationalization. Application of the system approach and functional and modular decomposition results in that systems become distributed, i.e., consist of a number of physically and/or logically distributed components that constitute the system to reach the common goal. Communication between such components is a crucial issue.

8.6.1 Distributed Control Systems

The instrumentation used to implement automatic process control has gone through an evolutionary process and still it is developing today. In the beginning, plants used to be local, large-case pneumatic controllers; later became miniaturized and centralized onto control panels and consoles. Their appearance changed very little when analog electronic instruments were used. The first application of process control computers resulted in a mix of the traditional analog and the newer direct digital control (DDC) equipment in the same control room. This mix of equipment was not only cumbersome but also rather inflexible because the changing of control configurations necessitated changes in the routing of wires.

Pneumatic systems have advantages over electronic instruments because of their safety in hazardous locations and the pneumatics is still widely used in the control valve portion of control loops. However, electronic systems are usually favored when computer compatibility, long transmission distances, or very cold ambient conditions are important factors, or long-term maintenance is to be minimized. A pneumatic system does not need additional hardware to operate an air-operated control valve, whereas a current to pressure (I/P) converter must be installed in an electronic loop. Because many transmitter outputs are inherently electronic, some pneumatic loops also require I/P converters for applications involving temperature, flow, etc. Electronic systems are superior for data loggers; computers distributed control, and programmable controller applications. Analog instrumentation is usually hard-wired and is dedicated to continuously performing a particular control function. Digital instrumentation is often shared and performs its control functions in a sequential manner.

In 1970, distributed control system was introduced and has become an accepted alternate to the other two forms of electronic control systems used for process control namely analog controllers and digital computers. As per this widely accepted Distributed Control System (DCS) concept, the control over the entire process is distributed, but it is not the case with the Direct Digital Control Systems. Distributed control system updates the status of a loop as often as every 200 ms or as infrequently every second. When analog instruments are used, their failure affects only one loop. However, in the case of DCS, when a system fails, it affects all controllers contained in that system. Such limitations as increased loop dead time and reliability are balanced against the many advantages of DCS, which include the economy inherent in sharing hardware among many loops, the flexibility and convenience of reconfiguration, the ability to handle complex mathematical relationships the ability to provide memory, and multicolor dynamic displays.

8.6.2 Computers in Industrial Control

A computer is a device that follows predetermined instructions to manipulate input data in order to produce new output data. When a computer is used as a part of a control system, instructions define what actions to be taken as the input data (both from plant as well as operator) changes, i.e., the operator's actions are part of input data. The output data are control actions to the plant and status display to the operator.

Computers used in industrial control must process the data at higher rates as control has to be performed in real time. Many emergency and alarm conditions require necessary actions that are to be taken in fraction of a second. An industrial process requires possibly hundreds of devices to be read in real time and signals sent to devices such as valves, motors and meters.

8.6.3 Applications of Computers in Process Industry

The term SCADA is often used to describe the applications of computers in process control, which rather nearly describe the functions that a computer can perform. If a computer to be used in industrial control, the areas where it can be applied firstly identified. Due to the complexity a large industrial control, the areas where it can be applied firstly identified. Due to the complexity a large industrial process can be simplified in to smaller subsystems, such as monitoring subsystems and closed loop control subsystem.

Monitoring Subsystems

Monitoring subsystems display the Process State to the operator and draw the attention to abnormal or fault conditions, which need attention. The plant conditions are measured by suitable sensors which are connected to input cards. The results of these parameters are displayed to the operator via indicators (for digital signals) or by meters and bar graphs for analog signals. The signals can also be checked for alarm conditions. A high temperature or a low liquid level could be typical analog alarm conditions. The operator could be informed of these, via warning lamps and an audible alarm. A monitoring system often keeps records of the consumption of energy and material for accountancy purposes, and produces an event/alarm log for historical maintenance analysis.

Closed Loop Control Subsystems

In many analog systems a variable such as temperature, flow or pressure is required to be kept automatically at some present value or made to follow some other signal. The block diagram representing such systems is shown in Fig. 8.18. Here a particular characteristic of the plant (e.g., temperature) denoted by PV (process variable) is required to be kept at a present value SV (set point value). PV is measured by suitable sensor and compared with the V to give an error signal.

This error signal is applied to a control algorithm; the output from the control algorithm is passed to an actuator, which affects the plant. For a temperature control the actuator could be a heater, for a flow control the

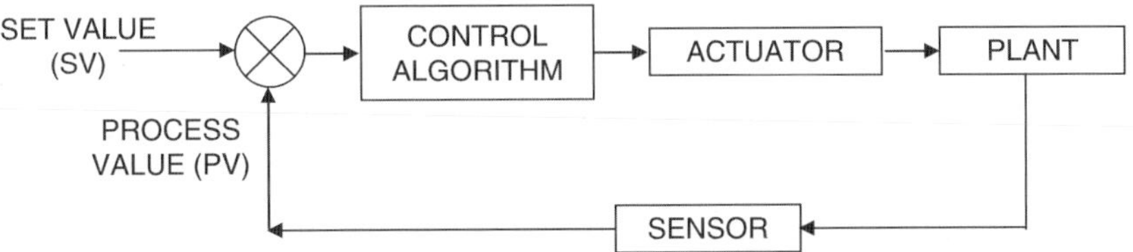

Fig. 8.18. Closed loop control system

actuator could be a flow control valve. The control algorithm will adjust the actuator until there is zero error, i.e., the process variable and the set point value are equal.

8.6.4 Direct Digital and Supervisory Control

If a computer is to be used for closed loop control there are two possible approaches.

(a) Direct Digital Control
(b) Supervisory Control

Direct Digital Control

In the early days of introduction of the Computers for process automation, a single big mainframe is used to control whole plant. This single computer is alone made responsible to carry out all the functions such as communicating with the field instruments, performing calculations and control operations which means computer itself can perform the control algorithm. This is known as Direct Digital Control (DDC). The block diagram is as shown in the Fig. 8.19.

The main problem with this type of configuration is that, if any problem occurs in the controller, the operation of the whole plant is stopped. Also as all the functions are to be performed by a single computer, the overloading is high and this overloading decrease the Mean Time Between Failures (MTBF) of the system, thus making it less reliable.

Supervisory Control

To overcome the difficulties of centralized control systems and still to enjoy the advantages offered by the digital technology, the concept of Distributed Control System came out. The basic concept of the Distributed Control System is to distribute the control system both functionally and geographically.

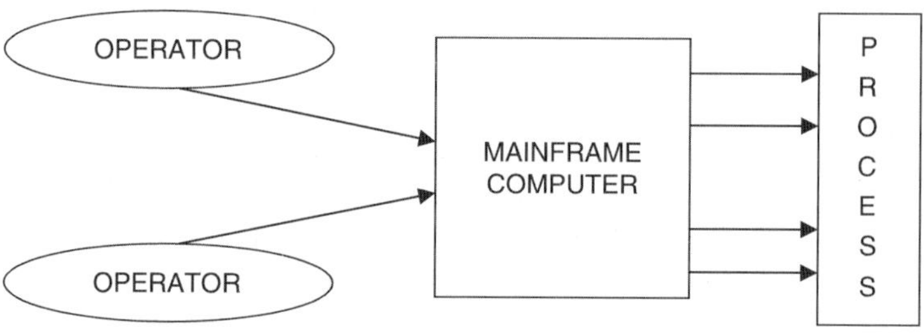

Fig. 8.19. Centralised Control System

To distribute the system functionally, it is required to divide the larger applications in to multiple smaller subsystems each of which carriers a portion of the application and allows smaller communication between the subsystems. Distribution of the system geographically means to allot a particular topography to a particular system as a whole. Today the distributed control system consists of large number of local controllers communicating with each other. They have hierarchic structure and have different processors for different functions. Sharing of the information and closed loop controllers providing the set points. This is known as supervisory control. The communication network in distributed environment is the key for system performance. The amount of information that could be processed depends upon the amount of the processing that could be done in the controllers themselves. The more the processing to be carried out on the process data the less the information that can be transmitted. This is the key feature of the Distributed Control system. Although more expensive supervisory control does remove the possibility of a total plant shutdown in the event of a computer failure.

Even processor based dedicated system such as PLC is designed around a subsystem of a plant, need for conventional computer is essential. The main advantage with the conventional computers is brute mathematical computing power, high speed and ease of connection to printers, keyboards and disc drives. If an application has few real I/O devices, which needs lot of mathematical operations to be performed, which requires several printers or graphics monitors, and where the security aspects are most important, a conventional computer is the optimal choice. The generalized distributed control system is shown in Fig. 8.20.

8.6.5 Architecture of Distributed Control Systems

The biggest disadvantage of the centralized computer control architecture is that the central processing unit represents a single point of failure that can

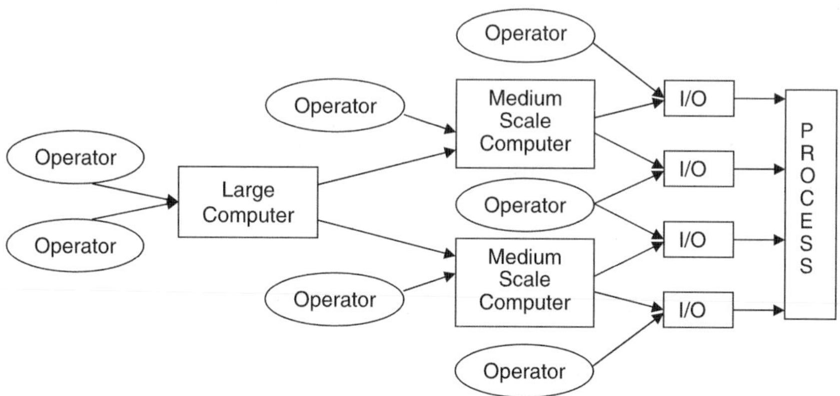

Fig. 8.20. Distributed Control Systems

shutdown the entire process. Since early industrial computer hardware was notoriously unreliable, two approaches were developed and have been used to solve the reliability problem; either a complete analog control system is used as a "HOT STAND-BY" to take over if the primary control computer fails. Either approach results in a significantly more expensive than analog control system that performs a comparable set of functions.

Another problem with its computer-based system is that software required to implement all of the functions is extremely complex, and requires a priesthood of computer experts to develop the system, start it up, and keep it running. Finally, the centralized system is limited in its capability to accommodate change and expansion. Once the loading on the computer approaches its limit, it becomes very difficult to add on to the system without a significant decrease in performance or increase in cost. Because of these problems, it became clear to both users and system designers that new architectural approach was needed. Control system engineers had been sketching concepts of Distributed systems composed of Digital control and communication elements. Unfortunately the technology to implement these concepts in a cost effective manner was not available. The architecture of the distributed system became practical only after the introduction of microprocessors. Supporting technology also became available, inexpensible solid state memories were developed to replace magnetic core memories; integrated circuit chip to implement the standard communication protocol were introduced; Display system technology flourished with the emergency of LED's and color CRT displays; in the software area, structure design techniques, modular software packages and new on-line diagnostics concepts were developed.

Result of this fortunate confluence of user needs and technological developments was the introduction or large number of distributed digital control system product lines. While each system has a unique structure and its own specialized features, the architectures of most of these systems can be described in context of generalized one shown in Fig. 8.21.

The devices in this architecture are grouped into three categories such as:

1. Devices that interface directly to the process to be controlled and monitored.
2. Devices that perform high-level human interfacing and computing functions.
3. Devices that provide the means of communication between the other devices.

A brief definition of each of these devices is as follows.

Local Control Unit (LCU)

The smallest collection of components in the system that can perform closed loop control is called local control unit. The LCU interfaces directly to the process.

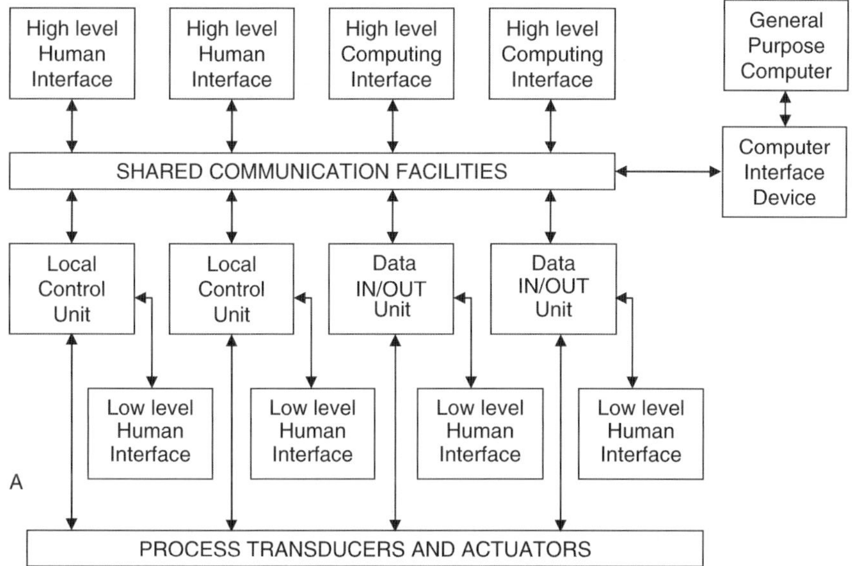

Fig. 8.21. Distributed Control System Architecture

Low-Level Human Interface (LLHI)

Low-level human interface is a device that allows the operator or instrument engineer to interact with the LCU using a direct connection. The LLHIs can also interface directly to the process. Operator oriented hardware at this level is called a "low-level_operator interface"; instrument engineer oriented hardware is called a "a low-level_engineering interface." This is referred to as Engineering Work Station.

Data Input/Output Module

A device that interfaces to the process solely for the purpose of acquiring or outputting data. Data input/output modules also perform control computations, referred to as I/O modules.

High-Level Human Interface (HLHI)

A collection of hardware that interfaces to other devices only one shared communication facilities. Operator oriented hardware at this level is called a "High-level operator interface"; Instrument engineer oriented hardware is called "High-level engineering interface." This is referred to as Human Interface Station (HIS).

High-Level Computing Device

High-level computing device is a microprocessor-based hardware that performs plant management functions traditionally performed by a plant computer. It interfaces to other devices only over shared communication facilities. This referred to as Field Control Station (FCS).

Computer Interface Device

A collection of hardware that allows an external general-purpose computer to interact with other devices in the DCS using the shared communication facilities.

Shared Communication Facilities

One or more levels of communication hardware and associated software that allow the sharing of data among all devices in the DCS. Shared communication facilities do not include dedicated communication channel between specific devices or between specific devices or between hardware elements within a device. This referred to as subsystem in CENTUM terminology.

8.6.6 Advantages of Distributed Control Systems

The major advantages of Distributed Control System are as follows.

Scalability & Expandability

Scalability refers to the ease with which a system can be sized for a spectrum of applications, ranging from small to large, and the ease with which elements can be added to the system after initial installations. The hybrid system is extremely modular, so it ranks high on both counts.

Control Capability

Control capability refers to the power and flexibility of the control algorithms that can be implemented by the system. The capability of the hybrid architecture is limited by the function available in the hardware module that makes up the system. To add a function involves both adding hardware and rewiring the system. This provides the full advantages of digital control; drift less set points and tuning parameters, availability of complex control algorithms, ability to change algorithms without changing hardware, remote and adaptive tuning capabilities.

Operator Interfacing Capability

The capability of the hardware provided to aid the operator in performing plant monitoring and control functions is known as operator interfacing capability. The operator interface in the hybrid system consists of conventional panel board instrumentation for normal control and monitoring function and a separate video display unit for supervisory control. The Video Display Units (VDU) are generally used as the primary operator for both the normal and supervisory control functions. The VDU provides significant benefits to the operator such as reduction in time needed to access control stations, flexibility of station grouping and graphic displays that mimic the process layout.

Integration of System Functions

The degree with which the various functional subsystems are designed to work with one another in an integrated fashion is known as integration of system functions. A high degree of integration minimizes the user problems in producing, interface, starting up and maintaining the system.

Significance of Single-Point Failure

The activity of the system's performance to a failure of any of its elements is known as single point failure. The failure of a hardware element in the computer can cause the system to stop performing completely unless a backup computer is used. Thus the system is relatively insensitive to single point failure due to the modularity of heir structure.

Installation Costs

An installation cost refers to the cost of system wiring, cost of control room and equipment room space needed to house the system. The distributed system reduces this cost further by using a communication system to replace the sensor wiring runs and by reducing the required equipment on space through the use of space efficient microprocessor based modules.

Maintainability

Maintainability refers to the ease with which the system can be kept running after installation. Low maintainability implies high maintenance costs including the cost of the spares, cost of process downtime when repairs are being made and personnel training cost. Since there are only few general-purpose control modules in the system, spare parts and personnel training requirements are minimal, the maintainability of the distributed system is excellent. Automatic online diagnostics are available to isolate failures to the module level. And the module replacement can be without disrupting the major portion of the process. Thus the distributed systems offer the user a tremendous amount of flexibility in the choice of control algorithm and location of the control equipments in the plant.

8.6.7 CORBA-Based Automation Systems

The Common Object Request Broker Architecture (CORBA) is a widely recognized by the control community object-oriented open-standard-based middleware for distributed systems integration. CORBA supports interoperability, portability and reusability of systems' components. There is a number of CORBA implementations and mappings to various programming languages that are available from different vendors. Also, its real-time and fault-tolerant specifications have been developed recently. Architecture of a typical CORBA-based system is presented in Fig. 8.22.

CORBA component's interface is described in a neutral Interface Definition Language (IDL). CORBA IDL is a declarative non-programming language; it does not provide any implementation details.

The methods that are specified in IDL can be implemented with any programming language such as C++ and Java using rules defined by OMG. The Object Request Broker (ORB) is an object bus that provides a transparent communication means for objects to send/receive requests/responds to/from other objects. The General Inter-ORB Protocol (GIOP) provides interoperability (a set of message formats) between ORB implementations

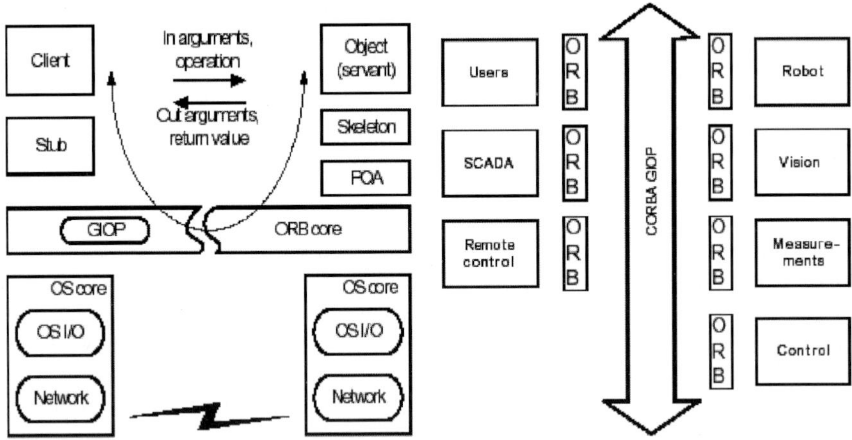

Fig. 8.22. CORBA-based architecture

from different vendors. The GIOP can be mapped onto different telecommunication protocols for transport protocol independence. Thus, the Internet Inter-ORB Protocol (IIOP) is a mapping of GIOP onto the TCP/IP protocol stack.

Distributed automation and control systems are considered from the implementation viewpoint. Theoretical and practical aspects of middleware in distributed automation systems and, in particular, the object oriented open-standard-based CORBA middleware are discussed. The usefulness of middleware for system integration and communication abstraction is shown on a practical example of the experimental robotic system. The main result is that middleware can significantly reduce development time, increase interoperability, portability and reusability of automation system's components and subsequently increase usability and competitiveness of a middleware-based automation system. Future work is related to the issues of hard real-time and reliable operations, exploitation of design patterns, context awareness, and mobile re-configurable automation distributed systems.

8.7 SCADA

Supervisory Control and Data Acquisition systems are basically Process Control Systems (PCS), specifically designed to automate systems such as traffic control, power grid management, waste processing, etc. Control systems are used at all levels of manufacturing and industrial processing. A manufacturing plant that employs robotic arms will have a control system to direct robotic arms and conveyor belts on the shop floor. It may use that same system for packaging the finished product and tracking inventory. It may also use a control system to monitor its distribution network. A chemical company will use control systems to monitor tank levels and to ensure that ingredients are mixed in the proper proportions.

A Las Vegas casino will use control systems to direct the spray from water fountains in coordination with the lights and music. Control systems are also used in the drilling and refining of oil and natural gas. They are used in the distribution of water and electricity by utility companies, and in the collection of wastewater and sewage. Virtually every sector of the economy employs control systems at all levels. The term "supervisory control and data acquisition" (SCADA), however, is generally accepted to mean the systems that control the distribution of critical infrastructure public utilities (water, sewer, electricity, and oil and gas). SCADA systems are still to come into widespread infrastructural use in India.

In this country they are being used primarily for automation in industrial production, and to some extent for specialized process control. Ranbaxy Labs and Voltas are two of the companies in India using SCADA systems for process control. However, they are increasingly common in the US, UK, Australia to name a few countries, where they are used in the control of infrastructural systems such as power, water and waste management, traffic control, etc. The economy and infrastructure of these countries is increasingly dependant on SCADA systems.

8.7.1 Architecture

SCADA systems are primarily control systems. A typical control system consists of one or more remote terminal units (RTU) connected to a variety of sensors and actuators, and relaying information to a master station. Figure 8.23 illustrates this generic, three tiered approach to control system design. Figure 8.24 shows a typical RTU.

Sensors and Actuators

Sensors perform measurement, and actuators perform control. Sensors get the data (supervision and data acquisition) and actuators perform actions dependent on this data (control). The processing and determination of what action to take, is done by the master control system (i.e., SCADA).

Remote Terminal Units (RTUs)

Advances in CPUs and the programming capabilities of RTUs have allowed for more sophisticated monitoring and control. Applications that had previously been programmed at the central master station can now be programmed at the RTU.

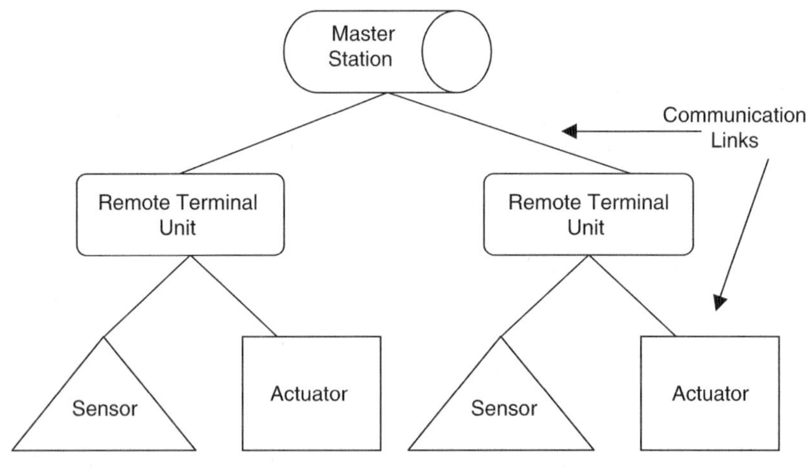

Fig. 8.23. SCADA architecture

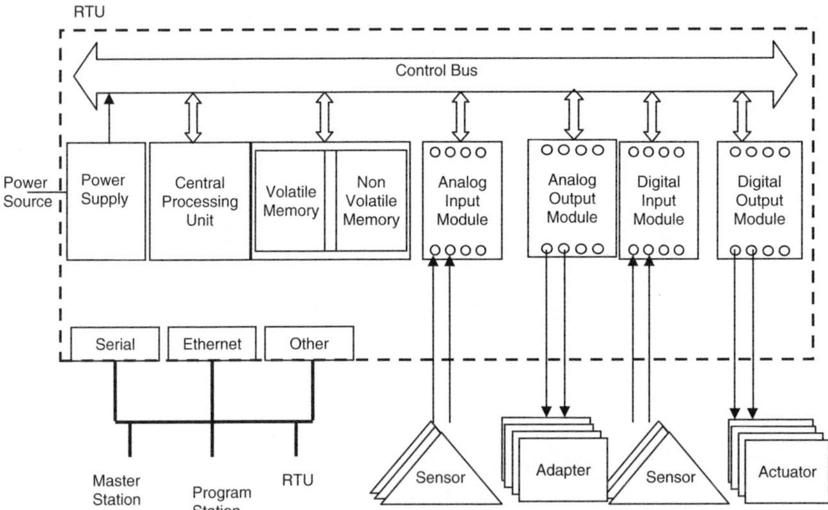

Fig. 8.24. SCADA registers

Programmable Logic Controllers

These modern RTUs typically use a ladder-logic approach to programming due to its similarity to standard electrical circuits. A RTU that employs this ladder logic programming is called a Programmable Logic Controller (PLC). PLCs are quickly becoming the standard in control systems.

Analog Input and Output Modules

The configuration of sensors and actuators determines the quantity and type of inputs and outputs on a PLC or RTU; depending on the model and manufacturer, modules can be designed solely for input, output, digital, analog, or any combination. An analog input module has a number of interfaces. Typical analog input modules have 8, 16, or 32 inputs. Analog output modules take digital values from the CPU and convert them to analog representations, which are then sent to the actuators. An output module usually has 8, 16, or 32 outputs, and typically offers 8 or 12 bits of resolution.

Digital Input and Output Modules

Digital input modules typically are used to indicate status and alarm signals. A specialized digital input module is used for counting pulses of voltage or current, rather than for strictly indicating "open" or "closed." This functionality, however, can also be implemented using standard input modules and functions found in the ladder-logic programming language of the PLC.

Master Station

Master stations have two main functions:

- Periodically obtain data from RTUs/PLCs (and other master or submaster stations)
- Control remote devices through the operator station Master stations consist of one or more personal computers (PC), which, although they can function in a multipurpose mode (email, word processing, etc.), are configured to be dedicated to master station duties. These duties include trending, alarm handling, logging and archiving, report generation, and facilitation of automation. These duties may be distributed across multiple PCs, either standalone or networked.

Communications Interfaces

Modern RTUs and PLCs (Programmable Logic Controllers) offer a wide variety of communications means, either built in directly or through a module. The following list represents a variety of transmission methods supported:

- RS-232/RS-442/RS-485
- Dialup telephone lines
- Dedicated telephone lines
- Microwave
- Satellite
- X.25
- Ethernet
- 802.11a/b/g
- Radio (VHF, UHF, etc.)

Design and Implementation Protocols

In SCADA Systems, the three major categories of protocols involve the specifications for design and manufacture of sensors and actuators, specifications for RTUs, and the specifications for communications between components of a control system. These can be segregated into three levels for a functional representation as shown in Fig. 8.25.

These can further be split into nine operational and management components as shown in Table 8.2.

Application Criticality

Since SCADA systems typically support vital distribution systems, an assessment of the criticality of that system will provide direction on its design and

structures are geared to that application. It is the favored standard in the United States for electrical power grid SCADA systems, but is not as popular in Europe.

DNP3

The second protocol specifically designed for SCADA communications is the Distributed Network protocol Version 3 (DNP3). Also created for the electrical industry, it has been adapted by other industry sectors and is the leading protocol employed in Europe for most SCADA applications.

HDLC

Several other SCADA standards exist, primarily High Level Data Link Control (HDLC) and Modbus. HDLC, defined by ISO for point-to-point and multipoint links, is also known as Synchronous Data Link Control (SDLC) and Advanced Data Communication Control Procedure (ADCCP). It is a bit-based protocol, the precursor to Ethernet, and is rapidly being replaced by DNP3, Industrial Ethernet2, and TCP/IP.

Modbus

Modbus is a relatively slow protocol that does not define interfaces, thus allowing users to choose between EIA-232, EIA-422, EIA-485, or 20 mA current loop. While slow, it is widely accepted and has become a *de-facto* standard-a recent survey indicated that 40% of industrial communication applications use Modbus.

Profibus

Profibus is a German standard that defines three types: Field Message Specification (FMS) for use in general data acquisition systems, Decentralized Peripherals (DP) for use when fast communication is required, and Process Automation (PA) for use when highly reliable and safe communication is required.

Foundation Fieldbus

Foundation Fieldbus is an extension to the 4–20mA standard to take advantage of digital technologies.

UCA

The Utility Communications Architecture (UCA) is a new initiative from the Electric Power Research Institute (EPRI) designed for the electrical industry.

It is more than just a protocol definition; it is a comprehensive set of standards designed to allow "plug and play" integration into systems, allowing manufacturers to design off-the-shelf compliant devices. IEEE assumed the UCA standards process in 1999 and has developed extensions for the water industry. Other industries are also examining UCA for suitability.

Hardware and Software (Operational Functions)

When discussing operational functions of an SCADA system, the hardware and software to be used must also be evaluated. In the context of standard SCADA systems, reliability, stability and safety are primary concerns. Adding a security perspective introduces the concept of assurance, or ensuring that the hardware and software have minimal exploitable flaws.

Users and Personnel Actions (Management Functions)

When discussing management functions, the users of the system and its components must also be evaluated. In addition, the automated decision-making that is programmed into the system must be evaluated in the same terms.

8.7.2 Security Concerns

SCADA systems are not designed with security in mind; rather the priority of developers has been reliability, availability, and speed. This does not mean they cannot be secured, however. The limitations of a system can be addresses by understanding the system's features, functions and capabilities. No inherent security is provided in these systems, since security is not a direct concern when the efficiency of the system is under consideration. This situation is acceptable as long as the systems are isolated from the outside world. However, in recent times, more and more of these systems are being exposed to open access, in order to promote intersystem communication and interaction.

Two recent trends' raising concerns are:

(a) Definition of standard interfaces and communication protocols in support of cross-vendor compatibility and modularity.
(b) Connection of nodes in an SCADA system to open networks such as the Internet.

While these phenomena have definitely brought about an increase in the efficiency of these systems, they have also caused them to inherit all the problems of common, networked information systems. The security of information, both against corruption and misuse, is now an increasing concern for these systems. This concern for security becomes even more magnified when these systems are deployed in key positions, where they are heavily depended upon for critical operations.

8.7.3 Analysis of the Vulnerabilities of SCADA Systems

To assist in determining optimal mitigation strategies, the vulnerabilities are grouped in the categories of

- Data
- Security Administration
- Architecture
- Network
- Platforms

Any given control system will usually exhibit a subset of these vulnerabilities, but may also have some unique additional problems.

Data

Sensitivity levels for control system data are usually not established. An essential characteristic of secure information systems is the identification and classification of data into categories of similar sensitivity. Absence of these fundamental distinctions makes it impractical and fruitless to identify where security precautions are appropriate (for example, which communication links to secure, databases requiring protection, etc).

Security Administration

Security administration is notoriously lax in the case of control systems, usually the result of poor legacy environment. The need to manage and administer security is usually overlooked, resulting in informal practices and inefficient management. As experience has proved, any system, which does not have well founded management and administrative policies, will eventually show vulnerabilities. This is the case with control systems as well.

Architecture

Architecturally, many control systems include centralized data storage and control. This results in the creation of a single-point-of failure. Occasionally, physical damage to infrastructure assets may be possible through permissible operation of control equipment. An effective control hierarchy would preclude this possibility. In addition to the above, many implementations of control systems have integrated in-house emergency services such as fire alarms, etc., into the control system itself. In view of the pathetic condition of the security of these systems, thoughtless addition of these services into the system adds to the complexity and further increases the vulnerability.

Networks

Vulnerabilities in control system networks depend on the type of system. Legacy implementations rely on proprietary protocols and low-bandwidth data channels. While there are fewer opportunities for disruptive behavior compared to newer networks, which closely resemble modern TCP/IP systems, problems are inherent because of the technology's age. Security is lamentable. This is due to the fact that these systems were designed in a time when error checking and integrity validation had not gained their present importance. In addition to this, accounting and logging are usually nonexistent, making it impossible to find the basis and reason for vulnerabilities. Configuration passwords are often simple and may be limited in effectiveness by the device itself. Wireless links are not secured. Networking equipment in these systems, particularly when physical access is presumed, is acutely vulnerable to attack.

Systems with contemporary technologies like Ethernet, routers, and firewalls have vulnerabilities that are more publicized than the vulnerabilities in the older networks. Little or no network restriction is implemented within the perimeter of the network, allowing 'telnet hopping' from innocuous network devices to sensitive utility equipment.

Two other factors contribute significantly to the vulnerability of control systems:

(1) The blind trust in the capability of PCS links to faithfully transmit data. The geographically sparse PCS network generally forces links of considerable span. These needs are filled by either cabled or wireless connections, which may be exclusively used by the PCS or shared. Shared links are more economically sensible, but many times the PCS systems at either end of the link are not adequately shielded from other entities using it. Furthermore, unsecured information on wireless and shared links is susceptible to eavesdropping or manipulation, and even long or unprotected unshared cable links may be vulnerable to a significant degree. E.g., if the master station and RTU have no security mechanism between them, an attacker could direct a malicious signal via the master station to the RTU and vice versa.
Recently a California based security firm, involved in vulnerability assessment of critical infrastructure systems, proved just this sort of vulnerability by accessing a remote substation of a large southwester United States utility company. They did this using a directional antenna and a wireless laptop from a vehicle parked in the vicinity of the substation 6.

(2) The connections between the PCS and external networks. An external network is any network that is not part of the PCS. Examples include interfaces to an administrative (non-automation) network or connections to other PCS systems for information transfer or mutual control.

> Often, interfaces to external systems assume that the outside network can be trusted, which leaves PCS security dependent on one or more organizations. This includes backdoor network access for strategic partners or IT consultants, which are not secured by adequate firewall measures, command logging, or privilege control.

With the world moving towards outsourcing, and strategic partnerships, security implementation suffers due to the absence of a common standard. Designers frequently omit to secure the backdoors left by them for easy tuning of a system, resulting in disaster at a later stage. Dial-up modem access is unencrypted, with a general lack of authentication practices. The data transfer that takes place over telephone lines, or wireless networks is usually either unencrypted, or encrypted with a weak algorithm, which does not take much effort to crack. The primary reason for this is a requirement to save time/resources on encryption. However, the result is that the signals can be easily analyzed and if so wished, modified by an attacker.

8.7.4 Security Recommendations

Computer security in the present context deals mainly with information stored on systems, or information in transit. This is not adequate for the protection of infrastructural systems such as SCADA networks Traditional information assurance does not adequately protect against cyber attacks on SCADA control systems in which the counter measures used may compromise the safety or operability of the SCADA control systems. The United States Federal Government has begun to lead in the protection of critical infrastructure but it will rest in the hands of the commercial industries to develop and implement an enterprise assurance policy to improve the SCADA control systems security posture. Several steps can be taken to assist in the improvement of the implementation and management of policies and procedures. Some of these are:

Implement Common Criteria Evaluations on SCADA Control Systems.

As standards and technology continued to change, the United States and Europe began working on standards for a common evaluation criteria for information security. The various evaluation criteria projects begun by the United States and Europe merged into a single International Common Criteria project ISO/IEC 15408 with the intent to standardize methods for evaluating information systems security. SCADA control systems products should be included and evaluated based on the Common Criteria standards to ensure the implementation of SCADA products does not compromise the security or safety of the critical infrastructure.

Adopt "Best Practices" and Procedures.

Most of the vulnerabilities of computer systems are well known and documented. Adopting "best practices" (i.e., implementing secured network devices and operating systems, and patch management) and procedures (i.e., backups) will allow administrators to protect systems not only from the cyber threats, but also normal system failures.

Isolate and Harden SCADA Networks.

Isolation of SCADA networks to a closed-loop network with limited and highly restrictive access from physical and electronic outside sources would help in mitigating the threat to them. If the connection of an SCADA network to the Internet or another open network is essential, appropriate buffers and checks should be placed between the layers. Segmented network topologies could increase the level of restrictive access and survivability. Utilization of authentication mechanisms such as passwords, tokens and biometrics could guard against unauthorized access. Enabling strong encryption for all data communications would further minimize the risk of a security breach.

Vulnerability and threat assessments should be performed regularly on current and newly implemented systems. Risk assessments should be conducted on each interconnection between the SCADA and corporate enterprise network. All unnecessary networks should be disconnected, especially if an open pathway to the Internet is formed. Unnecessary services that are not required to support the operation of the SCADA control systems should be removed or disabled.

Firewalls and intrusion detection systems should be implemented, to not only prevent entries but also monitor unintentional security breaches on the SCADA and corporate enterprise network. Detailed network knowledge should be restricted. Communicating IP addresses and DNS names is unnecessary and can be costly if in the wrong hands. Implementing single-sign-on procedures (through an administrative management portal) will pass authorized users to the command prompt of a device without knowledge of the IP address or password.

Limiting access privileges on a device and port level is advisable. There is no reason for PBX maintenance staff to access a data center database or for IT consultants to access all network devices. Implementing a virtual private network (VPN) for administrative channel access and partitioning dependent upon privileges provides additional levels of security.

In summation, it is easy to observe that SCADA technology holds a lot of promise for the future. The economic and performance advantages of this type of system are definitely attractive. However, since the vulnerabilities of current implementations are in proportion to the advantages, it is essential that measures be taken to mitigate the risk to current systems and to ensure that future systems are designed with sound policies and design.

Summary

The purpose of pulse codes along with its characteristics and types namely RZ, RB, NRZ, NRZI, and PE techniques were discussed. The various analog and digital modulation techniques like AM, FM, PM along with applications were also dealt in this chapter. The advanced version of RF network analyzer and trends in wireless communication has been discussed in detail. The essential and the various factors that govern the interfacing process of distributed automation and SCADA systems were also elaborated.

Review Questions

1. Write notes on pulse codes with neat sketches.
2. What are the primary requirements for modulation of signals?
3. List out the objectives, which are met by modulation.
4. Explain in detail about AM with sketches
5. Discuss about FM scheme with derivations.
6. Explain the PM method with neat diagrams and derivations.
7. Write the need for digital modulation.
8. Describe the types of digital modulation in detail.
9. Discuss the following: ASK, FSK, and PSK.
10. Write notes on MSK
11. Explain about QAM.
12. List out the applications of digital modulation.
13. Discuss about wireless communication and LANs.
14. Write the trends of wireless communication.
15. With neat sketches explain in detail about RF network analyser.
16. Explain in detail about the distributed automation systems architecture.
17. Write the functions of SCADA.
18. Explain in detail about the SCADA architecture and protocols.
19. What are pulse codes?
20. Compare and contrast analog and digital modulation.
21. What are the major essential elements in wireless communication?
22. What is meant by RF network analyzer what are its purposes?
23. What factors influence the success of distributed network
24. Explain SCADA in power systems with suitable example
25. What are the components of SCADA? Explain in detail.
26. Write the advantages and demerits of SCADA.

9

Current Trends in Instrumentation

Learning Objectives. On completion of this chapter the reader will have a knowledge on:

- Current Trends in Instrumentation
- Fiber Optic Instrumentation
- Fiber Optic Sensors such as Pressure Sensors, Voltage Sensor, Liquid Level Monitoring, Temperature Sensors, Stress Sensor
- Fiber Optic Gyroscope – Polarization Maintaining
- Advantages of Fiber Optics
- LASER Instrumentation
- Measurement of Velocity, Distance and Length using LASER
- LASER Heating, Welding, Melting and Trimming
- LASER Trimming and Melting
- Smart Intelligent Transducers
- Smart Transmitter with HART Communicator
- Computer Aided Software Engineering (CASE)

9.1 Introduction

Fiber-optic strands provide a significant advantage over standard copper transmission lines. The potential information-carrying capacity of a single fiber strand is much greater than a single copper wire. The carrier frequencies used in light waveguides are far greater than those deployed in copper lines; light is composed of different wavelengths each of which can carry a separate data channel. This section describes the fiber-optic sensors used in applications such as pressure measurement, voltage measurement, liquid level monitoring, temperature measurement, and stress analysis. The concepts of LASER instrumentation are also discussed in this chapter, which focuses on LASER trimming, welding and heating. The later sections of this chapter deal with the most recent developments in instrumentation such as Smart Transmitters and Use of CASE tools in Instrumentation.

9.2 Fiber-Optic Instrumentation

Optical fiber is the workhorse of any optical communication system. It is basically the conduit for the data and ends up linking all of the other components together. The light wave is transmitted by total internal reflection. A typical fiber consists of the following:

- The core, the actual conduit of the light
- The cladding, a glass material which, by doping, has achieved a certain index of refraction to cause total internal reflection
- Other layers to provide strength and protection from harsh conditions

The light travels down the fiber with minimal attenuation because of the difference in the indices of refraction between the core and the cladding, combined with the angle of reflection. Optical fiber is manufactured as either single mode or multimode. Multimode fiber (50 μm to a few mm) has a larger core diameter than a single mode fiber (4–9 μm). These measurements must be accurate because the diameter and concentricity of the core and cladding would affect such things as the modes of light that can travel in the optical fiber, the numerical aperture, and the coupling of light in and out of the fiber.

9.2.1 Fiber-Optic Sensors

Fiber-optic sensors are being used in a wide variety of applications. Sagnac interferometers, which can determine movement by measuring the shift in interference fringes from two counter propagating coherent beams in a ring, support fiber gyros on aircraft, missiles, rockets, and robots. Michelson interferometers, in which interference fringes indicate the length relationship between two legs of the interferometer, support strain and acoustics measurements for civil structures and underwater applications. Another type of fiber sensor, fiber grating sensors, is emerging as a potentially low-cost solution to a wide variety of point-sensor measurements such as axial strain and temperature, transverse strain, shear strain, moisture, pressure, acoustics, vibration, and chemical content. Applications areas for these sensors include aerospace and structural monitoring.

Over the past 20 years, two major product revolutions have taken place due to the growth of the optoelectronics and fiber-optic communications industries. The optoelectronics industry has brought about such products as compact disc players, laser printers, bar code scanners and laser pointers. The fiber-optic communication industry has literally revolutionized the telecommunication industry by providing higher performance, more reliable telecommunication links with ever decreasing bandwidth cost. This revolution is bringing about the benefits of high volume production to component users and a true information super highway built of glass.

In parallel with these developments fiber-optic sensor technology has been a major user of technology associated with the optoelectronic and fiber-optic communication industry. Many of the components associated with these industries were often developed for fiber-optic sensor applications. Fiber-optic sensor technology in turn has often been driven by the development and subsequent mass production of components to support these industries. As component prices have fallen and quality improvements have been made, the ability of fiber-optic sensors to displace traditional sensors for rotation, acceleration, electric and magnetic field measurement, temperature, pressure, acoustics, vibration, linear and angular position, strain, humidity, viscosity, chemical measurements, and a host of other sensor applications, has been enhanced.

In the early days of fiber-optic sensor technology most commercially successful fiber-optic sensors were squarely targeted at markets where existing sensor technology was marginal or in many cases nonexistent. The inherent advantages of fiber-optic sensors which include their ability to be lightweight, of very small size, passive, low power, resistant to electromagnetic interference (EMI), high sensitivity, wide bandwidth, and environmental ruggedness were heavily used to offset their major disadvantages of high cost and unfamiliarity to the end user.

The situation is changing over the recent years. Laser diodes that cost $3,000 in 1979 with lifetimes measured in hours now sell for a few dollars in small quantities, have reliability of tens of thousands of hours and are used widely in compact disc players, laser printers, laser pointers, and bar code readers. Single mode optical fiber that cost $20/m in 1979 now costs less than $0.10/m with vastly improved optical and mechanical properties. Integrated optical devices that were not available in usable form at that time are now commonly used to support production models of fiber-optic gyros. Also, they could drop dramatically in price in the future while offering ever more sophisticated optical circuits. As these trends continue, the opportunities for fiber-optic sensor designers to produce competitive products will increase and the technology can be expected to assume an ever more prominent position in the sensor marketplace. In the following sections the basic types of fiber-optic sensors that are being developed will be briefly reviewed followed by a discussion on these sensors and its applications.

Intensity-Based Fiber-Optic Sensors

Fiber-optic sensors are often loosely grouped into two basic classes referred to as extrinsic or hybrid fiber-optic sensors and intrinsic or all fiber sensors. Figure 9.1 illustrates the case of an extrinsic or hybrid fiber-optic sensor. These sensors consist of optical fibers that lead up to and out of a "black box" that modulates the light beam passing through it in response to an environmental effect.

In this case an optical fiber leads up to a "black box" which impresses information onto the light beam in response to an environmental effect. The information could be impressed in terms of intensity, phase, frequency, polarization, spectral content, or other methods. An optical fiber then carries

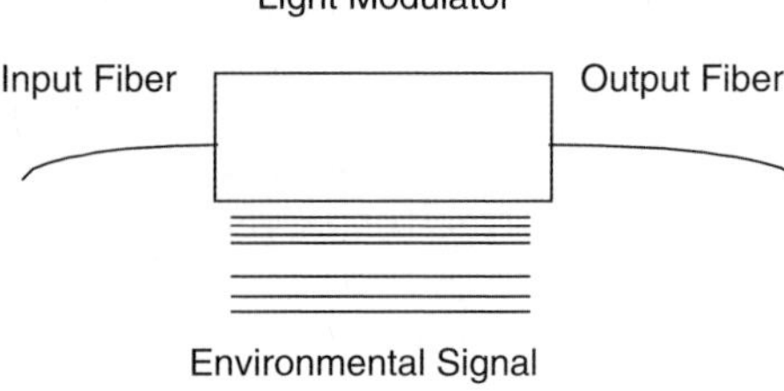

Fig. 9.1. Extrinsic fiber-optic sensors

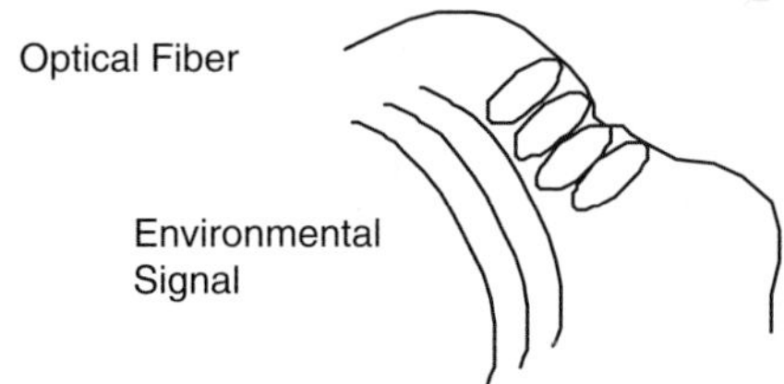

Fig. 9.2. Intrinsic fiber-optic sensors

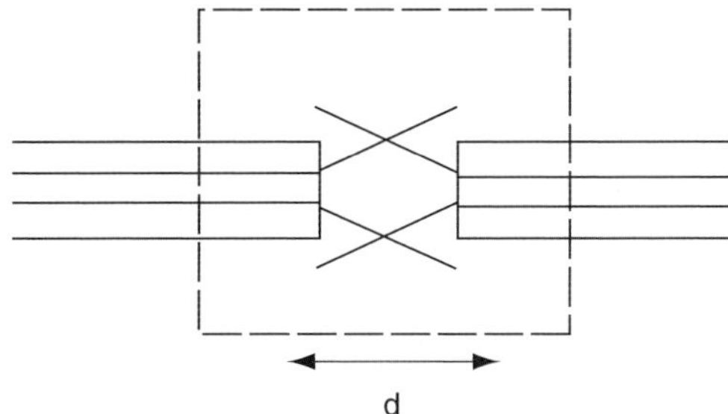

Fig. 9.3. Closure and vibration fiber-optic sensor

the light with the environmentally impressed information back to an optical and/or electronic processor.

In some cases the input optical fiber also acts as the output fiber. The intrinsic or all fiber sensor shown in Fig. 9.2 uses an optical fiber to carry the light beam and the environmental effect impresses information onto the light beam while it is in the fiber. These sensors rely on the light beam propagating through the optical fiber being modulated by the environmental effect either directly or through environmentally induced optical path length changes in the fiber itself. Each of these classes of fibers in turn has many subclasses with, in some cases sub-subclasses that consist of large numbers of fiber sensors.

In some respects the simplest type of fiber-optic sensor is the hybrid type that is based on intensity modulation. Figure 9.3 shows a simple closure or vibration sensor that consists of two optical fibers that are held in close proximity to each other. Light is injected into one of the optical fibers and

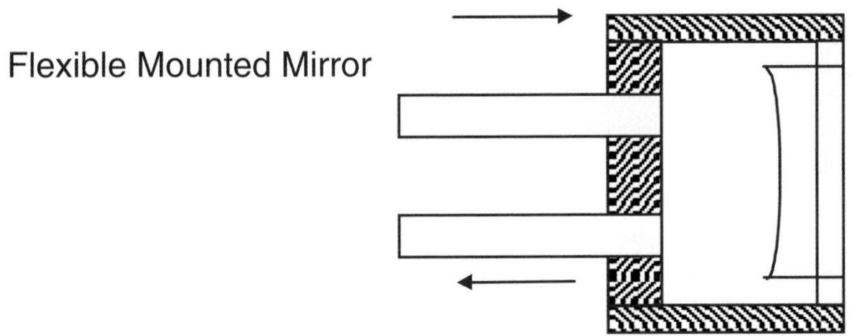

Fig. 9.4. Numerical aperture fiber sensor based on a flexible mirror can be used to measure small vibrations and displacements

when it exits the light expands into a cone of light whose angle depends on the difference between the index of refraction of the core and cladding of the optical fiber. The amount of light captured by the second optical fiber depends on its acceptance angle and the distance d between the optical fibers.

When the distance d is modulated, it in turn results in an intensity modulation of the light captured. A variation on this type of sensor is shown in Fig. 9.4. Here a mirror is used that is flexibly mounted to respond to an external effect such as pressure. As the mirror position shifts the effective separation between the optical fibers shift with a resultant intensity modulation.

These types of sensors are useful for such applications as door closures where a reflective strip, in combination with an optical fiber acting to input and catch the output reflected light, can be used. By arranging two optical fibers in line, a simple translation sensor can be configured. The output from the two detectors can be proportioned to determine the translational position of the input fiber.

9.2.2 Fiber-Optic Pressure Sensors

To fulfill the objective of providing robust and reliable fiber-optic pressure sensors capable of operating in harsh environments, this section presents the detailed research work on the design, modeling, implementation, analysis, and performance evaluation of the novel fiber-optic Self-Calibrated Interferometric/Intensity-Based (SCIIB) pressure sensor system. By self-referencing its two channels outputs, the developed SCIIB technology can fully compensate for the fluctuation of source power and the variations of fiber losses. Based on the SCIIB principle, both multimode and single-mode fiber-based SCIIB sensor systems were designed and successfully implemented.

To achieve all the potential advantages of the SCIIB technology, the novel controlled thermal bonding method was proposed, designed, and developed to fabricate high performance fiber-optic Fabry–Perot sensor probes with excellent mechanical strength and temperature stability. Mathematical models of the sensor in response to the pressure and temperature are studied to provide a guideline for optimal design of the sensor probe. The solid and detailed noise analysis is also presented to provide a better understanding of the performance limitation of the SCIIB system. Based on the system noise analysis results, optimization measures are proposed to improve the system performance.

Extensive experiments have also been conducted to systematically evaluate the performance of the instrumentation systems and the sensor probes. The major test results give us the confidence to believe that the development of the fiber-optic SCIIB pressure sensor system provides a reliable pressure measurement tool capable of operating in high pressure, high temperature harsh environments.

High accuracy fiber-optic pressure transducers and readouts have been developed by applying optical technology to the resonator-based Digiquartz Sensors. The Model 790 Pressure System consists of a Fiber-Optic Digiquartz Sensor connected solely by a pair of optical fibers to a local Interface/Display unit. The fiber links between the Sensor and the Interface/Display unit can be 500 m long (1,640 ft.). The down-link fiber provides an optical driving signal that excites a pair of quartz resonators inside the transducer. One resonator is load sensitive and vibrates at a frequency determined by the applied pressure.

The second resonator vibrates at a frequency that varies with the internal temperature of the transducer. Optical frequency signals from the resonators are transmitted back to the interface unit through the up-link fiber. The Model 790 measures the periods of the two up-link optical signals from the transducer, compensates for thermal effects, and calculates and displays fully temperature-corrected, linearized digital pressures. The Model 790 Interface/Display unit also provides a two-way RS-232 bus for configuration, control, and data acquisition.

The user may select or define engineering units, sample rate, baud rate, resolution, and single, continuous, or synchronized sampling commands. Over 30 absolute, gauge, and differential pressure transducers are available from 3 psid to 40,000 psia (276 MPa). Performance is comparable to the primary standards with typical resolution of 0.0001% and accuracy of 0.01% of full scale. High precision pressure measurements are possible even under difficult environmental conditions. The total optical isolation provides the important benefits of intrinsic safety, noise immunity, and long line-driving capability.

Application areas include aerospace, power plants, process control, energy exploration, pipeline monitoring, and custody transfer. All Digiquartz products come with a limited five year warranty with the first two years covered at 100%.

9.2.3 Fiber-Optic Voltage Sensor

Fiber-optic voltage and electric field sensors have attracted much interest in recent years as an alternative to traditional measurements in electric power systems. These sensors are based on the Pockels effect and do not contain any conductive elements. As a result, the sensing head of the sensor introduces slight perturbations in the measured electric field. The first fiber-optic voltage sensor was demonstrated many years ago. There are two obstacles that prevent the application of these sensors in industry. First, their sensitivity is somewhat low. The typical sensitivity of voltage sensors with a longitudinal modulation is between 0.039% per 1 Vrms and 0.051% per 1 Vrms. Second, the temperature stability of the entire sensor does not meet industrial requirements.

In general, attention is paid to the temperature stability of crystal parameters. There are other optical elements inside the sensor, such as quarter-wave plates, that are sensitive to temperature changes. The changes in birefringence with temperature in this element cause the strong variation of the sensor's sensitivity. It proposes a new configuration of a fiber-optic voltage sensor, based on the Bi12TiO20 crystal. Instead of a quarter-wave plate it used a glass back reflecting prism as a phase-retarding element. By using this prism, it demonstrated the good temperature stability of 61.5% from 220°C to 60°C and excellent sensitivity without the use of an additional temperature control channel.

Basic Principle of the Sensor

The optical scheme of the sensor is shown in Fig. 9.5. The light from an 850-nm light-emitting diode is launched into the sensor through a multimode fiber approximately 300–320 mm and Selfoc lens. The polarization state of the light after the addition of the polarizing prism is 45° to the axes of the electric field-induced birefringence of the crystal. When it applied an electric field, the original birefringence of the crystal became uniaxial because of the Pockels effect. The 5, 35, 32 mm crystal was aligned in such a way that the electric field and the optical path were normal to the crystal plane. The polarized light returned to the crystal after two reflections from the surface of the back

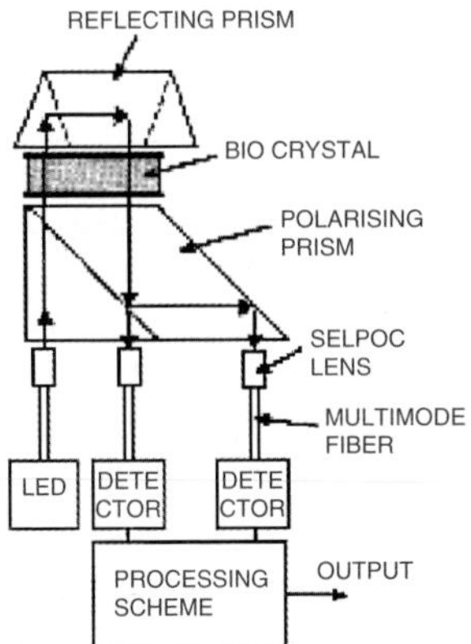

Fig. 9.5. Schematic of voltage sensor

reflecting prism. After the second pass through the crystal the beam meets the polarizing prism at the point where two orthogonal states of polarization separate into two channels.

The analyzer converts the polarization modulation into intensity modulation in two channels. Both intensity-modulated beams were launched into the fibers by use of Selfoc lenses. The vertical axes of the reflecting prism and the polarizing prism are aligned at the angle of 45° 6q, where q is the angle of polarization rotation that is due to the optical activity of the crystal. The plus sign or minus sign changes according to the crystal modification with respect to left-hand side or right-hand side rotation. On the other hand, this prism introduces a π/2 phase shift between the fast and the slow waves in the crystal, which permits one to use the best sensitivity of the sensor.

The angle of incidence and the refractive index of a prism can be used to determine a phase shift between S and P polarizations based on total internal reflection. If one selects the correct angle of incidence and refractive index, it is possible to obtain a desirable phase shift between S and P polarizations. To calculate the refractive index of the back reflecting prism. It used the phase shift D between S and P polarizations on the glass–air interface of the back reflecting prism under total internal reflection:

$$\tan(\Delta/2) = \frac{(n_1/n_0)[(n_1/n_0)^2 \sin^2\varphi - 1]^{1/2}}{\sin\varphi \tan\varphi}$$

where w is the angle of incidence, $n1$ and $n0$ are the refractive indices of the air and glass prisms, respectively. By using Eq. (1) and $\varphi = \pi/4$, It calculated $n1 = 1$, the refractive index of the back reflecting prism, as

$$n_0 = \left[\frac{2}{1 - \tan^2(\Delta/2)}\right]^{1/2}$$

The phase shift D should equal π/4 to obtain the circular polarization of the beam after two consecutive reflections from surfaces of the prism. From this condition the refractive index of the prism $n0$ was calculated to be 1.553773974.

9.2.4 Fiber-Optic Liquid Level Monitoring

Continuous monitoring of liquid level in oil tanks is important in the petroleum and chemical industries where the measured data is useful in process control loops. Numerous types of sensors with various principles have been used for liquid level measurement of oil tanks. Although most of them work well in practice, these sensors are limited by high cost or complicated configuration. Additionally, the atmosphere near the oil tanks is flammable and thus any electric spark is likely to cause severe disaster. Optical sensors have a history of more than 30 years and have been successfully used in certain areas such as underwater acoustic sensing, strain monitoring, chemical substance detecting, and so on.

These kinds of sensors have many advantages including intrinsical safety in explosive or fire-hazardous environments, electrical insulation, EMI resistance, resistance to corrosive fluids, small size and weight, fast response and potential low cost. They are very suited to measure physical parameters (pressure, temperature, density, etc.) in fields where safety is strictly required.

Principle of Operation

Liquid level is obtained through hydrostatic pressure measurement with the fiber-optical sensor. It is based on the principle that the hydrostatic pressure difference between the top and bottom of a column of liquid is related to the density of the liquid and the height of the column, i.e., p, g, H, where p, H, and g are pressure, liquid density, column height and the gravity acceleration, respectively. Using pressure measurement to determine liquid level is particularly useful for applications with foaming or bubbling liquid, where other level measurement technologies have difficulties. The sensor head is actually a pressure sensor based on the small displacement of a diaphragm and microbend effect of optical fibers.

It mainly consists of a sensing diaphragm with a hard center, a microbend modulator (a pair of tooth plates), sensing and reference fibers, adjusting bolts and stainless steel housing. The sensor is mounted at the bottom of an open liquid vessel and the sensing system configuration. Liquid pressure is converted into small displacement of movable tooth plate by the sensing diaphragm. The structure of fiber-optic sensor is shown in Fig. 9.6a. Then the sensing fiber between two tooth plates is bent and the through light intensity is reduced.

By detecting the output light power, the exerted pressure is obtained and the liquid level can be calculated with the specified density. The sensing diaphragm was directly fabricated instead of welded onto the steel housing

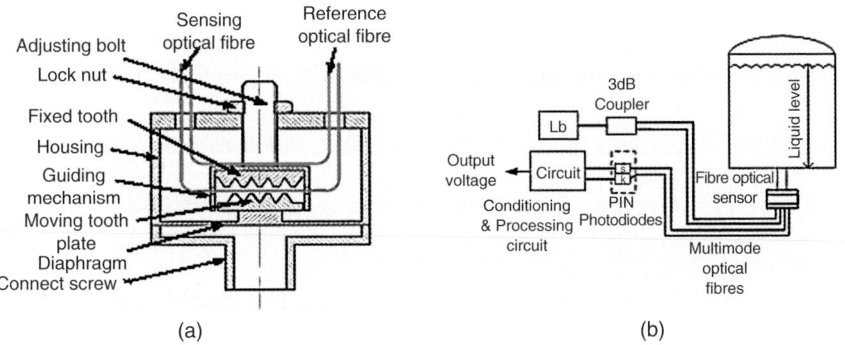

Fig. 9.6. (**a**) Structure of sensor (**b**) system configuration

to decrease additional stress caused by any temperature effects. Standard multimode fibers were used in order to lower the cost. Furthermore, two parallel optical fibers, one for sensing and the other for reference, were used to eliminate influence due to fluctuation of light source and other disturbance along the optical path as shown in Fig. 9.6b.

Structure Design and Theoretical Analysis

Microbend modulator Sensors based on microbend loss in optical fibers were first proposed and demonstrated in 1980. In addition to some general advantages mentioned microbend-based sensors have been demonstrated to have their own unique set of advantages including low parts count and low cost, easy mechanical assembly and fail-safe characteristics. The structure of Microbend modulator sensor is shown in Fig. 9.7a. According to optical wave theory, the attenuation coefficient can be deduced as follows q where K is a constant, and are the magnitude of microbend, the length of sensing optical fiber and the frequency of tooth spacing, respectively. It is the propagation constant difference between two adjacent propagated modes. It is dominated by measurands, by the fiber length just between tooth plates, and by the period of the tooth spacing as shown in Fig. 9.7b.

FEM Simulation

The analytical equation of diaphragm displacement is derived under many assumptions. Therefore, the deviation between design results and actual performance of sensor is inevitable. Fortunately, the finite element method (FEM) provides a powerful tool to precisely simulate the performance of designed sensor. The diaphragm was simulated using finite element analysis software ANSYS to accurately predict the displacement versus pressure and the stress distribution.

Generation, Detection, and Processing Unit

A packaged LD module used for standard communication is used here as the light source. Output light from the LD with wavelength of 1,310 nm is

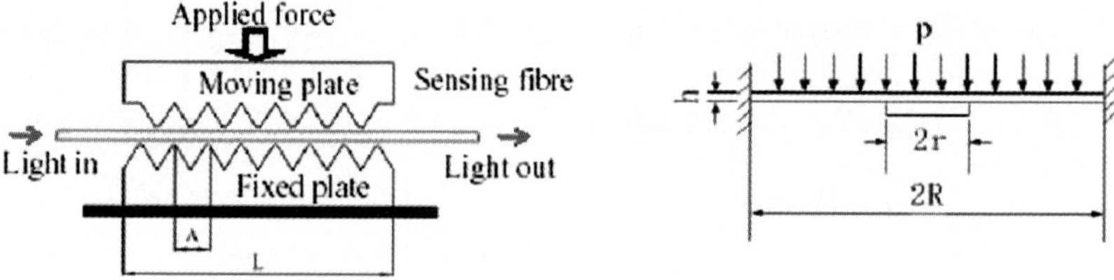

Fig. 9.7. (**a**) Structure of microbend modulator; (**b**) diaphragm with hard center

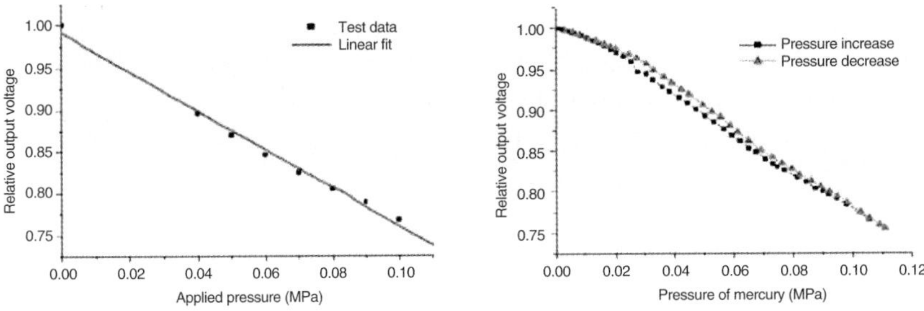

Fig. 9.8. Test result with: (**a**) piston gauge and (**b**) simulation setup

split into two channels by a 3 dB coupler and passes into two optical fibers. The light output is converted to electrical current by detecting photodiodes. The conditioning and processing circuits, which includes logarithmic amplifier, filter and so on, convert the current into output voltage, which indicates the variation of pressure.

Experimental Analysis

The manufactured fiber-optical sensor was first tested with a deadweight piston gauge. The curve of relative output voltage versus pressure is shown in Fig. 9.8. In addition, a simulating test setup was established. A vertically placed glass tube was connected to the fiber-optical sensor. Mercury was poured into the tube carefully to apply pressure to the sensor and the level was read from the scale etched along the tube. The decrease of pressure is implemented by discharging the mercury through the valve at the bottom of tube.

The pressure is calculated according to the height of mercury column (1 mmHg = 133.322 Pa). It can be seen that the results tested with two methods coincide very well. From results obtained by piston gauge, the sensitivity and nonlinearity can be calculated as $14.5\,\mu\text{V}\,\text{Pa}^{-1}$ and 0.08, respectively. To enhance the performance further, longer sensing fibers and an additional compensation circuit may be used.

9.2.5 Optical Fiber Temperature Sensors

There are many temperature dependent optical properties on which optical fiber temperature sensors are based. The required application of a temperature sensor will often determine one type being chosen over another. For example, if a relatively fast temperature measurement at a single location is desired, a point temperature sensor is suitable. For temperature monitoring of large or long structures such as tunnels or pipelines, a distributed temperature sensor is used, although the time of processing is normally slower. Other attributes of a temperature sensor, such as its temperature range, accuracy and cost, will also determine its applicability. A short review on well-regarded absorption and fluorescence-based techniques is presented in this section.

Absorption-Based Temperature Sensors

Some temperature sensing schemes are based on the temperature dependent optical absorption of a material. Two such sensing materials used are semiconductors and rare earth-doped fiber. An early extrinsic optical fiber sensor used a semiconductor material as the sensing element. A section of gallium arsenide was attached to the end of an optical fiber. Near ambient temperatures (300 K) the energy band gap of such materials is almost inversely proportional to temperature. That is, the wavelength relevant to the fundamental absorption edge moves toward longer wavelengths with increasing temperature.

The optical temperature measurement system used two light emitting diodes of different wavelengths ($\lambda 1$ and $\lambda 2$). The sensor's absorption of light at one wavelength ($\lambda 1$) changed as a function of temperature. The light of the second wavelength ($\lambda 2$) was used as a reference. An accuracy of $\pm 1°$C was achieved for the temperature range of -10 to 300° C. The sensor had a response time of 2 seconds. The semiconductor transmission as a function of wavelength and temperature is shown in Fig. 9.9. Rare earth-doped fibers, namely neodymium-doped fibers, have been used in temperature sensing schemes by different groups of workers. The schemes utilized the temperature sensitive absorption properties of the fiber.

The ratio of absorption intensities at two separated wavelengths was calculated to give the temperature. A schematic of the self-referencing system (Fig. 9.10a) used and the change in absorption with respect to temperature relative to 20 °C is shown in Fig. 9.10b. The sensing element used was 25 mm in length. An accuracy of $\pm 1°$C was achieved for the 20–200°C range of temperatures.

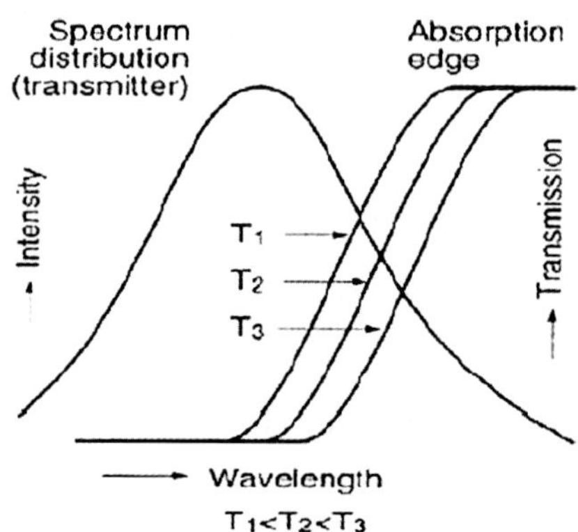

Fig. 9.9. Semiconductor transmission as a function of wavelength and temperature

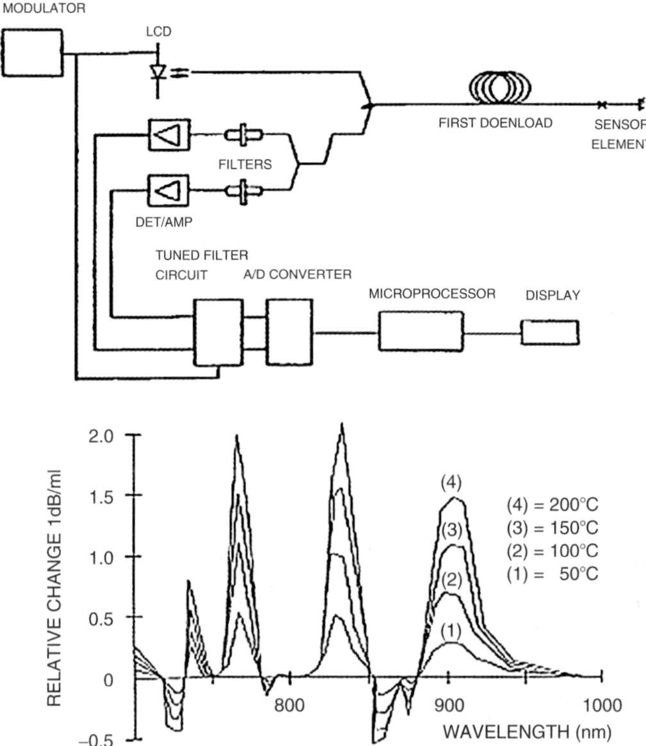

Fig. 9.10. (**a**) Schematic diagram of the sensor system. (**b**) The change in absorption with respect to temperature

If absolute absorption intensity at only a single wavelength is measured to provide the temperature, signal intensity fluctuations and connector losses can give inaccurate temperature readings as shown in Fig. 9.10b. The differential absorption technique overcomes these problems, since the ratio of absorption intensity at two separated wavelengths is calculated to provide the temperature information.

9.2.6 Fiber-Optic Stress Sensor

Measurements of seepage and movement are fundamental for dam surveillance, and are performed to various degrees at the majority of dams around the world. Recent developments in distributed measurements using optical fibers have radically improved the ability to measures those parameters. The key aspect of this optical technology is the ability to measure seepage and movements, at every meter along the dam. This proves invaluable information for effective overall dam surveillance. Temperature measurements for seepage monitoring is gaining rapid acceptance as a method for monitoring seepage in embankment dams.

The method has been used successfully especially in Germany and in Sweden, where Hydro Research has introduced and improved the method and the evaluation. Temperature measurements have successfully been used in about 30 Swedish dams since 1987, both for limited monitoring/investigations and for long-term monitoring.

The reliability of temperature measurements is well known, and the method is now also recommended in the Swedish Guidelines for dam monitoring. Until 2005, more than 20 dams, dykes and tailing dams have been equipped with temperature sensing fibers. This technology is being recognized as the most effective way to detect seepage flow changes with high sensitivity all along the entire dam.

Distributed Temperature and Strain Measurements

Distributed sensing takes advantage of the fact that the scattering characteristics of laser light, traveling down an optical fiber, vary with temperature and strain. The sensor consists of a length of standard telecoms optical fiber, normally housed in a protective cable. The measuring instrument uses a laser to send pulses of light into the sensing fiber. A detector measures the scattered light returning from the fiber as the pulse of light travels down its length. Measuring the change in power and frequency of the scattered light against time allows the instrument to calculate temperature and/or strain at all positions along the fiber.

The key feature is that the fiber itself is the sensor and it can be used to measure along its entire length. Sensor net, currently offers a range of measurement instruments, including the Sentinel DTS that can measure temperature to a resolution of 0.01°C over a distance of 10 km with a spatial resolution of just 1 m. This means that for a 10 km length of cable the user has the equivalent of 10,000 point sensors. Sensor net also provides the distributed temperature and strain sensor (DTSS), a preproduction instrument capable of measuring strain to a resolution of 10, independently of temperature, over a distance of 10 km again with a spatial resolution of 1 m.

To complement the instruments, Hydro Research and Sensor net have developed and tested an optical cable specifically for dam installations. The cable contains two single mode fibers and two multimode fibers, allowing connection to both the Sensor net Sentinel DTS, for very fine resolution temperature monitoring, and to the Sensor net DTSS, for monitoring of strain.

The cable is strong enough to withstand the high levels of compaction experienced during installation into a dam. The cable is designed to efficiently and reliably transfer strain from the cable jacket to the sensing fiber within, while protecting the fiber from the environment and rough handling.

Thermal Processes in Embankment Dams

The seepage monitoring method uses the seasonal temperature variations that occur in all surface water which causes a seasonal variation of the seepage water that passes through a dam (Fig. 9.11). The magnitude of this seasonal temperature variation can be measured in the dam and is correlated to the seepage flow through the dam. The basic approaches were described by Kappelmayer in 1957 and Johansson in 1991.

Temperature acts as a tracer with the seasonal temperature variation as source. Low seepage flows will not affect the thermal condition in the dam, and the temperature will remain almost constant. At increasing seepage flows the temperature in the dam will begin to vary seasonally. The amplitude of the variation depends on the seepage flow, on the seasonal variation at the inflow boundary, and on the distance from the boundary to the measuring point.

The seepage flow is normally small in embankment dams (the hydraulic conductivity is often less than 10–6 m s^{-1}) and the seasonal temperature variation in the upper part of the dam depends essentially on the air temperature at the surface. The influence from the air is however less than 1°C for depths in the dam body that exceed 10 m and is therefore negligible beyond such depths. This process must however be considered in small dams with heights less than about 20 m. In larger dams with heights above 100 m the geothermal flow must be considered, as well as the thermal stratification in the reservoir.

The thermohydraulic behavior of an embankment dam is complex. It includes such basic thermal processes as heat conduction (from the dam crest and from the foundation due to geothermal flow), advection and radiation. The first two processes are partly coupled to each other because viscosity and density of water are temperature dependent. The problem is further complicated by the variation in material properties in the dam, and the different conditions in the saturated and unsaturated parts of the dam. In order to analyze the problem, certain assumptions have generally to be made. The general problem can, however, be studied using coupled transport models.

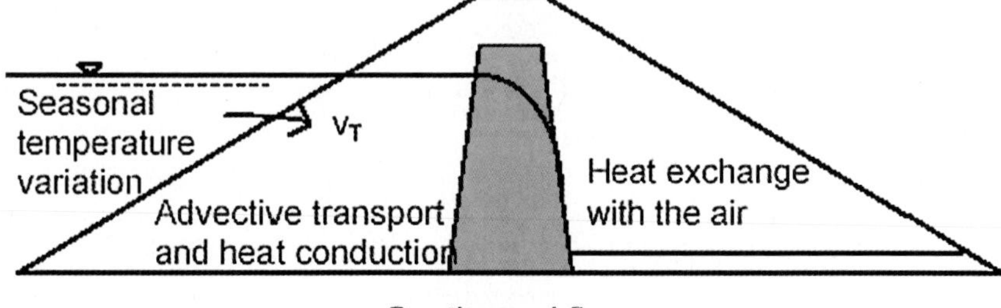

Fig. 9.11. Geothermal flow

Both heat conduction from the surface, and geothermal heat can be ignored at many applications, which simplifies the evaluation. In order to increase the evaluation possibility, especially at investigation of smaller dams a heat source could be added in combination with temperature monitoring. The seepage can then be calculated based on the temperature rise when heating up, and the thermal response when the heating is finished. This method will be valuable at lower dams or dykes where the temperature variation on the surface of the dam will disturb the temperature variation in the seepage area, but seems less appropriate for long term monitoring of high dams.

Generally, a constant temperature will be a sign of a small seepage, while large seasonal variations may be sign of significant seepage. At increasing seepage flows the temperature in the dam will be changed, and the seasonal variation will increase. This variation is dependent on seepage flow, the seasonal variation at the inflow boundary, and the distance from the boundary to the measuring point.

Measurements at Hylte Embankment Dam

Temperature measurements have been used in many dams and many examples can be found in the literature. Some interesting result from the Hylte dam is presented as an example in Fig. 9.12. Optical fiber cables were installed by the dam owner at the dam toe of the Hylte dam and dyke in late 2002. The purpose of this installation is primarily to identify areas with higher seepage or increasing seepage, but also try to estimate the seepage flow. The site consists of a 250 m dam with a maximum height of about 8 m and a 1.3 km dike with maximum height of about 8 m.

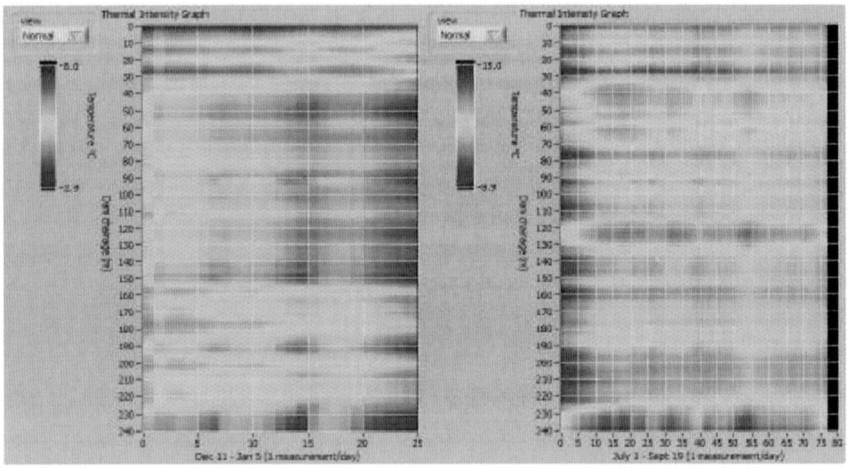

Fig. 9.12. Long term temperature measurements

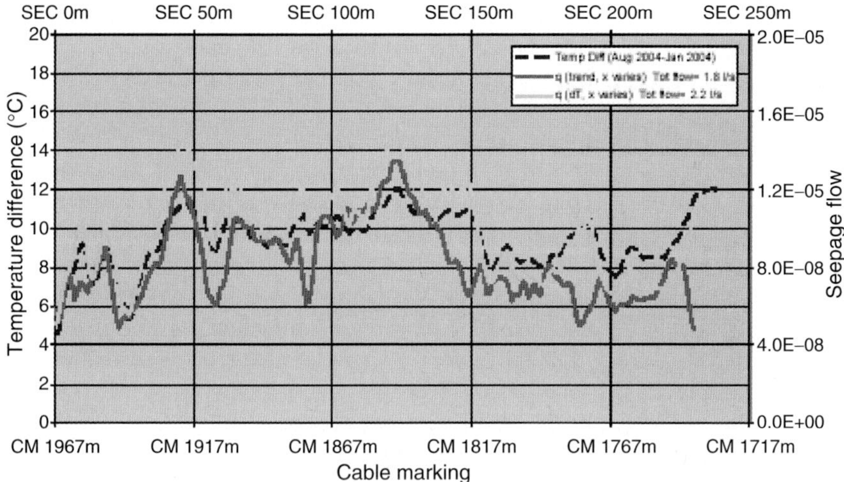

Fig. 9.13. Calculated seepage along the dam toe using two evaluation methods

The cable is placed at a depth of between 0.8 and 1 m, all along the dam toe, close to or just below the ground water level. Five measurements have been performed since 2003. Three of these have been done just for a few days, while two longer periods were measured during winter 2003–2004 and summer 2005. The result in Fig. 9.13 shows generally a higher temperature in the summer than in the winter. The fiber is placed closer to the surface at the right abutment (chainage 240), and the temperature is clearly affected by the air temperature. The fastest cooling sections during the winter period are observed at chainage 50, 90 145, and 190 m. These areas also correspond to those where the temperature is highest in the summer, which can be explained by higher seepage. Areas with anomalous seepage flow can then be identified.

The measurements can also be compared with calculated temperature for different seepage flow rates using the software DamTemp, which is based on the work by Johansson. The seepage flow through the Hylte dam was evaluated from the measured temperatures in two different ways. The first method calculates the maximum measured temperature difference during the measuring period. The seepage flow rate can then be estimated. The other method is based on the measured trend from the long-term measurements in 2003/2004.

The areas with highest seepage are the same as found from Fig. 9.13. The significant color changes seem more drastic than the calculated seepage differences, which are about ±50%. This indicates how sensitive the method can be with a proper installation. The total flow in the dam is estimated to about $2 - 1.8 \mathrm{l\,s^{-1}}$ for the first method and 2.2l/s for the second method. No seepage measurements are made on regular basis, but the calculated value agrees reasonable with single measurements. The absolute flow estimation is however less relevant, since the largest advantage with this method is the high ability to detect small seepage changes.

Bergeforsen Embankment Dam

The Bergeforsen power plant in the river Indalsalven was built in the early 1950s. The construction was complicated due to extreme geological situation. As a result an extensive monitoring program started including pressure sensors, inclinometers, strain gauges etc. However, no seepage measurements have been possible because a high downstream water level. A complementary installation was made in 2005, where 17 new standpipes were drilled from the downstream to the bedrock in order to measure the temperature in the seepage water (Fig. 9.14). Optical fibers were installed in the new standpipes, in five old standpipes, and also in the upstream water.

A Sensornet DTS unit was installed in September 2005, when also continuous measurements started. Initial measurements indicate excellent result which is necessarily to detect the small expected seasonal temperature changes, may be less than 1°C. One example is shown in Fig. 9.16, where the higher temperature in the air is seen, as well as the temperature in the water in the standpipes (with a temperature of about 6°C). Measurements is performed both at going down and up in each standpipe, which gives a symmetrical result. No significant temperature anomaly is observed, but the temperature seems to be slightly higher at the first standpipes close to the spillway, the first three in the left side in Fig. 9.14. A general increasing temperature with depth is seen for all standpipes. The initial temperature measurements indicate normal homogenous and normal seepage flow.

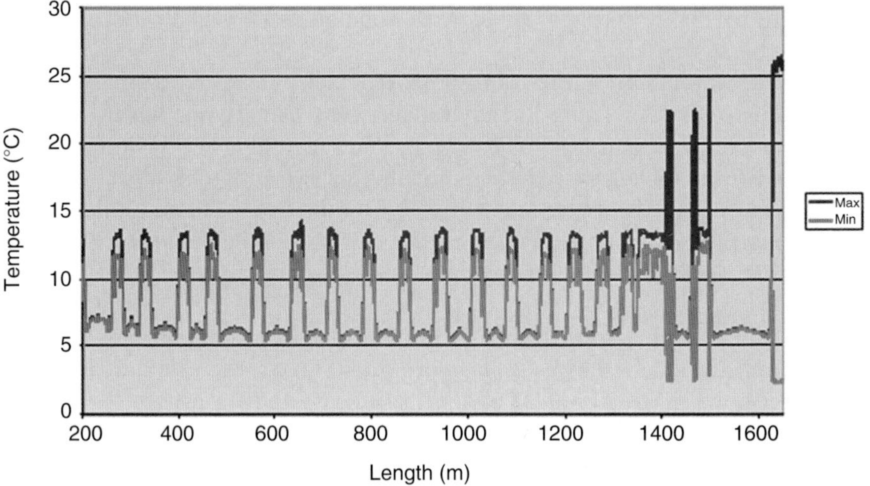

Fig. 9.14. Initial temperature measurement at Bergeforsen power plant

Strain Measurements at Ajaure Embankment Dam

In September 2004, the first distributed measurement of strain in an embankment dam was undertaken using the Sensornet DTSS within a research project funded by Elforsk. The system allows the strain and temperature to be measured simultaneously and independently at all points along the fiber. The measurement was carried out at Vattenfall Vattenkraft's Ajaure dam in Sweden, where a cable was installed at two levels in the crest as indicated in Fig. 9.15. The aim was to compare the measurements taken during September 2004, at full reservoir level, with measurements to be taken at low reservoir level in May 2005.

The result of the measurements showed no significant strain difference between the two measurements, indicating no significant localized movement of the dam as shown in Fig. 9.16. These initial measurements demonstrate the potential of distributed strain measurements, which is further described by Johansson and Watley 2005. Two more measurements will be performed before June 2006.

Whilst undertaking strain measurement at Ajaure, a series of tests were also performed to demonstrate the ability of the DTSS to determine the size of localized deflection in the dam which, could be detected by distributed strain measurements. Similar tests were also repeated in the Sensornet Laboratory. Those tests have demonstrated that a deflection of the dam by just 5 mm will be detectable. The current sensitivity of the system is suitable for detecting very small localized changes of the dam structure.

However, the sensitivity is set to improve with further development, providing enhanced temperature correction and hence much increased strain accuracy. Distributed optical sensing will be an essential tool for dam

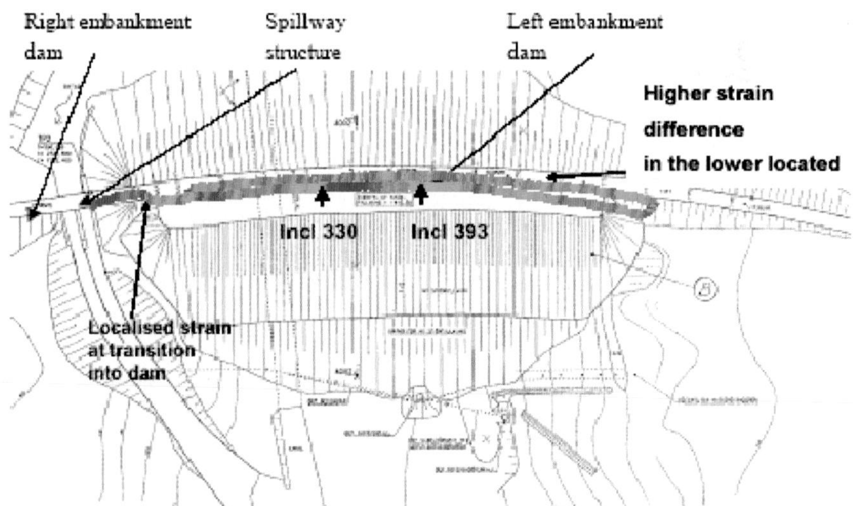

Fig. 9.15. Visualization of change in strain at Ajaure between May and September

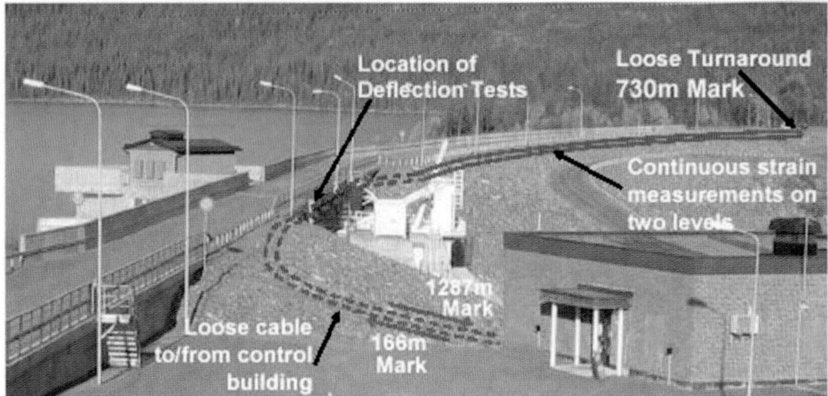

Fig. 9.16. Illustration of the sensing cable route at Ajaure dam

monitoring, providing location specific information which has not been available before. The distributed technology provides an excellent complement to conventional inclinometers.

9.2.7 Fiber-Optic Gyroscope: Polarization Maintaining

It is hard to imagine modern motor vehicle development without the use of gyroscope platforms as a means of vehicle motion analysis. The properties of today's inertial measuring systems that are used for motion analysis, stabilization and 3D topology surveys are the subject of this paper. As an example, this section will describe the motion analysis of motor vehicles (the so-called "elk test") that use the latest version of iMAR gyroscope platforms manufactured by iMAR GmbH, St. Ingbert (Germany). Precise sensor technologies for measuring even highly dynamic driving manoeuvres in combination with sophisticated mathematical procedures support the measuring engineer in the execution of even the most demanding measuring programs.

From Gyroscopes for Ships to Miniature Strap-Down Systems

For many years, inertial measuring systems have been used for navigation tasks in the air, at sea or on the ground. These systems measure the acceleration along the trajectory of a motor vehicle with the aid of sensor technology. The speed and position of a body can be determined by integration of its acceleration over time within an earth-fixed navigation coordinate system. The first inertial measuring systems were so-called north-seeking gyros. In 1913, the first passenger ship was equipped with an Anschutz gyro compass (named after its inventor), which comprised two heavy twin balls rotating in opposite directions.

Forty years later, in 1953, the first successful flight using an inertial navigation system in platform technology took place from Massachusetts to California. Almost 40 years later, in 1992, the iDIS-FV from iMAR was presented to the professional world as the first inertial measuring system in strap-down technology for vehicle motion analysis with fiber-optic gyroscope technology and a VME bus process computer. Four years later came the iDIS-FC, the new version came with an integrated microcontroller. Today, in 2002, the third generation of inertial measuring systems for vehicle motion analysis is being introduced to the market. It features half the weight of previous versions, a three-times higher accuracy, higher sampling rates with less signal delay and substantially enhanced functionality.

Inertial Measuring Technology

According to Newton's second axiom, the motion path of a body can be precisely determined if at all times all forces acting upon this body are known and if the starting position and initial speed are taken into consideration at the start of the observations. But in principle it is not possible to differentiate between inert and gravitational mass in a motion system. Basically, inertial measuring technology determines the forces acting on a body in motion according to Newton's second axiom using accelerometers to derive the motion from these data.

When navigating in a gravitational field, the value and direction of the gravity vector (gravity g $9.81\,\mathrm{m\,s^{-2}}$ in an earth-fixed measurement) in reference to the body in motion has to be known at all times. In a rotating reference system, the rotation rate vector also has to be known (earth rotation rate: $\mathrm{E} = 15.041^\circ\,\mathrm{h}^{-1}$). In terrestrial navigation, the acceleration of gravity is constantly comeasured by accelerometers as disturbance and has to be compensated for. This is achieved by constantly registering the alignment of the body in reference to earth's gravity field using fixed (or "strap-down") gyroscopes.

Consequently, an inertial measurement system in its simplest configuration consists of three orthogonally arranged accelerometers to determine the acceleration vector and of three orthogonally arranged gyroscopes to determine the rotation rate vector of the body, if motions in all six degrees of freedom of the body are permitted. As the orientation of the measuring system is to be referenced to an earth-fixed coordinate system – usually the local horizontal reference system – the earth's rotation rate vector also has to be compensated for in the rotation rate vector, measured by the gyroscopes.

The compensations are calculated using two simultaneous vector-valued differential equations. One can see its modular architecture, which allows the integration of customized features. Major design features of the iDIS-FMS are its high sampling rate of 400 Hz, the large measuring range of $\pm 500^\circ\,\mathrm{s}^{-1}$ and $\pm 10\,\mathrm{g}$, the internal or external GPS, an integrated odometer interface, the CAN bus and Ethernet interface as well as an open interface application software based on Windows. iDIS-FMS has been developed for various applications.

Devices of this series are used for vehicle motion analysis as well as the stabilization of antennas on ships, the guidance of aircraft and underwater vessels (drones and RPVs) and the determination of position and control of service robots (e.g., in pipelines). It complies with all environmental rules for such devices per the IP 67 system of protection and is operated with a voltage of between 10 and 34 V and a power consumption of only 30 W.

In accordance with the requirements of measuring field engineers, the following data are among the output resulting form internal real-time calculations:

- Useful rotation rate in vehicle-centered coordinate system (earth rotation rate compensation applied)
- Useful acceleration in vehicle-centered coordinate system (gravity compensation applied)
- Angular orientation in space (roll, pitch, yaw) of the vehicle in reference to its local, earth-fixed coordinate system (so-called Eulerian angles)
- Odometer-longitudinal velocity in vehicle's coordination system
- Position in local and world coordinate system (x/y/z and longitude/latitude/height)
- Output of measuring values centered on the vehicle's gravitational center or on any other point of the vehicle

Coordinate Systems

In strap-down technology several coordinate systems are differentiated. The four most important are:

1. Earth-fixed, global world coordinate system
2. Earth-fixed, local horizontal coordinate system
3. Sensor coordinate system
4. Vehicle coordinate system (also called body coordinate system)

In order to determine the global position of a vehicle, the WGS84 coordinate system supported by GPS (satellite-aided Global Positioning System) should be used, in which each position is precisely defined by longitude, latitude and height (world coordinate system). In order to determine metric distances, an earth-fixed local horizontal coordinate system in which the surface formed by the x- and y-axis of this coordinate system is tangential to earth's surface on the testing site should be used.

The longitude and latitude of the testing site define the origin of this earth-fixed coordinate system. For navigation algorithms, only the latitude is relevant, as the north–south axis through the earth's poles forms a symmetrical axis for the quasi-stationary navigation problem. By definition, the x-axis of the coordinate system is directed towards east, the y-axis towards north and its z-axis up (ENU).

All measured values refer to the sensor coordinate system and have to be transformed into a different coordinate system, which can be more easily interpreted. Besides the world or horizontal coordinate system, the vehicle coordinate system used in the automotive industry and defined in DIN 70 000 is such a system. The direction of its x-axis is the forward direction of the car, its y-axis goes to the left and its z-axis is positioned orthogonally on top of the two other axes in the upward direction. The transformation from the sensor or vehicle coordinate system into the earth fixed coordinate system is carried out internally in the system with so-called quaternia or direction cosine matrices.

The angular presentation is very efficient from the mathematical point of view; for the user it is of very little help, however. Therefore, Eulerian angles are usually used for presenting the position of a vehicle in space. These roll, pitch and yaw (RPY) angles are defined as follows. RPY angles according to DIN 70.000 (body frame):

Yaw Angle. “Psi” angle around the z-axis of the world coordinate system (upward direction). This is the first rotation required for the transformation from the world system into the sensor system.

Pitch angle. “Theta” angle around the y-axis of the world coordinate system rotated by. (angle around the y-axis of the sensor coordinate system for small lateral angles).

Roll angle. “Phi” angle around the x-axis of the sensor coordinate system (third rotation).

It is important to understand that the vehicle orientation in the world coordinate system is described by the RPY angles after three successive rotations with the standard world coordinate system (ENU: x east, y north, z up) as the starting point. After three successive rotations of the single axis, the spatial position of the vehicle is achieved. In this context, the fixed sequence of these rotations is essential. The measuring system of course also determines the angles in the case of simultaneous rotations around all axes by solving a simultaneous differential vector equation.

Motion Analysis Sensors

Only servo accelerometers are used as acceleration sensors for precise motion analysis. Their internal closed-loop structure renders excellent linearity in a measuring range, which covers more than 5 decades (± 20 g measuring range with a resolution of 10 µg) in these applications. They are shock resistant (100 g) and have proven their high reliability in military and civil applications over many years. Two types of sensors, based on two different effects, can be used as gyroscopes.

Usually, the speed is generated in the form of a harmonic oscillation with a piezo vibrating actuator or with capacitive acceleration. Today, the vibro gyroscope is frequently used in low-cost devices. In contrast to fiber-optic gyroscopes in these applications, noise, a large temperature-dependent drift (up to $3^\circ\,s^{-1}$ over temperature) and an obvious temperature-dependent scale factor with a small bandwidth are uncritical. Influences of translator acceleration, vibration or structure-borne noise can have a negative impact on these gyroscopes, as their measuring system is based on vibration and acceleration measurement. It comprises a light-emitting super luminescent diode (SLD), beam splitters, a modulator, the fiber coil which covers the effective surface A, and an interference detector.

Through the positioning of the beam splitters, the light emitted by the SLD with the wavelength split in two beams, both traveling along the coil in opposite directions. If the whole arrangement is rotated around the normal vector of the coil with the angular velocity "w," the path of one beam becomes shorter while the path of the other beam becomes longer. The resulting differences in lapse of time cause a phase shift between the two light waves. The interference detector recognizes this and measures the angular velocity. With feedback control of the measured phase shift to an optical phase-shift element (so that the sensor works internally on a constant phase), a so-called closed-loop optic gyroscope is the result.

Compared to mechanical or piezoelectric devices, optical gyroscopes have the advantage that the sensing element is not subject to any mechanical strain. It is largely insusceptible to vibration and acceleration. This explains why optic–optic gyroscopes are used for precision measuring systems. Single-axial optic–optic gyroscopes are frequently used by the automotive industry to determine the yaw rate.

Measuring Deviation and Measuring Results

When using inertial measuring technologies, one should be familiar with the basic measuring deviations of these systems. The gyro drift is the error angle per time unit of a gyroscope. If a gyroscope is at rest with an output offset of $3^\circ\,h^{-1}$ (rotation rate offset), then the gyro drift is 3° per hour or 1° in 20 min (without aiding). The random walk is a measuring deviation of the gyro, which is the consequence of integrating white random noise. The noise of the rotation rate can be calculated as 1-sigma-value in reference to an output bandwidth $f[\mathrm{Hz}]r[^\circ\,s^{-1}] = RW[^\circ\,\mathrm{sqrt}(h)^{-1}] \times \mathrm{sqrt}(f[\mathrm{Hz}]^{-1}) \times 60/3600$.

Looking closer at the drift of a pitch or roll gyroscope (usually y-axis or x-axis), one will notice that this drift causes the actual angles to be overlapped by an error angle which increases with the measuring time and drift. The precision of the attitude measurement in space (i.e., pitch and roll angle) is decisive for the precision of the gravity compensation in the measured acceleration signals. Therefore, the measuring deviation of the acceleration after gravity compensation increases with the duration of the measurement.

An angle error of 0.06° results in an acceleration error on the horizontal channels of approximately 1 mg. However, even a very slight gyro drift is in most cases not small enough to carry out longer measuring cycles and achieve the required measuring accuracy.

Therefore, iMAR's so-called adaptive aiding is used to eliminate the drift from the measuring results. In the case of land vehicles, external speed data are recorded optionally to correct the position angle. Using these data, the longitudinal and lateral acceleration can be estimated, which, together with the inertial measuring system, allow the correction of the position angle. If available, GPS is used for the drift compensation of the heading gyro.

These data are recorded with the iMAR platform iDIS-FMS, which is used by ZF Lemförder Fahrwerktechnik (a supplier of chassis components, modules and complete axle systems). In combination with the measured steering angle, they allow the user to assess, among other things, the lateral dynamic transmission behavior and the roll steer effect of a car. Additionally, all movements of the car body are recorded for use as evaluation criteria for vehicle dynamics and comfort.

iDRPOS: 3D Topology Surveying

The iDRPOS algorithm, which can be optionally integrated into the iDIS-FMS, uses a special Kalman filter for advanced aiding in addition to the general odometer aiding. Therefore, it can determine both driving dynamics data and topological path data. The measuring system also renders reliable position data even if the GPS is temporarily unavailable. There is no impact on the accuracy of measuring vehicle-dynamic related data such as rotation rates, accelerations or attitude, should the GPS be temporarily unavailable – quite in contrast to low-cost systems with micro-mechanical gyroscopes or low-cost optic-optic gyroscopes whose accuracy primarily depends on GPS satellite detection.

Due to the high accuracy of optic-optic gyroscopes and accelerometers, iDIS-FMS, suitable for the analysis of vehicle dynamics, use GPS only as an additional data source. They are also largely insusceptible to GPS interruptions. The quality of an inertial measuring system, however, also depends on its user interface. Therefore, when designing the new generation of these measuring devices, apart from the hardware a universal and modular interface was also developed which is used consistently in all iMAR measuring systems starting with version 2002.

It enables the user to modify parameters such as sampling rates, aiding criteria, threshold values, etc., to submit commands (e.g. Start/Stop of the alignment) and to output time-related or distance-related measuring data (e.g. attitude, acceleration, rotation rate, odometer speed, etc.). The open interface allows the user to integrate the measuring systems (with an optional DLL) directly into the user's own application. It is noted that the inertial algorithms run on the hardware (process computer) of the measuring system with a real-time operating system. Windows on an external laptop or PC serves only for user prompts and display purposes as well as for saving data.

9.2.8 Gratings in Fiber

UV light shining into the germanium-doped core of an optical fiber induces a local change in the index of refraction. One can create a grating in the core of the fiber by imaging an interference pattern created either holographically or by a phase mask through the side of the fiber. The resulting grating acts as a wavelength filter. The period of the induced index of refraction modulation in the optical fiber core determines the wavelengths that will be reflected. When the fiber grating is subject to strain, two effects cause the period of the fiber grating to change. One effect is that the overall length of the grating changes, and the other is that the index of refraction changes. The net result is that the overall period of the grating changes, resulting in a shift in the reflected wavelength, which provides a mechanism to measure strain. The wavelength shift in most silicon-based fibers is extremely repeatable and accurate, as the composition of the fibers is very nearly pure quartz, which provides excellent mechanical properties.

Fiber gratings are widely used to support wavelength division multiplexing (WDM) and dispersion compensation in the telecommunications industry. These volume applications have significantly reduced the price of fiber gratings and have triggered the production of fiber gratings with greatly improved reliability and performance. Because the long-term trend in optical networking is toward bringing fiber to the home with networks incorporating these devices, other fiber-grating users will continue to benefit from telecommunications development, which will improve the performance and lower the cost of these devices.

Each fiber grating may be made to reflect at a specific wavelength, so one can design a simple system that performs WDM along a single fiber line. This technique can readily be used to multiplex 10 or more fiber-grating sensors along a single fiber. By combining time-division multiplexing with WDM, one can support hundreds of fiber Bragg grating sensors in a single fiber.

Unique Capabilities

As a first example of the unique capabilities of fiber grating sensors, consider a multiaxis fiber grating strain sensor that offers the ability to measure 2-D and 3-D strain fields along a single fiber line as well as measuring transverse strain gradients and shear strain. Strain is the change in length divided by the original length of the object under test. It can be presented as a percentage or as a unit less number. A fiber grating written into polarization-preserving fiber, effectively acts as two separate gratings, one for each polarization because, light in each polarization experiences a different refractive index. This duality

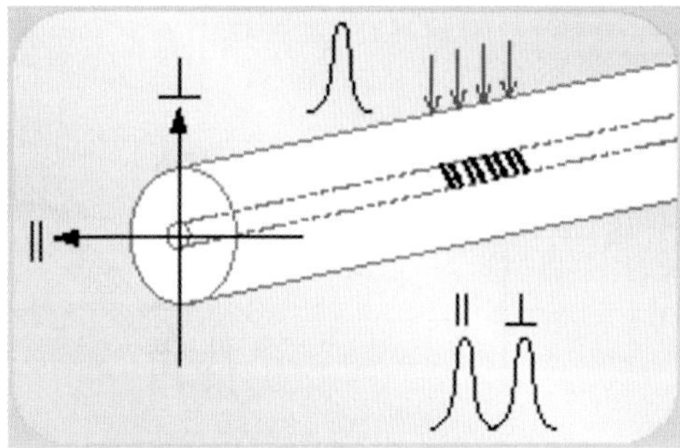

Fig. 9.17. Multiaxis fiber grating

allows users to quantitatively measure both transverse strain and axial strain. One can implement a series of similar sensors simply by writing fiber gratings tuned to different wavelengths at different points along the optical fiber.

When the fiber grating is illuminated by a broadband source such as an LED, the reflection spectrum shows two peaks as measured by an optical spectrometer or other spectral measurement device. The spacing between these peaks is a measure of transverse strain along the birefringent axes of the optical fiber. In the case of Fig. 9.17, the multiaxis fiber grating strain sensor is subject to uniform transverse strain along one of its axes. This results in two clear, distinct spectral peaks whose spectral separation is proportional to the transverse strain.

While fiber-grating strain sensors provide an absolute measurement of strain in the optical fiber, they are often embedded in materials that provide residual strains during manufacture. This residual strain can be measured, and for many applications the parameter of interest is the relative strain change after the structure has been fabricated, which involves recording the reflected wavelengths after the structure is fabricated and noting wavelength changes thereafter. By fastening this type of fiber-grating sensor onto a structure with an appropriate bonding material, the user can record an initial strain state.

Changes in the axial strain and transverse strain, then, may indicate changes in material properties, dimensions, pressure, or temperature of the structure. These indications of third-party damage and/or changes that affect structural integrity reference the baseline value. Blue Road Research (Gresham, OR) has typically achieved strain measurement accuracy of less than 1% for surface-mounted fiber-grating strain sensors, compared to their electrical equivalents. If the transverse strain along one of the axes is partially unloaded (as would be the case during an adhesive-bond failure or internal failure of a composite material), then transverse strain gradients occur that result in clear splitting of the appropriate spectral peak as shown in Fig. 9.18. The amplitude of the transverse strain is encoded in the spectral shift.

The length over which a specific load occurs can be determined by the amplitude intensity of the peak. The position of the load can be localized

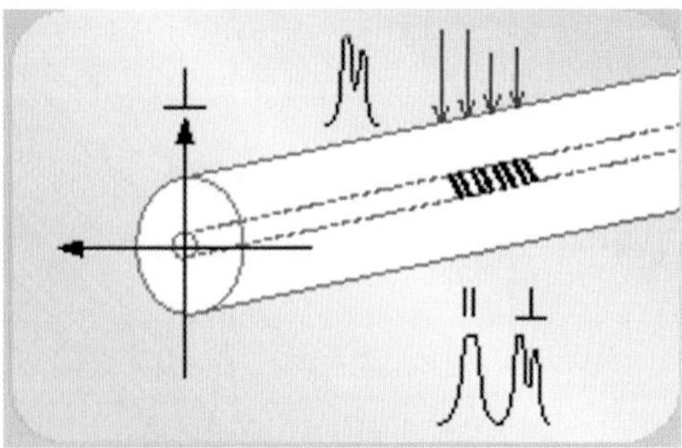

Fig. 9.18. Clear splitting

either by using short gauge length fiber-grating sensors whose dimensions are of the order of 5 mm or by using distributed techniques such as chirped fiber gratings in which the period of the fiber grating varies along its length, allowing position information to be wavelength-encoded.

A multiaxis fiber-grating strain sensor can also be used for testing joints by mounting the sensor in the edge of the adhesive bond, with its transverse axes aligned at 45° to the plane of the part in order to offer optimal response to shear. For zero load, the spectral response is simple. As the load increases, the spectral response becomes more complex, with peaks separating to show increased transverse forces or splitting to show bond failure. At higher loads, all the peaks move, which indicates axial strain. This application was tested by placing the ends of a test coupon containing a fiber-grating sensor in an Instron universal test machine and applying a load. As the test machine pulled the joint apart, shear force (transverse strain) increased, causing the principal peaks to move apart. When the joint began to fail, one of the principal peaks split, indicating that a portion of the bond was unloaded by approximately 600 microstrains.

The intensity of the two split peaks was approximately equal; indicating that half of the fiber-grating length had been unloaded. When the joint was loaded to higher levels, all the peaks moved, indicating axial strain. Thus, the application demonstrated simultaneously measuring transverse strain, transverse strain gradients, and axial strain interior to an adhesive bond. This technique offers unique capabilities for accurate diagnostics of previously.

9.2.9 Advantages of Fiber Optics

Fiber optic strands provide a significant advantage over standard copper transmission lines. The potential information-carrying capacity of a single fiber strand is much greater than a single copper wire. In addition, with modulation schemes, a single carrier frequency can carry numerous channels of information. The carrier frequencies used in light waveguides are far greater than those deployed in copper lines; light is composed of different wavelengths each of which can carry a separate data channel. Optical fiber is very light and has a very small diameter, typically smaller than the diameter of a human hair.

Application areas include:

- Aerospace
- Power plants
- Process control
- Energy exploration
- Pipeline monitoring
- Laboratory standards
- Process control systems

9.3 Laser Instrumentation

Laser instrumentation is a technology introduced in the control of processes for welding, heating, trimming, and melting. These advanced processes are extensively used in first world countries to extend available options in automation and increase productivity and profit. This new technology offers industry the opportunity to become world-class manufacturers that are globally competitive. This section describes laser instrumentation used in the measurement of velocity, distance and length along with the techniques such as welding, heating, trimming, and melting.

9.3.1 Measurement of Velocity, Distance, and Length

Laser Doppler velocimetry has become a very useful laboratory tool for the measurement of gas, liquid, and solid surface velocities. The majority of the commercial laser Doppler velocimeters are laboratory systems, which work well but require customized setups, which depend upon the application.

The number of possible uses for an industrial velocimeter is quite large. The primary applications thus far have been in the primary metals industries. A noncontact velocity gauge can assist in process control in many rolling mill applications. For example, in the aluminum industry large webs of sheet aluminum are cold-rolled in sizes that require more than one roller across the width of the sheet. If the rollers do not roll the material with the same tension, the resulting sheet will bow as it comes off the line. To control this type of process, a number of velocity measuring stations could be lined up across the web.

By continuously adjusting the tensions to keep the measured velocities at a predetermined value, the bow in the rolled sheet will be eliminated. In the steel industry, the biggest application is for measuring the length of hot steel for cut-to-length applications and the quality assurance on bar and structural steel rolling mills. Integrating the measured velocity easily derives length measurement. The requirements of a velocimeter to be used for such applications are very stringent. For the web measuring application, the velocity must be accurate to 0.1%, and the nearly specular surface of the aluminum causes a very strong Doppler signal. The steel mill applications require that length be measured to a similar accuracy, but the steel surface is very dark and the Doppler signal is considerably weaker.

Another problem with many steel mill applications is that the location of the part on the transport system is not well constrained, requiring the gauge to have both a large depth of field and large dynamic range. The cut-to-length applications require that the instrument be capable of handling both positive and negative velocities because of the manner of operation of the cutting station. The environmental problems can also be quite severe. Steam from water cooling of the transport system and scale on the hot steel must be considered. The high heat and steam can be handled by proper packaging, but the scale can cause dropouts in the return light signal, which requires special processing, particularly in length measurement applications.

Theory

This section presents the theory of velocimetry in a simple way with minimal mathematics. A laser Doppler velocimeter works on the principle that light scattered from a moving object is frequency shifted with respect to the incident light. If a collimated beam of light of wavelength X is incident on a moving surface (Fig. 9.19), the frequency or Doppler shift is given by AV $= (x)(d - as)v$, where *as* and d are the unit vectors of the incident and scattered light, and v is the velocity vector of the moving surface.

For industrial applications the velocity component parallel to the surface *vp* is of primary interest. To determine *vp* uniquely requires that there be no

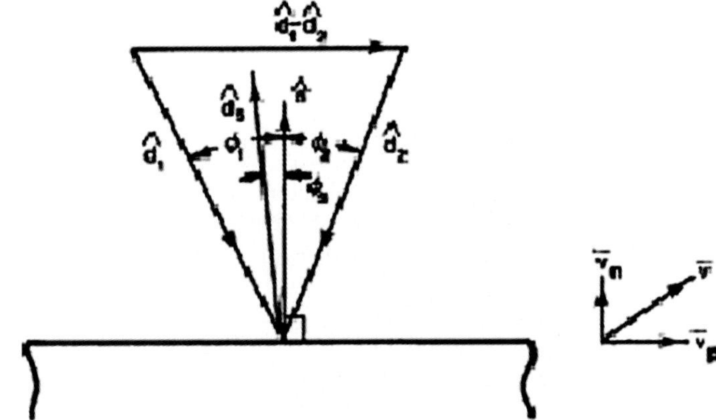

Fig. 9.19. Differential Doppler

motion of the surface along its normal direction t (i.e., vn = 0) and that the scattered light be collected over a small well-defined solid angle, necessarily resulting in a weak detected signal. Because of these constraints, a differential Doppler method is more suitable. The method requires two incident beams *a1* and *d2* as illustrated in Fig. 9.19. It is seen that all sensitivity to the viewing direction has dropped out. This is important in the design of the optical system because it allows a large solid angle of light to be collected without degrading the signal. In addition, if the angles of incidence are equal. The light incident on the detector consists of many waves originating from the scattering centers on the surface. Each wave is modulated at frequency v, but the phase between them is random.

It can be shown that the sum of n sinusoidal waves each of amplitude A produces a resultant sinusoidal wave of the same frequency. In addition to sensing the modulation due to the Doppler signal, the detector also senses overall intensity variations caused by the scattering characteristics of the surface. On most surfaces this will present a problem since the scattering characteristics of the surface may vary greatly. Much of the surface signal component can be removed using a high pass filter as long as the surface does not have a high frequency structure that creates a modulation near the Doppler frequency.

A method of eliminating the sensitivity to surface structure, as well as laser intensity fluctuations, is called the differencing differential Doppler technique. In this configuration, the two incident beams, d1 and d2, are orthogonally polarized, one parallel and one perpendicular to the plane of incidence. The scattered light is collected by a lens and directed through a polarizing beam splitter oriented at 45° to each polarization, causing half of the light from each polarization to be directed to separate detectors where interference occurs.

The signals from each detector are 180° out of phase so that subtracting the two signals enhances the Doppler component while canceling common intensity variations due to surface structure and laser instabilities. An example of the effectiveness of this technique is shown in Figs. 9.20 and 9.21 respectively. A signal of this quality is much easier to process and extract the Doppler frequency information. The use of the differencing differential Doppler method removes most of the noise in the frequency spectrum, but

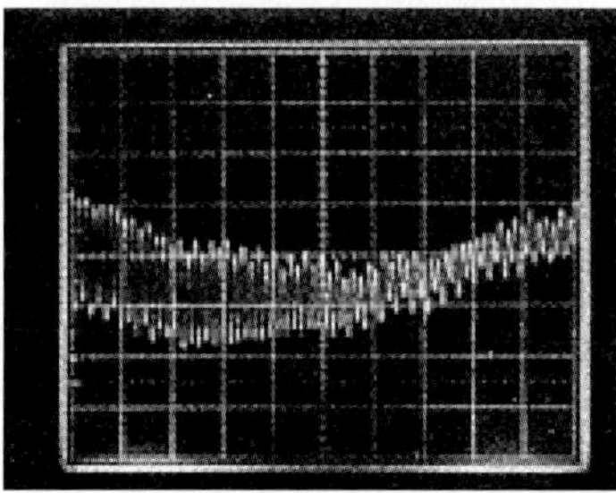

Fig. 9.20. Single detector differential Doppler

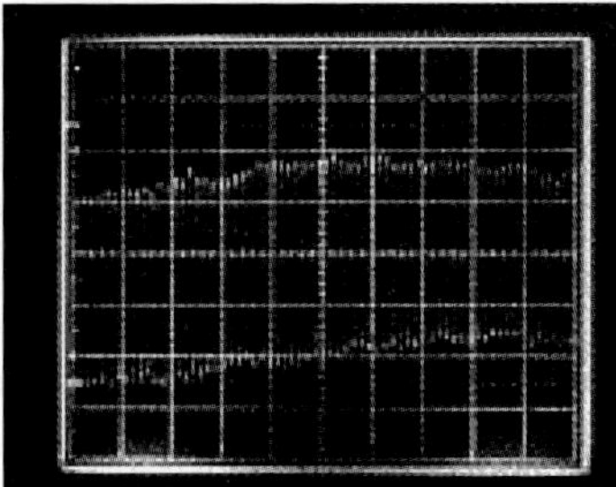

Fig. 9.21. Two detector differencing differential Doppler

there will always be some low frequency components as well as noise due to the 60-Hz line frequency and external disturbances. The presence of these low frequency components makes the measurement of velocities near zero a difficult task. The solution most commonly employed in laser Doppler systems is to introduce a frequency bias by frequency shifting one of the two incident beams.

The frequency shifting can be done in many ways depending on the magnitude desired. For high frequency shifts >10 MHz acousto-optic modulators are used, and for low frequency shifts (<1 kHz) a rotating halfwave plate can be used. For frequency shifts between these ranges, two acoustooptic modulators operating at slightly different frequencies are often used, although this is an expensive solution. The method used in this system is uniquely suited to the way data are taken, and it permits velocity measurements to be made for surfaces moving in either the forward or reverse direction.

Optical System

The optical system of the velocimeter has been designed to incorporate the differencing differential Doppler technique. The design also incorporates a frequency shifter as an option. A number of other features have been incorporated to improve the SNR and to make the instrument adaptable to a large range of installations. The optical schematic of the velocimeter is shown in Fig. 9.22. The light source is a 5-mW polarized He–Ne laser operating in a small number of longitudinal modes and the TEMOO transverse mode.

The output of the laser passes through lens *L1*, which controls the divergence of the laser and helps transfer the beam waist to the measuring volume. The next element in the optical train is the frequency shifter which takes the input beam polarized at 45° to the plane of the paper and divides it into a vertically and a horizontally polarized beam. The vertically polarized beam is then frequency shifted by −20 kHz, and the two beams are recombined. The polarizer and detector located directly behind the first mirror combine the two polarizations and detect the resulting beat frequency to monitor the frequency shift precisely.

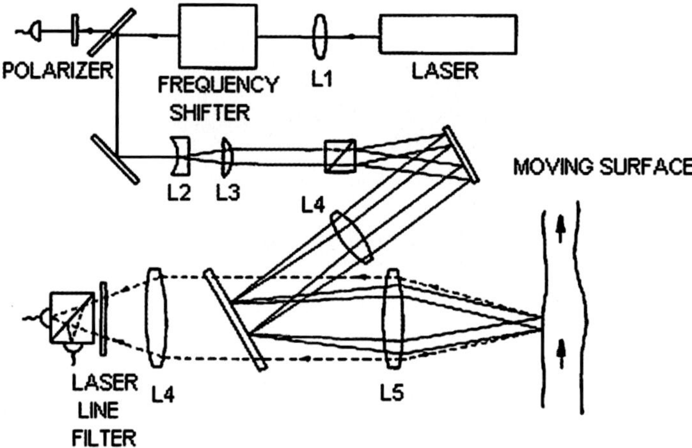

Fig. 9.22. Velocimeter optical System

The beam at this point is 0.8 mm in diameter. It is passed through beam shaping optics which create an 8×0.8 mm elliptical beam with its major axis in the direction of motion. The purpose of elongating the beam is twofold: the long dimension, which is in the direction of motion, narrows the frequency spectrum, and the short dimension improves the SNR. The beam then enters the Wollaston prism, which divides it into two beams of equal intensity, one horizontally polarized and one vertically polarized.

The two beams emerge from the prism with a fixed angle between them. Large angles are used for measuring slow speeds and small angles for high speeds, so that the resultant Doppler shifts are comparable, maximizing the dynamic range of the measurement. Wollaston prisms typically used have angular separations of 5, 1, and 0.25°. Lenses L4 and L5 form a 1.5:1 a focal beam expander which images the beam crossing point onto the moving surface. The light scattered by the moving surface is collected and focused onto the detectors by lenses L5 and L6. The fold mirror between lenses L5 and L6 is a long narrow strip of glass with a dielectric coating on one edge.

The mirror is only 3 mm wide and, therefore, only obstructs a very small percentage of the aperture. After passing through lens L6, the light is spectrally filtered by a 100-nm laser line filter to remove as much of the background radiation as possible. It would be optically more correct to place the laser filter in the quasi-collimated beam between the two lenses, but these lenses have a 100-mm clear aperture requiring a large and hence expensive filter. Since the required performance of the filter is not very stringent, the position shown is a more economical choice. Following the filter is a polarizing beam splitter oriented with its polarization axis at 45° to the axis of the incident polarizations.

This splits the light allowing an image of the moving surface to fall on each of the Doppler detectors. The optical system has also been designed with the ability to focus on surfaces varying in distance from 0.5 to 1.0 m. This was deemed a desirable feature because it allows the instrument to adapt easily to many different mounting arrangements. The focusing is accomplished by moving the Wollaston prism and lens L6 together in a fixed linear relationship. To move the measurement volume from 0.5 to 1.0 m, the Wollaston prism is moved to the right, and lens L6 is moved to the left. This optical system fits in a box 84 cm long X 16 cm wide X 23 cm high.

Performance

The velocimeter has been tested extensively both in the laboratory and in steel mills on red hot steel (9,000°C). The laboratory tests were designed to test both velocity and length measurements. The velocity measurements were done on a custom-designed motor-driven flywheel. The flywheel is 279 mm in diameter and is rotated at speeds from 60 to 3,400 rpm by a crystal controlled dc motor with <0.01% wow and flutter at 3,400 rpm and 0.5% wow and flutter at 60 rpm. The velocimeter is set up to measure the edge velocity of the wheel, which travels at 1–50 m s^{-1}. The flywheel calibrator was used for extensive characterization of the measurement of velocity.

To check accuracy and repeatability, the calibrator and the velocimeter are both connected to an HP-85 computer. The computer steps the calibrator through a sequence of speeds and averages 100 velocity readings from the velocimeter at each speed. This test is run for each velocity range more than once to test repeatability. The standard deviation that is printed with each data point is also important as it shows the number of samples that must be taken to obtain the desired repeatability.

The results of the test show that the particular instrument tested was calibrated 0.025% low on the 40 m s^{-1} range, 0.5% high on the 10 m s^{-1} range, and 0.3% high on the 2 m s^{-1} range. These calibrations were repeatable to within 0.1% from day to day and at different focal positions. A graph of the log of the standard deviation of the measurements versus the log of the velocity is shown in Fig. 9.23. The user may notice on the graph that for each range is a horizontal line called the bin limit. The bin limit is the level where the standard deviation of the velocity measurement becomes less than 1/3 times the frequency bin spacing in the FFT.

Below this limit the accuracy of measurement of a constant velocity object cannot be predicted by multiplying the standard deviation by the inverse of the square root of the number of data samples because the noise on the measurement is less than the bin spacing. Above this point the signal may be averaged to predictably improve repeatability. It must be pointed out that this is only for objects with very constant velocities. The entire 2 m s^{-1} range lies below the bin limit.

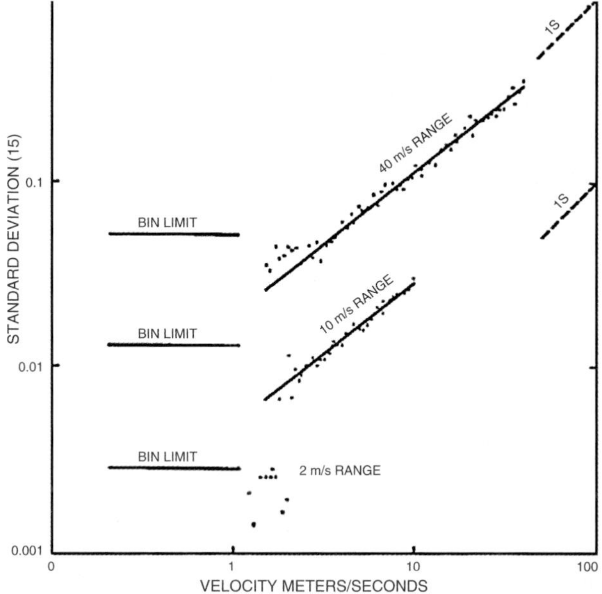

Fig. 9.23. Standard deviation of velocity measurements

This can be explained by noting that to produce a 200 kHz Doppler signal from a part moving $2\,\mathrm{m\,s^{-1}}$ there must be $100\,\mathrm{fringes\,mm^{-1}}$ on the part producing a total of 1,200 fringes over a 12 mm illuminated area. This large number of fringes produces a natural frequency peak width of 1,200 of the frequency being measured, well below the resolution of the FFT. Therefore, the peaks produced are so narrow that they do not produce a distribution of frequencies.

Setting the motor to a constant speed and noting any changes in the velocity output can also use the calibrator to check the instrument for drift. Such measurements show zero drift with time. This is to be expected since the only possible source of drift in the instrument is the crystal oscillator, which controls the data-sampling rate. These tests were repeated with the frequency shifter in operation, and the statistical results were unchanged.

The length measurement capability is tested by manually pulling a strip of Mylar tape through a gate in front of the velocimeter. The tape is 25 mm wide and 30.398 m long. Mylar was used because it is very resistant to stretching and, therefore, makes a repeatable sample for length measurement. The results of several measurements made in the $2\,\mathrm{m\,s^{-1}}$ range are shown in Table 9.1. From the velocity calibration it is known that the $2\,\mathrm{m\,s^{-1}}$ range is 0.3% high; if this is taken into account, the corrected measured mean is 30.461 m, which is within 0.2% of the actual length. This test shows that if the instrument is properly calibrated, the length measurement will be within the 0.3% required accuracy with a 3% variation of 0.1%.

Table 9.1. Length measurement data

Tape length (m)
30.543
30.553
30.565
30.554
30.536
30.560
Mean = 30.552 m
Standard deviation = 0.011 m

The instrument tests on line at steel mills were designed to check the accuracy and reliability of length measurements under harsh environmental conditions. To operate properly in a steel mill environment, the instrument must be able to handle high heat loads, scale on the surface of the part, steam, and lots of dirt. One steel mill test was on a continuous casting bar mill producing 5 m long X 172 mm diameter bars. The bars passed by the measuring station at a temperature of 9,000°C. Setting up the gauge was a very simple procedure. The gauge head was positioned alongside the roller conveyor immediately following the casting operation.

The head was installed 1 m from the conveyor with an air window and a Transit shield to protect the gauge from the heat. The unit was supported on a steel table with wood cribbing and wedges to adjust the height. The electronics enclosure was located on the floor −6 m away. The total installation time was −30 min. During operation of the mill, the castings were passing by the measurement head at a rate of one every 45 s at speeds between 0.75 and 1 m s^{-1}. The gauge was set on the 2 m s^{-1} range, and the resulting lengths were logged on a portable printer. To verify the length measurements being made by the velocimeter, four of the bars were taken off the line immediately following the measuring station while they were still hot.

These bars were then checked by tape measure with an accuracy of 12 mm. The resulting measurements are shown in Table 9.2. Taking into consideration the uncertainty in the tape measurements and the known 0.3% calibration error in this range, the data in Table 9.2 are consistent.

Measurement Setup

Typical cables were connected from the current source of an LDC-3916 Laser Diode Controller to a resistive load as shown in Fig. 9.24. A special adapter was made for the 9-pin connector on the LDC-3916, so voltages could be measured directly at the connector. The current source was set to drive a constant current through the load. Voltages were measured across the current

Table 9.2. Steel mill test data

Bar	Velocimeter (m)	Tape (m)	Delta (m)	Delta (%)
26	4.812	4.801	0.011	0.21
27	4.565	4.572	0.007	−0.17
30	4.438	4.420	0.018	0.40
40	4.444	4.432	0.012	0.27
Mean error in meters = 0.009 m				
Mean error in percent = 0.18%				

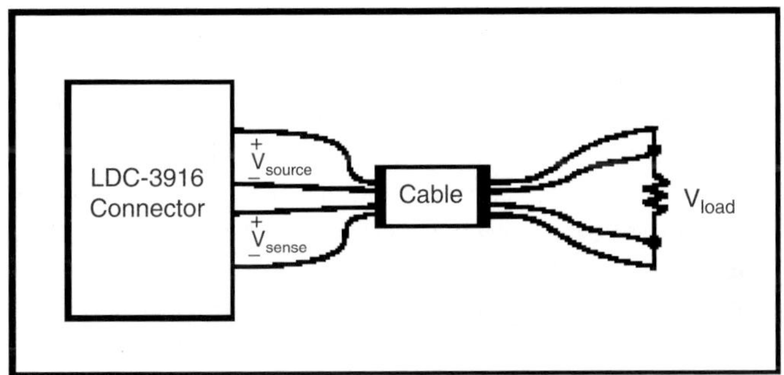

Fig. 9.24. Measurement setup

Table 9.3. Results from voltage measurement

	Current	Load Ω	Cable (ft.)	V source (V)	V load (V)	V sense (V)
Case 1	250 mA	10	6	2.66	2.50	2.50
Case 2	500 mA	10	6	5.26	4.99	4.99
Case 3	1.0 A	5	6	5.52	4.99	4.99
Case 4	250 mA	10	16	2.84	2.50	2.50
Case 5	500 mA	10	16	5.67	4.98	4.98
Case 6	900 mA	5	16	5.71	4.49	4.49

source (V source), resistive load (V load), and voltage sense connections on the LDC-3916 connector (V sense). Cable lengths, resistor values, and current source setpoints were changed to measure different conditions.

Results are shown in Table 9.3 for six different conditions. The four wire voltage sense of the LDC-3916 increases the accuracy of the voltage measurement across the load. Increased voltage measurement accuracy translates to better laser diode protection, particularly in test setups with high current drive, and/long cable runs between the controller and the laser diode.

9.3.2 LASER Heating, Welding, Melting, and Trimming

Laser welding is a new joining technology introduced to the South African industry. The process involves the joining of any weldable material using a laser beam. Laser welding is used in joint designs such as butt, T, lap, corner and edge joints. Joints are welded by means of butt, fillet, spot, and stake (continuous spot-type) welds. Weld joining comprises autogeneous, addition of cold wire, and addition of hot wire (hybrid welding) techniques. This advanced welding process is extensively used in first world countries to extend available options in automated welding and increase productivity and profit. It allows joints to be welded at very high speeds and at low heat-input rates, resulting in relatively small heat affected zones (HAZ) and minimal distortion. This new technology offers industry the opportunity to become world-class manufacturers that are globally competitive.

High-power near infrared diode lasers are of great interest to both industry and research institutions. Several different approaches are being developed for laser welding of plastics. The main principle now used to laser-weld plastics is known as "transmission welding." Laser welding for plastics is achieved through a process known as "Through Transmission Infrared" (TTI) as shown in Fig. 9.25. This process sends the intense light beam through one component (the transmitting part) to be absorbed by the other component (the absorptive part) at the weld interface while the parts are clamped together.

The energy created at the absorptive component causes molecular vibrations that heat the plastic. As the absorptive component starts to melt and expand, the heat and melted area are transferred to the other component. The end result is an intermolecular mixing of the two components. Laser welding depends on color, specifically on the color of the part to be welded. More scientifically, one material must transmit the laser light while the other absorbs it, and converts it to heat.

The basic thing is that the materials must be transmissive or absorptive only at the specific wavelength of the laser source used. Parts that appear to be black or some other color to the human eye can be either transmissive

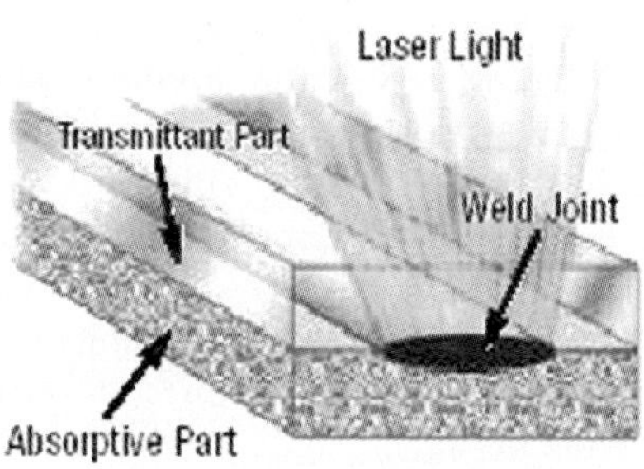

Fig. 9.25. Through Transmission Infrared (TTI) process

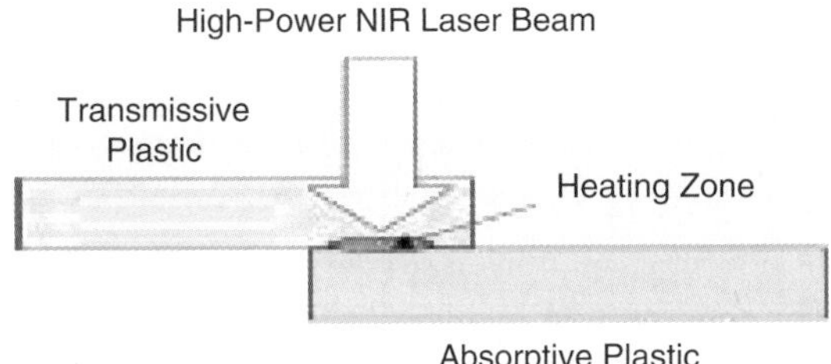

Fig. 9.26. Principle of Transmission Welding

or absorptive depending on the formulation of the pigment. Even joints that require optical clarity (clear-to-clear) can be achieved by the use of special coatings.

Laser welding of thermoplastics depends on the same rules of resin compatibility that the other processes do, but is more forgiving of resin chemistry or melt temperature differences than most other plastic welding processes. Contained weld joints have no flash or particulate outside the joint to cause problems.

Transmission welding in Fig. 9.26 demonstrates that precise, controllable heating and melting of low melting point thermoplastics can be produced at the interface between a transmissive and absorptive plastic.

Welding Trials

Relative quick transverse motion between a high-power diode laser beam and the heating zone can produce a strong welded seam at high coverage rates. This has been demonstrated with both conventional solid-state lasers and high-power diode lasers. Although some research dates back a number of years, no commercial industrial applications of the process have emerged until recently.

To enable laser transmission welding of plastics, three major elements must be considered:

1. Is the transmission of the transmitting layer and the absorption of the absorbing layer adequate at near infrared wavelengths?
2. Are the materials of the joint similar enough to make them weldable?
3. Does the laser provide enough energy to melt a large area at adequate speeds?

Each individual joint configuration and material combination requires a different average power for a particular weld speed at a particular width.

Even nylon and high density polyethylene (HDPE) have been laser welded, showing that it is only necessary for the polymer to transmit greater than 10% of the thermal energy in order to produce adequate melting at the joint interface for these 3 mm thick materials. To determine the mutual weldability

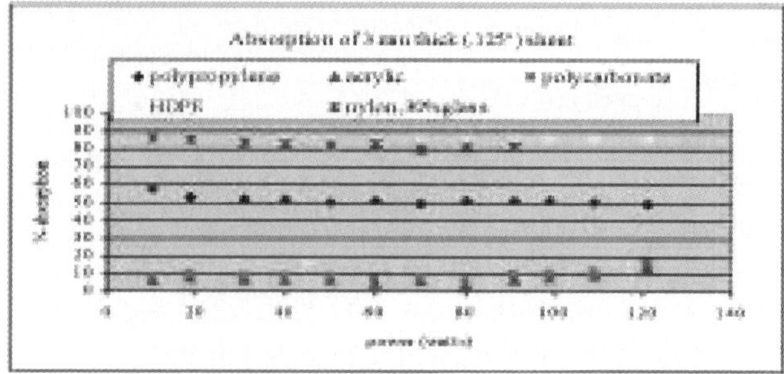

Fig. 9.27. Transmission of Major Groups of Polymers

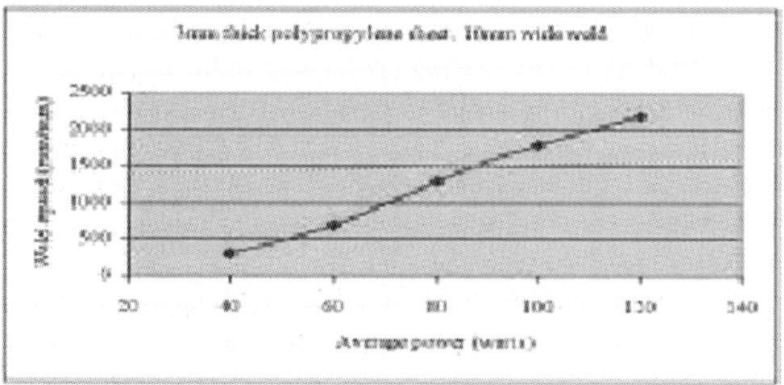

Fig. 9.28. Plastics welding performance using a 120 W FAP system

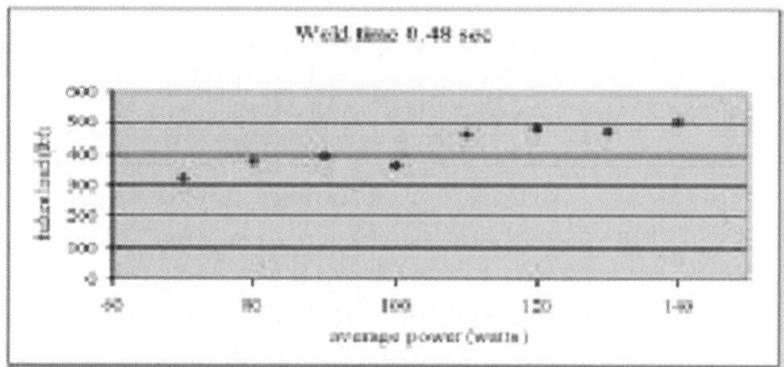

Fig. 9.29. Effect of average power on weld strength

of polymers, chemical similarity between the two components of the joint is essential. The transmission of polymers is shown in Fig. 9.27. However, it is relatively easy to determine laser weldability.

Welding of polypropylene homopolymer was chosen as the subject in this process because although it is widely used in a range of industries it is generally considered to be difficult to join. The performance of plastic welding using 120 W FAP system and the effect of average power as shown in Fig. 9.28 and Fig. 9.29, respectively.

Laser Technology

Diode lasers shown in Fig. 9.30 are essentially high-powered LEDs that operate at a specific wavelength. They are relatively compact, light, and cost-effective. Diode lasers are available in a wide range of power levels at both 808 and 940 nm wavelengths.

Nd:YAG lasers shown in Fig. 9.31 use a solid rod of rare earth material as the lasing medium. The longer-wavelength 1,064 nm light produced is useful in more applications than basic diode lasers, and a broader and higher range of power levels are available.

CO2 lasers shown in Fig. 9.32 are effective for welding thermoplastic films, and are available in very high power levels to accommodate very fast weld line speeds. They are also effective for cutting thermoplastics.

Direct beam delivery shown in Fig. 9.33 use optics that are mounted directly on the laser housing and fixed in focal length and beam position relative to the housing. They are used primarily in systems using the moveable part concept or with diode lasers mounted on robot arms.

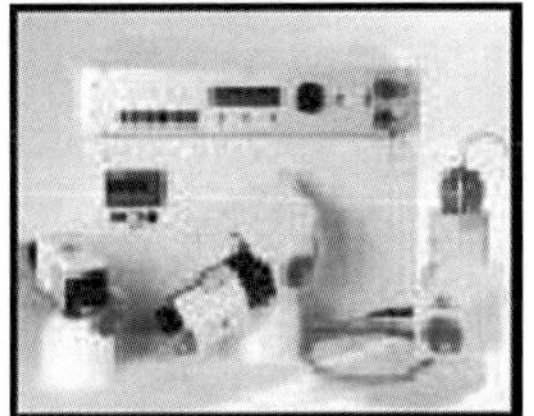

Fig. 9.30. Diode lasers

Fig. 9.31. Nd:YAG lasers

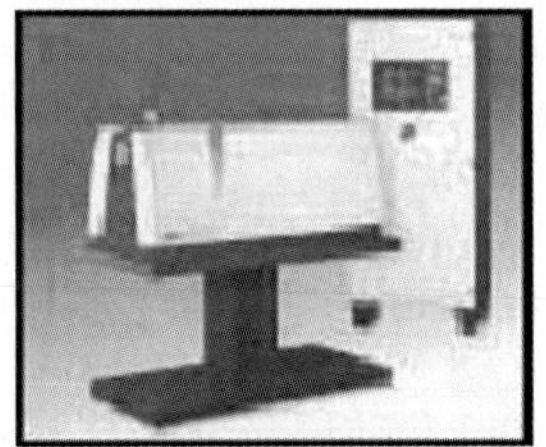

Fig. 9.32. CO2 lasers

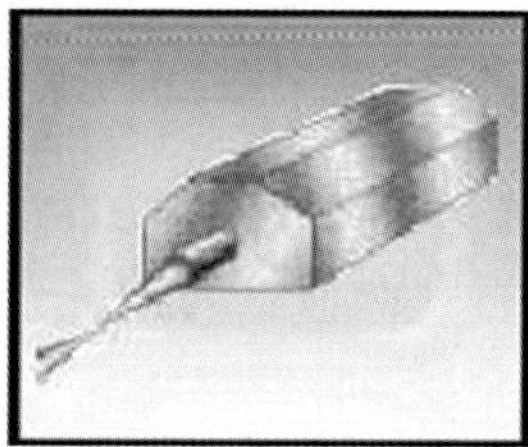

Fig. 9.33. Direct beam delivery

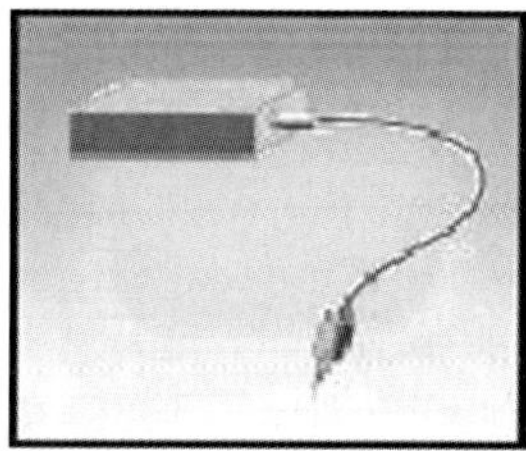

Fig. 9.34. Fiber delivery

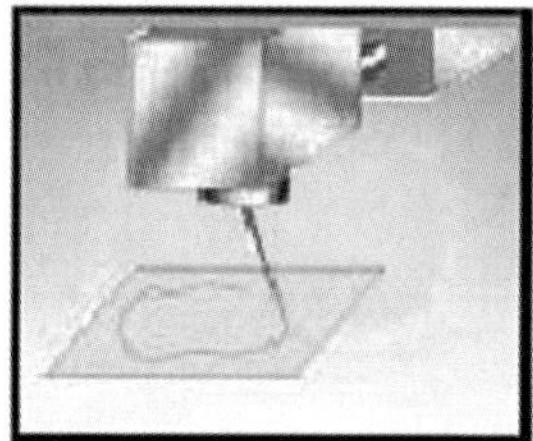

Fig. 9.35. Galvo heads

Fiber delivery shown in Fig. 9.34 is used to get the beam into a tight spot where the laser housing may not be able to go. It is useful for delivering an Nd:YAG or diode beam to a robot end effector or for fixed mounting in a small space within an automated machine.

Galvo heads shown in Fig. 9.35 can steer the laser beams at amazing speeds to allow for complex contour or quasisimultaneous operations. Galvo heads can also be mounted on moving machine elements or robots, and can also distribute a fiber-delivered beam.

Integration of the laser beam shown in Fig. 9.36 controls with the beam or part positioning system is simple with Dukane's laser welder software tools. Using sophisticated computer controls, it is ensured that each weld program results in a robust process and maximized part quality.

Beam splitting is useful for welding larger parts. A single higher-power laser beam can be split into two beams each at half the power of the original beam,

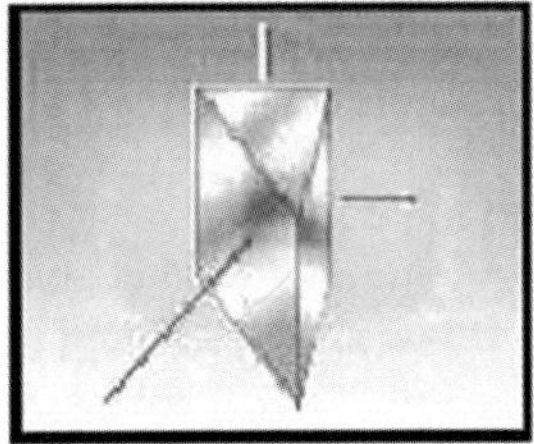

Fig. 9.36. Integration of the laser beam

and may be divided again if need be. This allows for multiple-part production from a single motion source, or attachment of multiple galvo heads to a single laser so quasisimultaneous welding can occur on a part larger than the working field of a single galvo head. Cost-per-unit-power drops as laser size increases, so beam-splitting is a good option for welding large parts quickly.

Positioning systems run the full spectrum from simple servo stages or galvo heads to robotic installations. *Enclosures* are required for laser systems in a factory work environment. All Dukane enclosures are Class I laser tight and utilize laser-safe glass viewing windows where appropriate. Guillotine doors are utilized for access to the fixturing area for operation, setup, or automated parts handling.

Advantages of Laser Welding

- The laser weld process can produce visually pleasing joints on parts with complex geometries, both very large and very small. The process is extremely precise and controllable. Full wall collapse welding in quasi-simultaneous mode even has some gap-filling capability.
- Laser welding is versatile and capable, and it probably costs less. The dedicated tool for laser welding is also relatively inexpensive.
- With laser welding, joints are dust-proof, humidity-proof, and crack resistant. They feature excellent aesthetics and can be made invisible.
- Laser welding process has it is Through Transmission Infrared Process
- Deep narrow welds are possible
- Metallurgical damage and severe grain growth is vastly reduced
- Microstructures are much finer, harder and more corrosion-resistant than other processes
- Component design opportunities are enhanced
- Welds can be made where access is limited and from one side only
- Welds can be made very close to heat-sensitive components
- High reproducible high-quality welds with relatively smooth surfaces can be obtained.

Table 9.4. Applications of LASER in industries

Industry sector	Application area
Aerospace	Weld-joining of airframe components
	Weld-joining of aero engine components
	Welding combustion chamber components
Automotive	Weld-joining of pressed sheet body parts
	Welding of tailored blanks (coated/uncoated)
	Welding of high strength steel
	Welding catalytic converters
Defence	Low-distortion welding of structures
Domestic products	Low-distortion welding of domestic appliances
	Welding of double glazing frames
	Electronic Weld-joining of electronic packages
Electronic	Weld-joining of electronic packages
Maintenance & repair	Weld repair of gas turbine blades & nozzle guide vanes (aero engine / power generation turbines)
	Sleeving technology (power generation repair)
	Weld rebuilding or cladding of worn surfaces
Oil & gas	Welding of thin-wall pipework
Packaging	Welding of thin sheet for can or tube applications
Power generation	Weld-joining of pipework & containers
Process plant	Weld-joining of tube-to-tube plate
Shipbuilding	Butt, fillet & stake welds

Typical Application Areas of Laser Welding

Laser transmission welding of 10 mm wide joint lines have been demonstrated using high power diode lasers. These optimized welds have been produced at speeds greater than 2,000 mm min^{-1}. The high power/low intensity beam produced by the 120 W FAP System has proven ideal for large area high quality welds. Some of the applications of Laser in industries are tabulated in Table 9.4

9.3.3 Laser Trimming and Melting

In the petroleum industry, perforating is a process of piercing the casing wall and the cement behind it to provide tunnels through which formation fluids may enter the wellbore. Current technology uses a perforating gun, or perforator to make these tunnels as shown in Fig. 9.37. The completion crew lowers the long cylindrical gun down the production casing or liner until it is opposite the reservoir zone. The bullets or special explosive charges carried by the perforator are aimed at the walls of the casing and shoot smooth, round holes in the casing and penetrate the rock as well.

The high velocity shaped charge jet creates high shock pressure of 1.5 million psi at tunnel entrance to 150,000 psi at the perforation tip, which com-

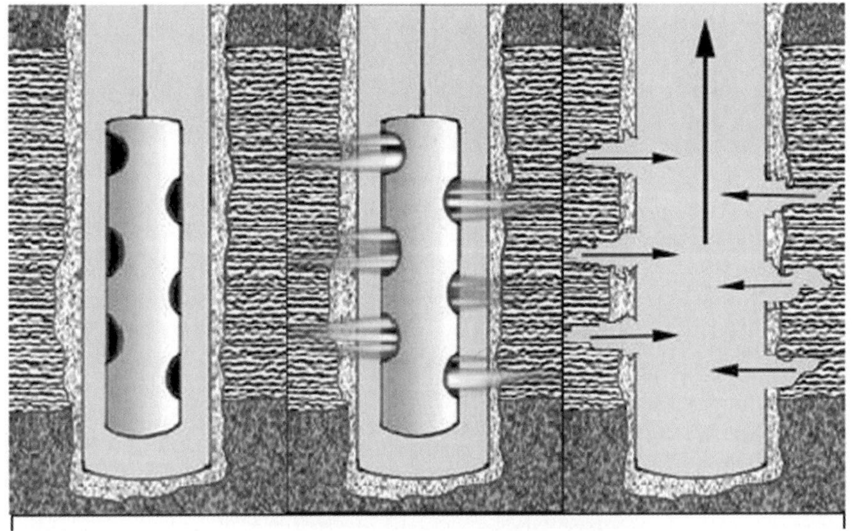

Fig. 9.37. Conventional perforation gun – lowered into a production zone

minutes the adjacent rock, fractures sand grains, fails inter grain cementation, debonds clay particles, and creates tunnels which are typically 0.25–0.4 in. in diameter and 6–12 in. in length. This mature technology has some disadvantages such as lack control of hole size and shape, and low flow performance due to reduction of permeability of perforated or crushed rock.

Permeability reduction factor values (permeability ratio of crushed zone to rock matrix) of 0.1–0.7 were reported in previous research works. Recent advances in high power laser technology provide a new tool to replace the current perforating gun for creating the holes. The laser perforator has the flexibility of drilling holes with different sizes and shapes. Previous laser rock tests also confirmed the increase of permeability of laser-drilled rocks.

Laser-induced permeability increase as high as 500–1,000% was reported on Berea gray sandstone rocks drilled by high power laser beams. The permeability increase is attributed to the micro cracks that were formed due to laser-induced clay dehydration and thermal cracking. Three drilling methods and their experimental results are presented in this section. The goal of the experiments was to drill a hole on rock as deep as possible using a series of laser bursts.

Method One: Fixed Beam

A fixed, defocused beam of 1-in. in diameter was fired on a rock with two 650 purging tubes in symmetrical configuration as shown in Fig. 9.38. The laser head can be lowered down to compensate the beam spot size change at the bottom of the hole as the hole gets deeper, but the available moving

Fig. 9.38. Method one – setup

distance is limited by the purging tubes first and then the laser head itself. In the experiment, the 1 in. defocused beam was pointed at a shale sample, 3 in. thick and 3 in. in diameter. The CO_2 laser power was 4,000 W and nitrogen flow rate from the two 650 purging tubes were set at 200 cubic foot per hour (cfh) each.

Method Two: Circular Motion Beam

In this method, the rock sample was moved circularly by the workstation under the fixed vertical beam and purge gas. This generated a relative circular motion of a defocused beam of 0.5 in. in diameter on the rock in a 0.5 in. diameter circle. A one inch diameter hole was formed by this circling beam after one revolution. A purging tube inside the hole circled together with the beam and was moved down after each revolution providing constant strong purging at the bottom of the hole as the hole got deeper. A 4 in. diameter by 6 in. thick limestone sample was lased by a circularly moved CO_2 beam. The beam power was 4000 watts and the gas flow rate was 300 cfh. The laser head was moved down 0.5 in. between bursts. One burst here is defined as one revolution that the beam rotates. The beam moved at a linear speed of 50 in. min^{-1}.

Method Three: Rotary Rock Method

In the third method the core rock sample is clamped by a rotary chuck and rotates around its own axis. The horizontal 0.5 in. diameter beam and 1/8 in. purging tube are positioned about $\frac{1}{4}$ in. away from the core axis and kept stationary for each lasing cycle, then are adjusted between the circles to keep the constant spot size and gas flow at the bottom of the hole as the hole gets deeper. A continuous wave CO_2 laser beam was used. The purging gas was nitrogen with flow rate 275 cfh. Two power levels, 4,000 and 2,500 W and four rotary speeds, 10,000° min^{-1}, 5,000° min^{-1}, 3,000° min^{-1}, and 2,000° min^{-1}, were tested.

9.4 Smart Instruments

This section focuses on the history of Smart Transducers and their standards along with the HART protocol which is a widely recognized standard in industry.

9.4.1 Smart Intelligent Transducers

The goals of the IEEE-P1451.2 Smart Transducer Interface provided to the user are summarized as:

- To select the transducer that is best suited to solve the measurement or control problem independently of the selected control network.
- To use the same transducers on multiple control networks.
- To select the control network that is best suited for the application without transducer compatibility constraints.
- To achieve automatic self-configuration when a transducer is connected to a network microprocessor.

At the core of the effort, Transducer Electronic Data Sheet (TEDS), which is a data structure stored in a small amount of nonvolatile memory, physically associated with the transducer was developed.

The TEDS is used to store parameters, which describe the transducer to the Network Capable Application Processor (NCAP), making self-identification of the transducer to a system possible. The working group has defined the contents of the TEDS and a digital hardware interface to access the TEDS, read sensors, and set actuators. The resulting hardware partition encapsulates the measurement aspects in a Smart Transducer Interface Module (STIM) on one side of the digital interface, and the application related aspects on the NCAP.

Background

Control networks provide many benefits for transducers, such as:

- Significant reduction of installation costs by eliminating long and large numbers of analog wires.
- Acceleration of control loop design cycles, reduction of commissioning time, and reduction of downtime.
- Dynamic configuration of measurement and control loops through software.
- Addition of intelligence by leveraging the microprocessors used for digital communication.

One major problem for analog transducer manufacturers is the large number of networks on the market today. It is currently too costly for many transducer manufacturers to make unique smart transducers tailored for each network on the market. In September 1993, the proposal of developing a smart sensor communication interface standard was accepted by IEEE-TC9.5 In March, 1994, the National Institute of Standards and Technology (NIST) and the Institute of Electrical and Electronics Engineers (IEEE), hosted a first workshop to discuss smart sensor interfaces and the possibility of developing a standard interface that would simplify connectivity of smart transducers to networks. Since then, a series of four more workshops have been held and two technical working groups formed in February, 1995:

- The P1451.1 working group concentrating on a common object model for smart transducers along with interface specifications to the model.
- The P1451.2 working group concentrating on defining the TEDS, the STIM, and the digital interface including connector pin allocation and a communication protocol between the STIM and the NCAP.

Technical Key Features

The STIM is shown here under the control of a network-connected microprocessor. In addition to their use in control networks, STIMs can be used with microprocessors in a variety of applications such as portable instruments and data acquisition cards. The STIM embodies specific unique features of this proposed standard, which are briefly described later.

Single General Purpose TEDS

The TEDS as presently defined supports a wide variety of transducers with a single general purpose TEDS structure. This approach makes the rest of the system easier to implement and the implementation scaleable. If specific fields are not required for a given transducer, these fields have zero width, saving the required memory.

Representation of Physical Units

The P1451.2 draft adopts a general method for describing physical units sensed or actuated by a transducer. A unit is represented as a product of the seven SI7 base units and the two SI supplementary units, each raised to a rational power. This structure encodes only the exponents; the product is implicit. The forms are for expressing "dimensionless" units such as strain (meters per meter) and concentration (moles per mole). The numerator and denominator units are identical, each being specified by subfields 2–10.

General Calibration Model

The P1451.2 draft provides a general model to optionally specify the transducer calibration. It is very flexible yet can collapse to an acceptable size for a simple linear relationship. The scheme showed in Fig. 9.39 supports a multivariable, piecewise polynomial with variable segment widths and variable segment offsets.

Triggering of Sensors and Actuators

The proposed digital interface has hardware trigger lines to allow the NCAP to initiate sensor measurements and actuator actions, and to allow the STIM to report the completion of the requested actions. The NCAP can trigger an individual channel, or all transducer channels at once. In the latter case, there are TEDS fields provided to specify timing offsets between the STIM's channels and to determine when each measurement or actuation has occurred

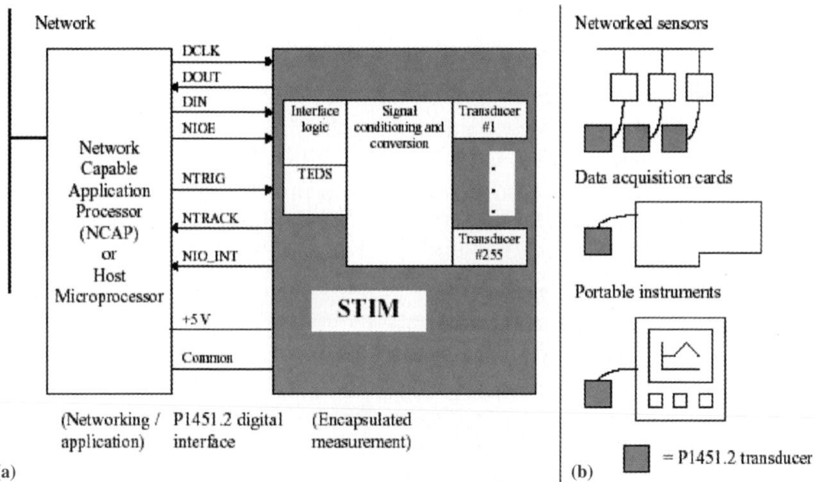

Fig. 9.39. Hardware partition proposed by P1451.2 and possible uses for the interface

relative to the single trigger acknowledgment. The draft proposes that the slowest channel be the reference channel and that all the offsets be specified relative to this channel.

Variable Transfer Rate Between Host and STIM

The hardware data clock line is driven by the NCAP. There is a field in the TEDS which specifies the maximum data transport rate that the STIM can support. This provides a flexible mechanism to match NCAPs and STIMs.

Support for Multivariable Transducers

P1451.2 includes support for multivariable transducers in a single STIM. An STIM may have up to 255 inputs or outputs allowing the creation of multivariable sensors, actuators, or combinations of several multivariable STIM.

STIM

The block diagram for an STIM, along with the interfaces between each module is shown in Fig. 9.40. A TEDS is incorporated into an STIM. In addition to the TEDS, the STIM contains logic to implement the P1451.2 interface, the transducer, and any necessary signal conditioning and signal conversion. Only interface "A" is defined by P1451.2. Interfaces "B," "C," "D," and "E" allow transducer manufacturers to continue to obtain competitive differentiation in areas of performance, quality, feature set, and cost by choosing how these interfaces are implemented. At the same time P1451.2 offers the opportunity to design transducers to a common interface between an STIM and NCAP enabling the use of the same transducer across many networks and applications.

The P1451.2 logic block may be implemented in several ways. The working group has now implemented STIMs using a field programmable gate array (FPGA) and a low-cost microcontroller to serve as the logic block. These methods demonstrate that P1451.2 STIMs can be built today using off-the-shelf parts. The microcontroller option provides the additional advantage of potentially combining all the logic, TEDS, and signal conversion into

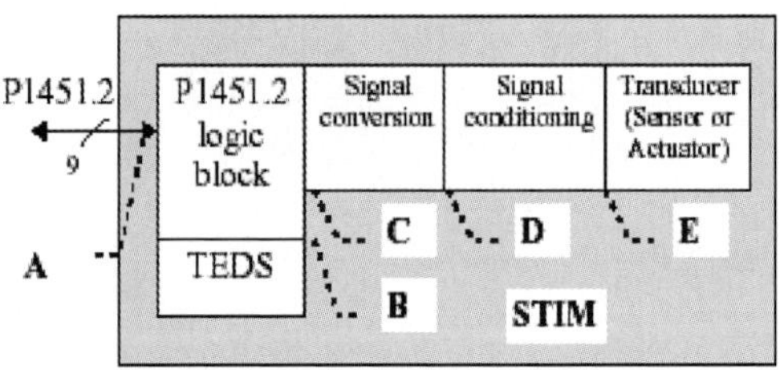

Fig. 9.40. Logic block diagram of STIM

(a) Temperature sensor STIM

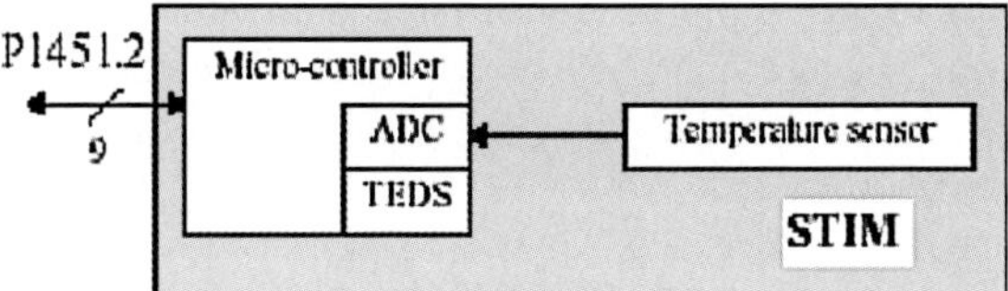

(b) Eight channel digital I/O STIM

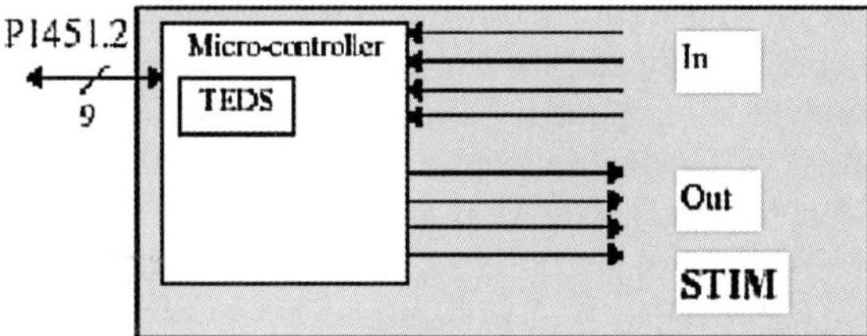

(c) Four channel sensor STIM

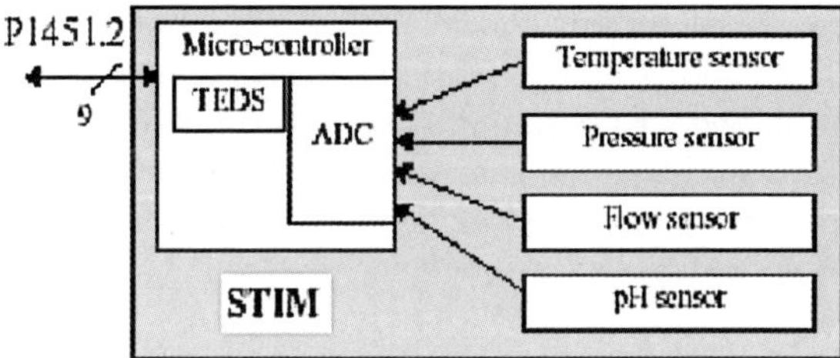

(d) Sensor and actuator STIM

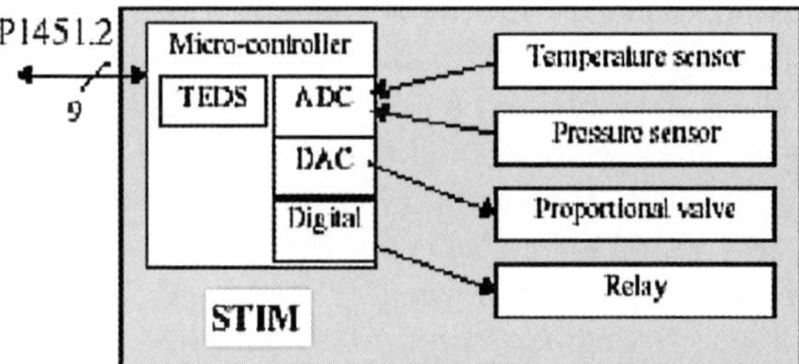

Fig. 9.41. STIM configuration

one integrated circuit, where the P1451.2 logic block is implemented using microcontroller firmware. Figure 9.41 shows four examples of STIM configurations using a low-cost microcontroller. These examples demonstrate the flexibility in STIM design provided by P1451.2.

The TEDS model in P1451.2 allows this STIM to be described as an eight-channel STIM or alternatively it could be described as a two-channel STIM with one input channel and one output channel each with a length of four. This flexibility in the model allows digital I/O modules with thousands of inputs/outputs to be implemented if such a product were needed. These four sensors could be measuring a process liquid. The code for implementing the control loop of the P1451.2 interface resides either in the NCAP or the microcontroller.

TEDS

The TEDS is one of the main technical innovations introduced in P1451.2. A TEDS, which carries information about the transducer and its performance, is not a new concept. Companies have been embedding data structures in memory associated with their products for many years. What is new is the general model of a transducer behind the P1451.2 TEDS, which supports a wide variety of sensors and actuators. The TEDS contains fields that fully describe the type, operation, and attributes of the transducer. If the transducer is moved to a new location, it is moved with the TEDS.

This way, the information necessary for using the transducer in a system is always present. The sections shown with dotted lines (calibration-TEDS, application specific-TEDS and extension-TEDS) are optional. The calibration specification in the TEDS permits the sensor manufacturer to describe a multidimensional calibration for each channel. To eliminate high order polynomials it is possible to specify a segmented calibration where each segment can have a variable width and offset. It is expected that a general correction engine will be present in the NCAP that understands this calibration scheme so that it can be run "blindly" no matter which transducer is attached.

It is a ceramic pressure sensor with an analog output between 0 to 5 V dc corresponding to 0–20,684,190 Pa (3,000 lb in.$^{-1}$) pressure input. The sensor has a 10 ms response time, no appreciable warm-up requirement and the maximum nonlinearity is measured to be 0.56% of Voltage supply. The primary components of the single channel STIM are a serial 12-bit analog-to-digital converter (ADC) for data conversion (75 ms conversion cycle), and an 8-bit PIC-type processor with 4 K by 12-bit on-chip EEPROM (8 MHz operation). Calibration is fixed and is specified using five equal segments with nonzero offsets for each segment. This allows first order calibration functions to be used to reduce the nonlinearity in the analog output (0.56% reduced to 0.03%) as shown in Fig. 9.42.

This specific TEDS implementation required the following amount of memory:

- Meta TEDS 366 bytes
- Channel TEDS 96 bytes
- Calibration TEDS 103 bytes

Out of the total 565 bytes, the manufacturer identification consumed 55 bytes, and product description consumed 205 bytes, for a total of 260 bytes or 46% of the TEDS.

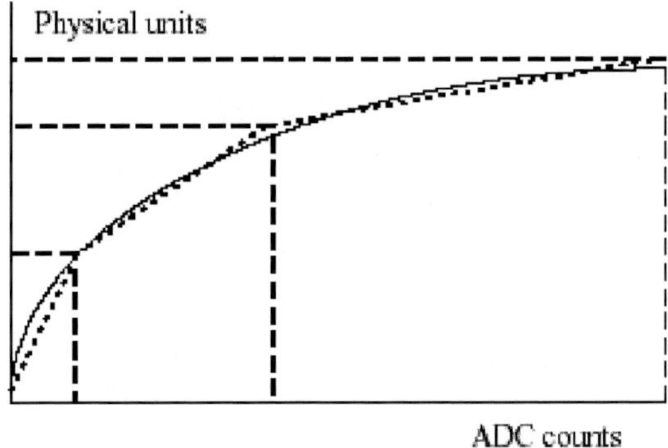

Fig. 9.42. Multisegment calibration curve

Digital Interface

Basic communications between the NCAP and STIM require four lines (DCLK, DOUT, DIN, and NIOE). DCLK is driven by the NCAP. The data transfers are based on SPI-like (serial peripheral interface), bit-transfer protocol. The NCAP drives the NTRIG line to initiate a measurement or action, and the STIM uses the NTRACK line to acknowledge that the requested function has been performed. The STIM can notify the NCAP of any exception conditions by use of the NIO_INT line.

The P1451.2 draft defines a set of status registers to support notification of standard exceptions such as hardware errors, busy channels, STIM power cycle, calibration failure, and self-test failure. A power supply of +5 V is also a part of the interface specification.

"Plug and Play" Operation

The operating mode for P1451.2 transducers is "plug-and-play" as indicated in Fig. 9.43. Since the TEDS reside with the transducer, one is able to move the transducer from one NCAP to another and simply plug it in to achieve self-identification.

The pressure STIM can be connected either to the same network or plugged into any of the NCAPs on the second network. "Plug & play" operation requires standardization of a connector. The P1451.2 draft proposes three types of connectors which have been defined for different environments where the standard may be used.

It is important to note that P1451.2 may be implemented as a single row of pins or wires connecting two circuit boards (an NCAP and an STIM) together. This implementation could be used internally on a product where the separation between NCAP and STIM is not visible to the user. A company may want to build several networked transducers and use a common internal interface (P1451.2) to join the transducer portion with the network/application portion

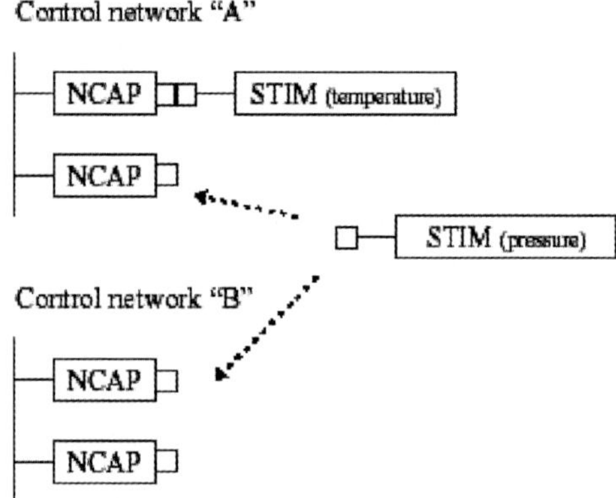

Fig. 9.43. Plug-and-play model of use

of the networked smart transducer. This may be as simple as firmware driving an eight bit microprocessor port or it may be hardware assisted. Figure 9.43 shows a very simple NCAP firmware block diagram with three blocks: STIM driver, application code and the network protocol stack. The STIM driver is composed of four major functions:

- "Bits and bytes" is responsible for getting data across the interface.
- "TEDS parser" has knowledge about the P1451.2 TEDS structure and assembles the data into meaningful pieces.
- "Correction engine" is the algorithm that converts raw readings from the STIM into units specified in the TEDS for sensors or units specified in the TEDS into STIM settings for actuators.
- "P1451.2 application programming interface (API) driver" provides access to TEDS blocks, sensor readings, actuator control, triggers and interrupt requests.

Most of the TEDS' contents need not be parsed and kept on the NCAP. This will depend on the measurement or control problem being solved and differentiation desired in NCAP products. If the application needs the contents of the TEDS for use in another part of the system, then the TEDS needs little parsing. If the NCAP will be making sensor measurements and sending sensor readings in engineering units, then the calibration factors must be parsed and kept on the NCAP for use by a correction engine.

In principle, only one STIM driver for each kind of NCAP is required to support P1451.2 as shown in Fig. 9.44. For each microprocessor family supporting this proposed standard there could be a single software driver for the P1451.2 interface, a single TEDS parser, and a single conversion engine. It is expected that the network providers will develop the drivers for the popular networks.

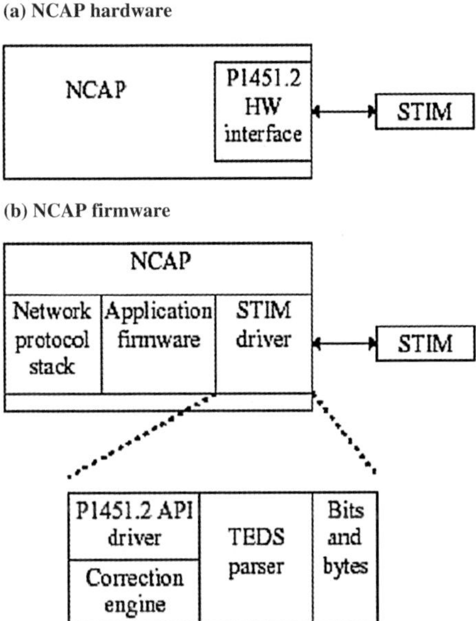

Fig. 9.44. NCAP support required for P1451.2

9.4.2 Smart Transmitter with HART Communicator

Highway Addressable Remote Transducer (HART) Field Communications Protocol is widely recognized as the industry standard for digitally enhanced 4–20 mA smart instrument communication. Use of the technology is growing rapidly, and today virtually all major global instrumentation suppliers offer products with HART communication. The HART protocol provides a uniquely backward compatible solution for smart instrument communication as both 4–20 mA analog and digital communication signals are transmitted simultaneously on the same wiring. HART provides many benefits promised by field bus, while retaining the compatibility and familiarity of existing 4–20 mA systems.

The HART Technology

The HART protocol makes use of the Bell 202 Frequency Shift Keying (FSK) standard to superimpose digital communication signals at a low level on top of the 4–20 mA. Since the digital FSK signal is phase continuous, it does not interfere with the 4–20 mA signal. A logical '1' is represented by a frequency of 1,200 Hz and a logical '0' is represented by a frequency of 2,200 Hz. The HART FSK signaling enables two-way digital communication and makes it possible for additional information beyond just the normal process variable to be communicated to or from a smart field instrument. The HART protocol communicates at 1200 bits per second without interrupting the 4–20 mA signal and allows a host application (master) to get two or more digital updates per second from a field device.

Flexible Application

HART is principally a master/slave protocol which means that a field (slave) device speaks only when spoken to by a master. Two masters (primary and secondary) can communicate with slave devices in a HART network. Secondary masters, such as handheld communicators, can be connected almost anywhere on the network and communicate with field devices without disturbing communication with the primary master. A primary master is typically a Distributed Control Systems (DCS), Programmable Logic Control (PLC), or computer-based central control or monitoring system.

The HART protocol can be used in various modes for communicating information to/from smart field instruments and central control or monitoring equipment. Digital master/slave communication simultaneous with the 4–20 mA analog signal is the most common. This mode, depicted in Fig. 9.44, allows digital information from the slave device to be updated twice per second in the master. The 4–20 mA analog signal is continuous and can still carry the primary variable for control.

Burst or Broadcast Mode

Burst is an optional communication mode (Fig. 9.45) which allows a single slave device to continuously broadcast a standard HART reply message. This mode frees the master from having to send repeated command requests to get updated process variable information.

The same HART reply message (PV or other) is continuously broadcast by the slave until the master instructs the slave to do otherwise. Data update rates of 3–4 per second are typical with 'burst' mode communication and will vary with the chosen command. Burst mode should be used only in single slave device networks as shown in Fig. 9.36.

Fig. 9.45. Master/slave mode or poll response model

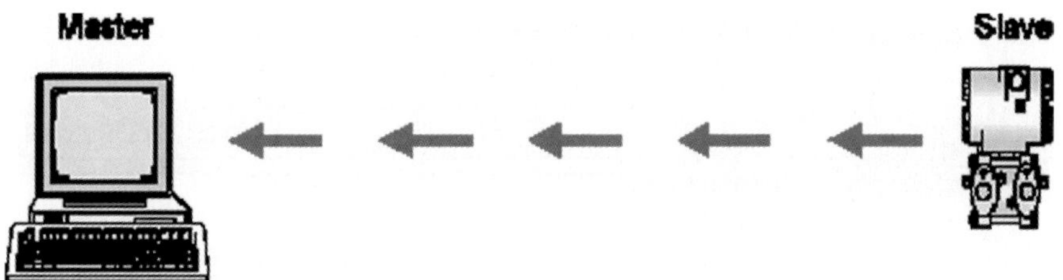

Fig. 9.46. Burst or broadcast model

The HART protocol also has the capability to connect multiple field devices on the same pair of wires in a multidrop network configuration as shown in Fig. 9.46. In multidrop applications, communication is limited to master/slave digital only. The current through each slave device is fixed at a minimum value to power the device (typically 4 mA) and no longer has any meaning relative to the process. From an installation perspective, the same wiring used for conventional 4–20 mA analog instruments carries the HART communication signals.

Allowable cable run lengths will vary with the type of cable and the devices connected, but in general up to 3,000 m for a single twisted pair cable with shield and 1,500 m for multiple twisted pair cables with a common shield. Unshielded cables can be used for short distances. Intrinsic safety barriers and isolators which pass the HART signals are readily available for use in hazardous areas.

Innovative Application Example

The power of the HART protocol is evident in the control diagram of Fig. 9.47. This innovative application uses the inherent feature of the HART protocol that both 4–20 mA analog and digital communication signals are transmitted simultaneously over the same wiring. In this application, the HART-compatible transmitter has an internal PID control capability. The device is configured such that the 4–20 mA loop current is proportional to the control output of the PID algorithm executing in the device (not the measured variable as in most transmitter applications). Since the 4–20 mA loop current is regulated by the PID control output, it is used to drive the valve position directly.

The control loop executes entirely in the field between the transmitter (with PID) and the control valve. The control action is continuous as the traditional 4–20 mA analog signal drives the valve. HART digital communication links the operator with the control loop to change set point, and read the primary variable, or valve position output as shown in Fig. 9.48. Substantial savings are possible in applications where this innovative control architecture is appropriate.

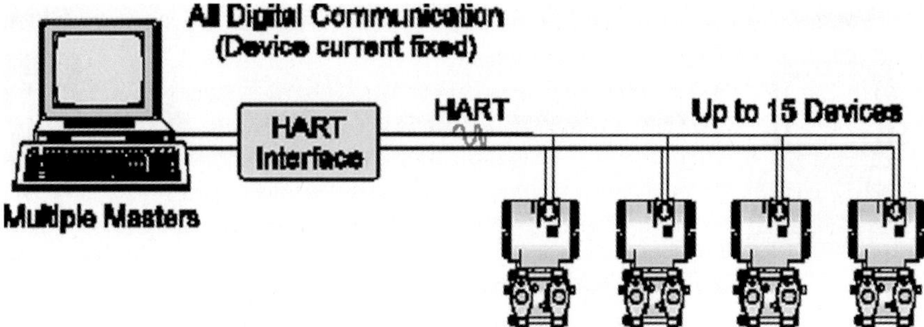

Fig. 9.47. HART field devices in a multidrop network

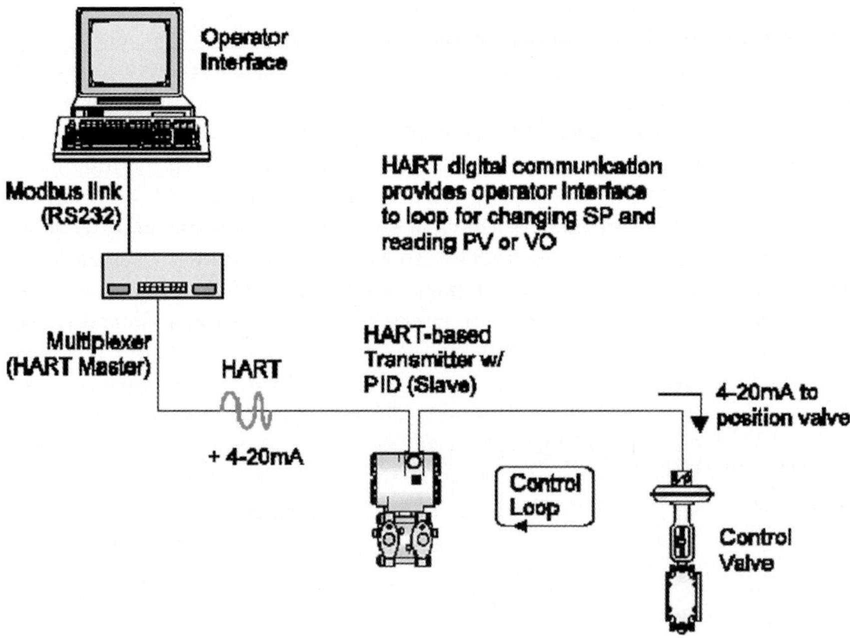

Fig. 9.48. HART instruments with cost effective control

Best Solution

The HART protocol provides users with the best solution and migration path for capturing the benefits of enhanced communication with smart instrumentation. No other communication technology can match the base of support or wide range of products that are available with HART today. The technology is easy to use and HART-compatible products are available from major instrumentation suppliers to address virtually all process measurement and control applications. The emergence of field bus will not displace HART in either existing or new production facilities.

HART provides users with many of the same benefits while retaining the compatibility and familiarity of existing 4–20 mA systems. HART allows the cost saving benefits of remote communication, flexible/accurate digital data transmission, field device diagnostics, and powerful multiparameter instruments to be captured without replacing entire systems.

Connection to current and future plant networks is assured by the digital communication capability and large installed base (more than 5,000,000 installations and growing rapidly).

9.5 Computer-Aided Software Engineering

Nowadays due to the increasing speed of changing market demands new products replace old ones much earlier than before, so the development of new products has to go faster. Thus the production lines have to be developed faster, too. A very important role in this development is software engineering because many production processes are 'computer aided', so software has to be designed for this production system. It seems very important to do the software engineering right and fast. In the past, software systems were built using traditional development techniques, which relied on hand-coding applications. Software engineers had to design software without help of computers, by programming each step at one time. This way is much too costly and time-consuming.

In order to speed up the process the bottlenecks in building software systems are to be found. This is hard to do because of the increasing role of computers in our society. Technology is developed further every day, so faster and bigger computers enter the scene. The software running on these computers can be more extensive because they can handle more information in the same time, so there is an increasing amount of data to go with it. Finding the right data out of this increasing amount of information is getting harder and harder, so finding the bottleneck is harder to do.

To speed up the software system building process, a new concept of designing software is introduced in the 1970s, called *Computer-Aided Software Engineering (CASE)*. This term is used for a new generation of tools that applies rigorous engineering principles to the development and analysis of software specifications. Simply, computers develop software for other computers in a fast way by using specific tools. When implanted in a Concurrent Engineering environment, this process is taking place while some parts of the developed software is running already.

It is a sort of on-line software engineering. There are a lot of problems with these kinds of systems because they are very complex, not easily maintainable and fragile. Some of the tools, which were very progressive, back then, are obsolete right now, so they have to be updated, which can be a problem because the software engineering process is fit around these tools. The tools developed right now are evolutional products out of earlier tools. The first tools, developed in the mid 1970s, were mainly introduced to automate the production and to maintenance structured diagrams. When this automation covers the complete life-cycle process, it also illustrates on *Integrated Computer Aided Software Engineering (I-CASE).*

Later on, integrated CASE (I-CASE) products were introduced. This was an important step because I-CASE products are capable of being used to generate entire applications from design specifications. Recently, CASE tools have entered a third phase: the introduction of new methodologies based on capabilities of I-CASE tools. These new methodologies utilize Rapid Prototyping techniques to develop applications faster, at lower cost and higher quality. By using Rapid Prototyping a prototype can be made fast, so the developed system can be tested more often in between the development-phases because it does not cost much time to create a prototype. Mistakes can be detected and corrected earlier this way. The earlier this can be done, the better because correcting these mistakes gets harder and more expensive when the system is developed further. So a lot of time and money can be saved using Rapid Prototyping.

As said above, a new set of tools is necessary. These tools should automate each phase of the life-cycle process and tie the application development more closely to the strategic operations of the business. A lot of different tools have been developed over the years and are being developed right now. There are so many tools, that the user could easily get confused. To survey all these CASE tools they are divided into the following categories:

1. *Information engineering-supporting products.* These are life-cycle processes, derived from the strategic plans of the enterprise and which provide a repository to create and maintain enterprise models, data models and process models.
2. *Structured diagramming-supporting products.* These are derived from several development methodologies like Gane–Sarson or Jackson. These products at least support data flow, control flow and entity flow, which are the three basic structured software engineering diagramming types.
3. *Structured development aids-providing products* These products are providing aids for a structured development of the process. These products are very suitable to be used by the system analysts, because they are helped very much by a structured process, because those can be analyzed faster and more accurately.
4. *Application-code-generating products.* These are products that generate application-code for a specific goal, set by the designer. Most of the products in this area are using a COBOL-generator, which is a tool that generates programming code in a specific language out of specifications set by the system-designer.

The heart of a well-designed I-CASE system is a repository, which is used as a knowledge base to store information about the organization, its structure, enterprise model, functions, procedures, data models, etc. The meaning represented by diagrams and their detail windows is stored in the repository. The repository steadily accumulates information relating to the planning, analysis, design, construction and maintenance of systems. In other words: *The repository is the heart of a CASE system.*

Two types of mechanisms have been used in CASE software to store design information:

1. A *dictionary*, which contains names and descriptions of data items, processes, etc.
2. A *repository*, which contains this dictionary information and a complete coded representation of plans, models and designs, with tools for cross-checking, correlation analysis and validation.

Before implanting CASE and designing tools, a series of steps should be followed:

1. Conduct a technology-impact study to determine how the basic business of the organization should change to maximize the opportunities presented by rapid technological change
2. Evaluate how the organization should be re-engineered to take advantage of new technology
3. Establish a program for replacing the old systems with the most effective new technology
4. Commit to an overall integrated architecture
5. Select a development methodology
6. Select a CASE-tool
7. Establish a culture of reusability
8. Strive for an environment of open interconnectivity and software portability across the entire enterprise
9. Establish intercorporate network links to most trading partners
10. Determine how to provide all knowledge to workers with a high level of computerized knowledge and processing power
11. Determine the changes in management-structure required to take full advantage of innovative systems, architectures, methodologies and tools

When an enterprise takes these actions ahead of its competition, it will gain a major competitive advantage. An enterprise should also be up-to-date because the rapid advances in computer technology allow the competition to get ahead of the user. Some significant trends in the development of new system environments include:

1. *Low-cost MIPS*, the price of fast processors is decreasing and even faster everyday
2. *Distributed computing environment*, end-users are moving towards a multilayered distributed computer architecture
3. *CASE and I-CASE tools*, highly integrated, repository-driven, computer-aided systems engineering tools are making it possible to generate application code directly from graphical specifications

4. *Forward/reverse engineering tools*, re-engineering tools enable analysts to convert low-level data definition and unstructured process code into standardized data elements and structured code.
5. *New development methodologies*, more efficient development life-cycle processes are making it possible to develop applications more rapidly and in closer coordination with end users.
6. *Growth of standards*, standards are emerging that will govern the future evolution of hardware and software.

9.5.1 The TEXspecTool for Computer-Aided Software Engineering

The tool can assist in the development of a broad range of software, but targets the development of software that implements mathematical models. The original application relates to the development of models of a repository for Canada's high-level nuclear waste, but is not limited to this use. TEXspec is particularly useful when documenting models and associated programs, which rely on mathematical notations to communicate the intent of the software.

Problem Definition

Canada has developed computer programs to model a deep geologic repository for used nuclear fuel. Regulators require these programs to be demonstrably high quality to support license applications. In 1999, the Canadian Standards Association (CSA) adopted standard N286.7 for the development of nuclear safety related computer programs, a scope that includes the AECL (Atomic Energy of Canada Limited) models. While the software development process used presently has been considered robust, it requires refinement in order to achieve compliance with this standard. The TEXspec project seeks to address the issue of compliance with CSA N286.7. The tool supports a compliant software development procedure, while imposing a minimum of additional overhead. While optimized to meet requirements associated with the modeling nuclear fuel waste, it is hoped that TEXspec will find more common usage.

Several commercial Computer-Aided Software Engineering (CASE) tools will support a robust software development methodology, but none provide support for the mathematical notations that are common in scientific models. TEXspec provides extensive support for this notation.

Software Development Methodology

Although Object Oriented (OO) analysis and design is appropriate for documenting many software applications, there are still applications for procedure-/flow-based software. In particular, some models, which are basically linear in structure, are best described using structured (non-OO) methodologies. Many scientific models have, to date, been described using a modified Yourdon/DeMarco methodology. Although OO methods would perhaps be more appropriate for some models, priority is given to the more common Yourdon/DeMarco analysis methodology. Products associated with this methodology are:

- Dataflow diagrams (DFDs)
- Process descriptions (mini-specs)
- Structure charts
- Subprogram design descriptions
- Data dictionary listings

DFDs and mini-specs comprise the requirements specification, while structure charts and subprogram design descriptions document the design. Data dictionary listings may be separated into requirements and design, or combined into a single product.

Requirements Specification

In Fig. 9.49, Diagram 0 shows the input and output "flows" to/from a single "process." This high level abstraction is intended to allow the reader to identify the functions of the complete system. Process 1 is broken into components in Diagram 1. The diagrams may be thought of as a hierarchy, with higher level diagrams having shorter process numbers.

The numbering convention of the diagrams and the processes allows the decomposition to be clearly seen. Once processes are decomposed to a point where they can be clearly specified in a short textual description, they appear on a DFD with a double circle, as Process 1.4 illustrates. Such "atomic" processes are associated with a mini-spec, rather than a lower level diagram. Data flows, like processes, can be broken up into constituent parts on lower diagrams. In Fig. 9.49, for example, 'Implements' on Diagram 0 becomes "Measuring Cup," "Bowl," "Oven," and "Pan" on Diagram 1. These flows are associated with multiple processes on Diagram 1, so they must be shown individually.

Mini-specs for each atomic process repeat some of the information on the DFD, and also detail the requirements for a low level process in any manner deemed suitable by the author. For scientific codes, mini-specs often make extensive use of diagrams and mathematical notations.

Design Specification

The boxes represent "subprograms" to be composed in a procedural programming language such as FORTRAN. The chart is intended to illustrate the nature of the interface between subprograms. The lines between the subprograms indicate a 'calling' relationship, with the subprogram, which is closer to the top of the structure chart invoking the lower.

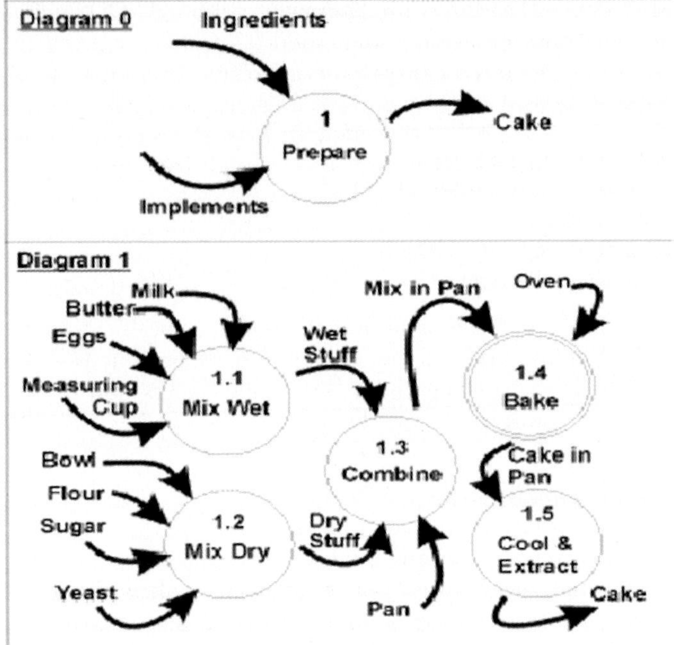

Fig. 9.49. Example dataflow diagrams (DFDs)

Data can be passed from one subprogram to another through an argument list, or through common storage that can be accessed from multiple subprograms. The interface variables can be input, output, or both, next to the variable name. Structure charts do not have a hierarchical organization paralleling DFDs. However, a large structure chart may span many pages using page connectors. Each subprogram must itself be documented. Subprogram design descriptions repeat some of the information on the structure chart, and document the algorithm and design details of the subprogram.

This may include material common with mini-specs, as the design reflects the requirements. In particular, many of the mathematical equations are referenced in both places.

Consistency Between Products

Experience at AECL has shown that lack of consistency between software products has been a major source of software defects. Commercial CASE tools have helped to reduce this inconsistency, but these tools all have difficulty in one or more critical areas:

- *Lack of support for scientific and mathematical notations.* The nature of scientific software demands that mathematical notations be permitted in specifications.

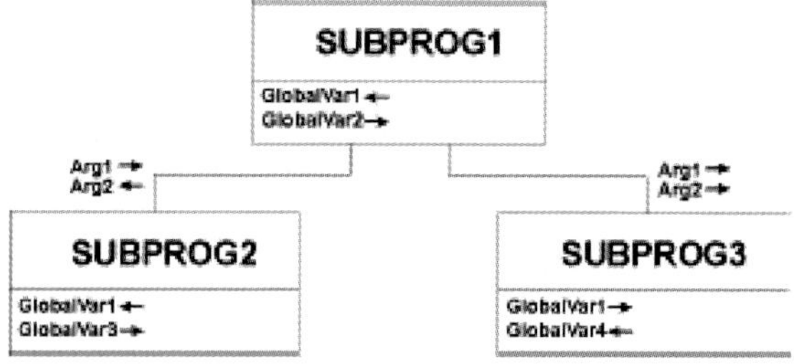

Fig. 9.50. Example structure chart

- *Insufficient accountability.* The principle of ownership and accountability for products is not strictly enforced. While a record of who updated products is often kept, the process control is typically inadequate.
- Assembling large products from smaller components is not adequately supported. Many defects originate as transcription errors between products. Mathematical equations are particularly susceptible.
- Insufficient consistency checking between products.

The TEXspec Solution TEXspec takes advantage of the plain text nature of LATEX input to permit processing and tracking of shared components. The main TEXspec processing is performed by modules which have been implemented in PERL, as indicated in Fig. 9.50 Graphical user interface (GUI) captures interactions with the user. Most of this interaction consists of displaying and manipulating "component" files, which form the inputs for the TEXspec scripts that select components and assemble them into products. These products are primarily LATEX input files, which can be postprocessed to produce output suitable for viewing, printing, or compiling.

While the GUI is a convenient way to construct components and initiate processing, it can be bypassed if required. The components can be generated by any means that can generate a plain text output file, including a text editor. More importantly, the processing can be controlled by any means that can initiate a process, with no requirement for interaction with a GUI. When processing many components, or when a log of processing is required, this "batch" style processing is a useful alternative. Neither the TEXspec scripts, nor the GUI can display or print the products. Figure 9.51 indicates that an intermediate script, which is intended to be edited by the user, initiates TEXspec to produce the product files, then controls postprocessing as appropriate. This flexibility allows the user to integrate TEXspec into existing procedures.

For example, if a static code analyzer such as Floppy is in use, it can be run automatically on code as it is generated. Interaction with a version control system might be desired, or the user may even wish to compile code as it is generated. Alternatively, processing that is not needed, can be removed, such as removing documentation generation (including LATEX processing) until the

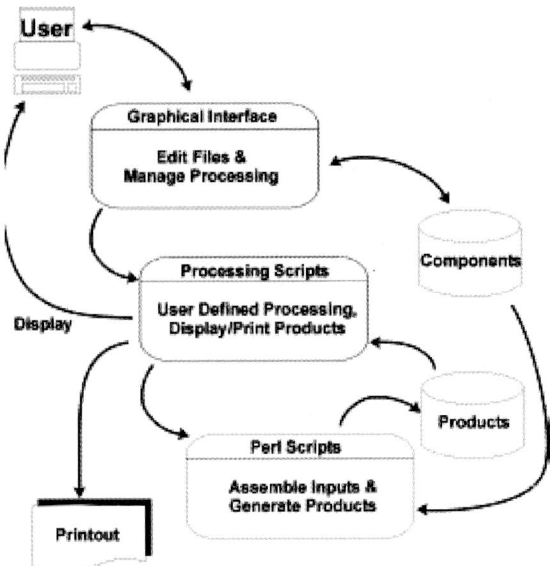

Fig. 9.51. TEXSpec architecture

code is stable. In order to support sharing of equations and data definitions, while tracking ownership and responsibility for content, TEXspec supports a fine granularity of components. Each TEXspec component is tracked independently by placing each in a unique file, which is mapped by the file name to the name of the component, and by the file name "extension" (in the tradition of MS-DOS or CP/M) to the type of component. TEXspec components, with associated file name extensions, are:

- Requirements data dictionary entries (.rdd)
- Design data dictionary entries (.ddd)
- Equations (.teq)
- Dataflow diagrams (.dfd)
- Mini-Specs (.ms)
- Structure charts (.sc)
- Subprogram design descriptions (.ds)
- Manuals (.tex).
- Dataflow Diagrams

TEXspec shows the details of composite dataflow decomposition explicitly on the DFD. for example, flow "conc-majoranions" (which would appear in Fig. 9.52) is shown as constituent components "CCL" and "CSUL." If a composite flow contains components which do not appear on the current diagram, they are shown in a regular font, while components that do appear on the current diagram are shown in a bold font. Consistency between DFDs is monitored by TEXspec. A warning message is generated for any inconsistency

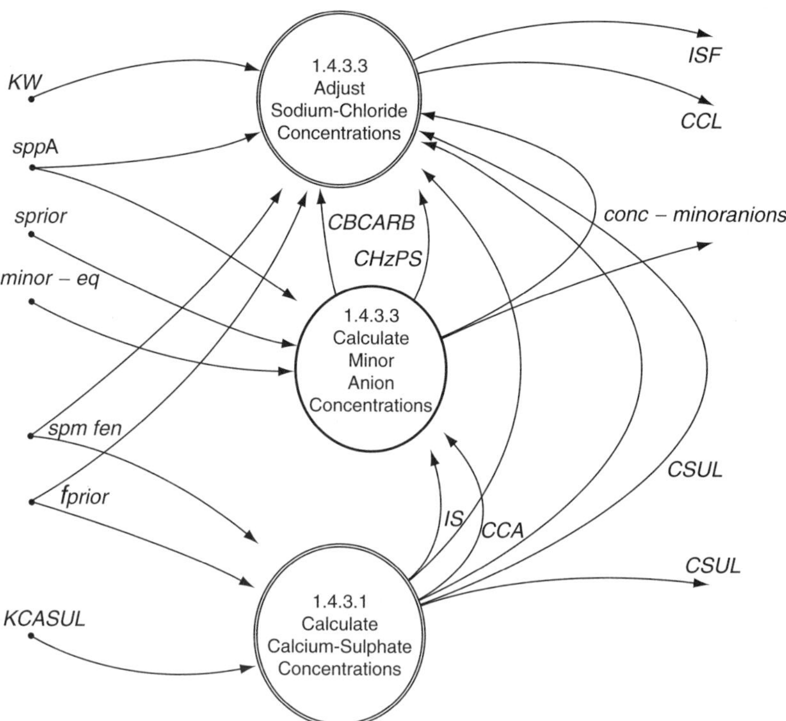

Fig. 9.52. TEXSpec dataflow diagram

between a DFD and it is parent. Labels can be shown with mathematical notation, rather than the plain text shown. Switching from plain text to mathematical labels is simple, since flows are taken from the requirements data dictionary (file name.rdd), which typically contains both a mathematical and a plain text label. Although this is an interesting capability, there has been little enthusiasm among users to take advantage of it.

The format of the header is common to all TEXspec products, detailing the project, the responsible author, implementer, and reviewer, along with an indication of the genealogy of the product (in very small type), which can be used to trace back the source of any defects. Atomic processes are not broken down into lower level DFDs, but are further specified using a mini-spec. This document is intended to be flexible in format, permitting the author freedom to communicate the intent of the process in whatever manner is most effective. The standard TEXspec header is generated for each mini-spec. The author must explicitly state input and output flows, which are presented in tabular format and verified for consistency with the DFD.

Equations appearing in mini-specs are often referenced in other documents. Authors are encouraged, but not required, to place each equation in a separate file (name.rdd), and reference that file from within the mini-spec. The equation can then be reused in subprogram design descriptions or manuals. At AECL, a commercial package is used to create equations that can be saved in LATEX format, which includes information encoded as LATEX comments, which permits reuse by word processors. By keeping an equation in a single file, available for reuse, transcription errors are reduced.

Structure Charts

TEXspec structure charts are produced using the xy-pic package under LATEX.

Subprograms can be grouped using color coded backgrounds. subprograms "SOURCE," "ZAPINT," and "REPFUN" are grouped with a yellow background indicating that they are library routines, not part of the software being documented.

Input and output variables are also color coded: green for input, red for output, and blue for both. This applies to both the arguments to a subprogram and to the common storage variables.

FORTRAN groups common storage variables into named blocks, which are indicated to the left of each variable name. Blocks are sorted alphabetically, and variables are sorted alphabetically within a block. For subprograms that are functions, rather than subroutines, an additional output variable is provided in the name of the subprogram itself. In this case the name of the subprogram appears in red.

Most of the information on the structure chart is extracted from the subprogram design description (subprogram.ds) file for each subprogram on the chart. Presentation details are contained in a structure chart file (name.sc). The structure chart file lists the subprograms to be placed at specified locations. It also indicates the calling relationship between the subprograms and the location of the connecting lines. TEXspec generates a warning message if the specified calling relationship is not consistent with the FORTRAN code. The description of the subprogram, and the argument list, are extracted from the subprogram design description file, and can be quite long. The structure chart may specify a maximum line length for these elements, and TEXspec inserts line breaks appropriately. The double box "INVTRY" at the bottom is a page connector to another structure chart. A page connector 'SIMALL' on another chart references this chart.

Subprogram Design Descriptions

TEXspec supports literate programming techniques through use of the LATEX package. Some preprocessing and post processing is required to achieve the desired products, but LATEX users will be immediately familiar with the format. Subprogram design descriptions can be long documents, but an abbreviated product. The TEXspec header is followed by a LATEX-style list of code blocks that comprise the subprogram. Code blocks to declare variables are replaced by a tabular form, which contains additional data dictionary information. Of particular interest is the 'Symbol', which is the mathematical notation for a variable. This allows variables to be traced through equations, making the relationship between code and requirements much easier to follow.

Also specified is the input/output status of each variable, which TEXspec validates against the contents of the code blocks. Several tables may be generated one for each subprogram calling arguments, common storage variables, local storage variables, and constants. After each table, the design may optionally specify preconditions and post conditions for the tabulated variables. Specifying valid ranges for data has proven to make testing much more accurate, and the process of specifying those ranges has identified many defects. Specifying preconditions and post conditions early in the design process is an effective and inexpensive quality control device.

Following the tables are the code blocks, each including commentary in LATEX format, which may feature equations, shared with mini-specs or other documents. Sharing equations ensures consistency. Consistency between subprogram design descriptions and compilable code is guaranteed, since LATEX extracts the code from the subprogram design description. Information for each subprogram on a structure chart is also extracted from the subprogram design description, ensuring that all design documentation is consistent. Data Dictionaries. TEXspec distinguishes between dictionary entries for requirements and design specification. Some design information is never applicable to requirements (e.g., a common storage block name). It is possible to have a close correlation between entries in the requirements and design dictionaries.

Design entries may optionally state a requirements dictionary entry, which is related. When this is done, fields in the design dictionary acquire default values equivalent to the requirements data dictionary. This is particularly useful to inherit the mathematical symbol and description. TEXspec produces a data dictionary listing which can show a cross reference of which products use which dictionary entries.

There is a considerable amount of data contained in many plain text files in a typical TEXspec documented project. To assist users, a GUI has been developed as a Java application to act as a front end to the process. While there is little new technology embedded in the GUI, it is interesting to note that the GUI, at over 20,000 lines of code, is much larger that the TEXspec scripts. Here, the structure chart 'SIMALL' is being edited in the upper window. A subprogram on the chart is being modified in the lower window. The GUI can open many windows, so it is contained within an application desktop, which produces only one icon on the user's desktop.

When editing a subprogram (module) on a structure chart, the user specifies the subprogram name, and sets a position in x,y coordinates. If the user wants to show an entry point on the chart to this subprogram (useful for charts with multiple entry points), then the location of the entry point must be specified. The location of the call string, and the maximum width of that string is also entered. Calls to other subprograms can be added from a dropdown list of all available modules. For each called subprogram, "waypoints" determine the shape of the line connecting the two boxes. Similar editing capability is provided for the overall structure chart in the upper window. Selecting a subprogram from the scroll list at the bottom of the upper window causes the lower window to appear.

Summary

This chapter focused on the current trends in instrumentation such as Fiber optic and LASER instrumentation. These sections dealt with the various types of sensors used for measurement and control in process industry. The key features of smart transducers and the HART communicator protocol are used in various modes for communicating information to/from smart field instruments and central control or monitoring equipment. The CASE tools are used to automate each phase of the life-cycle process and tie the application development more closely to the strategic operations of the business. A number of different tools have been developed over the years and are being developed right now.

Review Questions

1. Write short notes on fiber optic sensors.
2. Explain briefly about fiber optic pressure sensors and fiber optic voltage sensors.
3. Discuss briefly about optical fiber temperature sensors.
4. Mention the application areas of fiber optic instrumentation.
5. Explain in detail about laser instrumentation with neat sketches.
6. List the advantages of laser welding.
7. Mention the application areas of laser welding.
8. Explain the process of laser trimming and melting.
9. Discuss briefly about smart intelligent transducers.
10. Explain in detail about smart transmitter with HART communicator.
11. Write short notes on Computer-Aided Software Engineering.

10

VI Applications: Part I

Learning Objectives. On completion of this chapter the reader will have a knowledge on LabVIEW applications such as:

- Fiber-Optic Component Inspection using Integrated Vision and Motion Components
- Data Acquisition and User Interface of Beam Instrumentation System at SRRC
- VISCP: A Virtual Instrumentation and CAD Tool for Electronic Engineering
- Distributed Multiplatform Control System with LabVIEW
- The Virtual Instrument Control System
- Controller Design using the Maple Professional Math Toolbox For LabVIEW
- Embedding Remote Experimentation in Power Engineering Education
- Design of an Automatic System for the Electrical Quality Assurance
- Internet-Ready Power Network Analyzer for Power Quality Measurements and Application of Virtual Instrumentation in a Power Engineering Laboratory
- PC and LabVIEW-Based Robot Control System
- Mobile Robot Miniaturization: A Tool for Investigation in Control Algorithms
- A Steady-Hand Robotic System for Microsurgical Augmentation

10.1 Fiber-Optic Component Inspection Using Integrated Vision and Motion Components

Since the 1980s, fiber-optic technology has been used for long distance telephone systems. Now with the increasing need for bandwidth, fiber-optic systems have become in high demand. These optical systems, whether transmitting data across continents or providing bandwidth for large cities, are very complex and consist of dozens of components, such as dispersion compensators, collimators, gratings, and attenuators, across the whole transmission path. Manufacturers of these optical components are challenged with making amazingly small and sophisticated devices, which will become even more intricate in the future.

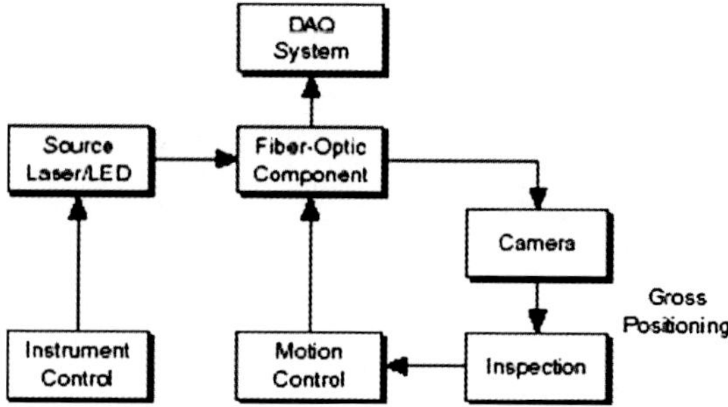

Fig. 10.1. Block diagram of important components of a generic fiber-optic inspection platform

In addition, these components are made from many different types of exotic materials (such as indium phosphide and gallium arsenide) instead of just silicon. The technologies involved in manufacturing these components are so new that most component manufacturers assemble their products by hand. In fact, Fortune magazine, in a report on JDS Uniphase in its September 2000 edition, stated that of JDS's 18,000 employees 12,000 are in manufacturing. Traditional, labor-intensive techniques cannot keep up with market demands that include tight design specifications, low cost, and high volume requirements. This section outlines common fiber inspection and measurement techniques and describes the equipment needed to create an automated optical component inspection system as shown in Fig. 10.1.

10.1.1 Fiber Basics

Optical fiber is the workhorse of any optical communication system. It basically is the conduit for the data and ends up linking all of the other components together. The light wave is transmitted by total internal reflection. A typical fiber consists of the following:

- The core, the actual conduit of the light
- The cladding, a glass material which, by doping, has achieved a certain index of refraction to cause total internal reflection
- Other layers to provide strength and protection from harsh conditions

The light travels down the fiber with minimal attenuation because of the difference in the indices of refraction between the core and the cladding, combined with the angle of reflection. Optical fiber is manufactured as either single mode or multimode. Multimode fiber (50 µm to a few mm) has a larger core diameter than a single mode fiber (4–9 µm).

These measurements must be accurate because the diameter and concentricity of the core and cladding would affect such things as the modes of light that can travel in the optical fiber, the numerical aperture, and the coupling of light in and out of the fiber.

10.1.2 Fiber-Optic Inspection Platform Overview

The approach to inspect the fiber is to use vision-guided motion control to perform gross alignment of the fiber and lens. The camera and inspection algorithms look at the core and cladding and perform vision measurements to determine if concentricity and diameters fall within tolerances. In addition, a National Instruments data acquisition (DAQ) board is used to measure the light signal attenuation through the fiber to determine characteristics such as optical loss. Figure 10.2 shows the front panel of the application.

10.1.3 Inspection Measurements

Typical inspection measurements are:

- Cladding diameters
- Core and noncircularities
- Core-cladding concentricity

These parameters can be measured using optical inspection techniques. Surface blemishes can also be detected using thresholding and blob analysis techniques.

10.1.4 Optical Inspection Overview

Using a CCD camera, the optical inspection system can measure the core and cladding diameter using edge detection crossing through the core and cladding boundaries; functions within National Instruments IMAQ Vision software can locate the edges. With IMAQ Vision, use the annulus region of interest tool to draw an inner and outer circle around a boundary as shown in Fig. 10.3. Line profiles are automatically placed perpendicular to the cladding boundary between the regions of interest. IMAQ Vision edge-detection functions precisely locate the boundary of cladding.

Once the cladding boundary is located, the center of the fiber, circularity, and diameter are all calculated using built-in IMAQ Vision functions. Example of the LabVIEW IMAQ Vision code is shown in Fig. 10.4. To automate the process completely, the user can use the IMAQ Vision pattern matching tools to locate the core and then automatically position the region of interest as shown in Fig. 10.5. Also, the user can use standard threshold and blob analysis techniques to locate and count the defects within the core.

10.1.5 Real Measurements

Image information can be misleading at times because these measurements are typically recorded in pixel values. For fiber-optic applications, it is important to record measurements in real world units such as microns. National Instruments IMAQ Vision contains easy-to-use functions to calibrate the data so that image output and measurements are done in microns rather than pixels.

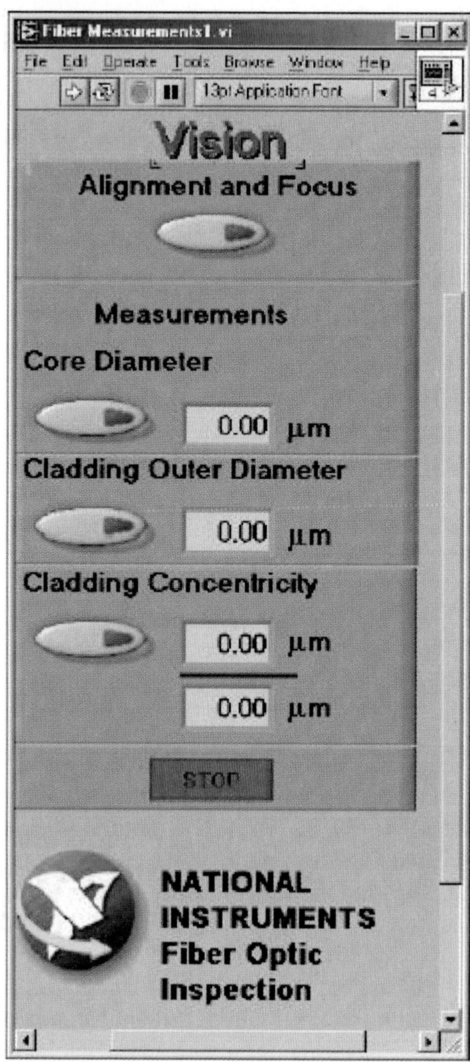

Fig. 10.2. LabVIEW front panel of fiber-optic inspection system for precise measurements of important fiber parameters

10.1.6 IMAQ Vision Functions

IMAQ Select Annulus

This VI is loaded from **Motion & Vision≫Machine Vision≫Select Region of Interest≫IMAQ Select Annulus.vi**. After the user specifies an annulus area in an image, IMAQ Select Annulus displays the image in the specified window, provides the annulus tool, and returns the coordinates of the annulus selected.

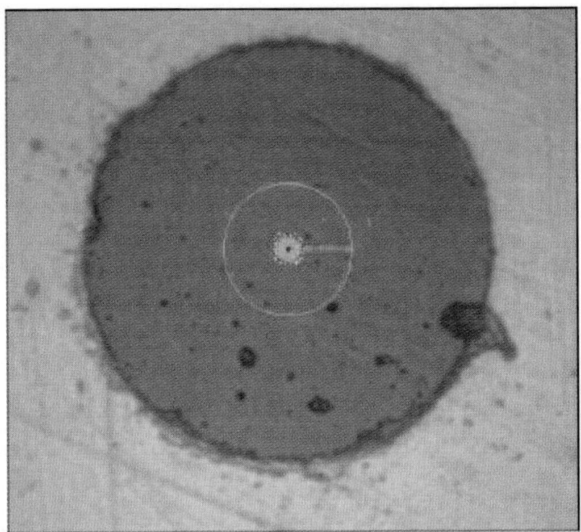

Fig. 10.3. IMAQ Vision functions draw two concentric circles and detect the edges where the core and cladding meet in this single-mode fiber

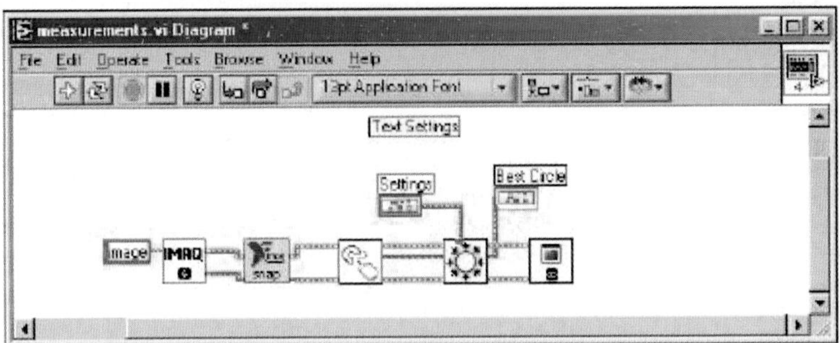

Fig. 10.4. Sample LabVIEW code for interactively measuring the core and cladding diameters

IMAQ Find Circular Edge

This VI is loaded from **Motion & Vision≫Machine Vision≫Locate Edges≫IMAQ Find Circular Edge.vi**. IMAQ Find Circular Edge locates a circular edge in a search area. It locates the intersection points between a set of search lines defined by a spoke and the edge of an object. The intersection points are determined based on their contrast and slope.

IMAQ Edge Tool

This VI is loaded from **Motion & Vision≫Machine Vision≫Caliper≫ IMAQ Edge Tool.vi**. IMAQ Edge Tool finds edges along a path defined in the image. Edges are determined based on their contrast, slope, and steepness.

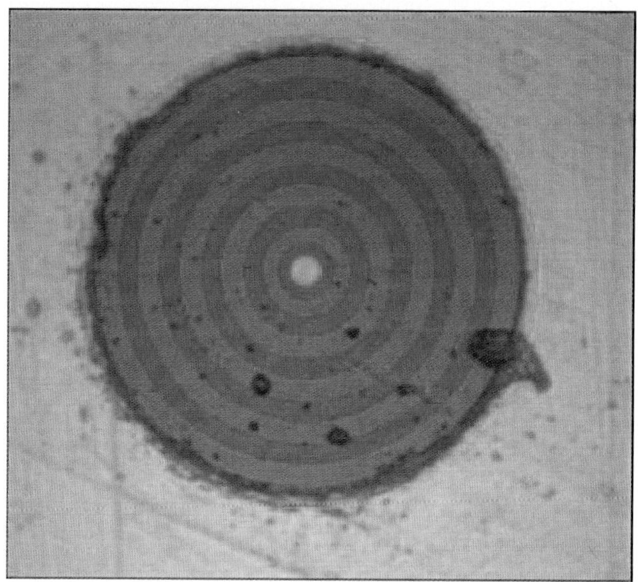

Fig. 10.5. IMAQ Vision functions draw concentric ROIs and in each concentric region

IMAQ Point Distances

This VI is loaded from **Motion & Vision≫Machine Vision≫Analytic Geometry≫IMAQ Point Distances.vi**. IMAQ Point Distances computes the distance between consecutive pairs of points.

IMAQ Convert Pixel to Real World

This VI is loaded from **Motion & Vision≫Vision Utilities≫Calibration≫ IMAQ Convert Pixel To Real World.vi.** Convert Pixel To Real World transforms pixel coordinates to real world coordinates, according to the calibration information contained in the image. Calibration information is attached to this image by IMAQ Learn Calibration Template, IMAQ Set Simple Calibration, or IMAQ Set Calibration Info.

10.1.7 Motion Control

Precision is the key to an inspection solution for optical components. When fiber measurements are in the micrometer range, it is imperative that the alignment be very accurate. National Instruments Motion Control Hardware and Software coupled with a third-party stage achieve the precise micrometer and nanometer accuracy necessary for fiber inspection or fiber alignment and bonding. This motion can all be controlled from the same interface where the visual measurements are being acquired.

Table 10.1. Fiber-optic components

Device	Company	Model
Instrumentation Chassis	National Instruments	PXI-1002
Instrumentation Controller	National Instruments	PXI-8156B
Image acquisition hardware	National Instruments	IMAQ PXI-1407
Camera	JAI	CV-M50
		OBJ 18 mm WD, 2.00X
Lens	Infinity	4x Adapter
		1.5x Main Body
Motion Control	National Instruments	PXI-7344
Stages	National Aperture	MM-3M-F-0.5
Amplifier	National Aperture	MC-3SA

Software Components

The basic software component is National Instruments LabVIEW Application Software, the development environment. Motion control software and IMAQ Vision software seamlessly integrate with LabVIEW to provide faster development time. Statistical Process Control software for LabVIEW provides the functionality to determine manufacturing yields, statistical trends, and process analysis.

Hardware Components

The hardware used in a fiber-optic components inspection system is listed in Table 10.1.

Automation through Integration

By tightly integrating different test and manufacturing functions and processes, optical component manufacturers can build highly automated systems that result in cost savings, shorter lead times, higher yield, and better quality. The main advantage of this integrated, automated approach lies in removing pressure from labor-intensive procedures. Productivity increases as a result of higher quality.

An open-ended system using National Instruments LabVIEW, NI-Motion, and IMAQ Vision software can easily be augmented with National Instruments image acquisition and motion control devices. The user can also add DAQ hardware coupled with a light-emitting source to perform beam profile analysis and quantify power dissipation. As new test or inspection requirements arise, the user can easily modify the system because a flexible open platform based on off-the-shelf components such as PXI and LabVIEW is used. Complete turnkey solutions for your fiber-optic inspection needs are available through local integrators.

10.2 Data Acquisition and User Interface of Beam Instrumentation System at SRRC

Data acquisition systems for the accelerator complex at Synchrotron Radiation Research Center (SRRC) are composed of various hardware and software components. Beam signals are processed by related processing electronics, and connect to control system by various type of interfaces in data acquisition front-ends. These front-ends include VME crates, personal computers, and instrument bus adapters. Fast Ethernet connected all elements together with control consoles. The user interface is running on the control console. Real-time data capture, display, and analysis are supported on the control console. Analysis tools based on Matlab scripts are adopted. Hardware and software implementation of the system is presented. User interface supports are also described.

10.2.1 Introduction to SRRC

The control system of SRRC is essential to operate the light source efficiently. The console level is composed of control console and the control server. Field level computer composed of more than 30 VME crates and several PCs for special devices. Both level computer systems are connected by a dedicated control fast Ethernet. Beam instrumentation system is composed of various beam monitors and associated processing electronics. The most important beam parameters include beam current monitor and closed orbit, beam profile, etc. Various software tools are integrated with the system and provide an efficient means of operation.

10.2.2 Outline of the Control and Beam Instrumentation System

The control system and the beam instrumentation system along with the user interface is discussed in this section.

SRRC Control System

The control system is a two-level hierarchy computer system. Upper layer computers include two process computers and many workstations and PCs. Database management, archive and various application programs are executed on the process computers. The main purpose of the workstations is for use as operation consoles. Bottom layer computers are distributed VME crate controllers that are in charge of the control and data acquisition for accelerator equipment.

Both computer layers are connected by a local area network. The software environment can be divided into four logical layers. They are device access layer, network access layer, database layer, and applications from bottom to top. The database plays a role as data exchange center for applications and subsystems of accelerator system. Most of the data is updated into database at ten times per second.

Beam Instrumentation System

The beam instrumentation system is composed of various monitors and supporting electronics. The system supports precision intensity measurements and lifetime calculations. All devices are controlled by VME based local controllers. The control console can access these devices through the user interface. The synchrotron radiation monitor is controlled by a PC and connected to control system via Ethernet. Beam diagnostic instrumentation systems provide the necessary electron beam parameters for commissioning, routine operations and beam physics studies.

User Interface

The Common Desktop Environment (CDE) is used for user interface development. It is an integrated graphical user interface for the SRRC control system, combining X Window System, OSF/Motif, and CDE technologies. Motif GUI Builder and Code Generator, UIM/X GUI builder is a used to generated various user interfaces.

It enables software developers to interactively create, modify, test, and generate code for the user interface portion of their applications. To satisfy various requirements, some applications are development in a LabVIEW based environment. For fast prototyping, users can customize user applications in the Malta environment. The control system supported various database access MEX files. User's can access the database directly in Matlab. For the IEEE-488 based interface, a fast Ethernet-based GPIB adapter was support. A GPIB/2 interface is also included to allow user access these instruments within Matlab.

10.2.3 Specific Applications

This section discusses the Business Process Management (BPM) system and the orbit display units along with the features of the BPM system such as Smartlink system.

BPM System

The BPM system consists of 57 BPMs equipped with switched electrode processing electronics. The data acquisition is done by a dedicated VME crate equipped with 128 16-bit ADC channels. The VME host sends raw data to the orbit feedback system via reflective memory. Averaged data are updated to the control database 10 times per second with a resolution of around 1μm.

Orbit Display Utilities

To provide better orbit observation, a Motif-based orbit display utility was developed. This utility provides basic orbit display and difference orbit display. Several display attributes can be selected by users, such as full scale range, persistent display mode enable/disable, average number, save, print, etc. The update rate of the display is up to 10 Hz. This is very useful for routine operation and various machine studies.

Turn-by-Turn BPM System

To support various beam physics studies, several BPMs are equipped with log-ratio processors for turn-by-turn beam position measurement. A multi-channel VME form factor digitizer with 12-bit ADC acquires turn-by-turn beam position. A server program running on a VME crate manages the data acquisition of beam position. A client program running on control console and Motif based GUI are used for the user interface.

SmartLink System to Acquire Beamline Data

To acquire data from a remote site, a SmartLink-based system was setup. The link used a private Ethernet as a field bus. The Ethernet is connected to a PMC Ethernet module installed on the VME host. The SmartLink data acquisition module is used to acquire data from the beamline monitor with high resolution, including photon flux (Io) monitor and photon BPM blades current. The update rate is about two times per second with 20-bit resolution. This slow update rate is the major disadvantage.

Synchrotron Radiation Monitor Interface

The synchrotron radiation monitor is used to measure beam profile. It consists of an optics and a high resolution CCD. To acquire profile information, a PC acquires the image from the CCD camera, analyses the profile and extracts profile parameters. The local display is also broadcast via facility-wide machine status CATV system. A server program running on PC serves the data request form control console. A client program running on the control console can access information of this PC by the help of LabVIEW program or Matlab MEX files running on the control console.

Gap Voltage Modulation Study Support

RF gap voltage modulation was adopted to relieve the effect of longitudinal coupled-bunch instability in routine operation of the storage ring at SRRC. Systematic measurements were done recently to investigate the mechanism

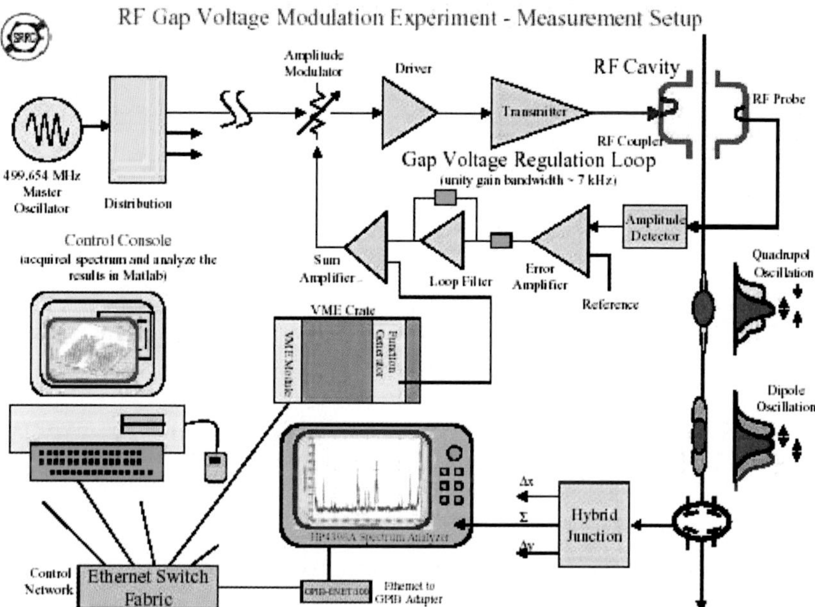

Fig. 10.6. Experimental setup for RF gap voltage modulation study

why RF gap voltage modulation can do this. The experimental setup is shown in Fig. 10.6. The gap voltage modulation frequency is about 50 kHz. The rolled-off frequency of the LLRF gap voltage regulation loop is about 7 kHz. A function generator in VME form factor generates the modulation sinusoidal wave.

This generator integrates with the control system to satisfy the requirement of routine operation. Frequency and amplitude can be adjusted on the control console. The modulation signal is injected after loop filter and added with a correction signal in the RF gap voltage regulation loop. HP4396A spectrum/network analyzer observes the beam spectrum form BPM sum signal.

National Instruments GPIB/ENET-100 controller connects the spectrum analyzer to the control network. The control console supports database accesses MEX-file that allows Matlab to read and set the modulation parameters. GPIB MEX-file allows Matlab to read and write the GPIB devices via a GPIB/ENET-100 controller. The experimental data can be acquired directly into Matlab. The gap voltage modulation is effective because the modulation frequency is far beyond the unity gain cutoff frequency of the gap voltage regulation loop. Modulation amplitude is about 10% of total gap voltage in routine operation.

The experiment's sequence is programmed by a simple Matlab script running on the control console that selects frequency scan range as well as modulation amplitude. The spectrum analyzer can select to measure upper or

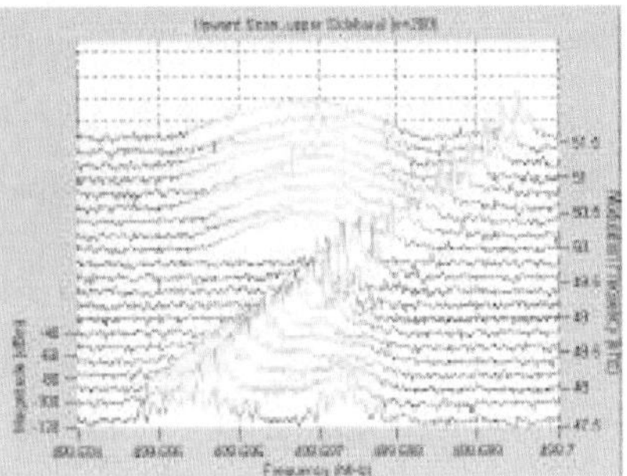

Fig. 10.7. Typical scan spectrum of RF gap voltage modulation study

lower sideband either 1 or 2 fs synchrotron oscillation frequency. Figure 10.7 shows the measured result and proves that the instability is suppressed near 50 kHz.

Booster Tune Measurement and Correction System

The electron beam is accelerated from 50 MeV to 1.5 GeV at the booster synchrotron with 50 ms ramping time. Tune is an important index indicated the tracking performance of the White circuit based power supply system. Tune information was obtained by the tune measurement system; it provides the x and y during energy ramping. Tune variation in the energy ramping process are correlated to the tracking performance of three families White circuit. The optimized lattice can be obtained by the help of measured tune for the booster synchrotron to get a better working point for efficient operation.

An application program was developed to automatically measure and correct the tune variation. The stored beam is excited by extraction kicker to perform damped betatron oscillation. Trigger timing and field strength of the kicker is set properly as the function of beam energy to ensure sufficient beam excitation and without killing the stored beam. Beam motion signals are picked up by stripline and processed by log-ratio BPM electronics. A transient digitizer in the VME crate records the betatron oscillation of the stored beam. Server programs running on the VME crate coordinate the process of data acquisition.

A client program running on control consoles is invoked by a Matlab script to perform data acquisition and analysis. Data analysis includes Fourier analysis, peak identification and visualization. The tune correction signal is generated and downloads to waveform a generator located in the VME crate. Figure 10.8 shows the system structure of the tune measurement and correction system. The Matlab script is used to generate a correction waveform by the measured tune and measured sensitivity matrix between tune and quadrupole setting. Tune variation during ramping can be reduced drastically by this feed-forward correction.

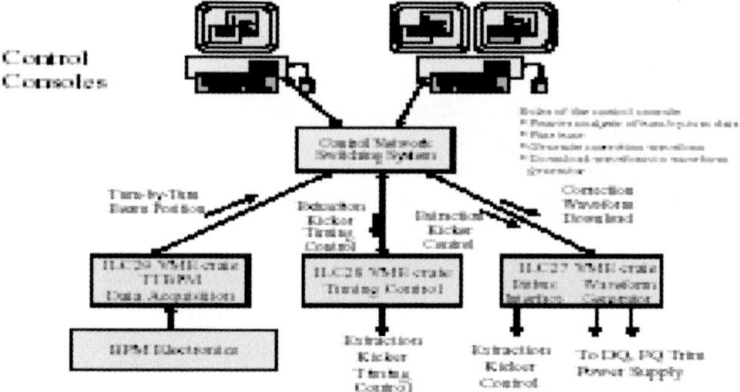

Fig. 10.8. Tune acquisition and correction system

10.3 VISCP: A Virtual Instrumentation and CAD Tool for Electronic Engineering Learning

The fall of prices of personal computers has made their introduction in educational laboratories economically feasible. Their use makes the teaching based on new design techniques such as Schematic Capture Programs or Simulators possible. On the other hand, Virtual Testing Instruments can be developed using additional hardware at low cost, thereby avoiding the need of expensive instrumentation. The use of commercial tools in educational applications has many inconveniences. The first one is that commercial tools are difficult to manage. The experience of using these tools in laboratory classes demonstrates that users need to spend a considerable amount of time in managing the program interface, thereby causing a reduction of the time that they dedicate to the circuits design. On the other hand, commercial tools are usually oriented towards professional applications and then, they do not provide help information about how to accomplish a design. Their help information is reduced to the program options. Also, there is not a commercial system that allows the design, simulation, and verification of electronic circuits within a unique environment. These drawbacks have been the motivating factors that lead to the development of an own system, called Virtual Instrumentation and Schematic Capture Program (VISCP), instead of using commercial tools for the practical teaching of electronics.

The principal aim of VISCP is to provide users with a simple and easy to use CAD tool instead of developing a system with the best performance. VISCP comprises of two main modules: a Schematic Capture Program (SCP) for circuit design, with a Netlist Generation Tool that interfaces with commercial simulators, and a Virtual Instrumentation (VI) module for test purpose. The block diagram of the system is shown in Fig. 10.9. The following sections describe the tools of the VISCP system and their operation, and the use of these tools in electronic design training.

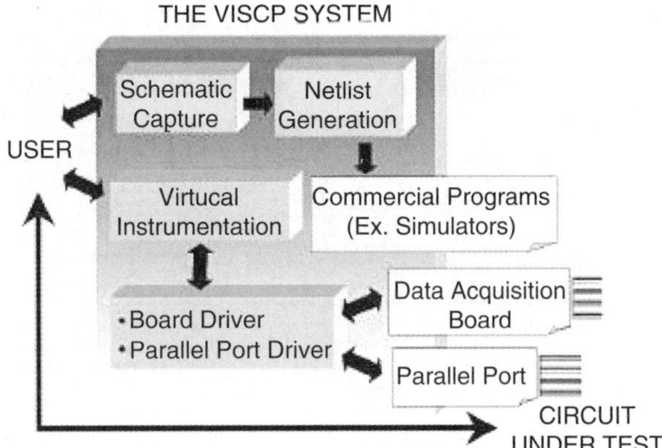

Fig. 10.9. Block diagram of the VISCP system

10.3.1 Schematic Capture Program

Recently, the development of professional CAD tools has improved the design task in electronics. At present, its use is indispensable in the case of complicated designs, so the laboratory teaching based on these new design techniques is mandatory. The users can design prototypes of systems with the help of a Schematic Capture Tool. Then, they can simulate its operation and ensure that the performance of the prototype is the desired one. If the results of the simulation are the expected ones, they can proceed to assemble and test the real circuit. If not, they must change their design and simulate its operation again. The simulation helps users to test and refine its designs before assembling the circuit with real devices.

The SCP of the VISCP system provides users with a useful and easy to use environment for the design of digital circuits. Its friendly and simple interface allows users to master the program performance in a short space of time. The program operates over Windows environment, where all programs present a similar graphic user interface that makes them easy to use. An Icon Bar with self-descriptive buttons and a Control Bar where help messages appear are provided to facilitate the use of the program options. Help on line is always available, and the component libraries have been improved in order to provide information to the user about the operation mode of the devices. Also, the SCP is oriented towards educational applications. It has been designed to help users in the design task by:

(1) Preventing typical mistakes in designs; for example, connections that can damage the devices in a real assembly are not permitted
(2) Avoiding graphical design errors by clearly showing the electrical connections between the system components
(3) Allowing users to link their designs with commercial simulators like PSpice and with PCB design programs, without the need of managing their configuration and input parameters.

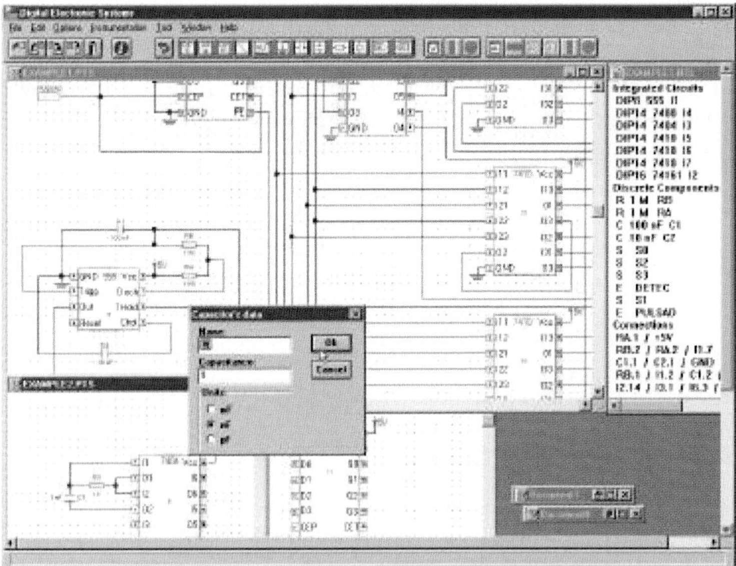

Fig. 10.10. Screen of the SCP of VISCP

These features are innovative with respect to commercial programs. On other hand, the SCP has the typical functions of the schematic capture programs. For example, a library with TTL standard digital circuits and discrete components like resistors, capacitors, and input or output ports is available. Any component can be selected, moved or erased, and their attributes can be modified. The program also facilitates the tracing of connection lines between components, thanks to the existence of a grid in the drawing area where all components are fitted. On the other hand, the simultaneous display of several windows representing different schematics is possible. In this way the user can, for example, compare several architectures for the same system. As opposed to traditional programs, the fundamental components of SCP are standard integrated circuits instead of logic gates. This characteristic ensures a direct relationship between the schematic and the real design that the user is going to assemble, making the understanding and test of the circuits operation easier. In Fig. 10.10 a screen of the SCP displaying several designs simultaneously is represented.

10.3.2 Netlist Generation Tool: Simulation

Within the SCP of VISCP, an option to generate the connections list of the circuit (netlist) is provided. So, once the user has designed the schematic of the circuit, he/she can proceed to its simulation using commercial programs (for example, PSPICE). The generated netlist contains all the information about the integrated circuits and components of the designed circuit, and their electrical connections. The adjustment of the netlist to a particular simulation program only requires the programming of a text format converter. A format converter/linker tool for the commercial simulator PSpice has been developed. It allows the simulation of the performance of a circuit inside the capture environment. In order to analyze the results of the simulation, the SCP system provides a tool that links with the Wave Form Analyzer PROBE of the software packet Design Center of MicroSim Corporation.

With the intention of avoiding mistakes in the simulation, the netlist is generated only when all the terminals of the devices are correctly connected (to other terminals, input or output ports or supply). If there are any mistakes, a message that indicates its origin is shown. The netlist will be represented in a text edition window.

10.3.3 Virtual Instrumentation

The testing of prototype systems is one of the main steps in the design of systems because its results can validate the operation of the systems or, on the contrary, can lead to the modification of some design aspects. Therefore, it is important that the users obtain the basis of the verification of systems at the same time as they learn to design them. How to test a system is a skill that can only be completely accomplished by means of experience and practice. This is because the test process is an interactive and sequential task and strongly depends on the particulars of each system. In the case of Digital Systems, the verification process is based on the measure of different signals when several series of stimulus are injected to the system. There are some basic theoretical rules about, for example, the suitable instrument to test some specific issues or the stimuli needed in order to detect some fixed failures. Nevertheless, the test strategy varies with the technology used, the final application of the system, the available points of measure, etc. Due to this, verification is always a difficult matter for users and, therefore, the learning of verification strategies must be one of the fundamental objectives of the laboratory classes.

Nowadays, the cost of the test process, in effort and time, is reduced by the use of professional measurement instruments. But the introduction of these instruments in laboratory classes in the first years of the degree presents some drawbacks. First, commercial instruments provide many features that are not required for educational purposes and consequently, the investment in instrumentation is not justified by the final use of the equipment. Second, the managing of commercial instruments is difficult because of their complex functions oriented toward the professional market. For this reason the instruments needed for educational purposes must be simple and interactive and its functionality must be adapted to the knowledge of the users. The Virtual Instrumentation technique provides an easy access to achieve these purposes. The virtual instrumentation technique is based on combining a computer with a simple and general data acquisition system in order to define a specific measuring instrument. In other words, the hardware of the system consists of the proper hardware of the computer plus a general-purpose data acquisition system. The software provides the functionality of the instrument. The main advantages of this approach are the following:

(1) *More efficient use of hardware resources.* The hardware is common to any instrument so the use of resources is more efficient. Using the same hardware, software modules can be developed in order to generate different instruments like an oscilloscope, a logic analyzer or a multimeter.
(2) *Adaptation of the control panel to the user's degree of knowledge.* The software determines the operation mode. In this way, the control panel can be adapted to the application, according to the requirements of the level of knowledge of the user. Besides, the users find the computer environment more familiar and then more manageable than the control panel of any traditional instrument.
(3) *Several applications at the same time.* Virtual Instruments use the resources of the computer like, for example, the graphic monitor to represent their results. Using actual multitasking operating systems (for example Windows95 for PCs), screens of different active programs can be represented simultaneously. In this way, the acquired data can be compared, for example, with simulation results easily. Moreover, data files can be shared between several applications and the users can analyze their results with datasheet programs, mathematical tools, etc.

10.3.4 User Interface

The Virtual Instrumentation technique permits the creation of configurable user interfaces. So, for example, during the first stages of learning, the controls can be reduced making the instrument programming easier. This means that the users can dedicate more time to the design and less to reading complex user guides. Also, the virtual instruments can take advantage of real windowed operating systems and the similar interface of its applications. Therefore if, for example, the users need to save some data recorded by the virtual instrument they know that they can find this function on the File Menu of the program, while in a traditional instrument this function depends on the manufacturer and the kind of instrument. Virtual Instruments developed in VISCP can be programmed graphically. For example, logical options are selected using ≪check-box≫ controls (see Fig. 10.11). Also, like in the schematic capture application, each control provides an online help describing its function.

Functionality

Users can create multiple instruments at the same time. The number of signals that can be displayed simultaneously is limited only by the data acquisition system. The same signal can be displayed by different instruments, assisting users to choose the best instrument for the visualization of a specific signal

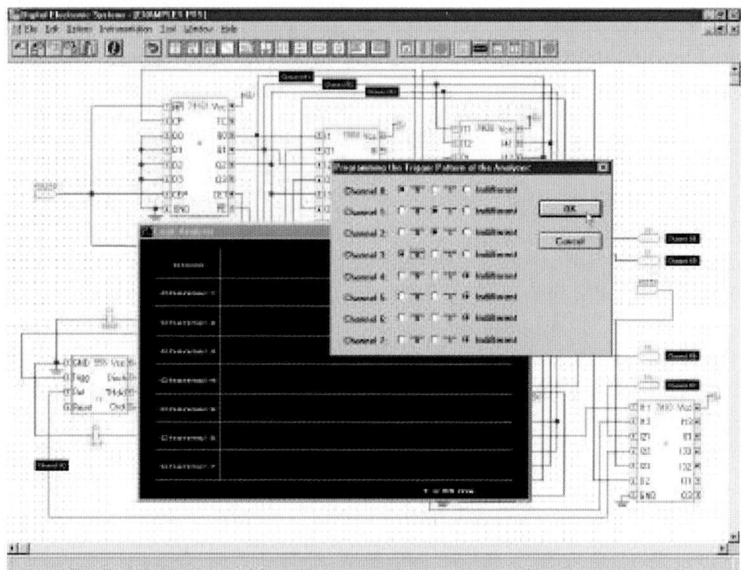

Fig. 10.11. Selection of the trigger pattern of the Logic Analyzer

according to its nature. Real measurements and simulation results can be directly compared because VISCP can display both simultaneously. Also, virtual instruments can be created into the schematic windows and a data channel can be linked with signals of the system. Therefore, the verification task can be made over the schematic design instead of over the prototype. Normally, schematics are better structured than real circuits so the error detection and therefore the test become easier.

10.3.5 Available Virtual Instruments

The software of the developed virtual instruments consists of two different kinds of modules, one that manages the data acquisition system (driver) and another that generates the proper virtual instruments. The driver has been developed using a low level programming language in order to optimize the data throughput. The driver module is unique and is shared by all the active instruments. The other modules have been developed using object-oriented languages. Common resources to different instruments, like for example, graphical displays are programmed only once by using the reusability characteristic of object oriented languages. Up to now, two virtual instruments for Digital System verification have been developed: a Voltmeter/Logic Probe and a Logic Analyzer. Nevertheless, new instruments can be created by only programming their specific characteristics. The Voltmeter/Logic-Probe is suitable for the test of combinational circuits while the Logic Analyzer facilitates the comprehension and test of sequential circuits. The instruments

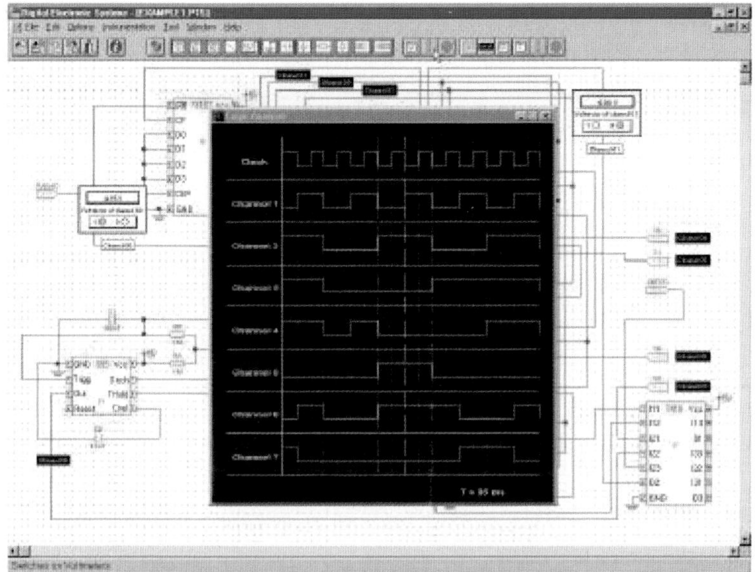

Fig. 10.12. Virtual Instrumentation of the VISCP System

are generated from the screen of the Schematics Capture and the user can associate the external test point and the corresponding node of the schematic. In this way, the results or waveforms are directly visualized on the schematic, thereby facilitating the test.

In Fig. 10.12 the schematic of a circuit and its test through a Logic Analyzer operating as a State Analyzer and Volmeters/Logic Probes is represented. The Digital Circuits test is completed with a Test Vector Generator that permits the verification without any external stimulus. In this way, the PC is the unique equipment required for the prototype test. Due to its educational orientation the Validation Tools for the system operation are included in VISCP. These tools use the acquired data as a source to generate a flow diagram describing the evolution of the system. For example, using a virtual logic analyzer and following some rules on signal acquisition the evolution of a finite state machine (FSM) can be obtained.

10.3.6 Hardware

Nowadays, VISCP works over PC platform and any data acquisition board can be used as interface to the external world. Only a small configuration must be done for each acquisition card. Another developed version of VISCP is oriented to distance learning and uses the parallel port of the PC as interface to external signals. In this way, only a reduced hardware must be added to the PC in order to operate like a virtual instrument. Obviously, this reduces the amount of signals to be observed or the acquisition rate of the system but this approach has the benefit that the users can test simple circuits at home without any specific equipment.

Conclusion

The professional tools of electronic systems design as well as the traditional instrumentation facilitate enormously the design, monitoring and prototypes test. However, their utilization in the first course of an Electronic Engineering Degree is not profitable because users never exploit all their possibilities. Furthermore, they have to devote a considerable amount of time to mastering these tools, reducing the time devoted to the learning of the fundamental design concepts. By contrast, the use of Computer-Assisted Learning tools, as the one presented in this paper, constitutes a most adequate solution. With their utilization, the learning process is improved, because they provide the same fundamental characteristics as the professional tools, and at the same time they avoid any drawbacks.

10.4 Distributed Multiplatform Control System with LabVIEW

The Optical Diagnostic Control System (ODCS) for the Tesla Test Facility is running on Macintosh personal computers. It is based on LabVIEW and interfaced with the machine control system through shared VME memory. The necessity for computers to upgrade and the need for a network access to all system functions, suggested a hardware and software upgrade of the system to include a network applications server. A VME CPU running UNIX is now the central component of the system. The LabVIEW VI Server feature allows the execution through the network of all the ODCS high level applications from any remote computer where LabVIEW and the ODCS software are installed.

10.4.1 Overview

At the Tesla Test Facility Linac, optical diagnostics have been extensively used since the beginning of operations. Optical Transition Radiation provides information about the beam: transverse shape, energy, divergence, bunch length, etc. As a consequence, a large number of optical diagnostic stations have been installed along the beam line and are extensively used for both beam transport optimization and accurate measurements. Every diagnostic station is equipped with one or more TV cameras and with stepper motors to insert or remove the radiator from the beam line and to control the position of lenses, TV cameras, and accessory equipment.

TV signals are collected from a video multiplexer whose output is connected to a frame grabber hosted in a dedicated Macintosh computer. Digitized images are uploaded into an RAM memory in a VME crate, where the ODCS runtime database is defined. Two other Macs are part of the system: one is the operator console, the other takes care of I/O modules executing commands sent by the operators. The large number of images being transferred between the three computers requires a wide bandwidth communication system: the initial solution was to use fiber-optic bus-to-bus adapters to connect the three Macs to the VME and let them communicate via the shared memory.

The integration of the ODCS into the machine control system was provided in the same way, with one DOOCS server CPU in the VME crate used to communicate with the other ODCS's computers through the shared memory.

Furthermore, new faster CPUs allow the user to replace two or more computers by a single one to simplify maintenance and obtain a more reliable system. The ODCS has been developed, since the beginning, with LabVIEW and the new features of version 5.1 of this software development tool are able to provide answers to the previous questions.

First of all, the LabVIEW code is portable to other platforms supported by this program, thus allowing to transport the developed software to another OS as simply as a file transfer. The choice was UNIX because of its powerful networking tools. Secondly, LabVIEW 5.1 embedded application server gives the possibility to run remote Virtual Instruments from a LabVIEW client application running on a remote computer. The next sections will illustrate in detail the new ODCS with its VIs server features and the possibility to run it on different OSs and different hardware.

10.4.2 The Software Structure

As mentioned earlier, a VME shared RAM memory hosts the core of the system: the ODCS database with online data and communication mailboxes. Since LabVIEW allows either to import compiled C-code subroutines or to directly call subroutines in a shared library, all the needed data structures were included in a standard C-structure that were used to map the 8Mbytes VME RAM memory. The C-code is mainly used to "place" the database structure over the VME RAM memory according to the address translation provided by the bus-to-VME adapters. After that, all the members of the structure, i.e., the whole ODCS database, are accessible from the pointer to the structure, as they are in the computer memory.

A number of LabVIEW Virtual Instruments, the low level ODCS Interface, have been created to interface subroutines to be used for R/W data from/to the ODCS database. All high-level applications (display, measurement, etc.) use the low level ODCS Interface Virtual Instruments to communicate with the database. Since different bus-to-VME adapters currently or earlier installed on the system use different ways to access the VME environment, these subroutines must be able to detect the specific hardware installed and process the VME access accordingly. A “virtual” VME database is also provided by the Low Level ODCS Interface by means of a LabVIEW global variable.

This feature proved to be very useful for off-line debugging and development. All of these different *operation modes* (virtual VME, VME Link Type1, VME Link Type2, etc.) can be selected at run-time according to the information provided by a configuration file that defines the environment in which the ODCS software will run. This conveniently simplifies the distribution of the new versions of the software, they being identical for all computers of the system, independently from the kind of connection to VME.

10.4.3 Software Portability

The C-code subroutines can be conveniently compiled as a *shared library* and loaded from LabVIEW using the Library Function tool. Since shared libraries are a mechanism common to MAC and Unix platforms, the porting of all the software initially developed to a SUN-Solaris has been straightforward. In fact, while the shared library source code only needs to be recompiled “as it is” in the Solaris environment, all the LabVIEW “code,” namely all programs written in LabVIEW graphic programming language, can directly be run in the UNIX environment.

The LabVIEW internal compiler will produce the runtime executable at the first run of any Virtual Instrument. It is worth mentioning that, since firmware was used to control the functionality of bus-to-VME adapters for the Macintosh, little extra work was required to create, for the Unix version, a library of *dummy* routines that complete the compilation of the source code. Clearly these *dummy* routines are never called in the UNIX environment but allow the user to maintain the compatibility of the shared library source code with both the Mac and the UNIX version.

10.4.4 The New ODCS with the LabVIEW VI Server ODCS on Unix

The earlier mentioned UNIX version of the ODCS software is now running on a Sparc VME computer with SunOS. Access to the VME environment is provided by the on-board VME interface and its driver’s library.

The use of this VME driver library has been embedded in the routines of the shared library as an extra operation mode and, in the same way, selected at runtime by means of the configuration file. Since the control room consoles are SUN workstations, operators can now open sessions on the ODCS Sparc CPU and run both diagnostic applications and accelerator control programs on the same console together. Furthermore, the high-level ODCS applications take advantage of the common environment with DOOCS to easily access

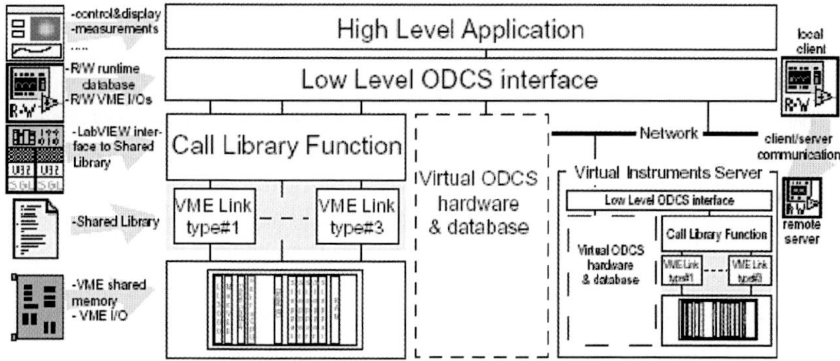

Fig. 10.13. The different operation modes for the low level ODCS Interface

its functionality and the Network File System. The modes of operation for the low level ODCS Interface is shown in Fig. 10.13.

This is convenient especially for measurement applications for which interaction with the accelerator (its magnetic elements, for instance) is needed. The Sparc CPU will run simultaneously both data display and measurements applications and applications to the execute commands sent by the operator. In the previous version those two features were provided by two separate Macs. The communication between partners is still provided by the mailbox system in the VME runtime database, to ensure compatibility with the Macintosh partners.

The ODCS Network Server

The LabVIEW application on the VME Sparc is also configured as a "Virtual Instrument Server." This feature of LabVIEW 5.1 allows to export a selection of Virtual Instruments that can be called and executed (locally, on the server machine) from a remote calling VI. The idea is to provide machines at remote networks, or that cannot be linked to VME with access to the VME runtime database, so that they become "active" partners of the ODCS. The list of VIs to be exported can be dynamically configured. The same is true for the IP numbers, or class of IP numbers, defining the list of clients authorized by the server. On the server side, once the server application is configured, one only needs to load the VIs in memory and they will be immediately available.

On the client side, a simple sequence of LabVIEW instructions will establish the communication with the server and ask it to run the selected VI with the given input parameters, receiving the output as result. The low level ODCS Interface Virtual Instruments running on the server do not need to be specially configured in order to be available for network calls. They will run in the operation mode defined by the hardware environment and will be available for both network clients and local high-level applications at the same time.

On the client environment (i.e., the remote ODCS console), one only needs to select the *network* operation mode for low level ODCS Interface and the R/W operation on the ODCS runtime database, or on VME I/O modules, will be automatically transferred to the server (which has direct access to the ODCS VME environment) to be executed locally. This system has been tested in a subnetwork controlled by a switching router. At the remote display, a 512×512 8-bit image takes approximately 300 ms to be downloaded from the server. This is roughly a factor too faster than with the previous system based on bus-to-bus adapter.

It is also planned to configure one of the system's Macs as a LabVIEW Virtual Instrument Server as well. Some high level applications use, for image analysis and manipulation, a set of tools provided by a library not included in the LabVIEW standard package common to all platforms. This so that, high level applications running on UNIX in the Solaris version, for which this LabVIEW library is not available, can nevertheless access the tools by running them, remotely, on the Macintosh server where they reside.

More complex configuration of the server can also be foreseen, for instance by creating a version of low level ODCS Interface not allowed to write the ODCS runtime database and VME I/O modules. The fully functional version could then be made available only to computers in the control room, to ensure safe operations, while the Read-only version could be made available to other selected networks for display only purposes.

A UNIX version of the ODCS software has been developed with a small amount of effort thanks to the portability of LabVIEW programs to other supported platforms. The system can thus be operated from any console of the TTF control system. The user also use the LabVIEW Virtual Instruments server to provide remote computers, running the ODCS software, with a way to run the Low Level ODCS Interface on the server and access the ODCS VME database, without any direct connection to the VME. A new version of ODCS software for LabVIEW on Linux will be also developed. It will be used for either desktop or VME embedded computer with Intel CPU.

10.5 The Virtual Instrument Control System

LabVIEW is a graphical programming language for data acquisition, data analysis and data presentation. It offers an easy and fast way to develop control systems and program instrumentation systems. Here a virtual instrumentation system is developed which will be used in the control system of BEPC or SSRF on the basis of LabVIEW. The main purpose of this virtual instrumentation system is to meet special needs for the control system of BEPC and SSRF. The virtual instrumentation system integrates many instruments and devices such as CAMAC, PLC, DAQ, and other devices with GPIB interface. Many applications have been designed and developed by using LabVIEW as interface.

The PC is used as a server that acquires data from CAMAC, PLC, DAQ, GPIB devices, and other data sources via the network. In addition, a command library to be used with PLCs has been implemented and stored in the userlib of LabVIEW. Users can build PLC monitor and control systems easily by using the programs in this library.

10.5.1 Introduction

There are many instruments and control devices in the control system and beam diagnostic system of BEPC. Most instruments can only provide special functions, at the same time some instruments cannot be accessed directly from control and beam diagnostic systems. In order to solve these problems, LabVIEW is chosen as the solution. The reasons for choosing LabVIEW is:

- Most instruments and devices used in the control and beam diagnostic system provide standard interfaces such as GPIB RS232, etc.
- LabVIEW includes many drivers for communication to other instruments, such as PLC, VXI, serial ports, etc.
- LabVIEW is a graphical programming software which provides a friendly man–machine interface, it is very easy to learn and efficient when used by developers.

This virtual instrumentation system includes applications which communicate from LabVIEW to the instruments and devices. A command library is developed to use LabVIEW with PLCs including almost all the commands of PLC devices by using LabVIEW as interface. This PLC command library contains many ICONs that have been put in the user library of LabVIEW. By using these ICONs users can build the PLC monitor and control system easily.

10.5.2 System structure

The virtual instrumentation system is divided into two levels as shown in Fig. 10.14.

First Level: The Interface Level

In this level, the server PC plays the key role. The server PC is the man–machine interface to the operator. All application programs run in the server PC. Three adapter cards are installed in the server PC. They are: GPIB adapter, DAQ adapter, and CAMAC adapter. All data acquired through these adapters are transferred to the server PC. Different application programs run in the server PC for CAMAC, PLC, DAQ, and instruments with standard GPIB interface. Each application program fulfills one task. For instance, the program in the PLC only communicates with a special PLC device.

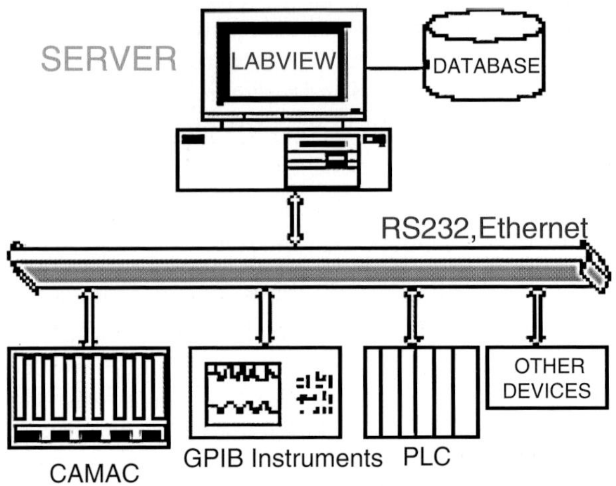

Fig. 10.14. Structure of virtual instrumentation system

All these LabVIEW application programs are called virtual instruments (VI) because their appearance and operation can imitate actual instruments. Each VI includes two parts: a front panel and a block diagram. The front panel is the user interface of the VI. It is combination of many controls and indicators. Controls simulate instrument input devices and supply data to the block diagram. Indicators simulate instrument output devices and display data acquired or generated by the block diagram. The block diagram is the graphical source code the VI. It integrates many objects which send or receive data, perform specific functions and control the flow execution.

The server PC also can receive and send data from and to other sources, such as EXCEL files and data via network. So far, we have developed the network communication VIs based on the TCP/IP protocol and VIs to read and write data from and to EXCEL files. The software items installed on the server PC are the following: Windows 95, LabVIEW 4.01, Visual C/C++2.0. All application programs are developed by using these software tools.

Second Level: The Hardware Level

This level includes CAMAC, PLC, DAQ, and instruments with GPIB interface. The data are acquired and sent to the server PC through adapters installed on the server PC. For CAMAC, we install a CCU-2-80B CAMAC crate controller in the CAMAC crate and an adapter in the server PC. All data are sent to CAMAC modules or received from CAMAC modules through this CCU-2-80B controller. For PLCs we use the RS232 protocol to communicate with PLC devices. The data from PLC devices are transferred through a serial port to the server PC. We also send and receive data from 390ADs (a programming digitizer with standard GPIB interface) and other DAQ devices.

10.5.3 Applications

LabVIEW-based CAMAC and PLC applications are discussed in detail in this section.

LabVIEW-CAMAC Application

CAMAC devices are widely used in the control and beam diagnostic systems. It has been the standard bus in the field of high energy physics. Although LabVIEW provides some drivers for CAMAC controllers, there are many CAMAC controllers which are not yet supported by LabVIEW, among them the CCU-2-80B controller which is now widely used in the control and beam diagnostic system of BEPC.

The CCU-2-80B is a CAMAC crate controller which was designed and developed in IHEP. It supports standard CAMAC modules. Because LabVIEW has not supplied a driver for the CCU-2-80B, we have developed our own driver which provides an interface to support the communication between LabVIEW and the CCU-2-80B controller. The driver is programmed using C/C++. It supports the basic CAMAC operations including reading, writing, and testing status bits. LabVIEW provides two modes to link C/C++ programs. One mode is CIN (Code Interface Node). Through CIN, users can directly call the C/C++ program within LabVIEW by using the special format offered by CIN. The other mode is Dynamic Link Libraries (DLL).

Users can call any DLL or shared library from LabVIEW. An application program has been developed which communicates with CAMAC devices by using the above two modes. For the DLL mode, we use Visual C/C++ to develop the C/C++ programs to support the basic CAMAC operations. Thereafter the program is changed into a DLL program which can be called by LabVIEW directly. For the CIN mode, we develop the C/C++ program using the specific format drafted by LabVIEW.

The following is the basic CIN format of LabVIEW:

```
/*
/* CIN source file
/*
# include "xtcode.h"
/* stubs for advanced CIN functions */
UseDefaultCINInit
UseDefaultCINDispose
UseDefaultCINAbort
UseDefaultCINLoad
UseDefaultCINUnload
UseDefaultCINSave
CIN MgErr CINRun(void var1);
```

```
CIN MgErr CINRun(void var1)
{
/* Enter your code here */
return noErr;
}
```

All graphic interface programs are developed in LabVIEW, however, all I/O programs are developed in Visual C/C++. All application programs will be used in the beam diagnostic system of BEPC.

LabVIEW-PLC Application

A programmable logic controller (PLC) is another important device in the control system. The command library for the PLC using LabVIEW was developed as shown in Fig. 10.15. It includes 34 VI. Each VI in the library has its own ICON and users can call this VI through its ICON which can be found the FUNCTIONS palette of LabVIEW. Because PLC devices which we use are linked to the server PC through a serial port, every VI of the command library is developed on the basis of the serial protocol. We use the basic serial subVI of LabVIEW for this command library. The basic serial subVI are Serial Port Read VI, Serial Port Write VI, Serial Port Init VI, and Serial Port Break VI. Although the PLC device which we use is the OMRON C200H, the command library can be used for any PLC device, which uses the serial port protocol, without any change. Nearly all PLC commands are included in our library.

Therefore users can easily and efficiently construct special PLC monitor or control systems by calling VIs of this PLC command library. The following is a sample of a VI in this library (see Fig. 10.15.). This VI is to execute a test command by responding to the test strings which are sent from the server to PLC devices.

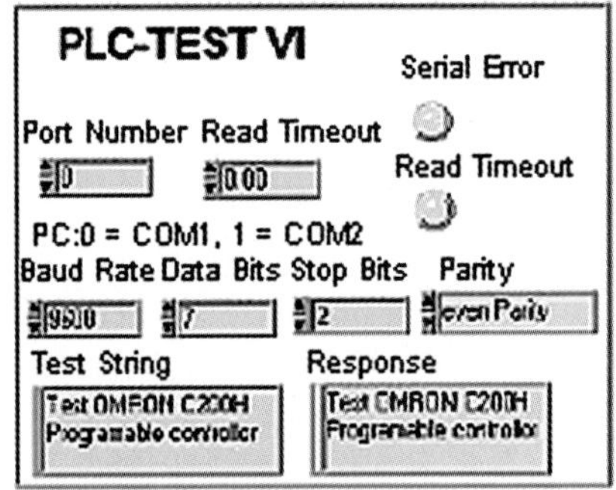

Fig. 10.15. LabVIEW-PLC application

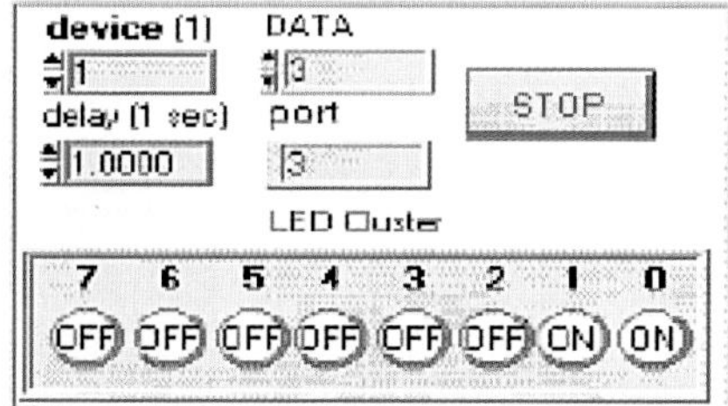

Fig. 10.16. LabVIEW-DAQ application

LabVIEW-DAQ Application

Data acquisition from DAQ adapters is another important task for which application programs had to be developed. The DAQ device used is the PC-DIO-96. It is a 96-bit parallel digital I/O board. It has been installed in the server PC. It contains four 24-bit programmable peripheral interfaces (PPIs). Each can be further divided into three 8-bit ports. The boards can be operated in either unidirectional or bi-directional mode. Digital signals acquired from data sources are transferred to the server PC through a 100-pin male ribbon connector. The following application, shown in Fig. 10.16 is a sample of the VI which sends data through the PC-DIO-96 board.

LabVIEW-GPIB Application

Most instruments used in the control and beam diagnostic systems provide a standard GPIB interface. In order to meet special needs of the beam diagnostic system of BEPC, we have implemented this application by using the 390AD which is a programming digitizer with standard GPIB interface. A GPIB adapter which is produced by NI corporation is installed in the server PC. On the server PC, users can operate the digitizer just the same as the real digitizer.

Our application does the following: data of channel 1 of the 390AD are acquired by the server PC through the GPIB adapter, then these data are processed using a series of mathematical algorithms in order to get the correct result needed by engineers. The calculated data are sent back to channel 2 of the 390AD and displayed on the server PC at the same time. Figure 10.17 is the front panel of the LabVIEW-390AD VI.

A virtual instrumentation system has been developed in the laboratory and LabVIEW-CAMAC application will be used in the beam diagnostic system. Because of the limited budget and resources, we have only developed CAMAC, PLC, GPIB, and DAQ applications by using LabVIEW as interface. In the near future, DATABASE will be installed in the server PC and all data acquired through instruments will be stored in a database. All applications send and receive data through this database. Also, VME is the next choice to link with LabVIEW.

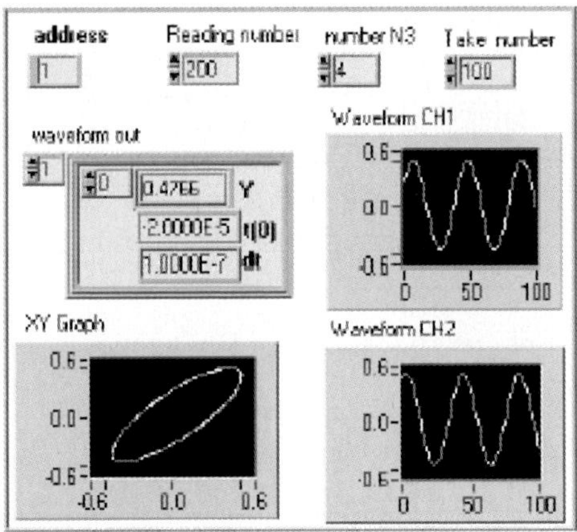

Fig. 10.17. Front panel of LabVIEW-390A

10.6 Controller Design Using the Maple Professional Math Toolbox for LabVIEW

This application demonstrates the usage of the *Maple Professional Math Toolbox for LabVIEW* to design and tune a Proportional–Integral–Derivative (PID) controller. This toolbox seamlessly integrates LabVIEW with Maple, allowing the user to develop applications that use both the data acquisition capabilities of LabVIEW, the sophisticated mathematical modeling of Maple. The following example application models an existing system and tunes the controller parameters. The final parameter values are automatically exported to the live system. This application illustrates the following features.

- Using Maple to derive a transfer function of the overall system
- Using LabVIEW to control a physical system
- Control engineering plots: step response, Bode plot, and root locus plot
- Powerful Maple symbolic capability to perform sophisticated "what if" analysis

This example in Fig. 10.18 illustrates the PID controller tuning for a two tank physical system. In this system, water flows through an input valve to Tank 1, and from this tank the water flows freely to Tank 2. The water flows freely from Tank 2. The input valve is controllable, while both of the output valves for tanks one and two are fixed in an open position. The goal is to control the position of the input valve to maintain the water level in Tank 2 at some preset level.

This system can be described by the following transfer function, where *H2(s)* corresponds to the height of tank 2 and *U(s)* is the input valve position.

$$\frac{H_2(s)}{U(s)} = \frac{R_2}{A_1R_1A_2R_2s^2 + (A_1R_1 + A_2R_2)s + 1}$$

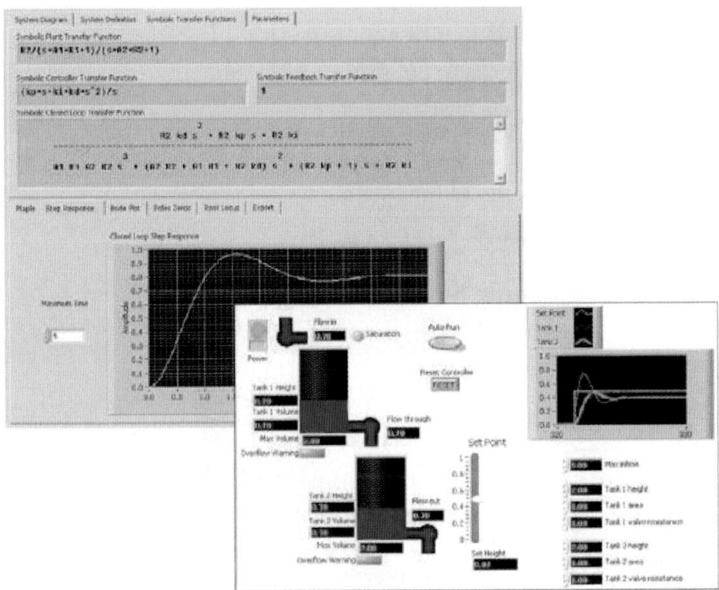

Fig. 10.18. Two tank system

In Fig. 10.19, *A1* and *A2* are the cross-sectional areas of Tanks 1 and 2 respectively, and the quantities *R1* and *R2* are the output valve resistances for Tanks 1 and 2. The goal is to tune a PID controller to maintain a given water level in Tank 2 and then deploy the controller parameters to a live system.

The described system is a live system controlled by LabVIEW. It contains an existing PID controller; however, the controller parameters are not optimal. The goal is to model the system with different controller parameters using Maple and then tune the parameters to improve the system performance. The new parameters are exported to the live system. In this example, a LabVIEW application controls the system and displays the various parameters and states of the system.

10.6.1 The Two Tank System

The system output (water height level of Tank 2) is shown in green in the upper-right portion of Fig. 10.20. This figure illustrates the response of the system to a step change of the Tank 2 desired height (set point) from zero to one half of the total height. From the figure, it is evident that the height of the water level in Tank 2 (green line) never reaches the desired height. It also exhibits a steady-state (settling) error. In addition, the height level first exceeds the target level, before it settles to a final value. The controller behavior is not satisfactory.

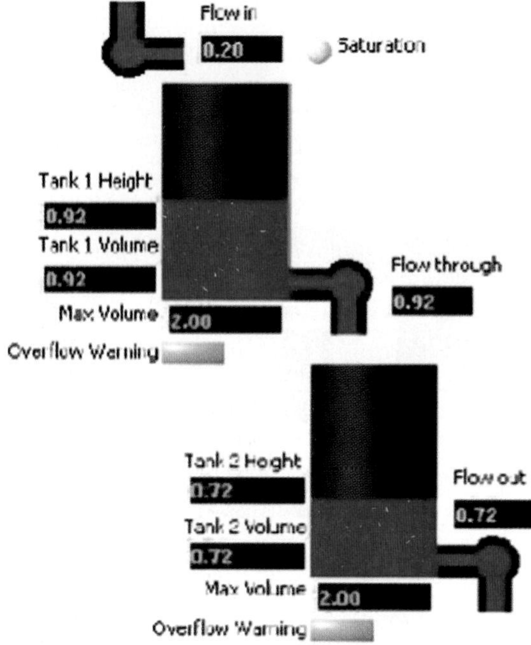

Fig. 10.19. Two tank system

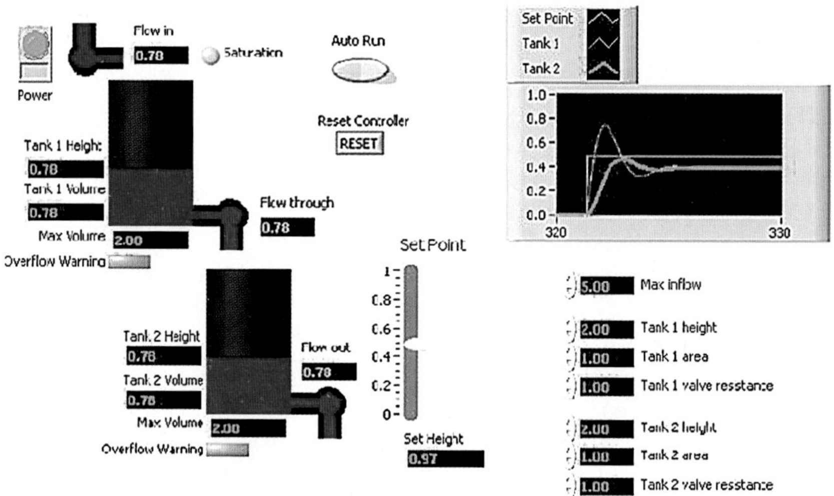

Fig. 10.20. Live system

10.6.2 Controller Parameter Tuning

Using the power of Maple, a simulation of the system is created within the LabVIEW application. Using this simulation, the system response is evaluated for different controller parameters. The user can enter arbitrary transfer functions for the plant as shown in Fig. 10.21(the two tank system) and controllers in both the forward and the feedback path.

Maple then formulates the transfer function for the entire closed-loop system shown in Fig. 10.22, from which the behavior of the entire system is simulated. Since Maple computes the overall system transfer function symbolically, the model and controller parameters shown in Fig. 10.23 are stored separately and are substituted only when needed. As a result, the user can investigate the sensitivity of the overall behavior to each of the parameters.

In addition to generating the transfer function of the system, Maple computes several commonly used control design and analysis plots, such as the step response, Bode plot, and root locus plot as illustrated in Figs. 10.24–10.26, respectively. In the case of the root locus plot, instead of using the default loop gain as the varying parameter, the user can select different parameters and the corresponding root locus plot for that parameter is generated automatically. This is a unique benefit that Maple offers, as a result of its symbolic

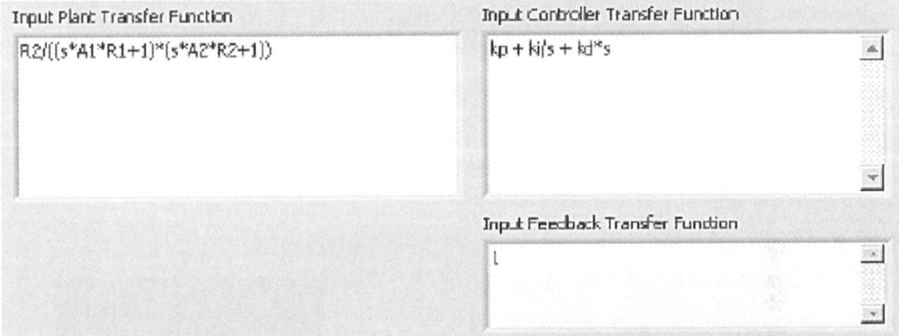

Fig. 10.21. Transfer functions for the controller, plant, and the feedback systems

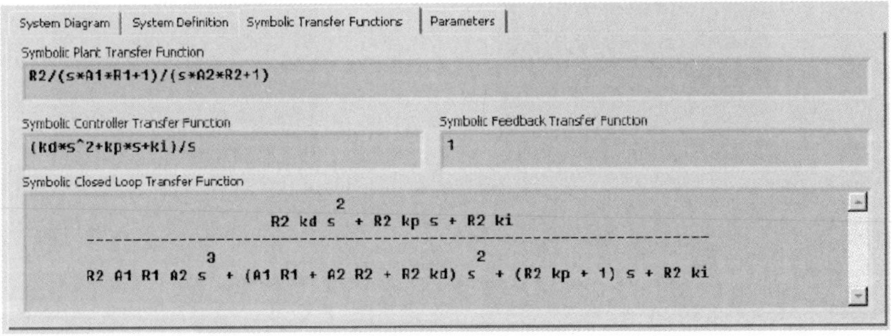

System Diagram | System Definition | Symbolic Transfer Functions | Parameters

Symbolic Plant Transfer Function

R2/(s*A1*R1+1)/(s*A2*R2+1)

Symbolic Controller Transfer Function

(kd*s^2+kp*s+ki)/s

Symbolic Feedback Transfer Function

1

Symbolic Closed Loop Transfer Function

```
                        2
              R2 kd s  + R2 kp s + R2 ki
---------------------------------------------------------
              3                            2
R2 A1 R1 A2 s  + (A1 R1 + A2 R2 + R2 kd) s  + (R2 kp + 1) s + R2 ki
```

Fig. 10.22. Transfer function for the entire system

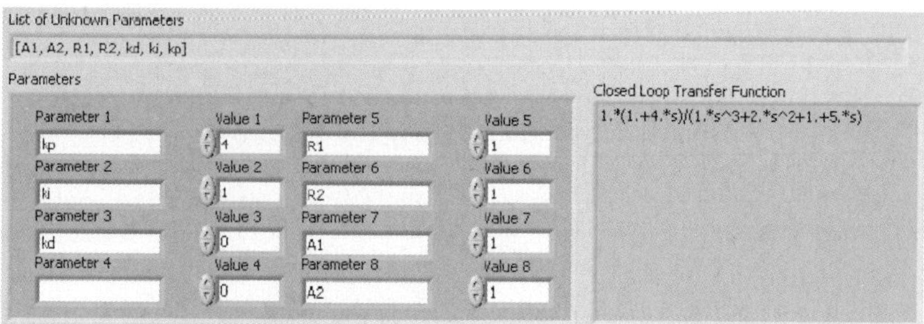

Fig. 10.23. List of the model and controller parameters

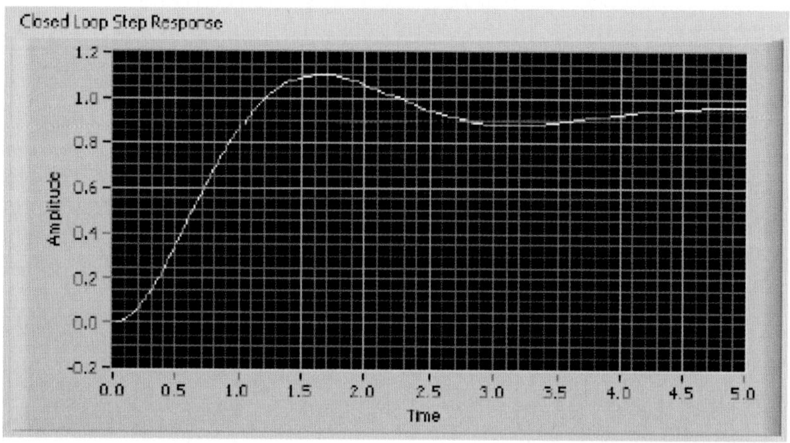

Fig. 10.24. Step response of the overall system

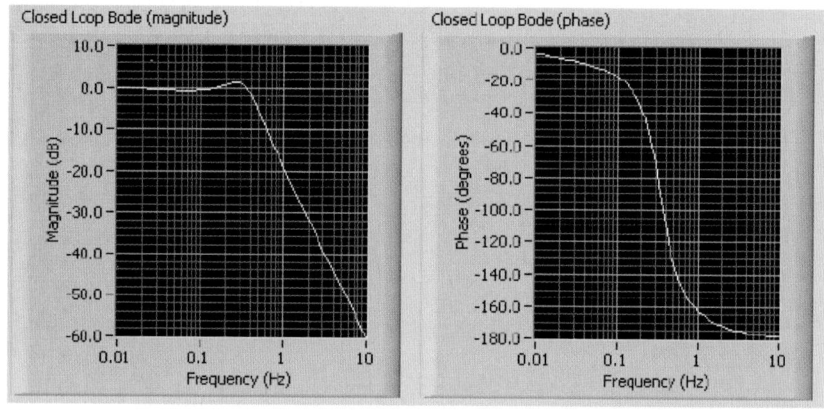

Fig. 10.25. Bode plot for the overall system

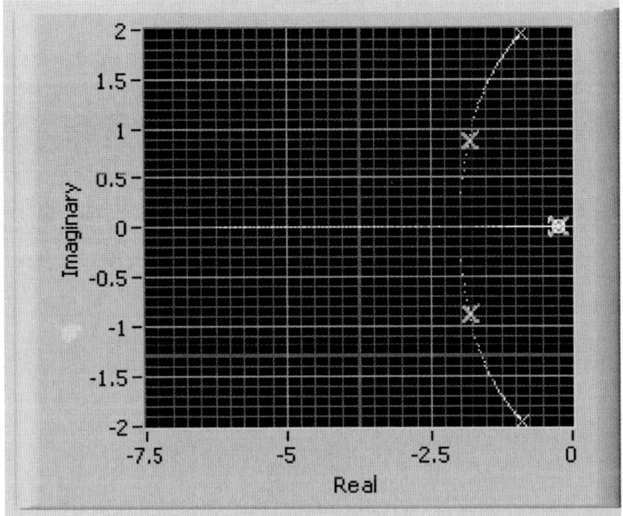

Fig. 10.26. Root locus plot for the entire system

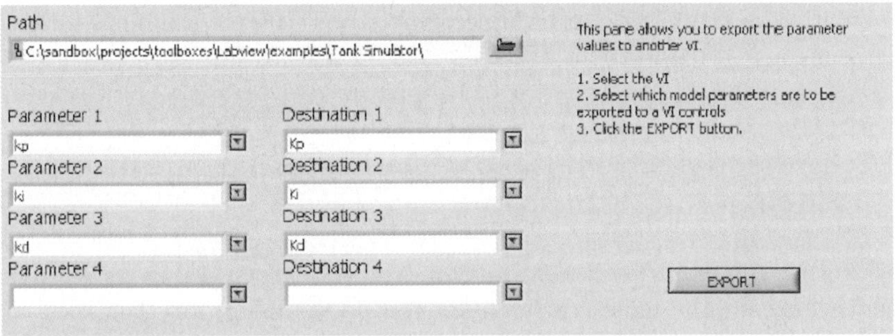

Path
C:\sandbox\projects\toolboxes\Labview\examples\Tank Simulator\

Parameter 1
kp
Parameter 2
ki
Parameter 3
kd
Parameter 4

Destination 1
Kp
Destination 2
Ki
Destination 3
Kd
Destination 4

This pane allows you to export the parameter values to another VI.

1. Select the VI
2. Select which model parameters are to be exported to a VI controls
3. Click the EXPORT button.

EXPORT

Fig. 10.27. The controller parameters are exported to a live VI

manipulation – it saves time by eliminating the need to manually re-derive the transfer function expression for each parameter value. Using these tools, the user can tune the parameter values for the PID controller until a satisfactory result is achieved.

10.6.3 Deployment of the Controller Parameters

The controller is designed in LabVIEW, and as a result the parameters are easily exported by writing them directly to the controller VI as shown in Fig. 10.27. The user need not stop the controller, read data files, or cut and paste the parameter values. The response is improved with the new controller parameters as shown in Fig. 10.28.

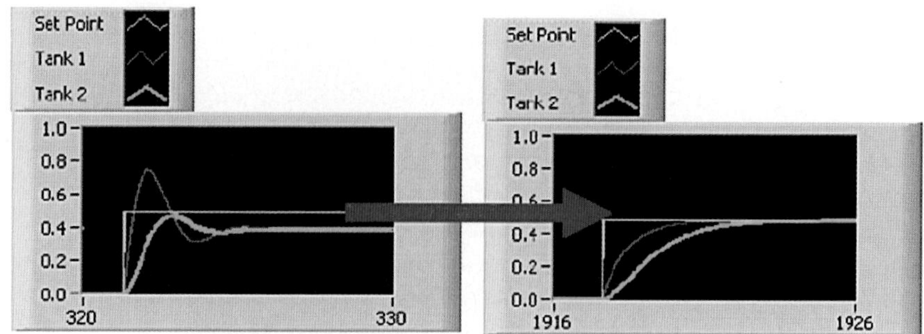

Fig. 10.28. The response of the system is improved with the new controller parameters

Conclusion

To develop and tune a PID controller, LabVIEW is used to implement a controller for the physical system, and Maple is used to model the system and design a controller. This controller is designed in the LabVIEW environment, making the export of the controller value to the real system effortless. The combination of LabVIEW and Maple offers a powerful environment for rapid development and deployment of control solutions.

10.7 Embedding Remote Experimentation in Power Engineering Education

Engineering education by its nature is a costly program in university environments. Perhaps the most costly component is the laboratory facilities, usually consisting of specialized equipment. Effective instruction of some topics in power engineering education requires experience with actual equipment, rather than small-scale replicas or simulation. In this application a new laboratory approach is described, as implemented in a virtual, Internet-based, experimentation platform.

This virtual laboratory (*VLab*) utilizes real equipment distributed among multiple universities from which remotely located users can perform experiments. The software solution is a multiuser, client–server architecture developed in the LabVIEW environment. Implementation details including video, chat, archiving, and the hardware and software platforms are also presented. An example presented herein is the study of current and voltage waveforms while controlling relays and low voltage contactors. The applications have been tested with teams enrolled in the electrical engineering department of Politehnica University of Bucharest and the power engineering program at Arizona State University.

10.7.1 Virtual Laboratories in Power Engineering

The dominance of the Internet is acknowledged in the development of information and communication technology that has made *Web-based distributed solutions* increasingly attractive. Apart from providing other services, the World Wide Web is being looked upon as an environment for hosting modeling and simulation applications. One of the major advantages such models provide is their ability to help users develop technical skills. The users manipulate virtual hardware (a simulator) to develop proficiency for operating the corresponding real world system. To preserve the advantage of the Internet, web-based laboratories require a tradeoff between simulation and actual operation of the laboratory equipment. A *laboratory* is commonly defined as:

- A place equipped for experimental study in a science or for testing and analysis; a place providing opportunity for experimentation, observation, or practice in a field of study
- An academic period set aside for laboratory work

A *virtual laboratory* is defined as an interactive environment for creating and conducting simulated experiments: a playground for experimentation. It consists of domain dependent simulation programs, experimental units called objects that encompass data files, tools that operate on these objects, and a reference book.

Consider the scientific meaning of *simulation* as "goal-directed experimentation with dynamic models." Also, denote a *virtual laboratory* as being based on the first definition of *laboratory* presented above, that is, a place providing opportunity for experimentation. In the work described below, a virtual laboratory becomes a virtual location inside the Web, having a distributed nature and a dynamic configuration. Laboratory equipment available in the web-ring can be shared together with experimental data and protocols. In the field of remote experiments used for distance learning purposes, one identifies two major solutions.

The first approach uses one or more data acquisition cards as a versatile interface between the physical phenomenon and the digital realm. The experiments are accessed either synchronously or asynchronously by at least one user via different software. A second pedagogical method uses standard digital instruments that can perform either standalone or connected to an external processing unit, usually a local personal computer (PC). The communication layer allows either serial or parallel data streaming and drastically determines the overall performance of the experiment. The first approach is more suitable for complex experimental setups and imposes special requirements in terms of security access.

The second method is mostly oriented toward realizing a more realistic software replica of the equipment itself. In this latter technique, a special concern is the equipment safety. Topics from distributed measurement systems field of study are very suitable for this second type of remote experiments.

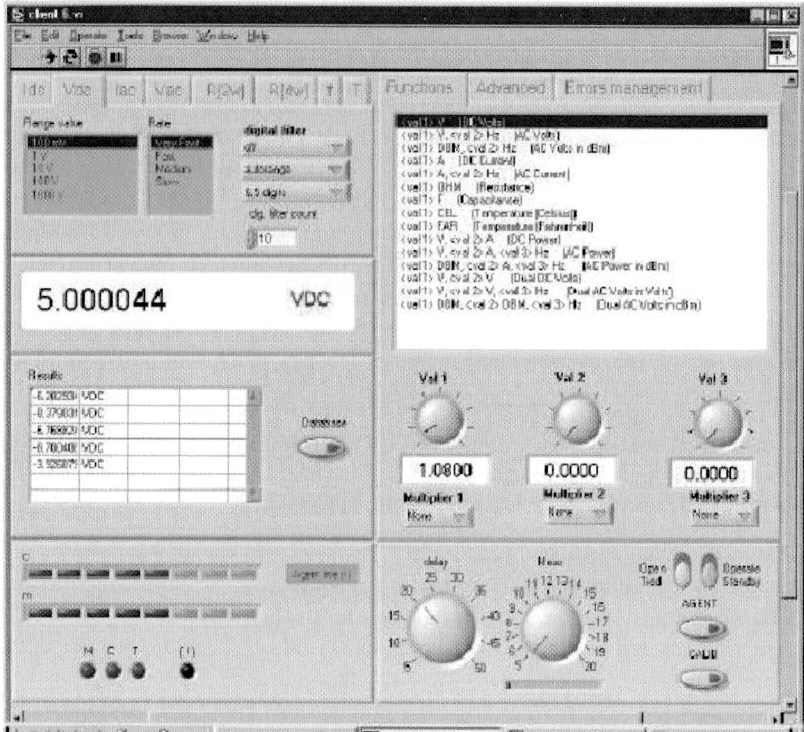

Fig. 10.29. An example of the front panel (client application) of the remote calibration process, using a step-by-step calibration procedure

Figure 10.29 shows the front panel of a client application as used in a Internet-based calibration of a digital voltmeter, e.g., the device under test (DUT), physically connected – at the remote location – to a calibrator unit (CLD), and a hardware platform connected to the Internet.

The equipment is located in a measurement laboratory at Politecnico di Milano, Italy. The actual calibration process is fully controlled from a web-location (this example is hosted on a computing platform at the Politehnica University of Bucharest, Romania).

10.7.2 Remote Experiments Over the Internet

In this section an experiment designed for users enrolled in electrical engineering is presented. The particular hardware employed here is organized around a data acquisition board. Ultimately, this online laboratory seeks to present fundamental topics in commutation such as synchronous contact closure in three-phase systems, closing time dispersion of a switches lot, relay

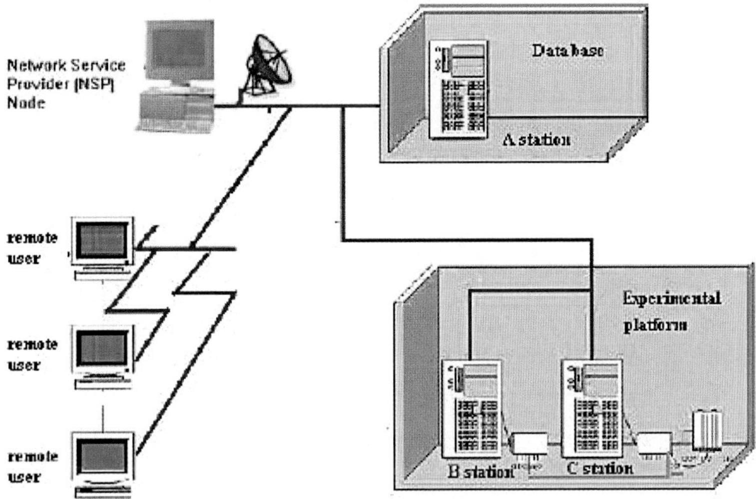

Fig. 10.30. General diagram of the VLab architecture

command, and electromagnetic noise measurements. The application is also intended to serve as online course support for the in-class lectures on measurements systems and other power engineering topics.

This virtual laboratory (VLab) utilizes equipment and software distributed between two universities from which remotely located users can perform experiments. The software solution is a multiuser, client–server architecture developed in the LabVIEW environment. The server application is unique to the communication interface and protocol (e.g., parallel, Centronics-based) between the computer and the specific laboratory hardware. Figure 10.30 shows the particular architecture of VLab. One of its nodes, VLab PUB, hosts a total of eight PC stations, and a network server, interconnected via a fiber-optic link with the main node of the Romanian Academic RoEduNet wide area network. Each remotely accessed device is connected to a PC running a specific server application.

Then the user is operating the equipment through a client application that can be formed by multiple client modules. All servers, including the one responsible for user authentication are running on PCs located in VLab PUB. It is not necessary for all server applications to be housed on the same machine; in fact, the bandwidth availability can be improved if the video server, for example, is running on a separate computer.

Figure 10.31 shows the schematic of the experimental setup in the case of three-phase contactors. The three voltages and three currents are acquired on six differential inputs of the data acquisition card (PCI-AT-E4 from National Instruments), while its maximum sampling rate (200 kHz) assures a maximum software delay between the input channels of at least 30 μs.

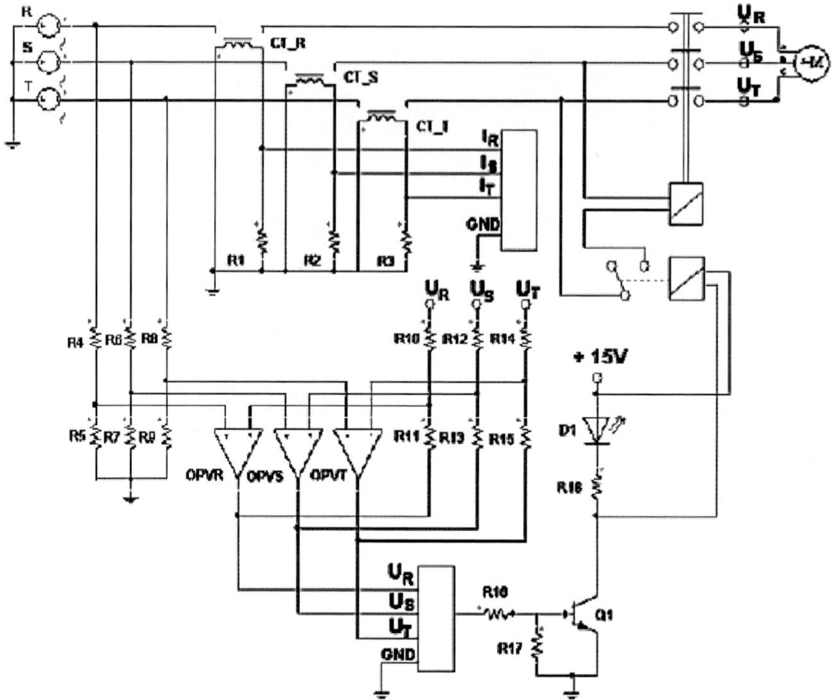

Fig. 10.31. Simple schematic for studying the three-phase contactor for motor-type load using a data acquisition card

This value is acceptable when considering the closing time of typical contactors. Users are required to save data and then process it in a specialized environment (e.g., MATLAB) to find the dependence of contacts nonsynchronicity from the load type. The control signal, sent from a DAQ digital output, is the time reference.

Figures 10.32 and 10.33 show typical waveforms as seen on the front panel of the client application. In this example, there are two online users, one of them becoming the master user after selecting the *Control* switch. All the actual selections made by the master user are visible on each user's front panel.

Figure 10.34 presents the schematic used for a synchronous controlled eight-relay bank using a parallel port with the corresponding voltages given in Fig. 10.35. The application is realized in the LabVIEW environment. A client of the LabVIEW application is embedded into another application, dedicated to the time dispersion study of the relay switches.

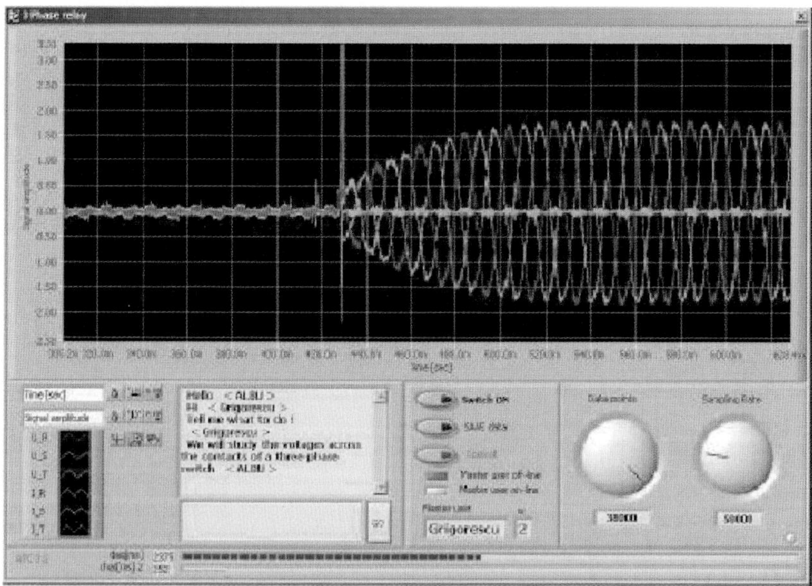

Fig. 10.32. Voltage and current waveforms at low-voltage contactor terminals as visualized at the remote-user location

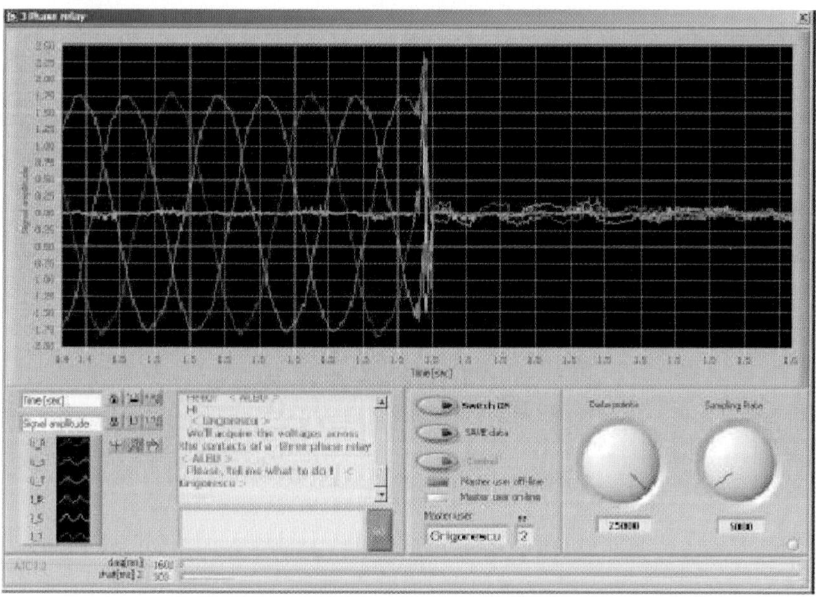

Fig. 10.33. Voltage and current waveforms at low-voltage contactor terminals as visualized at the remote-user location

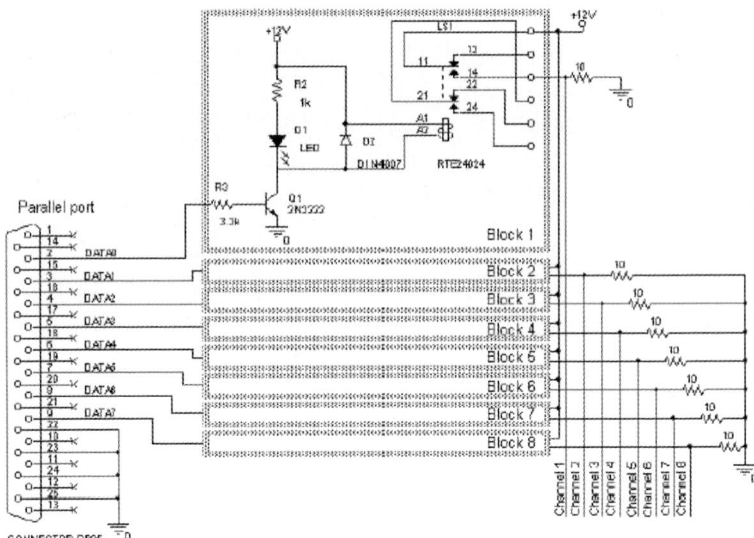

Fig. 10.34. Simple schematic for controlling a bank of eight relays from the computer parallel port

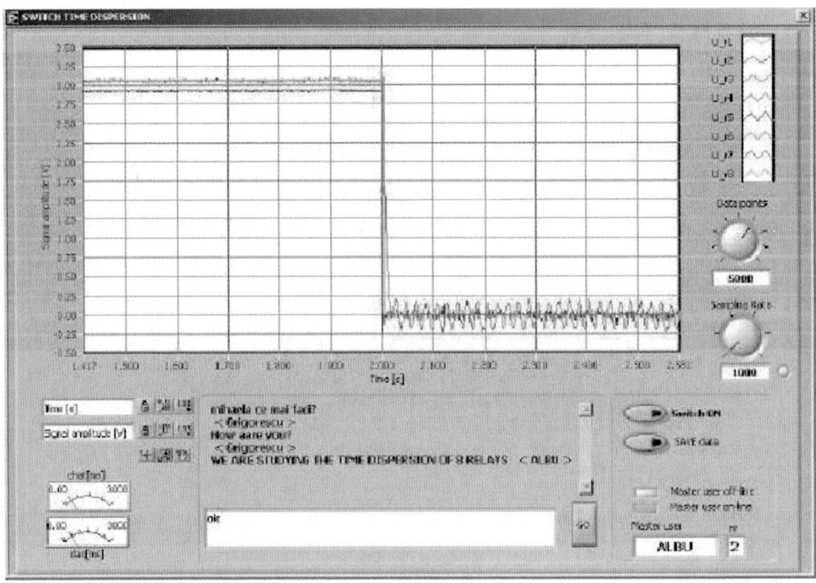

Fig. 10.35. Voltage across contacts of eight synchronously controlled relays

Conclusion

Designing a multiple-access experimentation platform is especially effective for power systems educational purposes. Users can practice utilizing the power devices in a tutorial mode, before they actually perform experiments. The equipment – usually sensitive to voltages and currents outside the rated range – is better protected from accidental misuse by implementing various levels of (inexpensive) software protection. Such an implementation allows real-time operation of distributed research teams. Similarly, the specific master (and client) software can be installed on the computing station of an industrial testing unit, which could then be exploited for research purposes. Overall, the educational exposure versus cost favors the use of remote electronic laboratory experiments in many circumstances. Some educators feel that virtual laboratories may not be as effective as manipulating real equipment.

For example, it can be argued that users do not attain respect for power measurements, and actual manipulation of test equipment afforded by hands-on experience; however, remotely located hardware may be superior to simulation. Further, users appeared to be attracted to computer use in this education application, thus taking advantage of the users' expertise in computer software applications. Often overlooked is the fact that, psychologically speaking, the remote user is shielded against the adverse consequences of the misconnection of equipment. Software protection interlocks prohibiting the user from making such mistakes increases the longevity of the equipment. For this reason, some institutions might consider initially utilizing the remote method to "train" the user before the use of the equipment for in-laboratory experiments. Another drawback to simulation is the fact that the user does not enter the actual laboratory, so there is less hardware troubleshooting (e.g., loose wiring or connections) whereas the (nonsimulation) approach taken here retains these possibilities but in a restricted form. One of the main advantages offered by VLab is that users from all over the world can use the equipment located in a particular laboratory. Each university that is part of the ring of virtual laboratory users can provide a different subset of experiments.

10.8 Design of an Automatic System for the Electrical Quality Assurance during the Assembly of the Electrical Circuits of the LHC

The Large Hadron Collider (LHC), constructed at CERN will bring into collision intense proton beams with a center of mass energy of 14 TeV and a nominal luminosity of 1034 cm–2 s–1. The magnet system required for this performance consists of 10,094 superconducting magnet units, most of which are operated at 1.9 K. During the assembly of the LHC machine one of the challenges will be the correct wiring of the 1,712 circuits with about 70,000 splices for the powering of the magnet units. At each interconnection plane between two cryomagnet assemblies, up to 72 junctions will be performed on electrical circuits with a current rating between 600 A and 13 kA. Considering the complexity of the electrical scheme, there is a considerable risk of wiring faults which will perturb LHC operation if they are not detected during the assembly phase in the tunnel. One of the most complex circuits is composed of

154 sextupole corrector magnets distributed along the full length of an entire arc composed of 23 cells.

Any electrical fault not detected during the assembly in the tunnel will be hardly discernible when the machine is at cryogenic conditions. As a consequence, repairs will be costly and time consuming. In order to ensure the proper assembly and functioning of the LHC machine, an electrical quality assurance (ELQA) plan has been established. This application describes the design of an automatic system for the ELQA activities to be applied during the assembly of the arc and dispersion suppressor zones of the LHC machine.

10.8.1 Methodology of Verification

The electrical powering layout and the verification procedures along with the parameters are explained in this section.

Electrical Powering Layout

The LHC machine layout comprises eight arcs and sixteen dispersion suppressor zones. The continuous cryostat includes one arc zone and two dispersion suppressor zones at each extremity, with a total length of 2.7 km. Six main bus-bars are used for the powering of the main dipoles and the quadrupole circuits rated for 13 kA. Twenty auxiliary bus-bars rated for 600 A supply the spool-piece correctors of the main dipole magnets. These 26 bus-bars will be joined at each interconnection plane between two cryomagnets. Further a cable with 42 auxiliary bus-bars feeds the corrector magnets associated with the main quadrupoles housed in the so-called short straight sections (SSS). The bus-bar cable is routed through the so-called line-N located outside the main magnet cold masses. Junctions of line-N cable segments will be made at each interconnection between a cryodipole and an SSS. The polarity of a magnet can be checked by means of voltage taps connected to the lead A of the magnet units. All the signals from the voltage taps are routed out of the cold mass locally via an instrumentation feed through system. Figure 10.36 shows the electrical routing of the bus-bars along an LHC cell.

Parameters to be Verified

To ensure the correct connection of bus-bars and in order to guarantee the right polarity of magnets powered in series, the following parameters are verified:

- Continuity of bus-bars and magnet interconnections.
- Authentication of magnet type by means of its measured ohmic resistance.
- Magnet polarity via the voltage taps on lead A.

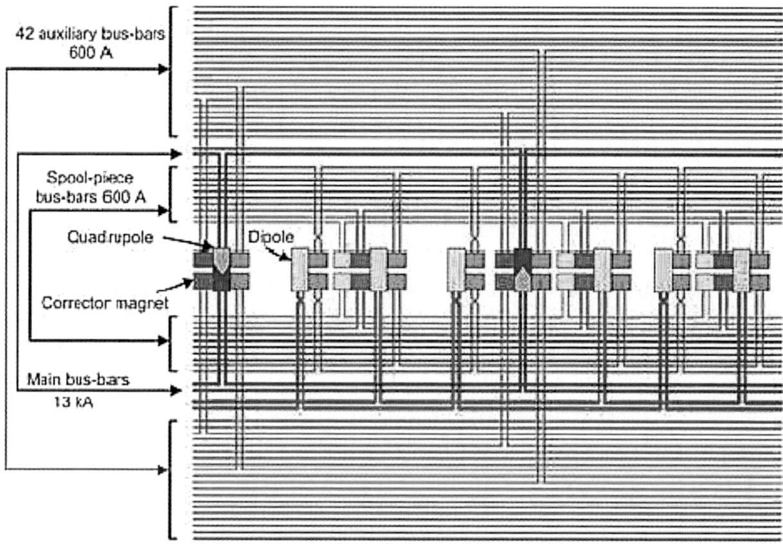

Fig. 10.36. Electrical powering diagram of one LHC cell

Test Procedures

The methodology applied for the continuity verification consists of feeding a stable DC current into a branch of the tested circuit. Voltage drops across precision resistors connected in series at both extremities of the branch confirm its continuity. The authentication and polarity of magnets connected in series within the branch are verified by measuring differential voltage drops between voltage taps at the magnet's lead A and the source and sink ends of the branch. The voltage measurements are automatically compared to known parameters stored in a database. In order to define a systematic method for the verification of the 70,000 splices all different electrical interconnection types and corresponding configurations have been determined by analyzing the data in the LHC reference database. This analysis lead to the definition of five interconnection types in the line-N circuits, six types for the circuits of the spool-pieces in the arc zone and six types for the circuits of the spool-pieces in the dispersion suppressor region. A test procedure has been developed and implemented for each interconnection type allowing its automatic verification. For the verification of the 42 auxiliary bus-bars in line-N, the access to three successive interconnection planes is required at the level of the so-called line-N interconnection board. Figure 10.37 shows the scheme for the verification of a magnet powered from the 42 auxiliary bus-bars cable. First the continuity of the circuit is assured by measuring voltage drops across the current reading resistors. Measured voltage drops between the allocated wire slots on the central interconnection board and the source and sink ends indicate the correct distribution of the wires. Finally the polarity and the type of the connected

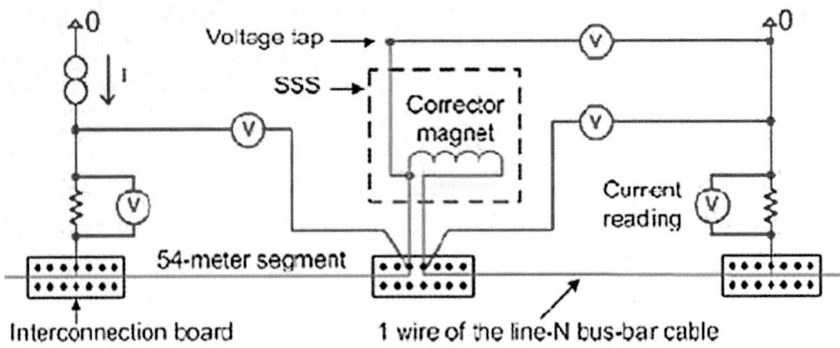

Fig. 10.37. Example of the verification of a circuit branch

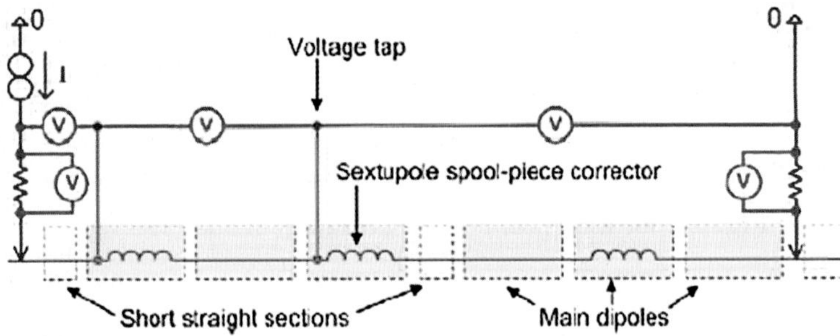

Fig. 10.38. Verification of the connection of two sextupole correctors powered via an auxiliary spool-piece cable

magnet are checked by measuring the expected voltage between the voltage tap attached to the magnet and the sink end.

The verification of the auxiliary spool-piece bus-bars only requires the access to the extremities of a cell. Figure 10.38 shows an example of the scheme for the verification of two sextupole spool-piece magnets attached to the main dipoles.

10.8.2 Technical Design

The design of a mobile system for the circuit verification in the LHC tunnel has been launched with the following technical and environmental constraints:

(1) Verification of a full cell.
(2) Qualification of all types of interconnections with a tool which can be independently operated by two persons.
(3) Hardware optimized for tunnel dimensions and storage underneath the cryomagnets.
(4) Fast connection of cables and connectors.
(5) Software for the automatic validation of measured data with respect to the LHC reference database.

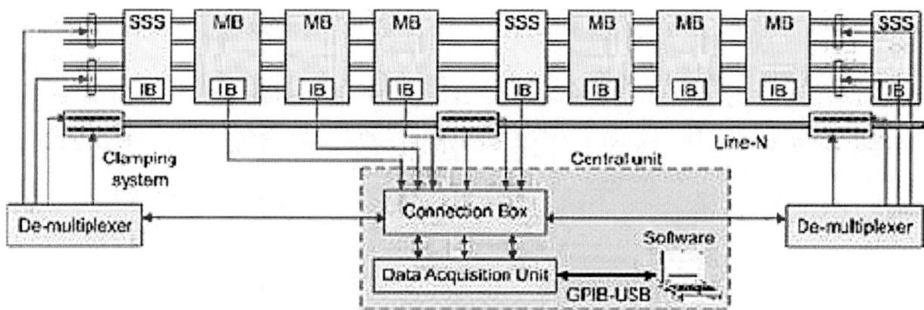

Fig. 10.39. Layout of the mobile system

The Mobile System

The mobile system shown in Fig. 10.39 is composed of a central unit to be placed at the center of the cell including a portable computer running the software application, of two demultiplexers positioned at the extremities of the cell, and the connectors and cables for the pick-up, routing and dispatching of the measurement and control signals.

The required signals dispersed along the 108 m long cell are collected in a connection box. All bus-bar signals are picked-up by connectors especially developed to ensure a fast plug-in and a reliable electrical contact. At the extremities of the cell two relay-based demultiplexers allow the selection of a couple of signals, among the 70 available bus-bars. The selection of the required channel is done via six digital lines driven from the central unit. A cable containing the current readings and the digital lines links the demultiplexer to the connection box. The signals coming from the central line-N interconnection board are directly routed to the connection box of the central unit. The voltage tap signals needed for the polarity checks are routed from each cryomagnet instrumentation interface box to the connection box via four dedicated cables.

The Data Acquisition System

The central unit contains a data acquisition system including a high precision DMM, two switching matrixes allowing the independent reading of the 3,160 voltage combinations generated from the 217 signals gathered along the cell, and an I/O card for the control of the whole system. In total 16 digital output lines are used in a sink mode to provide the control of the two demultiplexers. Five analog channels are used. Four provide the current reading and the fifth allows for the differential voltage reading of two signals out of the 3,160 combinations. The system is controlled by a LabVIEW based program with a GPIB/Universal Serial Bus (USB) interface using the standard protocol IEEE 488.2. The system allows fully automatic verification and it is adapted to any configuration of the electrical circuit under test.

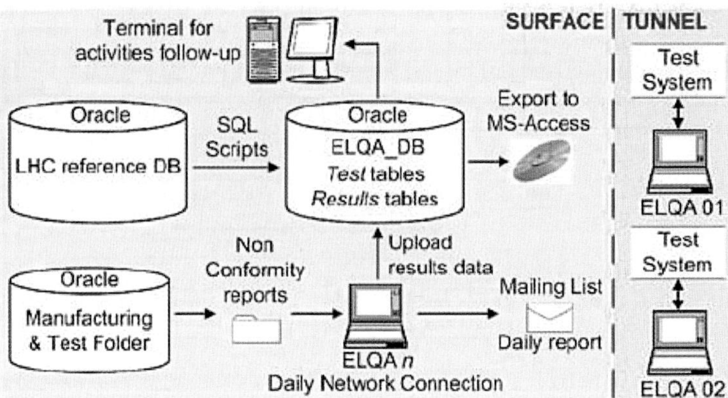

Fig. 10.40. Flow and management of data

Management and Transfer of Data

The control of the system is based on the LabVIEW application and an Oracle database containing the information needed to perform the electrical qualification and allowing the storage of the test results. Figure 10.40 shows the structure for the management and transfer of data.

The ELQA_DB database contains two parts, one containing the test tables and one containing the results tables. The test tables have been automatically generated by applying a package of PL/SQL scripts to the LHC reference database. This ensures that the latest version of the LHC machine parameters is used. The generated database contains all electrical interconnection data necessary to perform and validate the electrical verification. The Oracle database is duplicated into an MSAccess format which can be exploited on portable computers, in order to have a self-sufficient test system in the LHC tunnel.

The procedure to verify a cell starts with the download of all the existing non conformity reports from the LHC Manufacturing and Test Folder. Then, during the verification, the LabVIEW application points to the MSAccess database and automatically launches the tests for the particular cell to be verified. The measurement results are stored in the local database and a test report is locally generated. The collected information will afterwards be uploaded on the surface to the Oracle database for future processing. The test reports are automatically distributed to a mailing list. The LabVIEW application includes useful tools for easing the verification activities in the tunnel, e.g., the automatic calibration of the hardware and the manual diagnostic of faults. In the presence of non conformities the baseline parameters and criteria for the qualification of a cell can be modified by authorized specialists. For the follow up of the activities, specific LabVIEW applications pointing directly to the Oracle database allow to retrieve any type of acquired data and provide for a graphical user interface to coordinate the test teams following the up to six installation on fronts in the tunnel.

Test Bench and Surface Training

A test bench has been developed allowing the reproduction of any of the LHC cell electrical configurations. This system was essential for the validation of the test equipment as well as for the definition of the test procedures. It will serve for off-line diagnostic and simulation of particular faults and configurations and will be extensively exploited for the training program of the crews assigned to perform the verification in the tunnel.

Conclusion

The hard and software for the electrical quality assurance during assembly of the arcs of the LHC machine has been designed and validated. The equipment and procedures are now in place to perform the testing of the about 70,000 electrical interconnections of the superconducting magnets in the LHC.

10.9 Internet-Ready Power Network Analyzer for Power Quality Measurements and Monitoring

Problem Definition

This application aims at designing a PC-based power quality analyzer, including the set of instruments in one hardware system, capable of running all instruments in parallel. The software system based on modular concept for easy extensions in the future, the measurement and data processing algorithms according to the latest IEC and EN standards, and the remote-controlled instrument through Ethernet according to SCIPI standard are designed. In addition, two concepts are designed in one firmware – simple field instrument and distributed system.

The Solution

Developing a software application in National Instruments LabVIEW graphical development environment to act as the instrument firmware, providing the instrument GUI, and implementing the remote commands set running on the portable industrial PC. Eight instruments – FFT analyzer, power flow monitor, Flickermeter, EN50160 voltage monitor, transient recorder, symmetrical components analyzer of three-phase system, telegram monitor, and alarm monitor can run, capture data, and do all processing and data storage in parallel. Data is stored locally or in a central database, depending on instrument purpose. The postprocessing data module, developed in LabVIEW, for one instrument is processing local binary data. For the distributed system, the data post-processing module processes data from the SQL database using Microsoft Internet Explorer.

The Internet post-processing is based on ASP pages using ActiveX graphical elements from NI Measurement Studio. Both postprocessing modules are providing various data analyzing, data comparing, and data reporting. Power analyzers in the distributed system can be centrally online monitored via a supervising application developed in LabVIEW.

10.9.1 Computer-Based Power Analyzer

The faculty members of the Department of Electrical Measurements have been researching and developing electrical power quality for more than 11 years. During this period, the power quality analyzer has been developed jointly with the company ELCOM. The basis of the solution is the software application designed and developed within LabVIEW. From the very beginning, the application structure has been designed in such way to provide the open framework into which the user can easily implement new software modules. After having the solution for the field instrument, the distributed system concept is designed and implemented using an SQL database for data storage and Microsoft Internet Explorer for Internet-based postprocessing. For online remote supervising of distributed power analyzers, the remote commands tree is designed and implemented according to SCIPI. All of these significant extensions were done with minor changes in former LabVIEW code of BK500 power analyzer.

10.9.2 Instruments Implemented in the Analyzer

For all instruments the user can observe data on the instrument display and store them into the data files or SQL database. The most important instruments are described in more detail later.

FFT Analyzer

The first instrument from the virtual analyzers set is the Fast Fourier transform (FFT) which is an analyzer of harmonics and inter harmonics. The instrument can display the signal either in time domain or in frequency domain. For each measured signal (voltage and current of the power network line) the amplitude and phase spectrum is measured, and the FFT analyzer indicates the direction of the power flow for each particular harmonic. From Fig. 10.41 the user can see the amplitudes of the selected spectral lines.

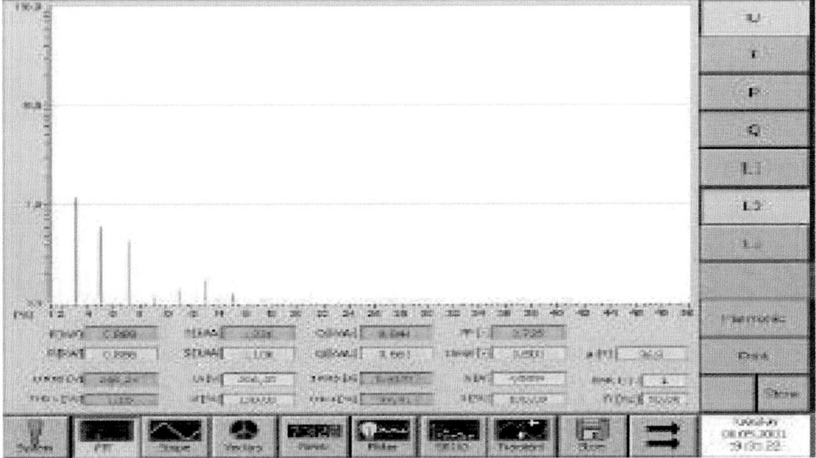

Fig. 10.41. FFT Analyzer front panel

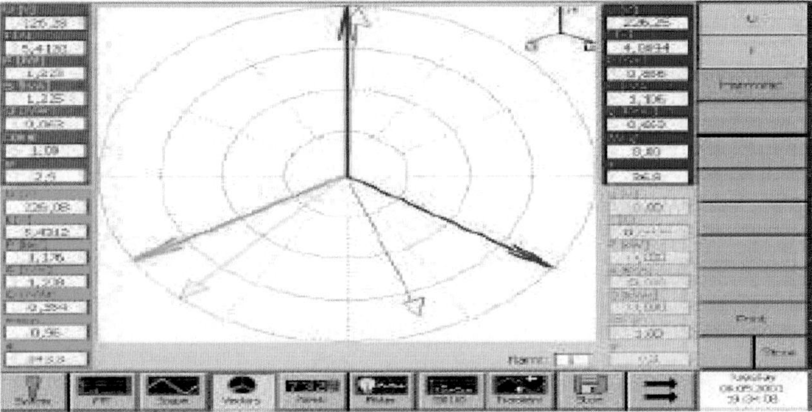

Fig. 10.42. Vector diagram panel

Power and Energy Monitor

The user can evaluate the main power and energy quantities calculated from the instantaneous values of the voltage and current signals using this instrument. The software implementation is designed in such way to allow the compensation of the errors caused by the nonlinearity and frequency dependent characteristics of the used transducers, if these characteristics are known for the particular transducer(s) used for the measurement. These corrections allow compensating for phase errors of power clamps that are very frequently used for current measurements in the power networks. The analyzer provides to the user the possibility of observing the phase conditions on the vector diagram continuously updated in real time as shown in Fig. 10.42.

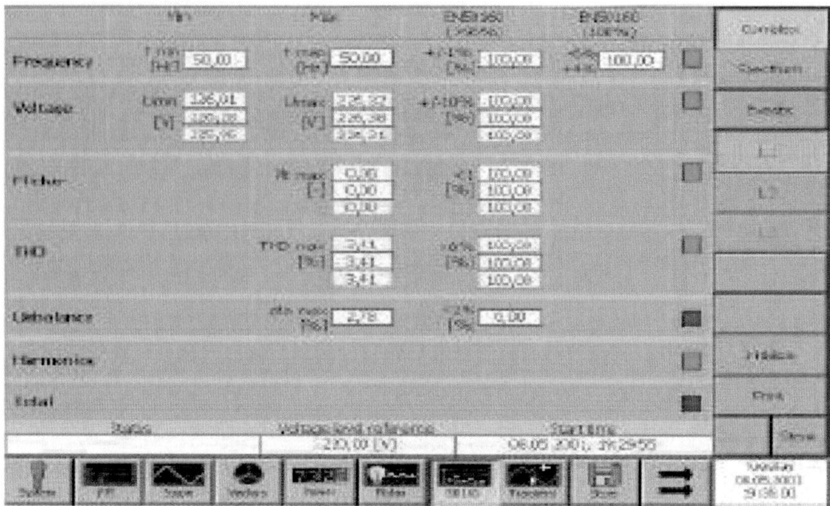

Fig. 10.43. Voltage Monitor front panel

Flickermeter

This instrument is designed for measuring the flicker effects of voltage signals according to IEC 868 recommendations. From the time domain of the voltage signals, the 10 min and 2 h flicker is being evaluated. Both evaluations are done each minute, minimizing the dependence between the instrument measurement start time and the particular time when the event affecting the level of the flicker occurs. Data are displayed on the display and can be also stored into the data files. Currently the research and development effort is focused into the modification of used algorithm in such way to allow measuring.

Voltage Monitor

This instrument analyzes the quality of the electrical energy according with EN 50160 standard. The user gets the complex overview of all quality parameters measured according with this standard. Some of the parameters are shown in Fig. 10.43. The user can individually set the levels for catching the voltage drop events. If the voltage drop occurs, the signal in time domain before and after the event is stored as well.

Transient Recorder

This instrument allows storing the voltage and current signals if the set trigger conditions are met. The user has a wide range of possibilities to set the trigger conditions for each particular power line network and combine them with logical AND and OR. There is also the possibility for setting the pre and post

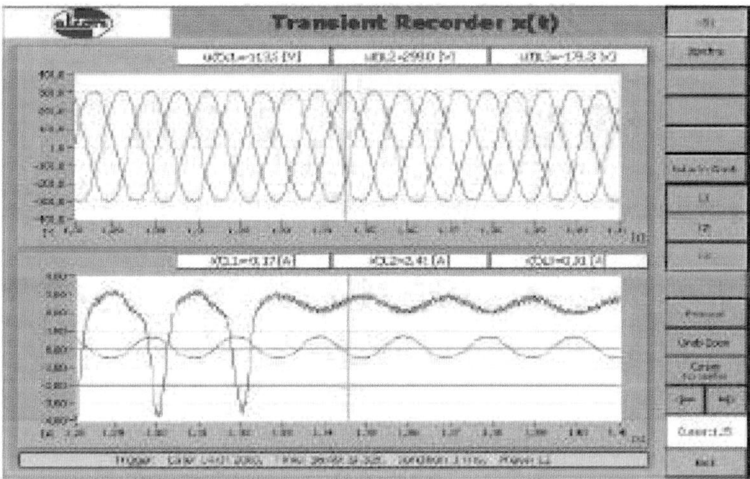

Fig. 10.44. Transient recorder postprocessing panel

trigger length within the time window. The postprocessing panel is shown in Fig. 10.44. Subsequently on the stored time window the instrument user can perform the off-line analysis.

10.9.3 Measured Data Analysis

Directly within the analyzer, the instrument user can perform all off-line analysis. With the developed post processing software module, users can process the measured data from any of BK550i instrument and print the protocols form the measurements. Also it is possible to export the measured data into ASCII files. Then it is possible to do any data processing within standard spreadsheet processors.

For the distributed system, a new revolutionary post-processing was implemented using Microsoft Internet Explorer and ActiveX graphical elements from NI Measurement Studio. With Internet Explorer and the ActiveX graphical elements from Measurement Studio, the user easily performed data analysis from anywhere on the Internet. The web-based postprocessing data histogram is shown in Fig. 10.45. The user could even compare data from different analyzers placed in different locations.

10.9.4 Supervising Module

For remote setup and online monitoring of the distributed system, the supervising software was designed and developed as a stand-alone application. With the supervising module, the user can setup ranges, constants, configure storing intervals. The user could monitor online up to four BK550i at the same time and display the most important screens.

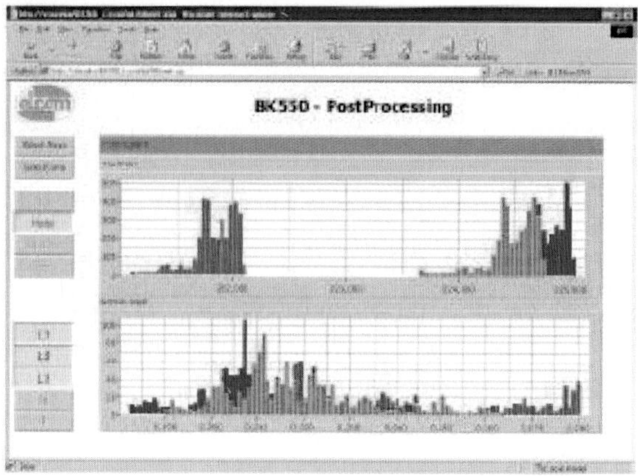

Fig. 10.45. Web-based postprocessing data histogram

10.9.5 Hardware Platforms for the Virtual Analyzer

Currently, the analyzer is implemented on two hardware platforms. One of the platform is a portable industrial computer. The computer is equipped with a plug-in data acquisition (DAQ) board, touch screen, and plug-in modules for signal conditioning. From the very beginning of the development process, high-quality DAQ plug-in boards from National Instruments were used. Currently, the PCI 6035 or PCI 6023E is part of the analyzer. Both boards provide 16 analog channels and the maximum sampling rate is $200\,\mathrm{kS\,s^{-1}}$. The DAQ process of instrument software uses sampling frequency $12.8\ \mathrm{kS\,s^{-1}}$ per each channel. Plug-in modules for signal conditioning are programmable through RS485 and are equipped with amplifiers/attenuators, galvanic isolation, anti-aliasing filters, and have very good linearity.

The second platform is a PC furnished with two USB ports that connect the variety of peripherals (keyboard, mouse, network etc.) as the hot plug-ins. Another component of the hardware is the module for signal conditioning and DAQ board. This configuration contains the touch screen as the standard way for operating the instrument. Using the keyboard or mouse is not intended in a case of standard usage.

10.9.6 Advantages of the Virtual Analyzer

The main advantage of the described solution is its flexibility. Using an all-purpose PC with hardware and software interfaces, plug-in DAQ board and signal conditioning is possible, implementing the powerful and user-friendly power quality analyzer. Using the development environment of graphical programming, dramatically simplifies the software maintenance and further development. With the application structure, further software modules are implemented, realizing new instruments without an impact on the current solution.

10.9.7 Future Vision

In addition to further extensions of the set of available instruments, extension can be made on the R&D effort on the innovation of the distributed system for monitoring of the electrical quality parameters base on the described analyzers. With the development of the hardware platform of real-time DAQ components of National Instruments and their compatibility with the graphical development environment of LabVIEW, the user can plan to run the DAQ core on the independent hardware with a real-time operating system (OS) and let only the time non-critical visualization run on the standard PC, while still using all the advantages of the graphical Windows OS features.

10.10 Application of Virtual Instrumentation in a Power Engineering Laboratory

In traditional power engineering laboratories, the data is collected using traditional analog voltmeters, ammeters, wattmeters, digital multimeters, and oscilloscopes. Users spend a large portion of time connecting the standard hardwired instrumentation, which allows them to measure only RMS voltage, RMS current, and real power. The quantities not measured, such as reactive power and phase angle, are calculated as part of the lab write-up using the power triangle and trigonometric identities. Since such quantities are calculated after the lab, users cannot observe the effect on these quantities due to changes of other experimental parameters. With the use of virtual instrumentation, users can see the realtime effects of an experimental variable on the real power, reactive power, and phasor quantities and the spectral representation of AC signals. For example, they can immediately see the result of loading a motor, changing a supply voltage, or changing a connection. Our laboratory has the capability of performing the actual experiments in the field of power engineering. In the study of electrical machines, computer simulation has been proposed by educators. However, many others agree that simulation is not a substitute for actual experiments. This application will emphasize some of the capabilities of utilizing virtual instrumentation to study single and three-phase transformers, DC generators, and synchronous and induction machines.

10.10.1 Lab Capabilities

Each computer is equipped with a digital acquisition (DAQ) board. There are eight measurement channels: four voltage and four current channels, which are connected to the digital acquisition board via conditioning and isolating circuitry. The DAQ card acquires data from any of the channels and passes it to LABVIEW, a National Instruments software product, for analysis and display. A LabVolt 8610 prime mover/dynamometer is placed at each station, enabling users to mechanically drive generators and to load motors. The prime mover/dynamometer has analog speed and torque outputs which are directly connected to the DAQ card.

The combination of digital acquisition, computer processing and display is referred to as "virtual instrumentation" (VI). The term "virtual" refers to the fact that the display and processing of information is controlled by software, and can easily be reconfigured to suit various purposes. It is important to emphasize that "virtual instruments" are real instruments: no simulation is involved. Real-world data is collected, processed, and displayed. Each VI consists of two pieces. The part of the virtual instrument that the user deals with the most is the "front panel." The front panel of the VI is just like the front panel of a standard hardware instrument: it may contain knobs, buttons, and switches for controlling the instrument. These controls are used to indicate instrument status while LEDs, meters, graphs, and numeric outputs are used to display the measured data. The other part of every VI is the "wiring diagram." This shows how the data is collected, processed, and routed. The wiring diagram is not just a visual representation of the computer program: it is the program. Programming in LABVIEW is entirely graphical. The OSU Power Engineering Lab has several experiments already set up, and the flexibility to create new ones.

10.10.2 Single and Three phase Transformers

To study the characteristics of a single-phase transformer, voltage is applied to the primary of a single-phase two-winding transformer. Using a current and a voltage channel, the input current, and voltage are measured. For small input voltages, the relation between flux and exciting current is linear. However, as the primary voltage increases, the nonlinear effects of hysteresis and saturation begin to become more evident, which are the cause of third and higher-order odd harmonics of the exciting current. Using the VI designed for this experiment the hysteresis curve obtained is shown in Fig. 10.46.

This curve is obtained by displaying core flux as a function of the exciting current. Core flux can be calculated as:

$$\phi(t) = \phi(0) + \frac{1}{N_2} \int_0^t v_2(\tau) d\tau,$$

where N_2 is the number of secondary turns and v_2 is secondary voltage. The capability of the VI to numerically integrate the secondary voltage has been used. To study the presence of harmonics in a three-phase wye–wye transformer and the effect grounding the neutral has on the harmonics, three single-phase transformers are connected in a wye–wye configuration as shown in Fig. 10.47. Connections are made by shrouded banana plugs, which help not only to provide a safe environment but also to save time. For example, it takes less than 2 min to make the connections of Fig. 10.46. By using the current

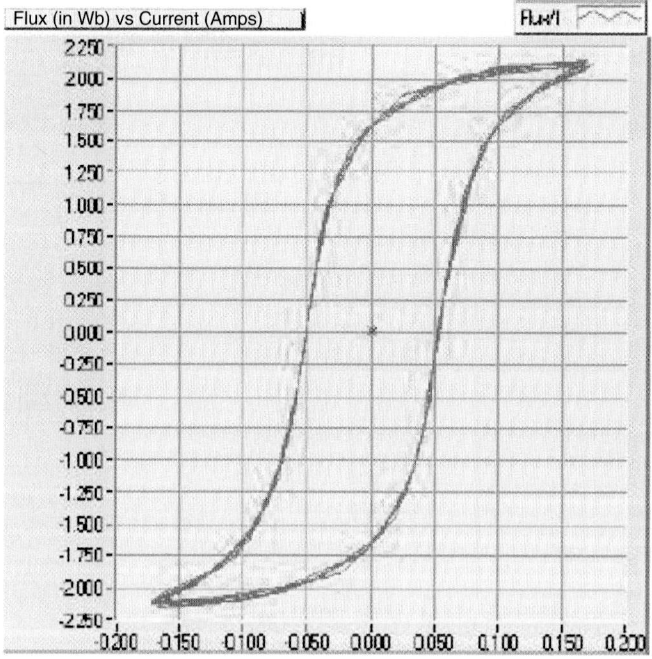

Fig. 10.46. Hysteresis curve of a 1ϕ transformer

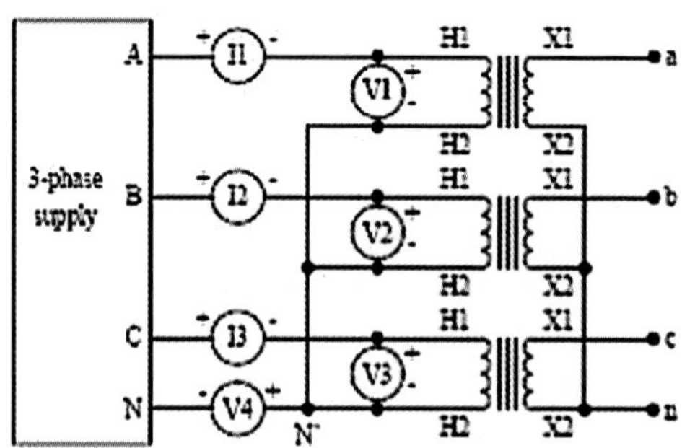

Fig. 10.47. Ungrounded Y–Y connections

and voltage waveforms, the VI can compute the full spectra of each. Fourier analysis is performed on each of the acquired waveforms in order to obtain different frequency components. The fourier analysis of input voltage V1 to the wye–wye transformer is shown in Fig. 10.48. One can see the large third, fifth, and other harmonics. For simplicity other voltages have been omitted. Users can position the cursors on the plots for accurate measurements. Here, the cursors show that the third harmonic in voltage spectrum is 4.55 dB below the 60 Hz fundamental (Fig. 10.48). However, when V4 is replaced by I4 (thus

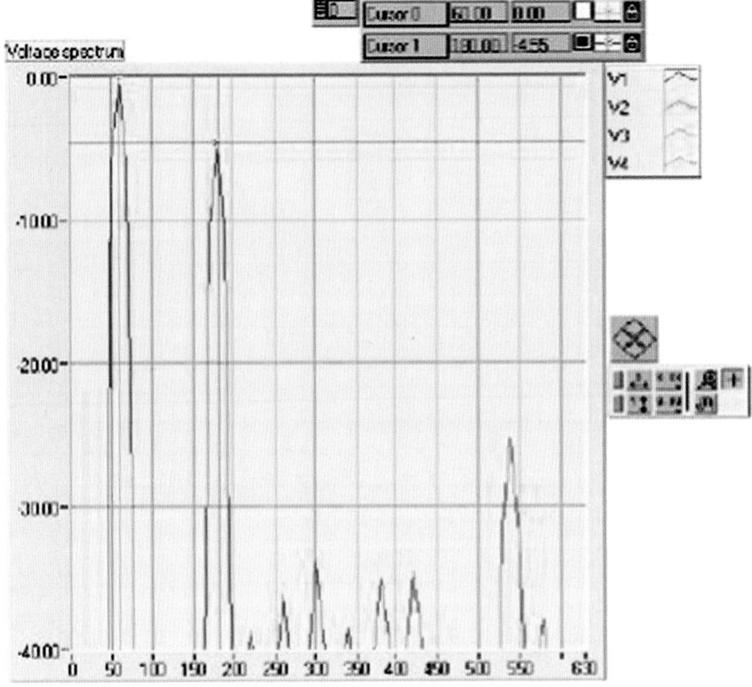

Fig. 10.48. Harmonics for ungrounded Y–Y 3ϕ transformer

providing a direct connection between N' and N), users can see that the third harmonic is 37.19 dB below the 60 Hz (graph not shown here). Therefore, users can visualize that the third harmonic is significantly reduced by connecting the neutral of wye–wye transformer to ground.

The ability to study the basic waveforms does not require virtual instrumentation. This is also possible by the oscilloscope. However, virtual instrumentation can provide other perspectives on the data which the oscilloscope cannot provide. For example, users can see the phasor diagram on the same VI program. Using the current and the voltage waveforms, the 60 Hz component is extracted and the results can be displayed on a polar plot as a phasor diagram. Such displays show the phase relations among the voltages and currents being measured. The phasor diagram, shown in Fig. 10.49, depicts line to neutral voltages and line currents of Y grounded Y 3ϕ transformer. Since the phasors are displaced by 120Æ, one can conclude that the connections are correct. Whereas, based on only RMS measurements of line to neutral voltages, one cannot tell whether the connections are correct or not. Thus, the phasor diagram helps users compare different voltages and currents.

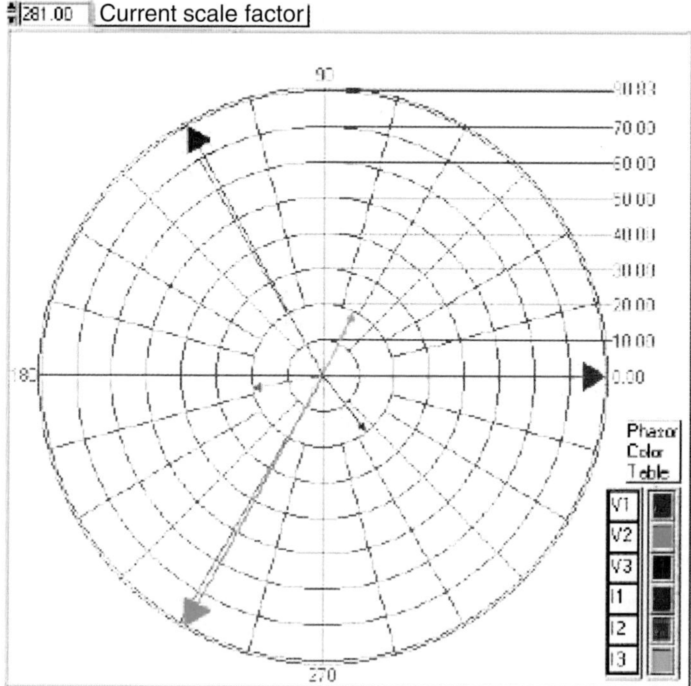

Fig. 10.49. Phasor diagram for Y–grounded Y 3ϕ transformer

10.10.3 DC Generator Characteristics

One of the VIs developed at OSU makes it very convenient to build a graph of one variable versus another. Users can record the data and plot all the corresponding points on the graph. Moreover, making multiple graphs on the same set of axes is possible, which is very useful when comparing characteristics of various machines. If a certain point is undesirable or recorded by mistake, then it can be removed by using the delete button. Users can delete an entire graph or multiple graphs at once. This application is used to plot the performance characteristics of various electrical machines. For example, the VI can be used to plot the magnetization curve of a DC generator, as shown in Fig. 10.50. This is a single graph of armature voltage (Y-axis, in volts) versus field current (X-axis, in amperes) at a constant speed of 1,725 rpm. This curve corresponds to the magnetization equation

$$E_a = K_a \phi_p \omega_{\mathrm{m}},$$

where E_a is the generated voltage, K_a is the machine constant, Φ_p is angular velocity, and ω_m is the flux per pole which depends on the field current.

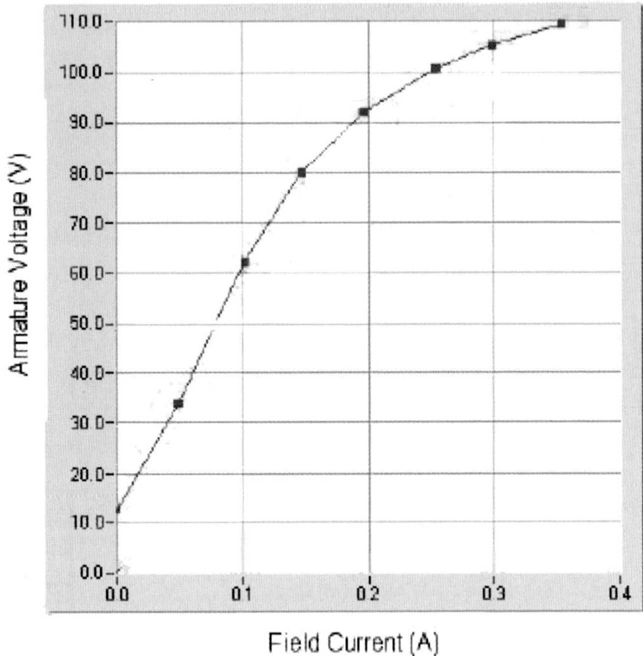

Fig. 10.50. Saturation curve of DC generator

The curve starts at approximately 12 V because of the residual magnetism in the machine. The characteristics of different DC generators in loaded condition are shown in Fig. 10.51, which depicts a set of three graphs of load voltage (Y-axis, in volts) versus armature current (X-axis, in amperes). The top-most curve represents a separately excited DC generator. The curve in center represents a cumulatively compounded long shunt DC generator and the third curve represents a self-excited DC generator. As expected, the separately excited DC generator has drooping characteristics because of the losses due to armature resistance, brush contact and armature reaction. Whereas, in the case of the self-excited shunt DC generator (without compounding), the field excitation is provided by the generator itself, so reduced terminal voltage will supply a reduced field current which causes further reduction in the voltage. In a cumulatively compounded generator the series and the shunt fields aid one another. As the load current increases, the current through series winding increases therefore series field strength increases which compensates the decrease in shunt field strength. As a result, the overall magnetic field strength increases, which is the cause of increase in the output voltage of the generator as compared to that of the self-excited case. Change in the output voltage from low load to high load is seen because magnetic field strengths of series and shunt fields do not compensate for each other completely.

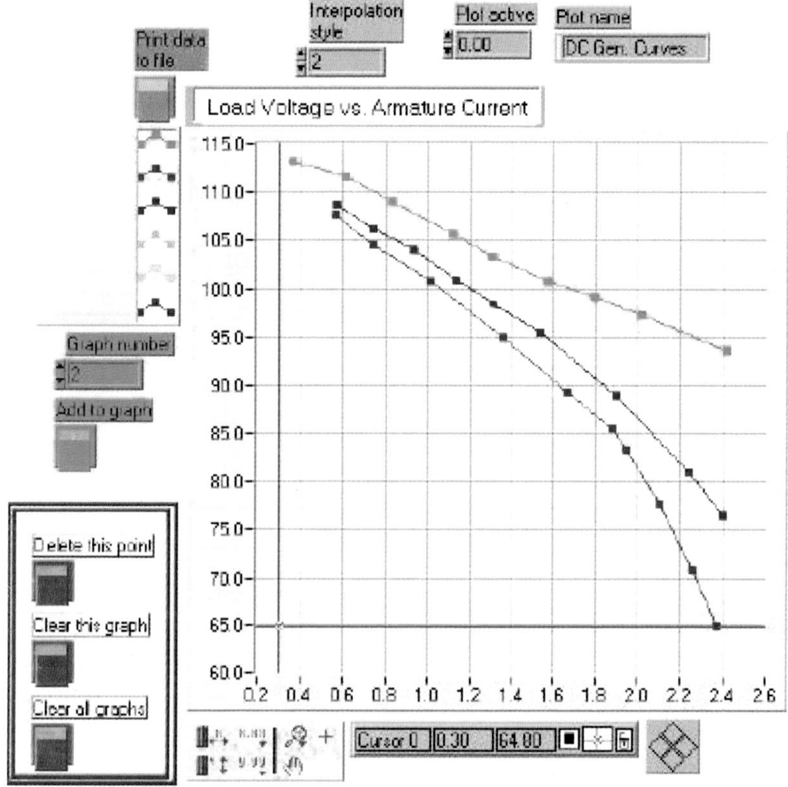

Fig. 10.51. DC generator performance

10.10.4 Synchronous Machine

A three-phase wound-rotor induction machine ("sel–syn") is used as a synchronous motor, as shown in Fig. 10.52. A start/run switch, when pressed, disconnects the excitation current and shorts all the three phases of the rotor so that the motor can be started as induction motor. After rated voltage is applied, the start/run switch is released allowing DC excitation to be supplied to the rotor which is synchronized with the stator MMF. Using the phasor diagram, the stator current and terminal voltage are displayed. By varying the excitation current, the magnitude and angle of the stator current is varied. From the phasor diagram, users can see that for low values of field current, the stator current lags the terminal voltage and for higher values it becomes leading. Users can plot the variation of stator current as the field current is changed i.e., Vcurve.

For different load levels, the recorded data provided the corresponding V-curves as shown in Fig. 10.53. A synchronized stroboscope is used to observe the change in the power angle as the power level is changed by connecting

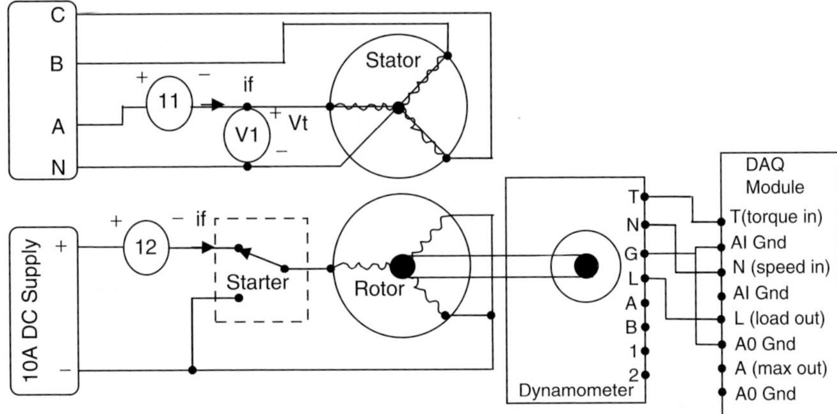

Fig. 10.52. Synchronous motor under load

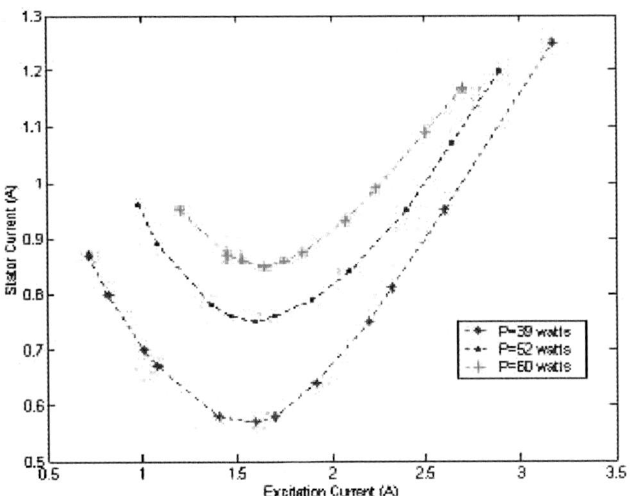

Fig. 10.53. V curves for a synchronous motor

it to the synchronous machine as shown in Fig. 10.54. This technique can be used to study the power angle characteristics of the synchronous machine. Using the prime mover to drive the machine as a generator, the power angle can be seen to be leading, and as the load is reduced and the 8610 is switched to dynamometer mode, the machine becomes a synchronous motor, with a correspondingly lagging power angle. To convert the synchronous motor into an unloaded synchronous generator, three-phase variac is disconnected from the stator and the prime mover is turned on to bring the machine to near 60 Hz. By varying the speed of the prime mover, the corresponding change in the generator voltage and frequency can be observed on the VI. Then, the

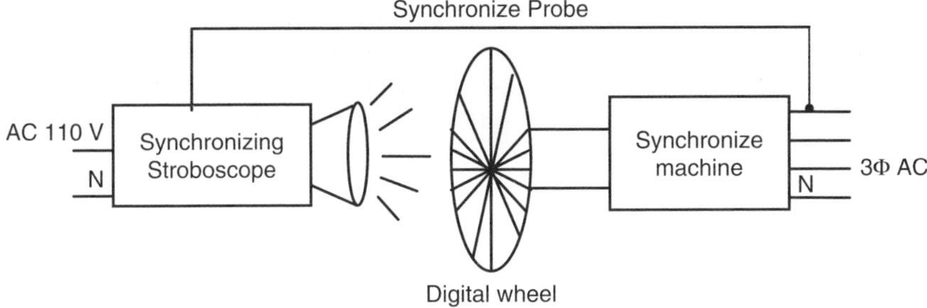

Fig. 10.54. Stroboscope used to measure power angle

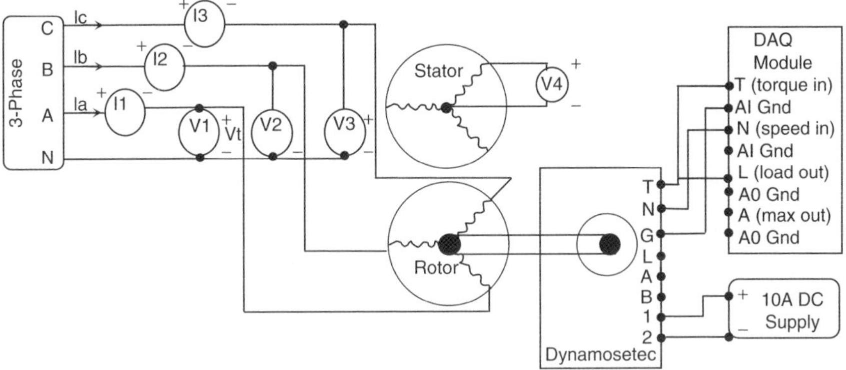

Fig. 10.55. Sel–syn machine as phase shifter and variable frequency source

speed is kept constant and the excitation is varied to observe the corresponding change in the terminal voltage. Using this data, users can plot the no-load saturation curve for the synchronous generator.

10.10.5 Induction Machine

The same machine as used in the synchronous machine experiment is used as a phase shifter and variable frequency source by connecting the three-phase supply to the rotor and moving it with the prime mover as shown in Fig. 10.55. When the shaft is rotated by hand, one can see the stator voltage phasor rotating on the phasor diagram. By observing the number of rotations completed by the shaft as the phasor rotates 360 electrical degrees, the number of poles of the machines can be determined. In our case, the stator voltage phasor completes 360Æ in half of a rotation of the shaft, which illustrates that the machine is a 4 pole machine. If the machine is rotated in the forward direction (using the prime mover) the stator voltage will have a higher frequency than 60 Hz, and conversely will have a lower frequency if the machine is rotated backwards. Fig. 10.56 shows that the frequency of stator voltage V4

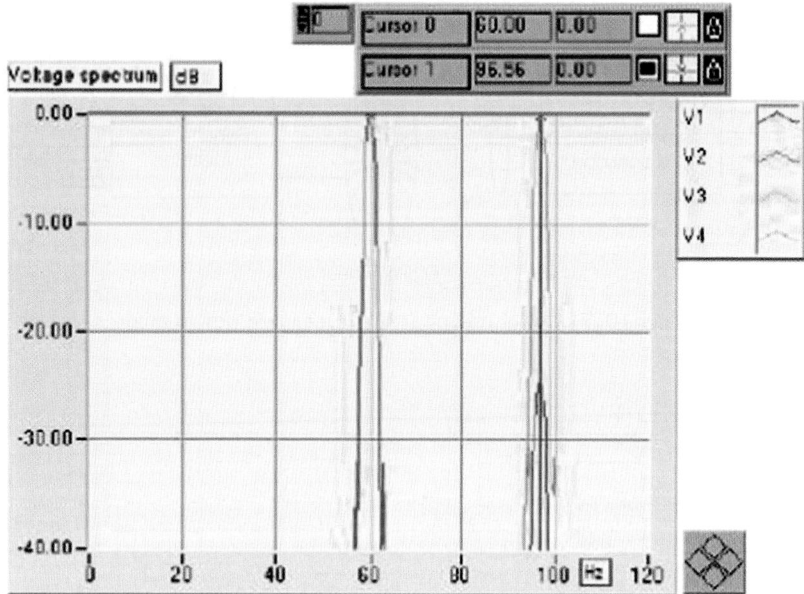

Fig. 10.56. Source and output voltage spectrum for frequency shifter

is 96.56 Hz, and that for rotor voltage V1 is 60 Hz. To convert the machine into an induction motor, all of the connections to the rotor are now connected to the stator, while the rotor windings are short circuited through a current channel to measure the rotor current. It takes approximately 20 s to convert the phase shifter into an induction motor, which can be subjected to several experiments.

Conclusion

The techniques and capabilities of an improved laboratory practice have developed and improved undergraduate experience for the users. The use of virtual instrumentation has not only provided a modern and interesting way for users to perform experiments but also has reduced the amount of time necessary to make connections. The real-time display of quantities, such as phasor voltages and currents, real and reactive power, and harmonic content, has made a great improvement in the users' ability to visualize these quantities and understand their relation to one another.

11

VI Applications: Part II

Learning Objectives. On completion of this chapter the reader will have a knowledge on LabVIEW applications such as:

- Implementation of a Virtual Factory Communication System
- Developing Remote Front Panel LabVIEW Applications
- Using the Timed Loop to Write Multirate Applications in LabVIEW
- Client-Server Applications in LabVIEW
- Web based MATLAB and Controller Design Learning with LabVIEW
- Neural Networks for Measurement and Instrumentation in Virtual Environments
- LabVIEW Interface for School-Network DAQ Card

11.1 Implementation of a Virtual Factory Communication System Using the Manufacturing Message Specification Standard

For interconnection purposes, a factory automation (FA) system can be combined with various sensors, controllers, and heterogeneous machines using a common message specification. In particular, interconnection of heterogenous machines through a common message specification promotes flexibility and interoperability. For this reason, the *manufacturing message specification* (MMS) standard have been developed. The standard specifies a set of communication primitives and communication protocols for the factory communication environment. In particular, the MMS standard specifies various functionalities of the different FA devices in a compatible way. Thus, users of MMS applications have to use functions from only one unique set of functionalities to operate various kinds of automation machines. Moreover, the different automation machines can communicate among themselves through the standard automation language. This enables transition from the traditional centralized control to the distributed control systems.

The MMS standard is composed of the common standards and the *MMS companion standard* (MMSCS). The MMSCS includes device-dependent specifications for a robot, numerical controller (NC), programmable logic controller (PLC), and process control (PC). In its initial stage, the MMS standard was developed for an application layer of the manufacturing automation protocol (MAP). Nowadays, MMS have been implemented on top of the TCP/IP protocol suite and is used as a reference model for industrial network or other message protocols, such as the home network protocol. In particular, MMS is implemented in a minimized form as the application layer for Profibus (Process Fieldbus) and Field Instrumentation Protocol (FIP).

A *virtual factory* environment can be used to examine the correctness of an implementation and as a test solution of an FA system prior to installing the system in a real factory communication environment. Furthermore, by using the virtual factory communication system, developing time and costs can be minimized. In addition, it can be used as a training tool of MMS users. In this article, the virtual factory communication system was designed and implemented with the use of the MMS standards and its CS part. Researches in the MMS technology include application of factory devices, middleware, and multimedia communication.

In this application, we describe the implementation of a virtual factory communication system using the MMS and the MMS-CS standards. In addition, we discuss here the implementation of an MMS Internet monitoring system (MIMS) for online monitoring of the developed virtual factory communication system.

11.1.1 MMS on Top of TCP/IP

Two types of protocol structures for the MMS communication on top of TCP/IP, referred to here as "MOTIP," have been studied. One method, which is based on Requests on Comments(RFC) 1006, uses the open systems interconnection (OSI) seven layer model. The other method is based on N578 that defines the direct mapping relation between the M-services and the TCP functions. In particular, we chose to use the N578 method in the implemented system because its simplicity. The differences between the structures of a Full-Map, Mini-Map, and MOTIP are presented in Fig. 11.1. In the case of Mini-MAP, IEEE 802.4 token-bus network is used instead of the Ethernet (IEEE 802.3) subnetwork. By Fig 11.1, Full-Map structure has similar scheme to MOTIP based on the RFC 1006 method, and Mini-Map structure is similar to and MOTIP based on the N578 method.

The implemented virtual factory communication system is based on the description of a practical test plant shown in Figs. 11.2 and 11.3. These figures show an overview (Fig. 11.2) and the layout (Fig. 11.3) of a practical test plant using the MOTIP-based program. MOTIP-based communication program have been tested with the Mini-Map and the Full-Map system using

1	MMS	ACSE	MMS	MMS
2	Presentation			
3	Session			
4	Transport			
5	Network			TCP
6	Data Link		Data Link	IP
7	Ethernet		Token Bus	Ethernet

Fig. 11.1. The various types of the MMS protocol structures (Full-Map, Mini-Map, and MOTIP)

Fig. 11.2. Overview photo of the referenced practical test plant

a developed gateway and the developed virtual factory communication system is a miniature of the test plant in Fig. 11.3.

M-Services and MMS Interface in MOTIP

The MMS functions convert a data control block (DCB) to the correspond parameters of the M services.

The protocol stack of MOTIP is requested at the MMS client side, the synchronous service or the asynchronous service can be used for communication. Using the synchronous service, the MMS client receives a confirmation

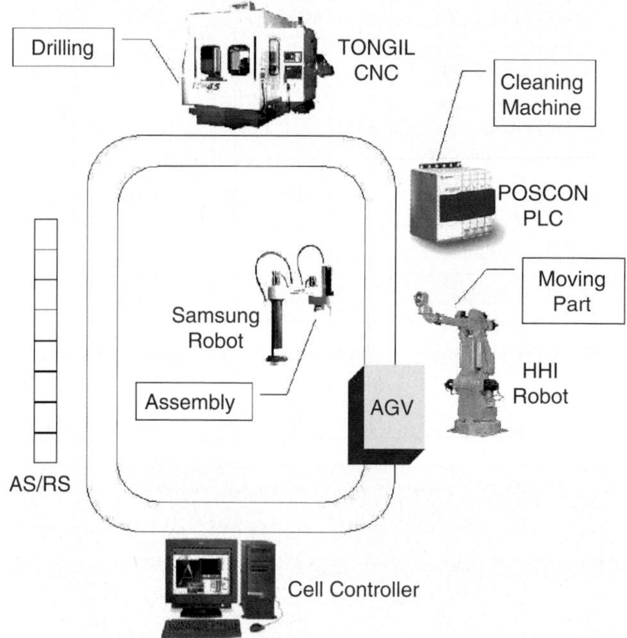

Fig. 11.3. Reference layout of the virtual factory

through a response from the MMS server. In the asynchronous service, the functions can be executed according to a received event through the response of MMS server. Most of implemented MMS services are executed by synchronous mode in the developed system

11.1.2 Virtual Factory Communication System

The developed virtual factory communication system was implemented with the Microsoft Visual C++ and OpenGL library. The virtual factory is embodied by integrating each virtual real machine (VRM), the MMS-CS program, and the MOTIP programs. Each VRMs can be communicated with through the developed MMS-CS program based on the SNU MMS library.

The MOTIP central server program (see Fig. 11.4) can be located in a control room of a factory plant. The MOTIP central server can monitor and command the MMS services to the networked manufacturing machines.

The implemented MMS services are as follows: virtual manufacturing device (VMD) service for the status acquisition of a networked machine, domain service for file transfer, program invocation (PI) service for the remote execution of a program, and miscellaneous services for reading and writing services of device status and operations. PI services includes creation, initialization, termination, and deletion of a remotely executed program. PI service can be carried out with other services. For example, the remote execution of

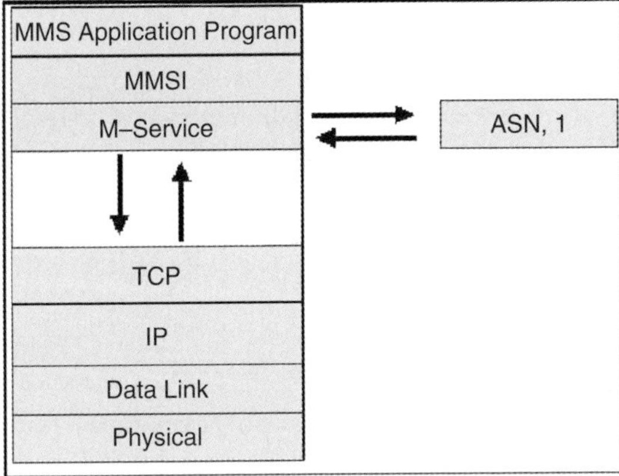

Fig. 11.4. The protocol stack of MOTIP

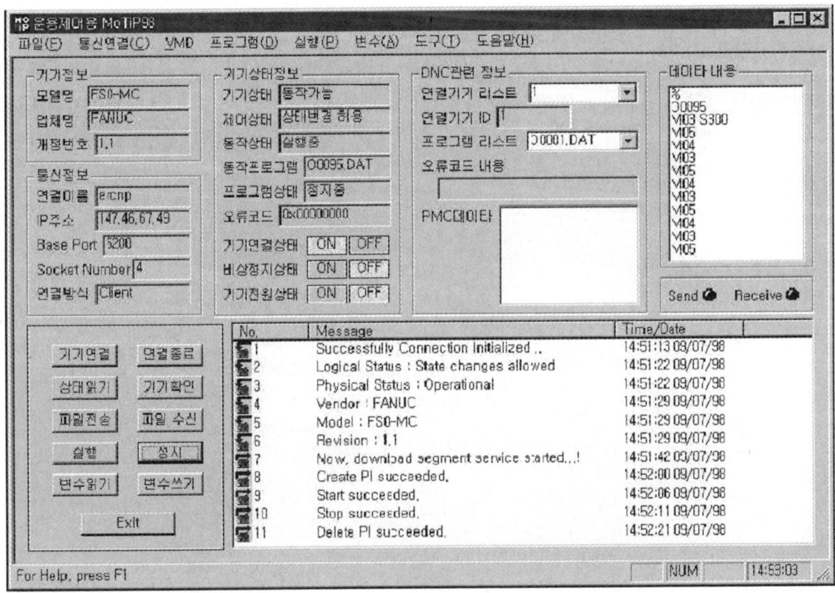

Fig. 11.5. Screenshot of the MOTIP-based central control program

specific program can be performed after downloading the status and segment service using VMD and Domain service. Figure 11.5 shows the execution of the VMD, the Domain, and the PI services in the MOTIP program.

MMS Companion Standard

The implementations of the MMS-CS VRMs are based on the MMS services specification. Specifically, the services of the implemented MMS-CS VRMS consist of: VMD, Domain, PI, and miscellaneous services.

The implemented services are as follows:

- VMD services: *Status*, *Identify*, and *Unsolicited Status*
- Domain service: *Initiate Download*, *Download Segment*, *Terminated Download*, *Initiated Upload Segment*, and *Terminate Upload*
- PI services: *Create*, *Start*, *Stop*, and *Delete*
- Variable service: *Read*, and *Write Semaphore*, *Operator*, *Event*, and *Journal* services have been implemented partially only.

VRM Using MMS-CS

The virtual factory communication system is operated by controlling each VRM with MOTIP-based communication programs. Figures 11.6 and 11.7 show reference models of practical machines.

The application program of Robot CS VRMs are modelled after a 6-axis robot made by HYUNDAI Heavy Inc. and the 4-axis scalar robot made by SAMSUNG Heavy Inc. The 6-axis robot is used for conveying the material with pallet throughout the implemented virtual system. The 4-axis robot takes part in a bolt assembling task of the overall process. NC-CS VRM is modelled

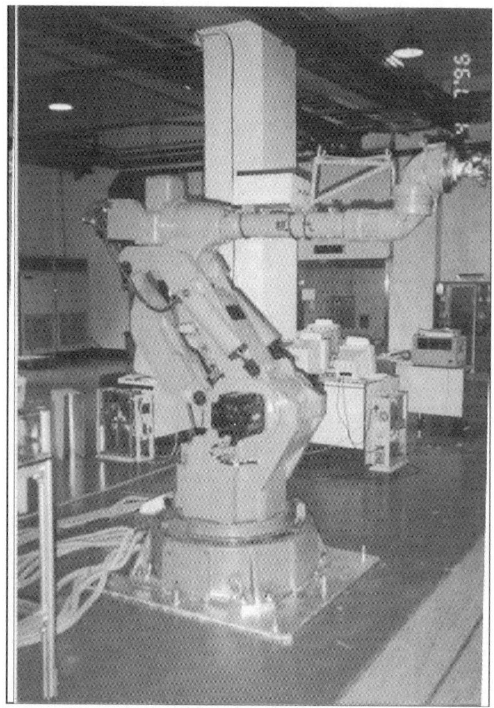

Fig. 11.6. The practical model of an 6-axis robot VRM

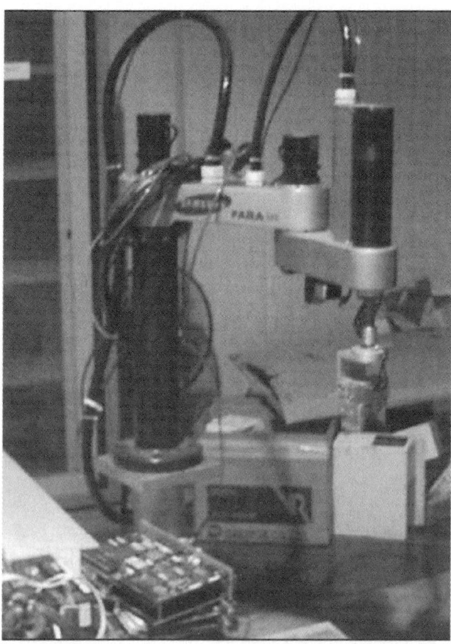

Fig. 11.7. The practical model of a 4-axis robot VRM

after an NC machine (TNV-40) made by TONGIL HEAVY Inc. NC-CS VRM is responsible for drilling a hole in the sample material.

The application program of PLC-CS VRM is modelled after a cleaning machine with an air compressor and clamps. It is controlled by PLC which is installed under the cleaning machine. PLC-CS VRM takes part in the cleaning works of assembled parts after the drilling process performed by the NC-CS VRM. The Process control VRM program can be scheduled for all the machines with the autonomous guided vehicle (AGV) movement. The CS application program includes all the VRMs, the common MMS, and MMSCS services. All machines are operated in accordance with their scheduled operation scenarios and the operation scenarios can be modified by the MMS user.

The test scenario begins with the AGV VRM idling status at station 4. The station 4 is starting and ending station in the developed virtual factory. AGV VRM loads a palette from station 4 (assuming that the AS/RS is aside) and move to station 1 for unloading the palette. In station 1, the NC VRM shown in Fig. 11.8, execute a hole drilling process in the material. After finishing the drilling process, AGV VRM as shown in Fig. 11.9 loads the palette and moves it from the station 1 to the station 2. In the station 2, the 6-axis robot VRM transfers the sample material from the palette to the clamps of the cleaning machine.

The PLC CS VRM starts to cleaning the sample material using an air compressor. Then after loading the palette, AGV moves from the station 2 to the station 3. The 4-axis robot assembles the bolt into a hole of the sample

Fig. 11.8. The practical model of an NC VRM

Fig. 11.9. The practical model of a Process Control VRM: AGV approaches Automated Storage and Retrival System (ASRS) to load a pallete

material at the station 3. After finishing the assembling process, the AGV moves from the station 3 to the station 4 and unload the palette at the station 4. The overall scenario can be scheduled by the MMS user. Screenshots of 6-axis and 4-axis robot VRM program are illustrated in Figs. 11.10 and 11.11 respectively.

11.1.3 MMS Internet Monitoring System

MIMS is designed for simultaneous monitoring and controlling the status of multiple clients. MIMS transfers the operational information of the virtual factory to an MMS administrator and external MMS users. Therefore, MIMS can concurrently monitor the status of each networked VRM and several MMS-CS

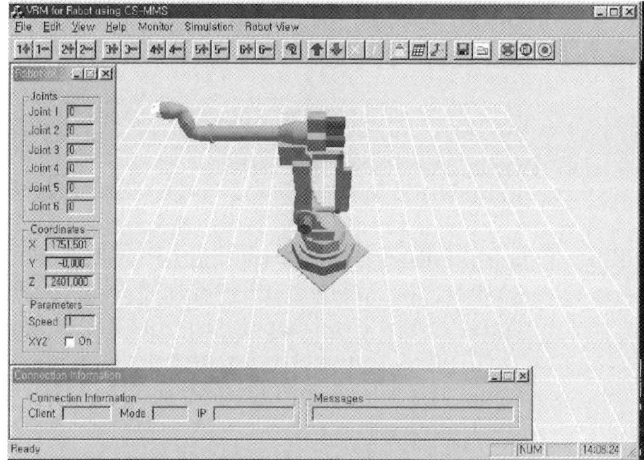

Fig. 11.10. Screenshot of 6-axis Robot VRM program

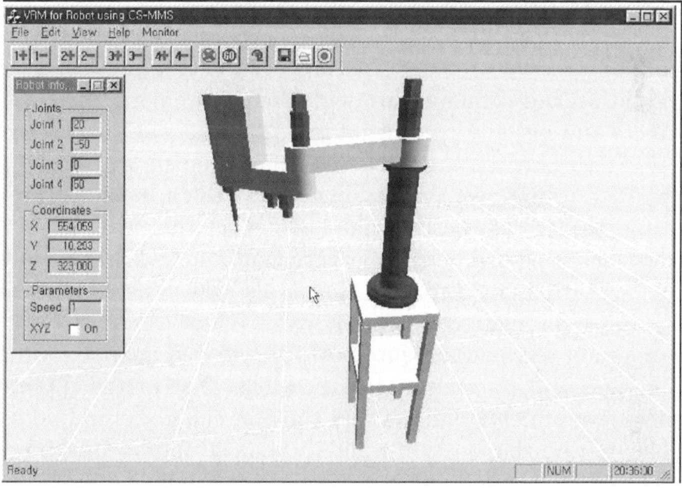

Fig. 11.11. Screenshot of 4-axis Robot VRM program

services. The MIMS is composed of MOTIP application program, the application program using MMS-CS, and each networked VRM. MIMS is designed for up to 100 multiple users. Whenever MMS clients uses the downloaded program based on the MMS-CS, MIMS can monitor the operation status through the user information and the communication data between MMS client and server.

The administrator can monitor all the messages from the multiple clients and check the problems in the networked system. The MMS user can then fix their technical problem through the online advisories from the MMS server administrator. Therefore, the developed system can serve as a training tool for MMS developers and users.

Conclusion

A virtual factory communication system using the general MMS, its CS part, and the robot VRMs, NC, PLC, and PC were designed and implemented. The applicability of the described-here virtual factory using the MMS and its CS part can be extended to the MMS implementation of a practical plant. In particular, the implemented system can be used as design guideline of an MMS system for practical plant. Furthermore, the developed virtual factory communication system and the VRMs can also be used for the verification of the operation of a real FA environment.

11.2 Developing Remote Front Panel LabVIEW Applications

The Web already has changed the way companies do business; they can quickly share information and communicate with their customers. They also can sell their products and visually present information in a way that never was possible before. With LabVIEW 6.0, National Instruments showed how the user can use the Web to improve the way engineers design, manufacture, and test products, and to decrease design time. The effect that Web-enabled distributed applications have had on the measurement and automation industry has made implementation not merely an advantage, but a necessity for any company that wants to remain competitive.

In the realm of test and measurement, distributed applications offer numerous advantages over conventional applications. Distributed applications can harness the power of networking so that the user can perform different aspects of the application in the location where it is most appropriate. The platform is fundamentally the same, but the deployment capabilities of the platform are enhanced radically. By using network technologies in the measurement solutions, the user can perform I/O on the production floor, deploy additional processing power for in-depth analysis in the control center, log and store postanalysis information to a corporate database, and display key information to clients around the world in a Web browser. The essential tool that is required to tie all of these pieces together is the software. LabVIEW provides a platform for designing the test system to take advantage of all the latest technologies while providing an environment that still is focused on quickly developing the application.

LabVIEW 6.1 introduces remote front panels, which can be used to view or control a VI in any Web browser. This feature greatly expands the application because an operator can run an application from anywhere. In addition, several users at different locations all can access the VI simultaneously. When developing a distributed application for the Web, the user must consider several factors involving bandwidth and network traffic. An application that will be controlled by many remote computers must be able to provide up-to-date information to all connected clients. This requirement is easy to meet if that information is simply a pass/fail status indicator, an output temperature level, or current engine speed. However, when the information required evolves into real-time charts, intensity graphs, and other data, the user must have high bandwidth. These types of data require that more data must be sent on each update and that the same amount of data must be sent to all connected clients. The easiest way to distribute large amounts of data on the Web is to use a high-end server with a large amount of bandwidth at its disposal. Because these resources are not always available, it is useful to examine other ways to optimize data transfer, such as the following:

- Reduce the amount of data sent
- Reduce the update rate of the data
- Minimize the amount of advanced communication, such as Property Nodes

11.2.1 Reducing the Amount of Data Sent

To reduce the amount of data sent, the unnecessary aspects of the application are eliminated. The Remote Panel Connection Manager is used to manage remote front panels. In LabVIEW, this tool can be opened by selecting **Tools≫Remote Panel Connection Manager**. The user can use the Remote Panel Connection Manager to perform a variety of tasks including viewing the number of computers connected to the Web Server, tracking the amount of data that is transferred across each connection, and displaying whether the connection is viewing or controlling the VI. Using this tool, the bandwidth requirements changes can also be viewed. For example, if the user has a temperature monitoring application, and the current temperature is published to only one client with a thermometer indicator, then the application requires only 1 or 2 kbytes of bandwidth (Fig. 11.12). If the user uses a graph that displays 1,000 points at a time on the front panel, the required bandwidth grows to more than 8 kbytes s^{-1} (Fig. 11.13). The waveform graph requires more data to be sent from the server to each client than a simple digital control. Therefore, it is more efficient to publish a single-point indicator instead of a multiple-point indicator such as a waveform graph, unless the application requires 1000 data points to be displayed simultaneously.

11.2.2 Reducing the Update Rate of the Data

Another way to reduce the bandwidth required by the application is to reduce the update rate of the data. If the user adds a 1 s delay to the VI, the required bandwidth drops from the 1 or 2 kbytes s^{-1} seen in Fig. 11.14 to 60 bytes/s. Reducing the update rate of front panel objects reduces the total bandwidth required by the application.

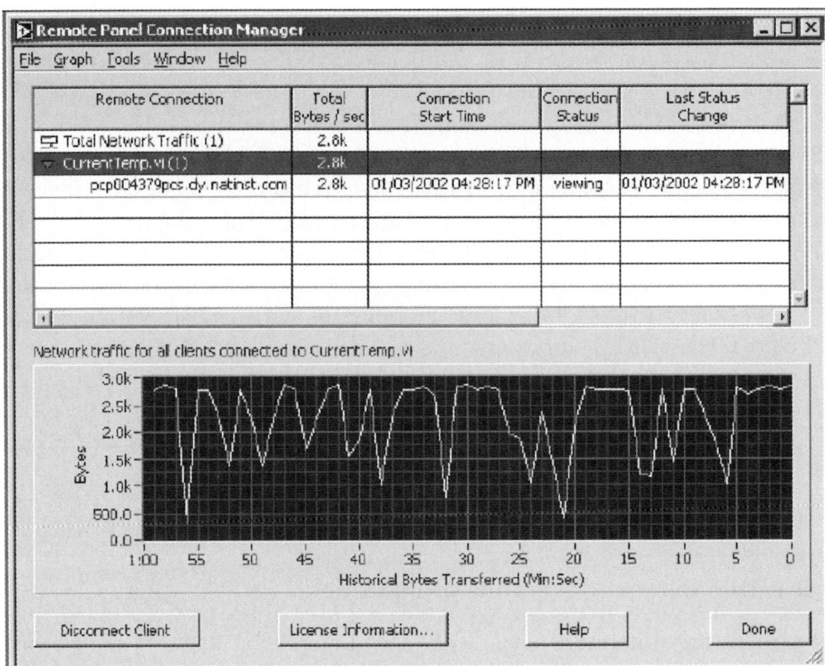

Fig. 11.12. Bandwidth required to serve a VI using digital controls connected to one subscriber

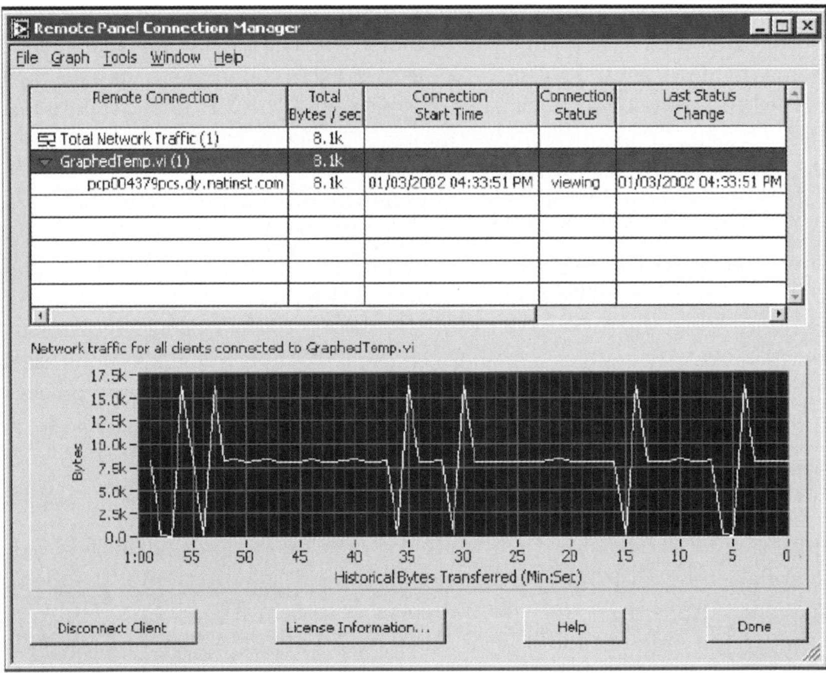

Fig. 11.13. Bandwidth required to serve a VI using a waveform graph control connected to one subscriber

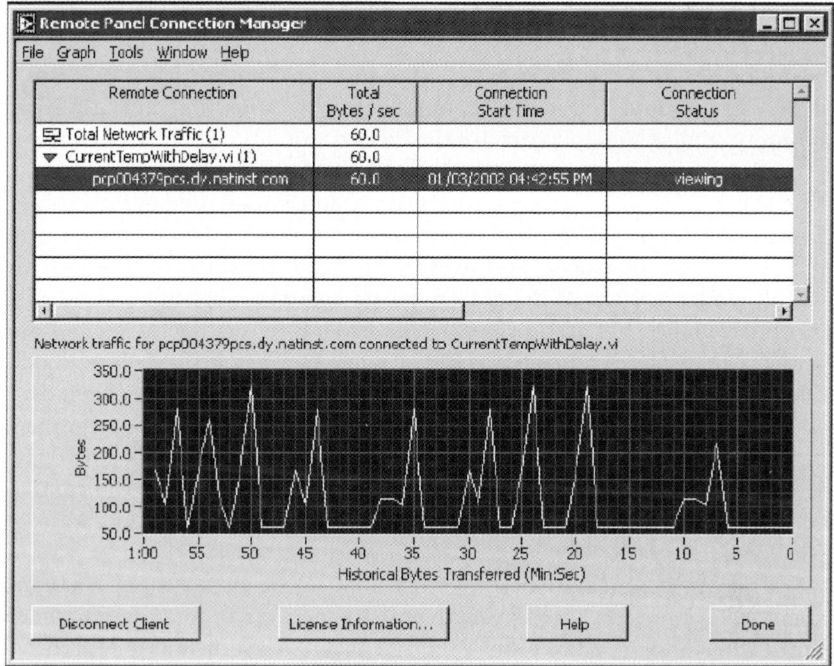

Fig. 11.14. Bandwidth required to serve the VI

11.2.3 Minimizing the Amount of Advanced Communication

Before the introduction of remote front panels, the user can control a LabVIEW application in other ways. For example, the user could use VI Server to control VIs on another computer. This control is made possible through a server/client architecture similar to the one used by remote front panels – the client requests action from the server, and the server responds to the request. Many aspects of LabVIEW use this architecture, including Property Nodes and events. Each instance of a Property Node or event requires a round trip between the server and client.

One trip is from the client to server requesting the property or event, and the other trip is from the server to the client sending the result of the request. Because of the nature of Property Nodes and events, using several Property Nodes or events in the VI results in several messages sent between the client and server. If the user controls a VI containing several Property Nodes or events remotely, the message queue on the server side runs the risk of overflowing. When the message buffer fills, the remote front panel server disconnects all clients connected to it, and the user may lose control of the application. Therefore, the application should be simple and the Property Nodes or events should be used only when necessary – to reduce the bandwidth required for the application.

11.2.4 Functionality to Avoid with Web Applications

It is important to note that not all standard application features translate well into distributed applications. Application features that are localized to the computer on which they run are the biggest potential pitfalls, because they are computer specific, and do not transmit on the Web. Two good examples of localized application features are ActiveX and system dialog boxes.

ActiveX is a powerful, useful communication and control technology. Any Web browser can download and execute ActiveX controls, after which they have full control of the Windows operating system. This full control yields great functionality but also presents a considerable security risk. To control this risk, Microsoft developed a registration system so browsers can identify and authenticate an ActiveX control before downloading it. After the browser downloads it, the user must register the control on the local system before it is executed. While it is not necessarily a problem to register an ActiveX control on a single computer, it can be difficult to ensure that the control is validated and registered by every computer that might view the application on the Web.

Another common component of most applications is a system dialog box. System dialog boxes are a good way to notify a user of a problem, communicate important results, or solicit input from the user. The Simple Error Handler VI even makes use of the system dialog box to manage errors. System dialog boxes are great tools, but they do not work well with distributed applications. The system dialog box remains on the local computer instead of being distributed with the application. In the following example, the VI is simple as shown in Fig. 11.15. It uses an Event structure to monitor the status of a Boolean control. When the user clicks a button, the VI displays a one-button system dialog box with the message Button Pressed.

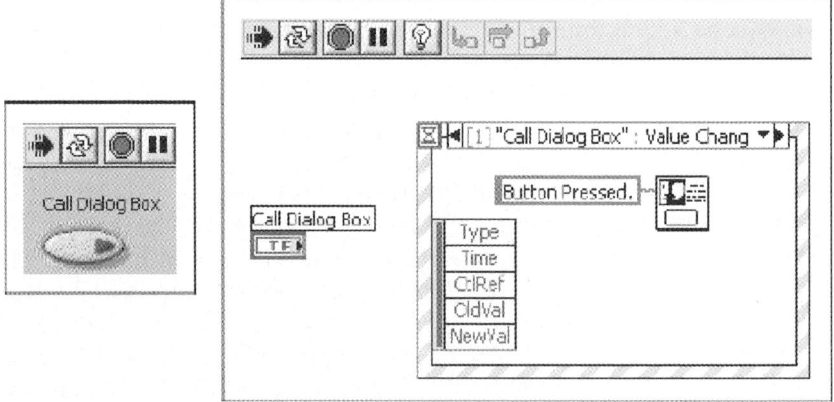

Fig. 11.15. Front panel and block diagram of a simple VI calling a system dialog box

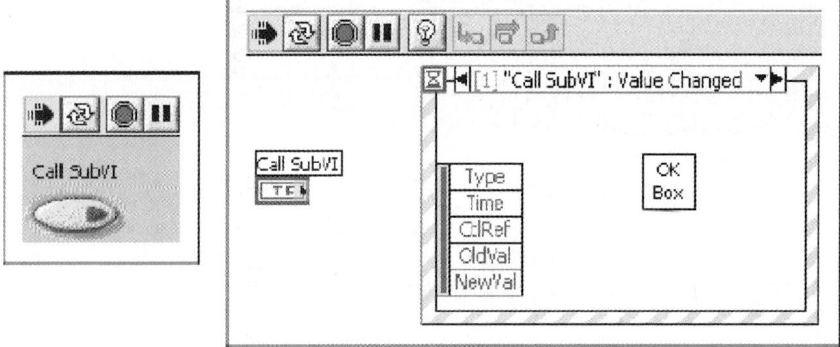

Fig. 11.16. Displaying a SubVI dialog box instead of a system dialog box for Web applications, consider eliminating dialog boxes completely, because they can distract the user

If the user distributes this VI remotely and a user on the Web clicks the button, a dialog box appears on the host computer only. Because that dialog box is a local system box, it does not appear to the user on the Web, but it prevents anyone from using the VI until it is closed. So the user on the Web cannot use the VI but is unaware that the problem is the open dialog box. Because of the localized nature of ActiveX and system dialog boxes, the user should not use these technologies in applications designed to be viewed and controlled remotely.

For remote front panels, the user can accomplish the tasks of system dialog boxes without any of the drawbacks. For example, the user can use a subVI for a dialog box instead of a system dialog box. Customize the **Window Appearance** settings in the **VI Properties** dialog box of the subVI to make the subVI behave identically to the system dialog box. Because the subVI is part of the LabVIEW application, the Web Server can serve it to remote clients as shown in Fig. 11.16.

An alternative way to present a user with information that might be displayed in a dialog box is to use a tab control. The user can design the application to run on one page of a tab control, and switch to another page when an error occurs or the user must make a decision. The user can use Property Nodes to disable the other pages of the tab control until the user acknowledges the error or makes the decision as shown in Fig. 11.17.

11.2.5 Security Issues

Besides bandwidth considerations, publishing a front panel on the Web presents some security concerns. If a VI is in edit mode, the user cannot view it remotely. Therefore, the user cannot modify a VI on the Web. However, anyone who requests control can control the VI. While the VI cannot be modified or damaged, the systems with which it interacts might be affected.

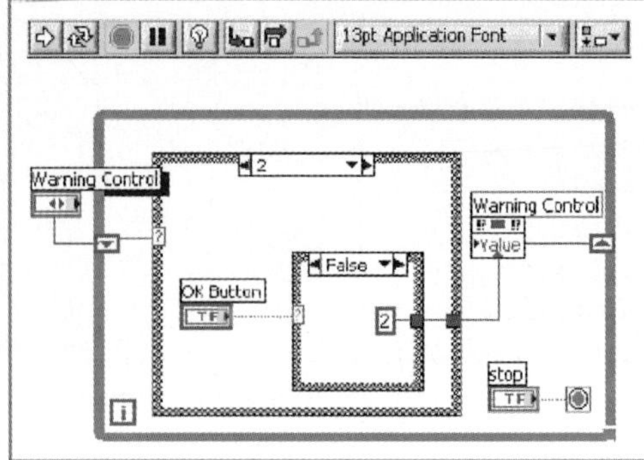

Fig. 11.17. Front panel and block diagram of a VI using a tab control to handle warnings

To prevent a user from gaining control of a VI, click the lower left corner of the front panel and select **Lock Control** from the shortcut menu. This option prevents a viewer from requesting control of the VI on the Web until the user unlocks it. Similarly, the user can regain control of a VI from any Web client.

To configure more Web Server security options, the user can select **Tools≫ Options**, select **Web Server: Configuration** from the top pull-down menu, and enter the **Root Directory**. After entering a root directory, Web clients have access only to files in that directory and its subdirectories. All other files on the computer are not accessible on the Web. Select **Web Server: Browser Access** to grant or deny viewing and/or controlling access to different IP addresses, so only known clients can view or run the application. Select **Web Server: Visible VIs** for control over which VIs in the Web Server directory are visible to a client, and to limit the time that a client can maintain control of a VI.

Alternatively, the user can program security into the VI itself. For example, the user can require a user to log in when the VI first runs, and disable all other aspects of the VI until the login is verified. To do so, place a login field on one page of a tab control, the other controls on subsequent pages, and enable the other pages after the user successfully logs in.

11.3 Using the Timed Loop to Write Multirate Applications in LabVIEW

This application describes the features of the Timed Loop and the method to use the Timed Loop to develop multi rate applications. The user can use the Timed Loop on Windows 2000/XP, PharLap ETS, and LabVIEW Real-Time Module on MacOS 10.x targeted to a supported real-time device.

11.3.1 Timed Loops

A Timed Loop executes an iteration of the loop at the period specified by the user. The Timed Loop is used when the user wants to develop VIs with multirate timing capabilities, precise timing, feedback on loop execution, timing characteristics that change dynamically, or several levels of execution priority. The VIs and functions used on the **Timed Loop** and the **DAQmx** palettes with the Timed Loop to develop VIs that control timing sources. The Timed Loop includes the (1) Input, (2) Left Data, (3) Right Data, and (4) Output nodes, as shown in the following illustration in Fig. 11.18.

The user can set configuration options of the Timed Loop by wiring values to the inputs of the Input Node, or using the **Loop Configuration** dialog box to enter values for the options. By default, the inputs of the Input Node appear as icons with the values specified in the **Loop Configuration** dialog box.

The Left Data Node of the Timed Loop provides timing and status information about the previous loop iteration, such as if the iteration executed late, the time the iteration actually started executing, and when the iteration should have executed. The user can wire data from the Left Data Node to the Right Data Node to configure future iterations of the Timed Loop. Right-click the border of the Timed Loop and select **Show Right Data Node** from the shortcut menu to display the Right Data Node. The user can also resize the Left Data and Right Data nodes.

The Output Node returns error information the Timed Loop receives in the **error in** input of the Input Node or error information the Timed Loop generates during an iteration. For example, if an external source aborts the execution of the Timed Loop or if more than 128 Timed Loops run at one time, the Output Node returns an error. The Output Node does not return error information from any sub diagram that executes within the Timed Loop.

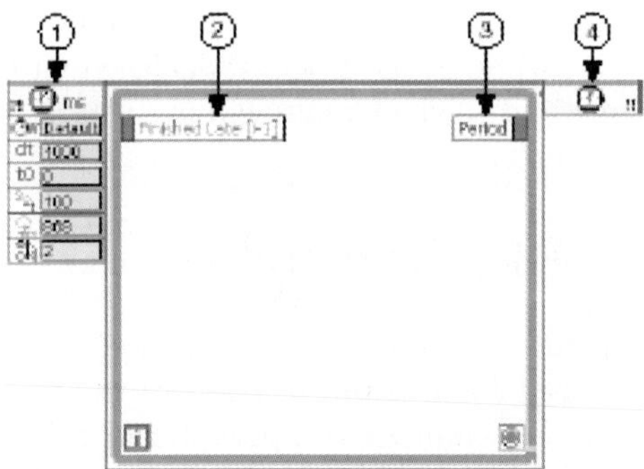

Fig. 11.18. Timed Loop

11.3.2 Configuring Timed Loops

The user can use the **Loop Configuration** dialog box to configure the Timed Loop execution. The **Loop Configuration** dialog box can be displayed by double-clicking the Input Node or right-clicking the Input Node and selecting **Configure Timed Loop** from the shortcut menu. The user can use this dialog box to specify a timing source, period, offset timing, and other options, or the user can use the dialog box to configure options so they appear as inputs on the Input Node. The user can place a checkmark in the **Use terminal** checkbox for any dialog box option the user wants to appear as an input on the Input Node. Any option configured in the **Loop Configuration** dialog box to appear as an input on the Input Node appears as an icon without the configuration values. To display the names of the inputs the user can Right-click the Input Node and select **Name Format≫Names** from the shortcut menu. Only the options selected **Use terminal** on the **Loop Configuration** dialog box appear as inputs when the user selects **Name Format≫Names** from the shortcut menu. The user can also use the Right Data Node to dynamically adjust the period, offset, priorities, and mode values for the Timed Loop. The updates take effect the next iteration of the loop.

11.3.3 Selecting Timing Sources

A timing source determines when a Timed Loop executes a loop iteration. By default, the Timed Loop uses the 1 kHz clock of the operating system as the timing source and can execute only once every 1 ms because that is the fastest speed at which the operating system timing source operates. If the system does not include a supported hardware device, the 1 kHz clock is the only timing source available. If the system does include a supported hardware device, the user can select from other timing sources, such as the 1 μs in CPU cycles available on some real-time hardware; or events, such as the rising edge of a DAQ-STC counter input or output; or the end-of-scan interrupt of an E-Series board AI engine.

The **Source type** listbox in the **Loop Configuration** dialog box is used to select a timing source or to the Create Timing Source VI to programmatically select a timing source.

11.3.4 Setting the Period and the Offset

The period is the length of time between loop executions. The offset is the length of time the Timed Loop waits to execute the iterations. The timing source determines the time unit of the period and the offset. If the timing source is a 1 kHz clock, the unit of time for the period and the offset is in milliseconds. If the timing source is a 1 MHz clock on a LabVIEW Real-Time target with a Pentium processor, the unit of time for the period and the offset

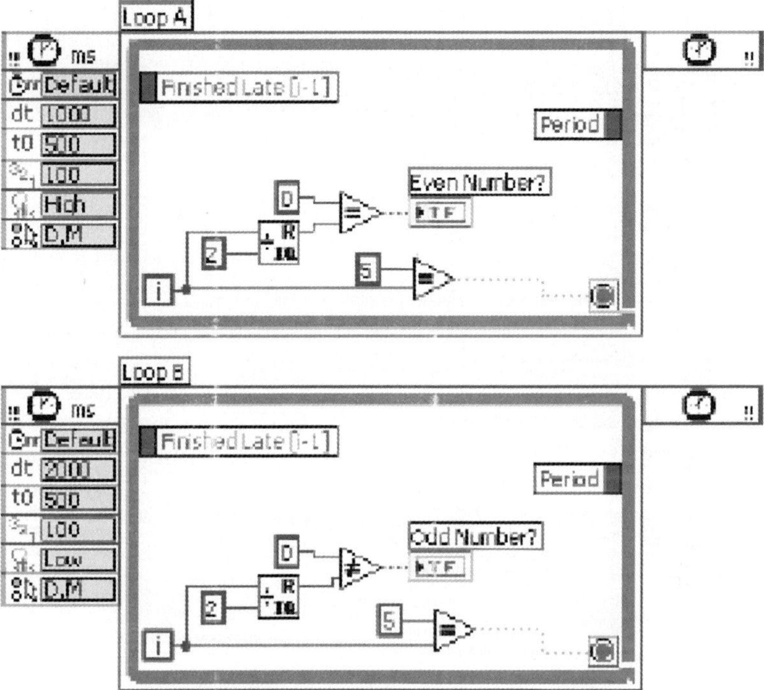

Fig. 11.19. Setting Timed Loop Period

is in microseconds. The time the first timing source starts determines the start time of the offset.

The Timed Loops in the following block diagram use the default 1 kHz timing source and have offset (**t0**) values of 500, so the Timed Loops wait 500 ms to execute their iterations. The period (**dt**) for Loop A is 1,000 ms, and the period for Loop B is 2,000 ms, which means Loop A executes every second and Loop B executes every 2 s. Both Timed Loops shown in Fig. 11.19 stops executing after six iterations. Loop A stops executing after 6 s because its period is 1 second, and Loop B stops executing after 12 s because its period is 2 s.

11.3.5 Setting Priorities

Each Timed Loop on the block diagram creates and runs in its own execution system that contains a single thread, so no parallel tasks can occur. The priority of a Timed Loop specifies when the loop executes on the block diagram relative to other Timed Loops. The priority setting of a Timed Loop is used to write applications with multiple tasks that can preempt each other in the same VI. The higher the value the user enters in **Priority** of the Timed Loop, the higher the priority the Timed Loop has relative to other Timed Loops on the

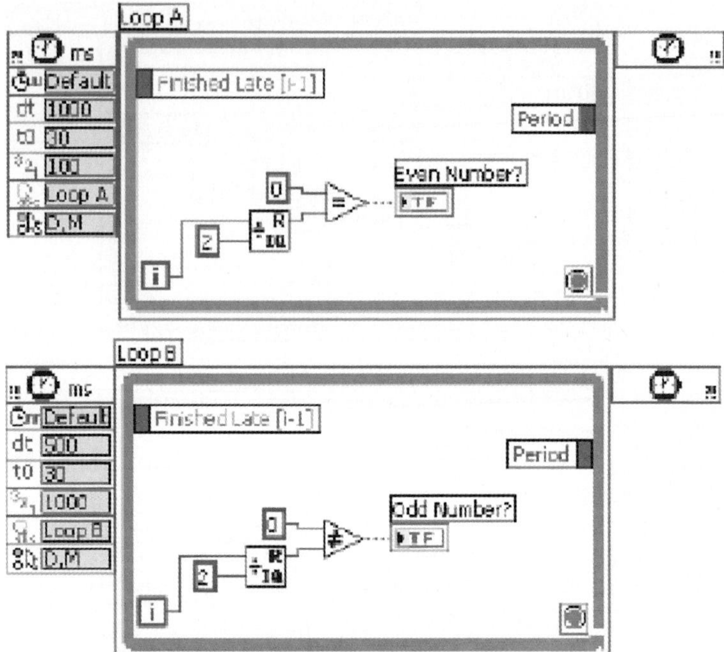

Fig. 11.20. Setting Timed-Loop Priority

block diagram in Fig. 11.20. The value entered in the Priority input must be a positive integer between 1 and 2,147,480,000. In the following block diagram, the Priority (**321**) value of Loop B (1,000) is higher than the priority of Loop A (100). Both loops are configured with an offset (**t0**) value of 30, so they execute at the same time. Because Loop B has a higher priority, it executes first.

If two Timed Loops have the same priority, the first Timed Loop that executes completes its execution before the other Timed Loop starts its execution. Timed Loops execute at a priority below the time-critical priority of any VI but above high priority, which means that Timed Loops execute in the data flow of a block diagram ahead of any VI not configured to run at a time-critical priority.

11.3.6 Naming Timed Loops

By default, LabVIEW automatically uniquely identifies each Timed Loop placed on the block diagram with a name, which appears in the **Loop name** text box of the **Loop Configuration** dialog box. The user can rename the Timed Loop by entering a name in this text box. The user can use the unique name of the Timed Loop with the VIs on the **Timed Loop** palette to programmatically stop the Timed Loop and to synchronize a group of Timed

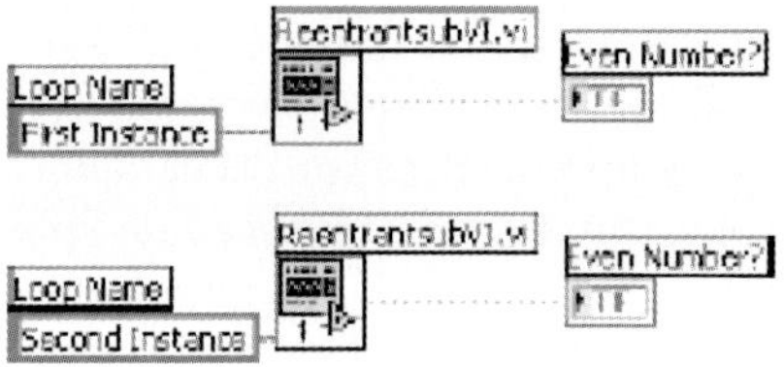

Fig. 11.21. Naming Timed Loops in Reentrant VIs

Loops to use the same start time. If a reentrant VI includes a Timed Loop and the user uses two or more instances of that reentrant VI as subVIs on a block diagram, then the user must programmatically change the name of the Timed Loop for each instance of the reentrant VI. It should be ensured that the reentrant VI which includes the Timed Loop has an input terminal on the connector pane connected to a string control wired to the **Loop name** input of the Timed Loop on the block diagram. On the block diagram where two or more instances of the reentrant VI are used as a subVI, wire unique string values to the **Loop name** input on the reentrant subVI to uniquely identify each Timed Loop within each instance of the reentrant subVI. For example, the following block diagram in Fig. 11.21 includes two instances of the same reentrant VI as subVIs. The **Loop name** string constants wired to the instances of the subVIs pass two different names to the Timed Loop in the block diagram of the reentrant VI each time the block diagram executes. If the Timed Loops were not programmatically renamed, LabVIEW would return an error.

11.3.7 Timed Loop Modes

Occasionally, an iteration of a Timed Loop might execute later than the time specified by the user. The mode of the Timed Loop determines the method the loop handles during any late executions. The options in the **Timed Loop action to take on late iterations** section of the **Loop Configuration** dialog box or the **Mode** input of the Input Node is used to specify the mode a Timed Loop uses to handle the late execution of a Timed Loop iteration.

The user can handle the late execution of a Timed Loop in the following ways:

- The LabVIEW Timed Loop Scheduler can align the execution with the original established schedule.
- The LabVIEW Timed Loop Scheduler can define a new schedule that starts at the current time.
- The Timed Loop can process the missed iterations.
- The Timed Loop can skip any missed iterations.

For example, if the user sets a Timed Loop with a period of 100 ms and an offset of 30 ms, the first loop iteration is expected to execute 30 ms after the first timing source starts running and in multiples of 100 ms after that at 130, 230, 330 ms, and so on. However, the first execution of the Timed Loop might occur after 240 ms have elapsed. Because other Timed Loops or hardware devices might already be running at the schedule specified, the user might want to align the late Timed Loop with the already running global schedule, which means the Timed Loop should align itself as quickly as possible with the schedule specified. In this case, the next Timed Loop iteration would run at 330 ms and continue to run in multiples of 100 at 430, 530 ms, and so on. If aligning the Timed Loop with other Timed Loops or other hardware devices is not important, the Timed Loop can run immediately and use the current time as its actual offset. In this case, the subsequent loop iterations would run at 240, 340, 440 ms, and so on.

If the Timed Loop is late, it might miss data other Timed Loops or hardware devices generate. For example, if the Timed Loop missed two iterations and some of the data from the current period, a buffer could hold the data from the missed iterations. The user may want the Timed Loop to process the missed data before it aligns with the schedule specified. However, a Timed Loop that processes the missed iterations causes jitter, which is the amount of time that a loop cycle time varies from the time specified. If the user does not want to process the missed data, the Timed Loop can ignore older data in the buffer the loop iterations missed and process only the latest data, such as the data available at the next period and the subsequent iterations.

11.3.8 Configuring Modes Using the Loop Configuration Dialog Box

By default, the Timed Loop discards any data generated during missed iterations and maintains the original schedule. For this mode setting, the icon for the **Mode** input of the Input Node appears with a D for discard and an M for maintain. The user can remove the checkmark from the **Discard missed periods** checkbox in the **Loop Configuration** dialog box to process data any missed or late loop iterations generate. For this mode setting, the icon for the **Mode** input of the Input Node includes a P for process. The user can also remove the checkmark from the **Maintain original phase** checkbox in the **Loop Configuration** dialog box to configure the Timed Loop to execute on a new schedule based on the first iteration of the Timed Loop.

11.3.9 Configuring Modes Using the Input Node

The **Mode** input of the Timed Loop Input Node is used to programmatically change the mode of the Timed Loop or maintain the current configuration of the modes configured in the **Loop Configuration** dialog box. The **Loop Configuration** dialog box is opened and a checkmark is placed in the **Use terminal** checkbox in the **Timed Loop action on late iterations** section. Close the dialog box, right-click the **Mode** input, and select

Create≫Constant or **Create≫Control** from the shortcut menu to create an enumerated constant or control to select a mode. Select **No Change** in the enumerated constant or control to maintain the current mode configuration or select the option to take on any late or missed iteration of the Timed Loop.

11.3.10 Changing Timed Loop Input Node Values Dynamically

The Left Data Node is used to acquire information about the execution of the Timed Loop, such as if the timing source executed late or if the offset or period values changed. The user can wire the values the Left Data Node returns to the Right Data Node or to nodes in the sub diagram within the Timed Loop. The Right Data Node is used to dynamically change the input values of the Timed Loop on the next loop iteration. The Timed Loop in the following block diagram shown in Fig. 11.22 runs 1 second (1,000 ms) longer each time the loop iterates until the loop has executed six times.

If the user dynamically changes the offset of the Timed Loop by wiring a value to the **Offset** input terminal of the Right Data Node, a mode with the **Mode** input terminal of the Right Data Node has to be specified. To set the mode, the user can right-click the **Mode** input terminal of the Right Data Node and select **Create≫Constant** or **Create≫Control** from the shortcut menu to create an enumerated constant or control to select a mode.

11.3.11 Aborting a Timed Loop Execution

The Stop Timed Loop VI is used to abort the execution of a Timed Loop programmatically. The user can specify the name of the Timed Loop that is to be aborted by wiring that name in a string constant or control to the

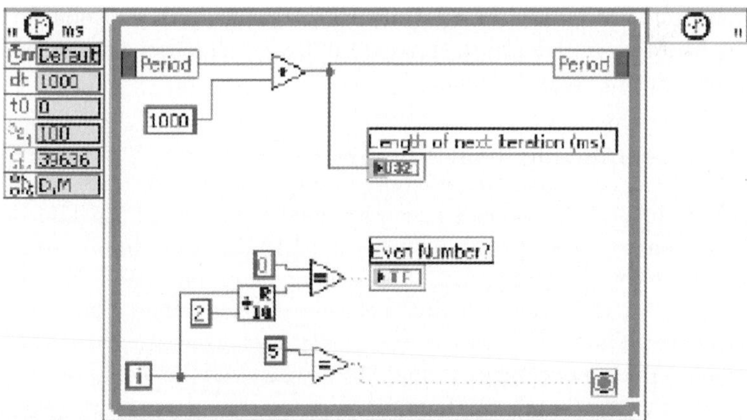

Fig. 11.22. Changing Timed Loop values dynamically using the Left Data and Right Data Nodes

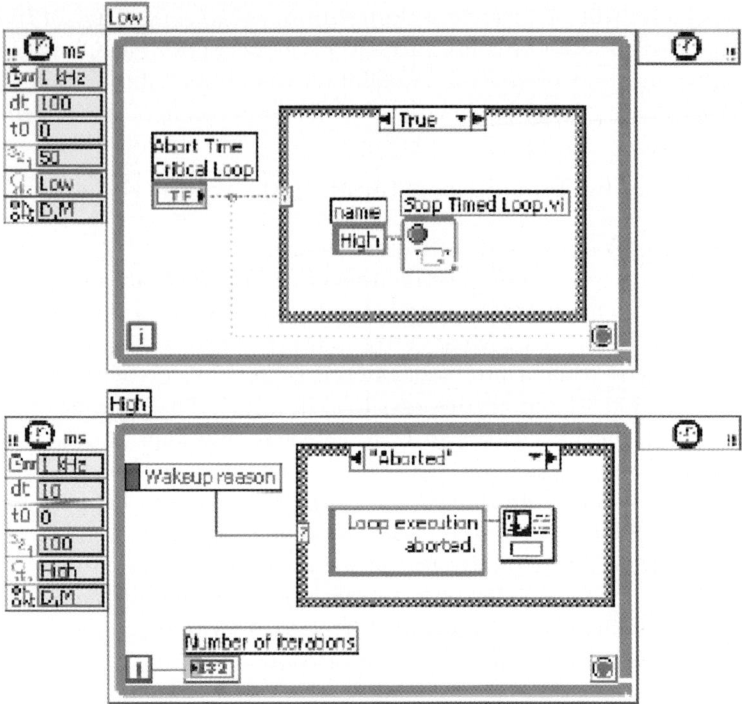

Fig. 11.23. Aborting Timed Loop Execution

name input of the Stop Timed Loop VI. For example, in the following block diagram in Fig. 11.23, the Low Timed Loop includes the Stop Timed Loop VI. The High Timed Loop runs and displays the number of iterations it has completed. If the user clicks the **Abort Time Critical Loop** button on the front panel, the **Wakeup reason** output of the Left Data Node returns a value of Aborted, a dialog box appears, and when the user clicks **OK** in the dialog box, the VI stops running.

11.3.12 Synchronizing Timed Loops

The Synchronize Timed Loop Starts VI is used to ensure all the Timed Loops on a block diagram use the same start time and the same timing source. For example, the user might have two Timed Loops and may want to ensure that they execute on the same schedule relative to each other. The user might want the first Timed Loop to execute first and generate data, then have the second Timed Loop process that data when the Timed Loop execution finishes. To ensure that both Timed Loops use the same start time as the basis for their execution, the user can create a Timed Loop group by wiring a name for the group to the **synchronization group name** input and wiring an array of Timed Loop names to the **loop names** input. The Synchronize

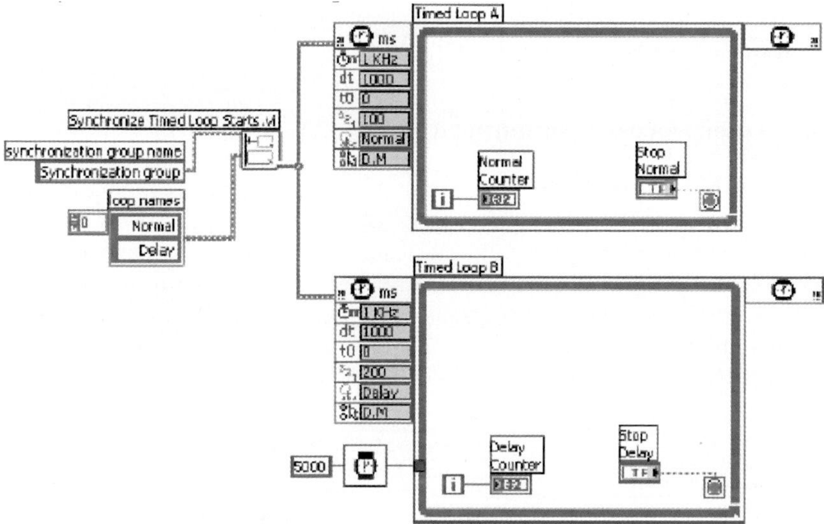

Fig. 11.24. Synchronizing Timed Loops

Timed Loop Starts VI in the following block diagram in Fig. 11.24 creates a synchronization group name and synchronizes Timed Loops A and B to use the same start time.

11.3.13 Timed Loop Execution Overview

The LabVIEW Timed Loop Scheduler, which is part of the LabVIEW Run-Time Engine, manages the execution of Timed Loops. Predefined events within the timing source represent a tick of the timing source. Based on the timing source type and hardware capability, all or some of the events generate interrupts that notify the LabVIEW Timed Loop Scheduler that an event or a timeout occurred. The LabVIEW Timed Loop Scheduler handles the interrupt, checks the list of Timed Loops associated with the timing source, and triggers a new iteration of a Timed Loop if necessary. The LabVIEW Timed Loop Scheduler uses the value of the **Priority** input of the Timed Loop to determine if the loop should preempt the currently running loop iteration. The scheduler also controls which Timed Loop continues its execution after the currently running Timed Loop finishes its iteration. Finally, the scheduler reconfigures the hardware component of the timing source for the next notification.

11.4 Client–Server Applications in LabVIEW

National Instruments LabVIEW is a very flexible programming environment for instrumentation development. In version 5.x LabVIEW has gained functionality to facilitate interprocess communication. In this application the method to build a client–server system is discussed. Here the server process runs asynchronously.

11.4.1 Interprocess Communication

Version 5.x of LabVIEW introduces a number of interprocess communication methods, which is discussed below. All of them, however, follow a similar pattern of use; creation of a refnum, writing or generation of an event, reading or waiting for event with a timeout, and deletion of refnum and freeing of resources. In the context we are discussing here, a process is an independently running LabVIEW routine, however, all such processes are launched from the same LabVIEW instance (either the development environment, or a runtime engine).

Notifiers

A notifier is a mechanism for passing transitory data from one process to another. The data contained in the notifier may be overwritten before the reader reads it. If there is more than one process waiting for a given notifier then they all receive the data sent. This process is somewhat analogous to sending data on a UDP broadcast, in that neither sender nor receiver should assume that all data that was sent was received. A simple example of a notifier is shown in Fig. 11.25. A notifier is well suited for transmitting informational messages about program state between processes.

Queues

Unlike notifiers, queues guarantee transmission of data from the sender to a receiver. Also, unlike notifiers, it is not possible to transmit information from one sender to multiple readers. The queue is somewhat analogous to a TCP-based protocol, in that information is reliably transferred from one sender to one reader, however, further control logic would be required to hold a two-way data exchange over a queue to prevent a process reading its own sent data. For our client–server application, the queue is the basis of the communication.

Semaphores and Rendezvous

Semaphores and rendezvous are used to control program flow. The semaphore is best suited to ensuring that only one process is carrying out a given task at any one time, for example accessing a piece of hardware. A rendezvous, on the other hand, is most appropriate for synchronizing separate processes, for example to synchronize a measurement with control of a piece of apparatus.

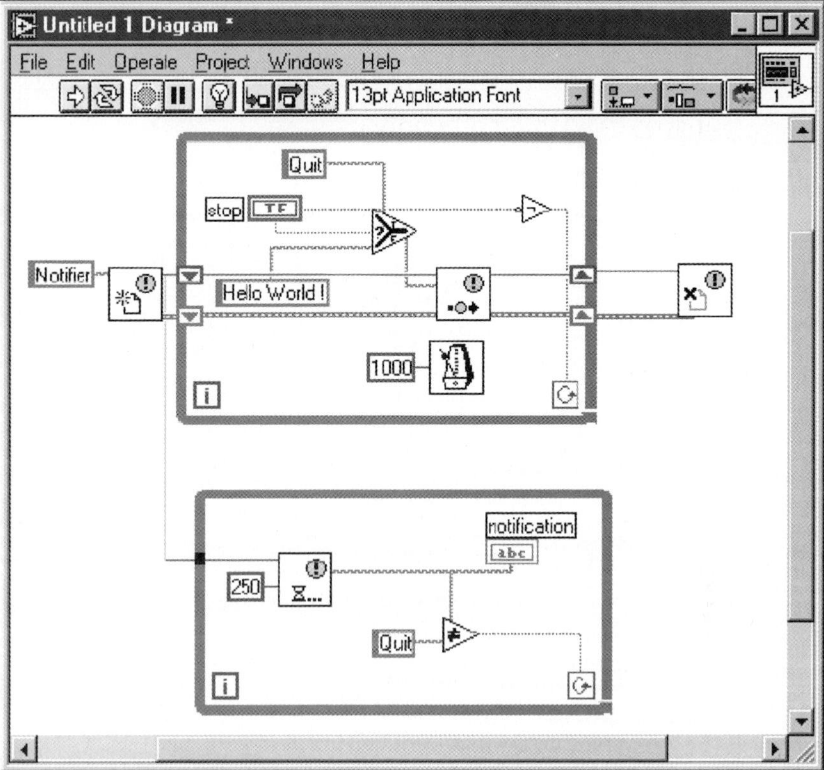

Fig. 11.25. A simple example of using a notifier to communicate between two parallel loops

11.4.2 A Simple Read-Only Server

The simplest form of server would be a process that only listened for incoming messages and then did whatever was appropriate based on the input.

Such a server shown in Fig. 11.26 would only create a suitable queue, and then attempt to read from the queue whilst running a while-loop. Adding a timeout to the read would allow the loop to cycle on a regular basis to poll a manual stop button. Clients would communicate with the server by sending simple string messages via the queue (which would have a well defined name). The obvious disadvantage with this system is that the communication process is only one-way. Even if the process does not inherently require to return data, the client has no way of knowing whether the server process is running and has correctly processed the data. For this reason, a two way communication process is required.

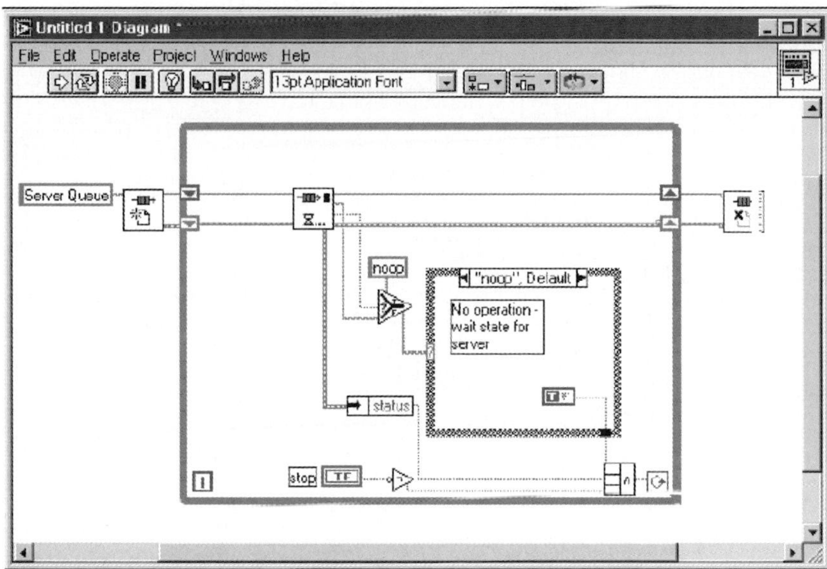

Fig. 11.26. A simple server which reads from a queue and executes commands

11.4.3 Two Way Communication: A Read–Write Server

As discussed above, sending two way data through a queue, whilst possible, does require additional flow-control logic. A more elegant method is to use two queues, one for upstream and one for downstream communication. It might seem, at first sight, that we could make do with two queues per server, so that the server would receive on one queue and return the results on the other. Figures 11.27 and 11.28 show the two-way communication of server and client, respectively. The problem with this is that it fails when more than one client is active. If one client sends a request to the server, and then starts to wait on the downstream queue, but before the server has finished processing the request, another client sends a request and starts waiting for its reply, then the two clients have entered a race to receive the answer from the server to client 1. In principle, we should expect that 50% of the time, client 2 will get client 1s data.

To circumvent this, we could tag the replies with an identification to allow clients to recognize their data, and to re-insert other data onto the queue. Such a solution, however, doesn't solve the underlying problem, and is likely to increase the entries being added and removed from the queue. Rather, we should have a separate queue for each client, so that the client only sees data intended for it. The process would now be; client creates a listening queue, sends command on server's queue including a reference to its listening queue. The server receives the command from its queue and writes its response to the client's queue. The client listens on its queue for the server. There is one twist to this scheme.

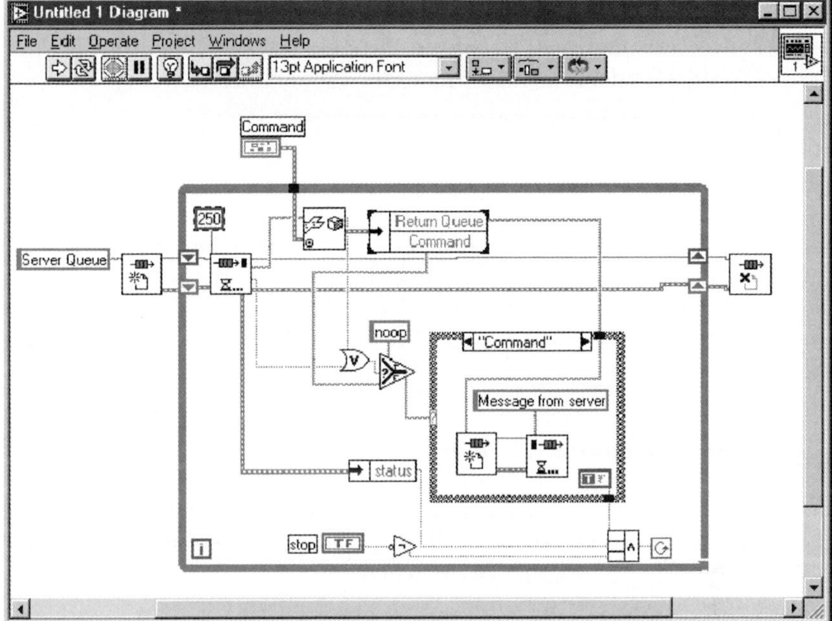

Fig. 11.27. A Read–Write Server

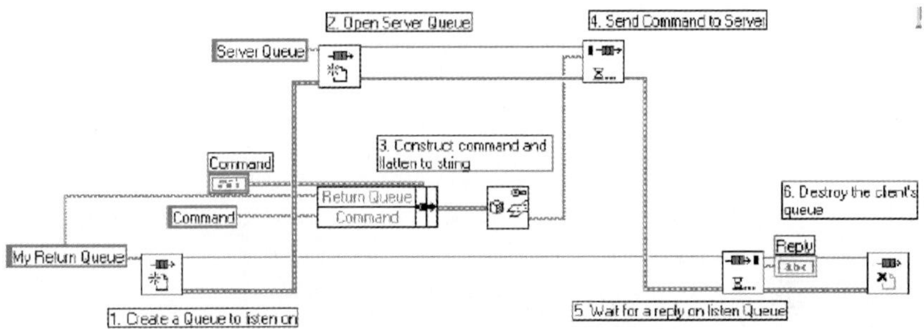

Fig. 11.28. A Read–Write Client

In order to send both a command and a return queue to the server, a cluster structure has been used. However, the queues only work with string data, so we have to flatten the cluster to a string before sending to the server and unflatten on receipt at the server. This scheme is perfectly adequate if the number of client calls is small. If the client calls the server many times, then the overhead of creating and destroying the queues gets excessive and memory consumption increases. The solution is to create a pool of queues which clients may use, and then use another queue to hold the names of theses queues.

Clients then remove a free client queue-name from this queue of queues, use it to receive a reply from the server, and then return the queue name to the queue of queues when finished. This will limit the number of simultaneous client calls to the number of queues in the pool, however, extra code can be added into the server or client to maintain the number of free queues in the pool within in certain constraints.

11.4.4 The VI-Reference Server Process

The VI-reference server shown in Fig. 11.29 follows the scheme for the Read–Write Server very closely. The main differences are that the command interpretation and processing and the replying to the client is all delegated to one sub-vi. In addition, the management of the pool of client queues is maintained in the server.

11.4.5 The VI-Reference Client

The Vi-Reference Client is actually a set of separate routines that each send one command to the server by calling a common sub-routine as shown in Fig. 11.30. The client also follows the scheme closely, using the pool of return queues to get a queue to listen on. We make use of some global variables to store static information such as the server queue name, timeout values and templates for the sent and returned data structures.

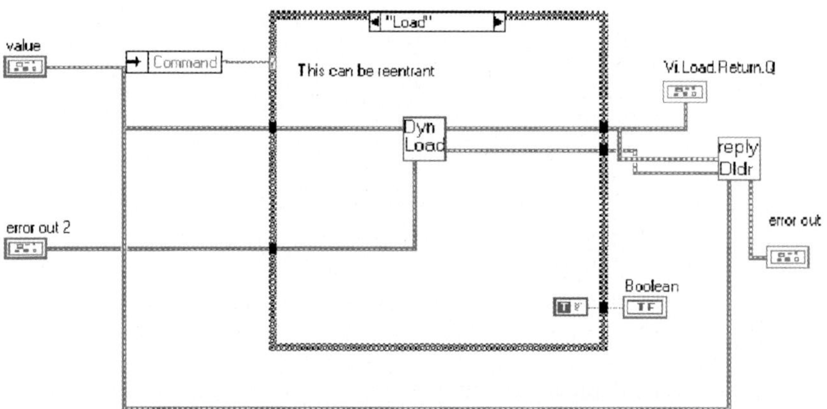

Fig. 11.29. The VI-reference server main process

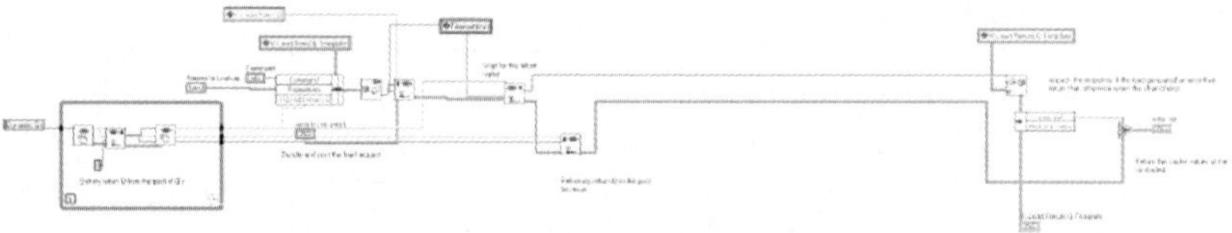

Fig. 11.30. The Client routine

11.4.6 Further Thoughts

Intermachine Client–Servers

In this application we have discussed only client–servers running within one instance of LabVIEW. Intermachine Client–Server setups are an example of the more general case of inter-LabVIEW client–servers. The approach used here will not work directly across LabVIEW instances, because the queue refnums are specific to a particular instance. One alternative approach would be to build a complete TCP/IP-based client–server system using a custom protocol. This would however add substantially to the complexity of the code. On Win32 platforms, the Data-Socket vis could be used to reduce the problem to a relatively simple case, in which the queue vis would be replaced with Data-socket VIs.

For a truly cross-platform approach, one could use the LabVIEW VI-server functions to invoke the create queue/read queue/write queue on the remote machine. A consistent standard to define on which machine the various queues were created and manipulated would be required. An obvious protocol would be to run the server queue on the server machine and the client queues on the client machine. The clients would then need to send not only the name of their queue but also their machine name and location of the queue VIs.

Persistent Client–Server Exchanges

The setup described here is orientated towards a single transaction, connectionless design – like for example http 1.0. The client makes a connection to the server, sends a single command and receives a single response. This system can process one client at a time, which is not a problem for the example given. If the server has a prolonged exchange with client, one would want the server to enable to process more than one client at a time. To adapt the system to work with multiple transactions per connection, one would have to allow the server to create a unique write queue for the client, and to pass this back to the client with the first response. Handling multiple clients takes rather more effort. If each transaction is short, then the server requires just one main loop than handles all available clients. If the transactions are lengthy, then passing both the up- and down-stream queues to a re-entrant VI that handles the transactions might be appropriate.

11.5 Web-Based Matlab and Controller Design Learning with LabVIEW

This application presents a web-based virtual laboratory for learning of MATLAB. Using this virtual laboratory, users could write their own MATLAB M-files and execute them using the MATLAB web server. In addition, web-based remote laboratory-experiment RC oscillator for learning of control design is presented. In RC oscillator experiment, interactivity between the controller design parameters in Bode plot and Root Locus with real experiment has been implemented in order to support full visualization of the controller design.

11.5.1 Introduction to Web-Based MATLAB

The Internet (Web) has become a widespread tool for teaching and learning. The Web enables more flexible delivery, distance education, new visualization possibilities, and cost reduction. Virtual and remote web-based laboratories have been developed to date. Remote web-based laboratories are based on sharing the same resources over the Internet. Therefore, remote laboratories are rarely opened to the public rather their use is limited to the assigned users.

The first one is a virtual laboratory that supports learning of MATLAB. In this virtual laboratory users can write their own MATLAB M-files and execute them using the MATLAB web server. The second web-based laboratory is a remote laboratory that supports the learning of control theory and control design. RC oscillator web-based remote control experiment is represented. For RC oscillator controller design, a visualization of controller design parameters is implemented interactively with the remote control experiment.

11.5.2 Learning of MATLAB

In engineering education, MATLAB is frequently one of the tools used to support support the learning of mathematics. Over the last five years, a hands-on introduction to the MATLAB program has been given to our undergraduate mechatronics and automation users. Lack of time and lab availability for practicing motivated us to develop a virtual lab, which enables users to learn MATLAB outside official lab time. MATLAB web server was designed for use with fixed MATLAB applications as proposed by MATLAB web server instructions. In such fixed MATLAB applications, the capabilities of the World Wide Web are utilized to send data to MATLAB server for computation and to display the results in a client Web browser.

The computation of MATLAB is defined in advance by selected application and can not be influenced by the remote user. Therefore, fixed MATLAB applications could not be used for learning of MATLAB, but only as a computation tool. Web-based learning of MATLAB is based on MATLAB programming capabilities. MATLAB enables writing of a series of MATLAB statements (a program) into a file with extension .m, and then executing them with a single command. Files with extension .m are called 'M-files'. A MATLAB application has been written, which enables transmission of the M-file over the web to the MATLAB web server, and a display of the results in a client Web browser. The m-file sent to the MATLAB web server should contain all data definitions, as well as computation commands. Users learning MATLAB can use any ASCII editor (Notepad or similar) to write the MATLAB m-file.

A standard web browser (Internet Explorer, Mozilla, etc.) is needed for learning of MATLAB over the web.

11.5.3 Learning of Controller Design

A remote lab at Faculty of Electrical Engineering and Computer Science, University of Maribor was established in order to support the learning of control. This remote lab provides various remote experiments such as cascade control of DC motor, two axes mechatronic device control, RC oscillator. The RC oscillator is a web-based version of interactive controller design and experiment. The objectives of WICDE are:

- To teach users control design,
- To minimize the gap between control theory and practice, by teaching control implementation,
- To show users how to learn by Web and how to use it and
- To support learning by doing.

The WICDE was designed to be available to a broad range of the users. Therefore, it was designed with minimum software requirements from the users prospective. To perform the WICDE experiment, a standard web browser (Internet Explorer, Mozilla, etc.) and 'LabVIEW Run Time Engine' are needed. Unfortunately, the assumption of minimum user's software requirements sets an undesirable limitation on the implementation of 'Learning through doing'.

This limitation means that users can not build their own experiment but could only vary the parameters of the already prepared experiment. Such a limitation is widely accepted for web-based experiments. Only a web-based experiment presented in assumes the MATLAB/Simulink software environment to be possessed by the users. Therefore, in the case of the web-based experiment, users could build their own experiment from home.

System Architecture of WICDE

WICDE is implemented using a DSP-2 learning module, and a breadboard with an RC circuit. The DSP-2 learning module is an 'embedded', light and small in volume DSP-based control system. It was developed at the Faculty of Electrical Engineering and Computer Science, University of Maribor. The DSP-2 learning module represents an open framework for rapid control prototyping (RCP) and rapid remote control experiment development presents the block scheme of the DSP-2 learning module-based remote laboratory.

A DSP-2 learning module, connected to a lab PC through the serial port, implements a control algorithm developed using Simulink, and through the analog and digital I/O signals, drive the real plant. LabVIEW virtual instrument and the LabVIEW server run on the same lab PC for the purpose of enabling remote control of the real plant. LabVIEW VI performs communication between the lab PC and the DSP-2 learning module, and enables DSP-2 data visualization and parameter tuning, while LabVIEW server enables remote operation of the LabVIEW VI. Remote users, connected to the server through the Internet, must have a "LabVIEW Run-Time Engine" installed on their personal computer in order to perform remote experiments. During remote experimentation, the remote user can adjust the controller parameters and send experimental results through email.

The DSP-2 learning module-based RCP system is based on two commercially available software packages such as MATLAB/Simulink and LabVIEW, and custom-made hardware like DSP-2 learning module. MATLAB, Simulink and Real-Time Workshop (RTW) are observed or controlled by using only the standard web browser.

RC Oscillator Controller Design and Experiment

RC oscillator exercise is described in this section and the tools for its solution are presented.

Task Description of the RC Oscillator Exercise

Build the RC oscillator shown in Fig. 11.31, which is based on the 3rd order RC circuit.

$$\mathrm{R1} = \mathrm{R2} = \mathrm{R3} = \mathrm{R} = 47\,\mathrm{ohm},$$
$$\mathrm{C1} = \mathrm{C2} = \mathrm{C3} = \mathrm{C} = 1\,\mu\mathrm{F}.$$

The objective is to build a mathematical model of the RC circuit and verify the model using the step and sinusoidal responses. Using Bode plot or Root locus methods, a P controller is designed with gain K_R for RC oscillator. RC oscillator is used for generation of sinusoidal signals, therefore, design gain K_R for the margin of select the step or the sinusoidal input to the RC circuit. When feedback control is selected another switch takes action.

This switch selects between the initial condition (IC) input and feedback with gain K_R for input into the RC circuit. Therefore two phases of the RC

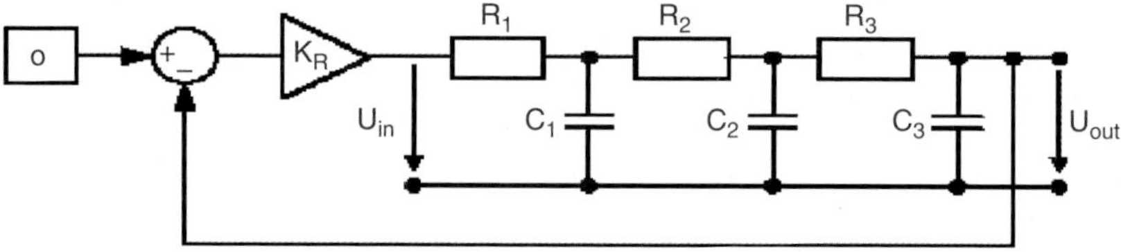

Fig. 11.31. RC oscillator control loop could be observed

oscillator response are observed. The capacitors of the RC circuits are charging to the value IC when IC input is selected. This phase is called the 'charging phase'. When feedback with gain K_R is selected the RC oscillator response is observed. This phase is called the "RC oscillator relaxation phase." Signals Uref, Uin, Uout1 and Uout2 are observed from the Simulink block scheme. Online tuning of the K_R (RC oscillator feedback gain), IC (the value to which capacitors should charge) and per the period of the pulse generator switching between the charging and the relaxation phase is possible. Due to delays appearing in Internet connections, delays also appear between the variation of the gain K_R and the interactive view of the RC oscillator response.

The RC oscillator Bode plot or Root Locus plot can be observed interactively with the RC oscillator response. Both mentioned plots are calculated by MATLAB on the basis of the RC circuit transfer function. Controller design parameters, such as phase margin and crossover frequency, are clearly marked in the Bode plot. Accordingly in Root Locus the actual roots of the control loop are clearly marked. Bode plot or Root Locus design is selected with the left button above the design diagrams in the LabVIEW front panel.

A MATLAB web server application has been written to enable users to write their own MATLAB M-files and execute them using MATLAB web server. In this way users could learn MATLAB over the web. In the future it is desirable to achieve web-based MATLAB/Simulink learning for users. For controller design, a visualization of controller design parameters interactively with the control loop step response is very important.

In MATLAB, the environment SISO tool offers such visualization capabilities but only for simulated control loops. In RC oscillator WICDE, experiment interactivity between the controller design parameters in Bode plot and Root Locus with real experiment has been established. More remote lab experiments with interactive controller design are planned for the future.

11.6 Neural Networks for Measurement and Instrumentation in Virtual Environments

Virtual Reality (VR) is a computer based mirror of the physically reality. Synthetic and sensor-based, computer representations of 3D objects, sounds and other physical reality manifestations are integrated in a multimedia *Virtual Environment* (VE), or *virtual world*, residing inside the computer. Virtual environments are dynamic representations where objects and phenomena are animated/programmed by scripts, by simulations of the laws of physics, or driven interactively directly by human operators and other real world objects and phenomena, Fig. 11.32. The original VR concept has evolved finding practical applications in a variety of domains such as the industrial design, multimedia communications, telerobotics, medicine, and entertainment. *Distributed Virtual Environments* (DVEs) run on several computers connected

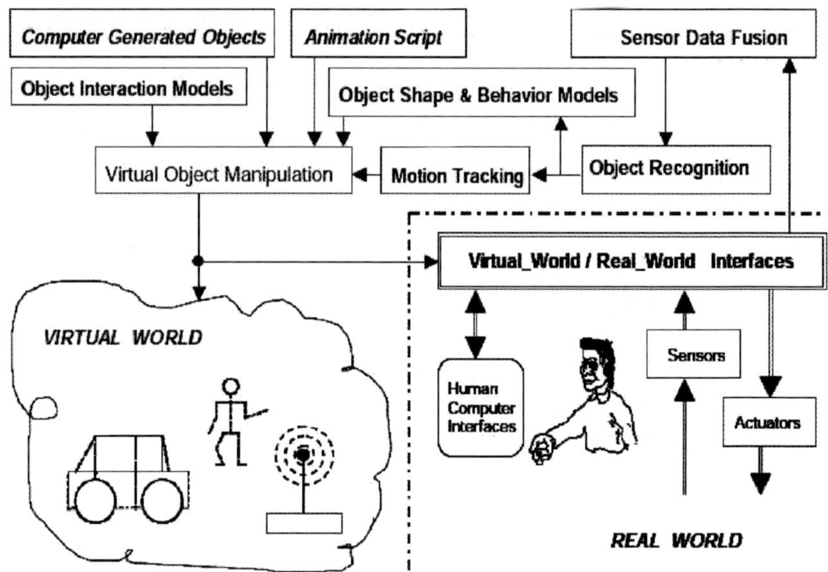

Fig. 11.32. Synthetic and sensor-based computer representations of natural objects and phenomena are integrated in a dynamic multimedia *Virtual Environment*

over a network allowing people to interact and collaborate in real time, sharing the same virtual worlds. Collaborative DVEs require a broad range of networking, database, graphics, world modeling, real-time processing and user interface capabilities.

Virtualized Reality^{TM} *Environment* (VRE), is a generalization of the essentially synthetic VE concept. While still being a computer-based world model, the VRE is a conformal representation of the mirrored real world based on sensor information about the real world objects and phenomena. *Augmented Reality* (AR) allows humans to combine their intrinsic reactive-behavior with higher-order world model representations of the immersive VRE systems. A *Human–Computer Interface* (HCI) should be able to couple the human operator and the VRE as transparently as possible. VRE allow for *nopenalty training* of the personnel in a variety industrial, transportation, military, and medical applications.

There are many applications such as remote sensing and telerobotics for hazardous environments requiring complex monitoring and intervention, which cannot be fully automated. A proper control of these operations cannot be accomplished without some AR telepresence capability allowing the human operator to experience the feeling that the user is virtually immersed in the working environment. In such cases, human operators and intelligent sensing and actuator systems are working together as *symbionts*, Fig. 11.33, each contributing the best of their specific abilities.

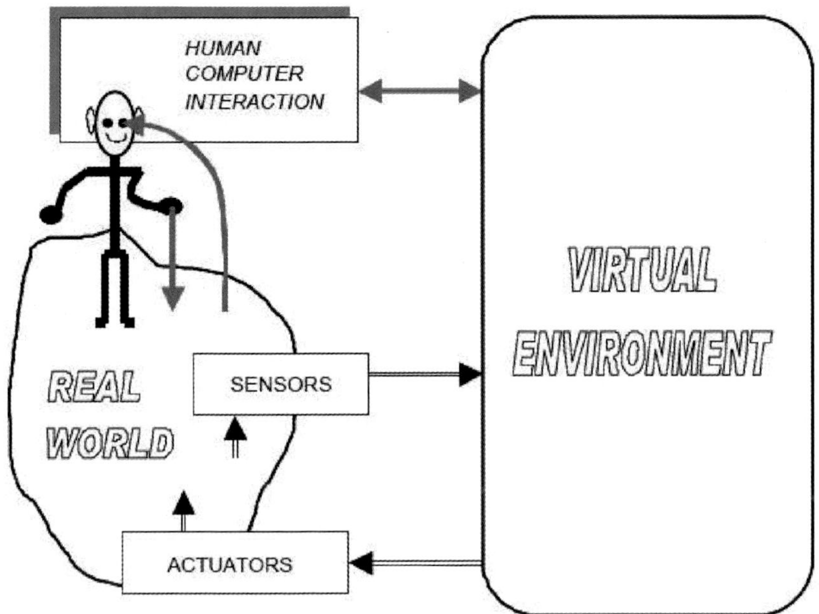

Fig. 11.33. Human–Computer–Real World interaction in the augmented VR

VR methods are also successfully used in the *concurrent engineering* design. The traditional approach to the product development is based on a two-step process consisting of a *Computer-Aided Design* (CAD) phase followed by a physical prototype-testing phase. The limitations of this approach are getting worse as the design paradigm shifts from a sequential domain-by-domain optimization to a multidomain concurrent design exercise. VR methods allow simulating the behavior of complex systems for a wide variety of initial conditions, excitations and systems configurations – often in a much shorter time than would be required to physically build and test a prototype experimentally. *Virtual Prototyping Environment* (VPE) design methods could be used to conduct interactive what-if experiments on a multidomain virtual workbench. This results in shorter product development process than the classical approach, which requires for a series of physical prototypes to be built and tested.

11.6.1 Modeling Natural Objects, Processes, and Behaviors for Real-Time Virtual Environment Applications

VREs and VPEs depend on the ability to develop and handle conformable (i.e., very close to the reality) models of the real world objects and phenomena. The *quality and the degree of the approximation* of these models can be determined only by validation against experimental measurements.

The convenience of a model is determined by its ability to allow for extensive parametric studies, in which independent model parameters can be modified over a specified range in order to gain a global understanding of the response. Advanced computation techniques are needed to reduce the execution time of the models used in interactive VPE applications when analysis is coupled with optimization, which may require hundreds of iterations. Model development problems are compounded by the fact that the physical systems often manifest behaviors that cannot be completely modeled by well-defined analytic techniques. Nonanalytical representations obtained by experimental measurements have to be used to complete the description of these systems.

Most of the object models used in VE are discrete. The objects are represented by a finite set of 3D sample points, or by a finite set of parametric curves, stored as *Look Up Tables* (LUTs). The fidelity of these discrete models is proportional with the cardinal of the finite set of samples or parametric curves. The size of the corresponding LUTs is not a matter of concern thanks to the relatively low cost of today's RAM circuits. However, the main drawback of these discrete models is the need for a supplementary time to calculate by interpolation the parameters of each point that is not a sample point. This will increase the response time of the models, which in turn will affect the real-time performance of the interactive VE. Higher efficiency models could be implemented using Neural Networks (NNs) that can learn nonlinear behaviors from a limited set of measurement data. Despite the fact that the training set is finite, the resulting *NN model has a continuous behavior* similar to that of an analog computer model. An analog computer *solves the* linear or nonlinear differential and/or integral equations representing mathematical model of a given physical process. The coefficients of these equations must be exactly known as they are used to program the coefficient-potentiometers of the analog computer's computing – elements (OpAmps). The analog computer does not follow a sequential computation procedure. All its computing elements perform simultaneously and continuously. Because of the difficulties inherent in analog differentiation, the equation is rearranged so that it can be solved by integration rather than differentiation. An NN does not require a prior mathematical model. A learning algorithm is used to adjust, sequentially by trail and error during the learning phase, the synaptic-weights of the neurons. Like the analog computer, the NN does not follow a sequential computation, all its neuron performing simultaneously and continuously. The neurons are also integrative-type processing elements.

The NN may take a relatively long time to learn the behavior of the system to be modeled, but this is not critical as it is done off-line. On the other hand, the recall phase, which is what actually counts in this type of interactive applications, is done in real-time. Due to their continuous memory, NNs are able to provide instantaneously an estimation of the output value for input values that were not part of the initial training set. Hardware NNs consisting of a collection of simple neuron circuits provide the massive computational parallelism allowing for even higher speed real-time models. The following section discusses NN techniques for real-time modeling and pattern recognition in VRE and VPE applications.

11.6.2 Hardware NN Architectures for Real-Time Modeling Applications

Random-pulse data representation was proposed by von Neuman in 1956 as a cybernetic model explaining how algebraic operations with analog variables can be performed by simple logical gates. Due to the simplicity of its circuitry, this data representation was used to build low cost instrumentation during the 60s, when the digital IC technology was still relatively expensive. There is a renewed interest in random pulse data systems as their high functional packing density makes them quite suitable for the VLSI implementation of NNs. Random-pulse data are produced by adding a uniformly distributed dither to an analog input signal V and then passing the result through a 1-bit quantizer. As variables are represented by the statistical averages of random pulse streams, the resulting data processing system has a better tolerance to noise than the classical deterministic systems. The digital technology used to implement these *random-pulse machines*, offers a number of advantages over the analog technology: modular and flexible design, higher internal noise immunity, simpler I/O interfaces. An important drawback is the relatively long time needed to get an acceptable accuracy for these statistical averages. However, the effects of this drawback can be mitigated by using a generalized multibit dithered quantization.

Generalized random data representations are produced by *multibit analog/ random-data conversion*, or dithered quantization, Fig. 11.34. The analog input V, supposed to have a relatively low variation rate, is mixed with an analog dither signal R uniformly distributed within a quantization interval, i.e. between $+\Delta/2$ and $-\Delta/2$. The resulting analog signal VR is quantified with a b-bit resolution and then sampled by a clock CLK to produce the random sequence VRP of b-bit data.

The ideal statistical average over an infinite number of samples of the random-data sequence VRP is

$$
\begin{aligned}
E[VRP] &= (k-1)p[(k-1.5)\Delta < VR < (k-0.5)\Delta] + kp[(k-0.5)\Delta \\
&\qquad < VR < (k+0.5)\Delta] \\
E[VRP] &= (k-1)\beta + k(1-\beta) = k - \beta
\end{aligned}
$$

The estimation accuracy of the recovered value for V depends on the quantization resolution Δ, the finite number of samples that are averaged, and on the statistical properties of the dither R. Because of the computational and functional similarity of a neuron and a correlator, it was found useful to consider Table 11.1 giving relative speed performance figures for correlators with different quantization levels.

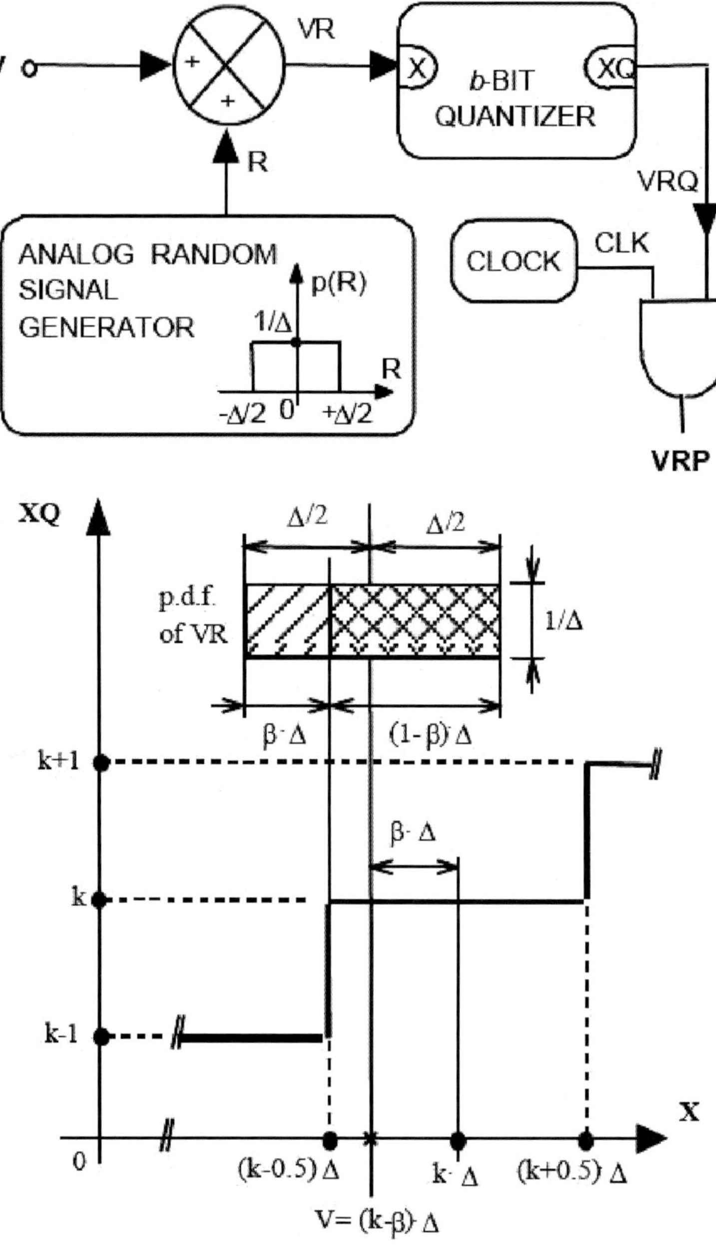

Fig. 11.34. Multibit analog/random-data conversion

Table 11.1. RMSE for different quantization levels

Quantization levels	Relative mean square error
2	72.23
3	5.75
4	2.75
...	...
8	1.23
...	...
analog	1

For instance, a basic 2-level (1-bit) random-pulse correlator will be 72.23 times slower than an ideal analog correlator calculating with the same accuracy the correlation function of two statistically independent Gaussian noise signals with amplitudes restricted within $\pm 3\sigma$. A 3-level (2-bit) correlator will be 5.75 times, and a 4-level (2-bit) correlator will be 2.75 times, slower than the analog correlator.

Based on these relative performance figures we have opted for a NN architecture using a 3-level generalized random-data representation, produced by a dithered 2-bit dead-zone quantizer. This gives, in our opinion, a good compromise between the processing speed and the circuit complexity.

Random-data/analog conversion allows to estimate the deterministic component V of the random-data sequence as an average V*N over the finite set of N random-data $\{\mathrm{VRP_i, i} = 1, 2, \ldots \mathrm{N}\}$. This can be done using a *moving average* algorithm:

$$V^*{}_N = \frac{1}{N}\sum_{i=1}^{N} VRP_i = \frac{1}{N}\left(\sum_{i=1}^{N} VRP_i + VRP_N\right)$$

$$V^*{}_N = V^*_{N-1} + \frac{VRP_N - VRP_\phi}{N}$$

While the classical averaging requires the addition of N data, this iterative algorithm requires only an addition and a subtraction. The price for this simplification is the need for a shift register storing the whole set of the most recent N random data. Figure 11.35 shows the mean square error of V*N, calculated over 256 samples, as function of the size of the moving average window in the case of the 1-bit and, respectively, 2-bit quantization.

The analog/random-data and random-data/digital conversions are illustrated in Fig. 11.36 showing a step-like analog signal x2 is converted to a sequence of random-pulses x2RQ which is then reconverted as a moving average over N = 16 random-pulses to produce the analog estimation MAVx2RQ.

One of the most attractive features of the random-data representation is that simple logical operations with individual pulses allow arithmetic operations with the analog variable represented by their respective random-pulse sequences to be carried out. This feature is still present in the case of low bit

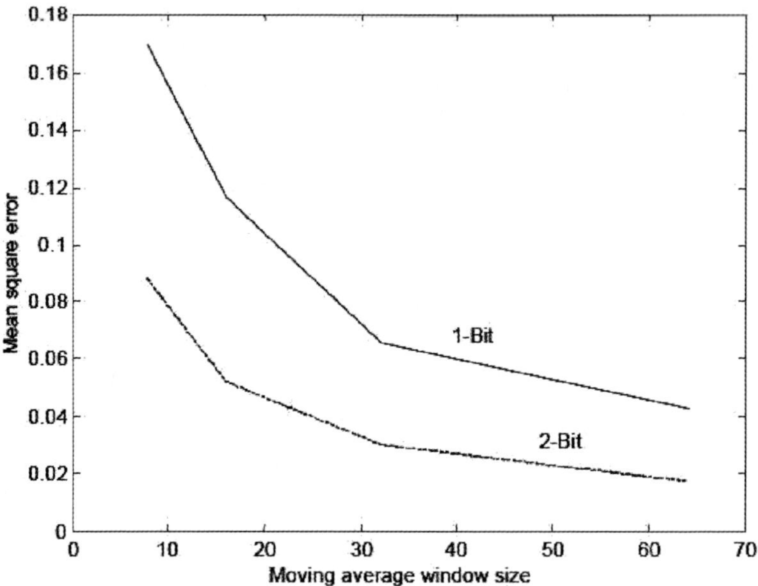

Fig. 11.35. Mean square errors of the moving average algorithm function of the size of the window

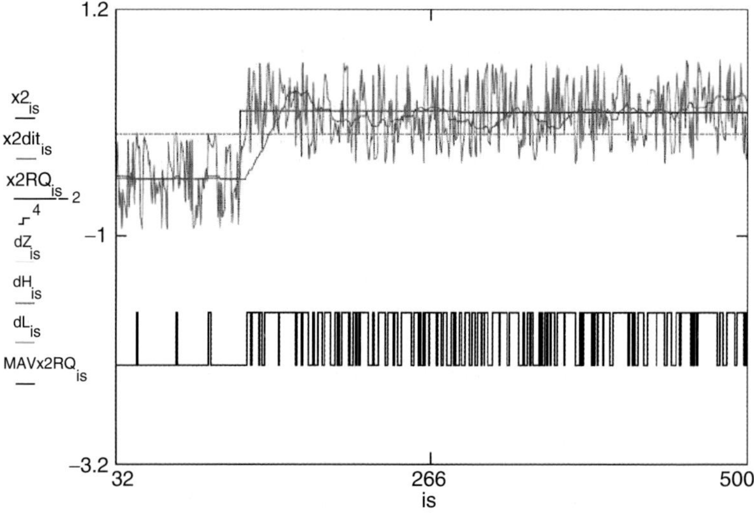

Fig. 11.36. Simulations illustrating the *analog/random -data* and *random-data/ digital* Conversions

random-data representations. The *arithmetic addition* of m signals $\{xi \,|\, .i = 1, 2, \ldots, m\}$ represented by their b-bit random data $\{Xi \,|\, .i = 1, 2, \ldots, m\}$ is be carried out, as shown in Fig. 11.37, by time multiplexing the randomly decimated incoming random-data streams. The decimation is controlled by uniformly distributed random signals $\{Si \mid i = 1, 2, \ldots, m\}$ with $p(Si) = 1/m$. This random sampling removes unwanted correlations between sequences with similar patterns. The random-data output sequence $Z = (X1 + \ldots + Xm)/m$ represents the resulting sum signal $z = x1 + \ldots + xm$.

The case of a 3-level unbiased random-data produced by a dithered 2-bit dead-zone quantizer is considered in this section. The truth table for the *multiplication* of two signed 2-bit random-data, $Z = X^*Y$ is as shown in Fig. 11.37.

Neural-network architecture using generalized random-data representation A NN hardware architecture has been developed based on the described 3-level 2-bit random data processing. Each synapse multiplies an incoming random data streams X_i, where $i = 1, 2, \ldots, m$, by a synaptic-stored weight w_{ij}, which is adjusted during the learning phase (Fig. 11.38). The positive-valued weights are considered *excitatory* and those with negative values are considered

X \ Y		0 *00*	1 *01*	-1 *10*
0	*00*	0 *00*	0 *00*	0 *00*
1	*01*	0 *00*	1 *01*	-1 *10*
-1	*10*	0 *00*	-1 *10*	1 *01*

Fig. 11.37. The resulting logic circuit for this 3-level 2-bit random data multiplier

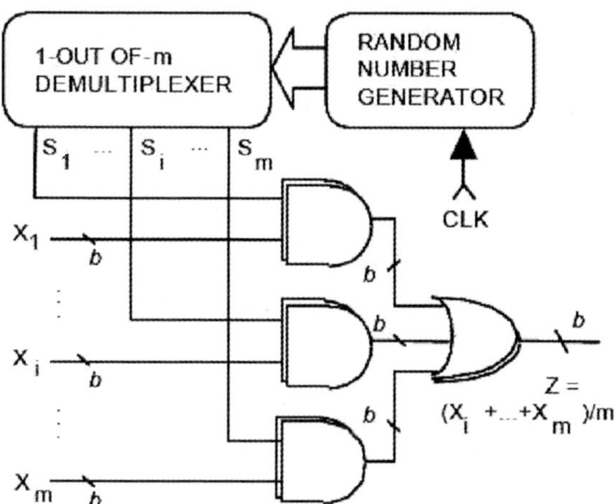

Fig. 11.38. Circuit performing the arithmetic addition of m signals represented by the b-bit random-data streams $X1, X2, \ldots, Xm$

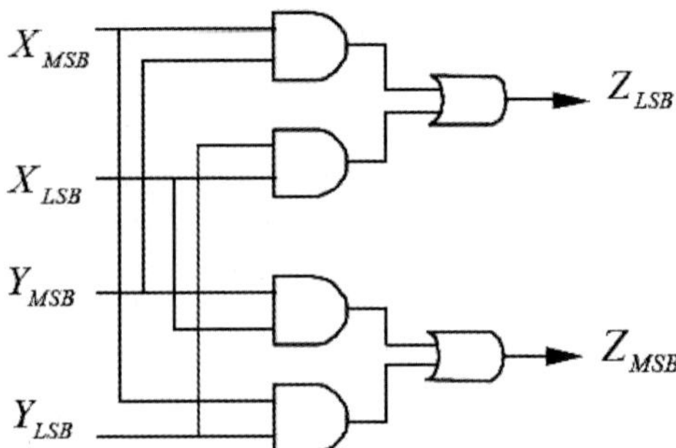

Fig. 11.39. Two-bit random-data multiplier

inhibitory. The neuron-body adds up the $\mathrm{DT}_{ij} = \mathrm{X}_i{}^*\mathrm{w}_{ij}$ signals from all the incoming postsynaptic channels. The results of this addition are then integrated by a moving average random-data/digital converter (Fig. 11.39). Since the neuron output will be used as a synaptic input to other neurons, a final *digital/random-data converter* stage is used to restore the randomness of Y_j (Fig. 11.40).

Using this 2-bit random-data neuron structure we implemented an autoassociative memory for pattern recognition applications, Fig. 11.41, which can learn input-pattern/target $\{\mathrm{Pq}, \mathrm{tq}\}$ associations:

$$\{P_1, t_1\}, \{P_2, t_2\}, \ldots. \{P_Q, t_Q\}$$

The NN is able to recognize any of the initially taught associations. If it receives an input $P = Pq$ then it produces an output $\mathrm{a} = \mathrm{t_q}$, for $q = 1, 2, \ldots, Q$. It can also recognize patterns corrupted by noise: i.e. if the input is changed $\mathrm{P} = \mathrm{P_q} + \delta$ the output will still be $\mathrm{a} = \mathrm{t_q}$.

Figure 11.42 shows as an example three training patterns, which represent the digits $\{0, 1, 2\}$ displayed as a 6×5 grid. Each white square is represented by a "-1," and each black square is represented by a "1." To create the input vectors, we scan each 6×5 grid one column at a time. The weight matrix in this case is

$$W = P_1 P_1^T + P_2 P_2^T + P_3 P_3^T$$

In addition to recognizing all the patterns of the initial training set, the autoassociative NN is also able to deal with up to 30% noise-corrupted patterns as illustrated in Fig. 11.43.

11.6.3 Case Study: NN Modeling of Electromagnetic Radiation for Virtual Prototyping Environments

The experimental VPE for *Electronic Design Automation* (EDA) developed at the University of Ottawa, provides the following interactive object specification and manipulation functions:

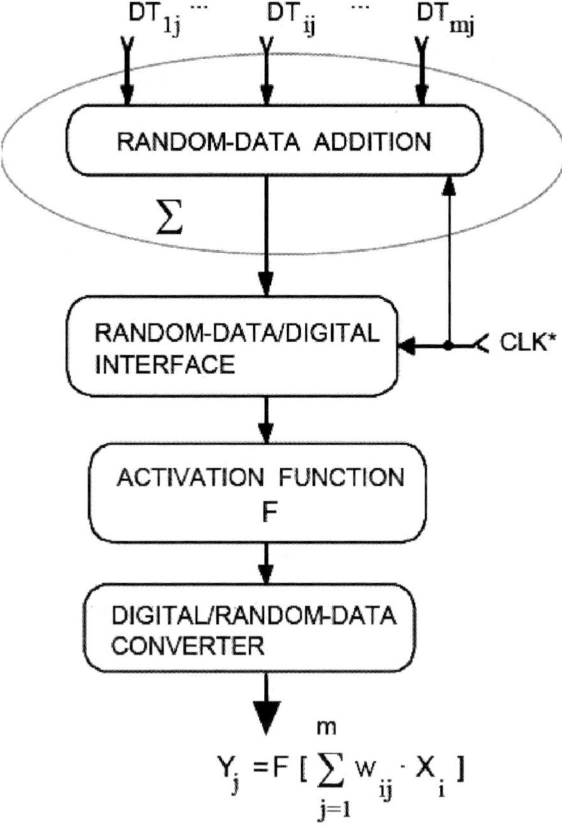

Fig. 11.40. Random-data implementation of the neuron body

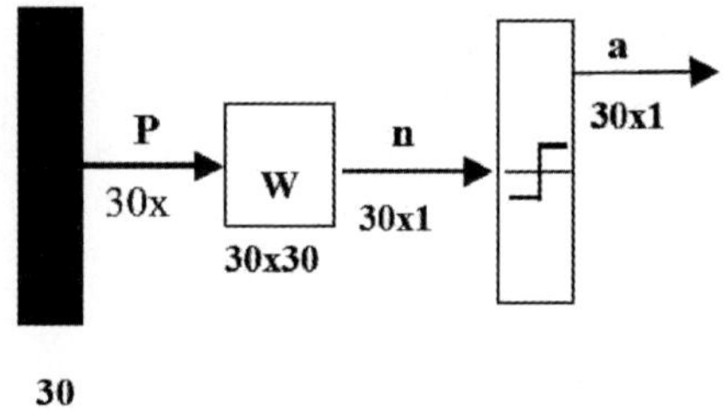

Fig. 11.41. Autoassociative memory NN architecture

(1) Updating geometric, electrical and material specifications of circuit components, as illustrated in Fig. 11.44
(2) 3D manipulation of the position, shape, and size of the circuit components and layout
(3) Accounting for 3D EM and thermal field effects in different regions of the complex electronic circuit

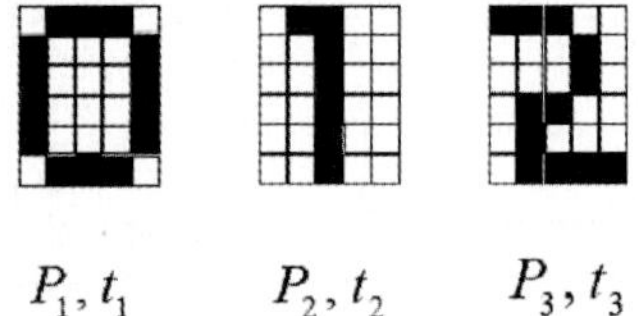

Fig. 11.42. Training set for the autoassociative NN

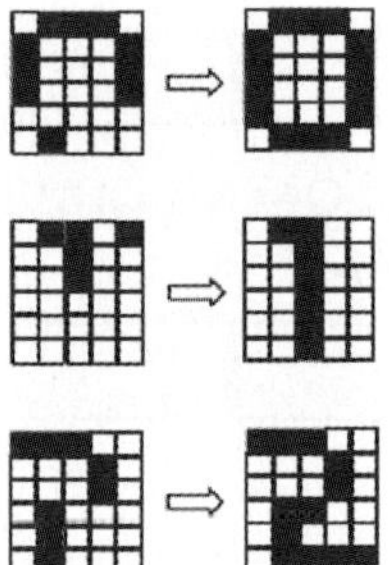

Fig. 11.43. Recovery of 30% occluded patterns by the autoassociative NN

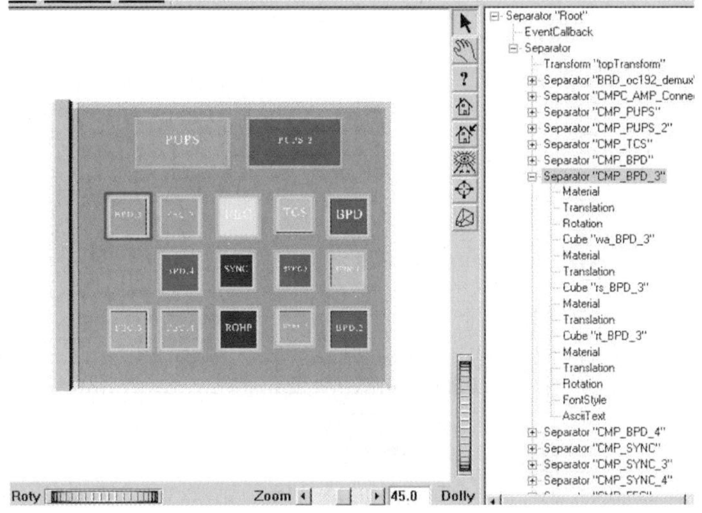

Fig. 11.44. Editing component properties in VPE

The VPE scenes are composed of multiple 3D objects: printed circuit boards (PCBs), electronic components, and connectors, Fig. 11.45. Any object in the VPE is characterized by its 3D geometric shape, material property and *safety-envelopes* defining the 3D geometric space points where the intensity of a given field radiated by that object becomes smaller than a user-specified threshold value. Each type of field (EM, thermal, etc.) will have its own safety-envelope, whereas the geometric safety-envelope is the object shape itself.

Fig. 11.45. 3D scene with a complex circuit assembly

Electronic components are placed on a PCB where they are interconnected according to functional CAD specifications and to design constraints taking in consideration the EM interference between the PCB layout components. However, this design phase does not take in consideration the interference due to the EM and thermal fields radiated by the integrated circuits and other electronic components. These problems are identified and ironed out during the prototyping phase, which may take more *what–if* iterations until an acceptable circuit placement and connection routing solution is found. Traditionally this phase involves building and testing a series of physical prototypes, which may take considerable time. Such a multidomain virtual workbench allows conducting more expediently, in a concurrent engineering manner, *what–if* circuit-placement experiments. The VPE is able to detect the collisions between the safety-envelope of the circuit currently manipulated by the manipulator dragger and the safety-envelopes of other objects in the scene. When a collision is detected, the manipulated circuit returns to its last position before the collision.

Virtual prototyping allows a designer to test the prototype's behavior under a wide variety of initial conditions, excitations and systems configurations. This results in shorter product development process than the classical approach, which requires for a series of physical prototypes to be built and tested, Fig. 11.46. A key requirement for any VPE is the development of conformable models for all the physical objects and phenomena involved in that experiment. NNs, which can incorporate both analytical representations

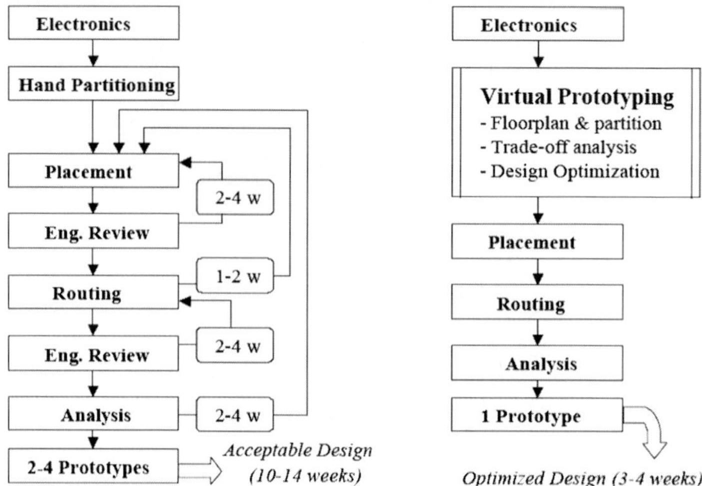

Fig. 11.46. Product design cycles for the traditional and respectively virtual prototyping

and descriptions captured by experimental measurements, provide convenient real-time models for the EM fields radiated by a variety of electronic components.

NN Modeling of EM fields

As a representative case, we consider the NN model of the 3D EM field radiated by a dielectric-ring resonator antenna, Fig. 11.47, with a ring radius a and a ring height d both with values from 4 to 8 mm in steps of 1 mm, and the dielectric constant of the ring ε_r with values from 20 to 50 in steps of 1.

The backpropagation NN using the Levenberg–Marquard algorithm consists of two input neurons, two hidden layers having 5 neurons with hyperbolic tangent activation function on each layer, and one output linear neuron. The training data set, Fig. 11.48, was obtained analytically by calculating far-field values in 3D space and frequency from near-field data using the finite element method combined with the method of integral absorbing boundary conditions. Each geometrical configuration was solved using the *Finite Element Method* (FEM) for each of the 31 dielectric constants and for 1,400 frequency steps from 2 to 16 GHz. The 200 epochs training took 55 s on a SPARC 10 workstation.

The resulting NN model, shown in Fig. 11.49, has a continuous (analog memory) behavior allowing it to render EM field values with a higher sampling resolution than that of the initial training data set. It took only 0.5 s on the same SPARC 10 workstation to render 5,000 points of the 3D EM field model.

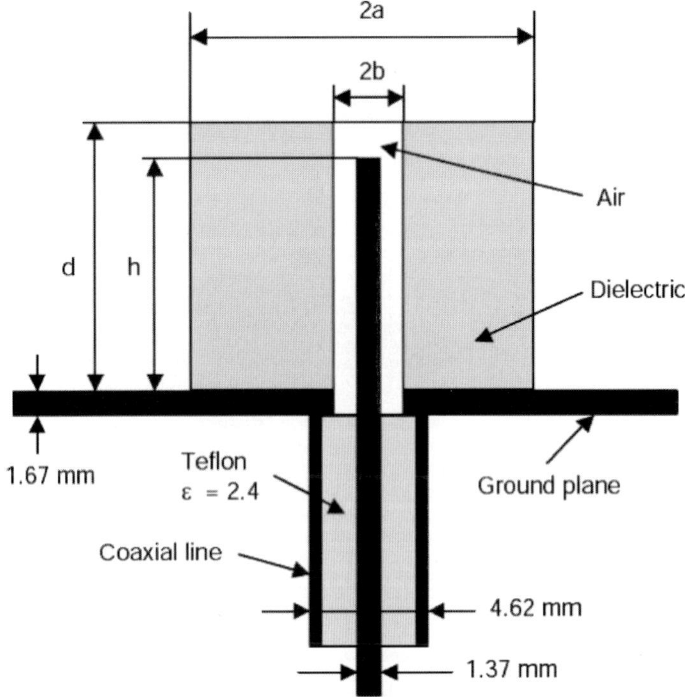

Fig. 11.47. The dielectric ring resonator antenna

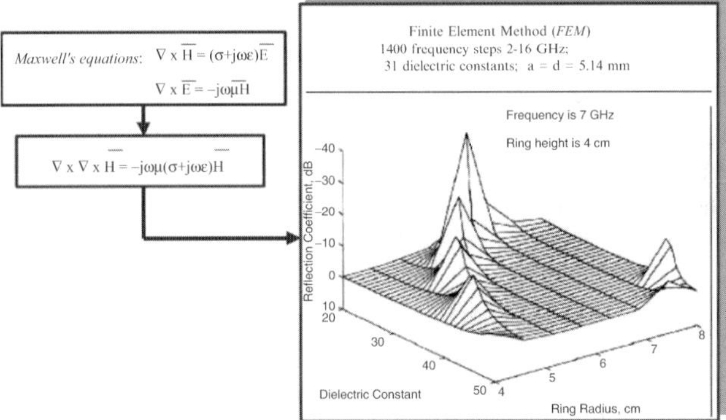

Fig. 11.48. The training data are obtained as analytically by calculating far-field values from near-field data using the finite element method

Model Validation

While the VPE idea is gaining wider acceptance, it also becomes apparent the need or calibration techniques able to validate the conformance with reality of the models incorporated in these new prototyping tools. Better experimental

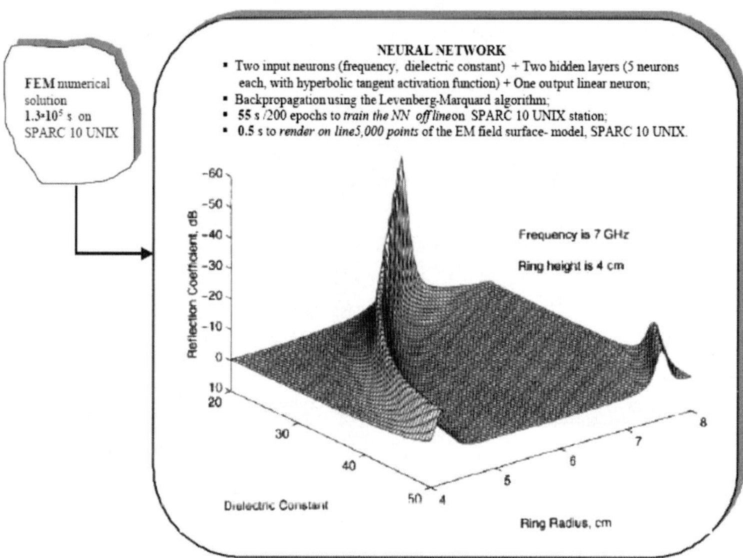

Fig. 11.49. NN model of the 3D EM field radiated by the dielectric-ring resonator antenna

test-beds and validation methodologies are needed to check the performance of the computer models against the ultimate standard, which is the physical reality. Some of the challenges to be faced are:

(1) The development of an experimental setup which should allow the desired manipulation of multidomain (geometric, mechanical, electric, thermal, material, etc.) design parameters
(2) The identification and measurement of multidomain phenomena which are considered to be behavioral characteristics for a given circuit design
(3) Finding the minimum set of experimental setups, cause–effect analytical/correlation methods, and calibration methodologies that provides a guaranty by interpolation (within acceptable error margins) the performance of the VPE computer models over wide ranges of multidomain design parameters.

The analysis in homogeneous space simplifies greatly the problem of calculating *Far Field* (FF) EM values form *Near Field* (NF) measurements. The radiating *Device Under Test* (DUT) is modeled as an array of short dipoles sitting on top of a table. The equation to solve for the electromagnetic fields is Helmholtz' wave equation:

$$\nabla^2 \vec{H} + k^2 \vec{H} = 0$$

in a homogeneous volume V bounded on one side by a surface where the magnetic field values of H are known through measurements and on the other side by the ground plane. An explicit solution allowing to evaluate the magnetic field H anywhere in the volume V from its field values and its derivatives on a surface $S1$ is given as

$$H(r') = \frac{1}{4\pi} \int_{s1} \left[G(r,r') \frac{\partial H(r)}{\partial n} - H(r) \frac{\partial G(r,r')}{\partial n} \right] dS_1$$

where $S1$ is the closed surface on which measurements are made, n is the normal to S1, and $G(r,r')$ is the free space Green's function. This algorithm is independent of the type of radiation. While it shares some sources of error with other transform algorithms, the integral transform employed here is more immune to aliasing errors than the FFT-based algorithms. Another advantage over conventional FFT transforms is that the far-field results are available everywhere and not only at discrete points.

The EM field measurement system, is shown in Fig. 11.50. It consists of a turning table with a highly conducting grounded surface on which the DUT is resting. The EM field probe can be positioned anywhere on a 90° arc of circle above the turning table. A special interface was developed for the control of the probe positioning and the collection of the measurement data via a spectrum analyzer. The turning table and the probe can be positioned as desired by steering them with position-served cables driven by motors placed outside an anechoic enclosure.

The probe positioning system and the steering cables are made out of nonmagnetic and nonconductive material in order to minimize disturbance of the DUT's fields. EM field measurements are taken on both hemispherical surfaces, providing data for the interpolative calculation of the derivative's

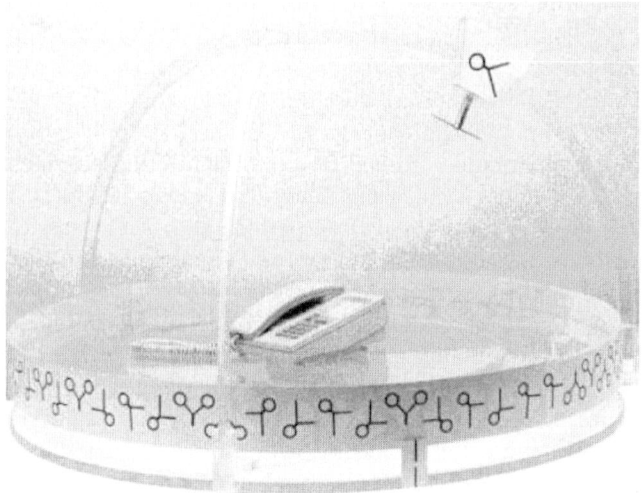

Fig. 11.50. Experimental setup for measuring the 3D EM near-field signature

variation on the surface S1. The surfaces are closed with their symmetric image halves. This is possible due to the presence of the ground plane. The actual angular positions of the table and that of the probe are measured using a video-camera placed outside the enclosure. The azimuth angle j is recovered by encoding the periphery of the turning table with the terms of a 63-bit pseudorandom binary sequence. This arrangement allows to completely identify the 3D position parameters of the EM probe while it scans the NF around the DUT.

Conclusion

VE technology has found applications in a variety of domains such as the industrial design, multimedia communications, telerobotics, medicine, and entertainment. *Virtualized Reality*™ environments, which provide more conformal representations of the mirrored physical world objects and phenomena, are valuable experimental tools for many industrial applications. VE efficiency depends on the ability to develop and handle conformable models of the physical objects and phenomena. As real world objects and phenomena more often than not manifest behaviors that cannot be completely modeled by analytic techniques, there is a need for nonanalytical models driven by experimental measurements. NNs, which are able to learn nonlinear behaviors from a limited set of measurement data, can provide efficient modeling solutions for many virtualized reality applications. Due to their continuous, analog-like, behavior, NNs are able to provide instantaneously an estimation of the output value for input values that were not part of the initial training set. Hardware NNs consisting of a collection of simple neuron circuits provide the massive computational parallelism allowing for even higher speed real-time models. As a representative case of the *Virtualized Reality*™ technology, the VPEs for EDA allow developers to interactively test complex electronic systems by running concurrently *what–if* multidomain (mechanical, electrical, thermal, etc.) experiments. For many an electromagnetic radiation problem it is not possible to derive an analytic solution. In this case, the only practical solution would be to use NN models trained by experimental measurement data. It is worthwhile to note that even when analytic solutions could be derived, the NN models still have better real time performance than classical numerical EM field methods. As the performance of the NN models can be decided only by a validation against experimental measurements, there is need for a concerted effort to characterize and catalog the EM radiated field signatures of all the integrated circuits and electronic components used for EDA.

Many real-world objects are characterized by a multitude of parameters of different nature. In order to capture this complexity, future efforts are need in order to develop composite models integrating 3D geometry, radiated fields, and other material properties. This could achieved by using a multisensor fusion system that integrates a variety of sensors covering all four phases in the perception process: far-away, near-to, touching, and manipulation.

11.7 LabVIEW Interface for School-Network DAQ Card

This application aims to create large-area extensive air shower (EAS) detector array for ultra-high energy (UHE) cosmic rays, by installing mini-arrays of scintillation counter detectors in secondary schools, in the Seattle, WA area. Data taken at individual school sites are shared via Internet connections and searched for multisite coincidence events. WALTA collaborated with CROP, a similar project based at the University of Nebraska at Lincoln, to salvage plastic scintillators, photomultiplier tubes (PMTs) and associated high voltage supplies from surplus equipment at the CASA detector site at Dugway, UT.

Individual detector stations each consist of four scintillation counter modules, front-end electronics, and a GPS receiver, as shown in Fig. 11.51. Preliminary training of secondary school teachers and users was conducted using obsolete NIM crates and fast electronics (discriminators, coincidence, and scaler modules) loaned from Fermilab. These modules are now being replaced by the new DAQ cards, which add GPS timing and a simple RS232 computer interface. The QuarkNet DAQ card provides a low-cost alternative to standard particle and nuclear physics fast pulse electronics modules.

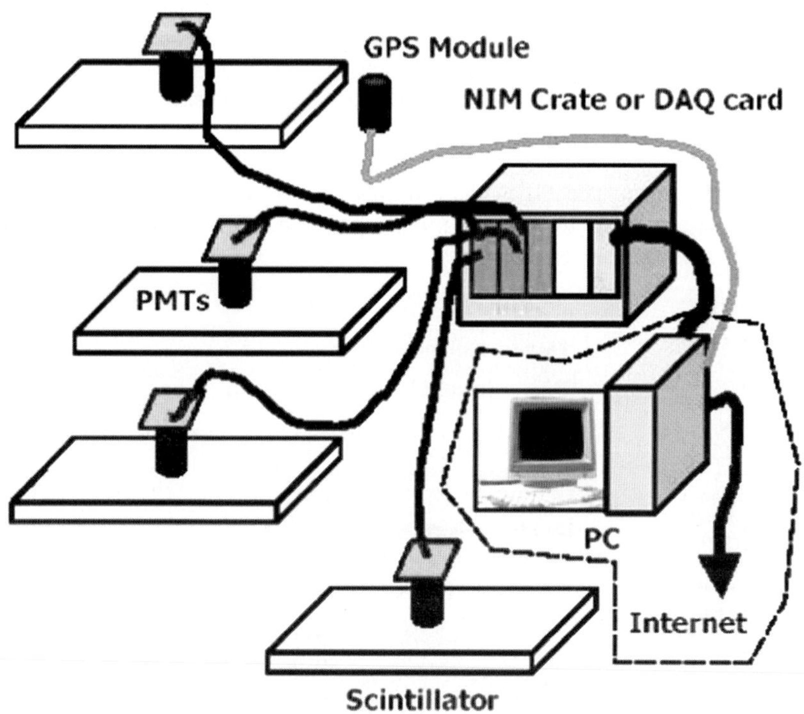

Fig. 11.51. School-network air shower detector stations. CROP and WALTA use PMTs and counters salvaged from the CASA experiment

The board, which can be produced at a cost of less than US$500, produces trigger time and pulse edge time data for 2 to 4-fold coincidence levels, via a universal RS232 serial port interface, usable with any PC. Details of the DAQ card have been presented elsewhere. The DAQ cards produce a stream of data in the form of ASCII text strings, in distinct formats for different functions. However, the highly compact data format, using hexadecimal coded data items without user-friendly prompts, makes use of the card via a simple terminal window rather challenging for typical users.

The user's host computer, a desktop PC which may have a Windows, Mac or Linux OS, is connected to the DAQ card via a standard RS232 serial cable. For schools that have relatively new Macintosh models, with USB ports only, a commercial USB-to-RS232 adapter can be inserted. The board and GPS unit are powered by a conventional 5VDC, 800 mA power adapter of the type available at consumer electronics stores.

Initial installations used NIM crates borrowed from Fermilab PREP, which will be replaced by the DAQ cards described in this application. The simplest way to operate the board is by opening a terminal window linked to the PC's serial port. Figure 11.52 shows a typical terminal screen during data-taking. Commands can be directly entered via the keyboard, and the resulting output viewed from the screen.

Typically, the user would enter commands to define the trigger logic level, and enable counting. At the end of the desired time interval, counting can be disabled and the trigger count read out. For each trigger, the display shows one or more lines of hexadecimal encoded data, providing time of trigger, pulse leading and trailing edge times relative to the trigger time, and information needed to determine the GPS time of the trigger to 24 ns precision. Other commands can be used to directly interrogate the GPS module, or act upon the counter registers, etc.

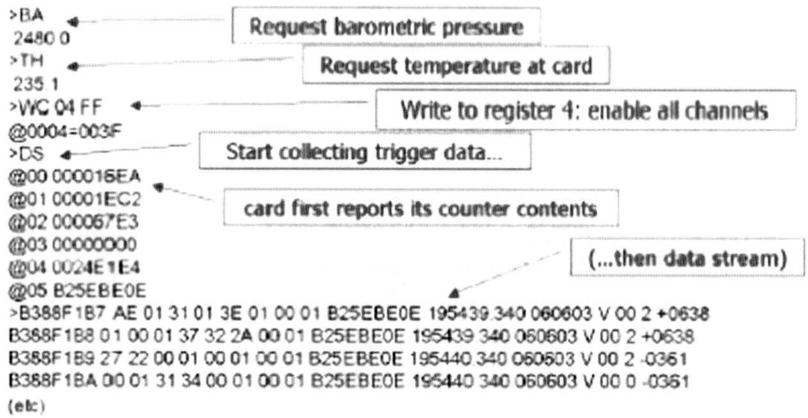

Fig. 11.52. Typical terminal window, showing raw data stream from the Quark Net DAQ card

Commands and responses illustrate reading the on-board barometer and temperature sensor, reading scalar contents, and the data stream produced while logging coincidences. These data serve as input to the LabVIEW module. Users may also prepare a script to implement a sequence of commands, or compile custom software to operate the card directly.

11.7.1 The WALTA LabVIEW Interface

In this application, an interface to the QuarkNet DAQ card using the National Instruments LabVIEW environment is developed. This interfacing environment is available at low cost to educational institutions, has a very large user base worldwide, and provides a simple graphical user toolkit with which it is possible to construct highly capable real-time programs. The large base of applications and program elements makes it easy for users to acquire help and training.

The QuarkNet LabVIEW interface software allows the user to send setup and operating commands to the card, and to log data of various types. Log files can be output in MS Excel compatible format, for direct handling by users who do not have programming skills. The GUI is divided into pages with different functions, for setup, housekeeping data, routine data-taking, etc. Figure 11.53 shows the configuration window, which can be opened from any tab of the GUI. Figures 11.54–11.56 show examples of the console, GPS record and event timing windows.

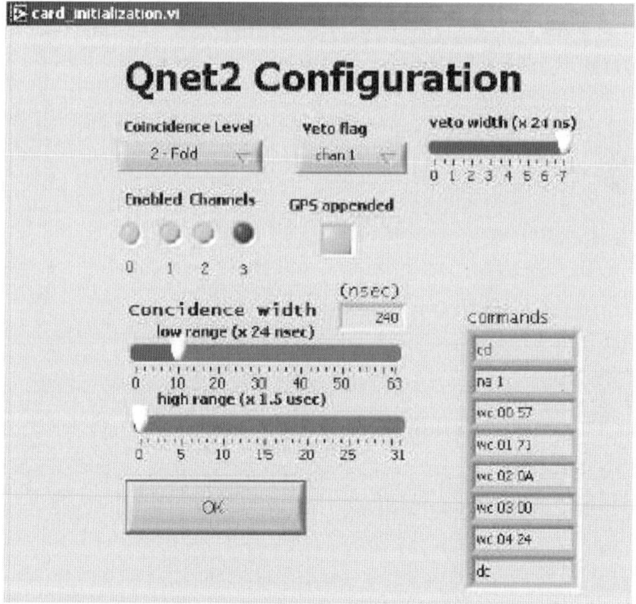

Fig. 11.53. DAQ card configuration window

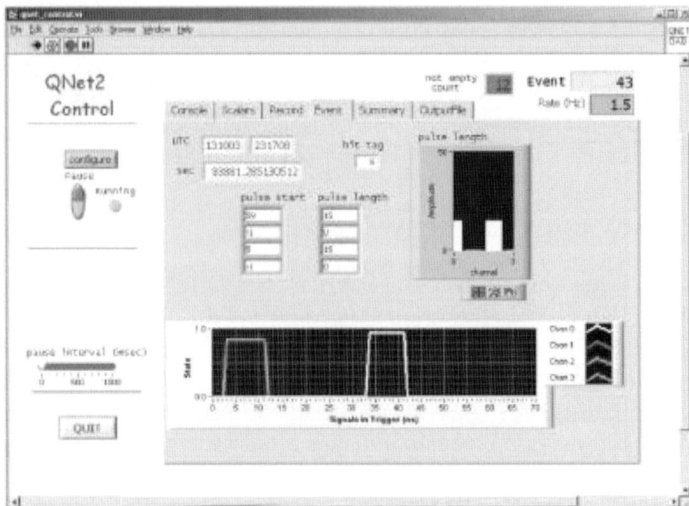

Fig. 11.54. Event timing data window

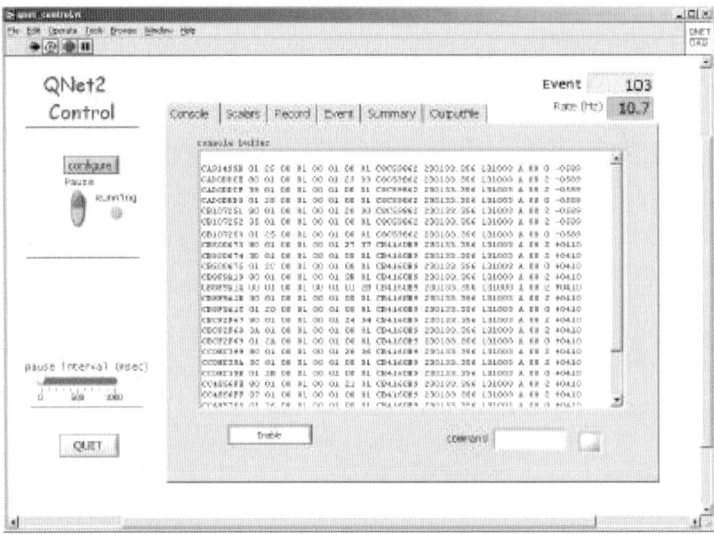

Fig. 11.55. "Console" data tab, showing raw hexadecimal data stream and allowing manual input of commands

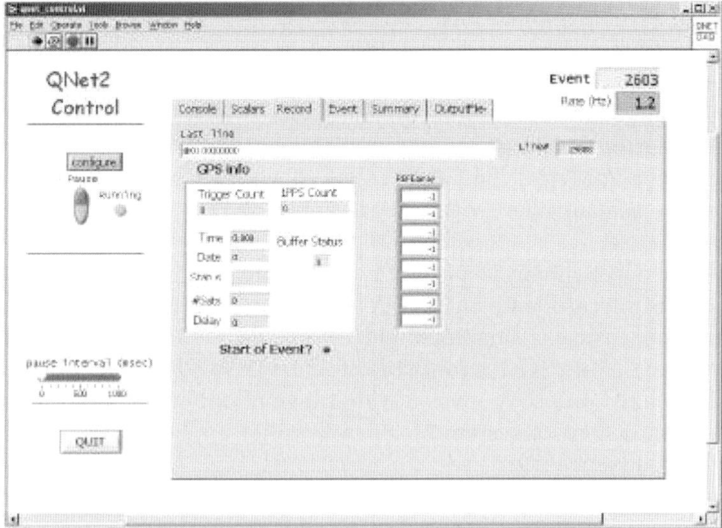

Fig. 11.56. GPS record data tab

11.8 PC and LabVIEW-Based Robot Control System

This application presents a new economical solution of robot control system for different sophisticated robot applications. The control system is based on PC controlled servomotor control card and an intelligent control software, which has been developed using high level graphical programming language (LabVIEW). The basic development is an interface software for making connection between the control card and LabVIEW. LabVIEW gives a wide range of opportunity of utilization of the developed control system at different robot applications. It shows a complete solution of robot control system for a ZIM 15 type 6-axis robot. The Digital Differential Analysis (DDA) method and the closed loop control system of the servomotor control card are described. The programming of the control card and the time optimal trajectory planning method are also presented.

11.8.1 Introduction to Robot Control System

The advent of new high-speed technology and the growing computer capacity provided realistic opportunity for new robot controls and realization of new methods of control theory. This technical improvement together with the need for high performance robots created faster, more accurate and more intelligent robots using new robot control devices, new drives, and advanced control algorithms. On the other hand – parallel with the need for high-performance robots – another need has appeared: the need for low cost robot control.

This application describes a new economical solution of robot control systems. The presented robot control system can be used for different sophisticated robot applications. The control system consists of a PC, two servomotor control cards, a digital I/O card, drivers of the servomotors of the robot and an intelligent control software, which has been developed using high level graphical programming language (LabVIEW). A complete solution of a robot control system is presented in this paper. The control system was developed for a ZIM 15 type 6-axis robot shown in Fig. 11.57, but any kind of robots can be controlled with the presented control system, which utilizes servomotors.

The basic development is an interface software for making connection between the control card and LabVIEW software. This interface software has been developed using C language and integrated with LabVIEW using the Code Interface Node (CIN) function of LabVIEW. The advantage of this approach is that LabVIEW gives a wide range of opportunities of utilization of the developed control system at different robot applications.

Only few advanced features are mentioned, e.g., easy use of the TCP-IP communication between the robots and the Flexible Manufacturing Cell (FMC), built-in functions allowing fast numerical calculations and easy implementation of different control methods and trajectory planning algorithms.

11.8.2 The Robot and the Control System

The robot control system was developed for ZIM 15-type 6-axis robot. Figure 11.57 shows the 6-axis robot and Fig. 11.58 describes the structure of the robot control system. Focusing on the actuators and sensors of the robot

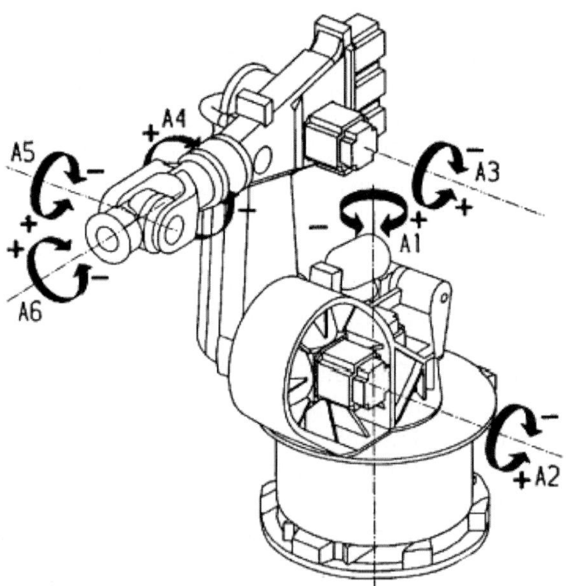

Fig. 11.57. 6-axis robot

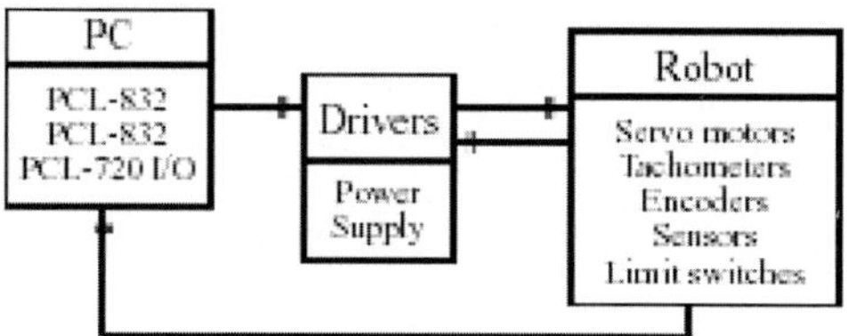

Fig. 11.58. The structure of the robot control system

hardware, each axis of the robot has its own servomotor, tacho-generator, encoder and two end-position sensors. It is important to mention that the sixth axis is not linked to the robot itself. The sixth axis is a rotating table located near the robot inside the robot workspace. The hardware of control system consists of a PC (an ordinary PC 486DX4/100), two PCL-832 servomotor control cards, a PCL-720 digital *I/O* card, six amplifier-drivers of the robot servomotors and a power supply.

11.8.3 PCL-832 Servomotor Control Card

Two servomotor control cards (PCL-832) are inserted into the control PC. Each card can control three axes simultaneously, and each axis has its own position control chip, allowing completely independent control. The card can supply a simulated tachometer output to the servomotor driver. This signal makes a tachometer unnecessary in some applications, reducing overall system costs. A special synchronization circuit synchronizes all the axes, called Digital Differential Analysis (DDA). In the described control system two PCL-832 cards are used to control all six axes simultaneously. The two PCL-832 servomotor control cards work in such a way that one of them is selected as the master card and the other as the slave card.

11.8.4 Digital Differential Analysis (DDA)

In order to obtain synchronization in a multijoint servomotor system, all axes should move simultaneously. The principle of DDA method is that all axes start sending position commands simultaneously (T1), and stop sending these commands simultaneously (T2) as well. The duration (T2–T1) is defined as one DDA cycle as shown in Fig. 11.59. The duration of the DDA cycle can be set (by the software) from 1 ms to 2 s as the application requires. For every pulse output the servomotor driver will advance the servomotor one step. One pulse therefore represents one position-command. In one DDA cycle the pulse number can be set from 0 to 4,095. The number of pulses in one DDA cycle represents the total position change possible by that DDA cycle. The continuous pulse sequence output to the servomotor driver ensures that a smooth position response is obtained. After the determination of the direction of the motion and the pulse number, the chip itself will control the motion.

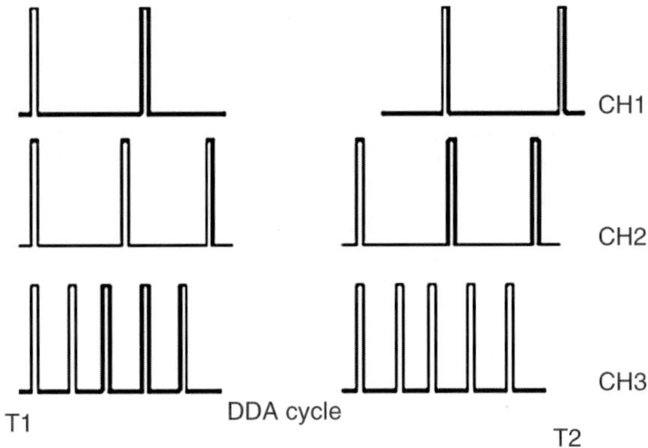

Fig. 11.59. DDA cycle

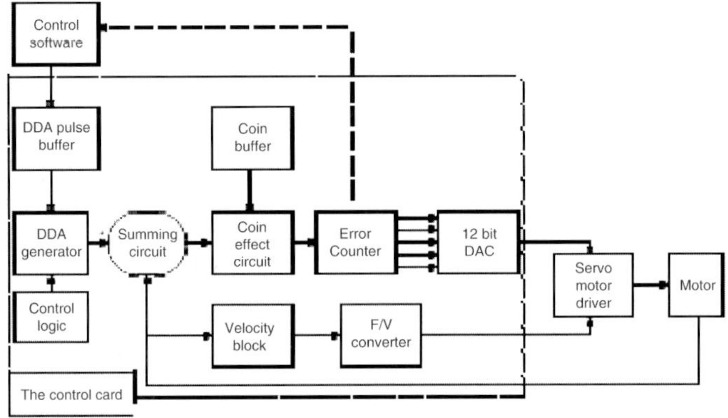

Fig. 11.60. The structure of closed-loop position control

11.8.5 Closed-Loop Position Control of the Control Card

The servomotor control card uses proportional closed-loop position control to obtain reliable and accurate results. It features an internal velocity feedback loop and an offset technique to compensate for steady-state error that is caused by using small values on P controller. This is a general control technique in robot control, called PD control. The functional block of the closed loop control of the servomotor control card is shown in Fig. 11.60. Physically the servomotor control card can be found inside the dash-dot area. The control software calculates the pulse numbers, then allocates these numbers into the DDA pulse buffer.

The DDA generates continuous command pulses, and these pulses are fed into a summing circuit, together with the pulses generated by the servomotor encoder device. The summing circuit determines the difference between both signals and feeds the result signal into the P pulse-offset controller.

This module has a programmable gain (Kp). This is the proportional gain of the control loop. After this calculation the signals' pulses are fed into the error counter, which drives the DAC chip in real-time. The velocity block is provided in the motion control chip. Its purpose is to add a velocity feedback loop in the whole system through a frequency-to-voltage (F/V) converter. The maximum input frequency to the F/V converter is 250 kHz. The simulated tachometer output gives −10 V to +10 V. This internal loop improves the motion dynamics of the servomotor system. This block provides the derivative part of the closed-loop control system. Similar development work is outlined earlier researches, where a LabVIEW-based control system for PUMA 560 is described.

11.8.6 Modified Closed-Loop Position Control of the Control Card

Basically the described servomotor control card utilizes a simple PD control as it is widely used in simple robot control. To improve the accuracy and the motion dynamics of the servomotor and to make the control system open for the user, another feedback is proposed by the author. In Fig. 11.60 the dashed lines show the new proposed feedback. Physically the control card utilizes two interrupts to make fast connection between the control card and the computer. One interrupt is used for the DDA pulses, and the other is used for the Error Counter. The Error Counter drives directly the DAC chip in real-time.

The information about the pulse numbers in the Error Counter is very useful and important for control purposes. The modified control algorithm can receive the real pulse numbers from the Error Counter using the second interrupt and a new subroutine of the control software. This new subroutine has been developed for the control system to make modification – regarding to the transient motion – in pulse numbers. The modification has to be done to less than one DDA cycle that is determined beforehand. Then the modified pulses are fed into the DDA generator. This real time operation is still under development.

11.8.7 Programming of the Control Card

The basic development of the control software involves an interface software for making connection between the control card and the LabVIEW software. This software contains the basic functions of driving the control card. The LabVIEW software has a Code Interface Node function (CIN), which enables to implement compiled C code into LabVIEW. Figure 11.61 presents the utilization of CIN in LabVIEW. Once the driver-software of the control card is implemented in LabVIEW, the user can build up several sophisticated control algorithms utilizing the advantages of LabVIEW programming. The control software consists of three main modules. The path-planning module is shown in Fig. 11.62, the trajectory-planning module is presented in Fig. 11.63 and the control module is given in Fig. 11.64. These figures are presented here only for demonstration, showing the form of the communications windows.

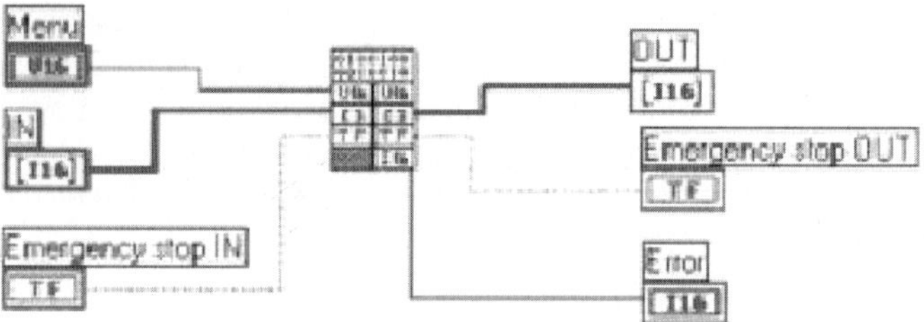

Fig. 11.61. The CIN function in LabVIEW

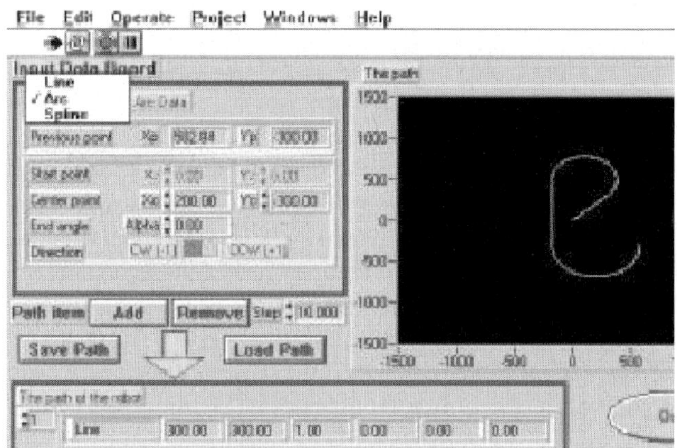

Fig. 11.62. The path planning module

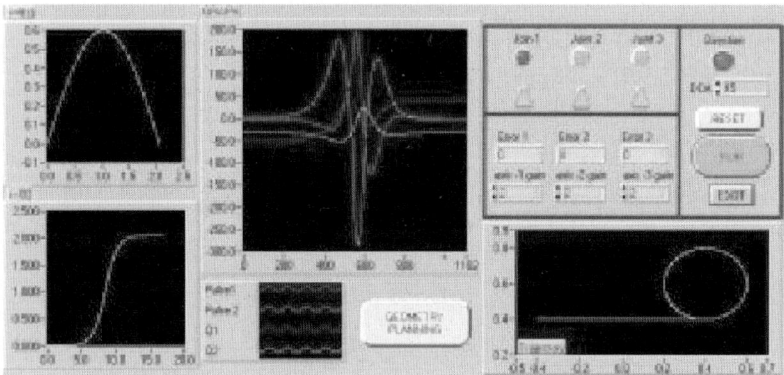

Fig. 11.63. The trajectory planning module

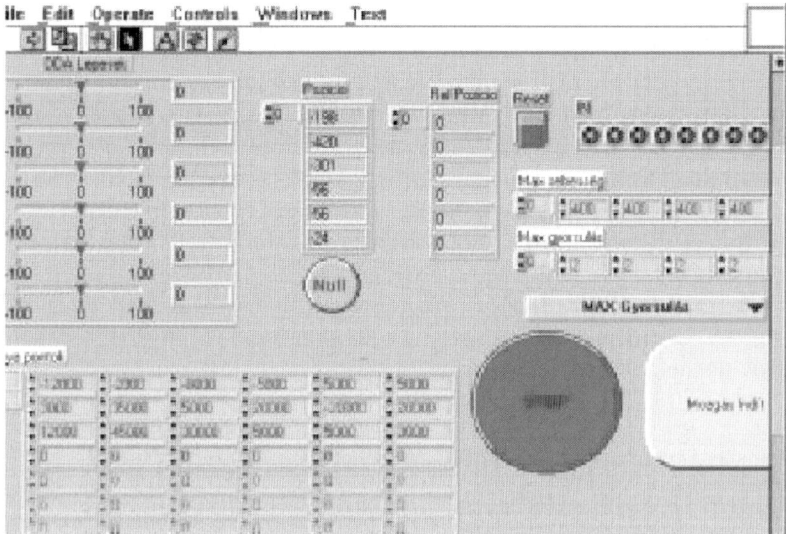

Fig. 11.64. The control panel of the control software

11.8.8 Optimal Cruising Trajectory Planning Method

The control system is capable of using several sophisticated trajectory planning methods. This application uses the time optimal trajectory planning method for CPC. The time optimal trajectory planning method is used for the cruising part of the constrained path, where the change of speed is slow in comparison with the transient motion. On the transient part of the path the robot uses the maximum acceleration to reach the desired speed value that the cruising trajectory planning method determines.

The minimum time cruising trajectory planning method is based on a simple idea. The method uses the maximum absolute value of the speed in every point of the path. This maximum is determined by the limit speed of the joints. Using the time optimal trajectory planning method the time history of the motion along the path for the cruising part is obtained. This is extremely useful for application planning because in a simple way a clear picture can be obtained about the motion features. It is also important that the generalization of the results is possible.

There is a question which is decisive for the problem such as the proportion, the role of the cruising and the transient motion in the motion as a whole. If the speed limits are low and acceleration is high, it is a good solution to use the minimum-time cruising trajectory planning as a basic procedure and complete it by the standard or special investigation of the transient part of the motion. For the transient part of the motion it turns out that using the parametric approach a similar procedure is possible as for the minimum-time cruising. For the transient part the absolute value of the acceleration in every point of the path should be maximized. The maximum is determined by the maximum of the possible force(torque) at one of the joints.

The presented robot control system utilizes a PC-based commercial servomotor control card and a LabVIEW-based intelligent control software (developed by the author). This implementation provides high robot performance at an affordable price. The basic development of the software enables the user to develop sophisticated control method for different robot applications with the utilization of the advantages of LabVIEW programming. The author proposed an additional feedback to improve the motion dynamics of the servomotors. The optimal cruising trajectory planning method is presented as one of the sophisticated trajectory planning methods that can be implemented using the control system.

11.9 Mobile Robot Miniaturization: A Tool for Investigation in Control Algorithms

Today the mobile robotics field receives great attention. There is a wide range of industrial applications of autonomous mobile robots, including robots for automatic floor cleaning in buildings and factories, for mobile surveillance systems, for transporting parts in factories without the need for fixed installations, and for fruit collection and harvesting. These mobile robot applications are beyond the reach of current technology and show the inadequacy of traditional design methodologies. Several new control approaches have been attempted to improve robot interaction with the real world aimed at the autonomous achievement of tasks. An example is the subsumption architecture proposed by Brooks which supports parallel processing and is modular as well as robust. This approach is one of the first solutions systematically implemented on real robots with success. Other researchers propose new computational approaches like fuzzy logic or artificial NNs.

The interest in mobile robots is not only directed toward industrial applications. Several biologists, psychologist and ethologists are interested in using mobile robots to validate control structures observed in the biological world. Franceschini uses a robot to validate the structure of the retina observed on a y, Beer to replicate the mechanism that coordinates leg movements in walking insects, Deneubourg to get a better understanding of collective behavior in ant colonies. All these research activities are based on mobile robot experimentation. A simpler way to validate control algorithms is to use simulations, but the simplifications involved are too important for the results to be conclusive. The control algorithm embedded in the robot must consider its morphology and the properties of the environment in which it operates. Real world features and anomalies are complex and difficult to modelize, implying that the experimentation of control algorithms through simulation can only be used in preliminary study but cannot prove the success of the control algorithm in the real world. The sole way to validate an algorithm to deal with these problems is to test it on a real robot. Many robots have been designed to perform experiments on control algorithms but only a few make cost effective experiments possible. Brooks has designed several robots with effective electronics and mechanics. The control algorithms are programmed in the subsumption behavioral language, taking into account software modularity, and real-time and parallel processing. Unfortunately, during experiments, only a few tools are available to improve the understanding of the control process. Moreover, the custom programming language makes code portability and algorithm difficult. Steels uses a video-camera to record robot actions during experiments but all the data concerning the robot control process are available only at the end of the experiment. Other platforms, such as the Nomad robot, make real-time interaction possible via a radio link, and have standard programming languages, but the size of the robot and the environment it requires make experimentation uncomfortable. The lack of a good experimentation mobile robot for single-robot experiments, means that it is impossible today to perform collective-behavior experiments. The programming environment and the real time visualization tools are totally insufficient for this purpose.

The development of the miniature mobile robot Khepera addresses the problems mentioned above. Its hardware is designed so that it is small enough for the operation of several at the same time and in small areas, for example on a desktop. Modularity allows new sensors and actuators to be easily designed and added to the basic structure. A versatile software structure is provided to help the user to debug the algorithms and to visualize the results.

11.9.1 Hardware

Miniaturization is an important challenge for industry: CD players, computers, video-cameras, watches and other consumer products need to implement many functionalities in a small volume. In the robotics field many applications need small actuators, small teleoperated machines or tiny autonomous robots. Dario gives a comprehensive description of the research field and of the available technology. In the Khepera design, miniaturization is the key factor in making cost-effective experimentations possible both for single or multiple robot configurations.

Generalities

The robot presented in this paper is only a first step in the direction of miniaturization. They measure no more than a few cubic centimeters, generate forces comparable to those applied by human operators and incorporate conventional miniature components. The next miniaturization step needs special fabrication technologies, today in development. The Khepera robot shown in Fig. 11.65 uses electronic technology available today: the new family of 683xx microcontrollers from Motorola makes the design of complete 32 bit machines

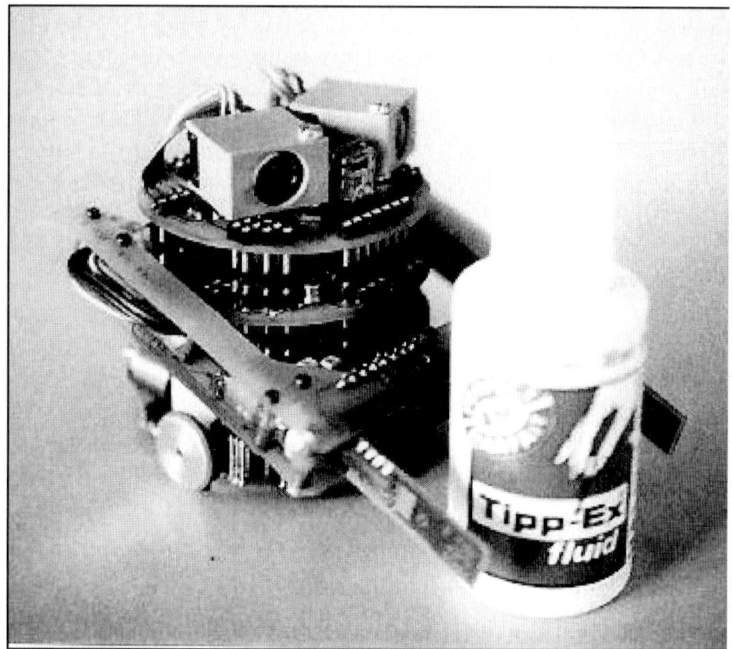

Fig. 11.65. The Khepera robot

extremely compact. Surface mounted devices (SMD) allow an important increase in component density on printed circuit boards. New compact sensors, including some signal preprocessing on the sensing chip, reduce the need of additional circuitry. Only the mechanical parts (wheels, gears, manipulator) are built expressly for Khepera, as well as the magnetic sensors for counting the wheel revolutions. The design of such miniaturized robots demands a great effort spanning several fields. The result is a complex compromise between functionalities to be implemented, available volume, current technology, power requirements, etc. Khepera is composed of two main boards (Fig. 11.66).

Application-specific extension turrets for vision, for interrobot communications, or which are equipped with manipulators can be directly controlled via the Khepera extension busses. Khepera can be powered by an external supply when connected for a long time to a visualization software tool; however, onboard accumulators provide Khepera with 30 min of autonomous power supply.

Distributed Processing

One of the most interesting features of Khepera is the possibility of connecting extensions on two different busses. One parallel bus is available to connect simple experimentation turrets. An alternative and more sophisticated interface scheme implements a small local communication network; this allows the

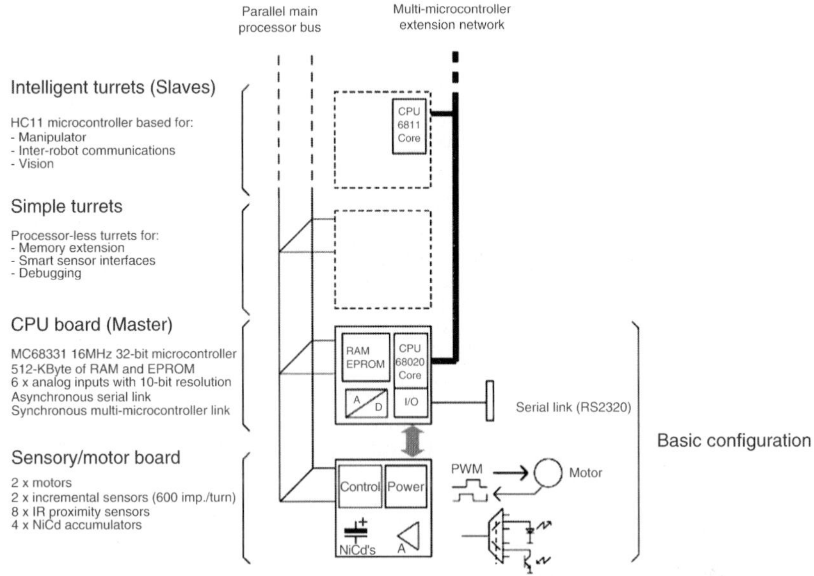

Fig. 11.66. Khepera hardware architecture

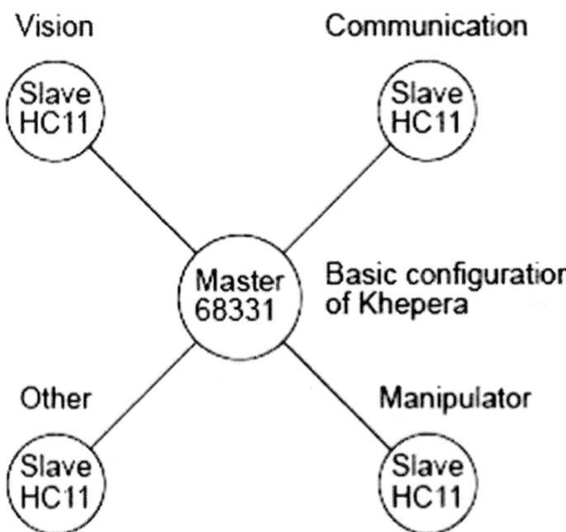

Fig. 11.67. Khepera communication network topology

connection of intelligent turrets (equipped with a local microcontroller) and the migration of conventional or neural preprocessing software layers closer to the sensors and actuators. This communication network (Fig. 11.67) uses a star topology; the main microcontroller of the robot acts as a master (at the center of the star). All the intelligent turrets are considered as slaves (on the periphery of the star) and use the communication network only when requested by the master. This topology makes it possible to implement distributed biological controls, such as arm movement coordination or feature extraction and preprocessing in the vision, as observed in a large number of insects. The multi-microcontroller approach allows the main microcontroller of Khepera to execute only high level algorithms; therefore attaining a simpler programming paradigm.

Basic Configuration

The new generation of Motorola microcontrollers and in particular the MC68331 makes it possible to build very powerful systems suitable for miniature neural control. Khepera takes advantage of all the microcontroller features to manage its vital functionality. The basic configuration of Khepera is composed of the CPU and of the sensory/motor boards.

The CPU board is a complete 32 bit machine including a 16 MHz microcontroller, system and user memory, analog inputs, extension busses, and a serial link allowing a connection to different host machines (terminals, visualization software tools, etc.). The microcontroller includes all the features needed for easy interfacing with memories, with I/O ports and with external interruptions. Moreover, the large number of timers and their ability to work in association with the I/O ports indicate that this device is the most important component in the design. The sensory/motor board includes two DC motors coupled with incremental sensors, eight analog infrared (IR) proximity sensors and on-board power supply. Each motor is powered by a 20 kHz pulse width modulation (PWM) signal coming from a dedicated unit of the microcontroller. These signals are boosted by complete four-quadrant NMOS H bridges. Incremental sensors are realized with magnetic sensors and provide quadrature signals with a resolution of 600 impulsions per wheel revolution. IR sensors are composed of an emitter and of an independent receiver. The dedicated electronic interface is built with multiplexers, sample/hold's and operational amplifiers. This allows the measurement of the absolute ambient light and the estimation, by reaction, of the relative position of an object from the robot.

Additional Turrets

To make experiments involving environment recognition, object detection, object capture, and object recognition possible, two intelligent turrets have been designed and built: one for stereoscopic vision, the other containing a manipulator. The stereoscopic vision turret employs two 64 pixel linear photoelement arrays and a dedicated optical element. The analog value of each pixel is coded on 16 gray levels. To obtain useable data under a wide spectrum of enlightenment conditions, an additional sensor is used to perform as an automatic iris: the integration time necessary for the photoelement arrays is controlled by intensity of the ambient light. Mondada proved the validity of this stereoscopic vision in robot navigation (spatial frequency filtering was used in obstacle detection and avoidance).

The manipulator turret makes Khepera capable of an interaction with objects of its environment. Two DC motors (Fig. 11.68) control the movements

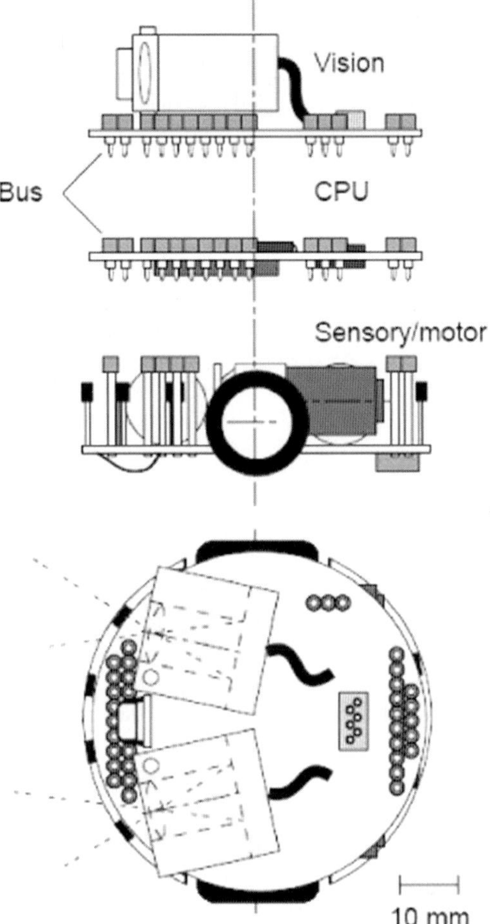

Fig. 11.68. Khepera physical structure: basic sensory/motor, CPU, and vision boards

of the manipulator (elevation and gripping). Different classes of objects can be detected by the gripper sensors which measure sizes and resistivities. Robots displaying collective behavior need means to perform interrobot communications and localization.

11.9.2 Software

Managing all the Khepera resources is a complex task. The large number of asynchronous events to control, and the necessity to share some critical interfaces led to the development of a complete low-level software organized as a collection of basic I/O primitives

Hierarchical Software Structure

The multi-microcontroller approach and the complex tasks to manage required a hierarchical approach to the software structure. The concept chosen applies when intelligent turrets (equipped with a microcontroller) are used. Two software structures are implemented: a single high-level application program and a number of stand-alone local processes (Fig. 11.69). Stand-alone local processes (e.g., for IR sensor sequencing, motion control, wheel incremental-sensor counting, etc.) are executed cyclically according to their own event timer and possibly in association with external interruptions. The high-level application software run the control algorithm and communicate with the stand-alone local processes via a mailbox mechanism. This decoupling of low- and highlevel tasks makes the development of complex applications quick and easy.

Control of Khepera

Experiments with Khepera are performed in two different ways: by running algorithms on autonomous robots or in connection with visualization software tools. As already mentioned, the details of the basic input/output activities are managed through a library of stand-alone processes. During the development, the standard RS232 link is used, through a generic high level protocol, to communicate with these processes from a workstation. The application software is therefore run on the workstation and calls to the basic primitives make it possible to monitor the robot activity possible. All standard and specialized visualization tools can be employed to simplify the control algorithm debugging.

Because the application software is written in standard C language, debugged algorithms can easily be converted to run on the Khepera CPU. Applications can be downloaded to Khepera and the robot becomes autonomous from the development environment.

11.9.3 Experimentation Environment

The quality of the measurements obtained in robot experiments and the efficiency of the whole experimentation process critically depends on the structure of the working environment. Tools currently available for simulation are far better developed than those used for experimenting with real robots. The real time interaction with the control process and the continuous visualization of the parameters make possible a faster and better understanding of the mechanisms involved. For these reasons, it is necessary to develop better visualization and interactive software tools adapted to the experimentation tasks. The simplest way to implement a comprehensive graphical interface is to use a scientific workstation. This must be connected to the robot to collect the data for display and to communicate the orders coming from the user.

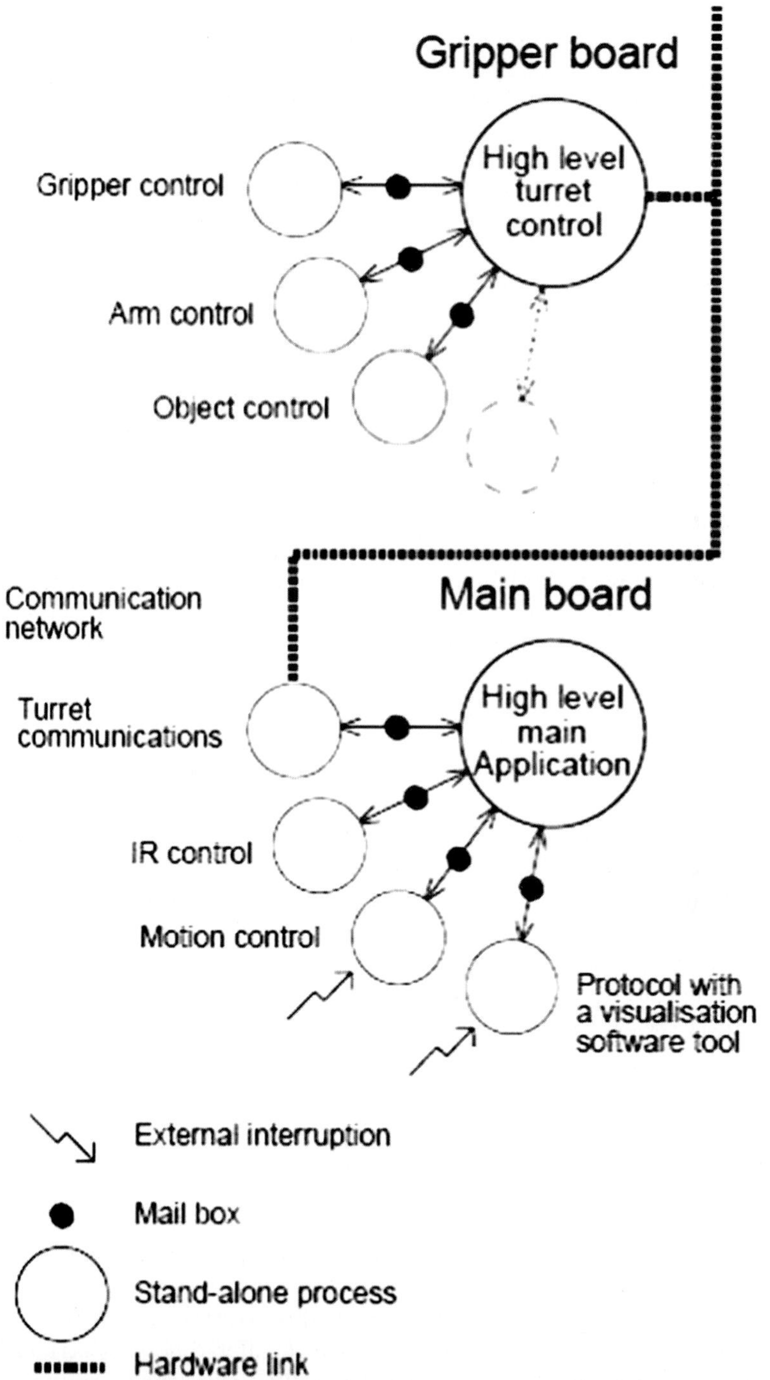

Fig. 11.69. Hierarchical software structure

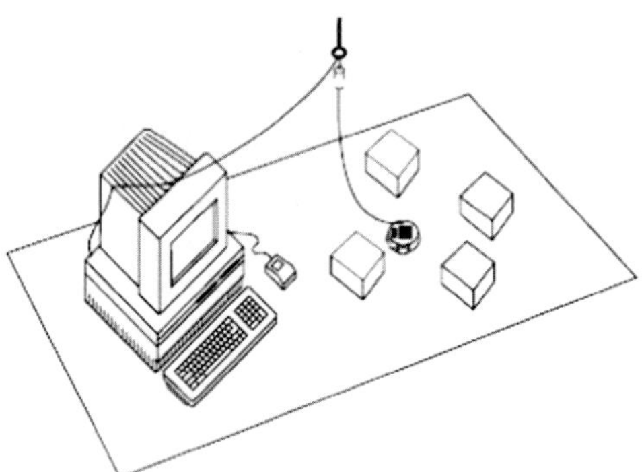

Fig. 11.70. Khepera experimentation environment

The physical arrangement of all elements involved in the experiment must be compact, allowing a complete and comfortable control. The entire configuration, including robot, environment, and workstation, is conveniently arranged on a normal table. In the arrangement shown in Fig. 11.70 the serial link cable does not disturb the movement of the robot.

A device to prevent the cable from rolling up is placed at mid-length on the serial cable. For experiments involving more than one robot, the wired serial link can no longer be used. Special radio communication modules are being developed for this purpose. This additional module will provide means to control several Khepera at the same time. With a wired or a radio serial link, the data flow between workstation and robot must be as little as possible. To minimize this flow without restricting user ability, the control algorithm runs on the workstation and communicates to the robot only the data concerning sensors and motors. This configuration is optimal when an important number of parameters must be displayed and controlled. Several programming and visualization tools are used to perform experiments with Khepera. Here, three programming styles will be presented: the first uses a classical programming language to build stand-alone applications, the second a complete graphical programming interface and the third is a compromise between the previous two, making the best of both.

Complete Applications

A programming possibility is to code the control algorithm and the user interface in a traditional way. This is a good choice for software engineers or researchers who have already developed a visualization and control interface, for instance in a simulation environment. It is often the case that, when a researcher starts to perform real robot experiments, a simulator has already been developed and used for preliminary studies. In this situations, the simulator can easily be adapted by replacing the simulated actions with calls to the interface with the real robot. This can usually be made without modifying the user interface. Some very interesting results have been achieved with this approach on the NNs simulator developed by Gaussier. The simulator is used as a tool to design neural networks for robot control. A real time visualization interface permits a variable step-by-step learning process on the robot. A similar experience in interfacing Khepera with a simulator is in progress using the simulator BugWorld, developed at the Institute for Informatik of the University of Zurich. In this case, the control interface will be complemented with a measurement system which enables the user to plot the trajectory of the robot in real time on the host screen.

LabVIEW

The software package LabVIEW is a commercial product from National Instruments and runs on several host machines (PC, Macintosh, or Sun workstations). LabVIEW has been developed as an environment for the design of virtual instruments (VI). Every VI comprises a control panel and an interconnection diagram. On the panel, the user can interactively specify graphical devices for input (e.g., sliding cursors, buttons, text controls) and for output (e.g., gauges, images, graphs). In the diagram, the user graphically enters the functionality of the VI. A library of standard functionalities is available to perform this task: icons performing numerical functions, string treatment, matrix computations, etc. can be linked together to design an algorithm. An icon can be associated with a complete VI and used hierarchically in the diagram of another instrument, thus allowing a modular approach. Moreover, modules can be written in standard programming languages, such as C or Pascal.

A sample experiment is shown in Fig. 11.71. The diagram represents the computation of a subsumption-based algorithm. Two modules, collide and turn wall, take inputs from the sensors (bottom left icon) and are connected

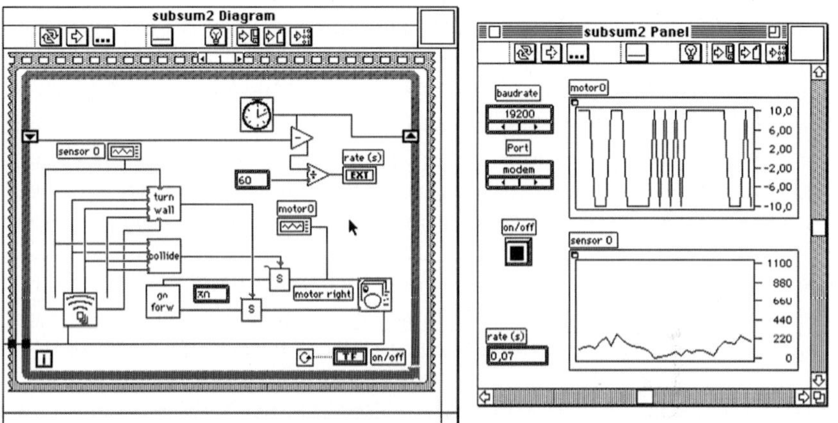

Fig. 11.71. LabVIEW display

to feed appropriate commands to the motors (right icon). The sensors and motor icons communicate with the robot through the wired serial link. The two graphs on the panel visualize the state of one motor and of one sensor. The modules in the top right part of the diagram evaluate the period required to recompute the algorithm; this is displayed at the bottom left of the panel. LabVIEW is an optimal tool for the design of experimentation environments without the use of programming languages. The complete graphical interface helps specifying the interaction items of the panel but becomes inefficient when designing complex control algorithms. In this case, it is more efficient to design modules using standard programming languages.

The only disadvantage of LabVIEW version 2 is that the display possibilities are somehow limited. The version 3 will provide more interactive capabilities and will be a better design tool for mobile robot experimentation. Due to code optimization and to improve performance, the software package is available only on SUN SparcStations. In the Grapher environment, an experiment is defined interconnecting modules that perform subtasks such as pure computation, visualization or control. The programming of these modules is done in C language. The interconnections are graphically specified by the user. The wiring diagram is on a single level, therefore preventing a hierarchical approach. For this reason, and to avoid over-complicated wiring schemes, the modules perform quite complex tasks. To facilitate the development, a large number of standard modules are available, making the best of the available hardware possibilities; the performance and flexibility of the visualization is particularly impressive. Comparing Grapher to LabVIEW, the former is less intuitive, needs some programming knowledge but make complex experimentation efficient. The experiment described in the next section illustrates this aspect.

11.9.4 Experimentation in Distributed Adaptive Control

As an example of the development techniques outlined above, the environment to evaluate a control architecture will be presented in this section. The control mechanism is developed according to the design methodology of distributed adaptive control. This approach is in turn derived from a distributed self-organizing model of the behavioral phenomenon of classical conditioning. The example involves an autonomous agent that can learn to avoid obstacles using collision and proximity sensors. The control structure consists of a neural net with three groups of neurons named Unconditioned Stimulus (US), Conditioned Stimulus (CS), and Motor actions (M). Neurons of the US group are directly connected with collision sensors, simulated here by the saturation of the proximity sensors. A prewired connection between the US and the M groups implements the robot basic task of avoiding obstacles at the time of a collision. Neurons of the CS group obtain their inputs from the proximity sensors. The learning is performed with an Hebbian rule on the connections between CS and US. The weights $K_{i,j}$ of these connections are updated according to

$$\Delta K_{i,j} = \frac{1}{N}(\eta . s_i s_j - \in .\bar{s} K_{i,j} \tag{11.1}$$

where N defines the number of units in the CS, η is the learning rate, ε the decay rate, and s the average activation in the group US.

This way, the robot can develop a conditional response, learning to avoid obstacles using the proximity sensors without producing collisions. During the experimentation it is interesting to observe the evolution of the learning process on the matrix K. The software environment used for this experiment is Grapher. Figure 11.72 shows the experiment display on a SUN Sparc Station. The principal window, on the top left, illustrates the functional diagram of the experiment: The DAC module (center top) performs the computation of the algorithm and interacts with the module Khepera to control the robot. Three other modules permit user interface. The panel module (top right) allows the user to control η and ε and the algorithm computation period by means of sliding cursors. The LabVIEW module displays the K matrix in the center bottom window. Finally, the c graph module displays the sensor state and the trajectory of the robot, as visible in the rightmost window. If the control algorithm C source with no display capabilities is available, the experimental setup can be designed in less than one day. The user gains complete control of the physical robot, the development environment and all the parameters of the algorithm in real time, thus obtaining an optimal visualization of the process.

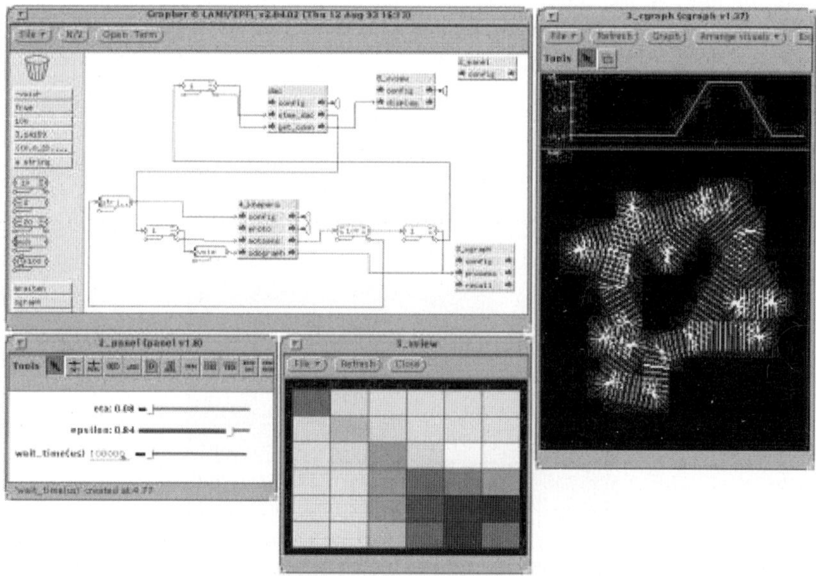

Fig. 11.72. Grapher display

Conclusion

The miniaturization of Khepera makes a compact and efficient experimentation environment possible. Associated with effective software tools, this robot is an optimal platform to test control algorithms. The modularity at the hardware, software and control tools levels gives to the user the necessary flexibility to perform accurate experiments quickly. An example of experimentation environment has been presented. The reduced size and cost of the miniature robots described make possible experimentation on collective behavior among groups of robots which is the main research activity in the near future.

11.10 A Steady-Hand Robotic System for Microsurgical Augmentation

This application reports a new robotic system developed to extend a human's ability to perform smallscale (submillimeter) manipulation tasks requiring human judgement, sensory integration and hand-eye coordination. Our novel approach, which we call "steady-hand" micromanipulation is for tools to be held simultaneously both by the operator's hand and a specially designed robot arm (Fig. 11.73). The robot's controller senses forces exerted by the operator on the tool and by the tool on the environment, and uses this information in various control modes to provide smooth, tremor-free precise positional control and force scaling. The result will be a manipulation system with the precision and sensitivity of a machine, but with the manipulative transparency and immediacy of handheld tools for tasks characterized by compliant or semi-rigid contacts with the environment. Humans possess superb manual dexterity, visual perception, and other sensory motor capabilities. We manipulate best at a "human scale" dictated by our physical size and manipulation capabilities

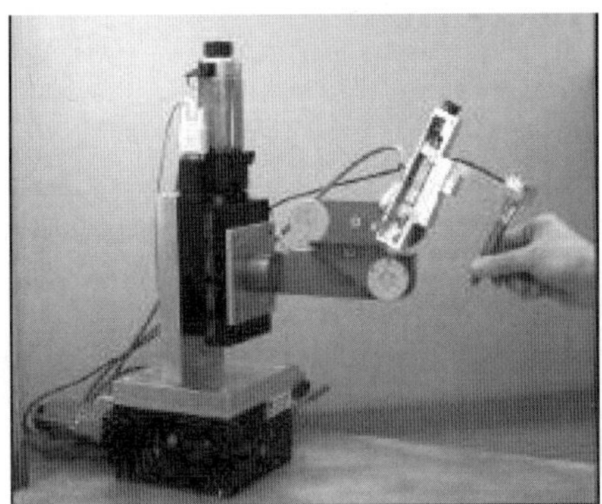

Fig. 11.73. The JHU "Steady-Hand" Robot for cooperative human–machine microsurgical manipulation

and roughly corresponding to the tasks routinely performed by our cave man ancestors.

Tasks that require very precise, controlled motions are difficult or impossible for most people. Further, humans work best in tasks that require relative positioning or alignment based on visual or tactile feedback. We do not come equipped with an innate ability to position or fabricate objects accurately relative to arbitrary measuring standards or to perform tasks based on nonhuman sensory feedback. For these tasks, we rely on machines. A good machine tool, for example, can routinely measure and fabricate parts to a precision of $\gg$2.5 mm ($\gg$0.001 in.). Fine-scale tasks such as microsurgery require both precise manipulation and human judgement. Other tasks may require combining precise manipulation with sources of information (assembly specifications, nonvisible-light images, etc.) that are not naturally available to a human. We thus have a choice: either automate the human judgement aspects of the task (difficult at best and often impossible) so that a machine can automatically perform the task or else find a way to use a machine to augment human manipulation capabilities while still exploiting the human's natural strengths. Most prior work on robotic micromanipulation systems has emphasized traditional master–slave and telerobotic manipulation. The approach might offer several advantages compared to these systems in the context of micromanipulation. These include: (1) simplicity; (2) potentially cheaper implementations; (3) a more direct coupling to the human's natural kinesthetic senses; (4) straightforward integration into existing application environment; and (5) greater "immediacy" for the human operator. The principal drawbacks are the loss of the ability to "scale" positional motions and the loss of the ability to manipulate objects remotely. These are certainly important abilities, but we believe there are many tasks in which they are not crucial and for which a simpler alternative is more attractive. These advantages are especially attractive in applications like microsurgery, where surgeon acceptance is crucial and where approaches that do not require a complete re-engineering of the surgical workstation are much easier to introduce into practice.

11.10.1 Robotically Assisted Micromanipulation

Mechanical systems have been developed which extend the capability of human operators using telerobotic principles including virtual training, manipulation of objects in hazardous environments, remote surgery, and microsurgery. In general, telerobotic devices rely on an operator commanding the motion of a robot using a secondary input device. The operator may reside in close proximity to the robot, observing its motions through a microscope as in microsurgery or may be many miles away as in space exploration. In both cases, the operator is an integral part of the system and has direct control over how the manipulator moves. An ideal teleoperated system would be transparent to the operator and give the impression of direct control. The input device manipulated by the operator may be either passive, such as a trackball, joystick or stylus, or made up of active devices such as motors. An active input device allows forces imposed on the robot to be measured, scaled and mimicked at the input device to be subsequently felt by the operator. Several systems have been developed for teleoperated microsurgery using a passive input device for operator control. Guerrouad and Vidal describe a system designed for ocular vitrectomy in which a mechanical manipulator was constructed of curved tracks to maintain a fixed center of rotation. A similar micromanipulator was used for acquiring physiological measurements in the eye using an electrode. While rigid mechanical constraints were suitable for the particular applications in which they were used, the design is not flexible enough for general purpose microsurgery and the tracks take up a great deal of space around the head. An ophthalmic surgery manipulator was designed for retinal vascular microsurgery and was capable of positioning instruments at the surface of the retina with submicron precision. While a useful experimental device, this system did not have sufficient range of motion to be useful for general-purpose microsurgery. Also, the lack of force sensing prevented the investigation of force/haptic interfaces in the performance of microsurgical tasks.

Many microsurgical devices are based on force-reflecting master–slave configurations. This paradigm allows an operator to grasp the master manipulator and apply forces. Forces measured on the master are scaled and reproduced at the slave and, if unobstructed, will cause the slave to move accordingly. Likewise, forces encountered by the slave are scaled and reflected back to the master. This configuration gallows position commands from the master to result in a reduced motion of the slave and for forces encountered by the slave to be amplified at the master. While a force reflecting master–slave microsurgical system provides the surgeon with increased precision and enhanced proportion, there are some drawbacks to such a design.

The primary disadvantage is with complexity and cost since two mechanical systems, one for the master and one for the slave, are required. Another problem with telesurgery in general is that the surgeon is not allowed to directly manipulate the instrument used for the microsurgical procedure. While physical separation is necessary for systems designed to perform remote surgery, it is not required during microsurgical procedures. In fact, surgeons are more likely to accept assistive devices if they are still allowed to directly manipulate the instruments.

Shared Autonomy and Cooperative Control

There is a large body of literature concerning provably stable control techniques for robots. Standard paradigms include (1) preprogrammed trajectory control of position and force; (2) fully autonomous robots; and (3) master–slave teleoperators. In our case, we are interested in developing provably stable controls for cases where both the robot and the human manipulate a single tool in contact with a compliant environment. The work most relevant to this includes that of Kazarooni who developed exoskeletons to amplify the strength of the human operator Kazarooni reports a linear systems analysis of the stability and robustness of cooperative human–robot manipulator control systems in which the manipulator scales-up the human operator's force input by a factor of ~10. The authors report a stability analysis of this closed-loop system (comprising a dynamical model of both the robot arm and the human arm) is complicated by the fact that precise mathematical plant models do not exist for either the hydraulically actuated robot and the operator's human arm. In consequence, in the authors perform a robustness analysis to develop stable robot force-control laws that accommodate wide variation in both human and robot arm dynamics. In contrast, we propose to address the control problem of cooperative human–robot manipulator systems in which the manipulator scales-down the human operator's force input by a factor of ~0.1. To achieve this scaling-down of human input we anticipate comparable (or greater) difficulties to arise from unknown human arm dynamics, we can construct the system using electrical motors (rather than hydraulic motors) for which accurate dynamical models are available.

A number of authors have investigated "shared autonomy" and for cooperative control of teleoperators, typically with space or other "remote" applications where time delays can affect task performance. There has also been some work on control of robots working cooperatively with humans to carry loads and do other gross motor tasks relevant for construction and similar applications. Within the area of surgery, we have long used "hands on" guiding of robots for positioning within the operating room hip replacement surgery system and in the JHU/IBM LARS system for endoscopic surgery). Davies, have combined hands-on guiding with position limits and have demonstrated 3 DOF machining of shapes in the end of a human tibia.

At JHU, we have been using the LARS robot to perform a variety of "steady hand" tasks combining hand guiding, active control, and safety constraints in neuro endoscopy and other areas. In one experiment using the LARS robot-assisted evacuation of simulated hematomas was found to take longer (6.0 min vs. 4.6 min) than freehand evacuation but was found to remove much less unintended material (1.5% vs. 15%). We have also made some preliminary experiments using the LARS for micromanipulation, although the compliance of the LARS upper linkage severely limits the benefit gained.

11.10.2 A Robotic System for Steady-Hand Micromanipulation

Design Goals

Cooperative micromanipulation requires capabilities not commonly found in conventional robots or teleoperator systems. Typically, these tasks will be performed by a human operator looking through a microscope while grasping a "handle" on the instrument or tool being used to perform the task. In the

Table 11.2. Steady-Hand robot design goals

Base (XYZ) assembly	(Off-the-shelf)
Work volume	100 mm × 100 mm × 100 mm
Top Speed	40 mm s^{-1}
Positioning resolution	≈2.5 μm (0.5 μm encoder res.)
RCM orientation assembly	(Custom)
Link length	100 mm
Range of motion	Continuous 360°
Top speed	180°/sec
Angular resolution	≈0.05° (0.01° encoder res.)
End-of-arm/guiding assembly	(Custom)
Range of motion	±2 cm; 360° continuous
Positioning resolution	0.5 μm; 0.1° (tentative)
Top speed	40 mm s^{-1}; 180° s^{-1}
Handle force resolution	0.01 N (tentative)

tasks that we are considering, we believe that motion "scaling" (in the sense that a 1 cm human hand motion might cause a 100 mm instrument motion) is much less important than smooth motion naturally aligned with the human's own kinesthetic senses. Pulling on the tool's handle should produce intuitively natural translation and orientation motions. Specific requirements required are included Table 11.2.

Positioning Performance

We are interested in manipulation tasks requiring very precise positional control, with controlled end-effector motion resolution on the order of 3–10 mm, when rotational motion is decoupled at the tool tip and 5–25 mm tip resolution when motion is decoupled about a fulcrum point 2 cm from the tool tip (i.e., when a point 2 cm from the tool tip remains fixed in space).

Safety

The strong preference is for relatively low power actuators with high-reduction, nonbackdrivable joints. Such systems are relatively easy to monitor, stop, and stay put once stopped.

Manipulation Forces

We are primarily interested in manipulation tasks with a reasonable degree of contact compliance between the tool and the environment being manipulated. In the case of microsurgery, this compliance is provided by the tissue being manipulated. The goal is moderate bandwidth (3–5 Hz) control and scaling of interaction forces, with tool tip forces ranging from $\gg$0.001 to $\gg$0.01 N, depending on specific application, and human interaction forces ranging from $\gg$0.01 to $\gg$3 N. We also wish to provide higher bandwidth sensing and haptic feedback of force discontinuities, and to explore the usefulness of such feedback in micromanipulation tasks.

System Design and Implementation

The steady-hand manipulator consists of several modular components:

(1) An off-the-shelf *XYZ translation assembly* composed of three standard motorized micrometer stages
(2) An *orientation assembly*, consisting of a custom-designed remote-center-of-motion (RCM) linkage providing two rotations about a "fulcrum" or remote motion center point located in free space approximately 100 mm from the robot
(3) A combined *end-of-arm motion and guiding assembly* providing one additional rotation about and displacement along a tool axis passing through the remote motion center. This subassembly also comprises a guiding handle with a 6 degree-of-freedom (DOF) force sensor and and a tool holder for mounting micromanipulator tool
(4) *Specialized instruments* held in the tool holder (e.g., microgrippers) with the ability to sense interaction forces between the tools and the environment being manipulated.

Base Translation Module

For expediency, we have employed a three-axis base translation module comprised of off-the-shelf motorized micrometer stages from New England Affiliated Technologies of Lawrence, MA. The XYZ translation assembly is formed by mounting a single axis Z-stage (NEAT: LM-400) orthogonal to a dual axis X-Y table (NEAT: XYR-6060). An axis consists of a crossed-roller way mounted table motivated by an encoded DC servomotor driven lead screw. Each axis has 100 mm of travel, can travel at speeds $>40\,\mathrm{mm\,s^{-1}}$ and has a positioning resolution of <2.5 mm (0.5 mm encoder resolution).

RCM: Remote-Center of Motion Module

The Remote Center of Motion (RCM) robot module (Fig. 11.74) is a compact robot for surgical applications that implements a fulcrum point located distal to the mechanism. The robot presents a compact design: it may be folded into a $171 \times 69 \times 52$ mm box and it weighs 1.6 kg. The robot can precisely orient an end-effector (i.e., surgical instrument) in space while maintaining the location of one of its points. This kinematic architecture makes it proper for laparoscopic applications as well as needle orientation in percutaneous procedures. RCM accommodates various end-effectors. We have used the RCM

Fig. 11.74. The Remote Center of Motion (RCM) module

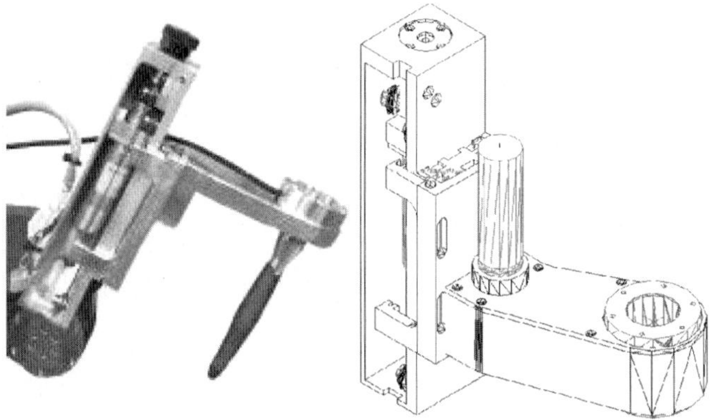

Fig. 11.75. Insertion end-effector with force sensor and guiding handle attached to rotation end-effector blank (*left*) and design drawing of rotation end-effector mounted on insertion stage (*right*)

in conjunction with the PAKY needle driver for performing image guided renal access. The robot has been successfully used at the Johns Hopkins Medical Institutions for nine surgical procedures.

Rotation/Insertion End-Effector Module

The insertion end-effector (Fig. 11.75) provides linear displacement along the tool axis passing through the remote motion center. The axis utilizes a two-stage telescoping crossed-roller slide mechanism driven via a cable by an encoded DC servomotor. The telescoping crossed-roller slides provide >150 mm of travel from a 70 mm closed slide length while maintaining high stiffness. The force transmission path consists of a low stretch nylon coated 304 stainless steel cable driven by a grooved drive pulley attached to the DC servomotor. The drive pulley always carries six wraps of cable to maintain good frictional contact and allow high repeatability. The insertion stage can travel at speeds of $\gg 30\,\mathrm{mm\,s^{-1}}$ and has a positioning resolution of $\gg 5$–10 mm (1.5 mm encoder resolution).

The rotation end-effector provides rotation about the tool axis and the mounting surface for the force sensor with guiding handle. The rotation stage is driven by a timing belt attached to a encoded DC servomotor. This axis is currently being fabricated and is nearing completion. It will provide a 360° continuous range of motion and is expected to travel at speeds of $\gg$120°–180°/sec with a positioning resolution of $\gg$0.05°–0.10° (0.01° encoder resolution).

Handle Force-Sensing Module

This module uses a small commercially available force sensor (Model: NANO-17 SI 12/0.12, ATI Industrial Automation, NC) to capture user forces. The 13–8 VAR stainless steel transducer (yield strength of 205 ksi) has a resolution of 0.025 N, 0.0625 N mm along the Z-axis and 0.0125 N, 0.0625 N mm in the X-Y axes. Force ranges of $\pm$22.5 N in the Z-axis and $\pm$12.5 N in the X-Y axes can be measured. The torque range is $\pm$125 N-mm. The force sensor has overload protection of 800 N in Z, 350 N in the X-Y axes and 2.5 N-m moment about any axis. The force sensor is 17 mm in diameter, 14.5 mm in height with mounting and tool adapter plates attached and weighs 9.4 grams. The force sensor is read using a 12-bit ISA bus F/T controller card with up to 7,800 Hz sampling rates. The force sensor is mounted on the instrument rotation stage with its z-axis parallel to the instrument insertion stage of the robot.

The rotation end-effector provides rotation about the tool axis and the mounting surface for the force sensor with guiding handle. The rotation stage is driven by a timing belt attached to a encoded DC servomotor. This axis is currently being fabricated and is nearing completion. It will provide a 360° continuous range of motion and is expected to travel at speeds of $\gg$120°–180°/sec with a positioning resolution of $\gg$0.05°–0.10° (0.01° encoder resolution).

Handle Force-Sensing Module

This module uses a small commercially available force sensor (Model: NANO-17 SI 12/0.12, ATI Industrial Automation, NC) to capture user forces. The 13–8 VAR stainless steel transducer (yield strength of 205 ksi) has a resolution of 0.025 N, 0.0625 N mm along the Z-axis and 0.0125 N, 0.0625 N mm in the X-Y axes. Force ranges of $\pm$22.5 N in the Z-axis and $\pm$12.5 N in the X-Y axes can be measured. The torque range is $\pm$125 N-mm. The force sensor has overload protection of 800 N in Z, 350 N in the X-Y axes and 2.5 N-m moment about any axis. The force sensor is 17 mm in diameter, 14.5 mm in height with mounting and tool adapter plates attached and weighs 9.4 g. The force sensor is read using a 12-bit ISA bus F/T controller card with up to 7,800 Hz sampling rates. The force sensor is mounted on the instrument rotation stage with its z-axis parallel to the instrument insertion stage of the robot.

Force-Sensing Microsurgical End-Effector Tool Module

A variety of surgical instruments such as pics, forceps, needle holders, and scissors are required during microsurgical procedures. To utilize the benefits offered by the cooperative control algorithms of steady-hand augmentation, these microsurgical tools must be equipped with sensitive, multidimensional force sensors. Our initial approach uses silicon strain gauges configured into bridges located within the surgical tool handle. The tool tip acts as a lever that imparts tourques upon the bridge during surgical manipulations.

Control System

The robot hardware control runs on a Pentium-II 450 MHz PC with Windows NT operating system. An 8-axis DSP series controller card (PCX/DSP, Motion Engineering Inc, CA) is used to control the robot. The card provides servocontrol using a 40 MHZ Analog Devices ADSP-2105 processor. It also has support for user digital and analog input, output lines. The PC also houses the ISA force sensor controller. A library of C + + classes has been developed for control purposes. This modular robot control(MRC) library provides Cartesian level control. It allows with multipriority clients and multiclient servers for distributed robot control transparent to the user application. It includes classes for kinematics, joint level control, command and command table management, sensor and peripheral support, and network support. Some exception and error handling is also built in. An array of sensors including serial and parallel ports, ATI force sensors, joysticks, digital buttons and foot pedals are supported. Support is also available for MEI motion controller cards and the proprietary LARS servocontroller. Although some of the MRC functionality is limited to WIN32 operating systems, most of the classes are operating system independent. A simple force controller has been implemented based on the MRC library. User forces at the tool are sensed using the force sensor, biased and resolved the coordinate frame of the robot for force – proportional velocity control. Both user forces and robot velocity are limited for safety. The base joints and the upper joints can be controlled independently by using a foot pedal. Control rates of over 1,000 Hz can be achieved using this controller. The force control can be used with the force sensor attached to the robot or positioned remotely (connected to another PC, networked to the robot controller). This simple control system is intended to allow us to test/refine the hardware. More sophisticated control will be used for actual applications.

11.10.3 Current Status

Mechanical integration of the steady-hand manipulation system is nearing completion. With the exception of the rotation axis, all axes are complete, integrated, and under computer control. The rotation axis fabrication is presently nearing completion in our machine shop. Experiments designed to evaluate the augmentation value of steady-hand manipulation are currently being performed. Initial indications are that the basic design assumptions of a stiff robot with force control are valid for surgical manipulations at a microscale.

Summary

The approach extends earlier work on cooperative manipulation to microsurgery and focuses on performance augmentation utilizing both force and position control. Our goal is to develop a manipulation system with the precision and sensitivity of a machine, but with the manipulative transparency and immediacy of hand-held tools for tasks characterized by compliant or semirigid contacts with the environment. The design is highly modular and represents one step in the evolution of a family of robotic surgical devices. Although the first focus is on retinal microsurgery, we believe that the proposed approach is more general. Other applications will include neuroendoscopy, ENT, and microvascular surgery.

A

LabVIEW Research Projects

The brief description of research projects collected from IEEE journals, Research Labs and Universities are given in this appendix.

A.1 An Optical Fibre Sensor Based on Neural Networks for Online Detection of Process Water Contamination

It is often required to know the contamination of process water systems from known contaminants e.g., oil, heavy metals, or particulate matter. Most of the governments are forcing rules to monitor the output water from all industrial plants before they may be discharged into existing steams or waterways. There is therefore a clear need for a sensor that is capable of effectively monitoring these output discharges. Optical fibers provide a means of constructing such sensors in an effective manner. Optical fibers are passive and therefore do not them selves contaminate the environment in which they are designed to work. The proposed sensor system is to be operated from and LED or broadband light source and the detector will monitor changes in the received light spectrum resulting from the interaction of the fibre sensor with its environment. The signals arising from these changes are often complex and require large amounts of processing power. In this proposed project it intended to use Artificial Neural Networks to perform the processing and extract the true sensor signal from the received signal.

A.2 An Intelligent Optical Fibre-Based Sensor System for Monitoring Food Quality

It is proposed to develop an intelligent instrument for quality monitoring and classification of food products in the meat processing and packaging industry. The instrument is intended for use as an effective means of testing food products for a wide range of parameters. Such parameters may range from simple external appearance e.g., color to degree of cooking internally in the case of pre-cooked products. The proposed instrument will incorporate two new technology areas namely Optical Fibre Sensors and Artificial Neural Networks and is intended for used on high volume food processing production lines.

A.3 Networking Automatic Test Equipment Environments

Automatic test equipment (ATE) is a term that, in its broadest meaning, indicates a generic system capable of performing measurements in an automatic or semiautomated (human-assisted) way. Years ago, this term was used specifically to refer to an automated measurement system employed to test the functionality of some electronic Device Under Test (DUT). Typical applications were in the manufacturing area, where ATE had a twofold nature: in-circuit testing and functional testing. For in-circuit testing, ATE often were stand-alone complex programmable machines, equipped with a bed-of-nails adapter specifically designed as a fixture to provide signal inputs and meaningful test-points of the DUT.

The test engineer had the responsibility of writing code that determined the exact sequence of stimulus signals, response measurements, and go/no-go decisions. For this aim, a switch matrix and the ATE itself were suitably controlled and coordinated by a workstation. For functional testing, ATE consisted of off-the-shelf instruments connected to the DUT by some kind of front-end adapter. In the latter case, most of the effort of the test engineer consisted of designing a program to control the various instruments to assess DUT performances. When planning the use of a dedicated testing machine as opposed to a test bench, other factors were taken into account: measurement speed, cost, and fault coverage.

A.4 Using LabVIEW to Prototype an Industrial-Quality Real-Time Solution for the Titan Outdoor 4WD Mobile Robot Controller

In the Titan project researchers applied a new approach to prototyping mobile robots by choosing tools which are commonly used by leading aerospace manufacturers and many other industries. Researchers have gained substantial experience when using the LabVIEW real-time programming environment coupled with the industrial quality data acquisition cards, both made by National Instruments. The methodology of virtual instruments software tools combined with the graphical programming environment was found to be very efficient for interactive cycles of design and testing, which are at the core of robotics prototyping.

A.5 Intelligent Material Handling: Development and Implementation of a Matrix-Based Discrete-Event Controller

A supervisory controller for discrete-event (DE) systems is presented that uses a novel matrix formulation. This matrix formulation makes it possible to directly write down the DE controller from standard manufacturing tools such as the bill of materials or the assembly tree. The matrices also make it straightforward to actually implement the DE controller on a manufacturing work cell for sequencing the jobs and assigning the resources. It is shown that the DE controller equations plus the Petri net marking transition equation together provide a complete dynamical description of a DE system. This means that a computer simulation can be performed to check the DE performance of the controller before it is implemented. In this research, the DE controller is implemented on an actual three-robot intelligent material handling cell at the Automation and Robotics Research Institute, University of Texas at Arlington, USA. Then, the actual implementation is performed and the simulated system produces commensurate results. The versatility of the system developed with this DE controller permits implementing different methodologies for conflict resolution, as well as optimization of the resource assignment and part throughput. Technical information given includes the development of the controller in LabVIEW and its simulation using MATLAB.

A.6 Curve Tracer with a Personal Computer and LabVIEW

The circuits and programs that conform the prototype of a curve tracer, operated under the control of a personal computer (PC) are presented. The prototype uses a data acquisition board that generates the signals applied to the base and collector terminals of the tracer. This board also carries out the necessary measurements to trace the characteristic curves of the device under test. It was designed an additional board with the appropriate circuit for the base and collector terminals. A program was written in the graphical language G of LabVIEW to program the complete operation of the tracer and to control the external circuits. The developed program presents the results in a graphic indicator as a family of curves, it can store in a file the parameters of the characterization process.

A.7 Secure Two-Way Transfer of Measurement Data

This research introduces a measurement system architecture, which has three main qualities: secure two-way transfer of measurement data, quick adaptability to different kinds of measurement tasks and strong data presentation capabilities through XML techniques. In the architecture, the researchers take advantage of well-tried technologies, like a commonly used visual programming language that offers predefined and adaptive measuring tools for managing measurement devices and tasks. XML is a widely adopted standard for a flexible text format and data exchange. It brings along a vast selection of readymade facilities for processing and transforming the content into any format desired. The researchers also propose a secure environment into the architecture, which can be accessed on demand using a wide range of terminal devices.

A.8 Development of a LabVIEW-Based Test Facility for Standalone PV Systems

To quantify the potential for performance improvement of a standalone photovoltaic (PV) system, a test facility has been installed. This research describes this development of a prototype standalone PV system. Essentially this entire system involves the integration of a Personal computer (PC), Data Acquisition (DAQ), a battery array and a solar array simulator (SAS) to create a standalone PV system and to test and simulate the system. This new system boasts of high accuracy measurements coupled with the commercial viability of low cost. The basic idea of this facility is that the SAS simulates solar power which is utilized to charge batteries. The information obtained by monitoring parameters, such as average battery's temperature, voltage and current is fed to the PC via the DAQ for analysis. This customized control interface has been developed by utilizing LabVIEW software, which forms the programming backbone of interinstrument communication through IEEE-GPIB bus.

The software created for this system is highly generic and can be used for other instances where different hardware is used. This project also discusses further research plan, in utilizing this standalone PV system to perform load analysis and batteries charging or discharging with the inputs to the SAS with actual meteorological data obtained from the Malaysian meteorological department.

A.9 Semantic Virtual Environments with Adaptive Multimodal Interfaces

Researchers present a system for real-time configuration of multimodal interfaces to Virtual Environments (VE). The flexibility of the tool is supported by a semantics-based representation of VEs. Semantic descriptors are used to define interaction devices and virtual entities under control. Researchers use portable (XML) descriptors to define the I/O channels of a variety of interaction devices. Semantic description of virtual objects turns them into reactive entities with whom the user can communicate in multiple ways. This article gives details on the semantics-based representation and presents some examples of multimodal interfaces created with our system, including gestures-based and PDA-based interfaces, amongst others.

A.10 A Dynamic Compilation Framework for Controlling Microprocessor Energy and Performance

Dynamic voltage and frequency scaling (DVFS) is an effective technique for controlling microprocessor energy and performance. Existing DVFS techniques are primarily based on hardware, OS time interrupts, or static-compiler techniques. However, substantially greater gains can be realized when control opportunities are also explored in a dynamic compilation environment. There are several advantages to deploying DVFS and managing energy/performance tradeoffs through the use of a dynamic compiler. Most importantly, dynamic compiler-driven DVFS is fine-grained, code-aware, and adaptive to the current microarchitecture environment. This project presents a design framework of the run-time DVFS optimizer in a general dynamic compilation system. A prototype of the DVFS optimizer is implemented and integrated into an industrial strength dynamic compilation system.

The obtained optimization system is deployed in a real hardware platform that directly measures CPU voltage and current for accurate power and energy readings. Experimental results, based on physical measurements for over 40 SPEC or Olden benchmarks, show that significant energy savings are achieved with little performance degradation. SPEC2K FP benchmarks benefit with energy savings of up to 70% (with 0.5% performance loss). In addition, SPEC2K INT show up to 44% energy savings (with 5% performance loss), SPEC95 FP save up to 64% (with 4.9% performance loss), and Olden save up to 61% (with 4.5% performance loss). On average, the technique leads to an energy delay product (EDP) improvement that is 3X–5X better than static voltage scaling, and is more than 2X (22% versus 9%) better than the reported DVFS results of prior static compiler work. While the proposed technique is an effective method for microprocessor voltage and frequency control, the design framework and methodology described in this paper have broader potential to address other energy and power issues such as di/dt and thermal control.

A.11 A Method to Record, Store, and Analyze Multiple Physiologic Pressure Recordings

To investigate the pathophysiology of syringomyelia, research scholars developed a system to simultaneously record arterial, central venous, airway, cervical subarachnoid, and intrathecal pressures. In addition, the proposed system recorded the electrocardiogram, and provided the means to analyze all pressures during individual and averaged cardiac cycles. During the study, flow magnetic resonance imaging and cardiac-gated ultrasound images were also used. To compare the pressure readings with these images the researchers needed to account for the low frequency change in the recorded pressures resulting from normal respiration. Research scholars therefore wrote a program to derive mean pressure waves during the cardiac cycle that averaged-out the pressure effects of respiratory cycle. Simultaneous cerebrospinal pressure, hemodynamic pressure, and electrocardiogram were plotted. This allowed to establish the simultaneous relationship of pressures, syrinx size, and cerebellar tonsil pulsation during the cardiac cycle. This system may prove valuable to other researchers investigating the interaction of multiple physiologic parameters during the cardiac cycle.

A.12 Characterization of a Pseudorandom Testing Technique for Analog and Mixed-Signal Built-in Self-Test

In this project, researchers characterize and evaluate the effectiveness of a pseudorandom-based implicit functional testing technique for analog and mixed-signal circuits. The analog test problem is transformed into the digital domain by embedding the device-under-test (DUT) between a digital-to-analog-converter and an analog-to-digital converter. The pseudo-random testing technique uses band-limited digital white noise (pseudorandom-patterns) as input stimulus. The signature is constructed by computing the cross-correlation between the digitized output response and the pseudorandom input sequence. The researchers have implemented a DSP-based hardware test bed to evaluate the effectiveness of the pseudorandom testing technique. The test results show that the user can achieve close to 100% yield and fault coverage by carefully selecting only two cross-correlation samples. Noise level and total harmonic distortion below 0.1% and 0.5%, respectively, do not affect the classification accuracy.

A.13 Power-Aware Network Swapping for Wireless Palmtop PCs

Virtual memory is considered to be an unlimited resource in desktop or notebook computers with high storage capabilities. However, in wireless mobile devices, like palmtops and personal digital assistants (PDAs), storage memory is limited or absent due to weight, size, and power constraints, so that swapping over remote memory devices can be considered as a viable alternative.

However, power-hungry wireless network interface cards (WNICs) may limit the battery lifetime and application performance if not efficiently exploited. In this project, the performance and energy of network swapping in comparison with swapping on local micro-drives and flash memories are studied. The results of extensive experiments conducted on different WNICs and local swapping devices, using both synthetic and natural benchmarks are reported. The study points out that remote swapping over power-manageable WNICs can be more efficient than local swapping, especially in bursty workload conditions. Such conditions can be forced where possible by reshaping swapping requests to increase energy efficiency and performance.

A.14 Reducing Jitter in Embedded Systems Employing a Time-Triggered Software Architecture and Dynamic Voltage Scaling

Earlier researches have demonstrated that use of an appropriate Dynamic Voltage Scaling (DVS) algorithm can lead to a substantial reduction in CPU power consumption in systems employing a time-triggered cooperative (TTC) scheduler. In this project, the impact of DVS on the levels of both clock and task jitter in TTC applications is discussed. The researchers describe a modified DVS algorithm (TTC-jDVS) which can be used where low jitter is an important design consideration. Then the effectiveness of the modified algorithm on a data set made up of artificial tasks and in a realistic case study are demonstrated.

A.15 End-to-End Testing for Boards and Systems Using Boundary Scan

ICs with IEEE 1149.1 Boundary Scan (BS) Architecture (a.k.a. JTAG) have been widely used in board level design to increase the testability. An end-to-end test methodology that utilizes BS architecture for testing boards and systems throughout the product life cycle is proposed. The proposed test methodology includes a programmable dynamic BS test architecture and a series of test modules that take advantage of the test architecture for complete fault coverage. Proposed design-for-testability (DFT) techniques guarantee the co-existence of BS testing with other system functions, such as in-system programming (ISP) and DSP JTAG emulation. At board level, programmable dynamic scan chains are used in a divide-and-conquer fashion to increase the flexibility in the development phase (or design verification testing, DVT). Besides, since the DFT techniques are programmable, they can be used as design-for-diagnosis (DFD) to increase diagnosis resolution during DVT. Address Scan Port (ASP) chips are used to enable multidrop test bus architecture for backplane testing as well as system embedded testing. Other advanced techniques, such as analog subsystem testing and board-level built-in self-test, as well as how to re-use BS architecture in in-circuit testing (ICT) and manufacture testing are also parts of the proposed methodology that takes advantage of BS architecture to provide full scale testing for systems.

A.16 An Approach to the Equivalent-Time Sampling Technique for Pulse Transient Measurements

An approach to the time-equivalent sampling technique in time-domain measurements of transient pulses is presented in this work. An application in fluorescence spectroscopy is mentioned and the sampling processes as well as the different sampling techniques are briefly studied. The approach makes use a Digital Signal Processor (DSP) and a Personal Computer (PC) to demonstrate the technique. The DSP communicates with the computer via a standard serial port to receive commands and to transmit data. In addition to the sampling technique, the DSP also applies a method of noise suppression by averaging the sampled digital signal. A control program was developed in the graphical language G of LabVIEW 6.1 to register and plot the time-resolved pulse. With this approach the users are able to resolve pulses with a time resolution of 16.66 ns and a bandwidth of 1.25 MHz.

A.17 Reactive Types for Dataflow-Oriented Software Architectures

Digital signal-processing (DSP) tools, such as Ptolemy, LabView and iConnect, allow application developers to assemble reactive systems by connecting pre-defined components in generalized dataflow graphs and by hierarchically building new components by encapsulating subgraphs. The researcher follow the literature in calling this approach dataflow-oriented development. The previous work has shown how a new process calculus, uniting ideas from previous systems within a compositional theory, can be formally shown to capture the properties of such systems. This project first re-casts the graphical dataflow-oriented style of design into an underlying textual architecture design language (ADL) and then shows how the previous modelling approach can be seen as a system of process-algebraic behavioral types for such a language, so that type-checking is the mechanism used to statically diagnose the reactivity of applications. The researchers show how both the existing notion of behavioral equivalence and a new behavioral pre-order are involved in this judgement.

A.18 Improving the Steering Efficiency of 1x4096 Opto-VLSI Processor Using Direct Power Measurement Method

This project reports the optimization of the steering efficiency of the 1-D Opto-VLSI processor using direct power measurement method for wavelengths in the near-IR and 632 nm. Highest improvement observed for the signal and inter port isolation is 8 and 12 dB, respectively. This improved performance of the processor is crucial to the realization of low crosstalk reconfigurable optical add/drop multiplexers (ROADM) using Opto-VLSI processors.

A.19 Experimental Studies of the 2.4-GHz IMS Wireless Indoor Channel

Experimental results from indoor measurements of the 2.4-GHz wireless channel are presented. These measurements include both quasi-static and time-varying channel conditions in a number of indoor environments. Four channel propagation characteristics of interest are evaluated. For the time-varying environment, the Doppler spread of the wireless channel is measured. For the quasi-static environment, mean excess delay, rms delay spread, and coherence bandwidth of the channel are determined. The effect each of these propagation characteristics has on multiple antenna wireless systems and their suitability for indoor use is discussed.

A.20 Virtual Instrument for Condition Monitoring of On-Load Tap Change

The Condition Monitoring of On-Load Tap Changers is very important because they have proved to be the elements with noticeable failures in a power transformer. This article describes the development of a portable virtual instrument for monitoring this kind of elements. The monitoring task is based in the measurement and analysis of the vibrations that a tap change produces. In contrast with other methods that can be used, this one has the advantage of being able to do continuous monitoring because the transformer can be operated online.

A.21 Toward Evolvable Hardware Chips: Experiments with a Programmable Transistor Array

Evolvable Hardware is reconfigurable hardware that self-configures under the control of an evolutionary algorithm. The search for a hardware configuration can be performed using software models or, faster and more accurate, directly in reconfigurable hardware. Several experiments have demonstrated the possibility to automatically synthesize both digital and analog circuits. The project introduces an approach to automated synthesis of CMOS circuits, based on evolution on a Programmable Transistor Array (PTA). The approach is illustrated with a software experiment showing evolutionary synthesis of a circuit with a desired DC characteristic. A hardware implementation of a test PTA chip is then described, and the same evolutionary experiment is performed on the chip demonstrating circuit synthesis/self-configuration directly in hardware.

A.22 Remote Data Acquisition, Control and Analysis Using LabVIEW Front Panel and Real-Time Engine

Students and faculty from South Carolina State University (SCSU) are collaborating with the staff at the Pisgah Astronomical Research Institute (PARI) to allow the SMILEY radio telescope to be accessed and controlled over the SCSU Network and the Internet. National Instruments (NI) LabVIEW software package has been used to design a virtual instrument (VI) for the system that has a front panel that will request a users name and password to either view and/or control the SMILEY. To publish the Web page, students used Hypertext Markup Language, Internet Protocol and a Uniform Resource Locator. The earlier versions of LabVIEW provided a number of VI options such as VI server, Data Socket Server, and Visual Basic and Active X to design and develop modules for remote data acquisition and control, usually in a local network. It also required that both the client and server must have LabVIEW and design and development of modules for remote control using the Internet was more involved. The recent LabVIEW 6.1 version introduced remote front panels to view and control a VI in any Web browser. The objective of this project is to discuss the design, development and testing of VI modules using LabVIEW front panel and real-time engine. This work was funded in part by a PAIR grant from NASA-MURED to SCSU under NCC 5-454.

B

LabVIEW Tools

The National Instrument's tools used to develop LabVIEW-based applications in various fields such as software management, electronics, data logging and supervisory control, image processing, motion control, and graphical system modeling are discussed in this appendix.

B.1 DIAdem

National Instruments' DIAdem is a software for managing, analyzing, and reporting data collected during data acquisition and/or generated during simulations. NI DIAdem is designed to meet the demands of today's testing environments, for which quick access to large volumes of scattered data, consistent reporting, and data visualization are required to make informed decisions. The DIAdem DataFinder, which is optimized for managing data from multiple sources and different formats, helps the user correlate data based on its descriptive attributes – functionality previously reserved for only those with a customized data management system. This tool is available at: http://www.ni.com/diadem/

B.2 Electronics Workbench

Electronics Workbench delivers the world's most widely used schematic capture and circuit simulation software with an installed base of more than 180,000 seats. The product portfolio includes software for schematic capture, circuit simulation (SPICE, VHDL, and patented cosimulation), PCB layout, and autorouting. Electronics Workbench software dramatically reduces time to market by helping the user to develop and create PCBs faster, more accurately, and with greater cost-effectiveness than comparable design suites. The popularity of circuit simulation continues to grow, and, at Electronics Workbench, simulation is the core of the technology. With Multisim, every circuit the user draw is automatically ready for simulation, so the user can begin testing at the earliest possible stage of the design flow – during schematic capture. Catch and correct errors almost before they happen. The tool is available at: http://sine.ni.com/nips/cds/view/p/lang/en/nid/202311

B.3 DSC Module

The National Instruments' LabVIEW Datalogging and Supervisory Control (DSC) Module is an extension of the LabVIEW graphical development environment that incorporates additional functionality for the rapid development of distributed measurement and control and high-channel-count monitoring applications. Whether the user needs to build a full-scale industrial automation and control system or just monitor a few dozen I/O points for historical collection, the NI LabVIEW DSC Module provides the tools to make the user more productive, including:

- Configuration-based monitoring and logging of system alarms and events
- One-touch data logging of networked I/O to a historical database
- Simplified networking of LabVIEW Real-Time and OPC-compliant devices
- Real-time and historical data trending
- Dialog-based front-panel security implementation

Further details on DSC module is obtained from: http://ni.com/labview/labviewdsc/

B.4 Vision Development Module

The National Instruments' LabVIEW Vision Development Module is for scientists, engineers, and technicians who are developing LabVIEW machine vision and scientific imaging applications. It includes IMAQ Vision, a library of powerful functions for vision processing, and Vision Assistant, an interactive environment for developers who need to prototype LabVIEW applications quickly without programming, or who need off-line inspection. Some of the features include:

- Critical distance measurement with gauging and caliper tools
- High-speed pattern matching
- Stream to disk with AVI support
- Grayscale, color, and binary image processing
- High-level machine vision, image processing functions, and display tools

National Instruments vision software products offer benefits for a breadth of machine vision applications. If the user wishes to build, benchmark, and deploy a vision application without programming, choose Vision Builder for Automated Inspection. If the user needs powerful programming libraries for the machine vision application, the Vision Development Module features hundreds of image processing and machine vision functions for LabVIEW, C/C++, Visual Basic, and .NET. Included in either software package is Vision Acquisition software, which acquires, saves, and displays images from thousands of different cameras, including IEEE 1394 (FireWire) cameras. Vision Acquisition software is also sold separately for security and monitoring applications. This module is available in: http://sine.ni.com/nips/cds/view/p/lang/en/nid/10419

B.5 FPGA

Using National Instruments' LabVIEW FPGA and reconfigurable I/O (RIO) hardware, the user can create custom I/O and control hardware without prior knowledge of traditional HDL languages or board-level hardware design. With the NI LabVIEW FPGA Module, the user can use LabVIEW graphical development to directly synthesize for FPGAs on NI RIO hardware. The user can also use reconfigurable FPGA technology for customizing measurement and control hardware with the same intuitive graphical programming as test and control systems. Additionally, the user can integrate the RIO hardware with the rest of LabVIEW for Windows or LabVIEW Real-Time application using the FPGA interface functions in LabVIEW. The user can create a LabVIEW FPGA block diagram, compile it, and download it to a RIO device. This code simultaneously executes a 32-bit counter and pulse generator and a custom control algorithm.

LabVIEW FPGA and RIO technology are ideal for applications requiring custom hardware. While testing requirements change, the user can simply modify the block diagram of the FPGA VI instead of scrapping the custom hardware the user built. Applications that often require custom hardware include:

- Custom timing and triggering routines
- Custom mix of analog, digital, counters, and triggers onto a single device
- High-channel counters, event detection, and time stamping
- Custom signal conditioning and parallel processing
- Digital communication protocols
- Rapid control prototyping
- Hardware-in-the-loop (HIL) simulation
- In-vehicle DAQ
- Machine control
- Machine condition monitoring

B.6 LabWindows/CVI

LabWindows/CVI is a proven test and measurement ANSI C development environment that greatly increases the productivity of engineers and scientists. Engineers and scientists use LabWindows/CVI to develop high performance, stable applications in the manufacturing test, military and aerospace, telecommunications, design validation, and automotive industries. LabWindows/CVI streamlines development in these areas with hardware configuration assistants, comprehensive debugging tools, and interactive execution capabilities developers can use to run functions at design time. Using built-in measurement libraries, developers can rapidly develop complex applications, such as multithreaded programs and ActiveX server/client programs. With the flexibility of the LabWindows/CVI environment, developers can optimize the data acquisition, analysis, and presentation components of their test and measurement applications. LabWindows/CVI 8.0 continues a history of innovation by offering the following features:

- Optimized Compiler Integration
- Improved System Deployment
- Call .NET Assemblies
- Directly Download LabWindows/CVI DLLs onto LabVIEW Real-Time Targets
- Native Tab Control
- More Than 80 New Analysis Functions
- IVI Custom Class Driver Wizard
- Updated Graph Control Features
- Include New Cell Types in the Table Control
- Custom Toolbar Enhancements
- Simplified Instrument I/O Assistant Code Generation
- Improved NI DIAdem File I/O Performance
- Source Code Browser Integration in the Workspace Window
- Ability to Add Custom Images to the Application Menu Bars
- Documentation Tooltips for User-Defined Functions

B.7 NI MATRIXx

National Instrument's MATRIXx is used for graphical system modeling and simulation in LabVIEW. The NI MATRIXx is used to:

- Build accurate, high-fidelity models and perform interactive simulations
- Analyze system models and build robust control algorithms
- Automatically document the system model and controller properties
- Generate readable, traceable, highly optimized C or Ada code
- Target the code to a real-time platform for control prototyping and hardware-in-the-loop test

For more than 20 years, engineers worldwide have relied on the MATRIXx product family for control design applications in automotive, aerospace and defense, process control, and academic environments. The MATRIXx suite of software includes:

SystemBuild. Graphical system modeling and simulation
Xmath. Interactive analysis, visualization, and control development
DocumentIt. Automatic documentation generation
AutoCode. Automatic embedded code generation for C and Ada
Target MATRIXx code to LabVIEW Real-Time and NI real-time hardware for rapid control prototyping and hardware-in-the-loop test.

B.8 Measurement Studio

National Instruments' Measurement Studio is an integrated suite of native measurement and automation controls, tools, and class libraries for Visual Studio .NET and Visual Studio 6.0. NI Measurement Studio dramatically reduces application development time with ActiveX and .NET controls, object-oriented measurement hardware interfaces, advanced analysis libraries, scientific user interface controls, measurement data networking, wizards, interactive code designers, and highly extensible classes.

Building on more than 16 years of measurement programming expertise, National Instruments created Measurement Studio with tools designed specifically for Microsoft Visual Basic, Visual C# .NET, and Visual C + + to bring the user simplified integration with data acquisition and instrument control interfaces for easier hardware integration. With Measurement Studio, the user uses the Visual Studio programming knowledge to get measurements started quickly and create a complete measurement solution – all the way from acquiring to analyzing to presenting.

Measurement Studio provides a collection of NET controls tightly integrated within the Visual Studio .NET 2003 environment that are designed specifically for engineers and scientists building virtual instrumentation systems. With Measurement Studio, the user can configure plug-in data acquisition devices, GPIB instruments, and serial devices from interactive wizards that also generate Visual Basic .NET or Visual C# .NET source code. With scientific user interface controls, the user can interactively configure graphs, knobs, meters, gauges, dials, tanks, thermometers, binary switches, and LEDs from property pages or collection editors. In addition, Measurement Studio delivers powerful networking components so the user can easily share live measurement data between applications via the Internet. The sources for Measurement Studio is obtained from: http://zone.ni.com/devzone/conceptd.nsf/webmain/

B.9 VI Logger

National Instruments' VI Logger Full is an easy-to-use yet very flexible tool specifically designed for the data logging applications. With intuitive dialog windows, the user can configure the logging task to easily acquire, log, view, and share the data. NI VI Logger is stand-alone, configuration-based data logging software – not a general-purpose application development environment. It is compatible with National Instruments plug-in data acquisition, Compact FieldPoint and FieldPoint products.

VI Logger Lite is data logging software with a subset of the features in VI Logger Full. Using VI Logger Lite helps the user to easily configure logging tasks, log data, export it to Microsoft Excel or NI DIAdem, and create LabVIEW example data logging programs. VI Logger Lite software is offered for FREE with the purchase of NI-DAQmx supported devices including M Series devices from: http://sine.ni.com/nips/cds/view/

B.10 Motion Control

The Windows motion control tools add powerfully integrated servo and stepper motor control to LabVIEW. NI offers VI libraries to control high-performance plug-in motion controller boards. Each VI library covers the full spectrum, from low-level motion control commands to high-level, integrated motion control on multiple synchronized axes. The VI libraries are separated into functional areas that provide initialization, individual control and programming commands, status reporting, and motion sequence flow control. High-level motion VIs provide simple access to the setup and initialization of each controller board and a hierarchy of single and multi-axis control VIs that access the power of the individual motion commands. Using NI Motion Assistant software (purchased separately) the user can generate new code based on these tools. This tool is obtained from: http://sine.ni.com/nips/cds/view/p/lang/en/nid/1397

B.11 TestStand

NI TestStand is a ready-to-run test management environment and framework designed to simplify the automation of the test and validation systems. TestStand is used to develop, manage, and execute test sequences.

These sequences integrate test modules written in any programming language. Sequences also specify execution flow, reporting, database logging and connectivity to other enterprise systems.

TestStand's architecture is open and provides all the necessary interfaces to fully customize the operator interface as well reporting, database logging, and management and execution environment. At the heart of TestStand is a high-speed, multithreaded execution engine which delivers the performance the user need, to meet the most rigorous test requirements, 24/7 and 365 days a year. This software tool is obtained from: http://zone.ni.com/devzone/conceptd.nsf/webmain/

B.12 SignalExpress

National Instruments' SignalExpress is interactive software designed to meet the needs of design and test engineers who need to quickly acquire and analyze electronic signals. SignalExpress simplifies the exploratory and automated measurement tasks by offering:

- A drag-and-drop, nonprogramming environment
- Rapid measurement configuration
- Integration of simulation data
- Extended functionality through NI LabVIEW

SignalExpress introduces an innovative approach to configuring measurements using intuitive drag-and-drop steps that do not require developing code. Unlike traditional benchtop measurement tools, SignalExpress combines the optimal balance of measurement functionality and ease of use to assist the user in streamlining a variety of applications:

- Design modeling
- Design verification
- Design characterization
- Device validation
- Automated test troubleshooting

SignalExpress tool is available at: http://ni.com/signalexpress/whatis.htm

C

Glossary

A

Absolute Accuracy

Used to describe the maximum or worst case error of a system when all potential system errors are considered. Absolute accuracy of a data acquisition system includes gain error, offset error, quantization error, noise, and nonlinearity.

Acquisition Rate

The rate at which samples are acquired. Typically described in terms of samples per second ($\mathrm{S\,s^{-1}}$), kilosamples per second ($\mathrm{kS\,s^{-1}}$), or Megasamples per second ($\mathrm{MS\,s^{-1}}$). Also may be referred to in terms of Hz, kHz, or MHz, where 1 Hz represents 1 sample per second.

ActiveX Control

A special function object that can readily be used by an OLE-enabled application, tool, or web browser. Examples include the functions in Measurement Computing and SoftWIRE products. ActiveX controls were replaced by .NET components and controls in version 7 of Visual Studio, which was released as Visual Studio .NET in 2001.

A/D

Analog-to-Digital (typically conversion).

ADC

Analog-to-Digital Converter. An electronic device that converts an analog input voltage into a digital value (discrete data values).

ADC Resolution

Resolution of the ADC, measured in bits. An ADC with 16 bits has a higher resolution and thus a higher degree of accuracy than a 12-bit ADC.

Analog Trigger

A trigger that is based on an analog threshold level. For example, the user may wish to start data acquisition scan when the input voltage first passes through 3.5 V. To accomplish this the user would set the analog trigger level to +3.5 V.

API

Application Programming Interface. Programming interface for controlling some software packages, such as Microsoft Visual SourceSafe.

Active Window

Window that is currently set to accept user input, usually the frontmost window. The title bar of an active window is highlighted. Make a window active by clicking it or by selecting it from the **Windows** menu.

AIGND

Analog input ground pin on a DAQ device.

Array

Ordered, indexed list of data elements of the same type.

Array Shell

Front panel object that houses an array. An array shell consists of an index display, a data object window, and an optional label. It can accept various data types.

Autoindexing

Capability of loop structures to disassemble and assemble arrays at their borders. As an array enters a loop with autoindexing enabled, the loop automatically disassembles it extracting scalars from 1D arrays, 1D arrays extracted from 2D arrays, and so on. Loops assemble data values into arrays as data values exit the loop in the reverse order.

Autoscaling

Ability of scales to adjust to the range of plotted values. On graph scales, autoscaling determines maximum and minimum scale values.

B

Block Diagram

A pictorial description or representation of a program or algorithm. In LabVIEW, the block diagram, which consists of executable icons called nodes and wires that carry data between the nodes, is the source code for the VI. The block diagram resides in the Diagram window of the VI.

Background

A DAQ system task (such as acquiring data) that occurs without interruption while another program routine is running.

Bandwidth, Small-Signal

A description of the highest frequency signal component that will pass through an amplifier and/or filter. Bandwidth is specified as the frequency where the attenuation is 3 dB.

Bandwidth, Large-Signal

Large signals are generally slew rate-limited before they are bandwidth-limited. Large signal bandwidth can be determined by the following equation: BW = Slew Rate/(2p × Vp) where Vp is the peak amplitude of the signal in question.

Bus Master

A type of a plug-in board or controller with the ability to take control of the computer's data bus and perform reads and/or writes without requiring the use of the host CPU.

Boolean Controls and Indicators

Front panel objects to manipulate and display Boolean (TRUE or FALSE) data.

Breakpoint

Pause in execution used for debugging.

Breakpoint Tool

Tool to set a breakpoint on a VI, node, or wire broken **Run** button Button that replaces the **Run** button when a VI cannot run because of errors.

Broken VI

VI that cannot run because of errors; signified by a broken arrow in the broken **Run** button.

Buffer

Temporary storage for acquired or generated data.

Bundle Function

Function that creates clusters from various types of elements.

Byte Stream File

File that stores data as a sequence of ASCII characters or bytes.

C

Conditional Terminal

The terminal of a While Loop that contains a Boolean value that determines whether the VI performs another iteration.

Control

Front panel object for entering data to a VI interactively or to a subVI programmatically.

Controls Palette

Palette containing front panel controls and indicators.

Cold-Junction

An unavoidable (and undesired) thermocouple junction that occurs where a thermocouple is connected to a measurement system.

Cold-Junction Compensation

A system developed to remove the error caused by the thermocouple's cold-junction.

Compact PCI

A bus configuration based on the popular PCI bus, but modified and enhanced for industrial applications.

Counter/Timer

A circuit or device that can be used to count or generate digital pulses or edges. Counter/timers are often used to set sample timing.

Calibration

A general term for any kind of alignment procedure or compensation routines to improve the accuracy of a sensor.

Capacitive Element

An MEMS structure designed to operate as a variable capacitor. Crossbow uses these to measure acceleration.

Caption Label

Label on a front panel object used to name the object in the user interface. The user can translate this label to other languages without affecting the block diagram source code.

Case

One subdiagram of a Case structure.

Case Structure

Conditional branching control structure that executes one of its subdiagrams based on the input to the Case structure. It is the combination of the IF, THEN, ELSE, and CASE statements in control flow languages.

Channel

Physical – a terminal or pin at which the user can measure or generate an analog or digital signal. A single physical channel can include more than one terminal, as in the case of a differential analog input channel or a digital port of eight lines. A counter also can be a physical channel, although the counter name is not the name of the terminal where the counter measures or generates the digital signal.

Channel Clock

Clock that controls the time interval between individual channel sampling within a scan. Products with simultaneous sampling do not have this clock.

Chart

2D display of one or more plots in which the display retains a history of previous data, up to a maximum that the user define. The chart receives the data and updates the display point by point or array by array, retaining a certain number of past points in a buffer for display purposes. See also scope chart, strip chart, and sweep chart.

CIN Source Code

Original, uncompiled text code. See also object code; Code Interface Node (CIN).

Class

A category containing properties, methods, and events. Classes are arranged in a hierarchy with each class inheriting the properties and methods associated with the class in the preceding level.

Clock

Hardware component that controls timing for reading from or writing to groups.

Cloning

To make a copy of a control or another object by clicking it while pressing the <<Ctrl>> key and dragging the copy to its new location. Mac OS Press the <<Option>> key. Linux Press the <<Alt>> key.

Cluster

A set of ordered, unindexed data elements of any data type, including numeric, Boolean, string, array, or cluster. The elements must be all controls or all indicators.

Cluster Shell

Front panel object that contains the elements of a cluster.

Coercion

Automatic conversion LabVIEW performs to change the numeric representation of a data element.

Coercion Dot

Appears on a block diagram node to alert the user that the user have wired data of two different numeric data types together. Also appears when the user wire any data type to a variant data type.

Color Copying Tool

Copies colors for pasting with the Coloring tool.

Coloring Tool

Tool to set foreground and background colors.

Common-Mode Voltage

Any voltage present at the instrumentation amplifier inputs with respect to amplifier ground.

Configuration Utility

Refers to Measurement & Automation Explorer on Windows and configuration utilities for the instrument on Mac OS and Linux.

Connector

Part of the VI or function node that contains input and output terminals. Data values pass to and from the node through a connector.

Connector Pane

Region in the upper right corner of a front panel or block diagram window that displays the VI terminal pattern. It defines the inputs and outputs the user can wire to a VI.

Constant

A terminal on the block diagram that supplies fixed data values to the block diagram. See also universal constant and user-defined constant.

Context Help Window

Window that displays basic information about LabVIEW objects when the user move the cursor over each object. Objects with context help information include VIs, functions, constants, structures, palettes, properties, methods, events, dialog box components, and items in the Project Explorer window.

Continuous Run Button

Icon that indicates a VI is set to execute repeatedly until the user stops it.

Control Flow

Programming system in which the sequential order of instructions determines execution order. Most text-based programming languages are control flow languages.

Conversion

Changing the type of a data element.

Coordinated Universal Time (UTC)

Time scale that is determined using highly precise atomic clocks. LabVIEW uses UTC as the standard for the internal representation of dates and times.

Current VI

VI whose front panel, block diagram, or Icon Editor is the active window.

D

D/A

Digital-to-Analog.

DAC or D/A Converter

Digital-to-Analog Converter. An electronic device that converts a digital number into an equivalent analog voltage or current.

DAQ

Common abbreviation for Data Acquisition.

DAQ Assistant

A graphical interface for configuring measurement tasks, channels, and scales.

DAQ Device

A device that acquires or generates data and can contain multiple channels and conversion devices. DAQ devices include plug-in drivers, PCMCIA cards, and DAQPad devices, which connect to a computer USB or 1394 (FireWire) port. SCXI modules are considered DAQ devices.

DAQ-STC

Data Acquisition System Timing Controller.

Data Acquisition (DAQ)

Acquiring and measuring analog or digital electrical signals from sensors, acquisition transducers, and test probes or fixtures. Generating analog or digital electrical signals.

Data Dependency

Condition in a dataflow programming language in which a node cannot execute until it receives data from another node.

Data Flow

Programming system consisting of executable nodes in which nodes execute only when they have received all the required input data and produce output automatically when they have executed. LabVIEW is a dataflow system. DAQ Channel Wizard Utility that guides the user through naming and configuring the DAQ analog and digital channels. DAQ Solution Wizard Utility that guides the user through specifying the DAQ application, from which it provides a custom DAQ solution.

Data Storage Formats

Arrangement and representation of data stored in memory.

Data Type

Format for information. In LabVIEW, acceptable data types for most VIs and functions are numeric, array, string, Boolean, path, refnum, enumerated type, waveform, and cluster.

Data Type Descriptor

Code that identifies data types; used in data storage and representation.

Datalog

To acquire data and simultaneously store it in a disk file. LabVIEW File I/O VIs and functions can log data.

Datalog File

File that stores data as a sequence of records of a single, arbitrary data type that the user specify while creating the file. Although all the records in a datalog file must be a single type, that type can be complex. For example, the user can specify that each record is a cluster that contains a string, a number, and an array.

dB or Decibel

A common unit used to express the logarithm of the ratio of two signal levels: $\mathrm{dB} = 20 \log 10\, \mathrm{V2/V1}$, for signal voltages V2 & V1 or $\mathrm{dB} = 10 \log 10\, \mathrm{P2/P1}$, for signal powers P2 and P1.

DDE

Dynamic Data Exchange. A software protocol in Microsoft Windows for inter-application communication. DDE allows a data acquisition application to share data real-time with Windows applications such as Microsoft Excel. DDE is an older protocol that has been replaced with COM and .Net protocols.

Device

An instrument or controller the user can access as a single entity that controls or monitors real-world I/O points. A device often is connected to a host computer through some type of communication network.

Device Number

Slot number or board ID number assigned to the device when the user configured it.

Differential Input

A differential analog input consists of two input terminals as well as a ground connection. The input measures the difference between the two inputs without (see CMRR) regard to ground potential variations. For example, the two inputs may be labeled High and Low, the difference between those two inputs is the measurement returned by the system. The third input, labeled Ground is a reference ground and must be connected to a reference ground at the signal source. A differential input can reject some signal difference between the Low input and the reference ground.

Digital Interface Card

Converts Crossbow analog sensor voltages to a digital signal and outputs in RS-232 format.

Digital Trigger

A trigger that is based on a standard digital threshold level.

DLL

Dynamic Link Library. A Windows-based file containing executable code that is run by other Windows applications or DLLs.

Drivers

Software that controls hardware devices, such as DAQ boards, GPIB. interface boards, PLCs, RTUs, and other I/O devices.

Dynamic Range

The ratio of the largest signal an input will measure to the smallest signal it can detect. Normally expressed in dB (20 log10 V2/V1).

E

Error Cluster

Consists of a Boolean status indicator, a numeric code indicator, and a string source indicator.

Error In

Error cluster that enters a VI.

Error Message

Indication of a software or hardware malfunction or of an unacceptable data entry attempt.

Empty Array

Array that has zero elements but has a defined data type. For example, an array that has a numeric control in its data display window but has no defined values for any element is an empty numeric array.

Enable Indexing

Option that allows the user to build a set of data to be released at the termination of a While Loop. With indexing disabled, a While Loop releases only the final data point generated within the loop.

End of File (EOF)

Character offset of the end of file relative to the beginning of the file. The EOF is the size of the file.

Error Out

The error cluster that leaves a VI.

Execution Highlighting

Debugging technique that animates VI execution to illustrate the dataflow in the VI.

Express VI

A subVI designed to aid in common measurement tasks. The user configure an Express VI using a configuration dialog box.

External Trigger

Voltage pulse from an external source that triggers an event, such as A/D conversion or a signal used to start or stop an event such as a scan of A/D channels. (Also, see Analog Trigger and Digital Trigger.)

Execution Highlighting

Feature that animates VI execution to illustrate the dataflow in a VI.

F

Flash ADC

An extremely high-speed analog-to-digital converter. The output code is determined in a single step by comparing the input signal to a series of references via a bank of comparators.

Floating

A condition in which there is no electrical connection between the ground of the signal source and the measurement system.

Fiber Optic Gyro

A high accuracy rate sensor that uses a laser and fiber optic ring to determine angular rate.

Flat Sequence Structure

Program control structure that executes its subdiagrams in numeric order. Use this structure to force nodes that are not data dependent to execute in the order the user want if flow-through parameters are not available. The Flat Sequence structure displays all the frames at once and executes the frames from left to right until the last frame executes.

Floating Signal Sources

Signal sources with voltage signals that are not connected to an absolute reference or system ground. Some common examples of floating signal sources are batteries, transformers, or thermocouples. Also called nonreferenced signal sources.

For Loop

Iterative loop structure that executes its subdiagram a set number of times. Equivalent to text-based code: For $i = 0$ to $n - 1$.

Foreground

An operation that is directly controlled by the active software application (see Background). When a foreground operation is running, no other foreground operations can be active in that window front panel. The interactive user interface of a VI. Modeled from the front panel of physical instruments, it is composed of switches, slides, meters, graphs, charts, gauges, LEDs, and other controls and indicators.

Formula Node

Node that executes equations the user enter as text. Especially useful for lengthy equations too cumbersome to build in block diagram form.

Frame

Subdiagram of a Flat or Stacked Sequence structure.

Front Panel

Interactive user interface of a VI. Front panel appearance imitates physical instruments, such as oscilloscopes and multimeters.

Function

Built-in execution element, comparable to an operator, function, or statement in a text-based programming language.

Functions Palette

Palette that contains VIs, functions, communication features, block diagram structures, and constants.

G

G

The graphical programming language used to develop LabVIEW applications.

Gain

A factor by which a signal is amplified, typically expressed in terms of "times" a number. Examples include X10 and X2 where the signal amplitude is multiplied by factors of 10 and 2, respectively.

General Purpose Interface Bus

GPIB. Synonymous with HP-IB. The standard bus used for controlling electronic instruments with a computer. Also called IEEE 488 bus because it is defined by ANSI/IEEE Standards 488-1978, 488.1-1987, and 488.2-1992.

Global Property

A VISA property whose value is the same for all sessions to the specified resource.

Global Variable

Accesses and passes data among several VIs on a block diagram.

Graph

2D display of one or more plots. A graph receives and plots data as a block.

Graph Control

Front panel object that displays data in a Cartesian plane.

Ground

A reference potential in an electrical system.

Grounded Signal Sources

Signal sources with voltage signals that are referenced to a system ground, such as a building ground. Also called referenced signal sources.

GPIB

General Purpose Interface Bus is the common name for the communications interface system defined in ANSI/IEEE Standard 488.1-1987 and ANSI/IEEE Standard 488.2-1987. Hewlett-Packard, the inventor of the bus, calls it the HP-IB.

Gyro

Gyroscope. A gyroscope uses angular momentum to fix an axis in space.

GUI

Graphical User Interface. A computer user interface in which the user interfaces with a computer via simple graphic displays or icons rather than text. GUIs often resemble common objects (e.g., stop signs) and are frequently developed to provide a "virtual" representation of a variety of instrumentation and data acquisition objects.

H

Handle

Pointer to a pointer to a block of memory that manages reference arrays and strings. An array of strings is a handle to a block of memory that contains handles to strings.

Handshaking

A type of protocol that makes it possible for two devices to synchronize operations.

Hardware Triggering

Form of triggering where the user set the start time of an acquisition and gather data at a known position in time relative to a trigger signal.

I

IEEE 488

See GPIB.

Impedance

The ratio of the voltage across a device or circuit to the current flowing in it. In AC circuits, the impedance takes into account the effects of capacitance and inductance. In most data acquisition specifications, the impedance listed is actually the DC impedance, which is the same as the resistance (in ohms).

Instrumentation Amplifier

An amplifying circuit whose output voltage with respect to ground is proportional to the difference between the voltages at its two inputs.

Input/Output

The transfer of data to or from a computer system involving communications channels, operator input devices, and/or data acquisition and control interfaces.

Integrating ADC

A slow but highly accurate and noise-immune analog-to-digital converter.

Internal Trigger

A condition or event that serves to start an operation or function. (See External Trigger.)

Isolation

Two circuits or devices are isolated when there is no electrical connection between them. Isolated circuits can be interfaced to each other via optical or electromagnetic circuits. A signal source is sometimes isolated from the measurement device in order to provide protection to the measurement device. IEEE 488.2 Institute of Electrical and Electronic Engineers Standard 488.2-1987, which defines the GPIB. indicator Front panel object that displays output. Instrument Wizard Utility that guides the user through communicating with instrument.

Indicator

Front panel object that displays output, such as a graph or LED.

Input Range

Difference between the maximum and minimum voltages an analog input channel can measure at a gain of the input range is a scalar value, not a pair of numbers. By itself, the input range does not uniquely determine the upper and lower voltage limits. An input range of 10 V could mean an upper limit of +10 V and a lower limit of 0 V or an upper limit of +5 V and a lower limit of −5 V. The combination of input range, polarity, and gain determines the input limits of an analog input channel. For some products, jumpers set the input range and polarity, although the user can program them for other products. Most products have programmable gains.

Instrument Driver

A set of high-level functions that control and communicate with instrument hardware in a system.

Isolation

Type of signal conditioning in which the user isolate the transducer signals from the computer for safety purposes. This protects the user and the computer from large voltage spikes and makes sure the measurements from the DAQ device are not affected by differences in ground potentials.

Iteration Terminal

Terminal of a For Loop or While Loop that contains the current number of completed iterations.

IVI

Interchangeable Virtual Instruments. A software standard for creating a common interface (API) to common test and measurement instruments.

IVI Driver

A driver written according to the IVI specification. The generic driver for a class of instruments (such as voltmeters) is called a class driver, whereas the driver for a specific instrument from a specific manufacturer is called a device-specific driver.

L

Label

Text object used to name or describe objects or regions on the front panel or block diagram.

Labeling Tool

Tool to create labels and enter text into text windows.

LabVIEW

Laboratory Virtual Instrument Engineering Workbench. LabVIEW is a graphical programming language written by National Instruments for the development of data acquisition software that uses icons instead of lines of text to create programs. X-view is written in the LabVIEW language.

Legend

Object a graph or chart owns to display the names and plot styles of plots on that graph or chart.

Linearity

The measure of a device's transfer function relative to a perfect Y = mX straight-line response.

Listener

A device in a GPIB system addressed by the controller for "listening" or receiving data. (See also Talker.)

LSB

Least Significant Bit.

Local Property

A VISA property whose value is unique to the session.

Local Variable

Variable that enables the user to read or write to one of the controls or indicators on the front panel of a VI.

M

Measurement & Automation Explorer

The standard National Instruments hardware configuration and diagnostic environment on Windows.

Measurement Device

A DAQ device such as the E Series multifunction I/O (MIO) device, the SCXI signal conditioning module, and the switch module.

Menu Bar

Horizontal bar that lists the names of the main menus of an application. The menu bar appears below the title bar of a window. Each application has a menu bar that is distinct for that application, although some menus and commands are common to many applications.

Module

A board assembly and its associated mechanical parts, front panel, optional shields, and so on. A module contains everything required to occupy one or more slots in a mainframe. SCXI and PXI devices are modules.

Multiplexer

A switch that allows one of multiple inputs to be selected and connected to a single output. Multiplexers are commonly used in DAQ products to allow a single A/D converter to acquire data from multiple analog input channels.

Mux

See Multiplexer.

Multiplexed Mode

SCXI operating mode in which analog input channels are multiplexed into one module output so that the cabled DAQ device can access the module's multiplexed output and the outputs on all other multiplexed modules in the chassis through the SCXI bus. Also called serial mode.

N

Name Label

Label of a front panel object used to name the object and as distinguish it from other objects. The label also appears on the block diagram terminal, local variables, and property nodes that are part of the object.

NaN

Digital display value for a floating-point representation of <Not A Number>. Typically the result of an undefined operation, such as log(−1).

NI-DAQ

Driver software included with all NI DAQ devices and signal conditioning components. NI-DAQ is an extensive library of VIs and functions the user can call from an application development environment (ADE), such as LabVIEW, to program an NI measurement device, such as the M Series multifunction I/O (MIO) DAQ devices, signal conditioning modules, and switch modules.

NI-DAQmx

The latest NI-DAQ driver with new VIs, functions, and development tools for controlling measurement devices. The advantages of NI-DAQmx over earlier versions of NI-DAQ include the DAQ Assistant for configuring channels and measurement tasks for a device for use in LabVIEW, LabWindows™/CVI™, and Measurement Studio; increased performance such as faster single-point analog I/O; NI-DAQmx simulation for most supported devices for testing and modifying applications without plugging in hardware; and a simpler, more intuitive API for creating DAQ applications using fewer functions and VIs than earlier versions of NI-DAQ.

Node

Execution element of a block diagram, such as a function, structure, or subVI. *See also* data flow, wire. Program execution element. Nodes are analogous to statements, operators, functions, and subroutines in text-based programming languages. On a block diagram, nodes include functions, structures, and subVIs.

Noise

Additional signals, coming from the product itself or other electronic equipment, that interfere with output signals the user are trying to measure.

Non-Displayable Characters

ASCII characters that cannot be displayed, such as null, backspace, tab, and so on.

Non-Referenced Single-Ended (NRSE) Measurement System

All measurements are made with respect to a common reference, but the voltage at this reference can vary with respect to the measurement system ground.

Numeric Controls and Indicators

Front panel objects to manipulate and display numeric data.

Nyquist Theorem

A sampling theory law that states that to create an accurate digital representation of a sampled waveform the user must sample the waveform, at least twice as fast as the highest frequency component contained in the waveform. Note that this is a minimum condition. In most applications, it is preferable to sample at a minimum of 3–4 times the highest expected frequency component.

O

Output Loading

The output characteristics of the device. We specify a minimum output equivalent resistance and maximum equivalent capacitance.

Object

Generic term for any item on the front panel or block diagram, including controls, indicators, nodes, wires, and imported pictures.

Object Code

Compiled version of source code. Object code is not standalone because the user must load it into LabVIEW to run it.

Object Shortcut Menu Tool

Tool to access a shortcut menu for an object.

Operating Tool

Tool to enter data into controls or to operate them.

One-Dimensional

Having one dimension, as in the case of an array that has only one row of elements.

P

Palette

Displays objects or tools the user can use to build the front panel or block diagram.

Panel Window

VI window that contains the front panel, the toolbar, and the icon and connector panes.

PC-Card

A credit-card-sized expansion card that fits in a PC-CARD slot, often referred to inaccurately as a PCMCIA card palette. A display of icons that represent possible options.

PCI

Peripheral Component Interconnect. An industry-standard, high-speed databus. A very high-performance expansion bus architecture developed by Intel to replace ISA and EISA. It has achieved widespread acceptance as a standard for PCs. It supports a theoretical maximum transfer rate of 132 $\mathrm{Mbytes\,s^{-1}}$.

PCMCIA

The Personal Computer Memory Card International Association. An international standards body and trade association that was founded in 1989 to establish standards for PC-CARD expansion cards used primarily in laptop computers. (See also PC-Card.)

Period

The period of a signal, most often measured from one zero crossing to the next zero crossing of the same slope. The period of a signal is the reciprocal of its frequency (in Hz). Period is designated by the symbol T.

Plug and Play

Describes plug-in boards that are fully configured in software, without the need for jumpers or switches on the boards.

Plot

Graphical representation of an array of data shown either on a graph or a chart.

Polymorphism

Ability of a node to automatically adjust to data of different representation, type, or structure.

Positioning Tool

Tool used to move, select, and resize objects.

Post Triggering

The technique used on a DAQ board to acquire a specified number of samples after trigger conditions have been met.

Probe

Debugging feature for checking intermediate values in a VI. PXI PCI eXtensions for Instrumentation.

Probe Tool

Tool to create probes on wires.

Port

A communications connection of one or more inputs on a computer. Common port types include RS-232 and USB.

Potentiometer

A continuously-adjustable variable resistor. They are used for adjustment of electrical circuits and as transducers for either linear or rotary position transmission.

Pretriggering

A technique used on a DAQ board in which a buffer is continuously filled with data. When the trigger occurs, the sample includes one buffer full of samples immediately prior to the trigger. For example, if a 1,000 sample pretrigger buffer is specified, and 20,000 posttrigger samples, the final sample set has 21,000 samples in it, 1,000 of which were taken prior to the trigger event.

Programmed I/O

A data transfer method where the data is read or written by the CPU.

Propagation Delay

The amount of time required for a signal or disturbance to pass through a circuit or process.

Project Explorer Window

Window in which the user can create and edit LabVIEW projects.

Project Library

A collection of VIs, type definitions, shared variables, palette menu files, and other files, including other project libraries.

Property Node

Sets or finds the properties of a VI or application.

PXI

PCI eXtensions for Instrumentation. A modular, computer-based instrumentation platform.

Q

Quantization Error

The uncertainty that is inherent when digitizing an analog value due to the finite resolution of the conversion process.

R

Rate Gyro

A special kind of gyroscope that measures rotation rate around a fixed axis.

Range

Region between the limits within which a quantity is measured, received, or transmitted. Expressed by stating the lower and upper range values.

Real Time

A method of operation in which data is processed as it is acquired instead of being accumulated and post-processed. Process control is generally done in real time where data analysis is not.

Relative Accuracy

A measure of accuracy (typically in LSBs) of an A/D converter. It includes all non-linearity and quantization errors. It does not include gain and offset errors of the measurement circuitry. As a measurement, it is the deviation of the measured data from the ideal data relative to a straight line drawn through the measured endpoints.

Referenced Single-Ended (RSE) Measurement System

All measurements are made with respect to a common reference or a ground. Also called a grounded measurement system.

Repeatability

The maximum variation between repeated measurements under the same conditions.

Resolution

The smallest increment that can be detected by a data acquisition or measurement system. Resolution is expressed in bits, in proportions, or in percent of full scale. For example, if a system has 12-bit resolution, it equals one part in 4,096 for resolution, or 0.0244% of full scale.

RS-232

Recommended Standard 232, a serial interface bus standard.

RS-485

Recommended Standard 485, a serial interface bus standard.

RTD

Resistance Temperature Detector. A sensor probe that measures temperature based on changes in resistance.

S

Sample Rate

The rate at which a signal or value is sampled. It is frequently expressed as samples per second ($\mathrm{S\,s^{-1}}$), kilosamples per second ($\mathrm{kS\,s^{-1}}$), or megasamples per second ($\mathrm{MS\,s^{-1}}$).

Scan

A scan is a group of channels sampled in sequence. Often, the sequence is repeated. For example, a scan of channels 0–3 samples those four channels. If more than four samples are requested, then the fifth sample will contain data from channel 0, and so on. Scans are generally sequential. Systems that have a channel/gain queue may be used to create scans that are not sequential.

Scope Chart

Numeric indicator modeled on the operation of an oscilloscope.

SCPI

Standard Commands for Programmable Instruments. An extension of the IEEE 488.2 standard that defines a standard programming command set and syntax for device-specific operations.

Scrolling Tool

Tool to move through windows.

SCXI

Signal Conditioning eXtensions for Instrumentation. The National Instruments product line for conditional low-level signals within an external chassis near sensors, so only high-level signals in a noisy environment are sent to DAQ devices.

SE

See Single-Ended Input.

Sensor

A device that responds to stimulus such as temperature, light, sound, pressure, motion, or flow and produces an output that can be measured to learn about the particular condition being monitored.

Sensitivity

The ratio of output volts to sensor output range (e.g., V/G.) Sensitivity allows the user to predict the sensor response to a sensor input.

Sequence Local

Terminal to pass data between the frames of a Stacked Sequence structure.

Sequence Structure

See Flat Sequence structure.

Settling Time

The time required for a voltage to stabilize at its final value (usually within a specified error range).

Shift Register

Optional mechanism in loop structures to pass the value of a variable from one iteration of a loop to a subsequent iteration. Shift registers are similar to static variables in text-based programming languages.

Shortcut Menu

Menu accessed by right-clicking an object. Menu items pertain to that object specifically.

S/H

Sample-and-Hold. A circuit that acquires and stores a signal (e.g., an analog voltage) on a capacitor or other storage element for a period of time.

Signal Conditioning

A catch all term for vibration filtering, signal amplification, etc. Crossbow products have built in signal conditioning that allow the sensors to be directly interfaced with data acquisition circuits.

Single-Ended Input

An analog input having an input terminal that is measured with respect to a common reference, usually analog ground. It has two input connections, one for the signal being measured and one for the common reference. In multiple input configurations, all signal inputs share the common reference. Input systems that use analog ground as the reference are called Referenced Single-Ended Inputs. Systems that allow an arbitrary reference within the input common-mode range are known as Non-Referenced Single-Ended Inputs. (See also Differential Inputs as a contrasting input type.)

Slew Rate

The specified (typically maximum) rate of change of a D/A converter or amplifier/buffer output. It is expressed in volts/microsecond.

SNR

Signal-to-Noise Ratio. The ratio of the overall signal level to the noise level, typically expressed in dB.

Software Trigger

An event that is started (triggered) based on software control.

Strain Gauge

A sensor with resistance that varies based on being either stretched or compressed. When attached to a solid object with known physical properties, the resultant deflection signal can be converted to units measuring force.

Strip Chart

Numeric plotting indicator modeled after a paper strip chart recorder, which scrolls as it plots data.

Structure

Program control element, such as a Flat Sequence structure, Stacked Sequence structure, Case structure, For Loop, or While Loop.

Subdiagram

Block diagram within the border of a structure.

Subpalette

A palette contained in an icon of another palette.

Subroutine

A group of software instructions separate from the main program that executes a specific function upon command and then returns control to the main program.

SubVI

VI used in the block diagram of another VI; comparable to a subroutine.

Supervisory Control

Control in which the control loops operate independently subject to intermittent corrective action.

Successive-Approximation A/D Converter

An ADC that sequentially compares a diminishing series of binary-weighted values generated by a D/A converter against the analog input.

Sweep Chart

Numeric indicator modeled on the operation of an oscilloscope. It is similar to a scope chart, except that a line sweeps across the display to separate old data from new data.

Synchronous

A timing configuration in which events occur in step with a reference clock or timer.

T

Talker

A GPIB device that sends data to one or more listeners on the bus.

Task

A collection of one or more channels, timing, triggering, and other properties in NI-DAQmx. A task represents a measurement or generation the user want to perform.

Task ID

Number LabVIEW generates to identify the task at hand for the NI-DAQ drive. The following table gives the function code definitions.

TCP/IP

Transmission Control Protocol/Internet Protocol. A standard format for transmitting data in packets from one computer to another. The two parts of TCP/IP are TCP, which deals with the construction of data packets, and IP, which routes them from computer to computer.

Function Code	I/O Operation
1	analog input
2	analog output
3	digital port I/O
4	digital group I/O
5	counter/timer I/O

Temperature Range

The temperatures between which the sensor will accurately operate.

Temperature Coefficient

The change of value or function corresponding with a change in temperature. This is often expressed as a percentage of reading per degree or in PPM (parts per million) per degree.

Terminal

Object or region on a node through which data passes.

Thermistor

A type of resistive temperature sensor. The thermistor resistance changes as a function of temperature.

Thermocouple

A temperature sensor made by fusing together dissimilar metals. The junction produces a small voltage (referred to as the Seebeck voltage) that varies as a function of temperature.

Tip Strip

A text banner that displays the name of an object, control, or terminal.

Tools Palette

Palette containing the tools the user can use to edit and debug front panel and block diagram objects.

Tunnel

Data entry or exit terminal on a structure.

Transducer

See Sensor.

Trigger

A signal that is used to start or stop an operation. Triggers can be an analog, digital, or software event.

Trigger Polarity

Trigger polarity defines whether the trigger occurs when the signal is rising in a positive direction or falling in a negative direction.

Ticks

Time in milliseconds required for the entire calculation.

Tool

Special cursor to perform specific operations.

Toolbar

Bar that contains command buttons to run and debug VIs.

Tools Palette

Palette that contains tools the user can use to edit and debug front panel and block diagram objects.

Top-Level VI

VI at the top of the VI hierarchy. This term distinguishes the VI from its subVIs.

Traditional NI-DAQ (Legacy)

An older driver with outdated APIs for developing data acquisition, instrumentation, and control applications for older National Instruments DAQ devices. The user should use Traditional NI-DAQ (Legacy) only in certain circumstances. Refer to the *NI-DAQ Readme* for more information about when to use Traditional NI-DAQ (Legacy), including a complete list of supported devices, operating systems, and application software and language versions.

Two-Dimensional

Having two dimensions, as in the case of an array that has several rows and columns.

U

Unevenly Sampled Data

Data that has been sampled with nonequal sampling intervals.

User-Defined Constant

Block diagram object that emits a value the user set.

Unipolar

A signal range from ground to a positive value (e.g., 0 to +5 V).

USB

Universal Serial Bus. A high-speed serial bus.

V

Vertical Gyro

A gyroscope whose spin axis is "stabilized" or gimbaled to remain vertical with respect to gravity. The vertical gyro is used to measure rotations away from vertical relative to the fixed earth reference frame.

Virtual Instrument

A combination of hardware and software elements that emulates a standalone instrument both in electrical function and in the computer screen representation.

Virtual Instrumentation

A program in the graphical programming language G that models the appearance and function of a physical instrument.

VISA

Virtual Instrument Software Architecture. A single interface library for controlling GPIB, VXI, RS-232, and other types of instruments.

VXI

VME eXtensions for Instrumentation (bus)

VI Server

Mechanism for controlling VIs and LabVIEW applications programmatically, locally and remotely.

W

Waveform

Multiple voltage readings taken at a specific sampling rate.

Waveform Chart

Indicator that plots data points at a certain rate.

While Loop

Loop structure that repeats a section of code until a condition occurs.

Wire

Data path between nodes. *See also* data flow.

Wire Branch

A section of wire that contains all the wire segments from one junction to another, from a terminal to the next junction, or from one terminal to another if no junctions exist in-between.

Wire Junction

The point where three or more wire segments join. wire segment A single, horizontal or vertical piece of wire.

Wiring Tool

Tool used to define data paths between terminals. Resembles a spool of wire.

X

X-Analyze

Software from Crossbow that interfaces to Crossbow sensor products and reads data through the serial port. X-Analyze allows the user to log data and print waveforms.

X-View

Demonstration software that reads data from Crossbow digital products through the PC serial port and displays the data.

Bibliography

1. J.S. Adam, E. Rosow, and T. Karselis, "Educational Applications Using Computerized Virtual Instrumentation", Presented at the Association for the Advancement of Medical Instrumentation in Philadelphia, PA, June 4, 1996
2. M. Akay and A. Marsh (Eds.), Information Technologies in Medicine, Volume 1, Medical Simulation and Education, Wiley-IEEE, Hokoken, NJ, 2001, ISBN: 0-471-38863-7
3. H. Hoffman, M. Murray, R. Curlee, and A. Fritchle, "Anatomic Visualizer: Teaching and Learning Anatomy with Virtual Reality", in M. Akay and A. Marsh (Eds.), Information Technologies in Medicine, Volume 1, Medical Simulation and Education, Wiley-IEEE, 2001, pp. 205–218
4. F. Amigoni, A. Brandolini, G.D'Antona, R. Ottoboni, and M. Somalvico, "Artificial Intelligence in Science of Measurements and the Evolution of the Measurements Instruments: A Perspective Conception", IEEE International Symposium on Virtual Environments and Measurement Systems VIMS'02, May 2002, paper VM-2015
5. M. Baszynski, C.R. Allen, Z. Moron, and P.A. Heasman, "3D Posture Measurement in Dental Applications", IEEE International Symposium on Virtual Environments and Measurement Systems VIMS'02, May 2002, paper VM-2017
6. A. Bernieri, G. Betta, A. Pietrosanto, and C. Sansone, "A Neural Network Approach to Instrument Fault Detection and Isolation", IEEE Transactions on Instrumentation and Measurement 44(3), 747–750, 1995
7. P. Bhagwat, "Bluetooth: Technology for Short-Range Wireless Apps", IEEE Internet Computing 5(3), 96–103, 2001
8. K.S. Bhaskar, J.K. Pecol, and J.L. Beug, "Virtual Instruments: Object Oriented Program Synthesis", Conference Proceedings on Object-Oriented Programming Systems, Languages and Applications, ACM, New York, NY, 1986, pp. 303–314
9. N.H.M. Caldwell, B.C. Breton, and D.M. Holburn, "Remote Instrument Diagnosis on the Internet", IEEE Intelligent Systems 2(3), 70–76, 1998
10. S. Chakravarthy, R. Sharma, and R. Kasturi, "Noncontact Level Sensing Technique Using Computer Vision", IEEE Transactions on Instrumentation and Measurement 51(2), 353–361, 2002
11. J. Charles, "Neural Interfaces Link the Mind and the Machine", Computer 32(1), 16–18, 1999

12. Object Management Group (OMG). The Common Object Request Broker: Architecture and Specification (CORBA), revision 2.0. Object Management Group (OMG), 2.0 ed
13. P. Daponte, L. Nigro, and F. Tisato, "Object Oriented Design of Measurement Systems", IEEE Instrumentation and Measurement Magazine 41(6), 874–880, 1992
14. B. Davie, V. Florance, A. Friede, J. Sheehan, and J.E. Sisk, "Bringing Health-Care Applications to the Internet", IEEE Internet Computing 5(3), 42–48, 2001
15. M. Dunbar, "Plug-and-Play Sensors in Wireless Networks", IEEE Instrumentation and Measurement Magazine 4(1), 19–23, 2001
16. J.F. Thayer, "Progress in Heart-Rate Variability Analysis", IEEE Engineering in Medicine and Biology Magazine 21(4), 22–23, 2002
17. E. Tatara and A. Cinar, "Interpreting ECG Data by Integrating Statistical and Artificial Intelligence Tools", IEEE Engineering in Medicine and Biology Magazine 21(1), 36–41, 2002
18. L. Ferrigno and A. Pietrosanto, "A Bluetooth-Based Proposal of Instrument Wireless Interface", IEEE International Symposium on Virtual Environments and Measurement Systems VIMS'02, May 2002, paper VM-2003
19. R.E. Filman, "Medicine on the Net: From Isolation to Universal Connectivity", IEEE Internet Computing 5(3), 40–41, 2001
20. J.P. Fisher, J. Martin, F.P. Day, E. Rosow, J. Adam, C. Chen, and L.D. Gillam, "Validation of a Less Invasive Method for Determining Preload Recruitable Stroke Work Derived with Echocardiographic Automatic Boundary Detection", Circulation 92, 1–278, 1995
21. G. Fortino and L. Nigro, "Development of Virtual Data Acquisition Systems Based on Multimedia Internet Working", Computer Standards and Interfaces 21(5), 429–440, 1999
22. M. Frigo and S.G. Johnson, "FFTW: An Adaptive Software Architecture for the FFT", ICASSP Conference Proceedings, Volume 3, 1998, pp. 1381–1384, also http://www.fftw.org/
23. L.A. Geddes and L.E. Baker, Principles of Applied Biomedical Instrumentation, 3rd ed., Wiley, New York, 1989
24. H. Goldberg, "What Is Virtual Instrumentation?", IEEE Instrumentation and Measurement Magazine 3(4), 10–13, 2000
25. K. Goldberg, "Introduction: The Unique Phenomenon of a Distance", in K. Goldberg (Ed.), The Robot in the Garden: Telerobotics and Telepistemology in the Age of the Internet, MIT, Cambridge, MA, 2000
26. D. Grimaldi, L. Nigro, and F. Pupo, "Java-Based Distributed Measurement Systems", IEEE Transactions on Instrumentation and Measurement 47(1), 100–103, 1998
27. D.L. Hudson and M.E. Cohen, Neural Networks and Artificial Intelligence for Biomedical Engineering, Wiley-IEEE, Piscataway, NJ, 1999, ISBN: 0-7803-3404-3
28. T.L. Huston and J.L. Huston, "Is Telemedicine A Practical Reality?", Communications of the ACM 43(6), 91–95, 2000
29. IEEE Standard for Medical Device Communications – Physical Layer Interface – Cable Connected, IEEE Std 1073.4.1, IEEE, 2000 edition, 2000, http://standards.ieee.org/catalog/olis/meddev.html

30. IEEE 1073 Family of Standards, http://standards.ieee.org/catalog/olis/meddev.html.
31. R.J. Kennelly, "The IEEE 1073 Standard for Medical Device Communications", IEEE Systems Readiness Technology Conference AUTOTESTCON '98, IEEE, 1998, pp. 335–336
32. J. Zidek, "Application of Graphical Programming and Benefit of Virtual Instrumentation in Teaching", International Network for Engineering Education and Research Conference – ICEE'99, 1999, Czech Republic, paper no. 277
33. E. Jovanov, J. Price, D. Raskovic, K. Kavi, T. Martin, and R. Adhami, "Wireless Personal Area Networks in Telemedical Environment", Third IEEE EMBS Information Technology Applications in Biomedicine – Workshop of the International Telemedical Information Society ITABITIS 2000, Arlington, Virginia, 2000, pp. 22–27
34. E. Jovanov, D. Starcevic, and V. Radivojevic, "Perceptualization of Biomedical Data", in M. Akay and A. Marsh (Eds.), Information Technologies in Medicine, Volume 1, Medical Simulation and Education, Wiley-IEEE, Hokoken, NJ, 2001, pp. 189–204
35. E. Jovanov, D. Raskovic, J. Price, A. Moore, J. Chapman, and A. Krishnamurthy, "Patient Monitoring Using Personal Area Networks of Wireless Intelligent Sensors", Thirty-Eighth Annual Rocky Mountain Bioengineering Symposium, Copper Mountain, Colorado, April 2001
36. E. Jovanov, D. Starcevic, A. Samardžic, A. Marsh, and Ž. Obrenovic, "EEG Analysis in a Telemedical Virtual World", Future Generation Computer Systems 15, 255–263, 1999
37. S.F. Khalid and J. Laney, Advanced Topics in LabWindows/CVI, Prentice-Hall, Englewood Cliffs, NJ, 2001
38. D.G. Kilman, D.W. Forslund, "An International Collaboratory Based on Virtual Patient Records", Communications of the ACM, 40(8), 111–117, 1997
39. W. Kozaczynski, G. Booch, "Component-Based Software Engineering", IEEE Software 15(5), 34–36, 1998
40. D.B. Lange, M. Ochima, "Seven Good Reasons for Mobile Agents", Communications of the ACM 42(3), 88–89, 1999
41. A.C. Lear, "XML Seen as Integral to Application Integration", IT Professional 1(4), 12–16, 1999
42. N. Leavitt, "Will WAP Deliver the Wireless Internet?", Computer 33(5), 16–20, 2000
43. X. Liu, G. Cheng, and J.X. Wu, "AI for Public Health: Self-Screening for Eye Diseases", IEEE Intelligent Systems 13(5), 28–35, 1998
44. W. Loob, "Virtual Instrumentation Boosts Capabilities and Reduces Costs", Medical Device & Diagnostic Industry Magazine, September 2000, pp. 10
45. B. Marovic and Z. Jovanovic, "Visualization of 3D Fields and Medical Data Using VRML", Future Generation Computer Systems 14 (1–2), 33–49, 1998
46. G.R. McMillan, R.G. Eggleston, and T.R. Anderson, "Non Conventional Controls", in G. Salvendy (Ed.), Handbook of Human Factors and Ergonomics, 2nd ed., Wiley, New York, 1997, pp. 729–771
47. P.T. Moran, G.F. Harris, K. Acharya, Z. Hongsheng, and J.J. Wertsch, "A Biofeedback Cane System: Instrumentation and Subject Application Results", IEEE Transactions on Rehabilitation Engineering 3(1), 132–138, 1995

48. F. Nebeker, "Golden Accomplishments in Biomedical Engineering", IEEE Engineering in Medicine and Biology Magazine 21(3), 17–47, 2002
49. National Instruments, "The Measurement Revolution, White Paper", September 1999, http://www.ni.com/revolution/learn.htm
50. Z. Obrenovic, D. Starcevic, E. Jovanov, and V. Radivojevic, "An Implementation of Real-time Monitoring and Analysis in Telemedicine", Third IEEE EMBS Information Technology Applications in Biomedicine Workshop of the International Telemedical Information Society ITABITIS 2000, Arlington, Virginia, 2000, pp. 74–78
51. Z. Obrenovic, D. Starcevic, E. Jovanov, and V. Radivojevic, "An Agent Based Framework for Virtual Medical Devices", Proceedings of The First International Joint Conference on Autonomous Agents and Multiagent Systems AAMAS 2002, ACM, New York, NY, USA, 2002, 659–660
52. J.B. Olansen and E. Rosow, Virtual Bio-Instrumentation: Biomedical, Clinical, and Healthcare Applications in LabVIEW, Prentice-Hall, Englewood Cliffs, 2001
53. S. Oviatt and P. Cohen, "Perceptual User Interfaces: Multimodal Interfaces that Process What Comes Naturally", Communications of the ACM 43(3), 45–53, 2000
54. S. Oviatt, "Ten Myths of Multimodal Interaction", Communications of the ACM 42(11), 74–81, 1999
55. M. Parvis and A. Vallan, "Medical Measurements and Uncertainties", IEEE Instrumentation and Measurement Magazine 5(2), 12–17, 2002
56. T.L. Huston and J.L. Huston, "Is Telemedicine A Practical Reality?", Communications of the ACM 43(6), 91–95, 2000
57. P. Pianegiani, D. Macii, and P. Carbone, "Open Distributed Control and Measurement System Based on an Abstract Client–Server Architecture", IEEE International Symposium on Virtual Environments and Measurement Systems VIMS'02, May 2002, paper VM-2007
58. C. Alippi, A. Ferrero, and V. Piuri, "Artificial Intelligence for Instruments and Measurement Applications", IEEE Instrumentation & Measurement Magazine 1(2), 9–17, 1998
59. A.A. Platonov and W. Winiecki, "Statistical Synthesis and Optimal Decomposition in Intelligent Monitoring Systems Design", Computer Standards & Interfaces 24(2), 101–110, 2002
60. C. Potter, R. Brady, P. Moran, C. Gregory, B. Carragher, N. Kisseberth, J. Lyding, and J. Lindquist, "EVAC: A Virtual Environment for Control of Remote Imaging Instrumentation", IEEE Computer Graphics and Applications 16(4), 62–66, 1996
61. Y. Qingping and C. Butler, "An Object-Oriented Model of Measurement Systems", IEEE Transactions on Instrumentation and Measurement 47(1), 104–107, 1998
62. H. Schumny, "Quality and Standardization Aspects of Digitalization", Computer Standards and Interfaces 22(4), 237–238, 2000
63. RMI: Remote Method Invocation Web Site. http://java.sun.com./products/jdk/rmi/index.html
64. R.A. Robb, Three-Dimensional Biomedical Imaging: Principles and Practice, Wiley, New York, 1994
65. R.A. Robb, Biomedical Imaging, Visualization, and Analysis, Wiley, New York, 1999

66. D.L. Rollins, C.R. Killingsworth, G.P. Walcott, R.K. Justice, R.E. Ideker, and W.M. Smith, "A Telemetry System for the Study of Spontaneous Cardiac Arrhythmias", IEEE Transactions on Biomedical Engineering 47(7), 887–892, 2000
67. A. Rosenbloom, "How the Virtual Inspires the Real", Communications of the ACM 45(7), 29–30, 2002
68. E. Rosow, J. Adam, F. Beatrice, K. Youngdahl, and R. Orlando, "A Virtual Instrumentation Evaluation System for Fiberoptic Endoscopes", Presented at the Society of American Gastrointestinal Endoscopic Surgeons (SAGES) March 21, 1997, San Diego, CA
69. Network Working Group Sun Microsystems. RPC: Remote Procedure Call Protocol Specification. RFC 1057, June 1988. http://info.internet.isi.edu:80/in-notes/rfc/files/rfc1057.txt
70. A. Sachenko, "Intelligent Data Acquisition and Advanced Computing Systems", Computer Standards and Interfaces 24(2), 97–100, 2002
71. L. Schwiebert, S.K.S. Gupta, J. Weinmann, "Research Challenges in Wireless Networks of Biomedical Sensors", Proceedings of the Seventh Annual International conference on Mobile Computing and Networking, ACM, New York, NY, 2001, pp. 151–165
72. F. Russo, "Fuzzy Systems in Instrumentation: Fuzzy Signal Processing", IEEE Transactions on Instrumentation and Measurement 45(2), 683–689, 1996
73. H.J.W. Spoelder, A.H. Ullings, and F.C.A. Groen, "Virtual Instrumentation: A Survey of Standards and Their Interrelation", Conference Record IMTC '97, Volume 1, 1997, pp. 676–681
74. P. Arpaia, "Introduction to the Special Issue on Digital Instruments Standardization", Computer Standards and Interfaces 22(2), 85–88, 2000
75. D. Starcevic, E. Jovanov, V. Radivojevic, Ž. Obrenovic, and A. Samardžic, "Virtual Medical Devices for Telemedical Applications", in P. Spasic, I. Milosavljevic, and M. Jancic-Zguricas (Eds.), "Telemedicine", Academy of Medical Sciences of Serbian Medical Association, Belgrade, Yugoslavia, 2000, pp. 218–244
76. M. Turk and G. Robertson, "Perceptual User Interfaces (Introduction)", Communications of the ACM 43(3), 33–35, 2000
77. F.M. Vos, G.L. van der Heijde, H.J.W. Spoelder, I.H.M van Stokkum, and F.C.A. Groen, "A New Instrument to Measure the Shape of the Cornea Based on Pseudorandom Color Coding", IEEE Transactions on Instrumentation and Measurement 46(4), 794–797, 1997
78. M.G. Vose and G. Williams, "Laboratory Virtual Instrument Engineering workbench", BYTE 11(9), 1986
79. The VRML Consortium Incorporated, Information Technology – Computer Graphics and Image Processing – The Virtual Reality Modeling Language (VRML) – Part 1: Functional Specification and UTF-8 Encoding. International Standard ISO/IEC 14772-1:1997, 1997, http://www.vrml.org/Specifications/VRML97/index.html
80. J.C. Waller and N. Foster, "Training Via the Web: A Virtual Instrument", Computers and Education 35(2), 161–167, 2000
81. Y.I. Wijata, D. Niehaus, and V.S. Frost, "A Scalable Agent-Based Network Measurement Infrastructure", IEEE Communications Magazine 38(9), 174–183, 2000

82. M.J. Wooldrige and N.R. Jennings, "Software Engineering with Agents: Pitfalls and Pratfalls", IEEE Internet Computing 3(3), 20–27, 1999
83. F. Zubillaga-Elorza and C.R. Allen, "A Java-Based Electronic Design Automation Toolkit for the Simulation and Testing of Instruments", Proceedings of the IEEE Instrumentation and Measurement Technology Conference IMTC '98, IEEE, 1998, pp. 304–309
84. F. Zubillaga-Elorza and C.R. Allen, "Virtual Instrument Toolkit: Rapid Prototyping on the Web", IEEE Internet Computing 3(5), 41–48, 1999
85. J. Bryzek, et al., "Common Communication Interfaces for Networked Smart Sensors and Actuators", SENSORS, Helmers Publishing, September, 1995, pp. 14–23
86. J. Bryzek, "Introduction to IEEE-P1451: The Emerging Hardware Independent Communication Standard for Smart Transducers", Proceedings of Eurosensors X, September, 1996, Leuven, Belgium
87. B. Hamilton, "A Compact Representation of Units", HP Laboratories Technical Report HPL-96-61, Hewlett-Packard Company, Palo Alto, California, May, 1996
88. B. Travis, "Smart-Sensor Standard Will Ease Networking Woes", EDN, Cahners Publishing, June 22, 1995, pp. 49–55
89. P. DuPuis, "Pressure Sensor Case Study #5", Fourth IEEE/NIST Workshop on Smart Transducer Interface Standards, Chicago, IL, September 14, 1995
90. J.C. Eidson and S.P. Woods, "A Research Prototype of a Networked Smart Sensor System", Proceedings Sensors Expo Boston, Helmers Publishing, May, 1995, pp. 223–232
91. R. West, "Automatic Calibration of Strain Gauges", SENSORS, Helmers Publishing, July, 1995, pp. 44–46
92. A.E. Brendel, "Sensor Stores Calibration Data in Nonvolatile ROM", Test and Measurement World, 32, 1996
93. "IEEE P1451.2 D2.01 IEEE Draft Standard for A Smart Transducer Interface for Sensors and Actuators – Transducer to Microprocessor Communication Protocols and Transducer Electronic Data Sheet (TEDS) Formats", Institute of Electrical and Electronics Engineers, August, 1996
94. J. Pinto, ISA paper #94-598, "Networked, Programmable, Intelligent I/O, The 'Truly' Distributed Control Revolution", Proceedings of the Industrial Computing Conference, Volume 4, Part 2, October 23, 1994, pp. 141–147
95. G. Tapperson, ISA paper #94-569, "Fieldbus: Migrating Control to Field Devices", Advances in Instrumentation and Control, 49(3), 1239–1252, 1994
96. J. Pinto, ISA paper #94-514, "Fieldbus- A Neutral Instrumentation Vendor's Perspective", Advances in Instrumentation and Control, 49(2), 669–675, 1994
97. S. Woods and K. Moore, "An Analysis of HP-IB Performance in Real Test Systems", Hewlett-Packard Laboratories Technical Report 93–20, August 1992
98. J.H. Saltzer, D.P. Reed, and D.D. Clark, "End-To-End Arguments in System Design", ACM Transactions on Computer Systems, 2(4), 277–288, 1984
99. S. Leichner, W. Kent, B. Hamilton, "Dimensioned Data", Hewlett-Packard Laboratories Private Communication, February 1995
100. Johnson, R.N. Building Plug-and-Play Networked Smart Transducers, Sensors 1997, 40–58
101. Eccles, L.H. A Brief Description of IEEE P1451.2, Proceedings of the Sensors Expo 81–90, 1997

102. A. Mason, N. Yazdi, A.V. Chavan, K. Najafi, and K.D. Wise, A Generic Multielement Microsystem for Portable Wireless Applications, Proceedings of the IEEE, 86(8), 1733–1746, 1998
103. J.M. Rabaey, M. Josie Ammer, J.L. da Silva Jr., D. Patel, Roundy, S. PicoRadio Supports AdHoc Ultra_low Power Wireless Networking, Computer Magazine, 2000
104. W. Ye, J. Heidenann, D. Estrin, An Energy-Efficient MAC Protocol for Wireless Sensor Networks, Usc/Isi Technical Report Isi-Tr-543, September, 2001
105. W.B. Heinzelman, Application-Specific Protocol Architectures for Wireless Networks. Ph.D Thesis, Massachusetts Institute of Technology, June, 2000
106. P. Rentala, R. Musunuri, S. Gandham, U. Saxena, Survey on Sensor Networks, http://citeseer.nj.nec.com/correct/479874, 2001
107. F. Primdahl, "The Fluxgate Magnetometer", Journal of Physics E 12, 214–252, 1979
108. T. Van Duzer and C.W. Turner, "Principles of Superconducting Devices and Circuits", Elsevier, NY, 1981, pp. 216–226
109. R.M.A. Azzam and N.M. Bashara, "Ellipsometry and Polarized Light", NH, Amsterdam, pp. 1–10, 1977
110. K.S. Lee, et al. "Fiber Optic Magnetostrictive Transducers for Magnetic Filed Sensing", Optical Engineering 43(12), 3577–3581, 1995
111. H. Okamura, "Fiber-Optic Magnetic Sensor Utilizing the Lorentzian Force", Journal of Lightwave Technology 8, 1558–1564, 1990
112. J.E. Lentz, et al. "Magnetic Materials Characterization Using a Fiber Optic Magnetometer", Journal of Application Physics 57, 3820–3822, 1984
113. G.W. Day and A.H. Ross, "Faraday Effect Sensors: The State of the Art", Proceedings of SPIE 985, Fiber Optic and Laser Sensors, 6, 138–150, 1980
114. R.B. Wagreich and C.C. Davis, "Magnetic Field Detection Enhancement in an External Cavity Fiber Fabry-Perot Sensor", Journal of Lightwave Technology, 14, 2246–2249, 1996
115. C.D. Butter and G.E. Hocker, "Fiber Optics Strain Gage", Applied Optics 17, 2867–2869, 1978
116. C.M. Dube, S. Thordarson, and K.H. Wansor, "Closed-Loop Fiber-Optic Magnetometer/Gradiometer", Proceedings of the SPIE 838, Fiber Optic and Sensors, 5, 17–27, 1988
117. D.M. Dagenais, F. Bucholtz, and K.P. Koo, "Elimination of Residual Signals and Reduction of Noise in Low-Frequency Magnetic Fiber Sensor", Applied Physics Letter, 53, 1417–1476, 1988
118. E. Udd, "Fiber Optic Sensors", Wiley, New York, 1991
119. J.H. Arthur, M.R. Sexton. LabVIEW Application: Energy Laboratory Upgrade. Proceedings of the 2002 American Society for Engineering Education (ASEE) Annual Conference & Exposition, Session 3233, 2002
120. R.H. Bishop. Learning with LabVIEW 6i. Prentice Hall, Upper Saddle River, NJ, 2001
121. N. Ertugrul. LabVIEW for Electric Circuits, Machines, Drives, and Laboratories. Prentice Hall PTR, Upper Saddle River, NJ, 2002
122. H. Franz. Use of LabVIEW Software for Virtual Instrumentation Technology. Proceedings of the 2003 American Society for Engineering Education (ASEE) Annual Conference & Exposition, 2003

123. R. Hennessey, H. Loya, B. Diong, R. Wicker. Using LabVIEW to Develop an Automated Control System. NI Instrumentation Newsletter, Special Academic Edition. Available Online http://www.nemesis-online.it/newsletters/Academic% 20Newsletter%201%202001.pdf
124. M.L. Higa, D.M. Tawy, S.M. Lord, An Introduction to LabVIEW Exercise for an Electronics Class. Proceedings of 32nd ASEE/IEEE Frontiers in Education Conference, Session T1D-13, 2002
125. C.D. Johnson, Process Control Instrumentation Technology, 7th ed. Prentice Hall, Upper Saddle River, NJ, 2003
126. N. Kiritsis, Y.W. Huang, and D. Ayrapetyan. A Multi-Purpose Vibration Experiment Using LabVIEW. Proceedings of the 2003 ASEE Annual Conference & Exposition. Session 1426, 2003
127. M. Kostic, Data Acquisition and Control Using LabVIEW™ Virtual Instrument for an Innovative Thermal Conductivity Apparatus. Proceedings of Virtual Instrumentation in Education 1997 Conference. June 12, 1997, MIT, 1997, pp. 131–136.
128. D. McDonald, A. Mahajan, Data Acquisition Systems: An Integral Part of Undergraduate Engineering and Technology Programs. Proceedings of the 1998 American Society for Engineering Education (ASEE) Annual Conference & Exposition. Session 1559, 1998
129. National Instruments LabVIEW Manual, National Instruments Corporate Headquarters, Austin, TX, 2002
130. NIWeek Papers. The Worldwide Conference on Measurement and Automation, 1998. Available online http://iawww3.epfl.ch/NIWeek98/pages/LabVIEW.htm
131. R. Pecen, M.D. Salim and M. Timmerman, A Hybrid Solar-Wind Power Generation System as an Instructional Resource for Industrial Technology Students. Journal of Industrial Technology, 16(3), 2000
132. R. Pecen and M. Timmerman, A Novel Power Quality Scheme to Improve a Utility Interface in a Small-Sized Hybrid Solar/Wind Generation Unit. Tenth International Power Quality Conference, 1999
133. R. Pecen, T. Hall, F. Chalkiadakis and A. Zora, Renewable Energy Based Capstone Design Applications for an Undergraduate Engineering Technology Curriculum. Proceedings of 33rd ASEE/IEEE 2003 Frontiers in Education Conference, Session S1E on Multidisciplinary Approaches, Boulder, CO, 2003
134. W.H. Rigby and T. Dalby, Computer Interfacing: A Practical Approach to Data Acquisition and Control. Prentice Hall, Englewood Cliffs, NJ, 1995
135. L. Sun, S. Lei, Y. Chen, H.T. Hsieh, and D. Peper, Environmental Data Management System – From Data Monitoring, Acquisition, Presentation to Analysis, 2002. Available Online http://www.unlv.edu/ Research_Centers/NCACM/HTML/research/daq/publicationdaqpdcs2002.pdf
136. J. Travis, LabVIEW for Everyone, 2nd Ed. Prentice Hall, Upper Saddle River, NJ, 1994. R. Waterbury, Software Trends Lead to Control Forefront. ISA Intech 41(6).
137. S. Wolf, and R.F.M. Smith, Student Reference Manual for Electronics Instrumentation Laboratories. Prentice Hall, Englewood Cliffs, NJ, 1990
138. W. Zabolotny, K. Pozniak, R. Romaniuk, T. Czarski, I. Kudla, K. Kierzkowski, T. Jezynski, A. Burghardt, and S. Simrock, "Distributed Embedded PC Based Control and Data Acquisition System or TESLA Cavity Controller and Simulator", Proceedings of SPIE 5484, 2004, pp. 223–230

139. K.T. Pozniak, M. Bartoszek, and M. Pietrusinski, "Internal Interface for RPC Muon Trigger Electronics at CMS Experiment", Proceedings of SPIE 5484, 2004, pp. 269–282.
140. T.S. Rappaport, Wireless Communications: Principles and Practice, Upper Saddle River, New Jersey, Prentice Hall, 2002
141. A.J. Viterbi, CDMA Principles of Spread Spectrum Communication, Addison-Wesley, Reading, MA, 1995
142. V.K. Garg, Wireless Network Evolution: 2G to 3G, Prentice Hall, Upper Saddle River, New Jersey, 2002
143. Bluetooth SIG Web Site: http://www.bluetooth.org
144. HiperLAN2 Global Forum Web Site: http://www.hiperlan2.com
145. HomeRF Web Site: http://www.homerf.org.
146. W.C.Y. Lee, Mobile Cellular Telecommunications Systems, McGraw Hill, New York, 1989
147. M. Yacoub, Foundations of Mobile Radio Engineering, CRC, Boca Raton, FL, 1993
148. IEEE, Wireless Media Access Control (MAC) and Physical Layer (PHY) Specification, P802.11D6.2, July 1998
149. Aly Elosery, Autonomous Power Control in CDMA Cellular Systems, PhD Dissertation, EECE Department, University of New Mexico, 2001
150. P. Gupta and P. Kumar, "The Capacity of Wireless Networks", IEEE Transactions on Information Theory, IT-46, May 2000
151. A.B. MacKenzie and S.B. Wicker, "Game Theory and the Design of Self-Configuring, Adaptive Wireless Networks", IEEE Communications Magazine, 39(11), 126–131, 2001
152. R. Ziemer, R. Peterson, and D. Borth, Introduction to Spread Spectrum Communications, Prentice-Hall, Upper Saddle River, NJ, 1995
153. W.C.Y. Lee, Digital Communications, 4th ed., New York, McGraw Hill, 2000
154. J.W. Norenberg, "Bridging Wireless Protocols", IEEE Communications Magazine, 39(11), 90–97, 2001
155. R.C. Petersen, P.A. Testagrossa, Radio-frequency Electromagnetic Fields Associated with Cellular-Radio Cell-Site Antennas, Bio-electromagnetics, 13, 527–542, 1992
156. B.S. Kaneshiro, Report on the Informational Workshop on Electric Magnetic Fields (EMFS) and Cellular Transceiver Facilities, Prepared for the California Public Utilities Commission, December 1993
157. T.S. Tenforde, Electromagnetic Fields and Carcinogenesis: An Analysis of Biological Mechanisms. Proceedings of the ICWCHR State of the Science Colloquium, Rome, Italy, November 13–15, 1995
158. N.A. Cridland, Electromagnetic Fields and Cancer: A Review of Relevant Cellular Studies, Rep. No. NRPB-R256, National Radiological Protection Board, Chilton, Didcot, Oxon, UK, 1993
159. D. Brusick, Genetic Effects of RFR, Proceedings of the ICWCHR State of the Science Colloquium, Rome, Italy, November 13–15, 1995
160. M.L. Meltz, Studies on Microwave Induction Genotoxicity: A Laboratory Report, Proceedings of the ICWCHR State of the Science Colloquium, Rome, Italy, November 13–15, 1995
161. J.L. Sagripanti, M.L. Swicord, C.C. Davis, Microwave Effects on Plasmid DNA, Radiation Research 110, 219–231, 1987

162. A. Maes, L. Vershaeve, A. Arroyo, C. DeWagter, L. Vercruyssen, In Vitro Cytogenetic Effects of 2450 MHz Waves on Human Peripheral Blood Lymphocytes, Bioelectromagnetics 14, 495–501, 1993
163. H. Lai, N. Singh, Acute Low Intensity Microwave Exposure Increases DNA Single-Strand Breaks in Rat Brain Cells, Bioelectromagnetics 16, 207–210, 1995
164. H. Lai, N. Singh, Single- and Double-strand DNA Breaks in Rat Brain Cells After Acute Exposure to Radiofrequency Electromagnetic Radiation, International Journal of Radiation Biology 16, 207–210, 69, 513–521, 1996
165. H. Lai, N. Singh, Melatonin and a Spin–Trap Compound Block Radiofrequency Electromagnetic Radiation-Induced DNA Strand Breaks in Rat Brain Cells, Bioelectromagnetics 18, 446–454, 1997
166. A.M. Lillienfeld, et al., Foreign Service Health Status Study – Evaluation of Health Status of Foreign Service and Other Employees from Selected Eastern European Posts, Final Report, Contract No. 6025-619073, Department of State, Washington, DC, 1978
167. S. Milham, Jr., Increased Mortality in Amateur Radio Operators Due to Lymphatic and Hematopoietic Malignancies, AJE, 127, 50–54, 1988
168. S. Szmigielski, Cancer Morbidity in Subjects Occupationally Exposed to High Frequency (Radiofrequency and Microwave) Electromagnetic Radiation, Science of Total Environment 108, 9–17, 1996
169. Environmental Epidemiology Program, Cancer Incidence in Census Tracts with Broadcasting Towers in Honolulu, Hawaii. Report to the City Council, City and County of Honolulu, Hawaii, State of Hawaii Department of Health, 1986
170. B. Hocking, I.R. Gordon, H.L. Grain, G.E. Hatfield, Cancer Incidence and Mortality and Proximity to TV Towers, Medical Journal of Australia 165, 601–605, 1996
171. H. Dolk, G. Shaddick, P. Walls, et al., Cancer Incidence Near Radio and Television Transmitters in Great Britain. Part I: Sutton Coldfield Transmitter, AJE 145, 1–9, 1997
172. H. Dolk, G. Shaddick, P. Walls, et al., Cancer Incidence Near Radio and Television Transmitters in Great Britain. Part II: All High Power Transmitters, AJE 145, 10–17, 1997
173. D. Hill, Epidemiological Studies of Radiofrequency Radiation Exposure, A Report to the California Public Utilities Commission, 1993
174. The ISIS ENGIN-X Engineering Diffractometer Instrument, http://www.isis.rl.ac.uk/engineering/
175. ISIS Second Generation Data Acquisition Electronics (DAE-II) – for Information Contact J.Norris@rl.ac.uk
176. The MXI-2 Interface from National Instruments, http://www.ni.com/
177. The LabVIEW Package from National Instruments, http://www.ni.com/labview/
178. The NeXus Data File Format, http://www.neutron.anl.gov/NeXus/
179. The Ray of Light Instrument Control Program, C.M. Moreton-Smith and A.M. Rice, http://www.isis.rl.ac.uk/computing/renaissance/index.htm.
180. *Open GENIE Reference Manual*, Technical Report RAL-TR-1999-031, Rutherford Appleton Laboratory, UK – http://www.isis.rl.ac.uk/OpenGENIE/
181. *Open GENIE – Analysis and Control*, arXiv paper cond-mat/0210442, NOBUGS2002/025 – see http://www.isis.rl.ac.uk/NOBUGS2002/
182. VNC Virtual Network Computing Software, AT & T Research Labs, Cambridge – http://www.uk.research.att.com/vnc/

183. NOAA-UMES Summer Camp, Available Online: http:/www.umes.edu/NOAA
184. D.G. Peterson, "Engineering Criteria 2000: A Bold New Change Agent", ASEE PRISM, September, 1997
185. A. Nagchaudhuri, "Introduction of Mechatronics in Pre-College Programs and Freshman Design Course in an Active and Cooperative Learning Framework", Thirty-First ASEE/IEEE Frontiers in Education Conference, Reno, NV, CD ROM, October 10–13, 2001
186. Beakman's Electric Motor, Available Online: http://fly.hiwaay.net/~palmer/motor.html
187. West Point Bridge Design Software, Available Online: http://bridgecontest.usma.edu/download.htm
188. D. Baum, Definitive Guide to Lego-Mindstorms, APRESS, 2000
189. F.G. Martin, "Robotic Explorations: A Hands-On Introduction to Engineering", Prentice Hall, Englewood Cliffs, NJ, 2001
190. J.R. Claycomb, K.E. Bassler, J.H. Miller, Jr., M. Nersesyan, and D. Luss. "Avalanche Behavior in the Dynamics of Chemical Reactions", Physical Review Letters 87, 178303-1-4, 2001
191. S.M. Lord, "An Innovative Multidisciplinary Elective On Optoelectronic Materials and Devices", Session 3a4, Proceedings of the 1995 Frontiers in Education Conference, Atlanta, GA, November 1995
192. In 1995, Funding Was Obtained from National Science Foundation Instrumentation & Laboratory Improvement (ILI) DUE grant #9552260 and the PA Equipment Grant Fund to Establish Bucknell's First Optoelectronics Laboratory.
193. C.J. Theesfeld and S.M. Lord, "Designing Optoelectronic Laboratories: A Unique Senior Design Opportunity", Session 7c2, Proceedings of the 1996 Frontiers in Education Conference, Salt Lake City, UT, November, 1996
194. Comprehensive Collaborative Framework (CHEF) Project Web Site, 2004. http://chefproject.org
195. "The MOST Experiment, July 30, 2003". NEESgrid Technical Report, 2003. http://www.neesgrid.org/documents/MOST_document_v1.0.pdf
196. R. Butler, D. Engert, I. Foster, C. Kesselman, S. Tuecke, J. Volmer, and V. Welch. "A National-Scale Authentication Infrastructure." IEEE Computer 33(12), 60–66, 2000
197. D. Gehrig, "Guide to the NEESgrid Reference Implementation." NEESgrid TR-2004-04, 2004 http://www.neesgrid.org/documents/TR_2004_04.pdf
198. I. Foster, C. Kesselman, G. Tsudik, and S. Tuecke, "A Security Architecture for Computational Grids." Fifth ACM Conference on Computer and Communications Security 83–91, 1998
199. W. Johnston, "Realtime Widely Distributed Instrumentation Systems." I. Foster and C. Kesselman (Eds.), The Grid: Blueprint for a New Computing Infrastructure, Morgan Kaufmann 75–103, 1999
200. D. Marcusiu, "NEESgrid Experiment-Based Deployment Strategy and Requirements", 2003 http://www.neesgrid.org/documents/NEESgrid_epd_strat_req_ v1.0.doc
201. M. Nakashima, H. Kato, and E. Takaoka, "Development of Real-Time Pseudo Dynamic Testing." Earthquake Engineering and Structural Dynamics 21, 79–92, 1999

202. Guo, T., D.Y. Chen, and F.C. Lee (1993) Separation of Common-Mode and Differential-Mode Conducted EMI. Proceeding of the Eleventh Annual VPEC Power Electronic Seminar, University of Virginia, Charlottesville, VA, USA, pp. 293–301
203. R. Zahavi, Enterprise Application Integration with CORBA, Wiley, 1999
204. R.J. Machado, J.M. Fernandes, A Petri Net Meta-Model to Develop Software Components for Embedded Systems, ACSD'01, IEEE CS, 2001, pp. 113–122
205. A. Guru, P. Savory, R. Williams, "A Web-Based Interface for Storing and Executing Simulation Models", in Proceedings of the 2000 Winter Simulation Conference, J.A. Joines, R.R. Barton, K. Kang, and P.A. Fishwick (Eds.), Department of Industrial and Management Systems Engineering, University of Nebraska, Lincoln, USA, 2000, pp. 1810–1814
206. Jorma Kyyra, "Challenges in the Teaching of Electrical Power Engineering–Motivation of Students and Change of Paradigms", in Proceedings of the International Conference on Engineering Education, ICEE2001, Oslo, August 2001, pp. 8D6-26–30
207. K. Holbert, M. Albu, "A Signal Processing Laboratory Employing On-Line Teaming for Remote Experiments", to be presented at the 2003 American Society for Engineering Education Annual Conference & Exposition, Nashville, TN, June 2003
208. T. Oeren, "Impact of Data on Simulation: from Early Practices to Federated and Agent-Directed Simulation", in Proceedings of EUROSIM 2001, Shaping Future with Simulation, A.W. Heemink, L. Dekker, H. de Swaan Arons, I. Smit, and Th.L. van Stijn (Eds.), Delft, NL, June 2001
209. M. Albu, A. Ferrero, F. Mihai, and S. Salicone, "Remote Calibration Using Mobile, Multi-Agent Technology", to be presented at the IEEE IMCT Conference, Vail, CO, 21–23 May 2003
210. P. Meadows, "Portable Electrical Stimulation Systems", IEEE AES Systems Magazine, October 1991, pp. 6–10
211. M. Ilic, D. Vasiljevic, and D.B. Popovic, "A Programmable Electronic Stimulator for FES Systems", IEEE Transactions Rehabilitation Engineering 2(4), 234–238, 1994
212. J.-P. Poindessault, C. Beauquin, and F. Gaillard, "Stimulation, Data Acquisition, Spike Detection and Time/Rate Analysis with a Graphical Programing System: An Application to Vision Studies", Journal of Neuroscience Methods pp. 225–235, 1995
213. M.A. Nordstrom, E.A. Mapletoft, and T.S. Miles, "Spike-Train Acquisition, Analysis and Real-Time Experimental Control Using a Graphical Programming Language", Journal of Neuroscience Methods 93–102, 1995
214. R. Muralidhar, S. Kaur, and M. Parashar: "An Architecture for Web-Based Interaction and Steering of Adaptive Parallel/Distributed Applications". In Proceedings of the European Conference on Parallel Processing, 2000, 1332—1339
215. C.C. Ko, B.M. Chen, S.Y.V. Ramakrishnan, C.D. Cheng, Y. Zhuang, and J. Chen, A Web-Based Virtual Laboratory on a Frequency Modulation Experiment, IEEE Transactions on Systems, Man, and Cybernetics 31(3), 295–303, 2001
216. S.H. Chen, R. Chen, V. Ramakrishnan, S.Y. Hu, Y. Zhuang, C.C. Ko, and B.M. Chen, Development of Remote Laboratory Experimentation Through

Internet. Proceedings of the 1999 IEEE Hong Kong Symposium on Robotics and Control Hong Kong, pp. 756–760, 1999

217. W. Blokland, "An Interface from LabVIEW to the Accelerator Controls Network", Proceedings of Accelerator Inst. Workshop, Berkeley, USA, 1992, pp. 320–329
218. A.A. Hahn and P. Hurh, "Results from an Imaging Beam Monitor in the Tevatron Using Synchrotron Light", HEACC'92, Hamburg, Germany, pp. 248–250, July 1992
219. E. Barsotti, "A Longitudinal Bunch Monitoring System Using LabVIEW and High-Speed Oscilloscopes", Accelerator Instrumentation Workshop, Vancouver, Canada, pp. 466–472, 1994
220. L. Catani, "Integration of a PC Based Experiment in a Network Based Control System: the TTF Optical Diagnostics Control System", PCaPAC98, Tsukuba (Japan).
221. R. Pascoe, "Introducing WWW Technology into Tertiary Teaching: A Personal Perspective", in Proceedings of North American Web Conference, Fredericton, NB, Canada, 1997. Available: http://www.uvm.edu/∼hag/naweb97/papers/pascoe.html
222. D. Tilbury, J. Luntz, and W. Messner, "Controls Education on the WWW: Tutorials for MATLAB and SIMULINK—Web Technology for Controls Education", in Proceedings of the American Control Conference, Philadelphia, PA, pp. 1304–1309, 1998
223. S.G. Crutchfield and W.J. Rugh, "Interactive Exercises and Demonstrations on the Web for Basic Signals, Systems, and Controls", in Proceedings of the 36th Conference Decision Control, San Diego, CA, pp. 3811–3815, 1997
224. P. Antsaklis et al., "Report on the NSF/CSS Workshop on New Directions in Control Engineering Education", IEEE Control System 19, 53–58, May 1999
225. B. Foss, T. Eikass, and M. Hovd, "Merging Physical Experiments Back into the Learning Arena", in Proceeding of the American Control Conference, Chicago, IL, pp. 2944–2948, 2000
226. A.C. Catlin et al. "The SoftLab experience: Building virtual laboratories for Computational Science". Available: http://www.cs.purdue.edu/research/cse/softlab/softlablabs/softlabframe work/softlab_report/report.html.
227. B. Aktan et al., "Distance Learning Applied to Control Engineering Laboratories", IEEE Transactions on Education 39, 320–326, 1996
228. T. Junge and C. Schmid, "Web-Based Remote Experimentation Using a Laboratory-Scale Optical Tracker", in Proceedings of the American Control Conference, Chicago, IL, pp. 2951–2954, 2000
229. C.C. Ko, "A Large Scale Web-Based Virtual Oscilloscope Laboratory Experiment", Proceedings of the IEEe Engineering Science and Education Journal 9(2), 69–76, 2000
230. J. Travis, Internet Applications in LabVIEW, Apr. 2000. Available: http://www.phptr.com/ptrbooks/ptr_0 130 141 445.html
231. L. Wells and J. Travis, LabVIEW for Everyone: Graphical Programming Made Even Easier, Prentice-Hall, Englewood Cliffs, NJ, 2000. Online Available: http://www.ni.com/reference/books/lv4every.htm
232. I.Y. Bar-Itzhack and Y. Vitek, The Enigma of False Bias Detection in a Strapdown System During Transfer Alignment, Journal of Guidance and Control 8, 175–180, 1985

233. H.K. Lee, J.G. Lee, G.I. Jee, Calibration of Measurement Delay in GPS/SDINS Hybrid Navigation, AIAA Journal of Guidance, Control, and Dynamics 25, 240–247, 2002
234. D. Goshen-Meskin, I.Y. Bar-Itzhack, Observability Analysis of Piece-Wise Constant Systems, Part II: Application to Inertial Navigation In-Flight Alignment, IEEE Transactions on Aerospace and Electronic Systems 28, 1068–1075, 1992
235. D. Knight, Achieving Modularity with Tightly Coupled GPS/INS, Proceedings of the IEEE PLANS '92, Monterey, CA, 426–432, 1992
236. M.K. Martin, B.C. Detterich, C-MIGITSII Design and Performance Boeing, C-MIGITS II Integrated GPS/INS User Guide, 1997
237. S. Dekey, Making Windows NT a Real-Time Solution – A Technical Overview, National Instruments. Available at: http://zone.ni.com/zone/jsp/zone.jsp.
238. N. Dorst, Using LabVIEW to Create Multithreaded VIs for Maximum Performance and Reliability, National Instruments. Available at: http://zone.ni.com/zone/jsp/zone.jsp.
239. J. Hill and D. Culler, "MICA: A Wireless Platform for Deeply Embedded Networks", IEEE Micro 22, 12–24, 2002
240. V. Shnayder, M. Hempstead, B.-r. Chen, G.W. Allen, and M. Welsh, "Simulating the Power Consumption of Large-Scale Sensor Network Application", in Proceedings of the Second ACM Conference on Embedded Networked Sensor Systems (SenSys'04), Baltimore, MD, USA, pp. 188–200, 2004
241. P. Levis, N. Lee, M. Welsh, and D. Culler, "TOSSIM: Accurate and Scalable Simulation of Entire TinyOS Applications", in First ACM Conference on Embedded Networked Sensor Systems (SenSys 2003), Los Angeles, CA, USA, pp. 126–137, 2003
242. A. Sinha and A. Chandrakasan, "Dynamic Power Management in Wireless Sensor Networks", IEEE Design and Test of Computers 18, 62–74, 2001
243. J. Polastre, R. Szewczyk, and D. Culler, "Telos: Enabling Ultra-Low Power Wireless Research", http://www.moteiv.com (January 2004)
244. D. Raskovic, "Energy-Efficient Hierarchical Processing in the Network of Wireless Intelligent Sensors (WISE)", Ph.D. Thesis, ECE Dept., University of Alabama in Huntsville, 2004
245. N. Chang, K. Kim, and H.G. Lee, "Cycle-Accurate Energy Measurement and Characterization With a Case Study of the ARM7TDMI", IEEE Transactions on Very Large Scale Integration (VLSI) Systems 10, 146–154, 2002
246. L. Day et al., "Automated Control and Real-Time Data Processing of Wire Scanner/Halo Scraper Measurements", Particle Accelerator Conference, 2001
247. J. Power et al., Experimental Physics and Industrial Control System [6] "Beam Position Monitor Systems for the SNS LINAC", Particle Accelerator Conference, 2003
248. J. Benford and J. Swegle, High Power Microwaves, Artech House, Boston, MA, 1992
249. C. Taylor and D. Giri, High-Power Microwave Systems and Effects, Taylor & Francis, Washington, DC, 1994
250. D. Price, J. Levine, and J. Benford, "Diode Plasma Effects on the Microwave Pulse Length from Relativistic Magnetrons", IEEE Transactions Plasma Science 26, 348, 1997

251. L. Moreland, E. Schamiloglu, R. Lemke, S. Korovin, V. Rostov, A. Roitman, K. Hendricks, and T. Spencer, "Efficiency Enhancement of High Power Vacuum BWO Using Nonuniform Slow Wave Structures", IEEE Transactions Plasma Science, 22(2), 554–565, 1994
252. L. Moreland, E. Schamiloglu, R. Lemke, A. Roitman, S. Korovin, and V. Rostov, "Enhanced Frequency Agility of High Power Relativistic Backward Wave Oscillators", IEEE Transactions Plasma Science 24(2), 852–858, 1996
253. C. Abdallah, W. Yang, E. Schamiloglu, and L. Moreland, "A neural Network Model of the Input/Output Characteristics of a High Power Backward-Wave Oscillator", IEEE Transactions Plasma Science 24(3), 879–883, 1996
254. C. Abdallah, V. Soualian, and E. Schamiloglu, "Towards "Smart Tubes" Using Iterative Learning Control", IEEE Transaction Plasma Science 26, 905–911, 1998
255. R. Smith and A. Packard, "Optimal Control of Perturbed Linear Static Systems", IEEE Transactons on Automatic Control 579–584, 1996
256. S. Hara, Y. Yamamot, T. Omata, and M. Nakano, "Repetitive Control System: A New Type Servo System for Periodic Exogenous Signals", IEEE Transactions on Automatic Control 33, 659–667, 1988
257. K.L. Moore, "Iterative Learning Control – An Expositionary Overview", Applied and Computational Controls, Signal Processing, and Circuits 1, 425–488, 1997
258. T.-Y. Kuc, K. Nam, and J.S. Lee, "An Iterative learning Control of Robot Manipulators", IEEE Transactions on Robotics and Automation 7, 835–842, 1991
259. S. Bennett, and S.R. Emge, "Fiber Optic Rate Gyro for Land Navigation and Platform Stabilization." Presented at Sensors Expo '94, Cleveland, OH, Septemper 2000, 1994
260. J. Borenstein, and J. Evans, "The OmniMate Mobile Robot – Design, Implementation, and Experimental Results." Proceedings of the 1997 IEEE International Conference on Robotics and Automation, Albuquerque, NM, April 21–27, pp. 3505–3510, 1997
261. J. Borenstein, B. Everett, and L. Feng, "Navigating Mobile Robots: Systems and Techniques." A.K. Peters, Wellesley, MA, 1996, ISBN 1-56881-058-X.
262. J. Borenstein, and L. Feng, "UMBmark: A Benchmark Test for Measuring Dead-Reckoning Errors in Mobile Robots", Presented at the 1995 SPIE Conference on Mobile Robots, Philadelphia, PA, October, 1995
263. H.R. Everett, "Sensors for Mobile Robots", A.K. Peters, Wellesley, MA, 1995. Komoriya, K. and Oyama, E., "Position Estimation of a mobile Robot Using Optical Fiber Gyroscope (OFG)", International Conference on Intelligent Robots and Systems (IROS '94). Munich, Germany, September 12–16, pp. 143–149, 1994
264. E. Udd (Ed.), Fiber Optic Sensors: An Introduction for Engineers and Scientists, Wiley, New York, 1991
265. J. Dakin and B. Culshaw, Optical Fiber Sensors: Principals and Components, Volume 1, Artech, Boston, 1988
266. B. Culshaw and J. Dakin, Optical Fiber Sensors: Systems and Applications, Volume 2, Artech, Norwood, 1989
267. T.G. Giallorenzi, J.A. Bucaro, A. Dandridge, G.H. Sigel, Jr., J.H. Cole, S.C. Rashleigh, and R.G. Priest, "Optical Fiber Sensor Technology", IEEE Journal of Quantum Electronics QE-18, 626, 1982

268. D.A. Krohn, Fiber Optic Sensors: Fundamental and Applications, Instrument Society of America, Research Triangle Park, North Carolina, 1988
269. N. Lagokos, L. Litovitz, P. Macedo, and R. Mohr, Multimode Optical Fiber Displacement Sensor, Applied Optics 20, 167, 1981
270. K. Fritsch, Digital Angular Position Sensor Using Wavelength Division Multiplexing, Proceedings of SPIE 1169, 453, 1989
271. K. Fritsch and G. Beheim, Wavelength Division Multiplexed Digital Optical Position Transducer, Optics Letters 11, 1, 1986
272. D. Varshneya and W.L. Glomb, Applications of Time and Wavelength Division Multiplexing to Digital Optical Code Plates, Proceedings of SPIE 838, 210, 1987
273. J.W. Snow, A Fiber Optic Fluid Level Sensor: Practical Considerations, Proceedings of SPIE 954, 88, 1983
274. T.E. Clark and M.W. Burrell, Thermally Switched Coupler, Proceedings of SPIE 986, 164, 1988
275. Y.F. Li and J.W. Lit, Temperature Effects of a Multimode Biconical Fiber Coupler, Applied Optics 25, 1765, 1986
276. Y. Murakami and S. Sudo, Coupling Characteristics Measurements Between Curved Waveguides Using a Two Core Fiber Coupler, Applies Optics 20, 417, 1981
277. D.A. Nolan, P.E. Blaszyk, and E. Udd, Optical Fibers, in Fiber Optic Sensors: An Introduction for Engineers and Scientists, in E. Udd (Ed.), Wiley, New York, 1991
278. J.W. Berthold, W.L. Ghering, and D. Varshneya, Design and Characterization of a High Temperature, Fiber Optic Pressure Transducer, IEEE Journal of Lightwave Technology LT-5, 1, 1987
279. D.R. Miers, D. Raj, and J.W. Berthold, Design and Characterization of Fiber-Optic Accelerometers, Proceedings of SPIE 838, 314, 1987
280. W.B. Spillman and R.L. Gravel, Moving Fiber Optic Hydrophone, Optics Letters 5, 30, 1980
281. E. Udd and P.M. Turek, Single Mode Fiber Optic Vibration Sensor, Proceedings of SPIE 566, 135, 1985
282. D.A. Christensen and J.T. Ives, Fiber optic Temperature Probe Using a Semiconductor Sensor, Proceedings of the NATO Advanced Studies Institute, Dordrecht, The Netherlands, p. 361, 1987
283. S.D. Schwab and R.L. Levy, In-Service Characterization of Composite Matrices with an Embedded Fluorescence Optrode Sensor, Proceedings of SPIE 1170, 230, 1989
284. R.B. Smith (Ed.), Selected Papers on Fiber Optic Gyroscopes, SPIE Milestone Series, MS 8, 1989
285. R.A. Lieberman (Ed.), Chemical, Biochemical, and Environmental Fiber Sensors V, Proceedings of SPIE, 1993
286. M.W. Sasnett and T.F. Johnston, Jr., "Beam Characterization and Measurement of Propagation Attributes", in Laser Beam Diagnostics: Proceedings of the SPIE 1414, Los Angeles, CA, January, 1991
287. D. Wright, et al., "Laser Beam Width, Divergence and Beam Oropagation Factor – an International Standardization Approach", Optical and Quantum Electronics 24, S993–S1000, 1992

288. J.G. Fujimoto, C. Pitris, S.A. Boppart, and M.E. Brezinski, "Optical Coherence Tomography: An Emerging Technology for Biomedical Imaging and Optical Biopsy", Neoplasia 2, 9–25, 2000
289. A. Molina, A.H. Al-Ashaab, T.I.A. Ellis, R.I.M. Young, and R. Bell, "A Review of Computer-Aided Simultaneous Engineering Systems", Research in Engineering Design 7(1), 38–63, 1995
290. D.C. Brown, "Intelligent Computer-Aided Design", Encyclopedia of Computer Science and Technology, 1998. J.G. Williams and K. Sochats (Eds.), Mc-Graw-Hill. Chou J.C.K. & Kamel M. 1988
291. J.C.K. Chou, M. Kamel, Quaternions Approach to Solve the Kinematic Equation of Rotation AaAx = AxAb of a Sensor Mounted Robotic Manipulator. Proceedings of the IEEE International Conference Robotics & Automation, Philadelphia, PA, pp. 656–662, 1988
292. H. Everett and L. Ong, Solving the Generalized Sensor-Mount Registration Problem. Proceedings of the ASME Winter Annual Mtg. DSC-Vol 29 (ASME) pp. 7–14, 1991
293. J.P. Huissoon and D.L. Strauss, Sensor Based Control of Industrial Robots. Proceedings of the IMC-12, Cork, Ireland, pp. 13–20, 1995
294. J. Weng, P. Cohen, and M. Herniou, Camera Calibration with Distortion Models and Accuracy Evaluation. IEEE Transactions on Pattern Analysis and Machine Intell 14(10), 965–980, 1992
295. R. Agah, J.A. Pearce, A.J. Welch, and M. Motamedi, "Rate Process Model for Arterial Tissue Thermal Damage: Implications on Vessel Photocoagulation", Lasers Surgery and Medicine, 15, pp. 176–184, 1994
296. B.L. Doyle, W.R. Wampler, and D.K Brice, "Temperature Dependence of H Saturation and Isotope Exchange" Journal of Nuclear Materials 103 and 104, 513–518, 1981
297. I. Foster, C. Kesselman, and S. Tuecke. The Nexus Approach to Integrating Multithreading and Communication. Journal of Parallel and Distributed Computing, 1996
298. E. Nahum, D. Yates, S. O'Malley, H. Orman, and R. Schroeppel, Parallelized Network Security Protocols, in Proceedings Symposium on Network and Distributed System Security, IEEE Computer Society, Los Alamitos, pp. 145–154, 1996
299. N. Venugopal. The Design, Implementation, and Evaluation of Cryptographic Distributed Applications: Secure PVM. Technical Report, University of Tennessee, Knoxville, TN, 1996
300. R. Pallas-Areny and J.G. Webster, Sensors and Signal Conditioning, Wiley, New York, 1991
301. J. Fraden, Handbook of Modern Sensors. Physics, Designs and Applications, AIP, Woodbury, NYN. White, 1997, "Intelligent Sensors" in Sensor Review 17(2), pp. 97–98
302. E.T. Powner and F. Yalcinkaya, "Intelligent Sensors: Structure and System", in Sensor Review 15(3), 31–34, 1995
303. "VSCADA" – Voltas Supervisory Control and Data Acquisition, http://www.voltasacnr.com/default.html.
304. Implementation Details Based on: "An Architectural Framework for Describing Supervisory Control and Data Acquisition (SCADA) Systems" Michael P. Ward, US Naval Postgraduate School, September 2004

305. "SCADA vs. the Hackers" Alan S. Brown, American Society of Mechanical Engineers, http://www.memagazine.org/backissues/dec02/features/scadavs/
306. Security Recommendations Based on: "SCADA Systems Security" Michael A. Young, SANS Institute, February, 2004
307. PCI Industrial Computer Manufacturers Group (PICMG): http://www.picmg.com PXI Systems Alliance (PXISA): http://www.pxisa.org.
308. VXI Plug & Play Systems Alliance: http://www.pnp.org.
309. John Pasquarette, Application Note 120 – Using IVI Drivers to Simulate Your Instrumentation Hardware in LabVIEW and LabWindows/CVI, National Instruments Corporation, Austin, 1998
310. E. Mandado, D. Valdés, M.J. Moure, and L. Rodríguez-Pardo, "VISCP: A PC-Based System for Practical Training of Digital Electronic Circuits Design and Verification", ALT-C 96 Association for Learning Technology (Third Annual Conference), Glasgow, Scotland, UK, 1996
311. L. Benetazzo, M. Bertocco, C. Narduzzi, "Networking Automatic Test Equipment Environments", IEEE Instrumentation and Measurement Magazine 8(1), 16–21, 2005
312. D. Ratner, P. Mapos Kerrow, "Using LabVIEW to Prototype an Industrial-Quality Real-Time Solution for the Titan Outdoor 4WD Mobile Robot Controller", Intelligent Robotics Lab., Wollongong University, NSW.
313. J. Mireles Jr., F.L. Lewis, "Intelligent Material Handling: Development and Implementation of Amatrix-Based Discrete-Event Controller", IEEE Transactions on Industrial Electronics 48(6), 1087–1097, 2001
314. L.M. Sheppard, "Visual Effects and Video Analysis Lead to Olympics Victories", IEEE Computer Graphics and Applications 26(2), 6–11, 2006